ANTARCTIC
RESEARCH
SERIES

American Geophysical Union

ANTARCTIC
RESEARCH
SERIES

American Geophysical Union

Volume 1 BIOLOGY OF THE ANTARCTIC SEAS
Milton O. Lee, *Editor*

Volume 2 ANTARCTIC SNOW AND ICE STUDIES
Malcom Mellor, *Editor*

Volume 3 POLYCHAETA ERRANTIA OF ANTARCTICA
Olga Hartman

Volume 4 GEOMAGNETISM AND AERONOMY
A. H. Waynick, *Editor*

Volume 5 BIOLOGY OF THE ANTARCTIC SEAS II
George A. Llano, *Editor*

Volume 6 GEOLOGY AND PALEONTOLOGY OF THE ANTARCTIC
Jarvis B. Hadley, *Editor*

Volume 7 POLYCHAETA MYZOSTOMIDAE AND SEDENTARIA OF ANTARCTICA
Olga Hartman

Volume 8 ANTARCTIC SOILS AND SOIL FORMING PROCESSES
J. C. F. Tedrow, *Editor*

Volume 9 STUDIES IN ANTARCTIC METEOROLOGY
Morton J. Rubin, *Editor*

Volume 10 ENTOMOLOGY OF ANTARCTICA
J. Linsley Gressitt, *Editor*

Volume 11 BIOLOGY OF THE ANTARCTIC SEAS III
Waldo L. Schmitt and George A. Llano, *Editors*

Volume 12 ANTARCTIC BIRD STUDIES
Oliver L. Austin, Jr., *Editor*

Volume 13 ANTARCTIC ASCIDIACEA
Patricia Kott

Volume 14 ANTARCTIC CIRRIPEDIA
William A. Newman and Arnold Ross

Volume 15 ANTARCTIC OCEANOLOGY
Joseph L. Reid, *Editor*

Volume 16 ANTARCTIC SNOW AND ICE STUDIES II
A. P. Crary, *Editor*

Volume 17 BIOLOGY OF THE ANTARCTIC SEAS IV
George A. Llano and I. Eugene Wallen, *Editors*

Biology of the Antarctic Seas IV

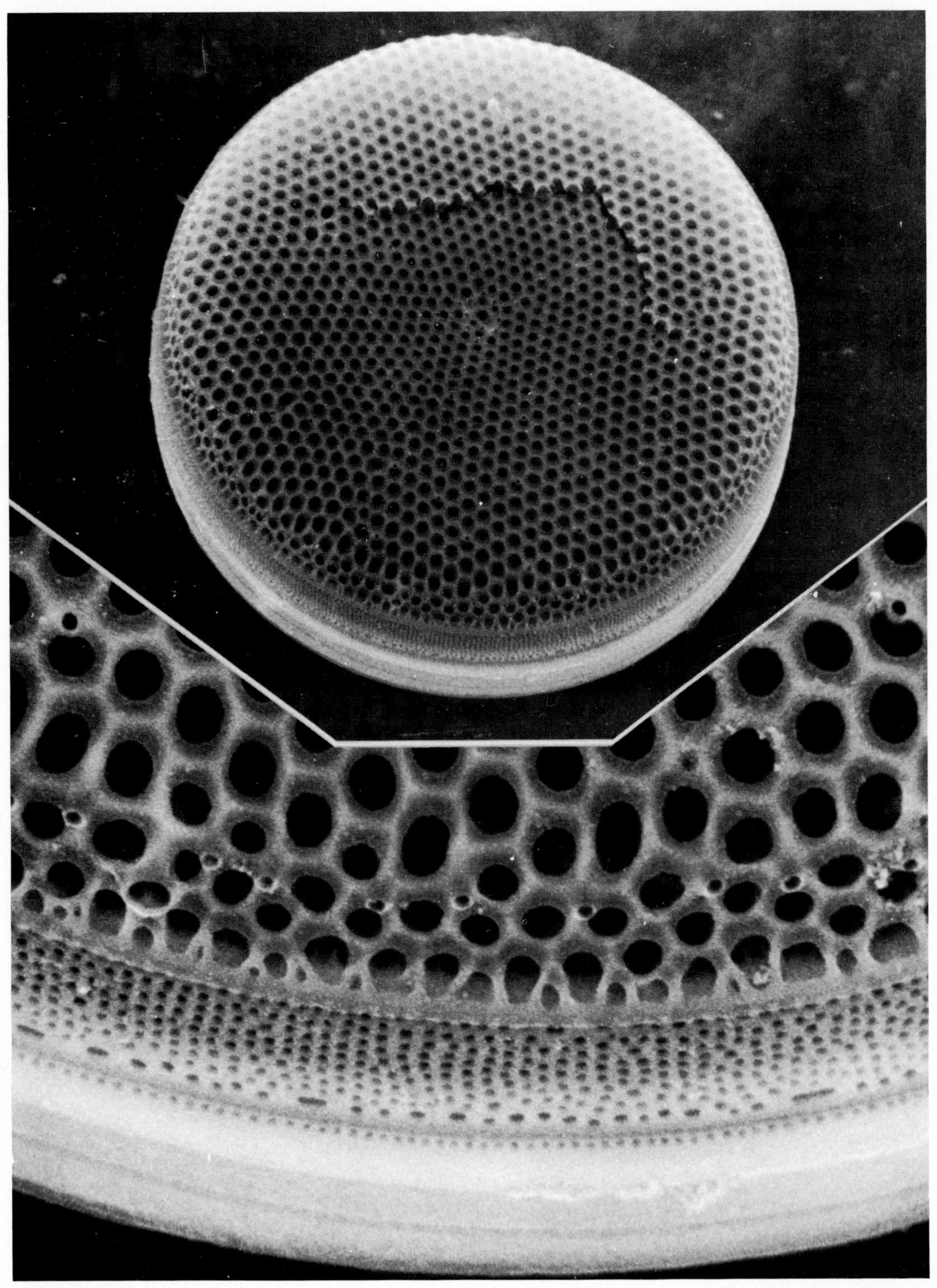

FRONTISPIECE

The marine diatom *Thalassiosira tumida* (Janisch) Hasle from an antarctic bloom, viewed in a scanning electron micrograph. (*Top*) The slightly broken valve is half of the silicious skeleton of the photosynthetic organism found in the icy waters off the Ronne Ice Shelf, ×2000. (*Bottom*) Enlargement of the same valve shows the hyaline girdle band and perforated intercalary band still attached to the valve itself, ×8000. See Figures 40a, 40b, and 46 of Hasle et al. (this volume pp. 329 and 331) and Figure 6 of El-Sayed (this volume, p. 311).

Volume 17 | ANTARCTIC RESEARCH SERIES

Biology of the Antarctic Seas IV

George A. Llano and I. Eugene Wallen, *Editors*

Published with the aid of a grant from the National Science Foundation

PUBLISHER

AMERICAN GEOPHYSICAL UNION

OF THE

National Academy of Sciences—National Research Council

December 3, 1971

Volume 17 | ANTARCTIC RESEARCH SERIES

BIOLOGY OF THE ANTARCTIC SEAS IV

GEORGE A. LLANO and I. EUGENE WALLEN, *Editors*

Suite 435, 2100 Pennsylvania Avenue, N.W.
Washington, D. C. 20037

Library of Congress Catalog Card No. 64-60030
International Standard Book No. 0-87590-117-4

List Price, $30.00

Printed by
THE HORN-SHAFER COMPANY
DIVISION OF
Geo. W. King Printing Co.
Baltimore, Maryland

THE ANTARCTIC RESEARCH SERIES

The Antarctic Research Series is designed to provide a medium for presenting authoritative reports on the extensive and detailed scientific research work being carried out in Antarctica. The series has been successful in eliciting contributions from leading research scientists engaged in antarctic investigations; it seeks to maintain high scientific and publication standards. The scientific editor for each volume is chosen from among recognized authorities in the discipline or theme it represents, as are the reviewers on whom the editor relies for advice.

Beginning with the scientific investigations carried out during the International Geophysical Year, reports of research results appearing in this series represent original contributions too lengthy or otherwise inappropriate for publication in the standard journals. In some cases an entire volume is devoted to a monograph. The material published is directed not only to specialists actively engaged in the work but to graduate students, to scientists in closely related fields, and to interested laymen versed in the biological and the physical sciences. Many of the earlier volumes are cohesive collections of papers grouped around a central theme. Future volumes may concern themselves with regional as well as disciplinary aspects, or with a comparison of antarctic phenomena with those of other regions of the globe. But the central theme of Antarctica will dominate.

In a sense, the series continues the tradition dating from the earliest days of geographic exploration and scientific expeditions—the tradition of the expeditionary volumes which set forth in detail everything that was seen and studied. This tradition is not necessarily outmoded, but in much of the present scientific work one expedition blends into the next, and it is no longer scientifically meaningful to separate them arbitrarily. Antarctic research has a large degree of coherence; it deserves the modern counterpart of the expeditionary volumes of past decades and centuries which the Antarctic Research Series provides.

With the aid of a grant from the National Science Foundation in 1962, the American Geophysical Union initiated the Antarctic Research Series and appointed a Board of Associate Editors to implement it. A supplemental grant received in 1966, the income from the sale of volumes in the series, and income from reprints and other sources have enabled the AGU to continue this series. The response of the scientific community and the favorable comments of reviewers cause the board to look forward with optimism to the continued success of this endeavor.

Morton J. Rubin
Chairman, Board of Associate Editors
Antarctic Research Series

PREFACE

Of the volumes currently available in the Antarctic Research Series, this volume is the fourth dealing with the biology of the antarctic seas. These collected papers comprise the results of original investigations, 11 of which are concerned mainly with the identification and distribution of marine plants and animals. In the first of these papers Stewart Springer gives a systematic appraisal of the five species of elasmobranch Rajidae from Antarctica, of which one represents a new and unique species. Heretofore one of the peculiarities of the antarctic ichthyological fauna has been the absence of sharks. In this very significant contribution, the author establishes the most southerly record for any member of the elasmobranchs. The second paper, by Patricia Kott, amplifies our systematic knowledge of the tunicates of the South Atlantic, South Pacific, and Indian oceans. It extends her monograph published as volume 13 of the Research Series under the title of *Antarctic Ascidiacea* and is based on collections made in the Antarctic through 1967; two new species are included. Additions and corrections to volume 13 are appended to this paper. John C. Markham reports on several lower chordates of the genus *Cephalodiscus* and discusses the systematics and distribution of the five species known from the Antarctic. The Deep Freeze materials examined in the course of this study were obtained through the U.S. Navy Hydrographic Office oceanographers from icebreakers assigned to task force 43 prior to and after the 1955–1959 International Geophysical Year and precede the National Science Foundation sponsored research now being conducted by the USNS *Eltanin* and the R/V *Hero* under the U.S. Antarctic Research Program.

Three papers dealing with crustaceans follow. The first, by John C. McCain and Scott Gray, Jr., lists 22 caprellid amphipods, of which a significant number are endemic to the Antarctic Peninsula, Scotia Ridge, and Tierra del Fuego regions; one-fourth of the species described are new to science. The second of these crustacean papers, by Gayle A. Heron and Thomas E. Bowman, deals with immature stages of copepods. It is based on several excellent collections by the USNS *Eltanin* with the aid of fine-mesh nets; one new species is described. The third crustacean paper, on benthic myodocopid ostracods by Louis S. Kornicker, reports on collections made in 1967 by James K. Lowry at Arthur Harbor, Palmer Station, in 1968 by V. A. Gallardo at Arturo Prat Station on Greenwich Island, and in 1963 by Waldo L. Schmitt from the South Shetland Islands and Palmer Archipelago. In addition to describing five new species, the author reviews the previous work done on this group as represented in Antarctica.

The seventh paper in this volume, the contribution by Joel W. Hedgpeth and John C. McCain, is a review of the deep-sea genus *Pantopipetta.* It includes a family placement and one new species. This group of pycnogonids is well represented in antarctic waters. Following this paper is the report by Robert B. Short on three new species of mesozoan parasites from cephalopods collected in New Zealand waters. In the study on distribution of recent benthonic Foraminifera, René Herb discusses the occurrence of arenaceous and calcareous forms in relation to the Antarctic Convergence. The more prominent species of the Drake Passage are amply illustrated.

Penetration of the normally inaccessible Weddell Sea by the USCGG *Glacier* during the 1967–1968 (Captain E. E. McCrory, Commanding Officer) International Weddell Sea Oceanographic Expeditions revealed an unexpectedly rich bloom of phytoplankton at very high latitudes. Chemical, physical, and biological observations by Sayed El-Sayed provided background information and samples that enabled phytoplankton specialists to contribute detailed studies on the morphologic variations of important antarctic diatom species. These two studies, the first on the *Fasciculati* group by Hasle, Heimdal, and Fryxell and the second by Fryxell and Hasle on *Corethron,* are enhanced by the outstanding photographs obtained by employing scanning electron microscope techniques that reveal details not discernible with light and transmission microscopes.

The analysis of zooplankton standing crop in both antarctic and subantarctic regions by Thomas Hopkins is an important contribution to antarctic plankton ecology. The data on biomass estimate are based on surveys made by the RSS *Discovery* and the USNS *Eltanin.* The information brought together in these last four contributions underscores the importance of these organisms as a dynamic ecological unit.

It is hoped that the continuing investigations of components of the populations of antarctic organisms will in time lead to the publication of monographs furthering the identification and distribution of antarctic and subantarctic species. Our better understanding of the antarctic ecosystems and their biota is dependent on the basic information that systematic research provides.

In this volume of marine biologic studies, we have annotated the listings in the table of contents to apprise the reader of the new species described by the respective authors and the number of other forms discussed in the primarily taxonomic papers.

George A. Llano and I. E. Wallen

CONTENTS

THREE SPECIES OF SKATES (RAJIDAE) FROM THE CONTINENTAL WATERS OF ANTARCTICA

STEWART SPRINGER

National Marine Fisheries Service Systematics Laboratory
U.S. Museum of Natural History, Washington, D.C. 20560

Abstract. Three recently collected skates (Rajidae) represent all three elasmobranch species now known from antarctic continental waters. One specimen, *Raja georgiana* Norman, 1938, taken from a depth of 1232 meters in the northern part of the Ross Sea, provides the southernmost record for an elasmobranch. The other two specimens, assigned to *Bathyraja* Ishiyama, 1958, on the basis of rostral and clasper characters, were taken near the continental shelf edge of the Antarctic Peninsula. One of them, *Bathyraja griseocauda* (Norman, 1937), has been reported earlier from the Antarctic Peninsula area. The other specimen is described as a new species, *Bathyraja maccaini.*

INTRODUCTION

A collection of 3 skates from off Antarctica, assembled through the coordinating efforts of Dr. George A. Llano, of the Office of Polar Programs, National Science Foundation, includes 3 species of unusual interest of which one proves to be new. These 3 specimens, together with 2 others previously reported by Bigelow and Schroeder [1965], make up the entire series of elasmobranchs now known from antarctic continental waters.

One of the 3 specimens, a newly hatched female, *Raja georgiana* Norman, 1938, from 1232 meters' depth in the northern part of the Ross Sea, represents the most southerly record for any species of elasmobranch. Only 5 additional specimens of *R. georgiana* have been recorded, all of them taken in shallower, subantarctic waters, in 179 to 825 meters, off South Georgia.

The second antarctic specimen reported here, *Bathyraja griseocauda* (Norman, 1937), is a sexually mature male from 94 meters off the Antarctic Peninsula. *B. griseocauda* was originally described from the female holotype, and 5 males from the subantarctic Patagonia-Falklands area and 2 additional specimens from the same area were later recorded by Hart [1946]. The species was redescribed in some detail by Bigelow and Schroeder [1965] from a mature male and an immature female taken off the Antarctic Peninsula. The 11 recorded specimens are from 94 to 585 meters' depth.

The third specimen on which I report is the holotype of the new species described below. It is a male of nearly mature size, as shown by the degree of clasper development, and was taken at 91 meters off the Antarctic Peninsula.

The presence of 3 species among only 5 specimens now known from antarctic continental waters suggests that the skate fauna is varied, but the small number of specimens does not mean that distribution is sparse. All of the skates were collected with trawls 10 feet or less across the mouth; if such nets are no more effective in antarctic waters than in temperate waters for catching bottom-dwelling fishes as large as most skates, we have no reason to assume that skates are especially rare around the antarctic land mass.

Before Bigelow and Schroeder's report [1965] of the presence of *Bathyraja griseocauda,* skates were known to be present on antarctic continental slopes solely from egg capsules. Dollo [1904] described *Raja arctowskii* entirely on the basis of 3 egg capsules, each about 60 mm in length, from 400 to 569 meters, at 70° to 71°S, in the Bellingshausen Sea. Norman [1938] referred to an egg capsule from 203 meters off Graham Land near Adelaide Island as *Raja* sp. The capsule was large (200 mm long) and contained an embryo 70 mm across the disc. An incomplete and thin-walled egg capsule, also about 200 mm long, was collected from 401 meters by the USNS *Eltanin* in the Ross Sea at 73°56′. In his general review of the antarctic fish fauna, Andriashev [1965] had only the records of egg capsules to indicate the presence of skates; but the two sizes, about 60 mm and about

200 mm, enabled him to conclude that more than one rajid species should be present.

Relatively great depths and low temperatures no doubt restrict the ranges of some demersal elasmobranchs, but it cannot be assumed without more evidence that populations of skates of the continental slope in Antarctica are isolated from populations of some subantarctic continental slopes. Although all rajids appear to be strictly demersal, some species are known from relatively deep water and some arctic species are known from very cold habitats. Furthermore, some skates of the northern hemisphere have a wide bathymetric range. Captures were recorded from 293 to 2393 meters for *Raja hyperborea* Collett and from 366 to 1907 meters for *R. jenseni* Bigelow and Schroeder. Maximum habitat depths are poorly known. For example, *R. abyssicola* Gilbert, 1895, is known from a single specimen taken off British Columbia from 2903 meters. Bigelow and Schroeder [1953] noted that *R. hyperborea* has been collected only from water somewhat below 0.0°C to 1.5°C and that *Breviraja spinicauda* has been recorded from water at 0.5°C to 3.8°C. That 2 of the 3 species now known in antarctic continental waters are identifiable with species described from the Patagonia-Falklands region is not surprising.

More than 100 species have been described in the genus *Raja*. Although a comprehensive study probably would find some of them invalid, the number no doubt would remain large. Furthermore, many new species of skates, especially from depths greater than 500 meters, have been discovered recently as a result of improved collecting techniques for deep-water species.

Most of the species of *Raja* were originally described solely on the basis of external characteristics. Bigelow and Schroeder [1948], however, set up the genus *Breviraja* as distinct from *Raja* chiefly on the structure of the rostral cartilage as shown in radiographs. Ishiyama [1958] greatly extended the use of internal characteristics and clasper characteristics in a thorough study of the systematics of rajids from Japanese waters. Some use of clasper structure has also been made in studies of limited scope (for example, Leigh-Sharpe [1924] and Hulley [1969]); but in the absence of a comprehensive review of rajids, diagnoses must be made primarily from external characteristics.

Insofar as feasible I follow the terminology, definition of structures, and directions for measurements and counts that were given by Hubbs and Ishiyama [1968] in their proposed methods for the study of skates. The diagnosis of each of the 3 species treated here, however, depends chiefly upon the traditional external characteristics, the only ones that appear in most of the descriptive accounts of skates. Such diagnoses are obviously unsatisfactory, especially for the antarctic species that are known to me from single specimens.

Two of the species are assigned to *Bathyraja* Ishiyama, 1958, on the basis of rostrum and clasper characteristics outlined by Ishiyama and Hubbs [1968]. The rostral structure of the third species places it in *Raja*. Since radiographs do not show details of the rostrum in any of the 3 antarctic specimens, the structural features were determined by partial dissection. Dissections were made from the ventral surface to avoid disturbing spines on the snout of one of the species.

Raja georgiana Norman, 1938

Figs. 1, 4

Raja georgiana Norman, 1938, p. 4, fig. 1.

Type locality. South Georgia; Bigelow and Schroeder, 1965, p. R38, fig. 1 (South Georgia).

Material examined. A young female (Los Angeles Co. Mus. 11407-1), total length 205 mm, disc width 148 mm; taken in a 10-foot Blake trawl, USNS *Eltanin* station 1935, from 1232 meters in the Ross Sea at 72°57′ to 72°51′S, 178°15′ to 178°18′E, January 30, 1967; estimated water temperature at point of collection about 0.7°C.

Diagnosis. The Ross Sea specimen falls in the genus *Raja*, as determined from the structure of its rostrum, which is stiff and stout basally and is flexible only near its tip. Also, the tips of the anterior pectoral rays do not reach a point as near the rostral appendix as they do in *Bathyraja* (Figure 4).

My identification of the specimen as *R. georgiana* is based chiefly on the number and distribution of spines and the presence of prickles over most of the dorsal surfaces. The present specimen has a patch of 5 small and inconspicuous spines that are definitely larger than prickles near the tip of the snout (6 are shown in the illustration of the young holotype [Norman, 1938, fig. 1]; Norman's description is otherwise in close agreement). In some skates, spines may be lost during growth or they may first develop after some growth or at maturity. They appear to remain essentially the same in number and distribution in *R. georgiana*. Bigelow and Schroeder [1965] reported

Fig. 1. *Raja georgiana,* a young female, disc width 148 mm, from the Ross Sea (drawing by Mildred H. Carrington).

only minor differences in spine characteristics in 2 young specimens, a half-grown male, and an adult male taken near South Georgia. The disappearance of the spines on the snout from older specimens seems to be the only important change in spine characteristics in *R. georgiana.*

The details of body proportions given by Norman [1938] for the young holotype of *R. georgiana* and by Bigelow and Schroeder [1965] for a half-grown male are not in close agreement with proportions of the Ross Sea specimen. The proportions in Norman's description, however, are obviously approximations. For example, he stated that the disc width is four-fifths of the total length; but from his measurements, the disc width may be calculated as 76% of total length as compared with 72% in the Ross Sea specimen. Norman gave snout length as one-seventh (14%) of disc width, but in the Ross Sea specimen the snout length is 21% of disc width. Such a large difference does not seem reasonable for skates of approximately the same size if they belong in the same species. The discrepancy may result from a combina-

TABLE 1. Proportional Dimensions of Three Antarctic Skates

Proportions are expressed as thousandths (‰) of disc width (first column) and total length (second column). Numbers in parentheses at the extreme left refer to numbered paragraphs in Hubbs and Ishiyama [1968], in which terms are defined.

	Raja georgiana		***Bathyraja griseocauda***		***Bathyraja maccaini***	
	‰ disc width	‰ total length	‰ disc width	‰ total length	‰ disc width	‰ total length
(1) Total length	1385	1000	1540	1000	1383	1000
(2) Disc width	1000	722	1000	650	1000	723
(3) Disc length	724	522	794	516	809	585
(4) Anterior projection	487	351	444	289	485	351
(7) Precaudal length	622	449	724	470	755	546
(8) Tail length	764	551	816	530	628	454
(9) Tail width at end pectoral	61	44	63	41	49	35
(10) Tail depth at end pectoral	47	34	33	22	28	20
(11) Tail width at origin first dorsal	30	22	29	19	27	19
(12) Tail depth at origin first dorsal	19	14	13	8	11	8
(14) First dorsal origin to tail tip	236	171	133	87	127	92
(15) Base first dorsal	64	46	54	35	40	29
(16) Between dorsal bases	0	0	6	4	4	3
(17) Base second dorsal	68	49	48	31	36	26
(19) Vertical height D_1	25	18	51	33	30	22
(20) Vertical height D_2	20	15	35	23	28	20
(22) Lateral fold length	595	429	686	445	560	405
(23) Lateral fold width	5	3	3	2	4	3
(25) Between front tips pectoral radials	166	120	48	31	32	23
(26) Cloaca to extended tip pelvic (anterior lobe)	223	161	244	159	215	155
(27) Pelvic fin length	186	134	206	134	183	132
(29) Clasper length	...	...	365	237	213	154
(31) Preocular length	186	134	197	128	202	146
(32) Preoral length (snout length)	209	151	190	124	202	146
(33) Prenarial length	159	115	143	93	149	108
(34) Internarial distance	142	102	105	68	89	65
(35) Nasal curtain length	74	54	67	43	66	48
(36) Nasal curtain width	23	17	22	14	19	14
(39) Mouth width	128	93	124	80	134	97
(40) Eyeball length	61	44	48	31	49	35
(41) Distance between orbits	81	59	57	37	66	48
(42) Between spiracles	115	83	111	72	111	80
(43) Spiracle length	29	21	43	28	45	32
(44) Distance between outer ends first gill slits	277	200	276	179	294	212
(45) Prefontanella length	...	...	...	...	160	115
(50) Between closest right and left scapular spines	135	98	...	...	149	108

tion of the approximate fractions used by Norman and the difference in definitions of snout length used by him and by me.

Bigelow and Schroeder [1965] gave a series of proportional dimensions only for a half-grown male, none of which are close to those for the Ross Sea specimen. The disc length in the half-grown male is 60.1% of total length, whereas it is 50.6% in the Ross Sea specimen. But Bigelow and Schroeder did give proportions for one set of dimensions in their juvenile specimens, disc length 50.0–51.8% of total length, which suggests to me that changes in proportion along the longitudinal axis are sufficiently large to influence all reported ratios, or at least those based on total length.

Norman [1938] stated that *R. georgiana* appeared to be most nearly related to *R. macloviana* Norman, 1938, from the Patagonian region and to *R. murrayi* Günther, 1880, from Kerguelen Island, but that it differed from both in having a triangle of large scapular spines. Both *R. macloviana* and *R. murrayi* are relatively small species; males mature at lengths of about 445 mm and about 425 mm, respectively, as compared to 860 mm for *R. georgiana.* Bigelow and Schroeder noted that claspers of the adult *R. georgi-*

ana are 'massive,' a term more applicable to claspers characteristic of *Raja* than to those characteristic of *Breviraja* or *Bathyraja*. In contrast, the claspers shown in Günther's illustration of an adult male of *R. murrayi* are relatively slender and cylindrical, with the general shape usual in described species of *Breviraja* or *Bathyraja*. The identification of antarctic rays should be much simplified if the characteristics that distinguish *Breviraja* and *Bathyraja* are more extensively used. Further study of larger specimens from antarctic continental waters is needed to verify my identification of the Ross Sea specimen as *R. georgiana*.

Description based solely on young female from Ross Sea. Total length 205 mm, greatest disc width 148 mm (for additional measurements, see Table 1; for shape and for color of dorsal surfaces and fins, see Figure 1). Prickles present over entire dorsal surface (including skin over eye) except extreme edges along anterior margins of disc and outer parts of pelvic fins; prickles distributed randomly except on posterior margins of pectorals where they tend to be in radiating rows, and along sides of tail, where they are in 4 or 5 longitudinal rows along each side of the central row of spines. A patch of 5 small spines near the tip of the snout, a large preocular spine, 2 large postocular spines, 3 large scapular spines forming a triangle in each scapular area, and a series of 24 large spines beginning with a nuchal spine and extending to the base of the first dorsal fin without interruption but becoming progressively smaller and closer together posteriorly. All spines and prickles on stellate bases, moderately slender with sharp tips slightly or moderately curved posteriorly, the larger spines ribbed, the longest spine 5 mm high. Ventral surfaces entirely smooth, the skin somewhat transparent, colorless except for darker color under posterior half of tail and along posterior pectoral margins where dorsal color shows through, yolk-sac scar plainly visible and not completely closed.

Total number of vertebrae 138 ± 4, trunk vertebrae 33, pre–dorsal-fin caudal vertebrae 64, vertebrae between first and second dorsal origins 11, vertebrae posterior to second dorsal origin 30 ± 4. Rows of upper-jaw teeth 35, rows of lower-jaw teeth 33. Number of turns in valvular intestine 11.

Bathyraja griseocauda (Norman, 1937)

Figs. 2 and 4

Raja griseocauda Norman, 1937, p. 8, fig. 9.
Type locality. Patagonia-Falklands region.

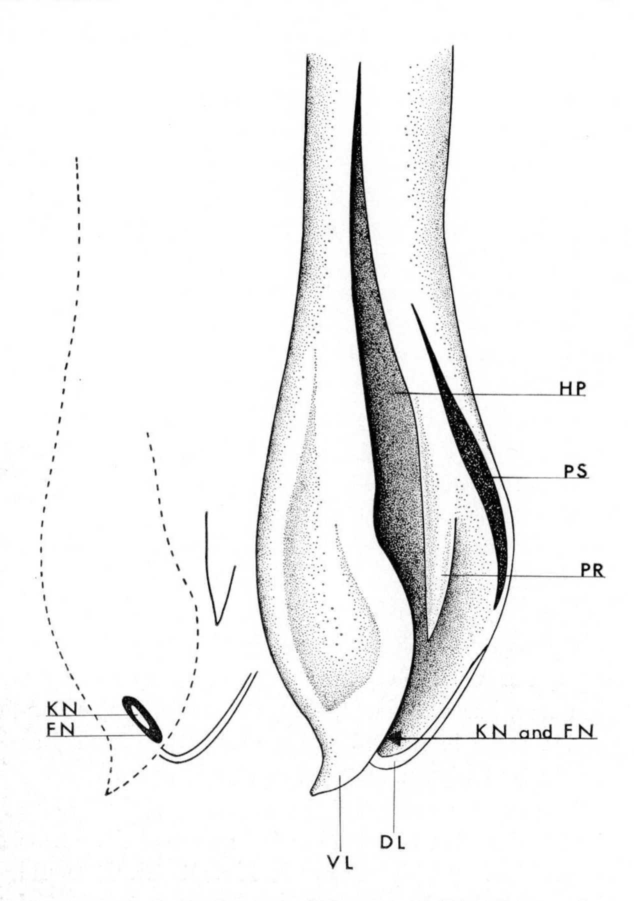

Fig. 2. Diagram of terminal portion of left clasper of a mature male *Bathyraja griseocauda*, disc width 315 mm, Antarctic Peninsula; HP, hypopyle; PS, pseudosiphon; PR, pseudorhipidion; FN, funnel; KN, knife; DL, dorsal lobe; VL, ventral lobe. The funnel and knife are hidden beneath the tip of the ventral lobe.

Breviraja griseocauda: Bigelow and Schroeder, 1965, p. R43, figs. 3 and 4 (off South Shetland Islands).

Material examined. One mature male, USNM 204703, total length 485 mm, disc width 315 mm; collected at R/V *Hero* Station 23, off Brabant Island north of Antarctic Peninsula at 64°12.1′S to 64°11′18″S, 62°39′36″W to 62°40′W, from about 94 meters, February 9, 1969; water temperature at 94 mm in area of capture 1.0°C.

Diagnosis. I regard *griseocauda* as belonging in *Bathyraja* Ishiyama, 1958, on the basis of characteristics outlined by Ishiyama and Hubbs [1968].

In both *Breviraja* Bigelow and Schroeder, 1948, and *Bathyraja*, the middle and anterior portion of the rostral projection is a slender and unsegmented cartilage so little calcified that it does not appear on any of the radiographs that I have seen. In both *Breviraja* and *Bathyraja* the tips of the anterior pectoral rays extend so far forward that they are nearly in contact with the rostral appendix (Figure 4), whereas in *Raja* a large gap separates the tips of the rays from the rostral appendix. Because the pectoral rays do not always appear on radiographs, as in the young specimen of *R. georgiana,* dissection may be required for generic diagnosis.

Ishiyama and Hubbs [1968] distinguished *Bathyraja* from *Breviraja* on the basis of combination of clasper and rostrum characteristics and found in a limited survey of species of rajids that *Bathyraja* included only Pacific species and *Breviraja* only Atlantic species. Their survey did not include antarctic or subantarctic species, which have been described only from external characteristics.

In rostrum characteristics, *griseocauda* meets criteria for *Bathyraja* in that it has small anterior notches (Figure 4) on the rostral appendix and that the slender rostral axis joins the appendix well back from the tip of the snout. The rostral appendix of *griseocauda,* however, has very large, thin, delicate wings, a condition more in line with that of *Breviraja colesi* as shown by Ishiyama and Hubbs [1968, figure 1B].

The clasper characteristics of *griseocauda* lead to its identification with *Bathyraja* in most of the details emphasized by Ishiyama and Hubbs [1968]. It should be noted that I follow the clasper terminology of Ishiyama [1958, pp. 200–204], which differs in some details from that used by Leigh-Sharpe [1924]. In *griseocauda,* a large and prominent pseudosiphon is present; the ventral terminal does not have a thin, sharp edge; a pseudorhipidion is present and is restricted to the central part of the terminal region. The *griseocauda* clasper differs from the clasper of *B. isotrachys,* the type species of *Bathyraja,* in having a funnel and knife hidden beneath the ventral lobe; but these structures are present in several species of *Bathyraja* from Japan [Ishiyama, 1958]. The strong, projecting, hooked ventral lobe of *griseocauda* may not be present in other *Bathyraja;* but a structure with some similarity is indicated in Leigh-Sharpe's figure for the clasper of *Raja murrayi* from Kerguelen Island [1924, figure 13].

Bathyraja griseocauda may be distinguished from other antarctic and subantarctic skates by external characteristics such as those used by Norman [1937] and Bigelow and Schroeder [1965]. The dorsal surfaces of *B. griseocauda,* except the pelvic fins and the anterior margin of the disc, are covered by prickles; but larger spines, except alar spines of mature males, are restricted to a single row of 19 to 27 stout, short spines along the midline of the tail extending from the region of origin of the pelvic fins to the first dorsal fin.

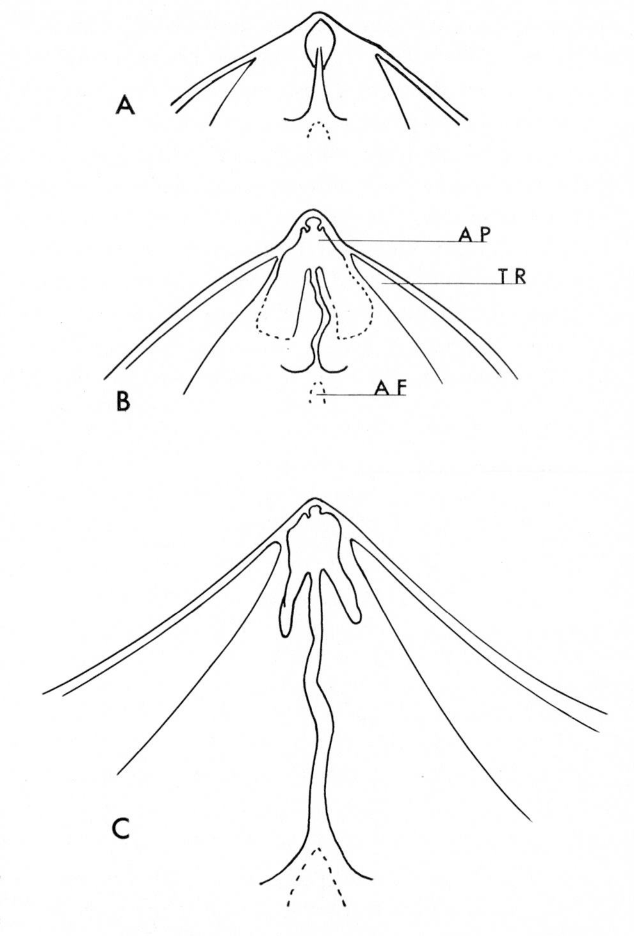

Fig. 4. Diagram of ventral aspect of rostral area of antarctic skates; A, young *Raja georgiana;* B, *Bathyraja griseocauda;* C, *Bathyraja maccaini;* AP, appendix of rostrum; TR, tips of pectoral rays; AF, anterior fontanelle (dorsal).

The absence of large spines on the disc should distinguish *Bathyraja griseocauda* from all other described antarctic rajid species except *Raja eatonii* and those species now usually assigned to the genus *Psammobatis.* That the absence of any spine on the disc is not always a reliable distinction is shown by Hureau's examination [1966] of a series of 35 females and 29 males of *Raja eatonii* Günther, 1876, from Kerguelen Island. *R. eatonii* was described originally from a Kerguelen specimen having a single large spine in the middle of the back (between the scapular regions). Hureau found, however, that the spines were absent from 13 females and 15 males in his series and that 2 females and 1 male each had 2 spines. My use of spine characteristics to separate *B. griseocauda* from *R. eatonii* thus is based on the number of large spines on the tail; 9 or 10 were noted in Günther's description and 8 to 12 were found in Hureau's series, as compared to 19 to 27 in *B. griseocauda.*

The proportional dimensions of the specimen of *B. griseocauda* at hand do not differ greatly from those given by Norman [1937] and Bigelow and Schroeder [1965] for the species.

B. griseocauda is not a large species, the two known mature males measuring 485 and 490 mm in length. The size of the mature female is probably somewhat greater; but it is more consistent with the production of egg capsules 60 mm long, such as were described under the name *Raja arctowskii* Dollo, 1904, than are the sizes of the other 2 larger species from antarctic continental waters, *R. georgiana* and *B. maccaini.* It is my view, however, that *Raja arctowskii* remains a nomen dubium.

Description of a mature male, USNM 204703. Total length 485 mm, greatest disc width 315 mm. For proportional dimensions see Table 1, for clasper structure Figure 2, for rostral structure Figure 4B, and for general shape Bigelow and Schroeder, 1965, figure 4.

Dorsal surfaces, except pelvic fins, anterior margins of disc, and region around alar hooks, covered with prickles; no enlarged spines on disc except 1 to 3 rows of alar hooks in 17 series; a single row of 20 short spines on large bases on the tail extending from the level of the pectoral-fin axilla to the first dorsal fin; ventral surfaces smooth.

First and second dorsal fins nearly equal in area, no space between dorsal-fin bases; anterior pelvic-fin lobe leglike, with 4 small projecting points along its posterior margin; narrow and inconspicuous fold of skin along each side of tail extending from the pelvic-fin axilla to the tail tip; claspers long and slender, their tips somewhat swollen and clublike; black curtains with scalloped edges over upper part of eyes; mouth with a small arch in the central portion but otherwise straight; teeth with pointed cusps; fringe of nasal curtain inconspicuous.

Dorsal surfaces dark brown or black, without markings; ventral surfaces yellowish except for some darker spots near base of tail.

Total number of vertebrae 134 ± 2, trunk vertebrae 30, pre–dorsal-fin caudal vertebrae 79, vertebrae between first and second dorsal-fin origins 11, vertebrae posterior to second dorsal origin 14 ± 2. Rows of upper-jaw teeth 27, rows of lower-jaw teeth 28. Number of turns in valvular intestine 8.

Bathyraja maccaini new species

Figs. 3, 4C

Holotype. An immature male, USNM 202702, total length 650 mm, greatest disc width 470 mm; collected at R/V *Hero* Station 28, cruise 69–1 off the south coast of Low Island on the north side of the Antarctic Peninsula, 63°25′30″S, 62°09′30″ to 62°05′36″W, from 91 meters, February 10, 1969; water temperature at 91 meters in area of capture 1.0°C.

Diagnosis. *Bathyraja maccaini* is a moderately large species of soft-nose skate known only from the immature male holotype. Rostrum and clasper characteristics place it in the genus *Bathyraja* Ishiyama, 1958, as further defined by Ishiyama and Hubbs [1968].

The extension of the rostrum anterior to its triangular base is very slender and flexible, with little, if any, calcification. The slender rostral axis broadens abruptly into an uncalcified appendix well back of the snout tip; the appendix extends nearly to the snout tip, and its wings are nearly in contact with the anterior pectoral radials. The shape of the appendix (Figure 4C) is similar to that shown for an adult male, *Bathyraja isotrachys,* by Ishiyama and Hubbs [1968, fig. 1].

The claspers of the holotype are not sufficiently developed to show details of the adult condition; they do show the presence of pseudosiphons, structures usually present in *Bathyraja* but not yet found in *Raja* or *Breviraja* [Ishiyama and Hubbs, 1968].

Raja doellojuradoi Pozzi, 1935, and *R. georgiana* belong in the genus *Raja* as restricted by Bigelow and Schroeder, 1948, and thus differ from *Bathyraja maccaini.*

B. maccaini may be distinguished from all other skates as described from the subantarctic and antarctic regions by the number and distribution of spines and prickles on the dorsal surfaces. Characteristics other than the number and distribution of spines do not appear consistently in descriptions. Although there are other differences that separate *B. maccaini* from each of the other species, no attempt is made here to set forth characteristics by which the species differs from skates beyond the limits of the subantarctic and antarctic region, on the grounds that this is impractical in the present state of published information on the numerous named species.

B. maccaini has one preocular, one postocular, and

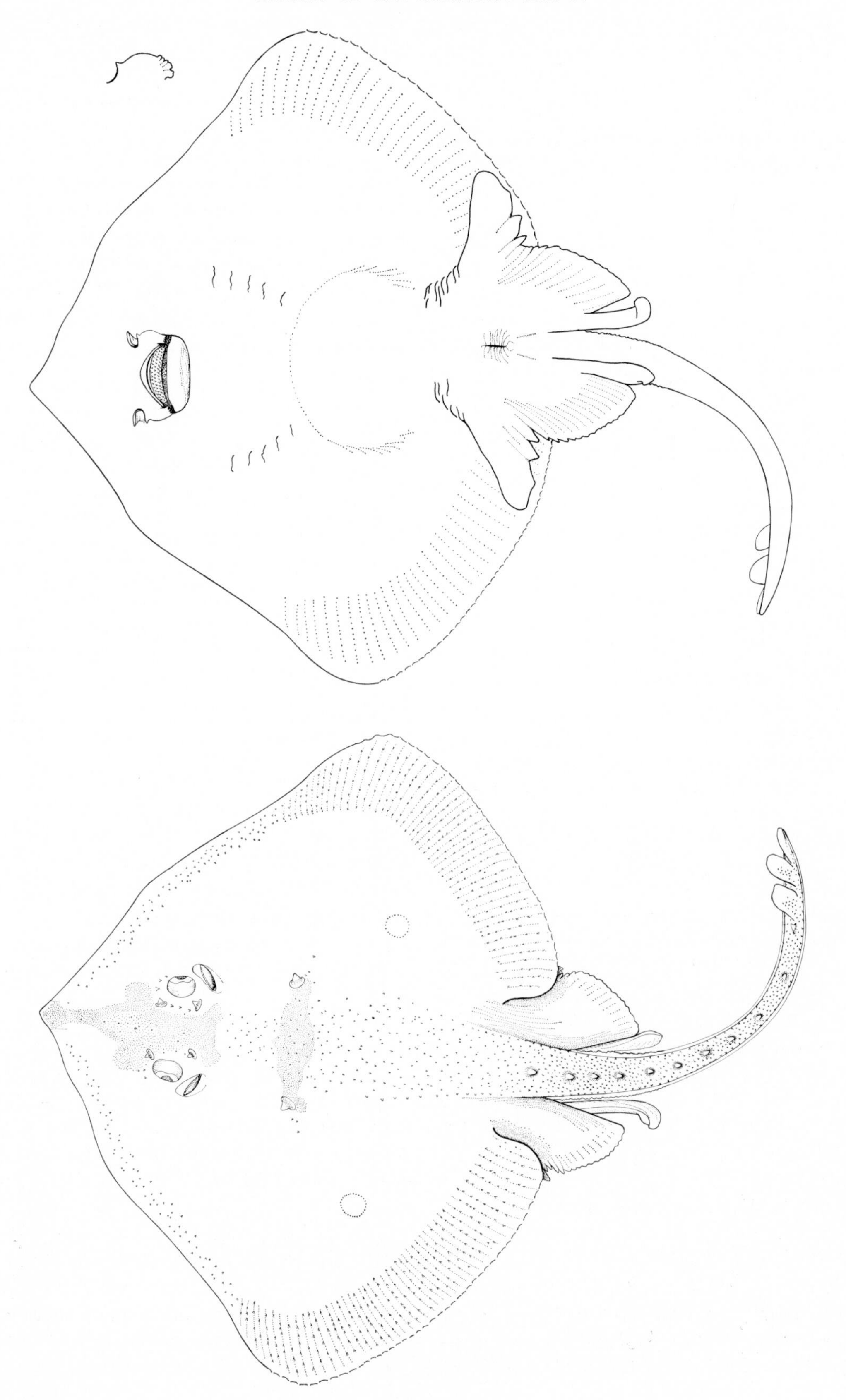

Fig. 3. *Bathyraja maccaini*, immature male holotype, disc width 470 mm, Antarctic Peninsula (drawing by Mildred H. Carrington).

one scapular spine on each side of the disc and a series of 11 spines on the midline of the tail extending from a point slightly posterior to the pectoral axilla to the first dorsal fin. Prickles, irregularly and somewhat sparsely spaced, are present in a broad band near the posterior pectoral margins and in a narrower band along the anterior pectoral margins. Prickles cover the dorsal surface of the tail and extend forward as a median band to the scapular region. They are present in reduced number on the head (Figure 3). Prickles are absent from the rest of the dorsal surface, the pelvic fins, the distal portions of the dorsal fins, most of the caudal fin, and a prominent skin fold along the sides of the tail.

Bases of spines and prickles are partially covered with skin, but both are ribbed basally. Prickles vary in size and shape; those of the malar area along the anterior pectoral margin are somewhat longer than the others and comparatively more slender. The holotype has no alar spines, but these probably should not be expected in an immature male.

In having ocular and scapular spines, *B. maccaini* differs from *Raja flavirostris* Philippi, 1892, *R. scaphiops* Norman, 1937, *R. albomaculata* Norman, 1937, *R. brachurops* Fowler, 1910, *R. eatonii*, Günther, 1876, and *Bathyraja griseocauda*. In lacking spines along the midline of the back anterior to the tail base, *B. maccaini* differs from *R. murrayi* Günther, 1880, *R. macloviana* Norman, 1937, *R. magellanica* Steindachner, 1903, and *R. multispinis* Norman, 1937.

The color of *B. maccaini* may be distinctive. The preserved holotype is gray-brown above, with well-marked patches of dark brown across the scapular area and on the head (Figure 3). Small and indistinct light spots are sparsely and irregularly distributed over the dorsal surfaces. These spots are not indicated in the figure, but 2 larger, indistinct, ocellus-like areas of lighter color are indicated by dotted lines.

The holotype of *B. maccaini* has a small space but no spine between the 2 dorsal fins. Several species *(Raja doellojuradoi, R. macloviana, R. magellanica, R. multispinis, R. scaphiops,* and *R. brachurops)* may have a spine or spines in the space between the dorsal-fin bases, but Norman [1937] used such phrases as 'a spine usually present,' suggesting that the characteristic is variable.

B. maccaini and *R. flavirostris* are readily separable by disc shape. *R. flavirostris* has a longer and more pointed snout. Anterior disc margins are concave in *R. flavirostris* and sinuous in *B. maccaini.*

Proportional dimensions appear to have little usefulness in diagnosis of single specimens. Although questions about the extent of allometry and the methods of calculation or measurement preclude detailed comparisons, proportional dimensions for the *B. maccaini* holotype are given in Table 1. The total number of defined vertebrae is 136 ± 2: 37 trunk vertebrae, 72 predorsal caudal vertebrae, 9 vertebrae between first and second dorsal-fin origins, and 18 ± 2 posterior to the origin of the second dorsal fin. Teeth of the upper jaw are in 29 rows, those of the lower jaw in 26 rows. The valvular intestine has 11 turns.

Bathyraja maccaini is named for Dr. John C. McCain, collector of the holotype.

REFERENCES

Andriashev, A. P.

1965 A general review of the antarctic fish fauna. Monogr. Biol., *15:* 491–550, figs. 1–8.

Bigelow, Henry B., and William C. Schroeder

1948 New genera and species of batoid fishes. J. Mar. Res., *7*(3) : 543–566, figs. 1–9.

1953 Fishes of the western North Atlantic. Mem. Sears Fdtn Mar. Res., no. 1, pt. 2, chap. 1: 1–514, figs. 1–117.

1965 Notes on a small collection of rajids from the subantarctic region. Limnol. Oceanogr., *10* (suppl.) : R34–R49, figs. 1–5.

Dollo, Louis

1904 Résultats du voyage du S. Y. Belgica. Zoologie. Poissons. 239 pp., figs. 1–6, pls. 1–12. Belgian government publication, Antwerp.

Fowler, Henry W.

1910 Notes on batoid fishes. Proc. Acad. Sci. Philad., *62:* 468–475, figs. 1–2.

Günther, Albert

1876 Remarks on fishes, with descriptions of new species in the British Museum, chiefly from southern seas. Ann. Mag. Nat. Hist., ser. 4, *17:* 389–402. [Description of *Raja eatonii* repeated with mention of one additional specimen, A. Günther 1879, *in* An account of the petrological, botanical and zoological collections made in Kerguelen's island and Rodriguez during the Transit of Venus Expeditions, in the years 1874–75. Phil. Trans. R. Soc. (London), *168:* 166.]

1880 Report on the shore fishes. Zool. Voy. H. M. S. *Challenger, 1*(6) : 1–82, pls. 1–32. Her Majesty's Stationery Office, London.

Hart, T. J.

1946 Report on trawling surveys on the Patagonian continental shelf. 'Discovery' Rep., *23:* 223–408.

Hubbs, Carl L., and Reizo Ishiyama

1968 Methods for the study and description of skates (Rajidae). Copeia, 1968, No. 3: 483–491, fig. 1.

Hulley, P. A.

1969 The relationship between *Raja miraletus* Linnaeus and *Raja ocellifera* Regan based on a study of the clasper. Ann. S. Afr. Mus., *52*(6) : 137–147, figs. 1–3.

Hureau, J. C.
1966 Nouvelle description de *Raja eatonii* Günther, 1879, Rajidae endémique des Îles Kerguelen (Antarctique). Bull. Mus. Natn. Hist. Nat., ser. 2, *38*(4): 396–399, fig. 1.

Ishiyama, Reizo
1958 Studies on the rajid fishes (Rajidae) found in the waters around Japan. J. Shimonoseki Coll. Fish., *7*(2–3): 193–394, figs. 1–86, pls. 1–3.

Ishiyama, Reizo, and Carl L. Hubbs
1968 *Bathyraja*, a genus of Pacific skates (Rajidae) regarded as phyletically distinct from the Atlantic *Breviraja*. Copeia, No. 2: 407–410, figs. 1–2.

Leigh-Sharpe, W. Harold
1924 The comparative morphology of the secondary sexual characters of elasmobranch fishes. The claspers, clasper siphons, and clasper glands. Memoir VII. J. Morph. Physiol., *39*(2): 567–577, figs. 1–15.

Norman, J. R.
1937 Coast fishes. Part II. The Patagonian region. 'Discovery' Rep., *16:* 1–150, figs. 1–76, pls. 1–5.
1938 Coast fishes. Part III. The antarctic zone. 'Discovery' Rep., *18:* 1–104, figs. 1–62, pl. 1.

Philippi, R. A.
1892 Algunos peces de Chile. An. Mus. Nac. Chile, sec. 1 (Zool.): 16 pp., 6 pls.

Pozzi, Aurelio J.
1935 Communicación preliminar sobre una nueva especie de 'raya' de la costa atlántica argentina, '*Raia doellojuradoi*' n. sp. Physis, B. Aires, *11*(40): 491–492.

Steindachner, Franz
1903 Die Fische der Sammlung Plate (Nachtrag). Zool. Jb. Suppl., *6:* 201–214.

ANTARCTIC ASCIDIACEA II

Patricia Kott

Zoology Department, University of Queensland, Australia

INTRODUCTION

Knowledge of the class Ascidiacea in Antarctica was considerably extended by large American collections made during the years 1947 to 1965 [Kott, 1969]. In the present work an account is given of the ascidians collected by the USNS *Eltanin* since 1965, together with some earlier collections not previously reported on.

Eighty species are recorded. Of these, 9 are endemic New Zealand species and 6 are widely ranging in the temperate zone of the southern hemisphere. Of the remaining 65 species taken south of 40°S, *Aplidium pellucidum* and *Molgula millari* are new to science. *Polyclinum sluiteri, Theodorella arenosa, Styela bythia,* and *Styela bathyphila* are recorded from the area for the first time. *Alloeocarpa incrustans, Alloeocarpa bridgesi, Cnemidocarpa ohlini, Bathyoncus enderbyanus, Bathyoncus mirabilis, Polycitor magalhaensis,* and *Aplidium stanleyi,* all rare species not previously in the American collections, are present in this new material. Relations between many of the species are clarified. *Bathyoncus herdmani* is shown to be synonymous with *B. mirabilis,* and *Molgula kerguelenensis* is shown to be a synonym of *M. euplicata.* Also, the genus *Askonides* Kott, 1962, is found to be not distinct from *Leptoclinides.*

In addition to the figures supplied for 2 new species and for those not previously recorded from the area, figures have been provided for species or certain diagnostic details that were not included in *Antarctic Ascidiacea* [Kott, 1969].

The total of 125 species previously recorded from the Antarctic [Kott, 1969] is consequently decreased by synonymy in 2 species, although 4 new records and 2 new species increase the number of antarctic ascidians now recorded to 129. Of these, 10 remain of doubtful identification. Of the total of 119 confirmed species now known from the geographical region south of 40°S, excluding endemic New Zealand and Australian species, 99 are present in the American collections housed in the Smithsonian Institution.

The geographical and vertical ranges of certain species have been extended slightly, but the new records generally fall within the ranges previously indicated. Exceptions concern the depth range of certain antarctic species *(Tylobranchion speciosum, Didemnum biglans, D. studeri,* and *Polysyncraton chondrilla)* that are now known to extend into deeper water than previously recorded for these species. This confirms and strengthens hypotheses concerning zoogeography and the origin and spread of antarctic ascidian fauna set forth by the author in *Antarctic Ascidiacea* [Kott, 1969].

STATION LIST

USS *Atka*

Sta. 4, Deep Freeze II, Ross Sea, Kainan Bay; 78°10′S, 162°22′W; January 20, 1957; 610 meters; bottom beam trawl.

Distaplia cylindrica, numerous colonies
Pyura georgiana, 2 specimens
Pyura discoveryi, 1 specimen
Molgula euplicata, 1 specimen

Sta. 24, Deep Freeze III, near Wilkes Station between Budd and Knox Coasts; 66°15′24″S, 110°28′48″E; January 23, 1958; 46 meters.

Distaplia cylindrica, 11 colonies
Aplidiopsis georgianum, numerous colonies
Styela nordenskjoldi, 1 specimen
Molgula pedunculata, 1 specimen

USS *Edisto*

Deep Freeze I, Kainan Bay, Ross Ice Shelf; 77°38′S, 163°11′W (approx.); January 29, 1956, collected by J. Q. Tierney; 644 meters.

Distaplia cylindrica, numerous colonies
Synoicium sp.?, 4 damaged colonies
Ascidia challengeri, 12 specimens
Caenagnesia bocki, 1 specimen
Pyura georgiana, 19 specimens
Pyura discoveryi, 6 specimens
Molgula pedunculata, 2 specimens

STA. 6, Ross Sea area, Robertson Bay (?); 73°19″S, 169°15′W; February 12, 1956; 100 meters; dredge.
Distaplia cylindrica, 1000+ colonies
Aplidium radiatum, 5 colonies
Synoicium adareanum, 34 heads
Didemnum biglans, 6 colonies
Cnemidocarpa verrucosa, 1 colony

STA. 14, SB-4, Weddell Sea (?); 71°35′S, 15° 19′W; January 21, 1959; depth ?.
Synoicium adareanum, 1 colony

STA. 38, TR 20, Port Lockroy, British I.G.Y. Base, Wiencke Island, Antarctic Peninsula; 64°50′S, 63°33′W; July 4, 1959, collected by J. Tyler; 104 meters.
Aplidium radiatum, 3 colonies (damaged)

USNS *Eltanin*

The *Eltanin* cruises for the greater part were conducted by the University of Southern California (USC); others were under the auspices of the Smithsonian Oceanographic Sorting Center (SOSC). Each organization maintained its own series of cruise and station numbers, each starting with 1.

The *Eltanin* stations referred to in this paper are listed chronologically as follows:

USC Cruises 6, 7, 9–12, 16, 19, Stations 339–1498.
SOSC Cruises 20, 21, Stations 91–203.
USC Cruises 22, 23, Stations 1500–1716.
SOSC Cruise 25, Stations 301–370.
USC Cruises 26, 27, Stations 1814–1978.

Cruise 6

STA. 339, Falkland Islands-Burdwood Bank; 53°05′ 04″S, 59°31′00″W; December 3, 1962; 512–586 meters; 40′ otter trawl.
Aplidium recumbens, 3 colonies
Didemnum studeri, 1 colony

STA. 432, South Shetland Islands; 62°52′03″S, 59°27′02″W; January 7, 1962; 935–884 meters; 5′ Blake trawl.
Tylobranchion speciosum, 20 colonies
Aplidium circumvolutum, 50 colonies
Caenagnesia bocki, 1 specimen
(See also Kott, 1969, p. 17)

STA. 436, South Shetland Islands; 63°14′S, 58° 45′W; January 8, 1963; 73 meters; 40′ otter trawl.
Distaplia cylindrica, debris only
Tylobranchion speciosum, 2 particles
Aplidium fuegiense, 7 colonies
Aplidium circumvolutum, 12 colonies
Synoicium adareanum, 8 colonies
Caenagnesia bocki, 1 specimen
Molgula pedunculata, 1 specimen
Molgula gigantea, 2 specimens
(See also Kott, 1969, p. 17)

Cruise 7

STA. 496, South Orkney Islands; 61°10′S, 45°09′ 30″W; February 20, 1963; 234–242 meters; rock dredge.
Aplidium radiatum, 4 colonies
(See also Kott, 1969, p. 18)

STA. 558, north of South Orkney Islands; 51°58′S, 56°38′W; March 14, 1963; 845–646 meters; 5′ Blake trawl.
Styela nordenskjoldi, 23 specimens
Eugyra kerguelenensis, 5 specimens

Cruise 9

STA. 671, South Georgia; 54°41′S, 38°38′W; August 23, 1963; 220–320 meters; 10′ Blake trawl.
Distaplia cylindrica, debris (zooids and larvae)
Aplidium stanleyi, 3 colonies
Molgula gigantea, 1 specimen

STA. 732, Scotia Sea area, north of South Georgia; 53°35′42″S, 36°50.8′W; September 12, 1963; 220–265 meters; 10′ Blake trawl.
Tylobranchion speciosum, 3 colonies
Didemnum biglans, 1 colony

STA. 740, south of Tierra del Fuego; 56°06′S, 66°19′W; September 18, 1963; 384–494 meters; 5′ Blake trawl.
Tylobranchion speciosum, 2 colonies
Aplidium fuegiense, several colonies
Aplidium variabile, 4 colonies
Aplidium recumbens, 16 colonies
Didemnum studeri, few colonies
Polysyncraton chondrilla, 20 colonies
Alloeocarpa incrustans, single specimen
Styelidae sp?
Pyura paessleri, 72 specimens
Bathypera splendens, 3 specimens
Molgula setigera, 6 specimens
Molgula malvinensis, 1 specimen

Cruise 10

STA. 786, north of Bellingshausen Sea; 63°01′S, 82°31′W; October 24, 1963; 4602 meters; Menzies trawl.
Bathyoncus enderbyanus, 1 specimen
Culeolus murrayi, 1 specimen

STA. 803, north of Bellingshausen Sea; 65°50′S, 82°40′W; October 28–29, 1963; 4328 meters; Menzies trawl.
Styela bathyphila, 1 specimen

Cruise 11

STA. 931, Amundsen Sea; 70°11′S, 106°38′W; January 20, 1964; 3495 meters; 5′ Blake trawl.

Polysyncraton chondrilla, single colony

STA. 945, Bellingshausen Sea; 67°55′S, 90°43′W; January 26, 1964; 4008 meters; 5′ Blake trawl.

Culeolus murrayi, 1 specimen

STA. 951, north of Bellingshausen Sea; 65°11′S, 86°52′W; January 29, 1964; 4529–4548 meters; 5′ Blake trawl.

? *Styela nordenskjoldi*, 1 specimen

STA. 954, north of Bellingshausen Sea; 63°02′S, 87°01′W; January 31–February 1, 1964; 4685 meters; 5′ Blake trawl.

Bathyoncus mirabilis, 2 specimens

STA. 958, off southern Chile; 52°56′S, 75°00′W; February 5, 1964; 92–101 meters; 5′ Blake trawl.

Aplidium fuegiense, 2 colonies
Aplidium irregulare, 5 colonies
Aplidium variabile, 1 colony
Aplidium pellucidum, 1 colony (new species)
Didemnum studeri, very numerous colonies
Alloeocarpa bridgesi, numerous specimens
Molgula pulchra, 2 specimens

STA. 959, off southern Chile; 52°55′S, 75°00′W; February 6, 1964; 92–101 meters; 5′ Blake trawl.

Sycozoa sigillinoides, 2 colonies
Alloeocarpa bridgesi, 9 specimens

STA. 969, Tierra del Fuego; 54°56′S, 65°03′W; February 10–11, 1964; 229–265 meters; 5′ Blake trawl.

Aplidium variabile, 1200 colonies
Pyura paessleri, 1 specimen
Molgula euplicata, 1 specimen

Cruise 12

STA. 991, north of South Shetland Islands; 60°57′S, 56°52′W; March 13, 1964; 2672–3020 meters; 5′ Blake trawl.

Styela nordenskjoldi, 4 specimens

STA. 993, north of South Shetland Islands; 61°25′S, 56°30′W; March 13, 1964; 300 meters; 10′ Blake trawl.

Styela nordenskjoldi, 1 specimen
Bathyoncus mirabilis, 2 specimens

STA. 1001, north of South Shetland Islands; 62°39′S, 54°46′W; March 15, 1964; 238 meters; Campbell grab.

Aplidium irregulare, 3 colonies
Pyura georgiana, 1 specimen

STA. 1003, between South Shetland Islands and South Orkney Islands; 62°41′S, 54°43′W; March 15, 1964; 210–220 meters; 10′ Blake trawl.

Polycitor magalhaensis, 6 colonies
Tylobranchion speciosum, 1 colony
Aplidium radiatum, 6 colonies
Aplidium circumvolutum, 3 colonies
Aplidium caeruleum, 15 colonies
Synoicium adareanum, 9 colonies
Polysyncraton chondrilla, 1 colony
Cnemidocarpa verrucosa, 3 specimens
Styela nordenskjoldi, 2 specimens
Pyura discoveryi, 1 specimen
Pyura obesa, 1 specimen (juvenile)
Molgula euplicata, 8 specimens
Molgula gigantea, 1 specimen
Eugyra kerguelenensis, 1 specimen
(See also Kott, 1969, p. 18)

STA. 1009, east of Antarctic Peninsula; 65°06′S, 52°00′W; March 17, 1964; 2818–2846 meters; 5′ Blake trawl.

? *Molguloides vitrea*, 1 specimen (empty test only)

STA. 1025, south of South Orkney Islands; 62°05′S, 40°44′W; March 24, 1964; 3250–3285 meters; 5′ Blake trawl.

Styela nordenskjoldi, 1 specimen

STA. 1058, east of South Orkney Islands; 59°50′S, 32°27′W; April 4, 1964; 650–659 meters; 5′ Blake trawl.

Distaplia cylindrica, 6 colonies

STA. 1073, east of South Orkney Islands; 60°33′S, 37°00′W; April 10, 1964; 1162–1226 meters; 5′ Blake trawl.

Didemnum biglans, several colonies

STA. 1078, east of South Orkney Islands; 61°27′S, 41°55′W; April 12, 1964; 604 meters; 5′ Blake trawl.

Corella eumyota, 40 specimens
Styela nordenskjoldi, 2 specimens
Molgula gigantea, 1 specimen

STA. 1079, east of South Orkney Islands; 61°26′S, 41°55′W; April 13, 1964; 593–598 meters; 5′ Blake trawl.

Aplidium fuegiense, 1 colony
Didemnum biglans, 1 colony
Corella eumyota, 30 specimens
Molgula gigantea, 3 specimens

STA. 1082, South Orkney Islands; 65°50′S, 42°55′W; April 14, 1964; 298–302 meters; 5′ Blake trawl.

Didemnum biglans, 4 small colonies
(See also Kott, 1969, p. 18)

Cruise 16

STA. 1417, west of Macquarie Island; 54°24′S, 159°01′E; February 10, 1965; 79–93 meters; 5′ Blake trawl.

Oligocarpa megalorchis, 12 specimens
(See also Kott, 1969)

STA. 1430, east of Stewart Island, on Campbell Plateau; 49°19′S, 171°36′E; February 22, 1965; 165–192 meters; 40′ otter trawl.

Styela nordenskjoldi, 5 specimens
(See also Kott, 1969)

STA. 1431, east of South Island, New Zealand; 45°37′S, 170°58′E; February 23, 1965; 51 meters; 40′ otter trawl.

Didemnum studeri, numerous colonies
Didemnum maculatum, numerous colonies
Didemnum lambitum, 3 colonies
Trididemnum natalense, numerous colonies
Leptoclinides rufus, numerous colonies
Corella eumyota, 2 specimens
Cnemidocarpa bicornuta, 2 specimens
Pyura cancellata, 2 specimens
Pyura picta, 1 specimen

Cruise 19

STA. 1498, east of North Island, New Zealand; 37°32′S, 178°42′E; August 31, 1965; 101 meters; 10′ otter trawl.

Aplidium variabile, 1 colony
?*Didemnum studeri,* fragments
Cnemidocarpa madagascariensis regalis, 2 specimens

Cruise 20

STA. 91, Auckland Island; about 36°51′S, 174°45′E; September 12, 1965; intertidal hand net.

Pyura subuculata, 4 specimens
Microcosmus kura, 30 specimens

STA. 107, southeastern corner of South-Australian basin; 51°06′S, 145°03′E; September 27, 1965; 4078–4146 meters; 10′ Blake trawl.

Styela bathyphila, 2 specimens

STA. 126, eastern corner of South-Indian basin; 58°06′S, 144°55′E; October 1, 1965; 3089–3164 meters; 10′ Blake trawl.

Molgula millari, 1 specimen (new species)

STA. 134, southwest of Macquarie Island; 59°48′S, 144°45′E; October 3, 1965; 3200–3259 meters; 5′ Blake trawl.

Molgula millari } (new species)
Pareugyrioides galatheae } 17 specimens

Cruise 21

STA. 188, Valparaiso, Chile (White Beach area); 415–1 Marsden Square; November 18, 1965; intertidal hand net.

Corella eumyota, 2 specimens

STA. 203, west of Peru-Chile trench; 33°45′S, 80°41′W; November 26, 1965; 79–91 meters; Blake trawl.

Corella eumyota, 1 specimen

Cruise 22

STA. 1500, east of Tierra del Fuego; 52°26′S, 68°35′W; January 19, 1966; 73–79 meters; 10′ Blake trawl.

Polyzoa opuntia, 3 colonies
Styela nordenskjoldi, 2 specimens
Pyura legumen, 1 specimen
Paramolgula gregaria, 1 specimen

STA. 1506, north of South Shetland Islands; 57°50′S, 56°51′W; January 24, 1966; 3788–3944 meters; 10′ Blake trawl.

Molguloides vitrea, 1 specimen

STA. 1509, north of South Shetland Islands; 58°54′S, 53°51′W; January 25–26, 1966; 3817–3931 meters; 10′ Blake trawl.

Culeolus murrayi, 3 specimens

STA. 1527, northwest of South Georgia; 51°06′S, 40°07′W; February 4–5, 1966; 3742–3806 meters; 10′ Blake trawl.

Styela nordenskjoldi, 1 specimen

STA. 1533, South Georgia, Bay of Isles; 54°00′S, 37°27′W; February 7, 1966; 3–6 meters; diving.

Polyzoa opuntia, 6 colonies
Molgula malvinensis, 1 specimen

STA. 1534, north of South Georgia; 53°50′S, 37°25′W; February 7, 1966; 271–276 meters; 5′ Blake trawl.

Molgula pedunculata, 2 specimens (test only)

STA. 1535, South Georgia; 53°51′S, 37°38′W; February 7, 1966; 97–101 meters; 5′ Blake trawl.

Aplidium fuegiense, 10 colonies
Aplidium stewartense, 2 colonies
Ascidia challengeri, 1 specimen
Theodorella arenosa, 50+ specimens
Cnemidocarpa verrucosa, 3 specimens
Pyura georgiana, 400 specimens
Pyura discoveryi, 2000 specimens
Molgula pedunculata, 5 specimens
Molgula malvinensis, 3 specimens
Molgula gigantea, 2 specimens

STA. 1536, west of South Georgia; 54°29′S, 39°

22′W; February 8, 1966; 659–686 meters; 5′ Blake trawl.
Didemnum studeri, single colony
STA. 1537, south of South Georgia; 55°01′S, 39° 55′W; February 8, 1966; 2886–3040 meters; 10′ Blake trawl.
Protoholozoa pedunculata, 1 specimen
STA. 1545, east of South Orkney Islands; 61°04′S, 39°55′W; February 11–12, 1966; 2355–2897 meters; 5′ Blake trawl.
Tylobranchion speciosum, 5 colonies
STA. 1553, southeast of South Orkney Islands; 62°09′S, 38°11′W; February 14, 1966; 3056–3459 meters; 5′ Blake trawl.
Styela nordenskjoldi, 2 specimens
STA. 1555, west of South Sandwich Islands; 60° 04′S, 35°59′W; February 15–16, 1966; 1976–2068 meters; 10′ Blake trawl.
Pareugyrioides galatheae, 2 specimens
STA. 1560, south of South Sandwich Islands; 59° 34′S, 27°18′W; February 18, 1966; 1190–1469 meters; 5′ Blake trawl.
Corynascidia suhmi, 1 specimen
STA. 1571, east of South Sandwich Islands; 54° 51′S, 14°54′W; February 28, 1966; 3947–4063 meters; 5′ Blake trawl.
? *Protoholozoa pedunculata,* stalk only
Bathyoncus enderbyanus, 1 specimen
Molgula millari, 1 specimen, new species
STA. 1578, northeast of South Sandwich Islands; 55°49′S, 22°11′W; March 4, 1966; 4236–4273 meters; 5′ Blake trawl.
Styela nordenskjoldi, 1 specimen
Pareugyrioides galatheae, 2 specimens
Fungulus cinereus, 10 specimens
STA. 1581, north of South Sandwich Islands; 56° 19′S, 27°29′W; March 6, 1966; 148–201 meters; 5′ Blake trawl.
Distaplia cylindrica, 6 colonies
Polycitor magalhaensis, 7 colonies
Tylobranchion speciosum, 1 particle
Aplidium recumbens, numerous colonies
Synoicium adareanum, 1 colony
Didemnum biglans, 1000+ colonies
Cnemidocarpa verrucosa, 5 specimens
STA. 1585, south of South Georgia; 56°11′S, 38° 36′W; March 9, 1966; 2869–3038 meters; 10′ Blake trawl.
Protoholozoa pedunculata, 1 colony
STA. 1593, Scotia Ridge, southeast of Falkland Islands; 54°43′S, 56°37′W; March 14, 1966; 339–357 meters; 5′ Blake trawl.
Diplosoma longinquum, 1 colony
Ascidia meridionalis, 1 specimen
Theodorella arenosa, 5 specimens
STA. 1595, southeast of Falkland Islands; 54°40′S, 57°05′W; March 14, 1966; 124–128 meters; 40′ otter trawl.
Didemnum studeri, several colonies

Cruise 23
STA. 1603, Tierra del Fuego; 53°51′S, 71°36′W; April 1, 1966; 256–269 meters; 5′ Blake trawl.
Aplidium fuegiense, 5 colonies
Didemnum studeri, many colonies
Ascidia meridionalis, 9 specimens
Cnemidocarpa ohlini, 6 specimens
Styela nordenskjoldi, 2 specimens
STA. 1604, Tierra del Fuego; 53°21′S, 73°02′W; April 1, 1966; 769–869 meters; 5′ Blake trawl.
Styela nordenskjoldi, 8 specimens
STA. 1605, southern Chile; 52°53′S, 74°05′W; April 1, 1966; 522–544 meters; 5′ Blake trawl.
Didemnum studeri, numerous colonies
Styela nordenskjoldi, 4 specimens
STA. 1621, Southeast Pacific basin; 61°27′S, 94° 58′W; April 10, 1966; 4419–4804 meters; 5′ Blake trawl.
? *Protoholozoa pedunculata,* 1 colony
Didemnum studeri, 2 colonies
STA. 1654, Southeast Pacific basin; 58°17′S, 107° 15′W; April 22–23, 1966; 4297–4667 meters; 10′ Blake trawl.
Styela sericata, 1 specimen
STA. 1660, Southeast Pacific basin; 61°31′S, 108° 00′W; April 25, 1966; 5042–5045 meters; 10′ Blake trawl.
Culeolus murrayi, 6 specimens
STA. 1668, Southeast Pacific basin; 63°53′S, 108° 39′W; April 27–28, 1966; 4930–4963 meters; 10′ Blake trawl.
Culeolus murrayi, 3 specimens
STA. 1673, Southeast Pacific basin; 64°08′S, 115° 17′W; April 30, 1966; 4866–4881 meters; 10′ Blake trawl.
Culeolus murrayi, 5 specimens
Fungulus cinereus, 2 specimens
STA. 1691, Albatross Cordillera; 53°56′S, 140° 19′W; May 14, 1966; 362–567 meters; 5′ Blake trawl.
Polyclinum sluiteri, 4 colonies
STA. 1716, off North Island, New Zealand; 37°35′S,

178°46′W; May 28, 1966; 128–146 meters; 5′ Blake trawl.

? *Didemnum* sp., juvenile, 2 colonies

Styela nordenskjoldi, 2 specimens

Cruise 25

STA. 301, Peru-Chile trench; 33°02′S, 72°57′W; September 25, 1966; 5249–5395 meters; 5′ Blake trawl.

? Phlebobranchia, empty test only

STA. 346, Southwest Pacific basin; 50°06′S, 127°31′W; October 25, 1966; 3914 meters; 10′ Blake trawl/Menzies trawl.

Styela sericata, 1 specimen

Styela bythia, 1 specimen

STA. 359, Southeast Pacific basin; 63°03′S, 128°12′W; November 3, 1966; 4682 meters; 5′ Blake trawl/Menzies trawl.

Protoholozoa pedunculata, 1 colony

? *Corynascidia suhmi*, 1 empty test

Culeolus murrayi, 2 specimens

STA. 364, Southwest Pacific basin; 56°17′S, 156°13′W; November 11, 1966; 3694 meters; 5′ Blake trawl/Menzies trawl.

Aplidium abyssum, 1 colony

STA. 366, Southwest Pacific basin; 49°21′S, 172°16′W; November 15, 1966; 5340 meters; 10′ Blake trawl/Menzies trawl.

Protoholozoa pedunculata, 1 colony

Corynascidia suhmi, 2 specimens

Culeolus murrayi, 4 specimens

Oligotrema psammites, 1 specimen

STA. 368, off east coast of New Zealand; 43°16′S, 175°23′E; November 19, 1966; 84 meters; 10′ Blake trawl.

Culeolus murrayi, 2 specimens (see last paragraph under 'Distribution,' p. 64)

STA. 369, off east coast of New Zealand; 43°17′S, 175°23′E; November 19, 1966; 95 meters; 10′ Blake trawl.

Didemnum lambitum, 1 colony

STA. 370, off east coast, South Island, New Zealand; 43°22′S, 175°20′E; November 19, 1966; 95 meters; 10′ Blake trawl.

Aplidium circumvolutum, fragments

Aplidium scabellum, 16 colonies and fragments

Didemnum studeri, 535 colonies (approx.)

Didemnum lambitum, 35 colonies

Leptoclinides rufus, 16 colonies

Corella eumyota, 1 specimen

Ascidia meridionalis, 6 specimens

Alloeocarpa affinis, 9 colonies

Okamia thilenii, 27 specimens (approx.)

Stolonica australis, 100 specimens (approx.)

Cnemidocarpa bicornuta, 10 specimens

Cnemidocarpa madagascariensis regalis, 13 specimens

Molgula sabulosa, 2 specimens

Cruise 26

STA. 1814, west of North Island, New Zealand; 38°58′S, 172°59′E; November 30, 1966; about 124–154 meters; 40′ otter trawl.

? *Didemnum maculatum*, 3 colonies

STA. 1814 bis, west of North Island, New Zealand; about 38°50′S, 172°48′E; November 30, 1966; about 154 meters; 40′ otter trawl. (Collector's notes for this entry, '1814 and 1815 inadvertently mixed. . . .' Location and depth here noted are an averaging of data of these two stations.)

Didemnum lambitum, 1 colony

STA. 1818, Tasman Sea; 40°15′S, 168°16′E; December 2, 1966; 913–915 meters; 10′ Blake trawl.

Pyura squamata, 2 specimens

Pyura subuculata, 2 specimens

STA. 1837, Tasman Sea; 45°38′S, 160°12′W; December 11, 1966; 4859–4868 meters; 5′ Blake trawl.

Bathyoncus enderbyanus, 1 specimen

Cruise 27

STA. 1978, south of Tasmania; 51°50′S, 150°27′E; February 21–22, 1967; 4213–4218 meters; 5′ Blake trawl.

Molgula millari, 1 specimen, new species

USS *Glacier*

Collected by W. L. Tressler

Deep Freeze II, 1½ miles west of Inaccessible Island, McMurdo Sound; 77°40′S, 166°14′E; November 4, 1956; orange peel grab.

Aplidium radiatum, 1 colony

USS *Staten Island*

Oceanographic cruise, 1962–1963; biological investigations; Antarctic Peninsula, January–March 1963; collection made by Waldo L. Schmitt with assistance of ship's personnel and survey team.

STA. 7/63, Antarctic Peninsula, Anvers Island, Arthur Harbor; 64°46′S, 64°04′W; January 25, 1963; 21–31 meters; dredged, stiff blue mud, clams, worm tubes, and amphipods.

Plate 1

Pyura georgiana with pycnogonid, *Ammothea gibbosa* (*Staten Island* Sta. 61/63; collected by Dr. Waldo L. Schmitt).

Aplidium circumvolutum, 1 colony
Synoicium adareanum, 1 colony
Eugyra kerguelenensis, 14 specimens
(See also Kott, 1969, p. 20)

STA. 24/63, Antarctic Peninsula, off Danco Coast, Paradise Harbor; 64°49′18″S, 62°51′W; February 4, 1963; 75 meters; dredge.

Aplidium radiatum, 2 specimens
Didemnum biglans, 3 large pieces
Pyura discoveryi, 7 specimens
Molgula gigantea, 1 specimen

STA. 61/63, South Shetland Islands, anchorage, False Bay, Livingston I.; 62°42′S, 60°22′W; February 25, 1963; 31 meters; dredge.

Synoicium adareanum, 4 colonies
Ascidia challengeri, 1 specimen
Pyura georgiana, 1 specimen (with pycnogonid predator, *Ammothea gibbosa*)
Pareugyrioides ärnbäckae, 3 specimens

STA. 66/63, Antarctic Peninsula, Port Lockroy, Wiencke Island; 64°49′05″S, 63°30′05″W; March 1, 1963; 62 meters.

Cystodytes antarcticus, 2 colonies
Tylobranchion speciosum, 1 colony
Aplidium circumvolutum, 1 colony
Synoicium adareanum, 3 colonies
Didemnum biglans, 1 colony
Polysyncraton chondrilla, 2 colonies
Styela wandeli, 1 specimen
(See also Kott, 1969, p. 21)

SYSTEMATIC DISCUSSION

Order ENTEROGONA

Suborder APLOUSOBRANCHIA Lahille

Family Clavelinidae Forbes and Hanley

Subfamily Holozoinae Berrill

Sycozoa sigillinoides Lesson

Sycozoa sigillinoides Lesson, 1830, p. 436.—Kott, 1969, p. 26, and synonymy.

New records. Off southern Chile: *Eltanin* Sta. 959 (92–101 meters).

Distribution. The present records fall well within the known vertical and horizontal range for this ubiquitous southern species.

Distaplia cylindrica (Lesson)

Figs. 1–3

Holozoa cylindrica Lesson, 1830, p. 439.
Distaplia cylindrica; Kott, 1969, p. 29, and synonymy.

New Records. Ross Sea: Deep Freeze I, *Edisto* Sta. ? (644 meters); Sta. 6 (100 meters); *Atka* Sta. 4 (610 meters). Between Budd and Knox Coasts: *Atka* Sta. 24 (46 meters). South Shetland Islands: *Eltanin* Sta. 436 (73 meters). South Orkney Islands: *Eltanin* Sta. 1058 (650–659 meters). South Georgia: *Eltanin* Sta. 671 (220–320 meters). North of South Sandwich Islands: *Eltanin* Sta. 1581 (148–201 meters).

Distribution. The present records fall within the previously known geographic range for the species, but the maximum recorded depth is extended from 439 to 659 meters.

Description. Only disintegrated colonies and larvae are available from *Eltanin* Station 671. Although some delicate test material and traces of a stronger outer cuticle are present in the debris, there is no indication of the form of the colony nor of the arrangement of zooids in it. Most of the zooids in the debris appear to be relatively immature, with a maximum length of 3 mm and a posterior abdominal stolon. The atrial opening persists as an anteriorly directed siphon with a circular sphincter. There are six lobes on both atrial and branchial apertures, but these are not readily apparent. Longitudinal muscles extend along the anterior aspect of the atrial siphon in the region homologous with the atrial lip in more mature zooids. There are about 20 very fine longitudinal muscle bands that anastomose on the thorax. More mature zooids in which the atrial aperture is expanded into a wide opening with a rounded anterior lip are also available among the debris but are less common. In all zooids there are 4 rows of about 22 long stigmata crossed by parastigmatic vessels. The stomach is of the usual form, tapering to the intestine, and lined with longitudinal striations. There is an oval gastrointestinal reservoir and a rosette of about 15–20 testis lobes in the loop of the gut. Colonies from *Eltanin* Sta. 1058 are typically long and narrow. Colonies from *Atka* Sta. 4 are rounded and irregular, with enormously branched root systems. These may represent regenerating colonies, since mature zooids are not present in the test, although there are developing zooids, the products of vegetative reproduction.

Some young colonies are available from *Edisto* Sta. 6 and from *Atka* Sta. 24 as small clavate lobes joined basally. Varying numbers of typical zooids, sometimes with an incipient brood pouch, are present, opening on the free end of the lobe. There is no differentiation between the head and stalk of the lobe. The stalk, which contains the posterior vascular stolons of the colony in an active state of vegetative reproduction, may extend for some distance along any surface to which the colony is fixed. In more mature colonies the stalk may expand across the surface of the substrate and increase in height to form investing colonies or sessile cushions that accommodate large numbers of zooids opening on the surface and with posterior abdominal vascular stolons curved into the basal half of the test. Occasionally the upright lobe expands to a rounded head on a short wide stalk and accommodates increasing numbers of zooids and zooid systems.

Larvae, present in the debris from *Eltanin* Sta. 671, are identical with those previously described and have four rows of the full number of stigmata (about 22), although there are no parastigmatic vessels. The larval tail is short and narrow and the larval test is clear, with adherent sand grains. The larvae appear to be in brood pouches, although in no case is more than a single larva present.

Remarks. The zooids of this species are distinguished by the parastigmatic vessels and by the relatively few and large larvae with a relatively short tail. The colony form varies, though the type of colonies present from *Edisto* Sta. 6 and *Atka* Sta. 24 indicates that these variations are largely due to age and can be affected by the shape of the substrate on which the colony develops. Accordingly, where large plane surfaces are available, the stalk of the colony extends

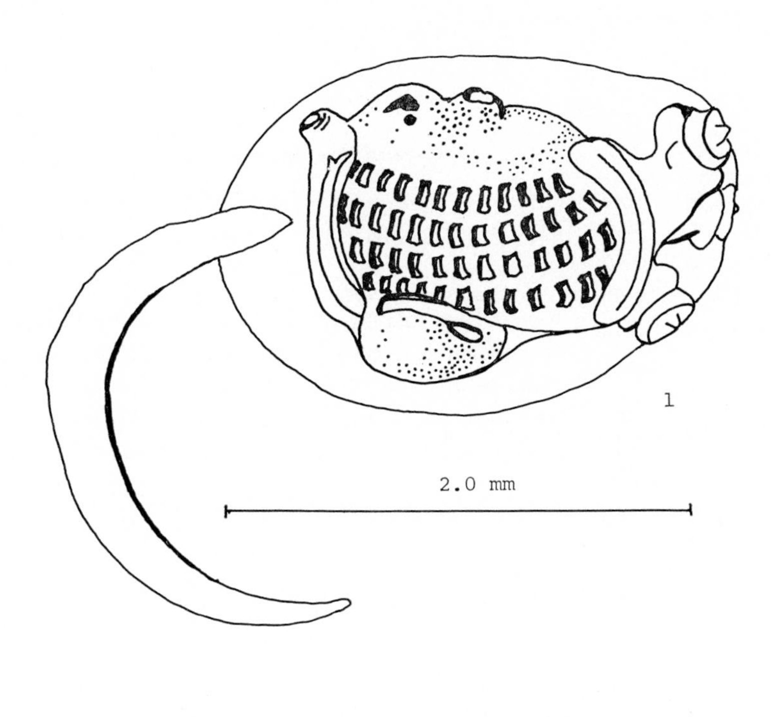

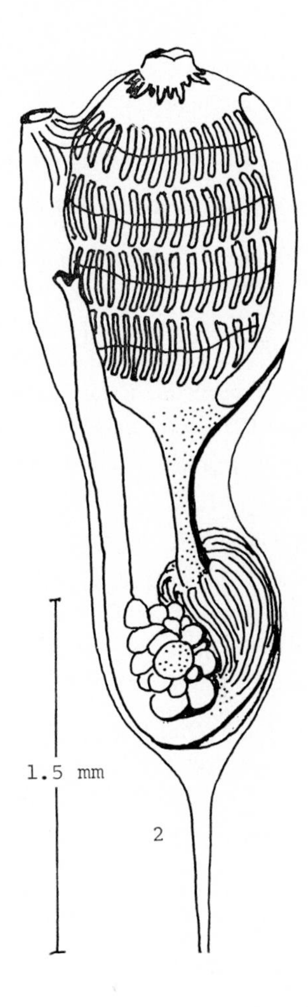

Distaplia cylindrica

1. Larva (from debris, *Eltanin* Sta. 671).
2. Young zooid (from debris, *Eltanin* Sta. 671).
3. Atrial opening of older zooid, showing muscles.

across this surface, and increasing numbers of zooids derived from vegetative reproduction cause the stalk to expand and form the base of a sessile cushion or investing sheet. Alternatively, where the surface is limited, the colony extends upward from the stalk to form the typical elongate cylinder previously described.

These investing colonies bear considerable resemblance to *Distaplia colligans* Sluiter [Kott, 1969], which is distinguished only by its darker purple-brown pigmentation, present even in preserved specimens.

The debris described above from South Georgia apparently represents a disintegrating colony of *D. cylindrica* from which larvae are released to metamorphose into adults. The presence of these young adult zooids in the debris with the mature larvae, together with the large size of the larva and relatively small tail, suggests that larvae are free-swimming for only a very limited time, if at all. However, both otolith and ocellus are present and may influence the selection of a site for settlement.

Protoholozoa pedunculata Kott

Protoholozoa pedunculata Kott, 1969, p. 35.

New records. South of South Georgia: *Eltanin* Sta. 1537 (2886–3040 meters); Sta. 1585 (2869–3038 meters). East of South Sandwich Islands: *Eltanin* Sta. 1571 (3947–4063 meters). Southeast Pacific basin: *Eltanin* Sta. 1621 (4419–4804 meters); Cruise 25, Sta. 359 (4682 meters). Southwest Pacific basin: *Eltanin* Cruise 25, Sta. 366 (5340 meters).

Distribution. Previous records have suggested that this species was confined to the Southeast Pacific basin and the deeper waters of the Scotia Sea. The previously known range of the species is therefore extended to the Southwest Pacific basin and possibly into deeper waters east of the Scotia Ridge.

Description. Zooids are present in colonies from *Eltanin* Stas. 1537, 1585, and 359 (Cruise 25). Only from the latter station, however, can the structure of the zooids be discerned, and here there are 3 transverse vessels instead of the 2 previously described for the species. The gonads in this specimen extend from the gut loop into the anterior part of the posterior abdominal extension. The stomach has internal longitudinal striations but no external folds.

A stalk only is available from Sta. 1571; and in the colony from Sta. 1621 all zooids are disintegrated and their structure is not discernible. In all colonies except that from Sta. 1537 the stalk is longer, with harder test than previously described for this species. The colony from Sta. 1585 is the largest, the stalk being 16 cm long and the maximum diameter of the head 3 cm. Larvae are present in colonies from Stas. 1585 and 359.

Remarks. While only 2 transverse branchial vessels have been previously identified in this species, there are 3 clearly apparent in the colony from Sta. 359. Although the latter condition could represent a different species or an individual variation in the same species, it is more likely to be the typical number of transverse vessels for the species. Owing to the poor condition of zooids when recovered from the depths in which they occur, the structure of the branchial sac is difficult to discern, and the third branchial vessel could have been overlooked. The absence of stomach folds in the present specimen may be due to its extended condition.

Family Polycitoridae Michaelsen

Cystodytes antarcticus Sluiter

Cystodites antarcticus Sluiter, 1912, p. 460.

Cystodites antarcticus; Kott, 1969, p. 37, and synonymy.

New record. Antarctic Peninsula, Port Lockroy, Wiencke Island: *Staten Island* Sta. 66/63 (62 meters). This record adds two colonies to the one colony previously reported from this station [Kott, 1961, p. 21].

Polycitor magalhaensis (Michaelsen)

Figs. 4–6

Paessleria magalhaensis Michaelsen, 1907, p. 69.

Sigillina (Paessleria) magalhaensis; Kott, 1969, p. 39, and synonymy.

New records. South Sandwich Islands: *Eltanin* Sta. 1581 (148–201 meters). Between South Shetland Islands and South Orkney Islands: *Eltanin* Sta. 1003 (210–220 meters).

Distribution. Previously known from a single specimen taken in the Strait of Magellan [Michaelsen, 1907]. The depth at the type locality is not given; it was probably less than that recorded above. The present specimens extend the range down to the Scotia Ridge.

Description. The colonies form large, rounded but irregular, and sometimes subdivided lobes of up to 4 cm in diameter and 5 cm in height, rising from a flattened spreading base; or they form cushionlike or clavate lobes. A heavy investment of sand in the surface layer of the test results in a hard, even surface and affects the color of the colonies, which are dark gray to black. Below the surface layer the test is less firm, gelatinous, and transparent, with sparsely distributed sand grains. There are whitish, opaque spherical bodies evenly distributed in the surface layer of the test and visible externally. The surface test is interrupted by the parallel compartments supporting the zooids. Basally and in the central core of the test, where the long compartments are less regular, the test has a more spongy consistency.

Zooids are arranged fairly densely, parallel to one another, and at right angles to the surface test for

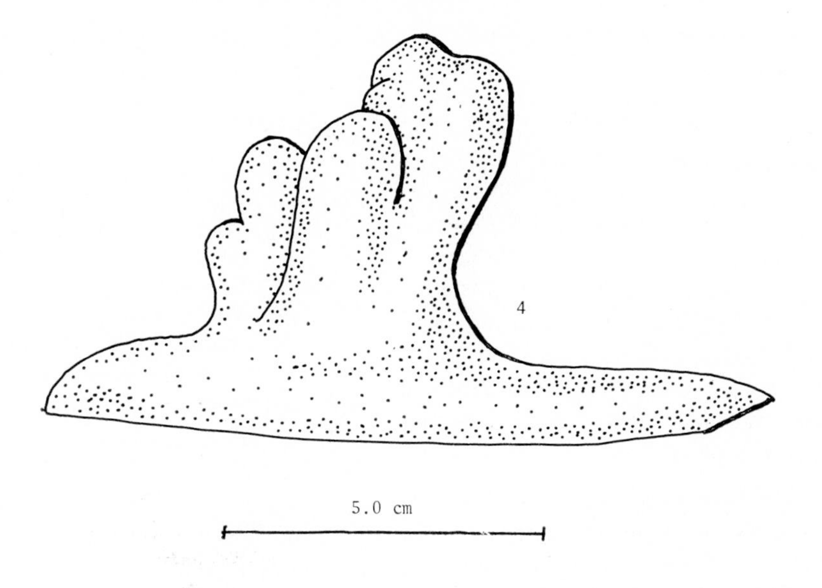

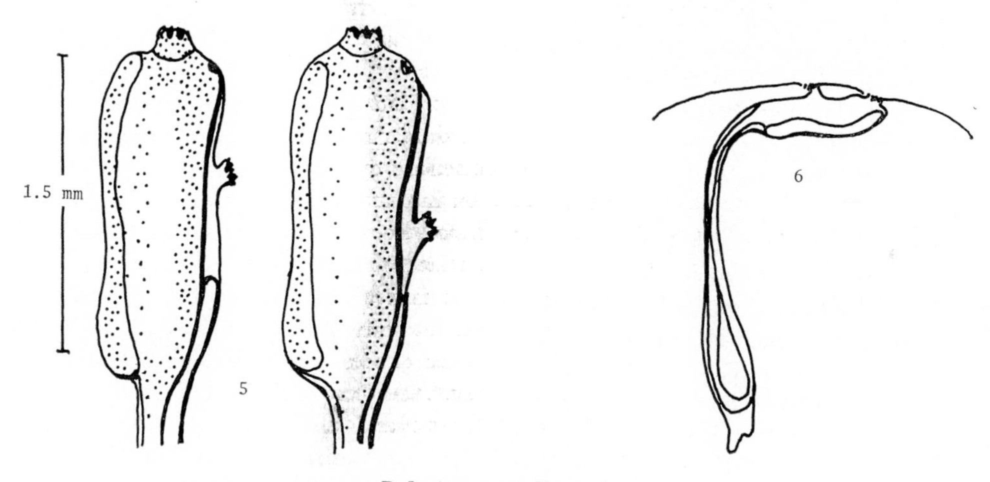

Polycitor magalhaensis

4. Colony (*Eltanin* Sta. 1581).
5. Thoraces of two zooids, showing alternative positions of atrial siphon.
6. Diagram of zooids opening to surface.

the greatest part of their length, although the thorax is bent at right angles to the rest of the zooid so that the dorsal border lies parallel to the surface of the colony. In the present colonies the zooids near the surface of the colony are contracted and filled with vegetative cells, and the structure of the zooids is discernible only from zooids in the basal test. These zooids are probably the result of vegetative reproduction from older zooids in the surface of the lobes and will probably later occupy new lobes developing from the basal spreading test. The thorax is very short, 1 to 1.5 mm long, while the long, narrow abdomen measures up to 2 cm. The branchial aperture is terminal and 6-lobed; and the atrial aperture, also bordered by indistinct lobes, is supported on a short siphon from the middorsal border of the thorax. Sometimes the atrial siphon projects at right angles to the length of the thorax, sometimes it is produced anteriorly, and sometimes the anterior lobes of its margin are slightly extended. These variations in the form and orientation of the atrial siphon are undoubtedly caused by the position of the thorax in relation to the curvature of the surface of the lobe. Both apertures open separately onto the surface of the colony, although these openings are inconspicuous and not evident on the surface. Fine longitudinal muscle bands extend down the thorax and along both sides of the abdomen. Circular muscle bands are present around the thorax and the anterior part of the abdomen. There are 14 to 18 rows of 12 very short rectangular stigmata. The gut loop is very long and narrow, extending the whole length of the abdomen. The esophagus expands into a stomach in the posterior part of the abdomen. No gonads are evident in the present colonies. There is no posterior abdominal vascular extension.

Remarks. Michaelsen [1907] described a single colony as the type species for the subgenus *Paessleria.* In Michaelsen's colony, however, the zooids were very contracted and, although he describes 3 rows of long narrow stigmata in the branchial sac, he qualifies this statement by saying that the inner details of the thorax are not certain. Van Name [1945] has also drawn attention to the fact that Michaelsen has not represented the stigmata in his figure [Michaelsen, 1907, pl. 3, fig. 11]. In view of the similarity in size and form of the colonies and the size, form, and relationships of different parts of the zooid, the present specimens are assigned to Michaelsen's species, which was described from a specimen in which the number of rows of stigmata was not discernible. The position and form of the atrial siphon is especially distinctive, and it seems unlikely that these records from adjacent geographic areas could represent 2 different species. Previously the only confirmed records for the family Polycitoridae in the Antarctic were the species *Cystodytes antarcticus* Sluiter and *Sigillina moebiusii* Hartmeyer [Kott, 1969]. Doubtful records for 2 other species, *Polycitor glareosus* (Sluiter) and ? *Polycitor clava* (Harant and Vernières) [Kott, 1969] do exist, but in neither case has the identification been confirmed. *Polycitor glareosus* is distinguished by a very short esophageal neck and is closely related to *Cystodytes. Polyclinum clava* Harant and Vernières, 1938, was assigned tentatively to the genus *Polycitor* [Kott, 1969] on the grounds of its similarity to *Polycitor giganteum* (Herdman), a circum-Australian species with a soft transparent test that distinguishes it from *P. magalhaensis.*

The spongy consistency of the basal part of the test traversed by canals for the zooids is most characteristic of the present species.

Family Polyclinidae Verrill

Subfamily Euherdmaniinae Seeliger

Tylobranchion speciosum Herdman

Figs. 7–9

Tylobranchion speciosum Herdman, 1886, p. 157.—Kott, 1969, p. 41, and synonymy.

New records. Antarctic Peninsula, Port Lockroy, Wiencke Island: *Staten Island* Sta. 66/63 (62 meters). South Shetland Islands: *Eltanin* Sta. 432 (935–884 meters); Sta. 436 (73 meters). Between South Shetland Islands and South Orkney Islands: *Eltanin* Sta. 1003 (210–220 meters). East of South Orkney Islands: *Eltanin* Sta. 1545 (2355–2897 meters). Scotia Sea area, north of South Georgia: *Eltanin* Sta. 732 (220–265 meters). North of South Sandwich Islands: *Eltanin* Sta. 1581 (148–201 meters). South of Tierra del Fuego: *Eltanin* Sta. 740 (384–494 meters).

Distribution. The previously recorded maximum depth for this species was 439 meters off the Knox Coast [Kott, 1969]. The new record from *Eltanin* Sta. 1545 considerably extends this depth range.

Description. The present colonies from Sta. 1545, in deeper waters than previously known for the species, are cylindrical lobes of 3 cm in length and 1.5 cm in diameter, sometimes extending from a common base. Zooids open around the border of the upper surface. There is some flattening of the lobes, but this is probably an artifact. The test is semitransparent and gelatinous. The stalk has a tough, opaque outer cuticle, which is absent in the region of the head. Zooids are about 2 cm long; the extended thorax and abdomen are each about 4 mm. The atrial aperture is from the anterodorsal corner of the body. On the thorax are 8 thick longitudinal muscle bands, the most ventral of which have branches extending across the ventral surface. These longitudinal muscles extend along both sides of the abdomen. The dorsal lamina has pronounced languets. There are about 25 rows of 12 to 20 stigmata. The transverse vessels support rounded papillae, sometimes biramous but often undivided. The papillae appear to correspond to each of the rectangular stigmata. The stomach is elliptical or rounded, according to the state of contraction of the abdomen; and there are traces of folds anteriorly where it joins the esophagus. The rectum extends almost to the base of the atrial opening, and the anus is bilabiate. The gonads are of typical form, clustered behind the gut loop. The colonies from *Eltanin* Sta. 432, also from

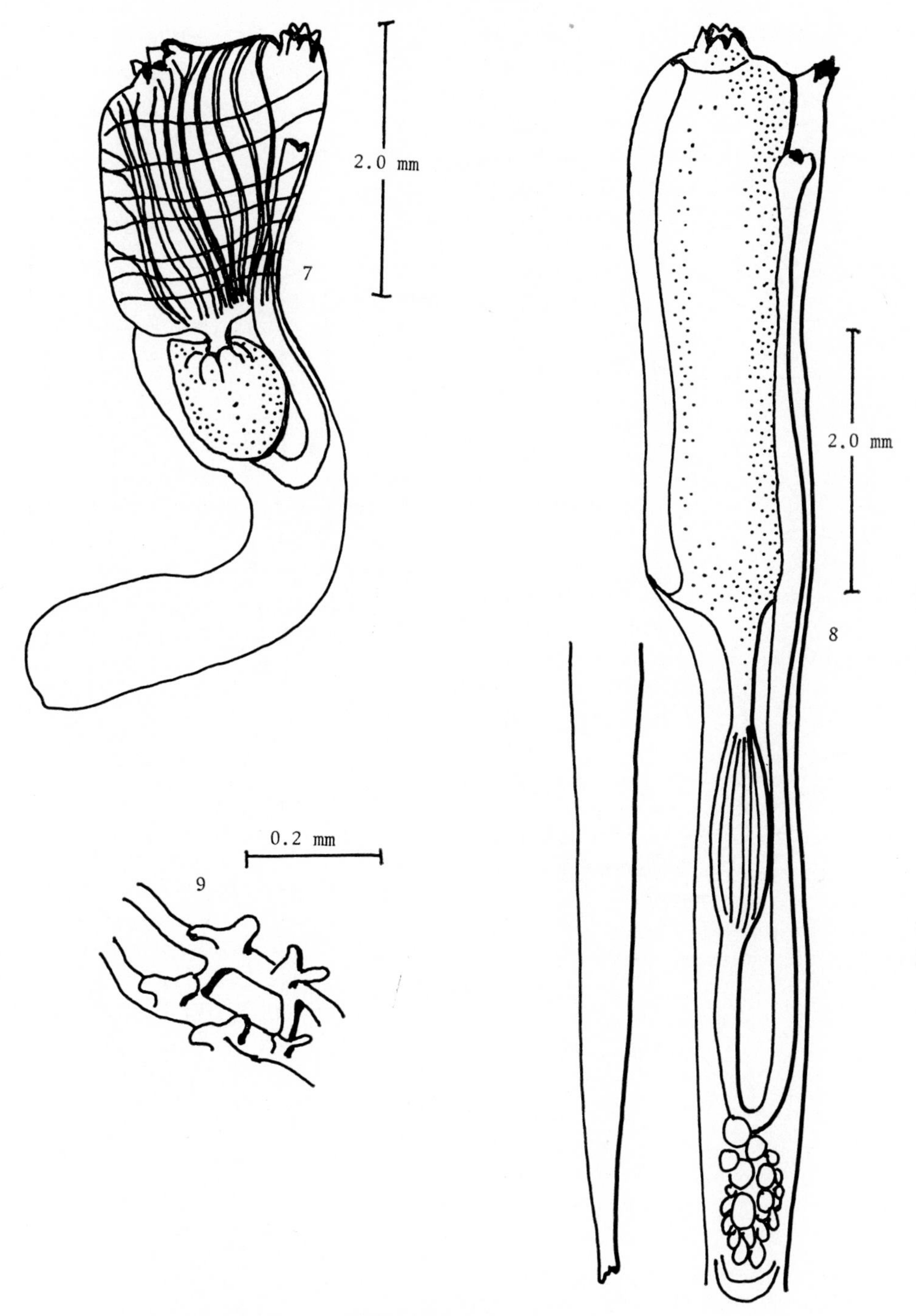

Tylobranchion speciosum

7. Thorax contracted, musculature showing (*Eltanin* Sta. 1545).
8. Proportions of extended zooid.
9. Portion of branchial sac.

deep water, are more regularly cylindrical than is general for this species. However, the free end of the head is very much damaged in all specimens, and only few zooids are present.

Remarks. Although these specimens have been taken from considerably deeper waters than previously recorded, they fall within the range of variation described for the species. The number of stigmata is

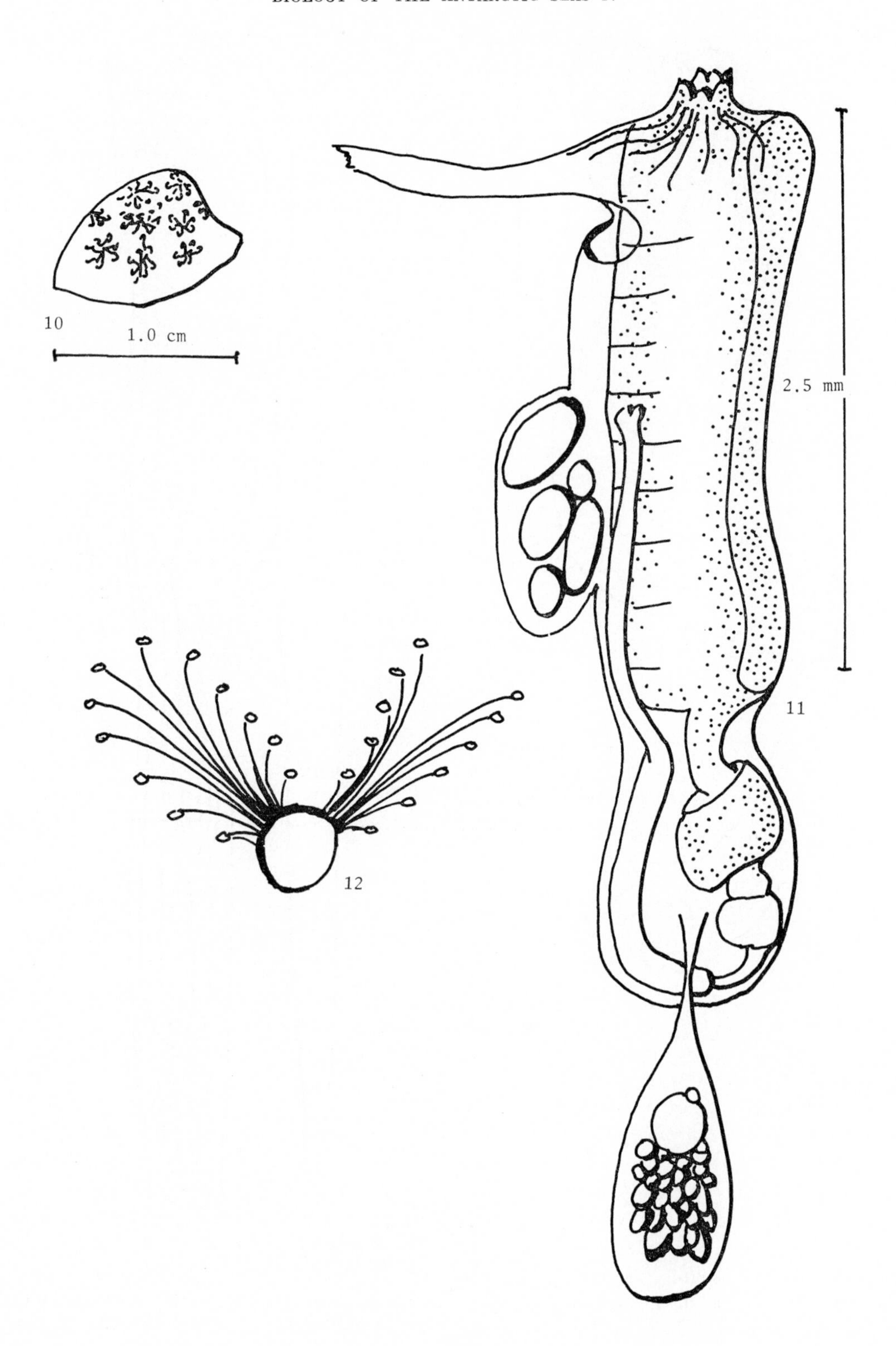

Polyclinum sluiteri

10. Colony.
11. Diagram showing arrangement of zooids around a common cloacal aperture.
12. Zooid.

slightly fewer than in species from more shallow waters, but this is not considered a characteristic sufficient to justify a separate species.

Subfamily POLYCLININAE Adams and Adams

Polyclinum sluiteri Brewin

Figs. 10–12

Polyclinum sluiteri Brewin, 1956, p. 124.

New record. Albatross Cordillera: *Eltanin* Sta. 1691 (362–567 meters).

Distribution. The species has previously been recorded only from the Chatham Islands in 129 meters [Brewin, 1956]. The present record suggests a contemporary or recent distribution along the Albatross Cordillera, through the Antarctic, and along the Macquarie ridge to New Zealand. Since the species has not been taken on the continental shelf of the antarctic continent, it is probable that its occurrence in widely separated locations further to the north represents relict populations.

Description. Sessile rounded colonies about 10 mm in diameter and 6 mm high. The test is very soft and semitransparent and has a tougher outer cuticle. The colonies are a light gray color when preserved in alcohol. There are two to three common cloacal apertures over the surface of the colony. These open into cloacal chambers from which canals radiate. Zooids are arranged along both sides of the cloacal canals, and their atrial lips are enormously extended along the roof of the canals almost to the cloacal openings. The atrial lips of zooids near the cloacal apertures are incorporated into the border of the opening. The length of the zooids is about 5 mm, of which the thorax is 2.5 mm and the abdomen and the saclike posterior abdomen are each 1 mm. There are 10 longitudinal muscles radiating from the branchial siphon and extending about halfway down on each side of the thorax. Muscles from the dorsal part of the branchial siphon extend along the atrial lip. The atrial opening is present in the anterior quarter of the dorsal surface. The ribbonlike atrial lip, varying in length, is longest where the zooids are more distant from the cloacal opening. It terminates in a crenellated tip. There are about 20 rows of about 20 rectangular stigmata. Rounded papillae are present on the transverse vessels, about 4 corresponding to each 3 stigmata. The stomach is rounded and smooth, and the rectum terminates about halfway up the thorax in a two-lipped anus. There is a large cluster of testis lobes posterior to the ovary in the saclike posterior abdomen, which is constricted off from the abdomen. About five developing embryos are present in a pouch from the posterior half of the thorax. This pouch appears to be a true brood pouch, accommodating a loop of the oviduct, rather than a simple expansion of the oviduct in the peribranchial cavity. Thus, in a well-developed brood pouch, the less well-developed embryos are present in the most proximal part of the oviduct, where it enters the pouch. Embryos of increasing maturity are present in the oviduct where it extends back into the pouch, before looping anteriorly. The most mature embryos present in the anterior or distal part of the oviduct are adjacent to the proximal portion, which enters the brood pouch and contains the least mature embryos. The oviduct then extends a short distance alongside the branchial sac.

Larvae. These are 0.5 mm with the usual three anterior papillae, four paired lateral ampullae, and paired clusters of vesicles dorsally and posteroventrally.

Remarks. *Polyclinum* species are very difficult to distinguish. They all have well-developed cloacal systems of a similar type, with only a few distinguishing characters available in the zooids. The species are often separated from one another on the basis of characteristics that in other genera are known to vary according to substrate and maturity. Other characteristics that have been used to characterize species are also not reliable. For instance, the extent to which the abdomen is oriented horizontally varies with the contraction of longitudinal muscles and the condition of the zooid in preservative. Similarly, the extent to which the ovary is embedded in the testis lobes appears to vary with the extent to which the testis is developed.

The present species closely resembles the ubiquitous *P. constellatum* from the Indian, Pacific, and Atlantic oceans and is distinguished from the latter species only by the two-lipped anal opening and the development of a brood pouch. The species resembles *Polyclinum marsupiale* Kott [1963] from southern and eastern Australia in the presence of a brood pouch, but it is distinguished by the shape of the colony, by the smaller zooid, and by the median ampullae in the larvae of the Australian species.

In addition to the present species, *P. cerebrale*, *P. michaelseni*, and *P. novaezelandiae* have been described from New Zealand waters. *Polyclinum cerebrale* Michaelsen, 1924, has zooids of similar size but is distinguished from the present species and others by the infolding of the surface of the test. *Polyclinum novaezelandiae* Brewin, 1958, can be distinguished generally by its larger zooids (12 mm long) and from the present species by the encrustation of sand in the surface test. *Polyclinum michaelseni* Brewin, 1956, is a stalked species with a single terminal cloacal opening.

The present specimens agree with Brewin's description of *Polyclinum sluiteri* in all characters with the exception only of the brood pouch, which Brewin has not described for her specimens, although the position and arrangement of the embryos in the mid-dorsal section of the peribranchial cavity suggest that they are retained there in the oviduct and that the development of the actual brood pouch is variable for this species. The situation is not altogether satisfactory, inasmuch as *P. michaelseni* and *P. novaezelandiae* are known only from single records and *P. sluiteri* and *P. cerebrale* have been taken only twice. Further collecting is needed to provide material with which to extend our knowledge of species variability within this genus, especially in New Zealand waters.

Aplidium abyssum Kott

Aplidium abyssum Kott, 1969, p. 47.

New record. Southwest Pacific basin: *Eltanin* cruise 25, Sta. 364 (3694 meters).

Distribution. This species is otherwise known only from the Peru-Chile trench. These two isolated occurrences may indicate a wide distribution in southern-hemisphere oceanic basins.

Description. The present colony is small and upright, with an irregular spreading base and a mushroom-like head. The base is extended into rootlike processes. The test is glassy and transparent and contains some enclosed Foraminifera shells, which are especially dense in the basal half of the test before it expands into the head. There is a large common cloacal opening in the middle of the upper surface of the head.

Zooids are small, and in these specimens there are no intact thoraces. There are four distinct stomach folds and a short posterior abdomen equal in length to the abdomen. Testis lobes are clustered in the anterior half of the posterior abdomen. The ovary is anterior to the testis lobes.

Single larvae are present in the test, but it is probable that these have remained after the disintegration of the thoraces. They are large (1.0 mm long) and possess an otolith but no ocellus. The usual three median papillae are present anteriorly; they are surrounded by ampullary vesicles, sometimes extending by narrow stalks from the body of the larva, although many of these vesicles are free in the larval test. These ampullary vesicles are often found in the family Polyclinidae. The ampullae, which are generally also present in larvae of this family, were not observed in the present species. The larvae are similar to but larger than those of *Aplidium variabile.*

Remarks. Although the form of the thorax could not be distinguished in the present specimen, the abdomen and posterior abdomen are identical with those previously described for this species. The shape of the colony differs from that of the type specimen from the Peru-Chile trench, but it is probable that this is a variable character in the species.

Aplidium fuegiense Cunningham

Aplidium fuegiense Cunningham, 1871, p. 66.—Kott, 1969, p. 47, and synonymy.

New records. Off Tierra del Fuego: *Eltanin* Sta. 740 (384–494 meters); Sta. 1603 (256–269 meters). Off southern Chile: *Eltanin* Sta. 958 (92–101 meters). South Georgia: *Eltanin* Sta. 1535 (97–101 meters). East of South Orkney Islands: *Eltanin* Sta. 1079 (593–598 meters). South Shetland Islands: *Eltanin* Sta. 436 (73 meters).

Distribution. The present record from the South Orkney Islands is only the second occasion on which the species has been taken south of South Georgia [Kott, 1969]. It is most commonly recorded from the Magellanic region and south to South Georgia and at Kerguelen.

Description. Generally the colonies in the present collection have a firm gelatinous test and small thin zooids that cross one another and have four folds in the stomach wall. Only in the spherical colonies from the South Shetland Islands and east of the South Orkney Islands is the test softer and less firm. Colonies from Tierra del Fuego are cylindrical, clavate, or long and pointed. The colonies from southern Chile are cushionlike and rounded, and those from South Georgia are globular. Small yellow spherical bodies are present in the surface layer of test of specimens from *Eltanin* Stas. 958 (southern Chile) and 1603 (Tierra del Fuego). Colonies from South

Georgia and southern Chile are encrusted with sand, while the other colonies in the collection have a smooth, naked surface.

Remarks. Colonies in the present collection show the same variation in form as previously noted for the species. The crowded small zooids, with four stomach folds, crossing one another in the test, and the small, often undivided atrial languets from the body wall anterior to the opening are characteristic.

Aplidium irregulare (Herdman)

Amaroucium irregulare Herdman, 1886, p. 223.
Aplidium irregulare; Kott, 1969, p. 50, and synonymy.

New records. Off southern Chile: *Eltanin* Sta. 958 (92–101 meters). North of South Shetland Islands: *Eltanin* Sta. 1001 (238 meters).

Distribution. These records are within the range previously recorded for the species.

Description. Colonies are rounded, flattened, or investing and up to 4 cm thick. In the best-developed investing colonies from Sta. 1001 the surface is raised into rounded lobes and the zooids curve up from the continuous basal test into the lobes and open around their upper surface where there are 2 to 3 common cloacal openings. In all colonies the circular systems developing into double row systems are more apparent than in *A. fuegiense.* There are always a large tripartite atrial lip from the upper border of the opening; 16 to 20 stigmata; and 6 stomach folds, sometimes irregular and slightly convoluted.

Larvae. Present in the colonies from *Eltanin* Sta. 958 are larvae as previously described with median and lateral paired ampullae and paired dorsal and ventral vesicles.

Remarks. This highly variable species is easily confused with *A. fuegiense,* which occurs in the same geographic area. Generally, *A. irregulare* is distinguished by a larger number of stomach folds, a larger number of stigmata in each row, more easily identified systems, larger atrial tongue, and by its more extensive, sometimes investing, colonies.

Aplidium variabile (Herdman)

Fig. 13

Amaroucium variabile Herdman, 1886, p. 216.
Aplidium variabile; Kott, 1969, p. 51, and synonymy.

New records. Off Tierra del Fuego: *Eltanin* Sta. 740 (384–494 meters); Sta. 969 (229–265 meters). Off southern Chile: *Eltanin* Sta. 958 (92–101 meters). East of North Island, New Zealand: *Eltanin* Sta. 1498 (101 meters).

Distribution. These records slightly extend the range previously recorded for this species. The specimens from *Eltanin* Sta. 740 represent the greatest depth from which the species has ever been taken. *Eltanin* Sta. 1498 is in the relatively shallow waters of the Chatham Rise, continuous with the continental shelf around the Chatham Islands, where Michaelsen has previously reported the species [Michaelsen, 1924]. Although this record represents the most northerly extent of the species, it falls within the circumsubantarctic range previously indicated.

Description. Colonies are investing clavate lobes from a common base. Zooids are parallel to one another and form circular systems around common cloacal opening. The test is transparent and fairly soft, and sometimes some sand is included, although there is never more than very sparse sand on the surface of the test. The thorax is narrow, and the stomach has 12 to 15 folds.

Remarks. The identification of this species with such highly variable colonies depends on the parallel arrangement of the zooids, the numbers of stigmata in each row, and especially the number of folds in the stomach.

Aplidium radiatum Sluiter

Fig. 14

Psammaplidium radiatum Sluiter, 1906, p. 25.
Aplidium radiatum; Kott, 1969, p. 54, and synonymy.

New records. Ross Sea: *Edisto* Sta. 6 (100 meters). McMurdo Sound, *Glacier,* Deep Freeze II (? meters). Antarctic Peninsula: *Edisto* Sta. 38 TR 20 (104 meters); *Staten Island* Sta. 24/63 (75 meters). Between South Shetland Islands and South Orkney Islands: *Eltanin* Sta. 1003 (210–220 meters). South Orkney Islands: *Eltanin* Sta. 496 (234–242 meters).

Distribution. The present records fall within the known range for this species. The record from Enderby Land [Kott, 1954] is the only one known from outside the western Antarctic.

Description. Colonies are rounded or irregular and lobed. They are fixed either by a small area of their basal surface or by a stalk, which may be only a short extension of the head from its basal surface or which may range up to 3 times the length of the head. Zooids are arranged in double rows radiating from conspicuous common cloacal openings about 2 cm apart over the upper surface of the lobe. Only in smaller colonies are single common cloacal openings present in the center of the upper surface. Zooids are fairly close together and appear superficially as circular areas in the slightly sand-encrusted surface test. The internal test is semitransparent, and sparse sand is included. Zooids extend radially in from the surface of the colony and down into the stalk where the amount of test between the zooids becomes less as they converge. The test of the lower part of the head and of the stalk is spongelike in consistency and is perforated by crowded canals or spaces to accommodate the zooids. This arrangement of the zooids accounts, to some extent, for the very soft nature of the internal test. In alcohol-fixed colonies the internal test is not quite so soft, although its spongelike nature is more easily apparent.

Colonies may be black to brownish gray where spherical black pigment cells in the zooids confer their color on the whole colony when it is viewed from the surface. When these pigment cells are not present, colonies appear buff-colored. Zooids have wide thoraces with 16 to 20 rows of 15 to 20 long rectangular stigmata, sometimes crossed by parastigmatic vessels. The atrial aperture opens directly into the common cloacal canal; and, when open, the muscular languet from the upper border is stretched across the top of the opening and only the tip of the median lobe is apparent. The stomach has 6 to 8 sometimes irregular and not well-defined folds.

Remarks. In earlier descriptions this species was characterized by spherical colonies with a single terminal common cloaca from which double rows of zooids radiate. It is apparent from the colonies in the present collection that the number of common cloacal openings increases with the size of the colony and that the regular rounded form of the head may be modified by development of lobes from the surface, each supporting one or more systems. The species is distinguished from others with similar zooids by their radial arrangement all around the head and projecting down into the stalk and by the soft or spongelike consistency of the internal test, increasing basally and reminiscent of the condition found in *Polycitor magalhaensis.*

Aplidium circumvolutum Sluiter

Fig. 15

Psammaplidium circumvolutum Sluiter, 1900, p. 14.
Aplidium circumvolutum; Kott, 1969, p. 57, and synonymy.

New records. Antarctic Peninsula: *Staten Island* Sta. 7/63 (21–31 meters); Sta. 66/63 (62 meters). South Shetland Islands: *Eltanin* Sta. 432 (935–884 meters); Sta. 436 (73 meters); Sta. 1003 (210–220 meters). Off east coast, South Island, New Zealand: *Eltanin* Cruise 25, Sta. 370 (95 meters).

Distribution. A wide circumpolar distribution in the Antarctic and Subantarctic is recorded for this species. These new records fall well within the known geographic and vertical range.

Description. Colonies are small and rounded (*Eltanin* Stas. 432, 1003) or form extensive investing sheaths always fixed by the greater part of their base. The surface test is impregnated by sand, and sand is also present throughout the test. The surface test is raised over the anterior end of each zooid. Irregular double-row systems of zooids are present, and conspicuous common cloacal openings are scattered over the surface. The surface of the test is depressed into furrows separating adjacent systems of zooids. Zooids are always placed vertically in the test and are parallel to one another. The posterior abdomen is rather short and is drawn up alongside the abdomen by contraction of the longitudinal muscle bands. The atrial lip is anterior to the border of the opening. It is broad and fleshy and has a tripartite tip, each lobe often irregular or subdivided. There are about 10 to 12 rows of stigmata in the present specimens and about 12 to 15 longitudinal muscle bands on the thorax, the most dorsal bands coalescing with the sphincter around the atrial aperture. The stomach is small and generally has 5 to 6 folds, although these are not present in the specimens from *Eltanin* Sta. 432. Typical larvae are present in colonies from Sta. 432.

Remarks. The form of the colonies and systems in this species is very variable. However, the larger investing sheets generally have more complicated double-row systems and may be older than the smaller colonies, in which the zooids appear to be arranged in circular systems around more prominent and more closely spaced common cloacal openings (*Eltanin* Stas. 432, 1003). The species may be identified, however, by the limited and even height of the colonies; the vertical and parallel arrangement of zooids opening to the upper surface but not on the sides of the colony; the broad, fleshy, muscular atrial lip from the body wall just above the atrial opening; and the bunched testis lobes in the posterior abdomen [Kott, 1969]. The larvae are also distinctive.

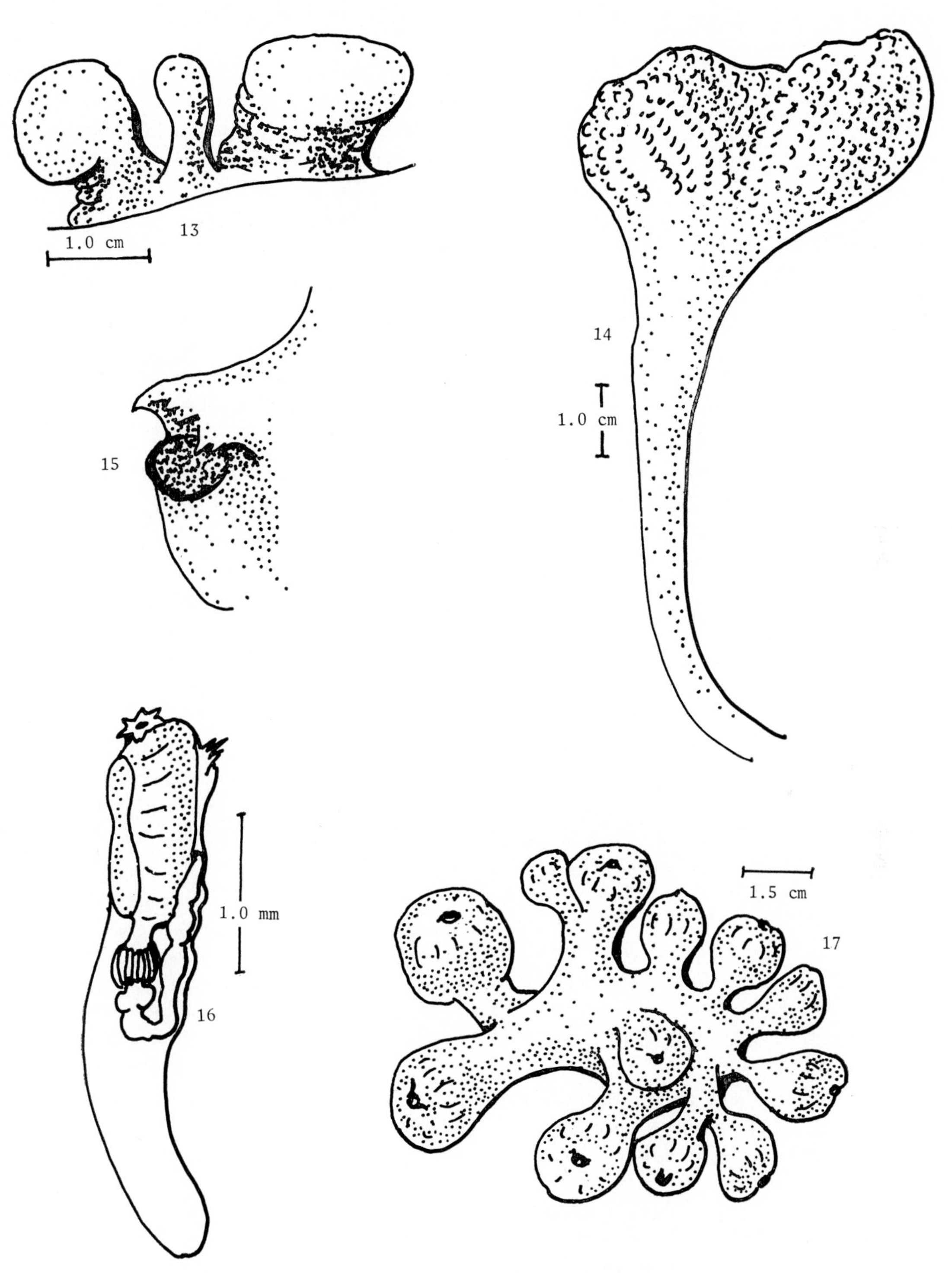

Aplidium variabile

13. Colony (*Eltanin* Sta. 969).

Aplidium radiatum

14. Long stalked colony (*Edisto* Sta. 6).

Aplidium circumvolutum

15. Atrial opening.

Aplidium stanleyi

16. Outline of contracted zooid (*Eltanin* Sta. 671).
17. Colony.

Kott [1969] has drawn attention to the similarity between this species and *Synoicium perreratum*. Although specimens from Sta. 432 in the present collection have no stomach folds, the zooids are still identified as *A. circumvolutum* by the larger number of muscle bands and their arrangement along the ventral surface of the abdomen and posterior abdomen.

Aplidium stanleyi Millar

Figs. 16–17

Aplidium stanleyi Millar, 1960, p. 41.—Kott, 1969, p. 58.

New record. South Georgia: *Eltanin* Sta. 671 (220–320 meters).

Distribution. Previously recorded from the Falkland Islands in 210–271 meters [Millar, 1960].

Description. Each colony in the present collection consists of about 12 small rounded lobes of 1–1.5 cm in diameter and up to 1.5 cm high, narrowing basally to join a central fleshy basal stolon continuous with the test of each lobe. The test throughout is firm, transparent, and free of sand; and there is no indication of attachment to the substrate. There is a single central common cloacal opening on each lobe surrounded by numerous zooids, although these are not in a single circle around the cloacal aperture. Zooids are short. The anterior border of the atrial opening is extended into a languet consisting of a median and 2 lateral pointed lobes. About 10 to 12 longitudinal muscle bands are present along both sides of the thorax; and these subdivide posteriorly into finer bands, forming a wide sheet along both sides of abdomen and posterior abdomen. There are about 10 rows of 15 stigmata and about 22 stomach folds. The posterior abdomen in the present specimens is short and filled with vegetative cells. The anal opening is 2-lipped.

Remarks. The colonies from the Falkland Islands [Millar, 1960] were less well developed than the present colonies. The species is distinguished, however, by the terminal cloacal openings, transparent test, basal fleshy stolon, lack of strong attachment to the substrate, and short zooids and number of stomach folds. The atrial languets in Millar's specimens were not well developed, but this may be due to the systems' being less well developed.

Aplidium stewartense (Michaelsen)

Macroclinum stewartense Michaelsen, 1924, p. 413.

Aplidium stewartense; Kott, 1969, p. 60, and synonymy.

New record. South Georgia: *Eltanin* Sta. 1535 (97–101 meters).

Distribution. The species has been taken previously from Stewart Island, Macquarie Island, and the Drake Passage. The present record extends its circumsubantarctic range to South Georgia. The maximum depth from which it has been taken is 115 meters.

Description. There are two small typical colonies in the present collection.

Remarks. The species is small, sandy, and inconspicuous and could easily be overlooked. This may explain the widely separated and limited records of its occurrence.

Aplidium recumbens (Herdman)

Amaroucium recumbens Herdman, 1886, p. 227.

Aplidium recumbens; Kott, 1969, p. 61, and synonymy.

New records. Falkland Islands–Burdwood Bank: *Eltanin* Sta. 339 (512–586 meters). South of Tierra del Fuego: *Eltanin* Sta. 740 (384–494 meters). North of South Sandwich Islands: *Eltanin* Sta. 1581 (148–201 meters).

Distribution. The present records confirm the circumsubantarctic distribution previously indicated. The northern limit of the species is off North Island, New Zealand, but it has not been taken south of the Antarctic Convergence.

Description. The colonies from *Eltanin* Sta. 740 are of characteristic fingerlike form. They are up to 6 cm high, 1 cm in diameter at the head end, and slightly reduced in diameter basally. The free end is slightly expanded into a rounded head onto which the zooids open in 2 to 3 circular systems each of 8 to 10 zooids surrounding a conspicuous common cloacal opening. The greater part of the length of the colony forms a transversely wrinkled stalk, and there is an incrustation of sand around the whole extent of the colonies. Colonies from the other stations, which may be younger, have lobes that are cylindrical and no more than 1 cm high. In these colonies sand is absent from the head of the lobe, and there are yellowish spherical marks in the surface test. In all zooids there are 12 longitudinal muscle bands on the thorax and 13 rows of about 15 stigmata in the branchial sac. The stomach is shield-shaped, but no folds are apparent.

Remarks. Specimens from the Falkland Islands are typical of the species; no new observations are available regarding the relationships of the species or of the significance of the absence of stomach folds. The zooids from the smaller colonies are identical; the colonies themselves, however, are more like those of *Aplidiopsis georgianum* and *Synoicium ramulosum.*

Aplidium caeruleum (Sluiter)

Figs. 18, 19

Amaroucium caeruleum Sluiter, 1906, p. 16.
Aplidium caeruleum; Kott, 1969, p. 62, and synonymy.

New record. Between South Shetland and South Orkney Islands: *Eltanin* Sta. 1003 (210–220 meters).

Distribution. The present record falls within the previously recorded vertical and horizontal ranges.

Description. The colonies from this station are small, with only a trace of the typical brilliant blue color around the atrial aperture of a single specimen. The outer layer of test is characteristically brittle with sand. Zooids are arranged in single systems around the upper surface of the head. The central common cloacal opening is conspicuous, but sand is not absent from the border of the opening. The atrial aperture is extended into a short siphon with a tridentate languet from the upper border of the opening. Parastigmatic vessels are present. There are 8 to 10 oblique stomach folds and, as is usual for the species, these are broken and interrupted along their length. The testis lobes are clustered in the posterior abdomen. Numerous eggs are present in the ovary, and there are embryos protruding from the posterodorsal corner of the thorax in contracted specimens.

Remarks. These specimens are typical of the species. Although the systems in these small colonies resemble, to some extent, those of *Aplidium recumbens,* they may be distinguished by their larger zooids, very soft internal test, and stomach folds.

Aplidium pellucidum new species

Fig. 20

Type locality. Off southern Chile: *Eltanin* Sta. 958 (92–101 meters). Holotype, colony (USNM 12012).

Description. The colony is rounded and gelatinous, investing a stalk (undetermined). Maximum thickness is 1 cm. The test is transparent and almost glassy, and the zooids are clearly visible. Zooids are arranged in more or less circular systems around common cloacal openings.

The zooids are long and slender; the thorax and abdomen are both 1.5 mm long, and the posterior abdomen is up to 6 mm long. The branchial aperture is 6-lobed and terminal, and the atrial aperture in the anterior one-third of the dorsal surface of the thorax has an anterior border that is produced into a languet with 3 terminal lobes. There are 8 longitudinal muscle bands on the thorax and 8 rows of about 10 rectangular stigmata. The stomach is large and rounded, with four not very distinct folds.

About 30 developing embryos are present in the peribranchial cavity. The larvae are small, with 3 anterior papillae, otolith and ocellus, 4 paired lateral ampullae, and corresponding median ampullae alternating with the papillae. There are also posteriorly projecting dorsal and ventral ampullae, each with median and paired lateral lobes.

Remarks. The larvae are similar to those of *A. fuegiense,* but the species may be distinguished by the nature of the colony and the consistency of the test. The soft gelatinous test and the larvae relate the species to *A. globosum* (Herdman) from Kerguelen. In the latter species, however, only 3 to 4 developing embryos are present in the peribranchial cavity, and the forms of the colonies differ. This single specimen undoubtedly represents a new species, distinguished by the transparent soft test, the large number of developing embryos in the peribranchial cavity, and the form of the colony.

Aplidium scabellum (Michaelsen)

Amaroucium scabellum Michaelsen, 1924, p. 374.—Brewin, 1956, p. 124.

New record. East of South Island, New Zealand: *Eltanin* Cruise 25, Sta. 370 (95 meters).

Distribution. The species is known only from New Zealand waters: North Island [Michaelsen, 1924]; Colville channel [Michaelsen, 1924; Brewin, 1956]; Chatham Island [Brewin, 1956]; Little Barrier Island [Michaelsen, 1924; Brewin, 1956]. It was previously known from 46 to 64 meters; the present record from 95 meters represents the greatest depth from which it has been taken.

Description. The colonies are top-shaped, sometimes joined basally, and often very irregular. Generally the zooids open only on the flat upper surface of the lobes, although there are several irregular lobes in the present collection where zooids open all around the surface. The test is heavily impregnated with sand, and the zooids extend vertically to the surface and parallel to one another in the upper part of the test, although basally they may cross one another.

Zooids are small. There are 16–19 rows of about 12 stigmata. Ten longitudinal muscle bands extend along both sides of the thorax and join into a single band along the ventral surface of the abdomen. On the posterior abdomen the longitudinal muscles separate into bands on both sides. The posterior abdomen therefore can be drawn up alongside the abdomen, although not to the extent that this occurs in *Aplidium circumvolutum*. The branchial siphon has the usual 6 lobes, and there are strong circular muscles for most of its length, forming a branchial sphincter. The atrial siphon is some distance along the dorsal surface opposite the third to fourth row of stigmata. It is small and cylindrical, directed anteriorly, and also has strong circular sphincter muscles. A single undivided and very short atrial lip extending from the body wall anterior to the atrial siphon appears to close down over the top of the atrial aperture. There are 5 distinct stomach folds. The posterior abdomen is only slightly longer than the abdomen.

Remarks. The colonies of this species collected by the *Eltanin* are extremely variable and, although the typical top-shaped colonies as described by Michaelsen and Brewin are present, there are other lobes that do not conform to their descriptions. Some of these colonies are easily confused with those of *Aplidium circumvolutum*, which have been taken also at this station. The zooids of the present species are, however, quite distinctive in the undivided condition of the small atrial lobe and its relation to the atrial siphon, as well as in the form of both the branchial and atrial siphons with their strong circular sphincter muscles.

Synoicium adareanum (Herdman)

Atopogaster elongata Herdman, 1902, p. 194.
Synoicium adareanum; Kott, 1969, p. 65, and synonymy.

New records. Ross Sea area: *Edisto* Sta. 6 (100 meters). Weddell Sea: *Edisto* Sta. 14, SB-4 (? depth). Antarctic Peninsula: *Staten Island* Sta. 7/63 (21–31 meters); Sta. 66/63 (62 meters). South Shetland Islands: *Staten Island* Sta. 61/63 (31 meters); *Eltanin* Sta. 436 (73 meters). Between South Shetland and South Orkney Islands: *Eltanin* Sta. 1003 (210–220 meters). North of South Sandwich Islands: *Eltanin* Sta. 1581 (148–201 meters).

Distribution. The present records fall within the vertical and horizontal range previously recorded for the species.

Description. Long and narrow, almost cylindrical, colonies are available from *Edisto* Sta. 6; the colonies from *Edisto* Sta. 14 SB-4 and *Eltanin* Sta. 1581 are large and rounded, while specimens from *Eltanin* Sta. 436 and *Staten Island* Sta. 7/63 and Sta. 66/63 are of variable form and size.

Remarks. The specimens in the present collection are typical of this species and show the great variation in size and form of colony previously described for the species.

Aplidiopsis georgianum (Sluiter)

Synoicium georgianum Sluiter, 1932, p. 11.
Aplidiopsis georgianum; Kott, 1969, p. 73, and synonymy.

New record. Near Wilkes Station, between Knox and Budd coasts: *Atka* Sta. 24 (46 meters).

Distribution. Previously a distribution mainly in the Magellanic region of the Subantarctic extending south to the Antarctic Peninsula was indicated. The present record from the antarctic continent considerably extends that range and suggests that the species may, in fact, have a circumpolar distribution around the antarctic continent.

Description. The majority of the present colonies are identical with those described by Millar [1960] from South Georgia: clavate colonies, joined basally with a semitransparent test and no differentiation between the stalk and the head. Some of the specimens, however, have a harder and less transparent test, especially on the stalk, and more crowded zooids. These resemble young colonies of *Aplidium recumbens* and *Synoicium adareanum*. In all colonies the zooids open onto the free upper surface of the lobe and form from 1 to 3 circular systems of up to 10 zooids around a central common cloacal opening. Zooids are fairly large and extend down into the stalk. There are about 11 rows of stigmata, and the stomach is large and roomy without folds. The gonads are bunched in the posterior abdomen. The atrial aperture is of typical form, the upper rim of the border of the opening being produced into a languet, while the opening itself is often produced toward the common cloaca.

Remarks. The form of the colonies, arrangement of zooids, musculature, atrial opening, stomach, and posterior abdomen distinguish this species. The relationship of the hard cylindrical colonies to the more translucent forms, both present from this station, is not clear. Both types have previously been identified with this species (Kott, 1969; Millar, 1960), and it is possible that they represent different stages of maturity. It is also significant that these cylindrical colonies are similar to young specimens of *Synoicium adareanum*, which many authors have not been able to distinguish from the present species. This may explain the lack of records around the antarctic continent. Unfortunately, no intermediate forms were present in this collection to establish any of these relationships.

Family Didemnidae Giard

Didemnum biglans (Sluiter)

Leptoclinum biglans Sluiter, 1906, p. 29.
Didemnum biglans; Kott, 1969, p. 75, and synonymy.

New records. Scotia Sea area, north of South Georgia: *Eltanin* Sta. 732 (220–265 meters). North of South Sandwich Islands: *Eltanin* Sta. 1581 (148–201 meters). South Orkney Islands: *Eltanin* Sta. 1073 (1162–1226 meters); Sta. 1079 (593–598 meters); Sta. 1082 (298–302 meters). Antarctic Peninsula: *Staten Island* Sta. 24/63 (75 meters); Sta. 66/63 (62 meters). Ross Sea, *Edisto* Sta. 6 (100 meters).

Distribution. The species was previously known to be circumantarctic in depths up to 600 meters. The new records extend the northern limit of the species to the South Sandwich Islands and extend the greatest recorded depth to 1226 meters off the antarctic continental shelf.

Description. Rounded lateral organs in this species are smaller than in *Didemnum studeri* and are found opposite the first row of stigmata in the thoracic wall. Burrlike spicules, previously described for the species, are also present in the specimens from off the South Orkney Islands. In the colony from Sta. 1082 the basal test is considerably thickened and projects up into the colony forming a central core with various foreign particles embedded in it.

Remarks. The species has not been taken north of the South Sandwich Islands, and it does not appear to occur with *D. studeri* off South Georgia. In view of its depth range, this distribution is surprising. It is possible that the species is comparatively recent and has developed in the Antarctic, isolated there by the deeper trenches in the Scotia Ridge. The species overlaps the range of the subantarctic species *D. studeri* between the South Orkney Islands [Kott, 1969, *Eltanin* Sta. 1082] and the South Sandwich Islands.

Didemnum studeri Hartmeyer

Fig. 21

Didemnum studeri Hartmeyer, 1911, p. 538.—Kott, 1969, p. 75, and synonymy.

New records. South Georgia: *Eltanin* Sta. 1536 (659–686 meters). South of Falkland Islands: *Eltanin* Sta. 339 (512–586 meters). Southeast of Falkland Islands: *Eltanin* Sta. 1595 (124–128 meters). South of Tierra del Fuego: *Eltanin* Sta. 740 (384–494 meters); Sta. 1603 (256–269 meters). Off southern Chile: *Eltanin* Sta. 958 (92–101 meters); Sta. 1605 (522–544 meters). New Zealand, east of South Island: *Eltanin* Sta. 1431 (51 meters); Cruise 25, Sta. 370 (95 meters). East of North Island: *Eltanin* Sta. 1498 (101 meters). Southeast Pacific basin: *Eltanin* Sta. 1621 (4419–4804 meters).

Distribution. The new records confirm the subantarctic distribution of this species, with its southern limits off the South Orkney Islands; extend the northern limits in the Magellanic region to southern Chile; and indicate that the species occurs at depths far greater than those previously known.

Description. The colonies are characteristic. In the specimen from off southern Chile the spicules are dense throughout, and large lateral organs occupy the thorax opposite the second and third rows of stigmata. Typical larvae are enclosed in the test. The larvae have a comparatively short tail. The fragments from *Eltanin* Sta. 1498 are small, the zooids are small, and the thoracic cloacal cavity is more limited than is usual for the species. No gonads were observed; consequently the identification of this material is doubtful. The specimens from east of the South Island have up to three testis lobes, large lateral organs, thick spicules throughout, and typical larvae 0.5 mm in length. The primary cloacal cavity extending between clumps of zooids is often very deep (*Eltanin* Sta. 1536), so that clumps of zooids extend in pillars between the surface and the basal test. The specimens from Sta. 1621 (4419–4804 meters) are small and rounded, with a maximum diameter of 5 mm. The zooids are in clumps and double rows surrounded by thoracic cloacal canals. The thoracic organ is of moderate size, and the spicules are moderately distributed throughout. The specimens are not distinguishable from the typical condition of the species.

Remarks. Although the variations in the depth of the primary cloacal canals also occur in *D. biglans,* the species are distinguished by the large lateral organs and by the even distribution of spicules in *D. studeri.* Although *D. studeri* is now known from depths of 4804 meters, it has not been taken south of the South Orkney Islands [Kott, 1969]. The occurrence of the species at 4804 meters is remarkable, in view of the absence of variation in the morphology of specimens from this depth. The larvae are similar to those of *D. lambitum* and *D. maculatum* but much smaller.

Didemnum maculatum (Nott)

Leptoclinum maculatum Nott, 1892, p. 316.
Leptoclinum psammatodes Sluiter, 1895, p. 171.
Didemnum psammatodes var. *maculatum* Michaelsen, 1924, p. 341.—Brewin, 1946, p. 97.
Didemnum psammatodes f. *maculatum* Kott, 1962, p. 325.

New records. New Zealand, east coast of South Island: *Eltanin* Sta. 1431 (51 meters). New Zealand, west coast of North Island: *Eltanin* Sta. 1814 (ranging from about 124 to 154 meters).

Distribution. Eastern coast of Australia from Coffs Harbour to Tasmania [Kott, 1962]. New Zealand: Hauraki Gulf, Stewart Island, Otago, Christchurch [Nott, 1892; Michaelsen, 1924; Brewin, 1946, 1950a, 1957, 1958].

Description. Extensive investing sheets on scallop shells, on debris, and on *Corella eumyota.* The spicules are thick, although they are less dense throughout and toward the base of the colony. Spicules also fill small pointed processes sometimes present on parts of the surface. Primary cloacal canals are deep and occupy the whole length of the zooid. Secondary canals are very shallow at about midthoracic level. The primary cloacal system is apparent from the surface in colonies as the surface is depressed into the deep canals, while the zooids are embedded in almost solid test in the polygonal areas marked out by the primary canals. The surface layer of test superficial to the spicules contains brown pigment and is sometimes rubbed off, causing the preserved colony to look rather dirty.

Zooids are of usual form, with a wide atrial opening, 4 rows of stigmata, a single undivided testis lobe, and 6½ coils of the vas deferens.

Larvae are plentiful, embedded in the basal test. They are 1.0 mm long and have well-developed otolith and ocellus, 4 paired lateral ampullae, and the usual 3 median papillae.

The colonies from Sta. 1814 are small, and the characteristic cloacal system is not developed. They do have 6½ coils of the vas deferens, a single testis follicle, and larvae identical with those of the present species.

Remarks. The species is distinguished by the very shallow secondary canals and deeper primary canals, the absence of spicules in the superficial layer of the test, and the sheetlike investing colonies. Michaelsen [1924], followed by Kott [1962], established several previously described species as variants of *Didemnum psammatodes* (Sluiter, 1895). Nott's species became *D. psammatodes* var. *maculatum.* However, *D. maculatum* (Nott) has priority over *D. psammatodes.* The present specimens represent the condition described for *D. psammatodes* var. *maculatum* [Michaelsen, 1924; Kott (as form), 1962]. There are no reliable distinguishing characteristics in the various forms or varieties that Michaelsen assigned to the species, and their ranges and relationships have never been clearly established.

Didemnum lambitum (Sluiter)

Didemnoides lambitum Sluiter, 1900, p. 18.
Leptoclinum jugosum Herdman and Riddell, 1913, p. 886 [part].
Didemnum lambitum Michaelsen, 1924, p. 352; Kott, 1954, p. 164; 1962, p. 317.

New records. New Zealand, west of North Island: *Eltanin* Stas. 1814–1815 ('about 154 meters'; cf. Station List, p. 000); New Zealand, east coast of South Island; *Eltanin* Sta. 1431 (51 meters); Cruise 25, Sta. 369 (95 meters); Sta. 370 (95 meters).

Distribution. The species is known from Australia and off both Tasmania [Kott, 1954] and New South Wales [Kott, 1962]; from New Zealand off Otago [new record; collected by R. Crump in 1969], the Chatham Islands and Waitangi [Sluiter, 1900; Michaelsen, 1924], and Stewart Island [new record; collected by E. Batham, 1969].

Description. Conical cylindrical fingerlike lobes solitary or joined basally. Each lobe has a terminal common cloaca, thoracic secondary cloacal canals, and a primary canal posterior to the zooids and surrounding a central core of test, although this is not always well developed. Spicules are dense just beneath the surface test but sparse in the central core. They are stellate and of varying sizes from 0.01 mm to 0.05 mm in diameter. There are 7½ coils of vas deferens and a single undivided testis lobe. Larvae are large (0.9 mm long) and have otolith, ocellus, 4 pairs of lateral ampullae, and the usual 3 median papillae.

Remarks. This appears to be a fairly common species off Portobello Marine Station, Otago Harbour (R. Crump, personal communication), although Brewin has not recorded the species in her extensive accounts of the New Zealand ascidian fauna. The colony form and cloacal systems are distinctive, and it is unlikely that the species has been overlooked, but it is possible that it does not extend into intertidal waters where most collecting has been done.

The species has a cloacal system similar to that of *Polysyncraton chondrilla;* but, in addition to distinctive gonads and larval forms, bladder cells are absent from the surface test. Spicules are practically absent from the central core of test in many specimens of *Didemnum lambitum,* but in specimens from New South Wales [Kott, 1962] spicules are often present throughout the test.

Polysyncraton chondrilla (Michaelsen)

Didemnum chondrilla Michaelsen, 1924, p. 344.
Polysyncraton chondrilla; Kott, 1969, p. 79, and synonymy.

New records. Bellingshausen Sea: *Eltanin* Sta. 931 (3495 meters). Antarctic Peninsula: *Staten Island* Sta. 66/63 (62 meters). Between South Shetland and South Orkney Islands: *Eltanin* Sta. 1003 (210–220 meters). South of Tierra del Fuego: *Eltanin* Sta. 740 (384–494 meters).

Distribution. The present records from the Magellanic region confirm the circumsubantarctic distribution for this species, which also has a wide distribution around the antarctic continental shelf and slope, where it has been taken in up to 935 meters. The present record in 3495 meters indicates a particularly extensive vertical range.

Description. Colonies from off Tierra del Fuego are typical, although small, and have larvae of characteristic form. The colonies from the Scotia Ridge and Antarctic Peninsula are typical and irregular.

The colony from Sta. 931 on the continental slope of the Southeast Pacific basin is less typical. It is 4 cm long, 0.8 cm wide, and 0.2 to 0.3 cm high in the center, with the edges spreading across the substrate. Spicules are fairly large and stellate and are present in the test just below the surface. As in more typical colonies, they are progressively less dense in deeper layers of the test. There is a characteristically extensive posterior abdominal cloacal cavity above a fairly thick basal test with shallow secondary canals at thoracic level. Zooids are of usual form with a wide atrial aperture; there are, however, only 6 large, rectangular stigmata in each of the 4 rows. Gonads are not apparent.

Remarks. The colony from Sta. 931 diverges from the typical condition of this species principally in the number of stigmata in each row (10 previously recorded for *P. chondrilla*). The spicules are also larger than is usual for the species, although their size is a variable characteristic, and similar spicules are known for specimens from Enderby Land [Kott, 1954, 1969]. Unfortunately, mature gonads, which could confirm the identity of the present specimens, were not located. Kott [1969] suggested that specimens from New South Wales comprise a distinct species, *P. jacksoni* (Herdman, 1886). The high number of testis lobes (8), large stellate spicules, and less well-developed cloacal systems of *P. jacksoni,* however, also occur in antarctic specimens of *P. chondrilla* [Kott, 1969]. The only morphological characteristic available to distinguish these species is the smaller number of arms (7) of the stellate spicules and the increased bifurcation of larval lateral ampullae in the Australian *P. jacksoni.*

The colonies bear some resemblance to those of *Didemnum lambitum,* and where testes are not available the species are distinguished only by the absence of a surface layer of bladder cells in *D. lambitum* and by the larvae [Kott, 1962].

Trididemnum natalense Michaelsen

Trididemnum natalense Michaelsen, 1920, p. 3.—Hastings, 1931, p. 92.—Kott, 1962, p. 278.
Trididemnum savignii; Hastings, 1931, p. 91.
? *Trididemnum planum* Sluiter, 1909, p. 42.
Trididemnum sluiteri Brewin, 1958, p. 445.

New record. New Zealand, east of South Island: *Eltanin* Sta. 1431 (51 meters).

Distribution. The species has been recorded from all around Australia [Hastings, 1931; Kott, 1962], from eastern Africa [Michaelsen, 1920], and from New Zealand [Brewin, 1958]. It apparently has a wide distribution in the southern hemisphere north of the Subtropical Convergence.

Description. Investing colonies with a thin basal membrane contain especially large spicules and an extensive cloacal system with thoracic secondary canals and posterior abdominal primary canals, although this cloacal system is not well developed in all

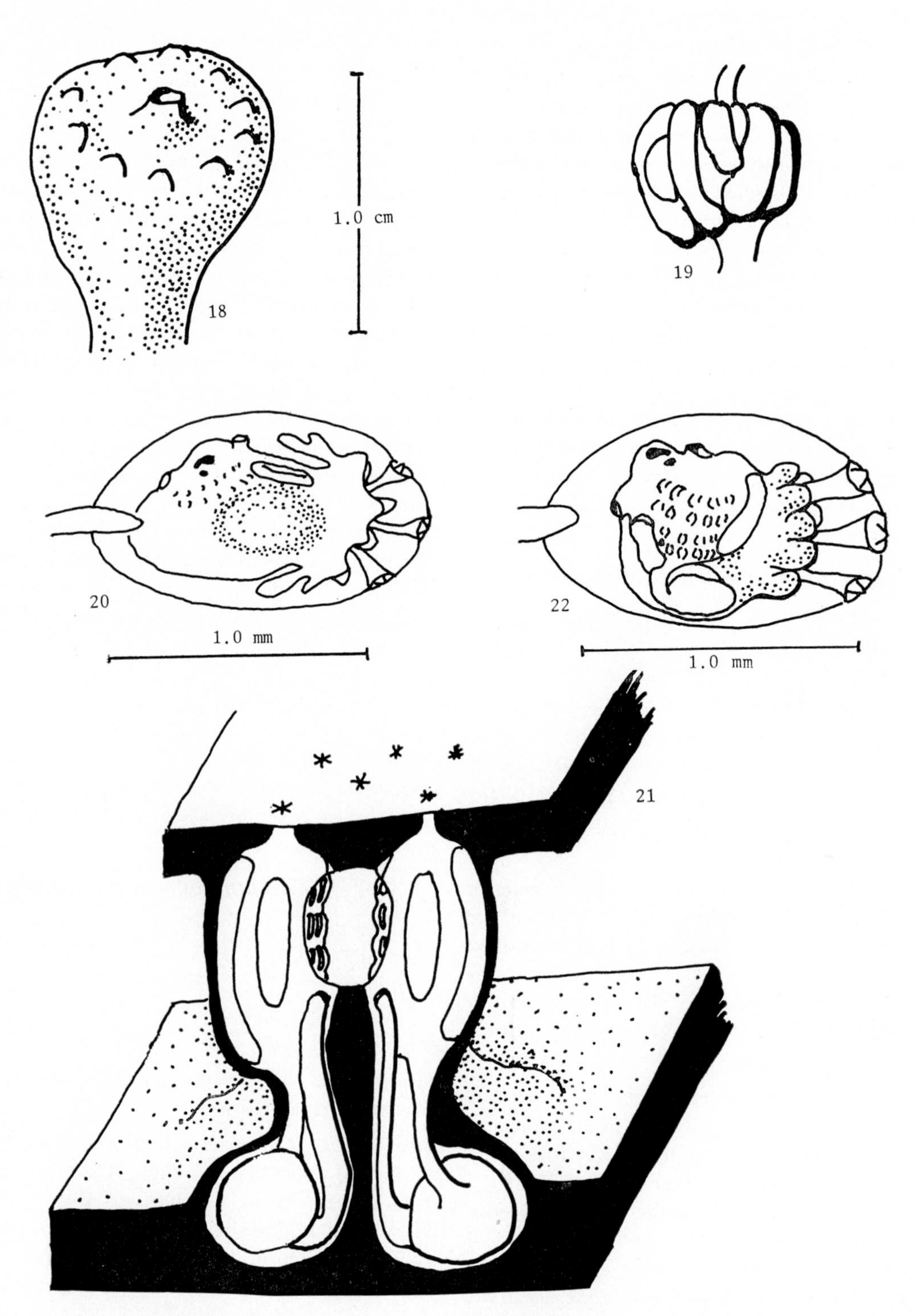

Aplidium caeruleum

18. Colony.
19. Stomach.

Aplidium pellucidum

20. Larva.

Didemnum studeri

21. Diagram of section through colony.

Leptoclinides rufus

22. Larva.

colonies and the primary canals may be limited to posterior thoracic canals. There is a thick layer of bladder cells in the surface test. There are also large stellate spicules around the anterior ends of the zooids. In some specimens there is an accumulation of pigment at the anterior end of the endostyle, and in one colony this is accumulated into a distinct pigment spot. The zooids are very small, with about 8 stigmata in each of the three rows; a posteriorly directed, fairly muscular atrial siphon; and large lateral organs. The gonads are not evident.

Remarks. The synonymy of *Trididemnum sluiteri* Brewin with the present species, previously suggested by Kott [1962], is confirmed by the present specimens from New Zealand, which have a variable accumulation of pigment at the anterior end of the endostyle as in *T. sluiteri,* a bladder cell layer superficially, large stellate spicules, and a posteriorly directed atrial siphon as in *T. natalense.* The species is distinguished from *Trididemnum auriculatum* only by the absence of the endostylar pigment cap, which has not yet been described for the Magellanic species. (See also Kott, 1969, p. 82, *Remarks.*)

Leptoclinides rufus (Sluiter)

Fig. 22

Polysyncraton rufus Sluiter, 1909, p. 72; 1913, p. 77.
Leptoclinides diemensis Michaelsen, 1924, p. 331.—Brewin, 1958a, p. 457.—Millar, 1960, p. 60.
Leptoclinides lissus Hastings, 1931, p. 93.
? *Leptoclinides sluiteri* Brewin, 1950b, p. 360.
Leptoclinides rufus; Tokioka, 1952, p. 92.—Kott, 1962, p. 286.
? *Leptoclinides auranticus* Brewin, 1956, p. 134.
? *Leptoclinides marmoreus* Brewin, 1956, p. 129.
? *Leptoclinides novaezelandiae* Brewin, 1958, p. 447.

New record. New Zealand, east of South Island: *Eltanin* Sta. 1431 (51 meters); Cruise 25, Sta. 370 (95 meters).

Distribution. Indonesia [Sluiter, 1909, 1913; Tokioka, 1952]; circum-Australia [Hastings, 1931; Kott, 1962]; New Zealand [Michaelsen, 1924; Brewin, 1950b, 1956, 1958, 1958a; Millar, 1960]. The species has a wide range in the temperate to tropic regions of Australasia on the continental shelf.

Description. Colonies are often extensive and always investing. They are fairly tough, with a thick superficial layer of bladder cells and an especially smooth surface. Beneath these there is a layer of fairly dense spicules that are only slightly reduced in density toward the base of the colony. Spicules form a distinct layer in the roof of the common cloacal cavity. The spicules are stellate and fairly large, although they vary considerably in size. There is an especially extensive posterior abdominal common cloacal cavity into which the zooids open directly. These openings are sometimes lined by spicules, indicating 5-lobed openings. Adjacent cloacal cavities are connected by canals.

Zooids are present in the surface layer of test and have 4 rows of 8 stigmata, a posteriorly directed and muscular atrial siphon, and a small oval lateral organ opposite the fourth row of stigmata. There are 6½ coils of vas deferens and 4 to 5 testis lobes.

Larvae, which are present in the basal test, are 0.75 mm long and almost as deep as their length. They have 4 pairs of short lateral ampullae and a single small median dorsal ampulla surrounding the 3 median papillae. An otolith and an ocellus are present. The larvae are identical with those previously described [Kott, 1962].

Remarks. There is no appreciable difference between *L. rufus* (Sluiter, 1909), *L. diemenensis* Michaelsen, 1924, *L. sluiteri* Brewin, 1950b, *L. marmoreus* Brewin, 1956, *L. auranticus* Brewin, 1956, and *L. novaezelandiae* Brewin, 1958. They all have fine longitudinal thoracic muscles, small lateral organs on the lower half of the thorax, a long branchial siphon, a long atrial siphon directed posteriorly, and a straight gut loop. There is always a superficial layer of bladder cells. There is a dense layer of stellate spicules up to 0.04 mm in diameter, and the density of spicules is reduced in the remainder of the test. The cloacal systems consist of posterior abdominal marginal canals and accessory canals traversing the colony. Millar [1960] has not described the cloacal system and 5-lobed atrial openings in his account of *L. diemenensis;* nevertheless, the agreement between his specimens and the present ones is so complete that they are undoubtedly conspecific. The extensive cloacal chambers of the present specimens probably represent a late stage in the development of the cloacal system, resulting from an increase in numbers of zooids and subsequent expansion of cloacal canals. Therefore these systems cannot be considered as a characteristic distinguishing either genera or species. Further, 5-lobed atrial apertures are only apparent and result from an appropriate arrangement of spicules in the vicinity of the openings. They do mark out the regions around the opening where longitudinal muscle fibers in the siphons are collected into bands. There are not, in fact, 5 lobes on the border of the siphon.

Therefore the genus *Askonides* Kott, 1962, is not distinct from *Leptoclinides*. *Leptoclinides imperfectus* (> *Askonides imperfectus* Kott, 1962) is closely related to the present species and is distinguished only by its larger spicules.

Diplosoma longinquum (Sluiter)

Figs. 23, 24

Leptoclinum longinquum Sluiter, 1912, p. 460.
Diplosoma longinquum; Kott, 1969, p. 83, and synonymy.

New record. Scotia Ridge, southeast of Falkland Islands: *Eltanin* Sta. 1593 (339–357 meters).

Distribution. This species has previously been recorded only from the Antarctic Peninsula. It is inconspicuous and easily overlooked. As it tolerates reasonably deep water, it could have a wide range in the Antarctic despite the present indication of a distribution limited to the Scotia Ridge.

Description. A single investing colony of typical form is available. The extended thorax with 4 rows of 12 rectangular stigmata is 2 mm long. The abdomen is 0.5 mm long. A typical hooked vas deferens curves around an undivided testis lobe. There is no accessory testis lobe, although the ovary is present in the position occupied by this accessory male follicle described by Sluiter [1912].

Larvae are present in the basal test. They are 1.6 mm long, with 4 pairs of fingerlike lateral ampullae, 3 median adhesive papillae, and an otolith, but no ocellus. There are up to 2 precocious buds present.

Remarks. Despite the general similarity of the colony and zooids to those of the ubiquitous *D. rayneri* Macdonald, the antarctic species is distinguished by the larger number of stigmata in each row, larger zooids, and a single undivided testis follicle. The larval form is larger in the present species and is further distinguished by the absence of an ocellus.

The species therefore is modified for an existence on the open sea bed rather than intertidal or shallow-water areas, where a well-developed larval ocellus aids in site selection.

Suborder PHLEBOBRANCHIA Lahille

Family Corellidae Lahille

Subfamily Corellinae Herdman

Corella eumyota Traustedt

Corella eumyota Traustedt, 1882, p. 271.—Kott, 1969, p. 84, and synonymy.

New records. Valparaiso, Chile: *Eltanin* Cruise 21, Sta. 188 (intertidal). West of Peru-Chile trench: *Eltanin* Cruise 21, Sta. 203 (79–91 meters). East of South Orkney Islands: *Eltanin* Sta. 1078 (604 meters); Sta. 1079 (593–598 meters). Off east coast, South Island, New Zealand: *Eltanin* Sta. 1431 (51 meters); Cruise 25, Sta. 370 (95 meters).

Distribution. These records fall within the limits of the depth and geographical ranges previously known for the species.

Description. There is little morphological variation in specimens of this species. The test is characteristically gelatinous, rather thin, semitransparent, and easily torn; and empty tests and individuals without the test are often present in the collections. Individuals are attached to rocks and especially shell fragments by part of or by all the left side. Apertures are sessile, although the atrial aperture on the dorsal border is directed upward and the terminal branchial aperture is directed horizontally. In larger specimens the branchial sac becomes enormously complicated, the stigmata form very numerous small spirals, and the supporting vessels are very much expanded to form a network on the outer surface.

Remarks. Specimens of this species were taken in far greater numbers off Macquarie Island (200+, Kott, 1969) than at the above stations, though they are here fairly well represented on the Scotia Ridge east of the South Orkney Islands.

Corynascidia suhmi Herdman

Corynascidia suhmi Herdman, 1882, p. 186.—Kott, 1969, p. 87, and synonymy.

New records. South of South Sandwich Islands: *Eltanin* Sta. 1560 (1190–1469 meters). Southeast Pacific basin: *Eltanin* Cruise 25, Sta. 359 (4682 metres). Southwest Pacific basin: *Eltanin* Cruise 25, Sta. 366 (5340 meters).

Distribution. The minimum depth previously recorded for this species is 1574 meters [Kott, 1969] off Macquarie Island. The new record (*Eltanin* Sta. 1560, 1190–1469 meters) extends its known range into shallower water, although there is a doubtful record of 146–174 meters from the Southeast Pacific basin.

Description. The present specimens are characteristically clavate, with thin, papery test. The specimen from *Eltanin* Sta. 1560 is fixed to a worm tube by fine hairlike rootlets on the base of the stalk.

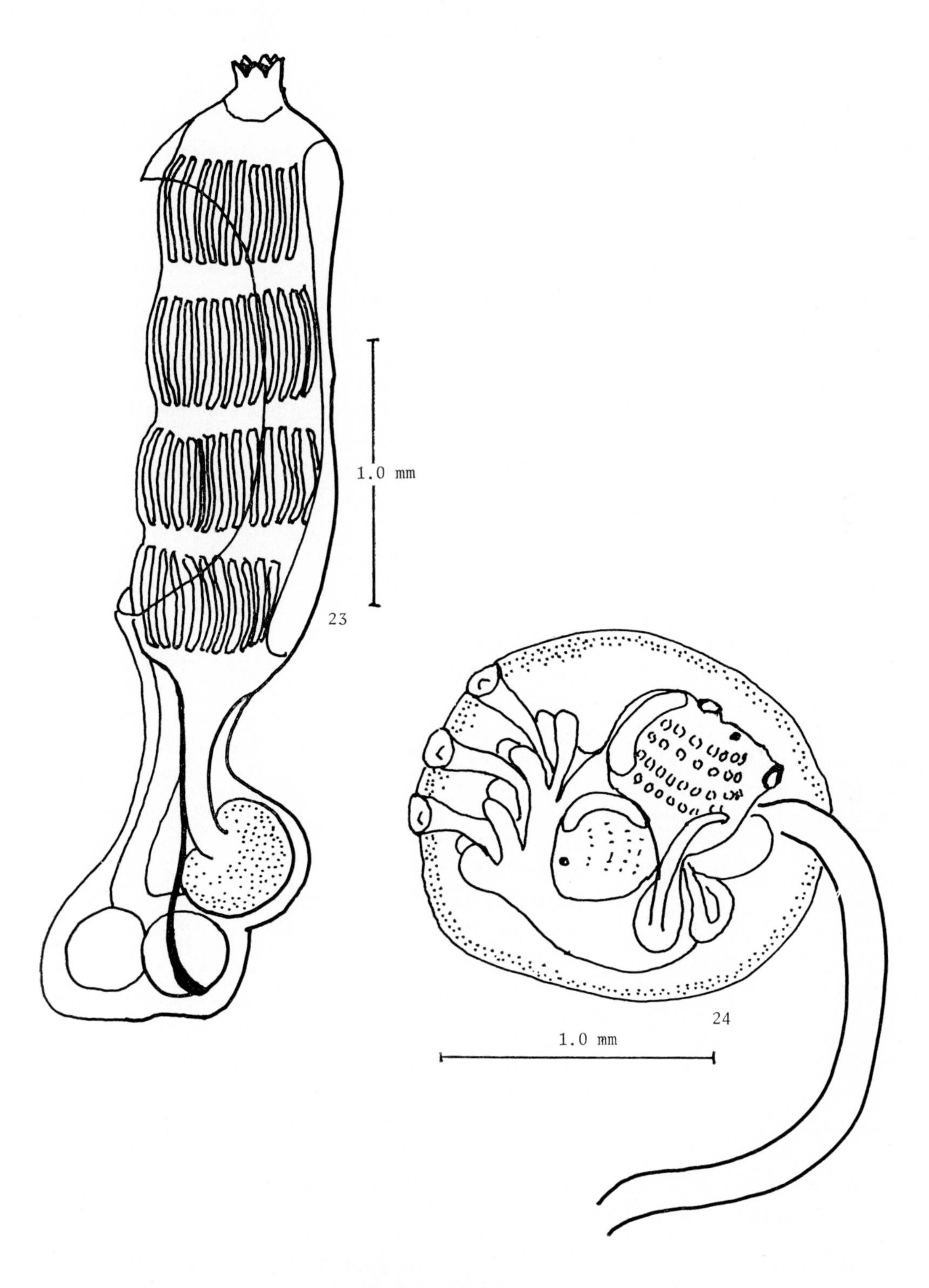

Diplosoma longinquum

23. Zooid.
24. Larva.

Family ASCIDIIDAE Herdman

Ascidia challengeri Herdman

Ascidia challengeri Herdman, 1882, p. 202.—Kott, 1969, p. 90, and synonymy.

New records. Ross Sea, Kainan Bay: Deep Freeze I, *Edisto* Sta. ? (644 meters). South Shetland Islands: *Staten Island* Sta. 61/63 (31 meters). South Georgia: *Eltanin* Sta. 1535 (97–101 meters).

Distribution. These records fall within the known geographical range. However, the collection from the Ross Sea is from a slightly greater depth than previously known.

Description. The specimens are typical.

Ascidia meridionalis Herdman

Ascidia meridionalis Herdman, 1880, p. 465.
Ascidia aspersa; Brewin, 1946, p. 106; 1950a, p. 344 [not *Ascidia aspersa* Mueller].
(For previous synonymy and literature see Kott, 1969, p. 92.)

New records. Tierra del Fuego: *Eltanin* Sta. 1603 (256–269 meters). Scotia Ridge, southeast of Falkland Islands: *Eltanin* Sta. 1593 (339–351 meters). Off east coast South Island, New Zealand: *Eltanin* Cruise 25, Sta. 370 (95 meters).

Distribution. New records from the Magellanic area of the Subantarctic fall within the previously known range. The placing of *Ascidia aspersa* of Brewin, from New Zealand waters, in synonymy with *A. meridionalis* extends the known range of the latter around the Antarctic and suggests a circumpolar distribution.

Description. The specimens are characteristic of the species. They are fixed by a large part of the left side. The apertures are sessile. Occasionally there are pointed papillae from parts of the test. There are a narrow prepharyngeal area behind the single ring of short and delicate tentacles and a simple U-shaped dorsal tubercle in the shallow peritubercular area. There are 5 to 6 stigmata in each branchial mesh, and intermediate papillae are associated with parastigmatic vessels. The branchial sac extends behind the gut loop.

Remarks. Externally this species resembles *Ascidiella aspersa* (Mueller, 1776); Kiaer, 1893; Millar, 1966a (> *Ascidia aspersa* Mueller), which is distinguished by the absence of internal papillae in the branchial sac. In the specimens from New Zealand waters, the branchial sac supports the internal branchial papillae characteristic of the genus *Ascidia* and cannot be separated from *Ascidia meridionalis* by any known character.

Family AGNESIIDAE Huntsman

Caenagnesia bocki Ärnbäck

Caenagnesia bocki Ärnbäck, 1938, p. 41.—Kott, 1969, p. 96, and synonymy.

New records. Ross Sea, Kainan Bay: Deep Freeze I, *Edisto* Sta. ? (644 meters). South Shetland Islands: *Eltanin* Sta. 432 (935–884 meters); Sta. 436 (73 meters).

Distribution. There have been, previously, isolated records of this species around the antarctic continent, especially on the continental slope. The present record from the Ross Sea confirms the circumantarctic distribution of the species.

Description. The present specimen from the Ross Sea is spherical and slightly sandy and has a more or less firm, gelatinous test, especially anteriorly. The apertures are depressed in a longitudinal area of thin test. The test is thinner posteriorly and has very fine, hairlike roots. The body musculature is as previously described [Kott, 1969a]. There are about 18 rows of infundibula. The specimens from the South Shetland Islands are small (1 cm to 1.5 cm in diameter), and anteriorly the apertures are surrounded by thicker test but are not protected by overlapping lips.

Order PLEUROGONA

Suborder STOLIDOBRANCHIA Lahille

Family STYELIDAE Sluiter

Subfamily POLYZOINAE Hartmeyer

Polyzoa opuntia Lesson

Fig. 25

Polyzoa opuntia Lesson, 1830, p. 437.—Kott, 1969, p. 100, and synonymy.

New records. East of Tierra del Fuego: *Eltanin* Sta. 1500 (73–79 meters). South Georgia: *Eltanin* Sta. 1533 (3–6 meters).

Distribution. A subantarctic species recorded from Kerguelen and Heard Islands. Most numerous records, however, are from the Magellanic region, from South Georgia, and on the Patagonian shelf, with a northern limit off northern Argentina. Although most records are from more shallow waters, the species has been taken from as much as 200 meters. It has not yet been found in New Zealand waters.

Description. The present specimens are dumbbell shaped, or long and rodlike, and apparently free, or flattened or clavate and fixed by a short stalk. The common test is leathery, rough, and wrinkled, and the zooids form small mounds on the surface. Zooids are 2 mm long when removed from the test. They are typical of the species, having 8 internal longitudinal vessels and about 4 stigmata per mesh, 14 internal stomach folds, and a curved pyloric caecum. There are simple endocarps between the esophagus and intestine and in the intestinal loop. There are also small endocarps scattered over the body wall. The anal border has 6 rounded lobes. There is a row of gonads either side of the endostyle: about 10 hermaphrodite polycarps on the right and 5 on the left. The gonads are sometimes missing from the left side of the body, or sometimes only male follicles are present on the left.

Remarks. The variation in the condition of gonads on the left side of the body is possibly related to the stage of maturity of these glands and draws attention to the fact that assumption of generic differences based on the presence of male and female components of these glands in this subfamily could be unreliable. The hermaphrodite gonads of *Theodorella* in the present collection are of identical form to those of the present species and, although other characters serve to distinguish the species, generic separation may not be justified.

Alloeocarpa incrustans (Herdman)

Figs. 26, 27

Synstyela incrustans Herdman, 1886, p. 342 (part; not Philippine specimens).
Alloeocarpa incrustans; Kott, 1969, p. 103, and synonymy.

New record. South of Tierra del Fuego: *Eltanin* Sta. 740 (384–494 meters).

Distribution. The species has been recorded only from the Falkland Islands and Tierra del Fuego. The present record represents the greatest depth from which the species is known and considerably extends its vertical range.

Description. The single specimen of this species from *Eltanin* Sta. 740 is low and rounded, with a spreading base containing club-shaped terminal vessels. Its maximum diameter is 3 mm. The test is thin and semitransparent, with brown pigment spots. The branchial tentacles are comparatively short and thick; the dorsal tubercle is in a shallow peritubercular area with a simple opening. The dorsal lamina is plain; there are about 30 internal longitudinal vessels on each side of the body and 1 to 2 oval stigmata per mesh dorsally, increasing in number ventrally. The gut occupies the posterior half of the left side of the body; the stomach narrows to the intestine and has 12 longitudinal folds expressed externally and a curved caecum. The intestine forms a simple loop with the stomach and esophagus, and it ends in a bilabiate anus. On the body wall are 3 to 4 endocarps. Anteriorly and to the right of the endostyle are 3 ovaries, and to the left of the endostyle in the middle of its length are 2 rounded, single-follicle testes with short ducts extending into the peribranchial cavity.

Remarks. The present specimen has more longitudinal vessels in the branchial sac and fewer stomach folds than is usual for this species. However, the position and nature of the gonads and gut are typical, and the condition here may represent a variation in the branchial sac found in individuals at the limits of their vertical range.

Both *A. capensis* Hartmeyer, 1912 [Millar, 1962] and *A. affinis* Bovien, 1922 have undivided testis follicles, although the South African species *(A. capensis)* has long male ducts. *A. affinis* is distinguished from *A. incrustans* only by the reduced small numbers of internal longitudinal vessels.

Although records of this species more often represent colonies, solitary individuals have been described *(A. emilionis* Michaelsen, 1900).

The undivided testis follicles of this and related species of *Alloeocarpa* resemble those of *Theodorella* spp., which are distinguished only by hermaphrodite gonads on the right side of the body (see Remarks under *Theodorella arenosa* below).

Alloeocarpa bridgesi Michaelsen

Figs. 28, 29

Alloeocarpa bridgesi Michaelsen, 1900, p. 41.—Kott, 1969, p. 104, and synonymy.

New records. Off southern Chile: *Eltanin* Stas. 958 and 959 (92–101 meters).

Distribution. The present records extend the known depth range of this species and extend its known geographic range to the north. It appears to be restricted, however, to the general area around Tierra del Fuego.

Description. Solitary and confluent individuals are joined by a thin spreading basal membrane. Their maximum diameter is 5 mm. They are attached to

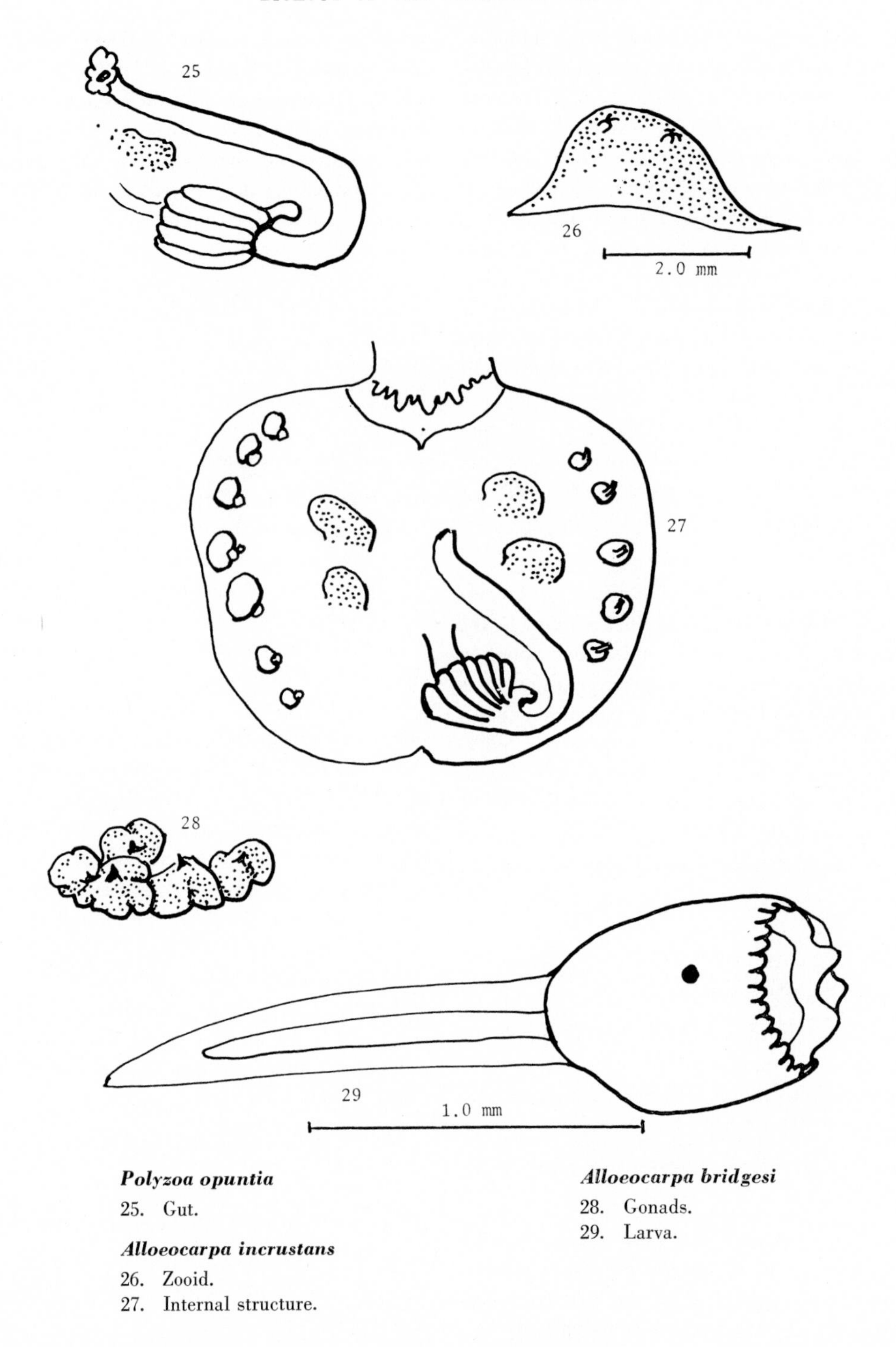

Polyzoa opuntia

25. Gut.

Alloeocarpa incrustans

26. Zooid.
27. Internal structure.

Alloeocarpa bridgesi

28. Gonads.
29. Larva.

coral particles and scallop shells, are dorsoventrally flattened, and have a transparent test containing brown pigment cells and elongate terminal ampullae in the spreading basal test. Both apertures are sessile on the dorsal surface. There are 7 longitudinal vessels on each side of the branchial sac. The dorsoventral flattening is so arranged that the endostyle runs around the right border of the body. The right side of the branchial sac occupies the raised, rounded upper surface, while the left side lies on the flat basal surface with the gut loop between the branchial sac and the body wall in the posterior half. The rectum extends over onto the upper surface to open at the base of the atrial siphon. There is an arc of about 5 testis follicles, adjacent to one another but not coalesced, around the left border of the body above the gut loop. Each follicle has a single short duct from its apex, but the wider basal part on the body wall is divided into 2 to 7 lobes by shallow to deeper indentations. The female gonads are arranged above the right border of the body, and developing embryos and larvae are present in the anteroventral part of the body, to the right of the endostyle. Larvae are identical with those of *Alloeocarpa incrustans,* with a single photolith obscured by pigment and anterior triradiate papillae surrounded by about 26 fingerlike ampullae.

Remarks. The species is distinguished from all others by the condition of the testis follicles, which represents a stage between the undivided follicles of *A. incrustans* and *A. affinis* and the coalesced and very lobed testis of *A. bacca* [Kott, 1969].

Alloeocarpa affinis Bovien

Fig. 30

Alloeocarpa affinis Bovien, 1922, p. 41.
Alloeocarpa minuta Brewin, 1951, p. 110; 1956, p. 123.

New record. Off South Island, New Zealand: *Eltanin* Cruise 25, Sta. 370 (95 meters).

Distribution. The species is known only from New Zealand waters: from Campbell Island [Bovien, 1922], Chatham Island [Brewin, 1956], Hauraki Gulf [Brewin, 1951]. It has been taken from intertidal waters, and the present record represents the greatest depth from which the species is known.

Description. The colonies comprise rounded, dorsoventrally flattened zooids enclosed in a continuous layer of test. The present specimens are found encrusting the mollusk homes of hermit crabs. Both apertures are sessile on the upper surface. There are 6 internal longitudinal vessels on each side of the body. The left side of the branchial sac extends across the base of the body. There are 4 to 8 stigmata per mesh. The gut loop is short and in the middle of the left or basal half of the body. The stomach is short, its walls are folded, and there is a small curved caecum. There are 4 to 5 small rounded ovaries on the right or upper side of the body wall and 2 single undivided testis follicles on the left side of the body wall above the gut loop. There is a single, short vas deferens from the center of each testis lobe.

Larvae are present in the peribranchial cavity on the right side of the body. They resemble those of other species of this subfamily. There are 3 triradiate anterior papillae and approximately 30 fingerlike ampullae forming a circle around the anterior part of the zooid. A single pigmented photolith beneath the ampullae is visible only from the front of the larva.

Remarks. The condition of the colony, gonads, and gut is identical with that described for *Alloeocarpa affinis* Bovien and *A. minuta* Brewin. Brewin [1951] distinguished her species from Bovien's by the comparatively small number of stigmata per mesh in each mesh of the branchial sac. The number of stigmata in the *Eltanin* specimens (4–8), however, is intermediate between the counts of Brewin (3–4) and Bovien (6–8). It appears, therefore, that the number of stigmata cannot be considered a distinguishing characteristic and that the species of Bovien and Brewin are synonymous.

Theodorella arenosa Michaelsen

Figs. 31a, 31b

Theodorella arenosa Michaelsen, 1922, p. 469.—Brewin, 1958, p. 449.
? *Theodorella torus* Michaelsen, 1922, p. 473.

New records. South Georgia: *Eltanin* Sta. 1535 (97–101 meters). Scotia Ridge, southeast of Falkland Islands: *Eltanin* Sta. 1593 (339–357 meters).

Distribution. The species has been previously taken only from Stewart Island, New Zealand [Michaelsen, 1922; Brewin, 1958], from waters up to 36.8 meters [Michaelsen, 1922]. The present records indicate a circumsubantarctic range out to the edge of the continental shelf.

Description. Sessile zooids joined by strands of basal common test extending across the surface of individuals of *Molgula gigantea* and *Pyura georgiana.* The younger zooids are dorsoventrally flattened, but more mature individuals are upright. Height of zooids is from 4 to 6 mm. Apertures are sessile at either end of the upper surface. The test is fairly thin and semitransparent in smaller individuals, becoming opaque and transversely wrinkled with adherent sand grains and mud as the individual matures.

There are 7–8 internal longitudinal vessels on either side of the branchial sac, 10 stigmata are present between the dorsal lamina and the first longitudinal vessel, and 6–7 stigmata per mesh are present in other parts of the branchial sac. Parastigmatic vessels are also present. The gut forms a simple loop in the posterior half of the left side of the body. The stomach is almost square, with about 15 broad, oblique folds and a curved caecum. There is a connective between the stomach and the intestine. The anus opening into the base of the atrial siphon is bilabiate. There are 10–11 hermaphrodite gonads around the ventral curve of the body to the right of the endostyle, and 5–8 undivided testis follicles to the left of the endostyle anterior to the gut loop. The gonoducts are all fairly short. Early embryos are present in the peribranchial cavity on the right.

Remarks. Both Michaelsen's and Brewin's specimens were found growing on stalks of *Pyura pachydermatina.* The present specimens on stalks and heads of *Molgula gigantea* and *Pyura georgiana* suggest that this location is advantageous in raising small species above the substrate. *Theodorella torus* Michaelsen, 1922, from the North Island in 3.7 to 14.7 meters, and *Theodorella stewartense* Michaelsen, 1922, from Stewart Island in 46 meters, are the only other species of the genus known. It is probable that *T. torus* is a synonym of the present species and that Michaelsen's colony represents a stage of development intermediate between the present specimens and the specimens from which Michaelsen described *T. arenosa,* in which the zooids become more confluent and the numbers of gonads increase. The large number of longitudinal vessels (14) distinguish *T. stewartense.*

The genus *Theodorella* appears to be intermediate between *Polyzoa,* with all hermaphrodite gonads, and *Alloeocarpa,* with dioecious gonads. *Theodorella arenosa, Polyzoa opuntia, Polyzoa reticulata,* and *Alloeocarpa affinis* have similar numbers of longitudinal vessels and stomach folds. Therefore when, as in present specimens of *Polyzoa opuntia,* the female component is missing from left polycarps, the species can be distinguished from *Theodorella arenosa* only by the form of the colonies. Similarly, if the female component from the right polycarps of *Theodorella arenosa* are not present, the species could be confused with *Alloeocarpa affinis. Alloeocarpa affinis, Theodorella arenosa,* and *Polyzoa opuntia* could therefore represent stages of increasing maturity in a single species. *Theodorella stewartense* is related to *Alloeocarpa incrustans* in a similar way. Further accounts of the range of variation and life history of colonies of one or more species should clarify the situation; but even if the species are shown to be distinct, their relations appear to be too close to justify generic separation.

Okamia thilenii (Michaelsen)

Metandrocarpa thilenii Michaelsen, 1922, p. 457.
Okamia thilenii; Brewin, 1948, p. 116; 1951, p. 104.
Okamia theilenii; Brewin, 1948, p. 123.

New record. Off east coast, South Island, New Zealand: *Eltanin* Cruise 25, Sta. 370 (95 meters).

Distribution. Known only from New Zealand waters: Tauranga, east coast, and New Plymouth, west coast of North Island [Michaelsen, 1922]; and Hauraki Gulf, east coast of North Island [Brewin, 1948].

Description. Zooids are rather elongate and crowded. They are joined by a basal membrane and by the test on the lower part of the body. The test is heavily impregnated with sand. Both apertures are on the narrow upper surface and are sessile. There are 10 internal longitudinal vessels on either side of the body. The stigmata are long and rectangular. The gut loop is simple, the stomach is almost square with distinct oblique folds, and the rectum extends well anterior to the base of the atrial opening. On the right side of the body there is a long arc of flask-shaped undivided testis follicles, each with a short duct from the narrow end of the flask directed toward the atrial opening. A similar, though not such a long, arc of testis follicles extends around the left side of the body anterior to the gut loop. A single row of ovaries extends along on the left of the ventral line posterior to the testis and ventral to the pole of the gut loop.

Remarks. These specimens conform completely with the descriptions previously given for the species. The gonads on the left in the present specimens exceed the number previously described, while those on the right are fewer. Brewin [1948] observed in her specimens that the number of gonads was less than those described by Michaelsen. The present observations confirm her suggestion that this is a variable character that cannot be used to distinguish separate species in this genus.

Stolonica australis Michaelsen

Stolonica australis Michaelsen, 1927, p. 202.—Michaelsen and Hartmeyer, 1928, p. 352.—Kott, 1952, p. 253.

New record. Off east coast, South Island, New Zealand: *Eltanin* Cruise 25, Sta. 370 (95 meters).

Distribution. This species has previously been recorded only from southwestern Australia and Tas-

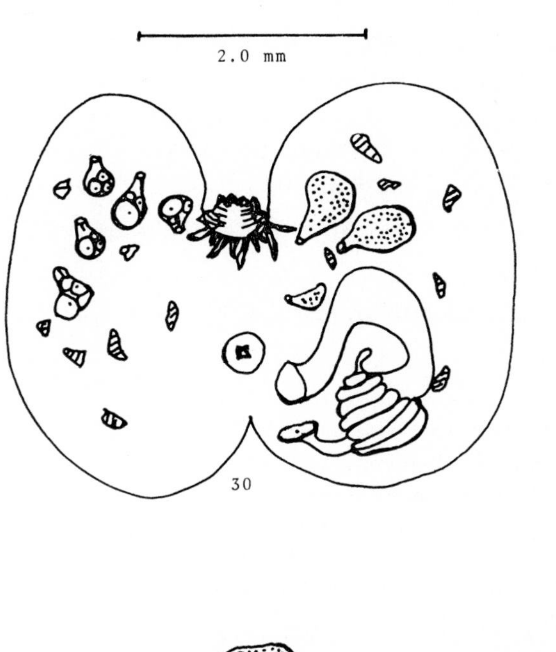

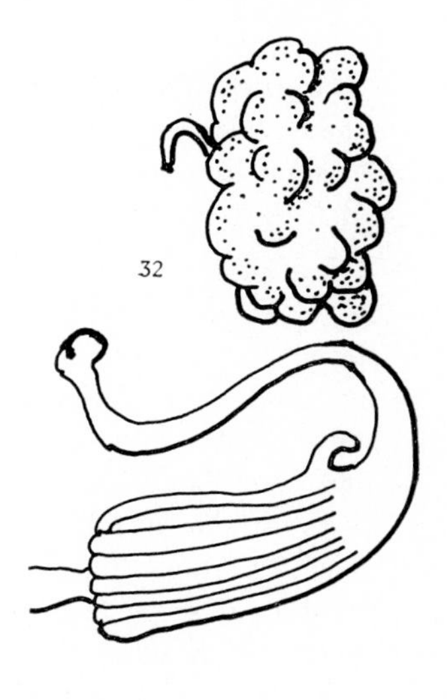

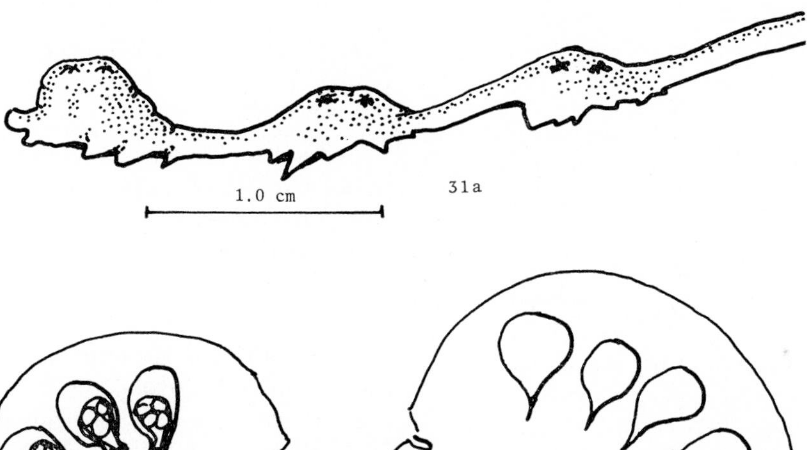

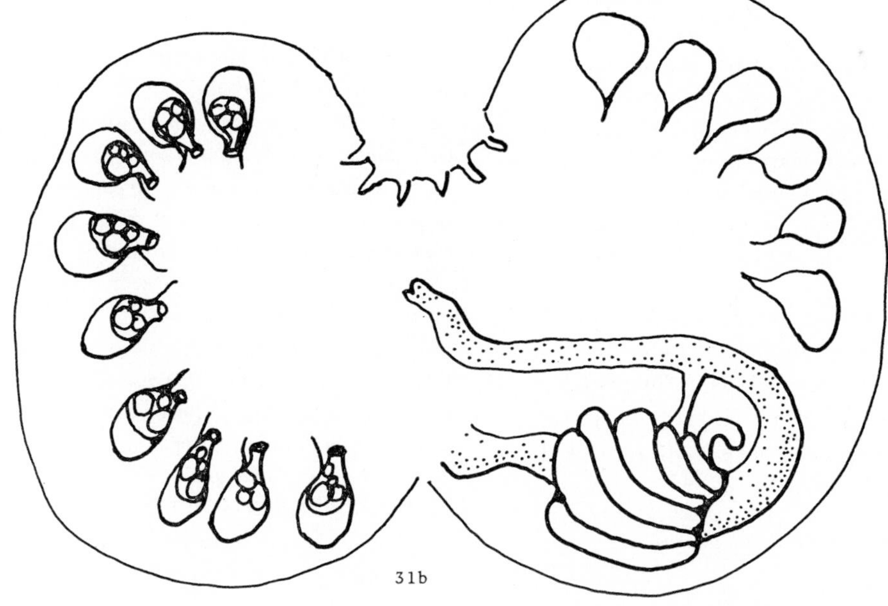

Alloeocarpa affinis

30. Internal structure (after Brewin, 1951, p. 110, fig. 6).

Theodorella arenosa

31a. Colony.
31b. Zooid.

Oligocarpa megalorchis

32. Left body wall.

mania. The present record extends its range to New Zealand and also represents the greatest depth from which the species has been taken.

Description. Zooids are small and rounded, joined by basal branching stolons. The test is heavily impregnated with sand. The branchial aperture is anterior and the atrial aperture is anterodorsal; both are sessile. There are 2 branchial folds on each side of the body with 7 to 10 internal longitudinal vessels on the folds and about 4 between folds, although between the most dorsal fold and the dorsal lamina there is only a single internal longitudinal vessel. There are 4 to 6 stigmata in each mesh. The gut loop is simple and the stomach oval with distinct folds. The gonads are present in rows around the ventral border of the body; on the left are found male and some hermaphrodite gonads, while on the right, male, female, and hermaphrodite gonads occur. An especially large clump of male glands is located just posterior to the atrial siphon on the right side of the body.

Remarks. These specimens are identical in all respects with those previously found around the coast of Australia except for the more plentiful gonads in the present specimens. This condition could be the result of age or seasonal differences and cannot be regarded as a distinguishing characteristic.

Oligocarpa megalorchis Hartmeyer

Fig. 32

Oligocarpa megalorchis Hartmeyer, 1911, p. 527.—Kott, 1969, p. 106, and synonymy.

New record. Macquarie Island: *Eltanin* Sta. 1417 (79–93 meters).

Distribution. The species has already been recorded from an adjacent location, *Eltanin* Sta. 1418 [Kott, 1969], and this new record falls within the known vertical and horizontal range.

Description. The present specimens conform with those previously described.

Subfamily STYELINAE Herdman

Cnemidocarpa verrucosa (Lesson)

Cnemidocarpa verrucosa Lesson, 1830, p. 151.—Kott, 1969, p. 107, and synonymy.

New records. Ross Sea: *Edisto* Sta. 6 (100 meters). Between South Shetland and South Orkney Islands: *Eltanin* Sta. 1003 (210–220 meters). South Sandwich Islands: *Eltanin* Sta. 1581 (148–201 meters). South Georgia: *Eltanin* Sta. 1535 (97–101 meters).

Distribution. The present records fall within the extensive range previously known for the species in the Antarctic and Subantarctic.

Description. As previously described [Kott, 1969, p. 107].

Cnemidocarpa bicornuta (Sluiter)

Fig. 33

Styela bicornuta Sluiter, 1900, p. 22.

Cnemidocarpa bicornuta; Michaelsen, 1922, p. 440.—Brewin, 1946, p. 117.

Cnemidocarpa bicornuata Brewin, 1948, p. 127; 1950, p. 58; 1956, pp. 122, 131; 1957, p. 577; 1958, p. 440.

Cnemidocarpa otagoensis Brewin, 1952, p. 457.

New records. New Zealand, east of South Island: *Eltanin* Sta. 1431 (51 meters); Cruise 25, Sta. 370 (95 meters).

Distribution. The species has been recorded from all around New Zealand, from the Chatham Islands, the Chatham Rise, and from Stewart Island. The present record from *Eltanin* Sta. 370 represents the maximum depth from which the species has been taken.

Description. The test is externally leathery and wrinkled, and the apertures are on prominent and transversely wrinkled siphons. The test is very tough, but thin, and the very muscular body wall adheres fairly closely to it. The large dorsal tubercle has a vesicular appearance, and the opening is horseshoe shaped, with both horns turned in. There are 4 branchial folds on each side of the body, with about 6 internal longitudinal vessels and about 15 to 20 on the folds. There are about 10 stigmata per mesh.

The gut forms a long, narrow, curved loop; the stomach is long with internal folds, and the anal border is bilabiate. There are two long, narrow gonads on each side of the body lying almost parallel to the gut loop. The testis follicles appear to project into the lateral aspect of the ovarian tube where it lies against the body wall, and they are enfolded by the ovary.

Remarks. Brewin [1952] established the species *Cnemidocarpa otagoensis* to accommodate specimens that had 'adipose' deposits in the body wall but that were otherwise indistinguishable from the present species. There are patches of different consistency in some of the present specimens corresponding to

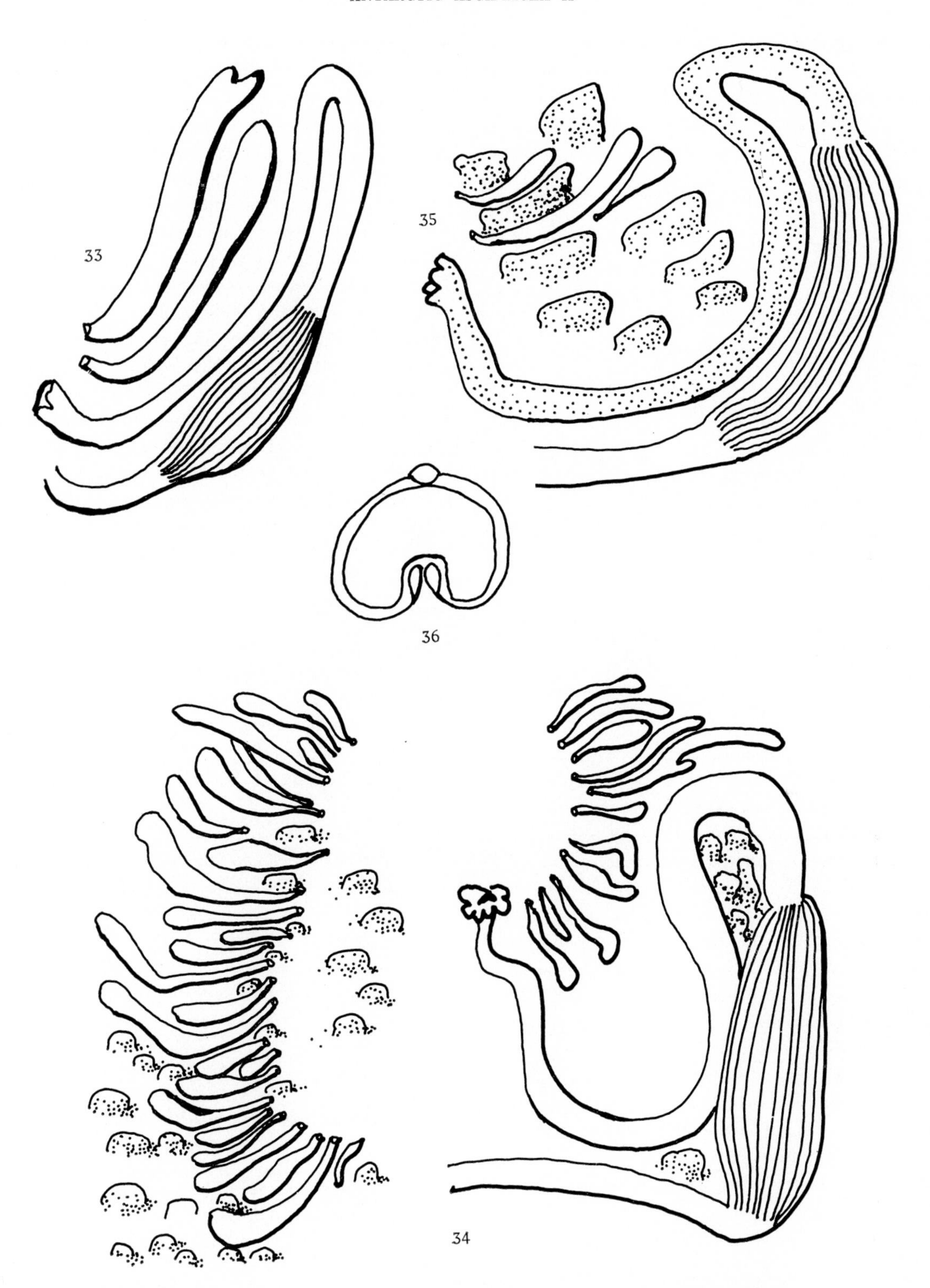

Cnemidocarpa bicornuta

33. Left side, gut and gonads.

Cnemidocarpa madagascariensis

34. Internal structure.

35. Left side, gut and gonads (gonads immature).

36. Section through gonad (diagrammatic), showing male gonads embedded in parietal surface of ovary with male ducts curving around ovary to join vas deferens on mesial surface.

the position in the body wall occupied by 'adipose' tissue in *C. otagoensis.* It is possible, therefore, that adipose tissue, as described by Brewin, is a seasonal occurrence associated with physiological changes in the species.

Cnemidocarpa madagascariensis regalis Michaelsen

Figs. 34–36

Cnemidocarpa madagascariensis var. *regalis* Michaelsen, 1922, p. 430.—Brewin, 1952, p. 455.

New records. New Zealand, east of North Island: *Eltanin* Sta. 1498 (101 meters). New Zealand, off east coast, South Island: *Eltanin* Cruise 25, Sta. 370 (95 meters).

Distribution. C. madagascariensis madagascariensis was first described from Madagascar [Hartmeyer, 1916]. Michaelsen [1922] and Brewin [1952] have taken a geographic subspecies, *C. madagascariensis regalis,* from the Three Kings Islands, New Zealand (120 meters), and off the Otago Peninsula (74–92 meters). The New Zealand records, although few, indicate a general range around these islands, possibly in deeper waters than are easily accessible from the shore.

Description. The specimens are up to 4 cm in diameter, laterally flattened, with sessile apertures. There is sometimes a solid stalk posteriorly of 1.2 cm in diameter and about 1 cm long. The test is tough, leathery, and hard; uneven and rough externally; and fairly thin. The body wall is always thin, with a strong, continuous layer of circular muscles, and is closely applied to the test. The inner lining of the siphons is raised into rounded projections. Branchial tentacles are not very long, and 16 of variable length alternate with rudimentary tentacles. There is a wide prebranchial area. A simple U-shaped opening, with both horns turned in, is present on the dorsal tubercle, which is set fairly deep in the peritubercular area. The dorsal lamina is wide and plain-edged. Branchial folds are tall and flat but do not overlap. Internal longitudinal vessels are arranged as follows:

E 7 (30) 10 (38) 9 (37) 11 (25) 5 DL

There are about 6 long rectangular stigmata per mesh between the folds, and these are crossed by parastigmatic vessels. The gut loop is very long, narrow, and deeply curved, and the rectum extends anteriorly to open in a 6-lobed anus. The esophagus is especially long and runs across the posterior end of the body. The long, narrow stomach with 30 longitudinal folds extends anteriorly at right angles to the esophagus. There is a gastrointestinal connective about one-third of the distance from the cardiac end of the stomach. Numerous small endocarps are present in the gut loop and scattered over the body wall, especially between and around the gonads.

The gonads, not longer than 1 cm, are very narrow. They may be in a single or double row in the middle of both sides of the body and are directed toward the atrial opening. They are very closely placed and are loosely attached to the body wall by a single row of long connectives from their lateral aspect. They are often overlapping, are sometimes branching, and are of typical cnemidocarp form, with testis follicles enveloped by the ovary on its lateral aspect and with testis ducts curving around to join the vas deferens on the mesial surface of the ovary. There are from 2 to 13 gonads on the left, in and anterior to the secondary gut loop. On the right there are 10 to 20 gonads.

Remarks. Only the position of the atrial aperture two-thirds of the body length from the branchial aperture distinguishes *C. madagascariensis madagascariensis* from *C. madagascariensis regalis.* Both subspecies have the characteristic long, cylindrical stomach and the numerous gonads by which the species is distinguished from *C. ohlini* (see p. 000) and *C. nisiotis* (Sluiter) [Brewin 1950a, 1950b]. The condition of the gonads suggests a phylogenetic link between *Cnemidocarpa* and those species of *Polycarpa* with numerous elongate gonads, as *Polycarpa circumarata* from Venezuela [Van Name, 1945].

Cnemidocarpa madagascariensis madagascariensis and *C. madagascariensis regalis,* together with *C. ohlini* from South America and *C. nisiotis,* also from New Zealand, suggest the circumpolar distribution of a common ancestor of these closely related forms.

Cnemidocarpa ohlini (Michaelsen)

Fig. 37

Styela ohlini Michaelsen, 1898, p. 366.—Kott, 1969, p. 122, and synonymy.

New record. Tierra del Fuego: *Eltanin* Sta. 1603 (256–269 meters).

Distribution. The only previous record of the species is from 27 meters, Strait of Magellan [Michaelsen, 1898]. It is a large, conspicuous species and apparently has a very restricted range.

Description. Specimens are flattened, dome-shaped, or upright, 2 cm in basal diameter, and 0.5–2 cm high. Apertures are sessile, 4-lobed, and close together on the upper surface. The test is thin, hard, and tough; it is white and papery, with a pearly luster, as described by Michaelsen. The body wall is thin and closely adherent to the test and has delicate, principally longitudinal muscle bands. There are about 30 simple branchial tentacles from the border of a pronounced velum. The dorsal tubercle has a simple C-shaped opening. The prepharyngeal area is narrow, and the dorsal lamina is broad and plain-edged. The branchial sac has 4 rounded folds on each side of the body, with tall internal longitudinal vessels close together on the folds according to the following branchial formula:

E 6 (14) 8 (18) 5 (25) 5 (16) 6 DL

The internal longitudinal vessels are convoluted, possibly because of contraction. There are 6 stigmata per mesh between the folds.

The gut loop is a simple, fairly narrow, short closed loop across the posterior end of the body. The stomach is short and oval with about 20 folds, and the anal border has about 6 rounded lobes. A single row of winding gonads is present on each side of the body, 7 on the right and 6 on the left anterior to the gut loop. These are directed toward the atrial opening, although the gonads terminate and their ducts open in the middle of the body wall well removed from the opening. The gonads consist of tubular ovary and testis follicles projecting into or enfolded by the ovary on its lateral aspect. The testis ducts extend around the ovary to join the vas deferens along its mesial surface. As the testis lobes increase in size, they expand further into the center of the ovary. The gonads are only lightly attached to the body wall and sometimes overlap one another. Occasionally they branch and appear to coalesce. Small endocarps are present on the body wall between the gonads.

Remarks. The gonads of *C. ohlini* clearly demonstrate that the species is correctly placed in the genus *Cnemidocarpa.* It is distinguished from *Styela* by the increasing envelopment by the ovary of the testis follicles as the latter increase in size. The present species is related to *C. madagascariensis madagascariensis* but has a shorter gut loop and fewer but longer gonads.

Cnemidocarpa nisiotis (Sluiter) [Brewin, 1950a, 1950b] is distinguished from *C. ohlini* by its thick, fleshy body wall, small number of gonads, larger stomach, longer and more curved gut loop, and smooth anal border. The gonads of both *C. ohlini* and *C. madagascariensis* and its subspecies represent a condition between species with few long gonads typical of the genus *Cnemidocarpa,* and species with numerous short gonads approaching a polycarplike form, as in *Polycarpa circumarata* from Venezuela (Van Name, 1945).

Styela nordenskjoldi Michaelsen

Figs. 38–41

Styela nordenskjoldi Michaelsen, 1898, p. 365.—Kott, 1969, p. 112, and synonymy.

New records. Knox Coast: *Atka* Sta. 24 (46 meters). North of South Shetland Islands: *Eltanin* Sta. 993 (300 meters); Sta. 991 (2672–3020 meters). Bellingshausen Sea: *Eltanin* Sta. 951 (4529–4548 meters). North of South Orkney Islands: *Eltanin* Sta. 558 (845–646 meters). Southeast of South Orkney Islands: *Eltanin* Sta. 1553 (3056–3459 meters). South of South Orkney Islands: *Eltanin* Sta. 1025 (3250–3285 meters). East of South Orkney Islands: *Eltanin* Sta. 1078 (604 meters). Between South Shetland and South Orkney Islands: *Eltanin* Sta. 1003 (210–220 meters). Northwest of South Georgia: *Eltanin* Sta. 1527 (3742–3806 meters). Northeast of South Sandwich Islands: *Eltanin* Sta. 1578 (4236–4273 meters). Tierra del Fuego: *Eltanin* Sta. 1500 (73–79 meters); Sta. 1603 (256–269 meters); Sta. 1604 (769–869 meters). Southern Chile: *Eltanin* Sta. 1605 (522–544 meters). East of Stewart Island: *Eltanin* Sta. 1430 (165–192 meters). Off North Island, New Zealand: *Eltanin* Sta. 1716 (128–146 meters).

Distribution. Widespread; see Kott, 1969.

Description. The same variation in size and form is observed in specimens in these collections as was previously described for the species [Kott, 1969]. Individuals, which range from 0.4 to 2.5 cm in maximum diameter, are stalked, sessile, or rooted and upright and are dome-shaped or flattened onto the substrate. The test is wrinkled or covered with small rounded papillae or is hard and scaly. Sometimes the test of sessile specimens is extended out into accessory rootlike extensions around the basal border, or there is a tuft of fine, tough hairs from the basal half of the test. The body wall is especially muscular with almost continuous outer circular muscles and inner longitudinal bands. Simple branchial tentacles alternate with rudimentary tentacles. With the contraction

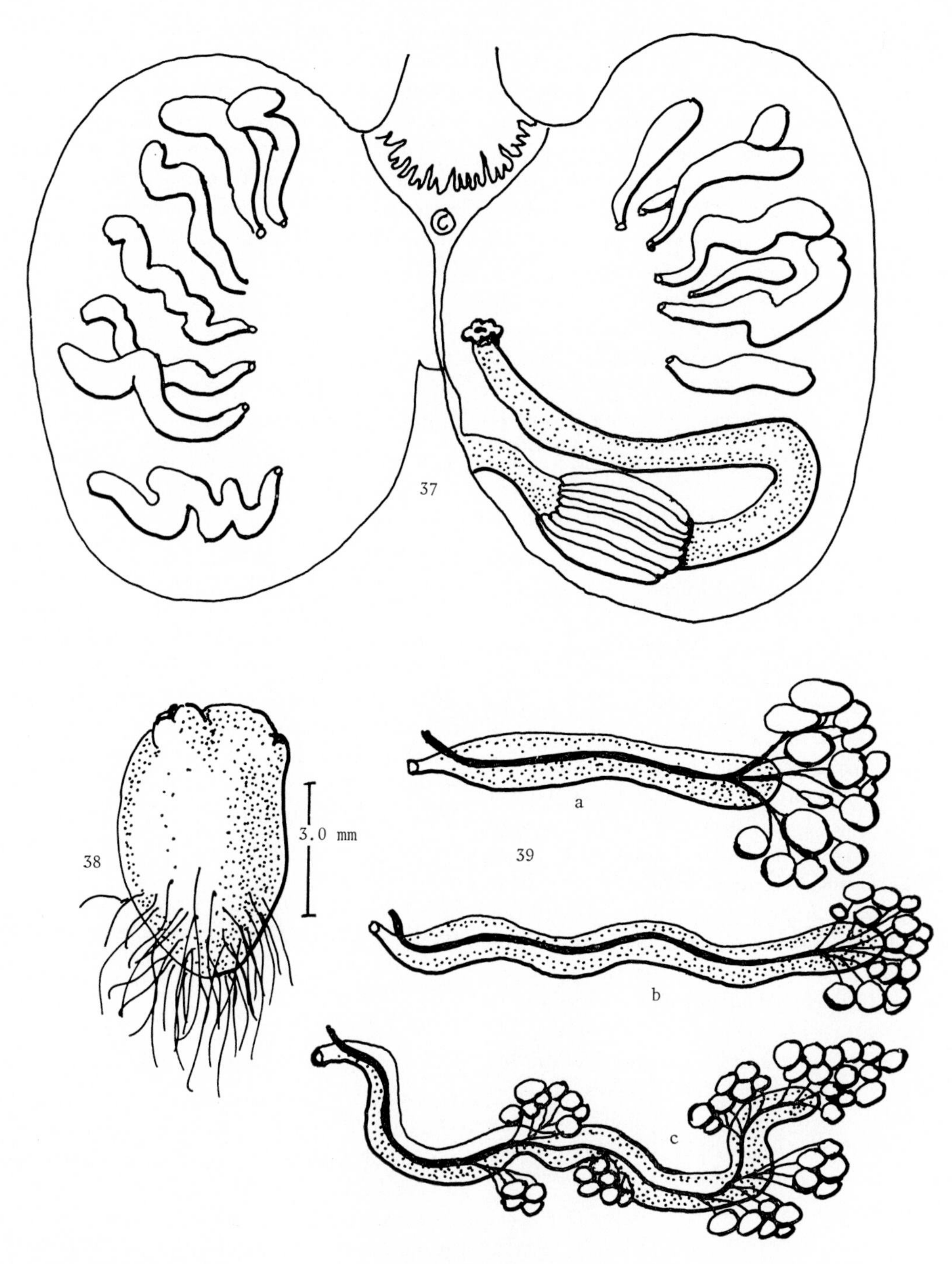

Cnemidocarpa ohlini

37. Internal structure.

Styela nordenskjoldi

38. Individual representation of rooted specimens taken at *Eltanin* Stas. 991, 1078, 1527.
39. Gonads from different specimens.

of the circular muscle at the base of the siphon the tentacles are supported on the edge of a fold of the inner siphonal wall, which projects into the lumen of the branchial opening. There is not a true velum between the branchial tentacles. The base of the tentacles extends anteriorly and posteriorly along the siphonal lining and across the prepharyngeal area respectively as fine ribs from the base of the tentacles.

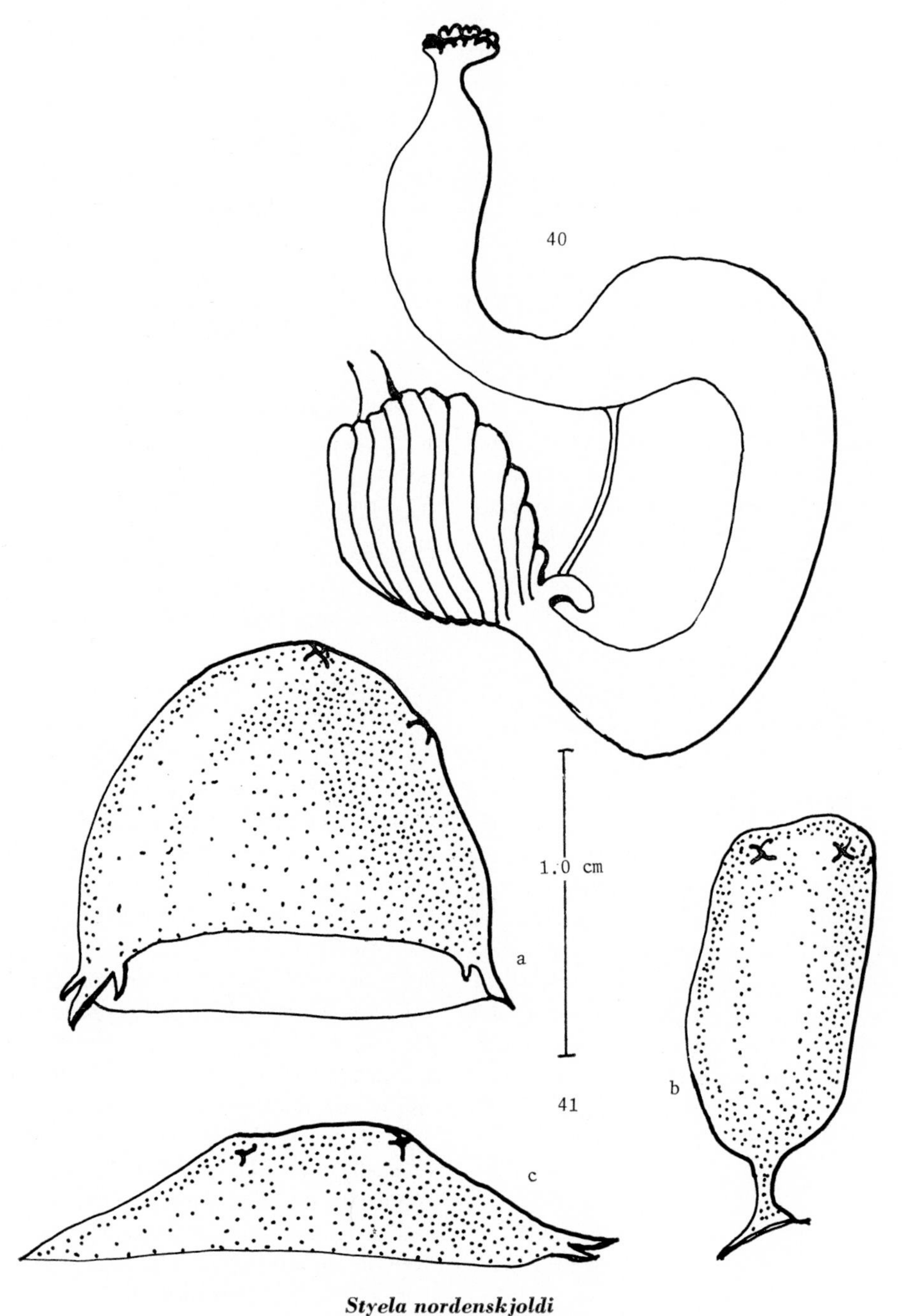

Styela nordenskjoldi

40. Gut (*Eltanin* Sta. 1078).
41. (a, b, c) External appearance of three specimens (*Eltanin* Sta. 558).

The 4 rounded, well-spaced folds in the branchial sac are sometimes suppressed altogether or fade out posteriorly. The internal longitudinal vessels are crowded, especially on the folds. The stomach may be voluminous and elongate, elliptical, or short and oval, with 10 to 20 internal folds. There is always a small, curved stomach caecum at the pyloric end of the stomach and a gastrointestinal connective. The intestinal loop may enclose an endocarp. The rectum is fairly long, and the anus has a lobed border. There are 1 or 2 long, sinuous or shorter, straight gonads on each side of the body. Testis follicles may lie against the body wall beneath the ovarian tube or be closely applied to the sides of the ovary; they may also be separated from the ovary and lie on the body wall around it, especially at its proximal end. The testis follicles are often discontinuous along the sides of the ovary or confined entirely to the distal end. The position of testis follicles in relation to the ovary depends on the extent to which they have proliferated (as previously described [Kott, 1969]). In some individuals the changing position of the testis lobes in relation to the ovarian tube can be observed where follicles are at different stages of development along the length of the ovarian tube. Where the 2 gonads are present, the gut loop may partly cover the posterior one on the left side.

Remarks. Specimens from *Eltanin* Sta. 558, all fixed on solitary corals, demonstrate almost the full range of external and internal variations in size and shape described above. Individuals with very fine hairs on the posterior half of the body, as in *C. barbata* Vinogradova, 1962, *C. bifurcata* Millar, 1964, and some specimens of *S. milleri* Ritter [see Van Name, 1945], are available from *Eltanin* Stas. 991, 1078, and 1527. Vinogradova [1962] has already noted the similarity between her *C. barbata* and *C. drygalski* (< *S. nordenskjoldi*). The branchial sac and gonads of *C. barbata*, *C. milleri*, and *C. bifurcata* also fall within the range of variations found in *S. nordenskjoldi* and observed in the present collection. Only *S. bifurcata* Millar, 1964, may be distinguished by the fine feltwork of hairs on the rest of the body. Therefore, although the test hairs of the specimens in this collection are sometimes finer and longer than those usually found in the species, this is shown to be a variable character, and the specimens appear to be correctly identified as *S. nordenskjoldi*. The suggested synonymy of *C. barbata* and *S. milleri* with *S. nordenskjoldi* [Kott, 1969] is also strongly indicated.

Styela wandeli (Sluiter)

Tethyum wandeli Sluiter, 1911, pp. 37, 38.
Styela wandeli; Kott, 1969, p. 117, and synonymy.

New record. Antarctic Peninsula: *Staten Island* Sta. 66/63 (62 meters).

Distribution. Limited area off the Antarctic Peninsula.

Description. As previously described, with a warty roughened test, paired ovarian tubes on both sides of the body, with very much proliferated and coalesced testis follicles on both sides of the body in a more or less continuous sheet around the proximal half of the ovaries and separated from them by endocarps.

Remarks. This new record of a specimen within the limited range indicated for this species confirms its distinguishing characteristics as reliable in separating it from the closely related species *S. nordenskjoldi.* Extensive proliferation of testis lobes in the latter species takes place across the proximal end of the ovary; but endocarps do not separate them from the ovary, and the continuous sheets of testis found in the present species do not occur in *S. nordenskjoldi.*

Styela bathyphila (Millar)

Figs. 42–44

Cnemidocarpa bathyphila Millar, 1955, p. 228.

New records. South Australian basin: *Eltanin* Cruise 20, Sta. 107 (4078–4146 meters). Bellingshausen Sea: *Eltanin* Sta. 803 (4328 meters).

Distribution. Having previously been recorded only from 01°03′–00°58′N, 18°40′–18°37′W, in 5250–5300 meters [Millar, 1955], this species can be expected to occur in the deeps of the Indian and Atlantic oceans at least as far north as the equator and probably into the North Atlantic.

Description. Specimens are ovoid and dorsoventrally flattened; maximum length is 0.8 cm to 1.5 cm. The test on the upper convex surface is whitish, thick, and tough, with a rough surface. The test on the undersurface, concave in the present specimens, is thinner but very tough and fixed to pebbles. Around the equatorial border the test is produced into short hairs or roots, undivided or with 2–4 branches from a common base. The apertures are sessile, 4-lobed, and well separated on the upper surface; the branchial aperture is almost terminal, while toward the opposite end of the body the atrial aperture is on the highest part of the upper surface, with the opening directed vertically upwards. The body wall is thin and closely applied to the test. There are sparse, fine, transverse and some longitudinal muscle bands on the left or base of these individuals where the dorsoventrally flattened body is folded around the endostyle. On the right or upper side of the body there is an almost continuous outer coat of transverse bands and inner longitudinal bands. There are 12 thin, simple tentacles joined basally by a broad tentacular velum. There are no rudimentary tentacles. The dorsal tubercle, in a very shallow peritubercular area, has a simple opening. In contracted specimens the prepharyngeal band supports a broad, thin, frilled velum, especially along either side of the dorsal tubercle. This disappears, however, when the branchial opening is

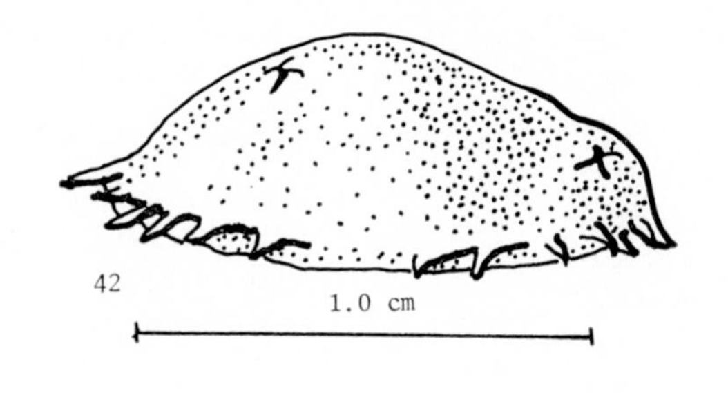

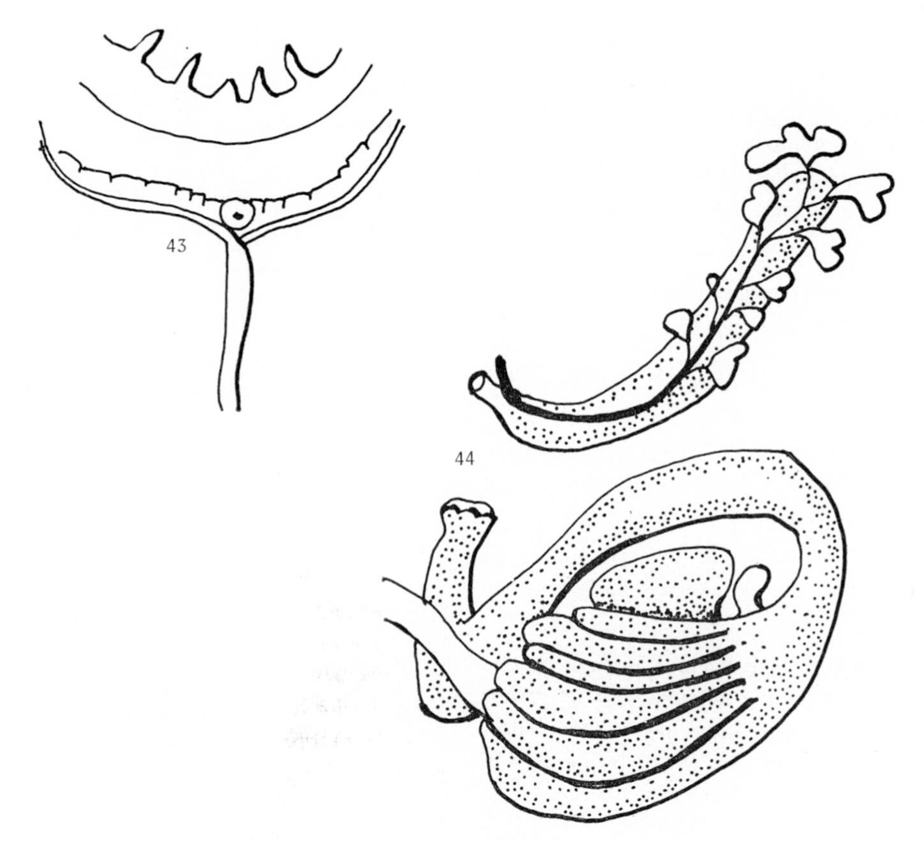

Styela bathyphila

42. External appearance.
43. Diagram showing tentacular velum, dorsal tubercle, and prepharyngeal velum.
44. Left inner body wall, showing gut and gonads.

extended. The dorsal lamina is a wide, plain membrane.

The branchial sac has about 30 delicate internal longitudinal vessels on each side of the body and is raised into 2–4 rounded but prominent folds by the expansion of transverse vessels into the folds. There are 1½–3 long, rectangular stigmata per mesh, crossed by parastigmatic vessels. The gut loop is small at the posterior end of the body. The esophagus is relatively long and the stomach pyriform, with 10 to 12 folds and a small caecum at the pyloric end. The intestine is short and encloses a narrow endocarp against the stomach before turning sharply anteriorly behind the esophagus to form a short rectum with lobed anal border. There are single gonads on each side of the body consisting of straight or curved ovarian tubes directed toward the ovary. Pyriform testis follicles lie beneath the ovary initially but extend out onto the body wall on each side of the ovary as the follicles proliferate into many lobes.

Remarks. The appearance of the surface test in this species resembles that of *S. nordenskjoldi,* and in certain individuals of the latter species the test in the equatorial region is produced into rootlike extensions, as in the present species. The gonads are identical in the two species. *S. bathyphila,* however, is distinguished by the position of the apertures, the branchial aperture being almost terminal and the atrial aperture near the opposite end of the upper surface. Differences in the gut are related to the position of the apertures, the long esophagus, the short rectum, and the restricted gut loop confined to the posterior end of the body, contrasting with the situation in *S. nordenskjoldi,* in which the apertures are close together and the gut loop is voluminous and extends anteriorly. The branchial tentacle, the velum, and the delicate internal longitudinal vessels further distinguish the present species.

Styela sericata Herdman

Figs. 45–47

Styela sericata Herdman, 1888, p. 153.—Kott, 1969, p. 122, and synonymy.

New records. Southeast Pacific basin: *Eltanin* Sta. 1654 (4297–4667 meters). Southwest Pacific basin: *Eltanin* Cruise 25, Sta. 346 (3914 meters).

Distribution. The species has a wide depth range down to 4820 meters in the deeper basins of all oceans.

Description. The present specimen, 1.5 cm high and 3.2 cm long, is typical of the species, with long, fine hairs all over the body except the areas immediately around the sessile apertures, which are terminal, and toward the opposite end of the upper surface. The body wall is thin and closely adherent to the test, with fine, sparse muscle bands crossing one another irregularly. The prebranchial area has small papillae, and about 25 branchial tentacles alternate with rudimentary tentacles. The dorsal tubercle has a simple opening. The peritubercular area is very shallow, and the usual V shape is partly obliterated by the anterior extension of the dorsal raphe. There is a V-shaped area of plain membrane in the anterior part of the branchial sac on either side of the dorsal lamina where it extends anteriorly almost to the dorsal tubercle. The most dorsal internal longitudinal vessels of the branchial sac terminate at the edge of this V-shaped membrane behind the dorsal tubercle and extend obliquely across the whole of the branchial sac, crossing the stigmata obliquely. The present specimen has 60 internal longitudinal vessels on each side of the branchial sac.

Remarks. The positions of the apertures and the related short posterior gut loop resemble those of *S. bathyphila,* although the gut loop of the latter is double whereas the rectum does not bend anteriorly in the present species. The present species is also distinguished by the oblique internal longitudinal vessels, the absence of branchial folds, the sparse body musculature, the fine long hairs covering the body, and the presence of rudimentary tentacles. The present specimen is the largest described and has an unusually large number of internal longitudinal vessels in the branchial sac.

Styela bythia Herdman

Styela bythia Herdman, 1881, p. 63; 1882, p. 151.
Cnemidocarpa bythia; Millar, 1959, p. 194.

New record. Southwest Pacific basin: *Eltanin* Cruise 25, Sta. 346 (3914 meters).

Distribution. The species has previously been taken from the South Australian basin [Herdman, 1881, 1882]; and from the Tasman Sea and Kermadec trench [Millar, 1959]. It can therefore be expected to have a wide distribution in the deeper ocean basins of the southern hemisphere.

Description. The present specimen is dome-shaped and 1 cm in diameter. The surface is without furrows or folds, and the test is covered with minute short papillae. Both apertures are sessile on the upper surface. The branchial tentacles are simple and extend from the edge of a pronounced branchial velum at the base of the branchial opening. The peritubercular area is very deep and encloses a circular opening at the base. The dorsal lamina is a broad membrane with long pointed languets from the free margin. There are 4 rounded branchial folds on each side of the body with up to 10 internal longitudinal vessels on all but the most ventral fold. There are 2 to 3 internal longitudinal vessels between the folds. The body wall, thin and closely adherent to the test, has very few muscles. Neither gut loop nor gonads were present in this specimen.

Remarks. The present specimen conforms entirely with previous descriptions of the species. The dorsal tubercle and the peritubercular area in particular are identical with the condition described and figured by Millar [1959].

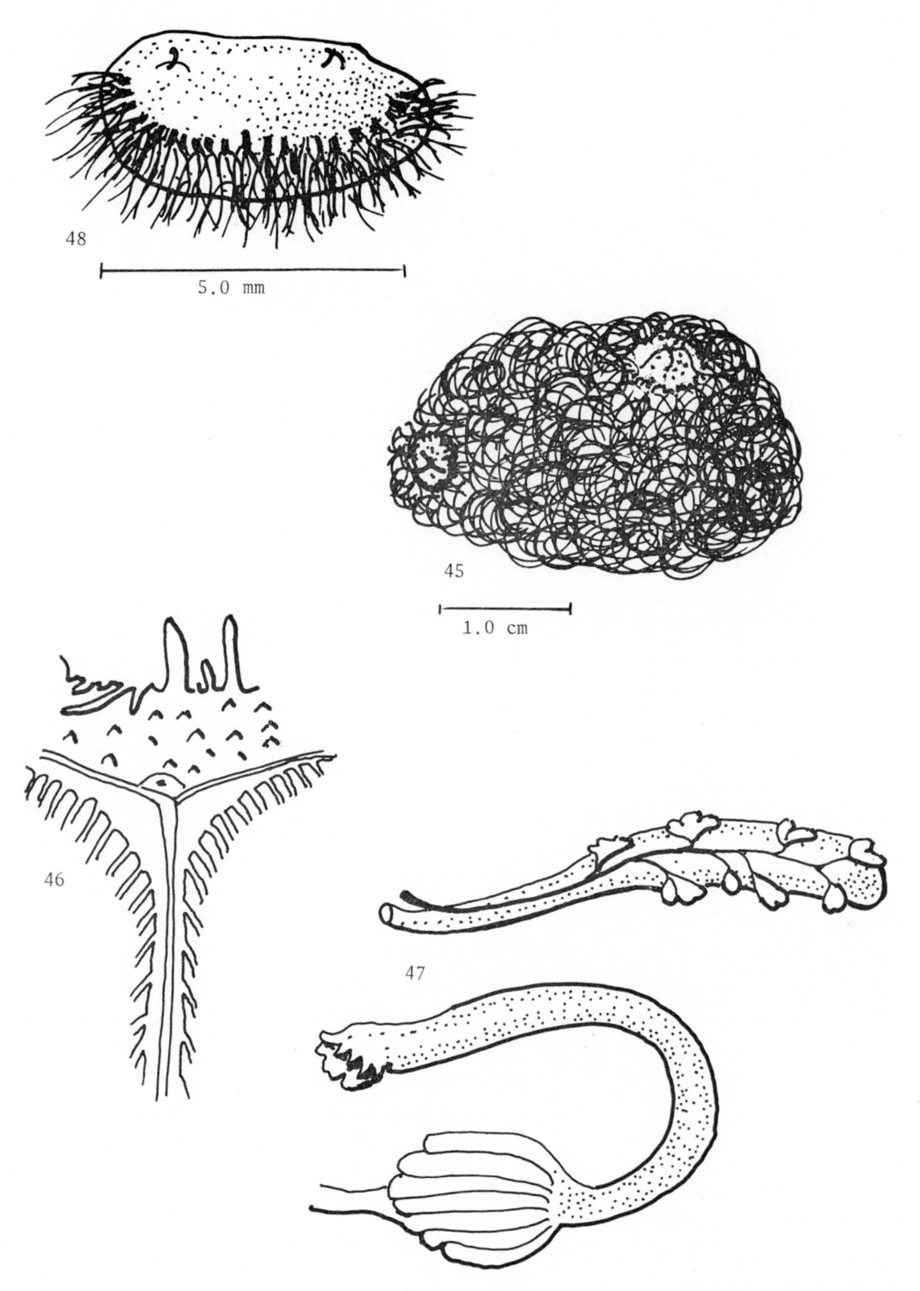

Styela sericata

45. External appearance.
46. Branchial tentacles, prepharyngeal papillae, dorsal tubercle, dorsal lamina, and origin of oblique internal longitudinal vessels.
47. Gut and gonads.

Bathyoncus enderbyanus

48. External appearance.

Bathyoncus enderbyanus Michaelsen

Fig. 48

Bathyoncus enderbyanus Michaelsen, 1904, p. 226.—Kott, 1969, p. 125, and synonymy.

New records. Bellingshausen Sea: *Eltanin* Sta. 786 (4602 meters). East of South Sandwich Islands: *Eltanin* Sta. 1571 (3947–4063 meters). Tasman Sea: *Eltanin* Sta. 1837 (4859–4868 meters).

Distribution. The present records confirm the distribution in the deeper waters of the southern Indian, Pacific, and Atlantic oceans.

Description. The specimens are typical of the species. In the single specimen from Sta. 1571 the filamentous processes arising from the papillalike trunks around the equatorial region of the body form a close feltwork coat externally, leaving a space next to the test.

Remarks. The fine feltwork of hair in one specimen of this collection resembles *Pyura tunica* Kott, 1969, in which test processes actually coalesce to form an outer coat, with a space between this and the test open to the outside around the siphons.

Bathyoncus mirabilis Herdman

Figs. 49–52

Bathyoncus mirabilis Herdman, 1882, p. 165.—Kott, 1969, p. 126.

Bathyoncus herdmani Michaelsen, 1904, p. 228.—Kott, 1969, p. 126.

Fungulus antarcticus Herdman, 1912, p. 308; 1915, p. 90.—Hartmeyer, 1912, pp. 374, 378.—Van Name, 1945, p. 367.

New records. Bellingshausen Sea: *Eltanin* Sta. 954 (4685 meters). North of South Shetland Islands: *Eltanin* Sta. 993 (300 meters).

Distribution. Previous records are from the Atlantic-Indian basin [Herdman, 1882, 1912; Michaelsen, 1904] in 2926–4636 meters. The present specimens extend the range to 300 meters on the Scotia Ridge and down to 4685 meters in the Southeast Pacific basin.

Description. Individuals have rounded to egg-shaped heads, narrowing into a short wide stalk about one-third the total height expanded basally over the substrate. The largest specimen available, from Sta. 954, 2.2 cm high, has a juvenile of 1.5 mm growing on the upper surface. The maximum diameter of the head region is about two-thirds of its length. The apertures are both sessile either side of the upper surface, although in the juvenile from Sta. 954 both apertures are on the same side of the head and the branchial aperture is produced toward the base, while the atrial aperture is near the upper surface. The branchial aperture is 4-lobed; the atrial aperture is also 4-lobed, but it is modified to present a transverse slitlike opening. The surface of the whitish test is fairly rough, with minute rounded or flattened papillae and some hairs. The specimens from Sta. 993 are black with encrusting and embedded sand. The thin body wall is closely adherent to the test, and fine muscle bands form a sparse network up to halfway down the body but converge posteriorly to enter the stalk. Strong circular muscles are present around the apertures. There is a circle of about 25 fairly long, simple branchial tentacles with a wide membranous vane along the posterior border, alternating with smaller tentacles. The dorsal tubercle has a simple C-shaped opening. There is a smooth-bordered dorsal lamina.

The branchial sac has 1 large, rounded fold near the dorsal lamina; 3 further vestigial folds are indicated by the close arrangement of longitudinal vessels held in place by intermediate transverse vessels [Herdman, 1882]. There is considerable variation in the numbers of internal longitudinal vessels forming the true or vestigial folds. For example:

Individual from Sta. 993:

3 (5) 3 (5) 3 (5) 2 (12) 3 DL 1 (14) 2 (4)
3 (5) 2 (5) 2

Individual from Sta. 954:

1 (4) 2 (2) 0 (3) 2 (8) DL 2 (8) 2 (2) 0 (3) 2 (4)

There are no stigmata, and the internal longitudinal vessels cross the transverse vessels to form longitudinal rectangular meshes occasionally crossed by accessory longitudinal connectives. The gut forms a long curved narrow loop posterodorsally on the left side of the branchial sac. The stomach, with about 30 folds, not all extending its whole length, is pyriform to long and elliptical, probably depending on the state of contraction of the body. There is a gastrointestinal connective and a vestigial caecum at the pyloric end. The anus is bilabiate. There is a single long, sinuous gonad of typical styelid form on each side of the body directed toward the atrial opening, the one on the left lying parallel with the gut loop. The testis follicles extend from beneath the ovary to lie alongside it as they increase in size. There is a large mushroom-shaped endocarp in the center of the body wall anterior to the gonad on each side of the body.

Remarks. These specimens agree with the description of the type [Herdman, 1882] in all respects except the stomach, which Herdman describes as being not of appreciably greater diameter than the rest of the gut. Variations in the shape of the stomach related to its state of contraction are observed in the present

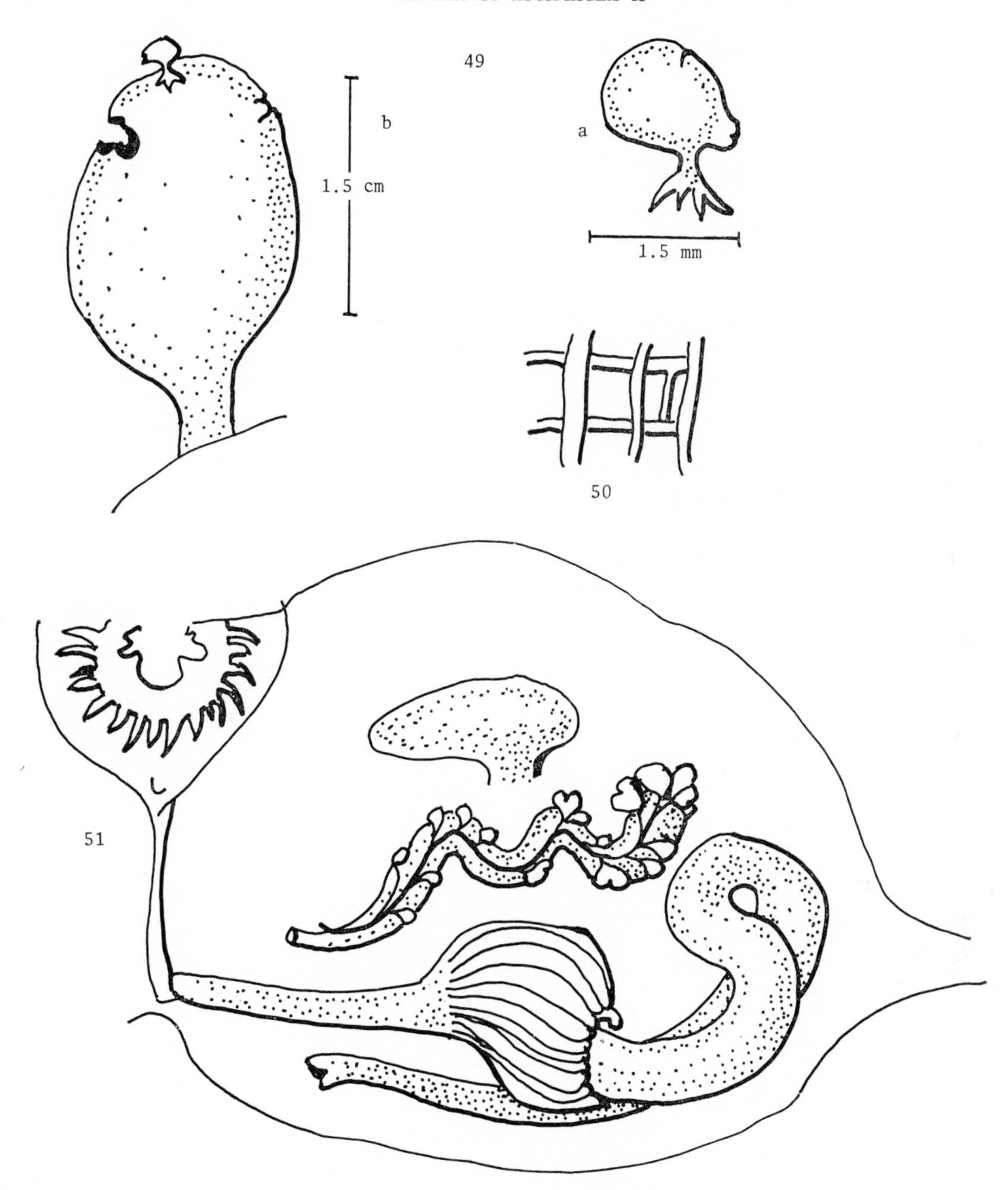

Bathyoncus mirabilis

49. Adult individual with juvenile fixed to test (*Eltanin* Sta. 954).

49a. Juvenile.

50. Portion of branchial sac.

51. Internal structure, left side of body (*Eltanin* Sta. 954). Stomach contracted.

specimens, and this phenomenon could explain the condition Herdman has described. Descriptions of specimens ascribed to the other synonyms indicated above agree with the present specimens and confirm the suggestion [Van Name, 1945; Kott, 1969] of the styelid nature of *Fungulus antarcticus* (< *B mirabilis*), despite the external appearance, which closely resembles *Fungulus cinereus* and demonstrates convergent evolution in these two deep-water genera.

Family PYURIDAE Hartmeyer

Pyura paessleri (Michaelsen)

Figs. 53, 54

Cynthia paessleri Michaelsen, 1900, p. 106.

Pyura paessleri; see Kott, 1969, p. 129, and synonymy.

New records. Tierra del Fuego: *Eltanin* Sta. 740 (384–494 meters); Sta. 969 (229–265 meters).

Distribution. This species, although common in the Magellanic province of the Subantarctic, has not been taken elsewhere. The present record from Sta. 740 represents the greatest depth from which the species has been taken.

Description. The specimens are typical, and in all of them the thin test is completely stiff, with adherent, often large particles of sand and shell. The ovarian tube of the styelid-type gonads regularly undulate, and the testis follicles are clustered around the outer curve of each undulation. Seminal ducts from the male follicles extend across the ovary to the vas deferens, which extends in a straight line on the mesial surface but does not follow the undulations of the ovary. It is probable that the vas deferens and the testis follicle ducts joining it do not increase in length to accommodate increases in length or width of the ovarian tube as it matures. Maturation of the ovary apparently causes an increase in the length of ovarian tube, which is accommodated by the development of an undulating or winding course along the axis formed by the vas deferens.

Remarks. The styelid gonad and test are distinctive.

Pyura georgiana (Michaelsen)

Boltenia georgiana Michaelsen, 1898, p. 364.

Pyura georgiana; Kott, 1969, p. 130, and synonymy.

New records. Ross Sea, Kainan Bay: *Atka* Sta. 4 (610 meters); Deep Freeze I, *Edisto* Sta. ? (644 meters). South Shetland Islands: *Eltanin* Sta. 1001 (238 meters); *Staten Island* Sta. 61/63 (31 meters). South Georgia: *Eltanin* Sta. 1535 (97–101 meters).

Distribution. The present records fall within the limits previously known.

Description. The very large numbers of specimens from South Georgia are all covered with needlelike spines with or without minute secondary branches. Minute spines are also present between the larger spines. The spines are thicker than usual and confer a furry texture to the surface of these individuals. The stalk of these specimens varies in length and thickness, although it rarely reaches the length found in specimens taken from deeper water. Parietal organs are present in most specimens, although no atrial organs were detected. The apertures are typically from either end of the upper surface.

Remarks. Millar used the form of test spines of specimens in the *Discovery* collections [Millar, 1960] to distinguish *P. georgiana* as an endemic South Georgian form from *Pyura bouvetensis,* which has a wider antarctic distribution. Kott [1969] found variability in test spines of individuals from all areas and was not able to confirm Millar's observations. The present collection from Sta. 1535, off South Georgia, supports the contention that the form of the test spines is not a reliable characteristic on which to base a specific distinction.

Attached by its proboscis to a specimen from *Staten Island* Sta. 61/63 is an immature pycnogonid identified by C. A. Child, Division of Crustacea, U.S. National Museum, as *Ammothea gibbosa* Möbius. The proboscis is inserted through the downwardly directed branchial aperture into the branchial sac. The tentacular ring and siphon are stretched around the posterior end of the pycnogonid proboscis, which extends along in the dorsal part of the branchial sac to a point about halfway along the dorsal lamina. Beyond this a circular section from the posterodorsal part of the branchial sac is absent, and the edges of the remaining part of the branchial sac are cut, frayed, and torn. The opening into the atrial siphon from the peribranchial sac is thus exposed. The gonads in this specimen are ripe, and it is probable that, in fact, the pycnogonid may have been feeding on the genital products of the ascidian after their release from the gonads. The gonads themselves, in this specimen, are not damaged.

Release of genital products in the Ascidiacea follows the intake of genital products of the same species into the branchial sac [Carlisle, 1951; Kott, 1969, p. 173], and it seems unlikely that the branchial aperture was completely occluded by the proboscis of the pycnogonid before this could happen. The initial release of sperm or eggs and subsequent cessation of ciliary feeding current may have provided the stimulus for the insertion of the pycnogonid proboscis into the branchial aperture. This, in turn, could have stimulated the muscular contraction of the siphonal muscles that closed the atrial aperture and firmly supported the pycnogonid in a position where it had access to gonadial products released into the closed ascidian.

Pyura legumen (Lesson)

Boltenia legumen Lesson, 1830, p. 433.

Pyura legumen; Kott, 1969, p. 133, and synonymy.

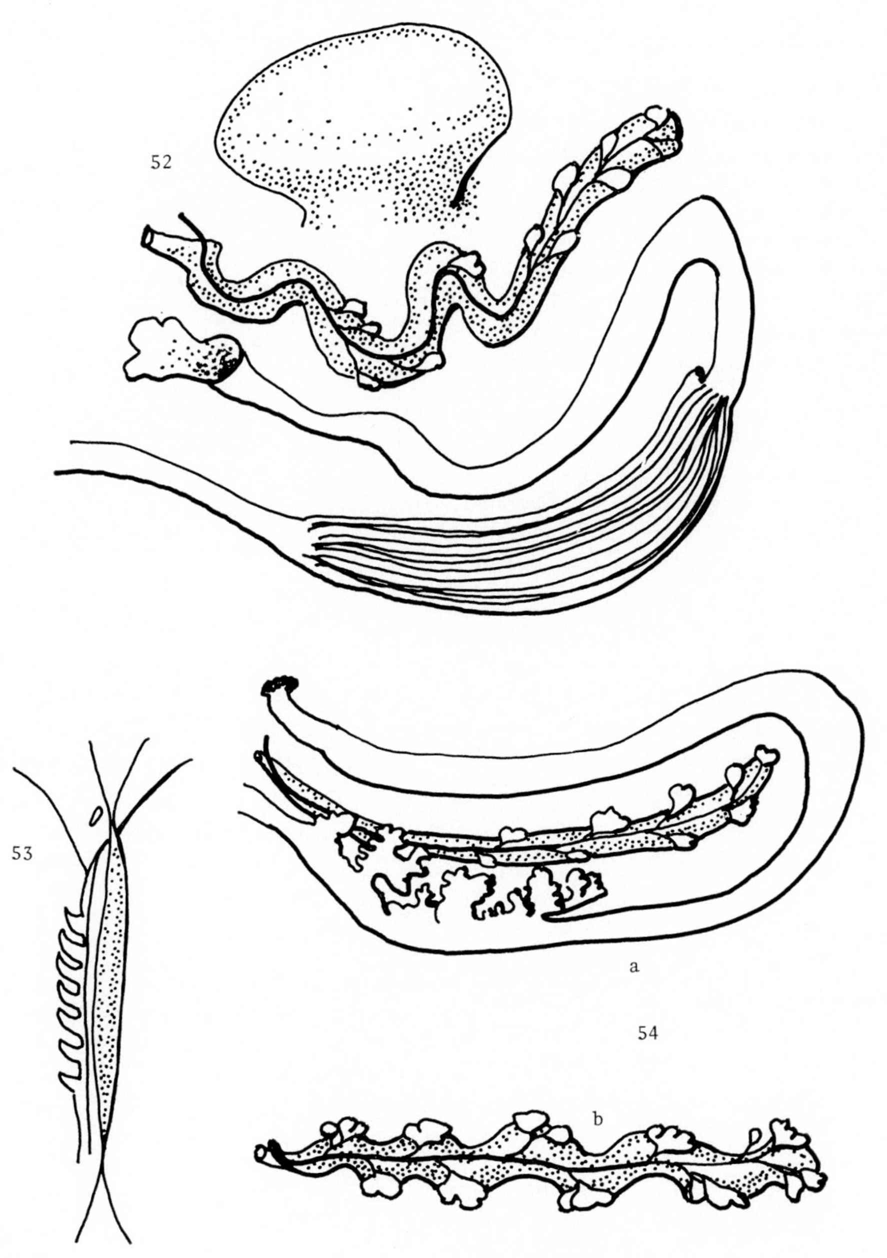

Bathyoncus mirabilis

52. Internal structure, left side of body (*Eltanin* Sta. 993). Stomach extended.

Pyura paessleri

53. Dorsal tubercle, long dorsal ganglion, and anterior part of dorsal lamina.

54a. Gut loop and gonad (*Eltanin* Sta. 740).

54b. Gonad (*Eltanin* Sta. 969).

New record. Tierra del Fuego: *Eltanin* Sta. 1500 (73–79 meters).

Distribution. Confined to the Magellanic province.

Description. A single small specimen with a short stalk and both apertures on one side of the elongate head. The test is whitish and leathery, with fairly sparse, minute, pointed needles. No parietal organs are developed, but atrial organs are present.

Remarks. The species is distinguished from *P. georgiana* by the elongate body with apertures on the side rather than on the upper surface and by the presence of atrial organs and absence of parietal organs.

Pyura squamata Hartmeyer

Pyura squamata Hartmeyer, 1909–1911, p. 1337.—Kott, 1969, p. 135, and synonymy.

New records. Tasman Sea: *Eltanin* Sta. 1818 (913–915 meters).

Distribution. This represents the most northerly record of this species, which is known from deep waters off the Antarctic Peninsula and Wilhelm II Coast. The species appears closely related to *P. lepidoderma* from shallow waters off eastern Australia and Japan [Kott, 1969].

Description. The two specimens in the present collection are very small, with maximum diameters of 0.3 and 0.6 cm. They are dorsoventrally flattened and fixed by the whole basal surface. Both apertures are sessile and at opposite ends of the upper surface. The test is hard, tough, and whitish internally, and sand grains are embedded in the surface. The polygonal scalelike areas are fairly small and obscured by some sand. There are 6 rounded branchial folds on each side of the body. Internal longitudinal vessels are crowded together on the folds, but there is only a single internal longitudinal vessel between the folds. The body wall has strong musculature.

The gut and gonads are as previously described for the species.

Remarks. These individuals are smaller than previously described, and the scalelike thickenings in the test are unusually small. However, the shape of the body, position of apertures, branchial sac, gonads, and gut are similar to those previously described. In *P. mariscata* [Rodrigues, 1966], a closely related if not identical species from 140 meters off Brazil, the greater part of the body has adherent sand, as in the present specimens, and the scalelike thickenings are not apparent. Their absence, especially in the present small specimens, does not necessarily indicate a distinct species.

Pyura discoveryi (Herdman)

Halocynthia discoveryi Herdman, 1910, p. 9.
Pyura discoveryi; Kott, 1969, p. 136, and synonymy.

New records. Ross Sea, Kainan Bay: *Atka* Sta. 4 (610 meters); Deep Freeze I, *Edisto* Sta. ? (644 meters). Antarctic Peninsula: *Staten Island* Sta. 24/63 (75 meters). Between South Shetland and South Orkney Islands: *Eltanin* Sta. 1003 (210–220 meters). South Georgia: *Eltanin* Sta. 1535 (97–101 meters).

Distribution. The present records fall within the known range of the species.

Description. The thick, hard, leathery, wrinkled test has scalelike thickenings. The specimens agree in all respects with those previously described. Individuals are often aggregated together in clumps.

Remarks. The scalelike thickenings in the test have not previously been described for this species but do in all cases exist, although the hard, leathery, wrinkled test has obscured them. They are similar to the scalelike thickenings in *P. squamata,* from which the present species is distinguished by the length of the siphons and the distance between the branchial folds. *Pyura subuculata* from New Zealand is also a closely related species, distinguished by the spines in the siphons, which have not been detected in *P. discoveryi.*

Pyura obesa Sluiter

Pyura obesa Sluiter, 1912, p. 454.—Kott, 1969, p. 138, and synonymy.

New record. The single juvenile present in a portion of the material from Sta. 1003 (210–220 meters) is the fourth known representative of this species. Two sizable, mature specimens from the same station were earlier discussed in *Antarctic Ascidiacea* [Kott, 1969].

Distribution. Limited to the South Shetland Islands and their immediate vicinity.

Description. Test is whitish, semitransparent, thick, and gelatinous, with randomly distributed black pigment spots associated with expanded terminal ampullae of test vessels. The outer surface is furrowed to form rounded elevations. The individual, 1 cm in length, is fixed along the ventral surface; and the sessile apertures are at either end of the upper surface. The 15 larger branchial tentacles, of variable length, have only very occasional rudimentary pinnate branches. There are 6 folds on each side of the body with internal longitudinal vessels arranged as follows:

E (3) 0 (8) 0 (8) 0 (12) 0 (6) 0 (6) 0 DL

There are 6 long rectangular stigmata per mesh between the folds. Parastigmatic vessels are present. The simple narrow gut loop and liver lobes are as previously described.

The gonads on the left are not developed. On the right, however, there is an ovarian tube with rudimentary testis follicles. The mature gonads of this species are separate polycarp sacs on either side of the central duct.

Remarks. The structure of the juvenile gonad on the right suggests the origin of this type of pyurid gonad: the straight styelid gonad is gathered into undulations as it increases in length while the vas deferens, extending straight along its upper surface, remains the same length. This is similar to the condition observed for *Pyura paessleri.* These undulations are eventually separated off as sacs on either side of the central ducts. The gelatinous test and expanded terminal ampullae of the test vessels resemble the conditions found in *Phallusia* spp.

Despite its lack of development, this juvenile is identified with *Pyura obesa* by the thick test and the form and course of the gut.

Pyura cancellata Brewin

Pyura cancellata Brewin, 1946, p. 121; 1948, p. 134; 1951, p. 104; 1956, p. 121; 1957, p. 578; 1958, p. 440.

New record. New Zealand, east of South Island: *Eltanin* Sta. 1431 (51 meters).

Distribution. Otago and Auckland harbors [Brewin, 1946, 1957]; Hauraki Gulf [Brewin, 1948, 1951]; Chatham Island [Brewin, 1956]; Stewart Island [Brewin, 1958], from the intertidal region near low water mark to 51 meters, the greatest depth at which the species is recorded.

Description. The 2 specimens in the present collection are irregular and sandy. The outer fibrous coat around the body is formed by fibrous extensions branching from papillalike trunks rising from the test. These fibrous extensions coalesce, involving sand particles. There is a space between this outer coat and the test, which is open to the exterior around the apertures and here and there where the sand and fiber coat is not complete. Worms, sand, etc., and water freely circulate between the test and the outer fibrous coat. The rampartlike extensions around the apertures are due to the forward extension of this outer coat.

The specimens conform in internal structure with those described by Brewin [1946].

Remarks. Pyura carnea Brewin, 1948, differs from the present species in the continuity of the outer coat with the test around the siphons, the greater number of internal longitudinal branchial vessels on each fold, the shorter dorsal ganglion, and the reduced number of gonad lobes. The largest recorded specimen of *P. cancellata* has a maximum dimension of 5.5 cm, while *P. carnea* extends up to 10.0 cm in length. Both have been taken in the Hauraki Gulf [Brewin, 1948].

Pyura tunica Kott, 1969, from the Knox Coast, has a test modified in a manner similar to the present species and is distinguished only by a wide gut loop and larger numbers of gonad sacs and internal longitudinal vessels than in New Zealand species.

Both *P. cancellata* and *P. tunica* are species from relatively shallow water, with a limited distribution in regions that are widely separated. They probably represent relict species. (See also Kott, 1969, p. 138, *Remarks.*)

Pyura picta Brewin

Figs. 55, 56

Pyura picta Brewin, 1950, p. 58; 1958, p. 440.

New record. New Zealand, east of South Island: *Eltanin* Sta. 1431 (51 meters).

Distribution. The species has previously been recorded from Oamaru, Foveaux Strait, and Pegasus Bay in the South Island to the Stewart Island region down to 92 meters [Brewin, 1950, 1958]. Its known range is therefore very limited.

Description. Individuals occur singly or in clumps. The branchial aperture is terminal, and the atrial aperture is on the dorsal surface. The test is very rough and irregular, with octagonal depressed areas marked off by sharp ridges. There are minute pointed scales (0.011 mm long, 0.006 mm wide) lining the siphons. Strong muscles radiate from the siphons, becoming more diffuse posteriorly. The branchial tentacles have short primary branches and very occasional minute secondary branches. The dorsal tubercle is large and the opening horseshoe-shaped, with the right horn turned in. Seven overlapping branchial folds are present on each side of the branchial sac. There are four stigmata per mesh between the folds. The internal longitudinal vessels are arranged as follows:

DL 1 (16) 1 (9) 1 (17) 1 (17) 1 (14)
1 (11) 1 (5) E

The gut forms a wide, open loop, with a large, lobed liver in the pyloric region. There are 6 well-defined anal lobes. The gonads, on each side of the body, are divided into 5 to 6 pairs of polycarp sacs either side of central ducts. The left gonad is in the gut loop.

Remarks. The gonads of this species are typical of many Antarctic-New Zealand species of the genus. In particular, the species resembles *P. subuculata*

(Sluiter), *P. squamata* (Hartmeyer), and *P. discoveryi* (Herdman). It is characterized by the distinctive sharp ridges of the test, which mark the surface into polygonal areas. It is further distinguished from *P. subuculata* by the absence of bushy tentacles and from *P. squamata* by the shape of the body and the position of the apertures.

Pyura subuculata (Sluiter)

Figs. 57, 58

Cynthia subuculata Sluiter, 1900, p. 27.
Pyura subuculata Michaelsen, 1922, p. 406.—Brewin, 1948, p. 117; 1950b, p. 354; 1951, p. 104; 1957, p. 578; 1958, p. 440; 1960, p. 119.

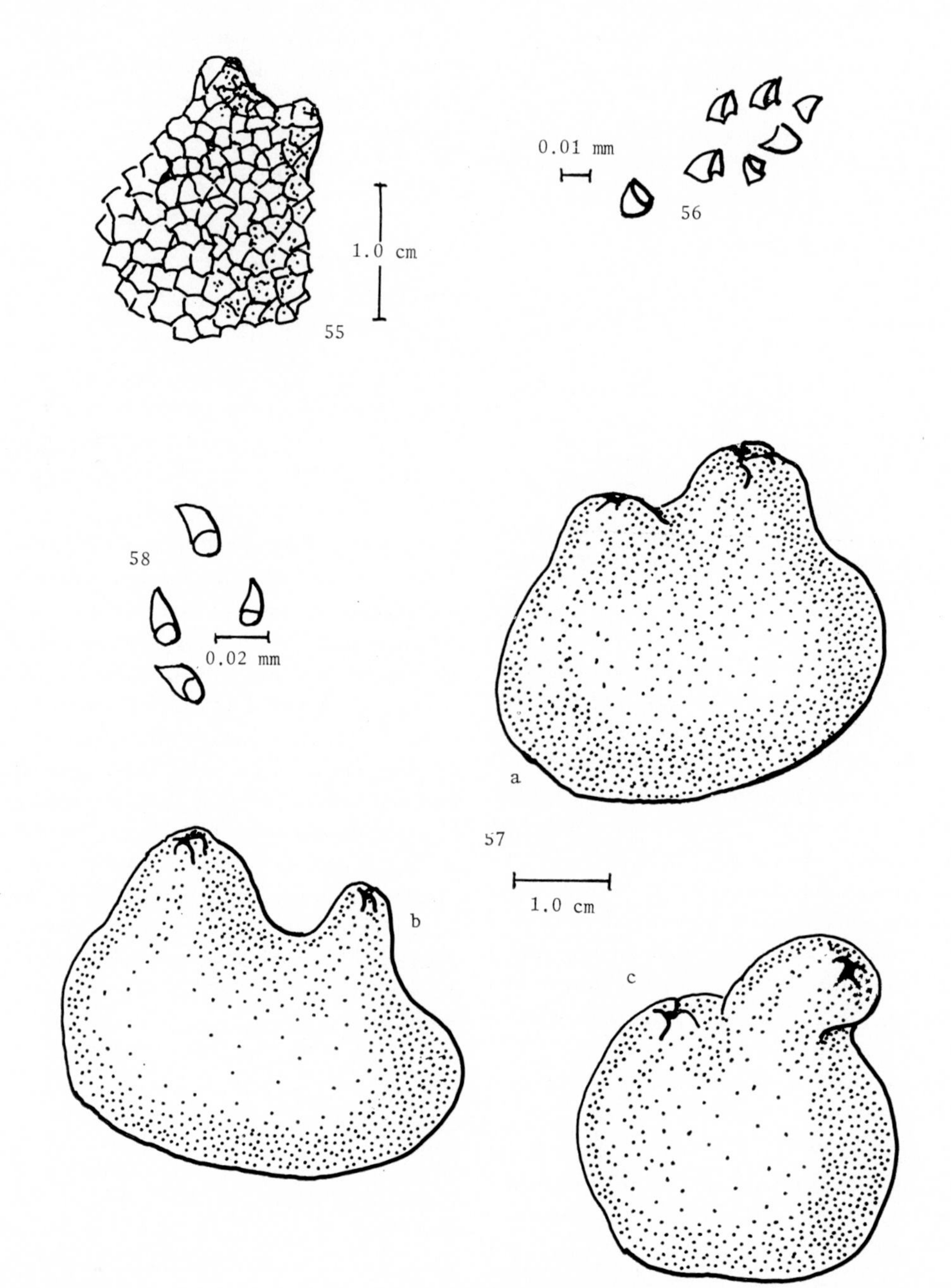

Pyura picta
55. External appearance.
56. Branchial siphonal spicules.

Pyura subuculata
57. (a, b, c) External appearance of three individuals.
58. Branchial siphonal spicules.

Pyura rugata Brewin, 1948, p. 136.
(Not *Pyura (Halocynthia) subuculata* var. *suteri* Michaelsen, 1908, p. 259; or *Pyura subuculata* Brewin, 1946, p. 119 [< *P. suteri* Michaelsen].)

New records. New Zealand, Auckland Island: *Eltanin* Cruise 20, Sta. 91 (intertidal). Tasman Sea: *Eltanin* Sta. 1818 (913–915 meters).

Distribution. New Zealand: French Pass; Sumner [Sluiter, 1900]; ? Cape Brett [Michaelsen, 1922]; Hauraki Gulf [Brewin, 1948, 1951, 1958]; Auckland [Brewin, 1957]; Great Barrier Island [Brewin, 1950b]; Cook Strait [Brewin, 1960]; Stewart Island [Michaelsen, 1922]. Previously, the greatest recorded depth for this species was 46 meters [Michaelsen, 1922].

Description. Specimens from Auckland Island taken in clumps with *Microcosmus kura* are rounded, up to 4.0 cm maximum in diameter, with apertures close together on the upper surface. The atrial aperture is almost sessile; the branchial aperture generally protrudes upward or anteriorly on a short siphon or is sometimes sessile. From *Eltanin* Sta. 1818 there are two separate individuals, each in the shape of a convex mound fixed by a flat base. The test is hard and rough externally and is sometimes produced irregularly, especially where individuals are crowded together in aggregates. Platelike thickenings with spherical inclusions are generally present and are most apparent where the test and outer cuticle are especially hardened. Minute conical spines, 0.015 mm long, line the branchial siphon. The branchial siphon is always fairly long internally; externally the length is not always so apparent. The body wall has musculature that is especially strong on the right side of the body, but not so strong elsewhere.

There are 12 bushy, compound branchial tentacles alternating with rudimentary ones. The dorsal tubercle has a horseshoe-shaped opening directed anteriorly, with both horns turned in. The dorsal lamina has short languets. In the Auckland Islands specimens there are 7 branchial folds on each side of the body. Internal longitudinal vessels are arranged as follows:

0 (8) 2 (11) 3 (15) 2 (15) 3 (15) 1 (12) 1 (8) DL

Only 6 branchial folds per side are present in the specimens from *Eltanin* Sta. 1818. There is a wide, open gut loop extending around the ventral aspect of the body with extensive arborescent liver lobes in the region of the stomach. The gonads on each side of the body are represented by 12 paired blocks, arranged on either side of a central duct. Each block is larger than the polycarplike sacs present in other species of this genus. The gonad on the left is in the gut loop.

Remarks. Pyura suteri, Michaelsen, is known from Christchurch (as *P. subuculata* var. *suteri* Michaelsen, 1908, and *P. suteri;* Brewin, 1950a, from Otago (as *P. subuculata;* Brewin, 1946, and *P. suteri;* Brewin, 1950), from Chatham Island (as *P. suteri;* Brewin, 1956) and from Stewart Island (as *P. suteri;* Brewin, 1958). Its range therefore overlaps the range of the present species. The specimens in the present collection from Auckland Island also overlap the range of the variable characters that Brewin [1948] had used to distinguish these two species; and it is apparent that there is a continuous gradation from the small '*suteri*' condition, with 7 to 18 gonad lobes, to the large '*subuculata*' condition, with 22 to 39 gonad lobes. In fact, sexual maturity is achieved in this and many other species of ascidian before the individual is fully grown, and the presence of embryos in the peribranchial cavity does not mean that maximum size has been reached. Siphonal spines, however, are larger in *P. suteri* (0.033 to 0.055 mm); and this provides the only reliable characteristic by which the two species can be distinguished from each other.

Pyura rugata from Auckland, taken in clumps with the present species and *Microcosmus kura* [Brewin, 1948], has siphonal spines 0.009 to 0.013 mm long, a length within the limits described for *P. subuculata* (0.010 to 0.015 mm); and there is no characteristic that could be used to distinguish these two species. The species most resembles *P. discoveryi* (Herdman) and is distinguished from it only by bushy branchial tentacles that are not present in *P. discoveryi.*

Microcosmus kura Brewin

Figs. 59–61

Microcosmus kura Brewin, 1948, p. 136; 1957, p. 578.

New record. Auckland Island: *Eltanin* Cruise 20, Sta. 91 (intertidal).

Distribution. New Zealand, Hauraki Gulf [Brewin, 1948, 1957]. The species is known only from the intertidal region and is taken always with *Pyura subuculata.*

Description. Up to 30 individuals from 0.5 to 2 cm in diameter are aggregated in clumps with *Pyura subuculata.* The test is tough and whitish and the surface rough and wrinkled, especially on the siphons.

The branchial siphon is directed upward and anteriorly, and the atrial siphon from about one-third of the distance along the dorsal surface is directed upward. The body wall is very muscular; there are outer longitudinal bands along the siphons and circular bands externally around the base of the siphons and around the body on the right. Posteriorly, longitudinal bands are also external. On the left the longitudinal bands from the branchial siphon extend beneath the circular bands at the base of the siphon and curve around the body over the circular bands at the base of the atrial siphon, under the circular bands at the base of the branchial siphon, and up along the right side of the branchial siphon. The longitudinal bands from the atrial siphon extend around the body in a similar fashion, deep in the branchial siphon musculature. There are also relatively sparse superficial bands around the middle of the body that branch off the longitudinal atrial bands as they sweep across the dorsal surface of the body. At the base of the long siphon there are 12 fairly long compound branchial tentacles alternating with shorter ones. The dorsal tubercle has a double spiral opening. There is a short prebranchial area. The dorsal lamina is a plain-edged membrane. There are 8 branchial folds on each side of the body, with 10 to 20 internal longitudinal vessels on the folds and 1 to 2 between. The gut forms a closed curved loop, open at the pole; and there are complicated liver lamellae in the stomach region. The anal border is smooth.

The gonads on each side of the body are embedded in endocarp material and are constricted into 3 consecutive lobes. The gonad on the left side of the body extends from the base of the atrial siphon across the descending limb of the intestine into the open pole of the gut loop.

In two very young specimens, 6 mm high, the test is rough and leathery except around the siphons, which are depressed into the body and surrounded by hard test at their base.

Remarks. The very rough test of this species is the only characteristic that can be used to distinguish it from other species of the genus.

Bathypera splendens Michaelsen

Bathypera splendens Michaelsen, 1904, p. 192.—Kott, 1969, p. 140, and synonymy.

New record. South of Tierra del Fuego: *Eltanin* Sta. 740 (384–494 meters).

Distribution. Records of this species have previously been confined to the antarctic continental shelf and north along the Scotia Ridge to the South Orkney Islands.

Description. The present specimens agree in all respects with those previously described.

Culeolus murrayi Herdman

Culeolus murrayi Herdman, 1881, p. 83.—Kott, 1969, p. 142, and synonymy.

New records. Bellingshausen Sea: *Eltanin* Sta. 786 (4602 meters); Sta. 945 (4008 meters). Southeast Pacific basin: *Eltanin* Sta. 1660 (5042–5045 meters); Sta. 1668 (4930–4963 meters); Sta. 1673 (4866–4881 meters); Cruise 25, Sta. 359 (4682 meters). North of South Shetland Islands: *Eltanin* Sta. 1509 (3817–3931 meters). Southwest Pacific basin: *Eltanin* Cruise 25, Sta. 366 (5340 meters). Off east coast, New Zealand: *Eltanin* Cruise 25, Sta. 368 (84 meters).

Distribution. The species is known from most of the deep basins of the world's oceans. The maximum depth previously recorded for it was 4804 meters (*Eltanin* Sta. 1150 [Kott, 1969]). This record, the three new records from the Southeast Pacific basin (*Eltanin* Stas. 1660, 1668, and 1673), and the new record from 5340 meters in the Southwest Pacific basin (*Eltanin* Cruise 25, Sta. 366) represent the five greatest depths from which the species has been taken.

The record of specimens from *Eltanin* Sta. 368 at 84 meters is believed to be an error. The shallowest depth from which the species has previously been recorded is 513 meters off the Aleutian Islands in the North Pacific. Though the specimens from Sta. 368 are in very damaged condition, they are identical in all ways with those taken in the same trawl (10′ Blake) at Sta. 366 in 5340 meters on November 15, 1966. One is inclined to believe that, although they were recorded from Sta. 368, they were in fact taken at Sta. 366, but were retained in the trawl by their long stalks to contaminate the collection from Sta. 368. Their condition would seem to bear this out.

Description. The specimen from *Eltanin* Sta. 786 is very small and probably young, with the stalk only about half the length of the head, which is only 2 mm long. The branchial aperture is directed toward the substrate, the atrial aperture is a transverse slit on the upper surface, and there is the usual semicircle of pointed papillae posteroventrally, enlarged in the dorsal midline. The remainder of the test is soft and

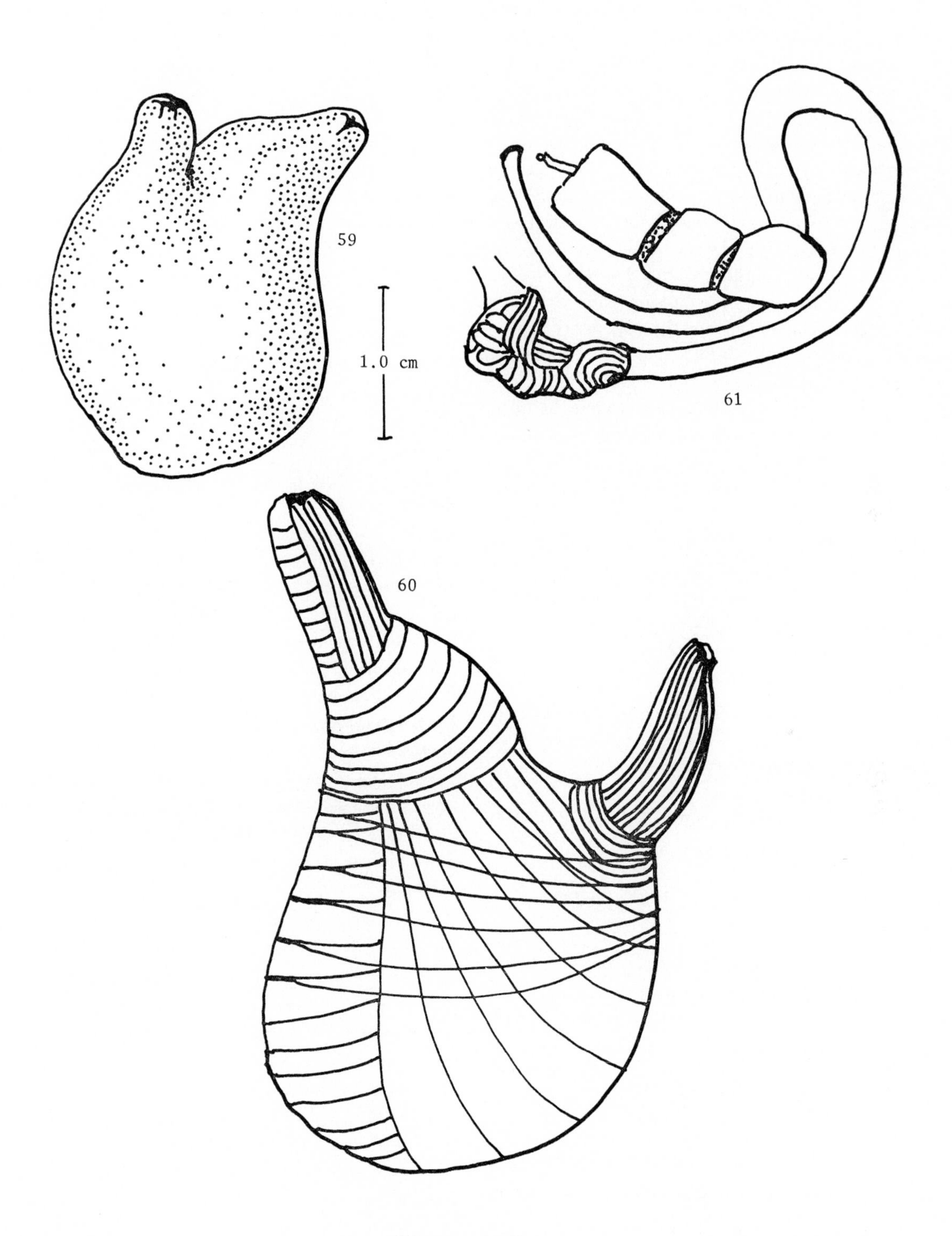

Microcosmus kura

59. External appearance.
60. Body musculature.
61. Internal structure of left side.

covered with minute rounded papillae. The base of the short stalk extends out over a calcareous fragment on which it is fixed. From *Eltanin* Stas. 1660, 1668, and 1673 there are specimens with heads 2–9 cm long with stalks up to 50 cm long. In all but specimens of about 2 cm, test papillae are reduced to scalelike, rounded structures that are generally depressed in the center to form a concavity, and there do not appear to be even minute points in the center of the scales, although they are present between the test scales in parts of the test of the larger individuals. In the smallest individuals, papillae with a hardened central point, rather than scales, are present over the body, close together; no points from the test between were observed. The conical papillae appear to lose their points and flatten into scales as the individual grows. There are the usual ring of tubercles around the upper posterodorsal aspect of the body, sometimes extending a little anteriorly down the ventral line. These tubercles are covered with scalelike papillae more or less similar to those over the rest of the body. Only in 2 individuals from *Eltanin* Sta. 1509, in which the test is whitish and semitransparent, are there no atrial tubercles. There is, instead, a continuous ridge of thickened test in the semicircle usually occupied by the tubercles. In a third individual from the same station tubercles are present.

Remarks. The specimens in the collection show the variation in the structure of the surface test previously documented for this species [Kott, 1969].

Family Molgulidae Lacaze-Duthiers

Molgula pedunculata Herdman

Molgula pedunculata Herdman, 1881, p. 234.—Kott, 1969, p. 145, and synonymy.

New records. Ross Sea, Kainan Bay: Deep Freeze I, *Edisto* Sta. ? (644 meters). Between Budd and Knox coasts: *Atka* Sta. 24 (46 meters). South Shetland Islands: *Eltanin* Sta. 436 (73 meters). North of South Georgia: *Eltanin* Sta. 1534 (271–276 meters). South Georgia: *Eltanin* Sta. 1535 (97–101 meters).

Distribution. The species, which is known mainly from the antarctic continental shelf, has its northern limit at South Georgia in waters down to 400 meters.

Description. The specimens in the present collection are small, covered with dense flexible hairs, generally mingled with sand and mud. The hairs are especially long posteriorly but shorter and thicker around the apertures, which are fairly close together on the upper surface. The arrangement of test hairs on the empty tests from *Eltanin* Sta. 1009 is typical.

The species is fairly constant, with up to 14 internal longitudinal vessels on the branchial folds; minute tertiary branches on the branchial tentacles; a thick body wall; a fairly long, narrow curved kidney; a long, narrow, deeply curved gut loop; and gonads generally inclined from the curve of the kidney and the gut loop toward the atrial opening. Very occasionally they form a more acute angle with the anterior pole of the gut loop and kidney. The testis lobes are especially small and very branched and may cover the whole mesial aspect of the ovary. The testis duct, obscured by the testis lobes, is difficult to detect.

Remarks. There is a close relationship between this species and *Molgula euplicata* and *M. gigantea* (see below). All have a relatively short dorsal lamina and a narrow curved gut loop and kidney. Their testis lobes are all unusually small, tending to cover the ovary on the mesial surface; and all have male ducts that open on the surface of the ovary and not with the oviduct near the atrial opening. The species are readily distinguished by the loss of hairs and complicated branchial sac in *M. gigantea* and by the reduction in branches of branchial tentacles and the excessive development of the ovary in *M. euplicata.* In *M. pedunculata* the testis lobes are especially well developed on the mesial surface of the ovary, obscuring the testis ducts. In *M. gigantea* and *M. euplicata* the gonads extend almost parallel to the gut loop and kidney, while in *M. pedunculata* the gonads are more or less at right angles to the axis of kidney and gut loop.

It is likely that the difference in the morphology of these closely related species might also be reflected in their life history. Thus *M. euplicata* might be oviparous and *M. pedunculata* and *M. gigantea* viviparous. *M. pedunculata* is probably successful in open conditions where sufficient concentrations of sperm are necessary for adequate fertilization. *M. euplicata* satisfies the requirement for oviparous species in more protected environments with provision for production of large numbers of eggs. *M. gigantea* probably flourishes in unexposed conditions, in view of the relatively limited development of gonads, and, if populating the open sea bed, probably has a short-lived larva. At present, larvae are not known for any of these species.

Molgula setigera Ärnbäck

Molgula setigera Ärnbäck, 1938, p. 7.—Kott, 1969, p. 147, and synonymy.

New record. South of Tierra del Fuego: *Eltanin* Sta. 740 (384–494 meters).

Distribution. The present record falls within the known geographic range of this subantarctic species, whose southern limit is South Georgia. The known depth is increased to 494 meters off Tierra del Fuego.

Description. Test hairs around the apertures are long and stiff. The hairs on the rest of the body are also long but more sparse. The apertures are well separated and are located at either end of the upper surface; the dorsal lamina is long. The kidney is short, oval, and not very curved. The gut loop is narrow and curved, but not deeply so. It extends around the posterior aspect of the body, but not anteriorly. The stomach is expanded into lobed glandular liver pockets, diverticula, and longitudinal furrows at its cardiac end. The gonads of this species lie close and parallel to the short oval kidney and the gut loop. The testis follicles are subdivided into small lobes that extend around the border of the tubular ovary but never over the mesial surface of the ovary.

Remarks. This species may be distinguished by the stiff bristles around the apertures (though these are not always present); the position of the gonads in relation to the oval kidney; the narrow but short gut loop; the position of the testis follicles around, but never covering the ovary; the position of the apertures; and the long dorsal lamina. Otherwise the glandular lobes and diverticula resemble those of other species of the genus.

Molgula malvinensis Ärnbäck

Molgula malvinensis Ärnbäck, 1938, p. 5.—Kott, 1969, p. 149, and synonymy.

New records. South of Tierra del Fuego: *Eltanin* Sta. 740 (384–494 meters). South Georgia: *Eltanin* Sta. 1533 (3–6 meters); Sta. 1535 (97–101 meters).

Distribution. The present records fall within the range of this widely distributed antarctic species.

Description. Oval to rounded specimens with hair-like extensions from the test, which is sometimes thicker and tougher than previously described. There is a thick incrustation of sand, pebbles, and shell fragments. Sessile apertures are fairly close together. The body wall has the usual well-developed musculature. There are no internal longitudinal vessels between the folds. The gonads curve around the top of the kidney and the gently curved gut loop. The stomach is short, with about 8 regular longitudinal glandular folds. The testis follicles are interrupted around the border of the ovarian tube. They are not very much divided, and the lobes remain fairly large.

Remarks. This is a rather robust species; the course of the gonads and the form of the testis lobes are characteristic.

Molgula pulchra Michaelsen

Molgula pulchra Michaelsen, 1900, p. 128.—Kott, 1969, p. 150, and synonymy.

New record. Off southern Chile: *Eltanin* Sta. 958 (92–101 meters).

Distribution. The present record confirms this as a subantarctic form.

Description. As previously described. Tailed larvae are present in the peribranchial cavity. Larvae have an otolith but no ocellus. One of the two specimens from the present collection is fixed to a shell; the other is free.

Remarks. The larvae of this free-living viviparous species are adapted for life on the open sea bed by loss of the light-sensitive organ [Berrill, 1955; Kott, 1969].

Molgula euplicata Herdman

Figs. 62–66

Molgula euplicata Herdman, 1923, p. 15.—Kott, 1954, p. 132; 1969, p. 153.

Molgula lutulenta Herdman, 1923, p. 14 (not *Caesira lutulenta* Van Name, 1912, p. 468; or *Molgula lutulenta* Van Name, 1945, p. 397).

Molgula kerguelenensis Kott, 1954, p. 137; 1969, p. 153.

New records. Ross Sea, Kainan Bay: *Atka* Sta. 4 (610 meters). Tierra del Fuego: *Eltanin* Sta. 969 (229–265 meters). Between South Shetland and South Orkney Islands: *Eltanin* Sta. 1003 (210–220 meters).

Distribution. This species had previously been considered an antarctic form. The present synonymy extends its range into both the Magellanic and Kerguelen provinces of the Subantarctic. Its distribution therefore resembles that of other antarctic species of this genus (*M. pedunculata, M. malvinensis*) that do not appear to be confined by the Antarctic Convergence.

Description. Almost spherical specimens up to 5 cm in diameter. The test is rigid but very thin and covered with a thick coating of short hairs in which mud and sand are mixed to form a felt work or rigid firm coating. The hairs may be absent from a small area around the inconspicuous apertures, which are fairly close together on the upper surface and are each surrounded by small areas of naked and delicate test. The hairs surrounding this area are longer than those from other parts of the test and, in the specimen from *Eltanin* Sta. 969, are fused together to form straight tubes extending out from the apertures at their base. These tubes are not siphons; and it appears that the individuals may be settled in the substrate, while the hairs around the apertures extend upward and fuse into cylinders, giving clear passage for streams to and from the apertures.

The body wall is thin and not very closely adherent to the test. There are strong, dense longitudinal and circular muscles around each aperture; and these anastomose with one another in the anterior part of the body but fade out posteriorly. There are 15 branchial tentacles with short and sparse primary branches and sometimes minute secondary branches. There is a thin, broad dorsal lamina about 5 mm wide (in a specimen 5 cm in diameter) with a minutely serrated border. The dorsal tubercle is large, with a C- or E-shaped slit, which may be inclined to the right or anteriorly. The branchial sac has 7 to 9 folds on each side. There are 4 to 5 internal longitudinal vessels per fold in a specimen 3 cm in diameter and up to 18 internal longitudinal vessels in a specimen 5 cm in diameter. In young specimens with few internal longitudinal vessels these are confined to the ventral side of the folds. Occasionally there is a single internal longitudinal vessel between the folds. The transverse vessels in older specimens become especially wide and anastomose between the folds to form an almost continuous membrane with small, sometimes slitlike openings in which the stigmata are often lost. Behind the folds the anastomosing transverse vessels leave an opening into the peribranchial cavity opposite each infundibulum. The stigmata form very tight spiral infundibula, which subdivide into 2 in the apex of the fold. The gut forms a long, narrow loop, deeply curved, which appears to become more straight with an increase in body size. The stomach wall is extended into glandular lobed diverticula with sometimes longitudinal glandular ridges at the cardiac end of the stomach, although these seem to be broken up into lobes with growth. The kidney is fairly long and gently curved. The gonads extend parallel and close to the kidney and the gut loop and then curve dorsally toward the atrial aperture; or they may curve dorsally, then anteriorly and dorsally again. The curve of the ovary is obscured when the gonad is embedded in the body wall, which appears to increase in thickness with age, or when the outlines of the gonad are obscured by proliferation of testis follicles. The origin of the oviduct is thus obscured, and it may appear to extend dorsally from a point some distance along the dorsal length of the ovary. In specimens 3 cm in diameter from *Eltanin* Stas. 969 and 1003 the gonad on the left is enclosed in the secondary gut loop, although the gonad on the right is oriented in the usual manner parallel to the kidney and curved dorsally in its distal extent. The left gonad in the large specimen from *Atka* Sta. 4 is duplicated. In this specimen the gut and gonad were so deeply embedded in the body wall that dissection was necessary to determine their course, and only the apertures of the gonoducts can be seen where they open on the inner surface of the body wall. Male ducts occasionally extend across the body wall for some distance before opening into the peribranchial cavity. More often male ducts are apparent crossing the surface of the ovary to open in many short common ducts that increase in number with the size of the individual. The testis follicles surround the ovarian tube on all aspects, although generally not so thickly on the mesial surface. The testis follicles proliferate immensely into minute lobes.

Remarks. The species is very closely related to *M. pedunculata* and *M. gigantea,* which it resembles in the minutely proliferated testis lobes that extend onto the mesial surface of the ovary and very often completely obscure it, and in the long, narrow curved kidney, often deeply curved gut loop, and short dorsal lamina. It may be distinguished from both by the thicker coating of hairs and the less bushy branchial tentacles. The orientation of the gonads in relation to the gut and kidney is usually distinct, although the left gonad in the deeply curved secondary gut loop sometimes resembles that of *M. pedunculata.* The present species, however, has more numerous male ducts opening on the surface

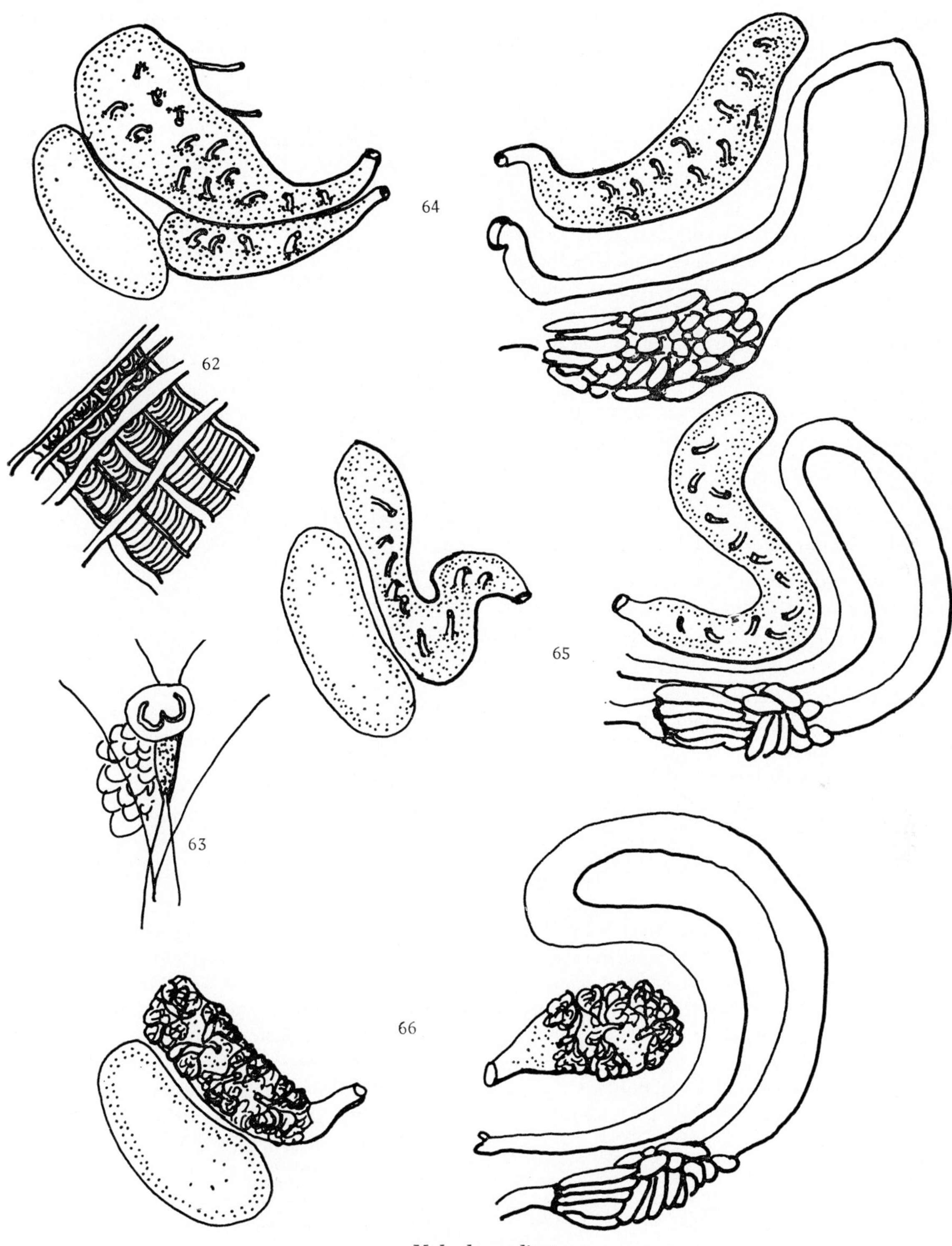

Molgula euplicata

62. Section of branchial sac (*Eltanin* Sta. 969).
63. Dorsal tubercle, gland and ganglion.
64. Gonads, kidney, gut loop (*Atka* Sta. 4).
65. Gonads, kidney, gut loop (*Staten Island* Sta. 57/63).
66. Gonads, kidney, gut loop (*Eltanin* Sta. 969).

of the ovary. If these, however, are obscured by the proliferation of testis follicles on the mesial surface of the ovary, as in *M. pedunculata,* the orientation and shape of the right gonad provides the main distinguishing characteristic for the species. Many of the variations observed in this species are associated with growth: the dorsal tubercle, C-shaped in an individual of 3 cm, is a more complicated E shape in an individual of 5 cm; and similar differences are observed in the shape and size of the gonads, the thickness of the body wall, the number of testis ducts opening into the peribranchial cavity, the complexity of the spiraling stigmata, the number of internal longitudinal vessels, and the number of branchial folds. The distinctions between this species and *M. pedunculata* are often obscured; and it seems probable, since their range overlaps, that there may have been some confusion in their identification.

Molgula gigantea (Herdman)

Ascopera gigantea Herdman, 1881, p. 238.
Molgula gigantea; Kott, 1969, p. 155, and synonymy.

New records. Antarctic Peninsula: *Staten Island* Sta. 24/63 (75 meters). South Shetland Islands: *Eltanin* Sta. 436 (73 meters). Between South Shetland Islands and South Orkney Islands, *Eltanin* Sta. 1003 (210–220 meters). South Orkney Islands: *Eltanin* Sta. 1078 (604 meters); Sta. 1079 (593–598 meters). South Georgia: *Eltanin* Sta. 671 (220–320 meters); Sta. 1535 (97–101 meters).

Distribution. Within the limits previously recorded for the species.

Remarks. All specimens have the minutely papillated, thin, paperlike transparent test and hollow stalk by means of which the small specimens of the species may be recognized before the characteristic proliferation of the branchial sac is evident.

Molgula millari new species

Figs. 67–71

? *Molgula immunda* f. *monocarpa* Millar, 1959, p. 201.

Type locality. South Indian basin: *Eltanin* Cruise 20, Sta. 126 (3089–3164 meters). Holotype, USNM 12013.

Additional records. South of Tasmania: *Eltanin* Sta. 1978 (4213–4218 meters). Southwest of Macquarie Island: *Eltanin* Cruise 20, Sta. 134 (3200–3259 meters). East of South Sandwich Islands: *Eltanin* Sta. 1571 (3947–4063 meters).

Distribution. Present in 3089 to 4410 meters in the deeper basins to the north of the antarctic continent. It has not yet been recorded from the Southeast Pacific basin, although it could be expected to occur there.

Description. Single specimens are available from each station. They are almost spherical, from 0.7 to 2 cm in diameter. The test is covered with very fine hairs, which form an almost feltlike protective coat around the body in which foraminiferan skeletons are enmeshed. The apertures are both sessile, opening at either side of the upper surface. In mature specimens the test on the upper surface above the apertures is thickened to form a domelike structure. The anterior and posterior borders of this thickened dome project over the atrial and branchial apertures respectively. The openings, when open, are consequently directed laterally from opposite sides of the body; and when the muscles of the body wall contract, the dome is closed down onto the surface test below the apertures, thus covering them. There is an area of very thin naked test around the apertures. The test around the upper border of the branchial aperture is produced into 6 hollow, pointed, fingerlike lobes, while around the upper border of the atrial aperture there are 7 such lobes. Rudimentary lobes are present on the lower border of both the branchial and the atrial apertures.

The body wall is inserted into the outer rim of the protective dome and extends into the lobes around the apertures. There is an almost continuous superficial layer of circular muscle bands around the siphons, extending about halfway down the body. Longitudinal bands radiate from the apertures, where some extend from the large lobes on their upper borders. These longitudinal bands break up into fibers over the ventral surface of the body wall. The siphon-protecting mechanism described above is unusual, but depends, as in Agnesiidae and Corellidae, on localized thickening of the test associated with special development of body musculature.

There is a branchial siphonal velum about halfway along the pretentacular region. There are about 12 large, bushy branchial tentacles with primary and secondary branches; these alternate with smaller tentacles. The small, circular dorsal tubercle with a simple opening is deep in the apex of a very long peritubercular area extending about halfway along

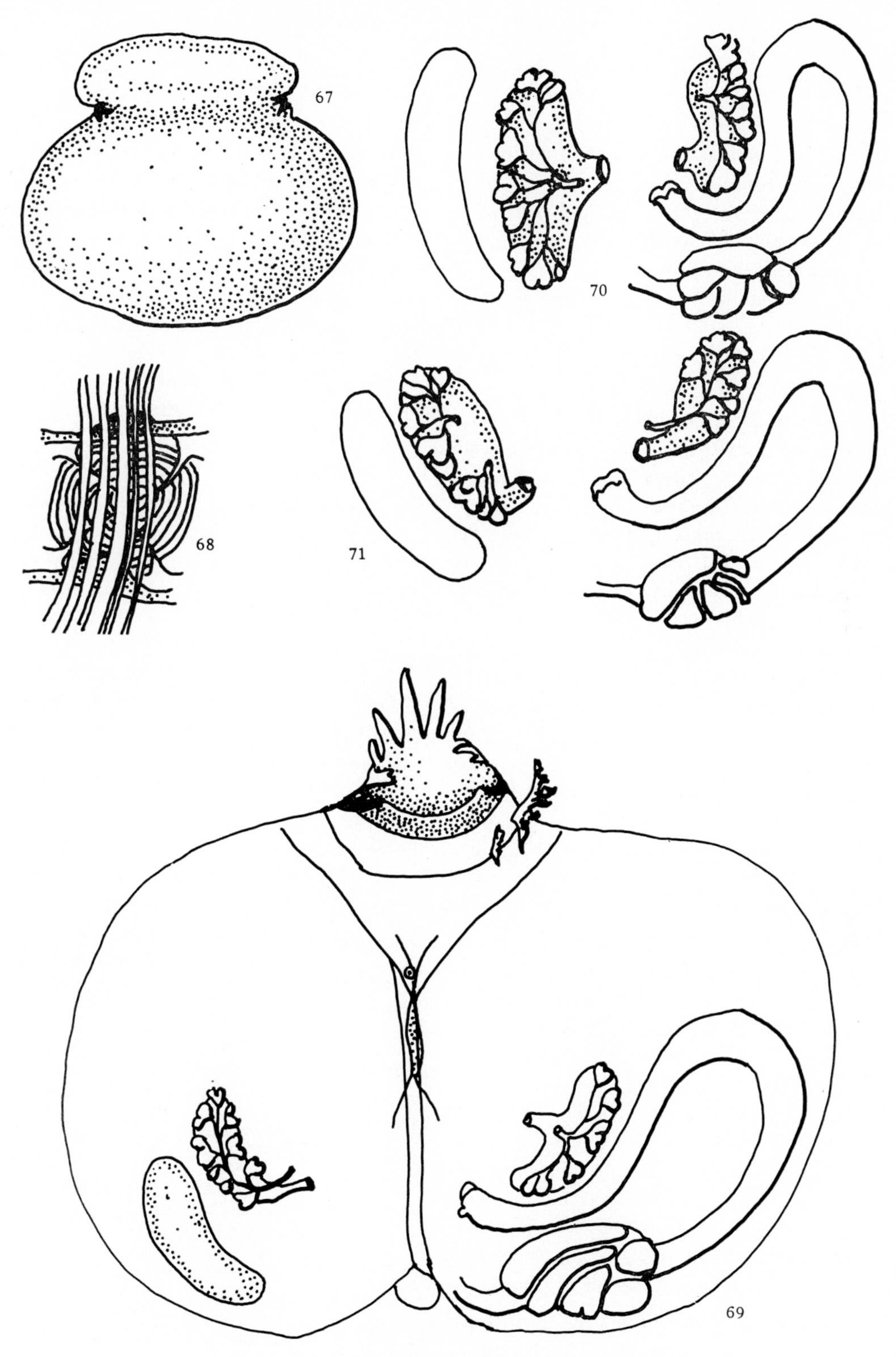

Molgula millari

67. External appearance.
68. Portion of branchial sac.
69. Internal structure (*Eltanin* Sta. 126).
70. Kidney, gut, gonads (*Eltanin* Sta. 134).
71. Kidney, gut, gonads (*Eltanin* Sta. 1571).

the dorsal surface. The dorsal lamina is a wide broad membrane. There are 7 branchial folds on the left and 6 on the right. These folds are formed by the internal longitudinal vessels supported by connectives from the transverse vessels, and by intermediate transverse vessels confined to the folds. There are no internal longitudinal vessels between the folds, and the stigmata do not project into the folds at all. Internal longitudinal vessels are arranged as follows:

E (3) (4) (6) (7) (6) (6) DL
(6) (6) (7) (8) (7) (5) (3)

The stigmata form flat cobweblike spirals of about 5 to 6 turns in rows of 6 beneath the folds. Sometimes interstitial spirals are present.

The gut forms a rather broad but closed simple loop. The stomach is short and very wide. The stomach wall is extended into wide, longitudinal, oblique glandular furrows which are broken up into large, rounded diverticula as the animal grows. The kidney is broad and slightly curved in the right posteroventral corner of the body. There are sausage-shaped, oval to rounded gonads on each side parallel and dorsal to the kidney and gut loop. Large, pear-shaped testis follicles are present around the whole border of the ovary or confined to its dorsal border. Testis ducts join on the mesial surface of the ovary and open by one or more common openings along its surface or at the base of the oviduct at the distal end of the ovary. The gonoducts are directed toward the atrial opening.

Remarks. The test hairs, the dorsal lamina, the short stomach with broad glandular folds, and the form of the branchial folds all resemble these organs in *Molguloides* spp., in particular *M. immunda* (Hartmeyer) [Kott, 1969]. *M. millari,* however, is distinguished by the position of the left gonad outside the gut loop; the straight gonads, in contrast to the coiled organs in *M. immunda;* and the thickened cap of test protecting the apertures, with the regularly radiating longitudinal muscle bands and thin but evenly spaced circular muscle fibers contrasting with the sparse and irregular musculature of *M. immunda.*

Molgula immunda f. *monocarpa* [Millar, 1959], from 4410 meters in the Kermadec trench, differs from the present specimens only in the absence of the gonad on the left side of the body. The form of the gonad on the right is similar to the gonads of the present species, which it also resembles in the development of pointed lobes on the upper borders of the apertures, in the characteristically thickened and expanded dome protecting the apertures, and in the form of the branchial folds. Millar's specimen may therefore be conspecific with the present specimen, but its designation as type specimen of a taxon elevated to species rank is not desirable in view of the probably abnormal absence of the left gonad.

It has therefore seemed more appropriate to describe a new species and designate the type from specimens in the present collection. *Molgula immunda* f. *monocarpa* [Millar, 1959] may subsequently be shown to be a synonym of this new species, *Molgula millari,* which has been named for Dr. R. H. Millar.

Molgula sabulosa (Quoy and Gaimard)

Ascidia sabulosa Quoy and Gaimard, 1834, p. 613.
Molgula amokurae Bovien, 1922, p. 34 [part].
Molgula sabulosa; Kott, 1952, p. 298 [part].—Millar, 1966, p. 374.—Hartmeyer and Michaelsen, 1928, p. 449, and synonymy.

New record. East of South Island, New Zealand: *Eltanin* Cruise 25, Sta. 370 (95 meters).

Distribution. From various localities around the coast of Australia, from Sharks Bay in western Australia to Bowen in Queensland; and in Malaya [see Michaelsen, 1928; Kott, 1952]. The single record of *M. amokurae* [Bovien, 1922] from Port Ross, 1.84 meters, is the only known occurrence in New Zealand, but it is probable that other specimens of this sandy, inconspicuous species have been overlooked.

Description. The specimen is small, rounded, and heavily invested with sand. The test is thin. The body wall is fairly muscular. Both siphons are sessile and on the upper surface. The dorsal tubercular opening is C-shaped and directed posteriorly. There are 7 folds on each side of the body, with 6 internal longitudinal vessels on each fold and none between the folds. A simple, long gut loop curves around the ventral border of the left side of the body. On the right side of the body the kidney is gently curved, and the gonad is placed with its long axis at right angles to the length of the kidney. On the left side of the body the gonad is in the secondary gut loop, and is directed toward the atrial opening. The ovaries are short and flask-shaped; the testis follicles form an arc around the proximal end of the ovary where it lies against the concavity of the kidney on the right side of the body and the gut loop on the left.

Remarks. The specimens described as *Molgula amokurae* [Bovien, 1922 (part)] are identical with the present specimen in every respect; nor can they be distinguished from the Australian species *M. sabulosa,* which has a wide distribution from the Malayan Peninsula and around the coast of Australia. Although the present specimen has a dorsal tubercular opening that differs from the typical condition found in the Australian specimens, Bovien did describe a specimen with the horizontal S-shaped opening that has been considered characteristic for the species. It is most probable, therefore, that *M. sabulosa* has a wide distribution in shallower waters of the southern hemisphere from tropical to temperate regions.

Molguloides vitrea (Sluiter)

Molgula vitrea Sluiter, 1904, p. 119.
Molguloides vitrea; Kott, 1969, p. 159, and synonymy.

New records. North of South Shetland Islands: *Eltanin* Sta. 1506 (3788–3944 meters). East of Antarctic Peninsula: ? *Eltanin* Sta. 1009 (2818–2846 meters).

Distribution. The present record confirms the circumantarctic range of this species, which is also taken off Indonesia and the Philippines [Kott, 1969].

Description. The specimen, 3 cm in diameter, is soft and spherical, with fine hairs from the surface of the semitransparent test. The test is thicker around the siphons, where the hairs are especially dense but shorter than on the rest of the body. The body musculature is fairly sparse and irregular. The branchial tentacles are four times branched and very bushy. There are 6 to 7 branchial folds. Infundibula with about 12 spirals project into the folds. The gut loop is wide and closed, and the stomach plications are rather broad. On the left side the gonad is coiled in the gut loop; the testis follicles are arranged around the outer curve of the ovary and open by several separate short ducts on the mesial surface of the ovary.

Remarks. There is some variation in the condition of the stomach observed in this species. It is sometimes short and broad with broad glandular folds as in *Molgula millari* new species. When extended, however, the stomach is longer and of lesser diameter, and the folds appear narrower than in *M. millari.* There is also some variation in the length and density of test hairs. The species is distinguished from *M. immunda* by the projection of infundibula into the folds and the presence of testis follicles along the length of the ovary, rather than at the proximal end only.

Eugyra kerguelenensis Herdman

Fig. 72

Eugyra kerguelenensis Herdman, 1881, p. 237.—Kott, 1969, p. 159, and synonymy.

New records. Antarctic Peninsula: *Staten Island* Sta. 7/63 (21–31 meters). Between South Shetland and South Orkney Islands: *Eltanin* Sta. 1003 (210–220 meters). North of South Orkney Islands: *Eltanin* Sta. 558 (845–646 meters).

Distribution. The present records fall within the known range for this species.

Description. Very delicate, laterally flattened specimens covered with very fine hairs and mud, which form a sometimes brittle protective coat through which the naked siphons project. Siphons are fairly close together in a naked area on the upper surface over which the surrounding test may fold when the siphons are withdrawn. Very short transverse muscle bands curve around the right and left sides of the dorsal and ventral borders of the body, and short longitudinal muscle bands radiate from the siphons. The 7 internal longitudinal vessels covering 7 rows of infundibula are each crossed by 4 radial vessels. The left gonad is in the primary gut loop, which is deeply curved. The stomach has glandular elevations. The anal border is lobed. Gonads consist of a tubular ovary with pyriform testis follicles around the ovary and opening by many separate short common ducts on the mesial surface of the ovary. In the specimen from Sta. 1003 the testis lobes are confined to the proximal end of the ovary, where they open by a single short duct.

Remarks. Apart from variations in the distribution of the testis, follicles, this species is very constant. It is interesting to observe that, as in Agnesiidae, in association with the lateral flattening of the body and the development of a mechanism whereby folds of test close across the withdrawn apertures, the body musculature is modified into short radiating longitudinal bands and rows of short transverse bands around the borders of the body. In association with the branchial sac and the position of the gonads, the body musculature is characteristic of this species.

Pareugyrioides galatheae (Millar)

Figs. 73, 74

Molgula galatheae Millar, 1959, p. 202.—Kott, 1969, p. 161.

New records. Southwest of Macquarie Island: *Eltanin* Cruise 20, Sta. 134 (3200–3259 meters). West of South Sandwich Islands: *Eltanin* Sta. 1555 (1976–2068 meters). Northeast of South Sandwich Islands: *Eltanin* Sta. 1578 (4236–4273 meters).

Distribution. These represent the most extensive southern records of the species, which has been previously recorded from the Guinea basin and the Southeast Pacific basin. The known depth range is extended into more shallow waters by the record from Sta. 1555.

Description. Spherical specimens of 1 to 2 cm in diameter are covered with a very fine feltwork of hairs with adherent foraminiferan skeletons. The test is otherwise thin and transparent. The sessile apertures are at either end of the upper surface directed away from one another. The body wall is thin, with longitudinal bands radiating from the siphons. Fine outer bands extend across the dorsal surface and around the base of the siphons, while some extend across the sides of the body to form a very irregular network posteriorly. Eight long branchial tentacles have small pinnate branches with minute secondary branches from either side of the anterior border or stem of the tentacle. These large tentacles are expanded posteriorly along the whole of their length into a broad membranous flange, which is fixed basally to the posterior surface of a flat tentacular velum and across the prepharyngeal area to the base of a narrow prepharyngeal velum. The larger tentacles alternate with 3 minute rudimentary ones. The dorsal tubercle is small and upright, with a simple slit. The dorsal lamina is a wide, smooth, double-edged membrane. In the branchial sac both the 5 transverse and 5 internal longitudinal vessels are especially high, although the longitudinal vessels are reduced as they accommodate and support the cone of infundibula and preserve the three-dimensional condition of the branchial sac. About 10 fine radial vessels with small, randomly distributed papillalike expansions cross each infundibulum. There are about 20 spirals in each infundibulum.

The gut forms an open loop and the rectum extends anteriorly for a short distance. There are about 15 longitudinal glandular folds in the stomach, which gradually narrows into the intestine. The kidney is long, narrow, and very much curved. The gonads are short and flask-shaped, anterodorsal to the kidney and pole of the gut loop. They have lobed testis follicles around the proximal border of the ovary. Testis ducts join to a short common opening on the surface of the ovary. The oviduct is very short.

Remarks. Superficially the species resembles *Molgula euplicata, M. millari, Molguloides* spp., and *Eugyra kerguelenensis* in the protective coat of test hairs intermingled with sand, shell, etc. It is further related to *M. millari* and *Molguloides* spp. in the position of the apertures, but it is distinguished by the tentacular velum, the stomach folds, and the absence of branchial sac folds. *Pareugyrioides galatheae* and *Molgula millari* are present together from *Eltanin* Cruise 20, Sta. 134, and demonstrate convergence in their external appearance.

Pareugyrioides ärnbäckae (Millar)

Eugyra ärnbäckae Millar, 1960, p. 144.
Pareugyrioides ärnbäckae; Kott, 1969, p. 161.

New record. South Shetland Islands: *Staten Island* Sta. 61/63 (31 meters).

Distribution. The present specimens of this very constant species fall within the range previously recorded.

Paramolgula gregaria (Lesson)

Cynthia gregaria Lesson, 1830, p. 435.
Paramolgula gregaria; Kott, 1969, p. 164, and synonymy.

New records. East of Tierra del Fuego: *Eltanin* Sta. 1500 (73–79 meters).

Distribution. The present record falls within the range previously recorded for this species, which is limited to the Magellanic province.

Description. The single specimen is large and typical of the species: it has a very modified branchial sac, short dorsal lamina, and bushy tentacles.

Fungulus cinereus Herdman

Fungulus cinereus Herdman, 1882, p. 127.—Kott, 1969, p. 167, and synonymy.

New records. Southeast Pacific basin: *Eltanin* Sta. 1673 (4866–4881 meters). Northeast of South Sandwich Islands: *Eltanin* Sta. 1578 (4236–4273 meters).

Distribution. These records fall within the known range of the species.

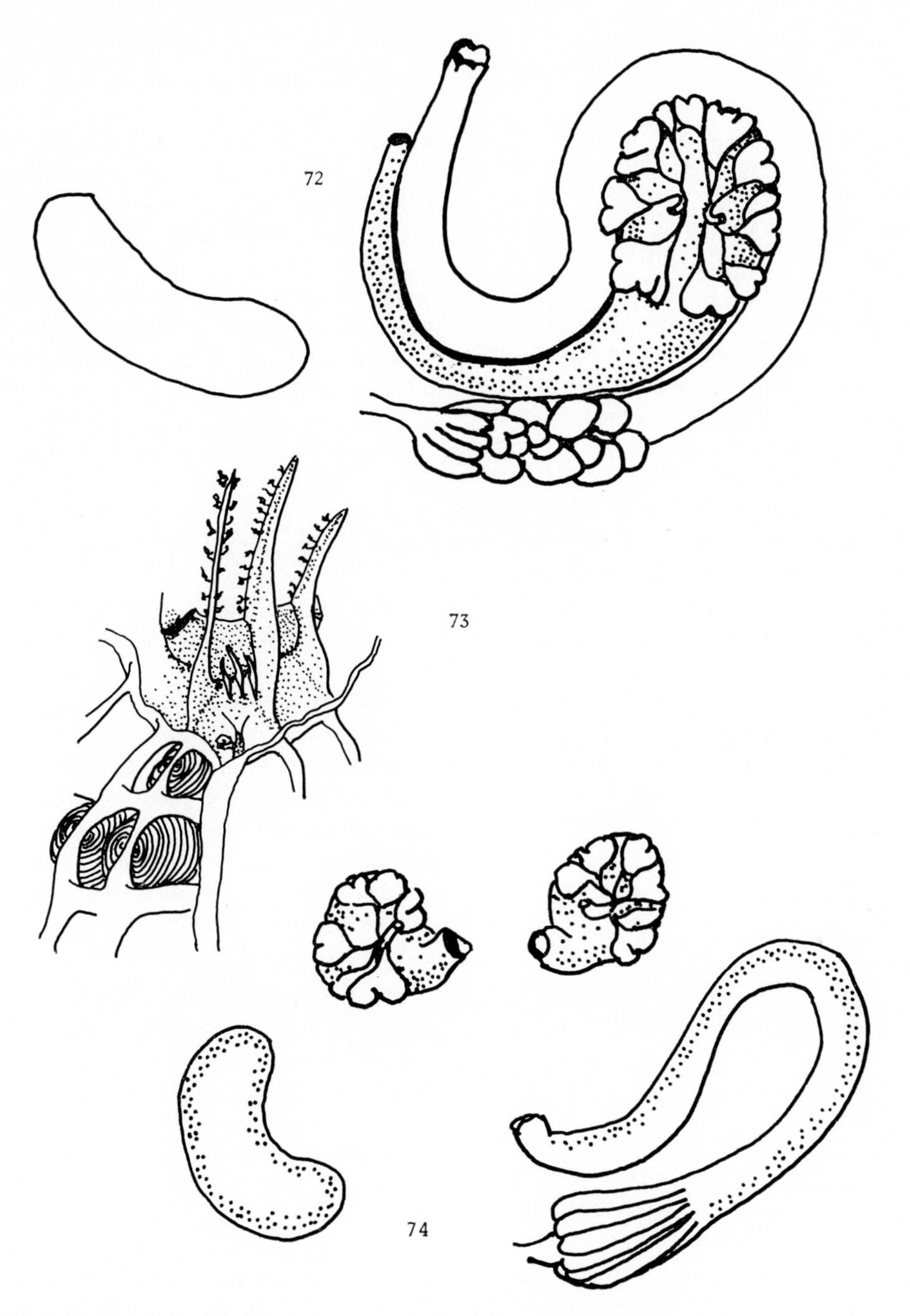

Eugyra kerguelenensis

72. Gut loop and gonad on left side of body.

Pareugyrioides galatheae

73. Branchial tentacles, dorsal tubercle, and part of branchial sac.
74. Internal structure.

Description. The present specimens are as previously described, with a rounded head on a fleshy, bulbous stalk and a thin, papery test with pointed conical papillae or minute hairs. The longitudinal muscles from the stalk terminate in the body wall where it is attached to the test dorsally just anterior to, or below, the branchial opening. The gut loop and the kidney vary in length and do not appear to increase with the size of the body. The kidney is sometimes quite short and oval. The stomach has long, convoluted internal glandular plications. The gonads, on each side of the body, are complicated; and the ovarian tube apparently consists of several separate compartments. Testis follicles are compacted into the ovarian tube in an undulating line down the length of the ovary and between the ovarian compartments. This condition appears to have developed from an undulating ovarian tube with testis follicles along either side. As the ovary develops, the outline of its winding course is obscured; and its compact oval shape with straight sides is thus secondarily acquired. The arrangement of testis follicles suggests the original course of the ovary, although they appear to be embedded in the ovarian tissue. There is a terminal oviducal opening, and there are secondary female ducts opening along the length of the ovary. These probably arise following occlusion of the tube at different points along its length, possibly by development of the testis follicles enclosed in the inside curve of each loop. Testis follicles are especially numerous and compact around the proximal end of the ovary, but they do not appear to be developed on the outer aspect of the more distal part of the ovary. Male ducts join a vas deferens on the mesial surface of the ovary. The vas deferens also extends in an undulating course along the ovary to open at the base of the terminal oviduct.

Oligotrema psammites Bourne

Oligotrema psammites Bourne, 1903, p. 233.—Kott, 1969, p. 168, and synonymy.

Hexacrobylus indicus; Oka, 1915, p. 8.—Monniot and Monniot, 1970, p. 334.

New record. Southwest Pacific basin: *Eltanin* Cruise 25, Sta. 366 (5340 meters).

Distribution. The species has been taken from widely separated localities in the deeper basins of the world. The present record represents the greatest depth from which the species is known.

Description. The present specimen is small. The test is semitransparent and covered with minute papillae, and it is quite firm. Internally the arrangement of muscles is identical with that previously described for the species. The specimen is highly contracted, and no attempt was made to dissect it to determine variations in the internal structure.

DISTRIBUTION

The zoogeographic boundaries indicated by the distribution of species of Ascidiacea in Antarctica [Kott, 1969] are confirmed by the additional records above.

Dominant Species of the Antarctic Ascidian Fauna

Ascidiacea appear to comprise a significant fraction of the antarctic benthos, although no accurate assessment of their biomass has yet been attempted. The pattern of distribution of dominant species supports the suggested explanation for the distribution of species in Antarctica [Kott, 1969]. Thus, species well adapted for survival in the Antarctic flourish as dominant species in the continental subprovince, although they may also occur as less important components of other populations further to the north. Similarly, other antarctic species are more successful and occur as dominants in the less inclement and more ecologically suitable regions of the South Georgian subprovince; and although they do occur in the continental subprovince, possibly through their ability to inhabit deeper waters, they are not present as dominant species.

Subantarctic species do not extend south of South Georgia and are apparently isolated north of the Subantarctic Convergence.

Those species that are sometimes dominant in ascidian populations south of the Subtropical Convergence are listed in Table 1.

The relative composition of the ascidian fauna at different locations also confirms the zoogeographic boundaries the author has based on information about the distribution of species studied [Kott, 1969]. The following groups of species can thus be identified from the patterns of their distribution:

(1) *Tylobranchion speciosum, Synoicium adareanum, Ascidia challengeri* [Kott, 1969, Table 4], *Sycozoa georgiana, Distaplia cylindrica, Caenagnesia bocki,* and *Aplidium radiatum* [Kott, 1969, Table 5] are, at different times, the dominant species in the ascidian fauna from stations around the antarctic continental shelf generally as far north as the South Shetland Islands. Somewhere along the Scotia Ridge, however, owing to zoogeographic or ecological factors, the composition of the collections changes, and, although these species have often been recorded from stations north of the South Shetland Islands, they are never the dominant form in those locations. The area from which they have been recorded in large numbers appears to coincide with the extent of the continental subprovince and with the southern extent of their range.

(2) *Molgula pedunculata, Pyura georgiana,* and *Molgula gigantea* [Kott, 1969, Table 4], *Pareugyrioides ärnbäckae, Didemnum biglans, Pyura dis-*

TABLE 1. Dominant Antarctic Species

(X, present; XX, dominant)

Information is derived from collections made by the RRS *Discovery* and reported on by Millar [1960]; and from various American collections listed by Kott in 1969 and in the present paper.

Species	Ross Sea	Victoria Quadrant	Weddell Sea	Bellings-hausen Sea	Antarctic Peninsula	S. Shetland Is.	S. Sandwich Is.	S. Orkney Is.	S. Georgia	Drake Passage	Burdwood Bank	Falkland Is.	Tierra del Fuego
Tylobranchion speciosum	XX	X	X		X	X	X	X	X			X	
Sycozoa georgiana	XX	XX	XX	XX		X			X			X	
Distaplia cylindrica	XX	X	X		XX	X	X	XX	X	X			
Synoicium adareanum	XX	X	X		X	XX	X		X				
Ascidia challengeri	XX	X	X		XX	XX		X	X	X		X	
Pareugyrioides ärnbäckae	X	X	X	XX		X	XX						
Cnemidocarpa verrucosa	XX	X	X	X	XX	XX	X	X	X		X	XX	
Aplidium radiatum	XX	X	XX		XX	XX		X	X				
Caenagnesia bocki		XX	X	XX		X			X				
Didemnum biglans	X				XX	XX	XX						
Molgula pedunculata	X	X	X		X	X		X	XX				
Pyura discoveryi	X	X	X		X	XX		XX	XX				
Pyura georgiana	X	X	X		XX	XX		XX	XX				
Pyura setosa	X	X			XX	XX		X					X
Molgula gigantea	X				XX	XX	XX	X	XX				
Corella eumyota	X	X	X		X	X		XX				X	X
Sycozoa sigillinoides		X			X	X			XX			XX	XX
Aplidium variabile									X	XX			XX
Didemnum studeri									X	XX	XX	X	XX
Paramolgula gregaria									X		XX	XX	XX
Pyura paessleri										XX	X	X	XX
Molgula setigera									X	XX		X	X

coveryi, and *Pyura setosa* [Kott, 1969, Table 5] are often the dominant forms off the Antarctic Peninsula and on the Scotia Ridge as far north as the South Shetland Islands, South Sandwich Islands, or South Georgia; their area of dominance coincides with the area of the Georgian subprovince. Species of this group are dominant in the northern extent of their range, although all seven have been recorded at stations around the whole antarctic continent.

(3) Of subantarctic species [Kott, 1969, Tables 6, 7] only *Aplidium variabile* occurs in large numbers as far south as South Georgia, while *Didemnum studeri, Paramolgula gregaria, Pyura paessleri,* and *Molgula setigera* may be dominant at stations from the Drake Passage to Tierra del Fuego but are never recorded south of South Georgia. Thus, the areas from which they may be taken as the dominant forms coincide with their range.

(4) Both *Corella eumyota* and *Sycozoa sigillinoides* have been regarded as species with a wide distribution in the antarctic region. They are, however, more often the dominant species north of the Antarctic Peninsula than around the antarctic continent. *Corella eumyota,* especially, is found in large concentrations off the South Orkney Islands, and it has already been observed [Kott, 1969] to form remarkable concentrations off the east coast of New Zealand and around Macquarie Island.

(5) *Cnemidocarpa verrucosa* appears to be the only species with a wide range in the whole region that does occur as a dominant species at locations in both the antarctic and subantarctic subregions.

(6) There are undoubtedly ecological factors that influence the distribution of ascidian species and that might explain the occurrence of *Eugyra kerguelenensis* in Leith Harbor, South Georgia, and Arthur Harbor, Antarctic Peninsula.

The New Zealand Fauna

The high degree of endemism in the New Zealand fauna is indicative of a long isolation. The relationships of many of these New Zealand species are with the antarctic fauna rather than with the fauna in other parts of the southern temperate area, although the New Zealand fauna is characterized by a remarkable species diversity in genera that in Antarctica are very stable. In this regard the Didemnidae, as well as *Cnemidocarpa*, in the family Styelidae, and *Pyura*, in the family Pyuridae, are represented by numerous species in the New Zealand fauna. *Pyura cancellata* and the antarctic *Pyura tunica* are closely related. *Pyura subuculata* and *Pyura picta* are not readily distinguished from *Pyura discoveryi*. *Trididemnum natalense* and *Microcosmus kura*, however, bear close relationships to species in Australia and South Africa.

The Chatham Islands fauna has been regarded by Brewin [1956] as a 'convergence fauna with species from North Island and the South Island and Stewart Island area.' Only those subantarctic species with an especially wide range to the north have been recorded from Stewart Island, the Chatham Islands, or both (*Corella eumyota, Sycozoa sigillinoides, Molgula sluiteri*). In addition to some endemic species, the ascidian fauna of these islands is predominantly New Zealand in character [Brewin, 1956, 1957]. The Stewart Island fauna, however, has the closer relationship to the circumsubantarctic fauna indicated by the endemic species *Pareugyrioides filholi*, which is closely related to the Magellanic *Paramolgula gregaria*, and by *Theodorella* spp., previously thought to be endemic to Stewart Island but now recorded also from the Magellanic province. These species, however, must be regarded as relicts of antarctic species. The large areas of *Corella eumyota* off Macquarie Island (*Eltanin* Stas. 1417, 1418) disappear between Macquarie Island and Auckland Island and may indicate the northern limit of the subantarctic faunal region. The fauna at Auckland Island is definitely New Zealand in its affinities, and consequently Stewart Island is within the New Zealand faunal area.

Circumpolar Fauna

Further attention is drawn to the circumsubantarctic distribution of faunal elements by new records or by synonymy in the following cases:

(1) The genus *Theodorella*, previously regarded as endemic to Stewart Island, has been found to occur at South Georgia.

(2) The synonymy of *Ascidia aspersa;* Brewin 1946, 1950, from New Zealand, with *Ascidia meridionalis*, an antarctic species extending into the Magellanic region, has been established.

(3) *Polyclinum sluiteri*, previously regarded as a New Zealand species, is now recorded from the Albatross cordillera.

(4) *Molgula kerguelenensis*, recorded from Kerguelen, is shown to be synonymous with *M. euplicata* from the Antarctic Peninsula and Scotia Ridge. The latter species demonstrates not only circumsubantarctic distribution but also the wide distribution from the Antarctic and into the Subantarctic that is characteristic of many species of this genus.

(5) The similarity between *Cnemidocarpa madagascariensis*, recorded from New Zealand and Madagascar, and *Cnemidocarpa ohlini*, which has been taken only off Tierra del Fuego, is also striking. Although these are undoubtedly distinct species, the inference of a common ancestor previously distributed around the antarctic continent cannot be avoided.

PRESERVATION OF ASCIDIANS

Facilities and time permitting, it is desirable to narcotize ascidians with menthol or chloral hydrate before fixation. This is not always possible, however, since they are very sensitive to disturbance and the conditions necessary to obtain the animals in an expanded state are not often available. Generally, therefore, specimens must be killed and preserved without narcotization.

Specimens are adequately killed and fixed in 10% neutral formalin. Shorter-term storage may be in 4–5% formalin, but for long storage collections should be transferred to 70% alcohol. All ascidians lose their characteristic pigmentation when placed in preservative, so it is most desirable that their coloration be noted or photographed in the field in the fresh state before fixation.

Specimens should not be crowded, and no container should be more than two-thirds full. When preservative is added, a small space should be left to take care of expansion. If alcohol is used, it should be renewed after a day or two and brought up to 70% strength. In pouring off the preservative, care should be taken to retain sediment or debris in the bottom of the container, for it often harbors small organisms attached to or commensal with the host ascidians.

Acknowledgments. Investigations contributing to knowledge of the Antarctic and its invertebrate fauna have, for many years, been stimulated by the enthusiasm and initiative of Dr. Waldo L. Schmitt of the Smithsonian Institution. The association with Dr. Schmitt that I have enjoyed over the past four years during the preparation of works on the antarctic Ascidiacea has been especially rewarding.

Miss Lucile McCain, also of the Smithsonian Institution, has contributed immeasurably to the value of these works on Ascidiacea by her careful editorial checking for consistency and accuracy in the manuscripts and in the references cited.

The present work was supported by National Science Foundation funds administered through the Smithsonian Institution, Washington, D. C. I am indebted to Dr. I. E. Wallen for the allocation of these funds.

REFERENCES

Ärnbäck-Christie-Linde, Augusta
1938 Ascidiacea, Part 1. *In* Further Zool. Results Swedish Antarctic Exped. 1901–1903, *3*(4): 1–54, figs. 1–11, pls. 1–4.

Berrill, N. J.
1955 The origin of vertebrates. Clarendon, Oxford. viii + 257 pp., figs. 1–31.

Bovien, P.
1922 Ascidiae from the Auckland and Campbell Islands (Holosomatous forms). *In* Papers from Dr. Th. Mortensen's Pacific Expedition 1914–1916, part 4. Vidensk. Meddr dansk naturh. Foren., *73:* 33–47, figs. 1–5.

Brewin, B. I.
1946 Ascidians in the vicinity of the Portobello Marine Biological Station, Otago Harbour. Trans. R. Soc. N. Z., *76*(2): 87–131, figs. 1–19, pls. 2–5.
1948 Ascidians of the Hauraki Gulf, part 1. Trans. R. Soc. N. Z., 77(1): 115–138, figs. 1–9, pl. 9.
1950 Ascidians from Otago coastal waters. Trans. R. Soc. N. Z., *78*(1): 54–63, figs. 1–5.
1950a Ascidians of New Zealand. Part 4. Ascidians in the vicinity of Christchurch. Trans. R. Soc. N. Z., *78*(2–3): 344–353, figs. 1–5.
1950b Ascidians of New Zealand. Part 5. Ascidians from the east coast of Great Barrier Island. Trans. R. Soc. N. Z., *78*(2–3): 354–362, figs. 1–7.
1951 Ascidians of New Zealand. Part 6. Ascidians of the Hauraki Gulf, part 2. Trans. R. Soc. N. Z., *79*(1): 104–113, figs. 1–7.
1952 Ascidians of New Zealand. Part 7. Ascidians from Otago coastal waters, part 2. Trans. R. Soc. N. Z., *79*(3–4): 452–458, figs. 1–5.
1952a Ascidians of New Zealand. Part 8. Ascidians of the East Cape Region. Trans. R. Soc. N. Z., *80*(2): 187–195, figs. 1–3, pl. 37.
1956 Ascidians from the Chatham Islands and the Chatham Rise. Trans. R. Soc. N. Z., *84*(1): 121–137, figs. 1–4.
1957 Ascidians of New Zealand. Part 10. Ascidians from North Auckland. Trans. R. Soc. N. Z., *84*(3): 577–580, figs. 1–3.
1958 Ascidians of New Zealand. Part 11. Ascidians of the Stewart Island region. Trans. R. Soc. N. Z., *85*(3): 439–453, figs. 1–3.
1958a Ascidians of New Zealand. Part 12. Ascidians of the Hauraki Gulf, part 3. Trans. R. Soc. N. Z., *85*(3): 455–458, fig. 1.
1960 Ascidians of New Zealand. Part 13. Ascidians of the Cook Strait region. Trans. R. Soc. N. Z., *88*(1): 119–120.

Carlisle, D. B.
1951 On the hormonal and neural control of the release of gametes in ascidians. J. Exp. Biol., *28*(4): 463–471, 1 fig.

Cunningham, R. O.
1871 Notes on the natural history of the Strait of Magellan and west coast of Patagonia made during the voyage of H.M.S. *Nassau* in the years 1866, 67, 68, and 69. xvi + 517 pp., 21 pls. (col.), 1 map, Edinburgh.

Harant, H., and P. Vernières
1938 Ascidiae compositae. Scient. Rep. Australas. Antarct. Exped. 1911–1914, ser. C, *3*(5): 1–13, pl. 14.

Hartmeyer, R.
1909–1911 Ascidien (continuation of work by O. Seliger). *In* H. G. Bronn, Klassen und Ordnungen des Tier-Reichs, Leipzig, vol. 3, Suppl., parts. 81–98: 1281–1773, figs. 1–43 (of which figs. 6–43 are maps). C. F. Winter, Leipzig. (The species included are noted in Arch. Naturgesch., *6*(1): 3–27, 1911 (1912), under 'Tunikata für 1910,' grouped both systematically and faunistically, by A. Schepotieff.)
1911 Die Ascidien der deutschen Südpolar-Expedition (1901–1903), *12*(4, Zool.): 403–606, figs. 1–14, pls. 45–57.
1912 Die Ascidien der deutschen Tiefsee-Expedition. Wiss. Ergebn. dt. Tiefsee-Exped. 'Valdivia' 1898–1899, *16* (3): 225–392, figs. 1–10, pls. 37–46.
1916 Neue und alte Styeliden aus der Sammlung des Berliner Museums. Mitt. Zool. Mus. Berl., *8*(2): 203–230.

Hartmeyer, R., and W. Michaelsen
1928 Ascidiae Diktyobranchiae und Ptychobranchiae. *In* W. Michaelsen, and R. Hartmeyer, Fauna Südwest-Austr., *5*(6): 249–460, figs. 1–61.

Hastings, A. B.
1931 Tunicata. Scient. Rep. Gt Barrier Reef Exped., *4*(3): 69–109.

Herdman, W. A.
1880 Preliminary report on the Tunicata of the *Challenger* expedition. Proc. R. Soc. Edinb., *10:* part 1, Ascidiadae, 458–472; Ascidiadae and part 2, Clavelinidae, 714–726.
1881 Preliminary report on the Tunicata of the *Challenger* expedition. Proc. R. Soc. Edinb., *11:* part 3, Cynthiadae, 52–88, 1 fig.; part 4, Molgulidae, 233–240.
1882 Report on the Tunicata collected during the voyage of H.M.S. *Challenger* during the years 1873–1876. Part 1. Ascidiae simplices. *In* C. W. Thompson and J. Murray, Report on the scientific results of the voyage of H.M.S. *Challenger* during the years 1873–76. Zoology, *6*(17): 1–296, figs. 1–23, pls. 1–37.
1886 Report on the Tunicata collected during the voyage of H.M.S. *Challenger* during the years 1873–76. Part 2. Ascidiae compositae. *In* C. W. Thompson, and J. Murray, Report on the scientific results of the voyage of H.M.S. *Challenger* during the years 1873–76. Zoology, *14*(38): 1–432, figs. 1–15, pls. 1–49, 1

chart, 2 appendices: A. Supplementary report upon the Ascidiae simplices; B. Description of a new species of *Psammaplidium*.

1888 Report upon the Tunicata collected during the voyage of H.M.S. *Challenger* during the years 1873–76. Part 3. The Ascidiae Salpiformes. The Thaliacea. The Larvacea. *In* C. W. Thompson and J. Murray, Report on the scientific results of the voyage of H.M.S. *Challenger* during the years 1873–76. Zoology, *27* (76): 1–166, figs. 1–28, pls. 1–11, 2 appendices: A. Descriptions of two new species of simple ascidians. B. Descriptions of the dorsal tubercle of a large species of Ascidia from Kerguelen Island.

1902 Tunicata. *In* Report on the collections of natural history made in the Antarctic regions during the voyage of the *Southern Cross*. 190–200, pls. 19–23. London.

1910 Tunicata. *In* National Antarctic Expedition (S. S. *Discovery*) 1901–1904. Nat. Hist., *5:* 1–26, figs. 1, 2, pls. 1–7. London.

1912 The Tunicata of the Scottish National Antarctic Expedition, 1902–1904. Trans. R. Soc. Edinb. *48*(2): 305–320, 1 pl. (Reprinted, under same title, 1915, *In* W. S. Bruce, Rep. Scient. Results Voyage S. Y. *Scotia*, 1902–1904, *4*(7): 83–102, 1 pl.).

1915 See 1912 entry.

1923 Ascidiae simplices. Scient. Rep. Australas. Antarct. Exped. 1911–1914, ser. C, *3*(3): 1–35, pls. 8–13.

Herdman, W. A., and W. Riddell

1913 The Tunicata of the 'Thetis' Expedition. *In* Sci. Results Trawl Exped. 'Thetis,' part 17. Mem. Aust. Mus., *4:* 873–889, pls. 90, 91.

Kiaer, J.

1893 Oversigt over Norges Ascidiae simplices. Forh. Videnskselsk. Krist., *1*(9): 1–105, pls. 1–4. (On pp. 24 and 25 is given a full synonymy for *Ascidiella aspera* (Mueller). A section entitled 'English descriptions of the new and unknown species' appears on pp. 88–103.)

Kott, Patricia

1952 The ascidians of Australia. Part 1. Stolidobranchiata Lahille and Phlebobranchiata Lahille. Aust. J. Mar. Freshwat. Res., *3*(3): 205–333, figs. 1–183.

1954 Tunicata. Ascidians. Rep. BANZ Antarct. Res. Exped. 1929–1931, ser. B, *1*(4): 121–182, figs. 1–68.

1962 The Ascidians of Australia. Part 3. Aplousobranchiata Lahille: Didemnidae Giard. Aust. J. Mar. Freshwat. Res., *13*(3): 265–334, figs. 1–50.

1963 The ascidians of Australia. Part 4. Aplousobranchiata Lahille: Polyclinidae Verrill (continued). Aust. J. Mar. Freshwat. Res., *14*(1): 70–118, figs. 1–26.

1969 Antarctic Ascidiacea. A monographic account of the known species based on specimens collected under U. S. government auspices 1947 to 1963. Antarct. Res. Ser., *13:* i–xv, 1–239, figs. 1–242, pls. 1–3, map, tables 1–9. AGU, Washington, D. C.

1969a A review of the family Agnesiidae Huntsman, 1912 with particular reference to *Agnesia glaciata* Michaelsen, 1898. Proc. Linn. Soc. N.S.W., *93*(3): 444–456, figs. 1–7.

Lesson, R. P.

1830 Zoologie. *In* Voyage autour du monde sur la corvette *La Coquille* pendant 1822–1825, *2*(1): 1–471, pls. 1–16. Paris. (Tunicates, pp. 256–279, 433–440, pls. 4–9, 13.) (See Hopkinson, 1913, A Bibliography of the Tunicata. 121. Ray Soc. Publs.)

Michaelsen, W.

1898 Vorläufige Mitteilung über einige Tunicaten aus dem Magalhaensischen Gebiet, sowie von Süd-Georgien. Zool. Anz., 21: 363–371.

1900 Die holosomen Ascidien des magalhaensisch-südgeorgischen Gebietes. Zoologica, Stuttg., *12*(31): 1–148, 1 fig., pls. 1–3.

1904 Die stolidobranchiaten Ascidien der deutschen Tiefsee-Expedition. Wiss. Ergebn. dt. Tiefsee-Exped. Valdivia, *7*(2): 181–260, pls. 10–13.

1907 Tunicaten. *In* Ergebnisse der Hamburger Magalhaensischen Sammelreise 1892/93. Hamburg, 1: 1–84, pls. 1–3.

1908 Die Pyuriden (Halocynthiiden) des Naturhistorischen Museums zu Hamburg. Jb. Hamb. Wiss. Anst., *25*(2): 227–287, pls. 1, 2.

1920 Die krikobranchen Ascidien des westlichen Indischen Ozeans: Didemniden. Jb. Hamb. Wiss. Anst., *37:* 1–74, figs. 1–6, pls. 1, 2.

1922 Ascidiae Ptychobranchiae und Diktyobranchiae von Neuseeland und den Chatham-Inseln. (Papers from Dr. Th. Mortensen's Pacific Expedition 1914–16. No. 11.) Vidensk. Meddr Dansk Naturh. Foren., *73:* 359–498, figs. 1–35.

1924 Ascidiae Krikobranchiae von Neuseeland, den Chatham- und den Auckland-Inseln. Vidensk. Meddr Dansk Naturh. Foren., *77:* 263–434, figs. 1–30.

1927 Einige neue westaustralische Ptychobranchiate Ascidien. Zool. Anz., *71:* 193–203.

1928 See R. Hartmeyer and W. Michaelsen.

Millar, R. H.

1955 Ascidiacea. Rep. Swed. Deep Sea Exped. Zoology, *2*(18): 223–236, figs. 2–7.

1959 Ascidiacea. Galathea Rep., *1:* 189–209, pl. 1.

1960 Ascidiacea. 'Discovery' Rep., *30:* 1–160, figs. 1–67, pls. 1–6.

1964 Ascidiacea: Additional material. Galathea Rep., *7:* 59–62, figs. 1–4, pl. 1.

1966 Port Phillip Survey 1957–1963. Ascidiacea. Mem. Natn. Mus. Vict., no. 27: 357–375, figs. 1–11.

1966a Tunicata. Ascidiacea. Martine Invertebrates of Scandinavia, *1:* 1–123, figs. 1–86. Scandinavian University Books, Oslo.

Monniot, C., and Françoise Monniot

1970 Les ascidies des grandes profondeurs recoltées par les navires *Atlantis*, *Atlantis II*, et *Chain* (2ème note). Deep Sea Res., *17:* 317–336, figs. 1–13.

Nott, J. T.

1892 On the composite ascidians of the North Shore Reef. Trans. Proc. N. Z. Inst., *24:* 305–334, pls. 24–30.

Oka, A.

1915 Report upon the Tunicata in the collection of the Indian Museum. Mem. Indian Mus., *6:* 1–33, pls. 1–5.

Quoy, J. R. C., and J. P. Gaimard
1834–1835 Voyage de découvertes de l'Astrolabe pendant les années 1826–1829. Zoology, *3:* 1–952. [Ascidia, *3*(2): 603–626, pls. 87–92, 1835.]

Rodrigues, S. de A.
1966 Notes on Brazilian ascidians, part 1 (in English). Papéis Dept Zool., São Paulo, *19*(8): 95–115, figs. 1–42.

Sluiter, C. P.
1895 Tunicaten. *In* R. Semon, Zoologische Forschungsreisen in Australien und den malayischen Archipel. Denkschr. Med.-Naturw. Ges., Jena, *8:* 163–186, pls. 6–10.
1900 Tunicaten aus dem Stillen Ozean. Ergebnisse einer Reise nach dem Pacific. (Schauinsland 1896–1897.) Zool. Jb. Systematik, *13:* 1–35, pls. 1–6.
1904 Die Tunicaten der Siboga-Expedition. Abt. 1. Die socialen und holosomen Ascidien. Siboga-Exped., Monogr. 56a: 1–126, pls. 1–15.
1906 Tuniciers. *In* Expédition Antarctique Française (1903–1905). 1–48, figs. 1–10, pls. 1–5. Masson, Paris.
1909 Die Tunicaten der Siboga-Expedition. Abt. 2. Die merosomen Ascidien (Krikobranchia excl. Clavelinidae). Siboga-Exped., Monogr. 56b: 1–112, 2 figs., pls. 1–8.
1911 Une nouvelle espèce de *Tethyum* (*Styela*) provenant de l'expédition antarctique française (1903–1905) commandée par le Dr. J. Charcot. Bull. Mus. Hist. Nat., Paris, *17*(1): 37–38.
1912 Les Ascidiens de l'expédition antarctique française du *Pourquoi-Pas?* commandée par le Dr. Charcot, 1908–1909. Bull. Mus. Hist. Nat., Paris, *18*(7): 452–460.
1913 Ascidien von den Aru-Inseln. Abh. Senckenb. Naturforsch. Ges., *35*(1): 65–78, pls. 5, 6.
1932 Die von Dr. L. Kohl-Larsen gesammelten Ascidien von Süd-Georgien und der Stewart-Insel. Senckenbergiana, *14*(1/2): 1–19, figs. 1–12.

Tokioka, T.
1952 Ascidians collected by Messrs. Renzi Wada and Seizi Wada from the pearl-oyster bed in the Arafura Sea in 1940. Publs Seto Mar. Biol. Lab., *2*(2): 91–142, figs. 1–29.

Traustedt, M. P. A.
1882 Vestindiske Ascidiae simplices. Første Afd. (Phallusiadae). Vidensk. Meddr Dansk Naturh. Foren., ann. 1881: 257–288, 1 fig., pls. 4, 5.

Van Name, W. G.
1912 Simple ascidians of the coasts of New England and neighboring British provinces. Proc. Boston Soc. Nat. Hist., *34*(13): 439–619, text figs. 1–43, pls. 43–73.
1945 The North and South American ascidians. Bull. Am. Mus. Nat. Hist., *84*: i–vii, 1–476, figs. 1–327, pls. 1–31.

Vinogradova, N. G.
1962 Ascidiae simplices of the Indian part of the Antarctic. Biological Results of the Soviet Antarctic Expedition (1955–1958). Part 1. Explorations of the fauna of the seas. (In Russian.) Akad. Nauk. SSSR, Zool. Inst., *1*(9): 195–215, figs. 1–5.

ASCIDIAN SPECIES DISCUSSED IN TEXT

CORRIGENDA

Volume 13 of the Antarctic Research Series, *Antarctic Ascidiacea*

p. xv Under 'Ascidian Fauna,' after 'Nature of the Fauna,' add 'Chart ....196, 197.'

p. 16 Under 'Sta. 339' insert after position, 'December 3, 1962;.'

p. 57 Column 2, line 3, under '*Description:*' insert semicolon after 'investing.'

p. 107 Under '*Cnemidocarpa verrucosa* (Lesson)' insert 'Frontispiece' after 'Text-figs. 147–149.'

p. 115 Table 2, under 'Species,' '*oblogna*' to read '*oblonga.*'

p. 127 In 'Key to Antarctic Species of Pyura,' entry for *P. tunica*, add '(Knox Coast)' after 'p. 137.'

p. 135 Pagination lacking.

p. 151 Column 1, line 5, under heading, '*Molgula confluxa* (Sluiter),' add '[part]' after 'p. 35.'

p. 163 Column 1, line 10 above *Pareugyrioides filholi* (Pizon) to read 'of the oviduct supplementing the 2 or 3 short common.'

p. 168 Column 2, line 8, after '1913a, p. 6' insert '1915, p. 8.'

Column 2, line 17, after 'Indian Ocean' change '4158' to '3518.'

Column 2, line 18, after '1913a' add '1915.'

p. 175 Column 1, last line under *Pelonaia corrugata*, change 'viviparous' to 'oviparous.'

p. 176 Column 1, last paragraph, first line, add 'aplousobranch' after 'antarctic.'

p. 194 Column 2, line 5 from bottom, add '(p. 198)' after 'below.'

p. 211 Column 1, line 4, change 'six' to 'seven.'

p. 216 Column 1, line 1, change '1882' to '1822.'

p. 232 Index entry 'Podoclavella' has been repeated. Delete second entry and in first entry add '11' after '6.'

p. 238 Column 2, insert 'Ova, see eggs' in alphabetical order.

Column 2, line 26, entry 'phototactic,' change '*337*' to '174.'

THE SPECIES OF *CEPHALODISCUS* COLLECTED DURING OPERATION DEEP FREEZE, 1956–1959

JOHN C. MARKHAM

Rosenstiel School of Marine and Atmospheric Science
University of Miami, Miami, Florida 33149

Abstract. The specimens of *Cephalodiscus* M'Intosh collected during Operation Deep Freeze (1956–1959) belong to five previously described species: *C. nigrescens, C. solidus, C. densus, C. fumosus,* and *C. hodgsoni.* The geographical ranges of three of these species are extended, and the bathymetric range of the genus is extended to 1554 meters. The variation of *C. fumosus,* recorded for the second time, is discussed.

INTRODUCTION

The material on which this report is based is in the collections of the United States National Museum, Smithsonian Institution. It was collected in the vicinity of Antarctica over a period of slightly more than three years (February 1956 to April 1959) and during the course of Operation Deep Freeze operations, conducted by the United States Navy. They were devoted primarily to nonbiological investigations, so the collections are much less complete than they might otherwise have been. The bulk of the material is from Deep Freeze III and IV (1958 and 1959), with a few samples from Deep Freeze I and II (1956 and 1957).

Throughout this report, general remarks have been made on the species dealt with only where there seemed a need to supplement previously published accounts. Nearly all the drawings and photographs are of coenecial features, because nothing essentially new was discovered regarding the zooids.

(The author of the genus *Cephalodiscus,* M'Intosh, 1882, at a somewhat later date spelled his name McIntosh. His works are indexed under McIntosh in the Catalogue of the British Museum (Natural History), vol. 3, letters L–O, 1910.)

HISTORICAL RÉSUMÉ

The genus *Cephalodiscus* was established by M'Intosh [1882] to include a colonial, tube-inhabiting animal taken by HMS *Challenger* in the Strait of Magellan, which he named *Cephalodiscus dodecalophus.* M'Intosh [1887] stressed the strong affinities of *Cephalodiscus* with the colonial, tubiferous *Rhabdopleura normani* Allman, which had been found earlier in the North Sea. *Rhabdopleura* was at that time placed in the phylum Polyzoa, an assemblage of animal groups containing the ectoprocts, endoprocts, and other superficially similar forms now regarded as members of separate phyla. Accordingly, M'Intosh [1887] assigned *Cephalodiscus* to the Polyzoa. Harmer [1887], however, recorded clear affinities of this animal with the enteropneust *Balanoglossus.* These include the division of the body into three distinct regions, the presence of a notochord and dorsal nervous system, and the existence of paired gill-slits. He therefore proposed placing *Cephalodiscus,* and probably *Rhabdopleura,* in the Hemichordata.

No new records for *Cephalodiscus* were reported until the *Siboga* Expedition (1899–1900) discovered three new rather shallow-water species in the East Indies [Harmer, 1903], which Harmer [1905] named *Cephalodiscus gracilis, C. sibogae,* and *C. levinseni.* Shortly thereafter, the Swedish South Polar Expedition of 1901–1903 [Andersson, 1903] turned up *C. dodecalophus* again near the type locality. During this same expedition, some different forms were found near the Antarctic Peninsula, which Andersson [1907] described as the new species *C. aequatus, C. inaequatus, C. solidus, C. densus,* and *C. rarus.* Ridewood [1921] later discovered that the *Challenger* had taken *C. densus* in 1874, but the specimens had remained unrecognized.

During this same period, the British National Antarctic Expedition of 1901–1904 procured two new antarctic species, which Lankester [1905] and Ridewood [1907] described as *C. nigrescens* and *C.*

hodgsoni, respectively. It later became evident that *C. nigrescens* had been collected before any other species, the *Erebus* and *Terror* having taken it in 1841 or 1842 [Ridewood, 1912]. Also, Ridewood [1906] found a temperate-water species, *C. gilchristi*, off the coast of South Africa. Gravier [1912a, 1913] described the antarctic species *C. anderssoni*, while Harmer and Ridewood [1913] recorded *C. agglutinans* from near the Falkland Islands. Further collections and studies produced evidence that *C. rarus* [Ridewood, 1918b] and *C. anderssoni* [Ridewood, 1920] were synonyms of *C. densus*, and that *C. aequatus* [Johnston and Muirhead, 1951] and *C. inaequatus* [Harmer and Ridewood, 1913] were identical with *C. hodgsoni*. Harmer [1905], in describing *C. gracilis* and *C. sibogae*, stated that they might be opposite sexes of the same species. Since neither species has been found again, the matter remains unresolved.

Schepotieff [1909] reported a fourth tropical species, *C. indicus*, from near India and Ceylon; and Ridewood [1918a] found another southern temperate species, *C. evansi*, in the vicinity of northern New Zealand. More recently, John [1931] described a new antarctic species, *C. fumosus*, and a subantarctic one, *C. kempi*. Johnston and Muirhead [1951] reported the first discovery of the genus near Australia as *C. australiensis*, and Bayer [1962] found fragments of the first species recognized from the tropical Atlantic, *C. atlanticus*.

The total number of species described to date is thus 20, of which 4 have been synonymized with others, leaving 16 recognized species of *Cephalodiscus*.

STATION LIST

In the following list, stations from which specimens of *Cephalodiscus* were taken are listed under the names of the participating ships, arranged alphabetically; the annual Deep Freeze operations are arranged alphabetically by ship.

USS *Atka*

Deep Freeze III, December 1957–February 1958.

STA. 1A, Ross Sea, McMurdo Sound; 77°38.8′S, 165°52.7′E; December 20, 1957; 549 meters; orange peel bottom grab; collected by L. W. Wilson.

Cephalodiscus densus
Cephalodiscus hodgsoni form B

STA. 29, Knox Coast, Vincennes Bay, Wilkes Land; 66°17′35″S, 110°18′40″E; January 27, 1958; 135 meters, rocky bottom; orange peel bottom grab; collected by L. W. Wilson.

Cephalodiscus densus
Cephalodiscus hodgsoni form B

USS *Burton Island*

Deep Freeze III, December 1957–February 1958.

STA. 3, Ross Sea, east of Cape Hallett; 72°08′S, 172°10′E; January 13, 1958; 434 meters, hard gravelly bottom, −1.69°C; *Challenger*-type trawl; collected by R. B. Starr.

Cephalodiscus solidus

STA. 5, Davis Sea, near Mirnyy, Queen Mary Coast; 66°33′S, 93°01′E; January 29, 1958; 80 meters, hard rocky bottom, −1.53°C; *Challenger*-type trawl; collected by R. B. Starr.

Cephalodiscus densus

USS *Edisto*

Deep Freeze I, January–February 1956.

STA. ED-8, Ross Sea, McMurdo Sound; 77°26′S, 169°30′E; February 18, 1956; 321 meters; dredge; collected by J. Q. Tierney.

Cephalodiscus densus
Cephalodiscus hodgsoni form A
Cephalodiscus hodgsoni form B

Deep Freeze IV, January–April 1959.

STA. ED-14, TD-2, Weddell Sea, west of Cape Norvegia; 71°50′S, 15°50′W; January 18, 1959; 1000–1100 meters; triangular dredge; collected by J. Tyler.

Cephalodiscus densus

STA. ED-15, TD-3, Weddell Sea, west of Cape Norvegia; 71°55′S, 15°35′W; January 23, 1959; 1280 meters; bottom trawl; collected by J. Tyler.

Cephalodiscus densus

STA. ED-18, TR-4, Weddell Sea, west of Cape Norvegia; 71°40′S, 15°35′W; January 25, 1959; 1554 meters; bottom trawl; collected by J. Tyler.

Cephalodiscus hodgsoni form A

STA. ED-20, TR-5, Weddell Sea, off Vahsel Bay (Duke Ernst Bay); 77°40′S, 35°30′W; January 28, 1959; 384 meters; bottom trawl; collected by J. Tyler.

Cephalodiscus densus
Cephalodiscus fumosus
Cephalodiscus nigrescens

STA. ED-20, TR-6, Weddell Sea, off Vahsel Bay; 77°40′S, 35°30′W; January 28, 1959; 394 meters; bottom trawl; collected by J. Tyler.

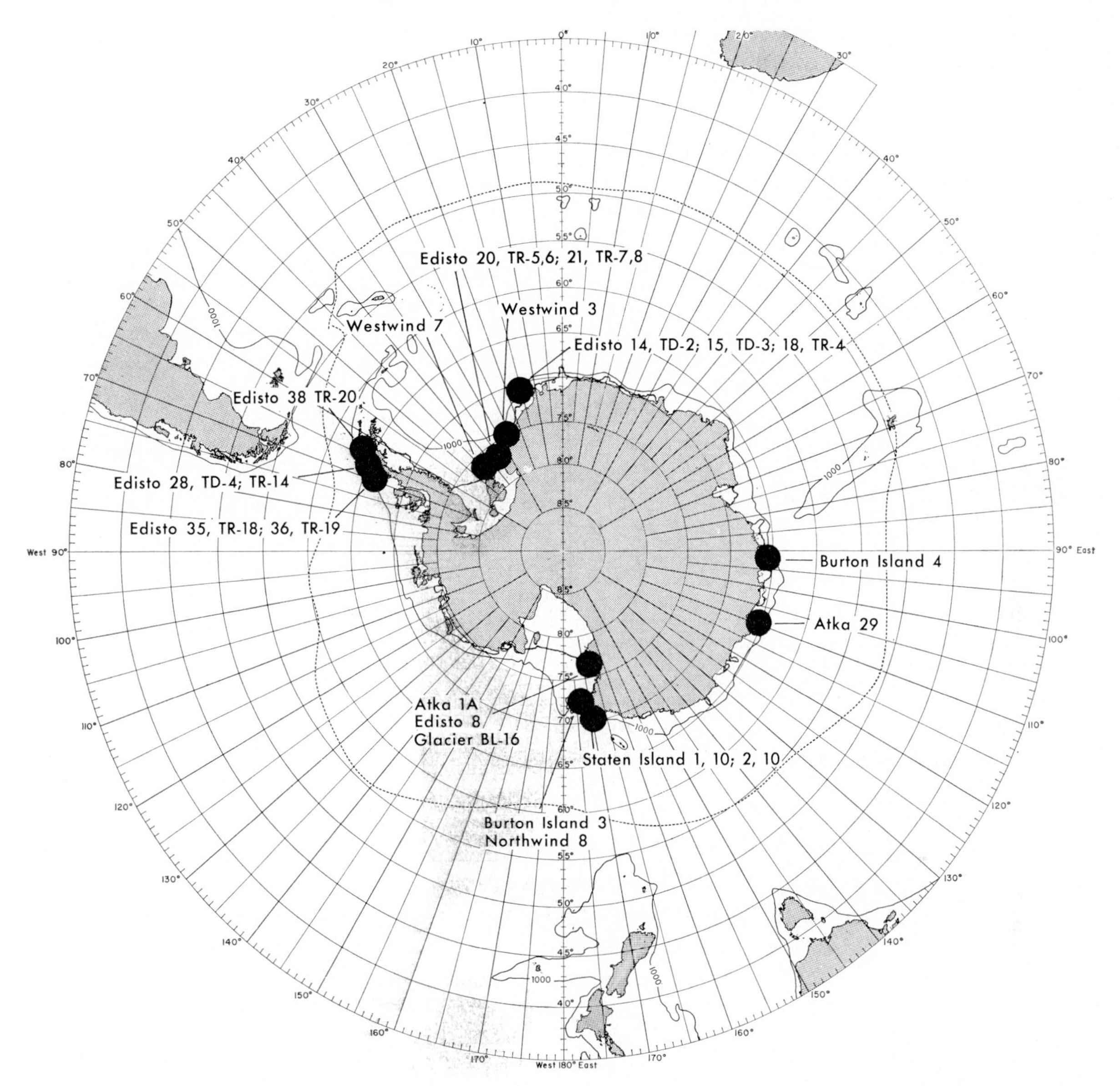

Fig. 1. Map of Antarctica showing locations of Deep Freeze stations from which specimens of *Cephalodiscus* were obtained.

Cephalodiscus densus
Cephalodiscus fumosus

Sta. ED-21, TR-7, Weddell Sea, off Vahsel Bay; 77°40′S, 35°30′W; January 30, 1959; 412 meters; bottom trawl; collected by J. Tyler.

Cephalodiscus densus
Cephalodiscus fumosus
Cephalodiscus nigrescens

Sta. ED-21, TR-8, Weddell Sea, off Vahsel Bay; 77°40′S, 35°30′W; January 30, 1959; 412 meters; bottom trawl; collected by J. Tyler.

Cephalodiscus densus
Cephalodiscus fumosus

Sta. ED-28, TD-4, Antarctic Peninsula, off Hugo Island; 65°08′S, 66°04′W; March 22, 1959; 135 meters; triangular dredge; collected by J. Tyler.

Cephalodiscus densus

Sta. ED-28, TR-14, Antarctic Peninsula, off Hugo Island; 65°08′S, 66°04′W; March 22, 1959; 130 meters; triangular dredge; collected by J. Tyler.

Cephalodiscus nigrescens

Sta. ED-35, TR-18, Antarctic Peninsula, off Lavoisier Island; 65°58′S, 66°51′W; April 4, 1959; 154 meters; bottom trawl; collected by J. Tyler.

Cephalodiscus nigrescens

STA. ED-36, TR-19, Antarctic Peninsula, off Lavoisier Island; 65°39'S, 66°19'W; April 5, 1959; 307 meters; bottom trawl; collected by J. Tyler.

Cephalodiscus nigrescens

STA. ED-38, TR-20, Antarctic Peninsula, off southwest coast of Wiencke Island; 64°50'S, 63°33'W; April 7, 1959; 134 meters; bottom trawl; collected by J. Tyler.

Cephalodiscus hodgsoni form A

USCG *Glacier*

Deep Freeze III, December 1957–February 1958.

STA. BL-16, Ross Sea, 2 miles northeast of Marble Point, near Cape Bernacchi; 77°26'S, 163°50'E; February 9, 1958; 142 meters; glacial moraine bottom; beam trawl; collected by W. H. Littlewood.

Cephalodiscus hodgsoni form B

Cephalodiscus nigrescens

USCG *Northwind*

Deep Freeze IV, January–April 1959.

STA. 8, Ross Sea, Moubray Bay, off Cape Hallett; 72°16'40"S, 170°18'E; January 12, 1959; 135 meters, rocky bottom; triangular dredge; collected by L. Wilson.

Cephalodiscus hodgsoni form A

USS *Staten Island*

Deep Freeze IV, January–April 1959.

STA. SI-1, 10, Victoria Land, Robertson Bay, west of Cape Adare; 71°27.5'S, 169°55.5'E; January 23, 1959; 439 meters, coarse volcanic sand, −1.30°C; beam trawl; collected by R. B. Starr and R. G. Miller.

Cephalodiscus hodgsoni form A

Cephalodiscus hodgsoni form B

STA. SI-2, 10, Victoria Land, Robertson Bay, west of Cape Adare; 71°21'30"S, 170°05'E; January 24, 1959; 128 meters, volcanic gravel and hard mud, −1.30°C; *Challenger*-type trawl; collected by R. B. Starr.

Cephalodiscus hodgsoni form A

Cephalodiscus hodgsoni form B

USCG *Westwind*

Deep Freeze III, December 1957–February 1958.

STA. 3, Weddell Sea, off McDonald Ice Rumples, Coats Land; 75°31'S, 26°43'W; January 8, 1958; 210 meters; triangular dredge; collected by J. Q. Tierney.

Cephalodiscus densus

STA. 7 (No. 91), Weddell Sea, off Filchner Ice Shelf, Vahsel Bay; 77°39'S, 44°50'W; January 16, 1958; 256 meters, sandy bottom; triangular dredge; collected by J. Q. Tierney.

Cephalodiscus nigrescens

SYSTEMATIC DISCUSSION

As mentioned in the section 'Historical Résumé,' the genus *Cephalodiscus* has been placed within the hemichordates. This group, Hemichordata, is variously considered a separate phylum [Hyman, 1959] or a subphylum of the phylum Chordata [Rothschild, 1965].

The Hemichordata contains three classes, Pterobranchia (erected to include *Rhabdopleura* and *Cephalodiscus* and considered by Barrington [1965] to be the most primitive hemichordates), Enteropneusta (including the various 'acorn worms' such as *Balanoglossus* and *Saccoglossus*), and Planctosphaeroidea (known only from some pelagic larvae) [Hyman, 1959; Rothschild, 1965]. In addition, the long-extinct graptolites are considered by some to be close relatives of the pterobranchs [Barrington, 1965; Bulman, 1955; Schepotieff, 1909], although Hyman [1959] considers the evidence for this relationship to be insufficient.

Kozlowski [1966] has more recently presented carefully detailed evidence that he believes indicates that the graptolites and pterobranchs were indeed closely related.

The class Pterobranchia is defined thus by Hyman: 'The Pterobranchia are small hemichordates with or without gill slits, with two or more tentaculated arms borne by the mesosome, and with recurved digestive tract, that live as aggregations or colonies housed in an externally secreted encasement' [Hyman, 1959, p. 155].

Within the Pterobranchia, there are two divisions. In one are those pterobranchs the members of whose colonies are not organically united, *Cephalodiscus* and the more recently discovered *Atubaria*. *Atubaria heterolopha* Sato, the only known member of the latter genus, is very similar in appearance and internal structure to *Cephalodiscus*, differing mainly in that it lacks a tube and its anus is posterior to the ovary [Komai, 1949]; further, in the few specimens that have been taken, no budding of new individuals was observed, although this is very common in *Cephalodiscus*.

The other division contains *Rhabdopleura*, in which the individuals are organically united. Hyman [1959] calls these divisions orders, namely Cephalodiscida and Rhabdopleurida, but recognizes no families. On the other hand, Van der Horst and Helmcke [1956], who call Pterobranchia an order but acknowledge that it might well deserve the rank of class, call these divisions families, namely Cephalodiscidae and Rhabdopleuridae. In a third major review of the group, Dawydoff [1948] presents yet another alternative, calling the Pterobranchia a class with the two above orders, each containing a single family, those already mentioned. Bulman [1955] recognizes a second family in the Cephalodiscida, Eocephalodiscidae, which was erected to include the Ordlovician fossil *Eocephalodiscus,* which is extinct.

The most conspicuous feature of a *Cephalodiscus* colony is the protective structure that its members secrete. Ridewood [1907], in observing that this structure commonly has been called the 'coenecium' (see glossary for definitions of terms applied to *Cephalodiscus*), considered that term inappropriate for the cephalodiscid structure because it is not homologous with the coenecium of ectoprocts; so he designated it the 'tubarium.' The term 'coenecium' has remained in use, however, and even Ridewood himself [1918a and later] subsequently used it.

Although the texture of the coenecium would lead one to describe it as chitinous, its actual chemical composition is uncertain. Andersson [1907] described a number of chemical degradation experiments on the coenecial material of *C. inaequatus* (<*C. hodgsoni*) that led him to conclude that it was neither chitin nor tunicin, while Rudall [1955] found that the coenecium of *Rhabdopleura normani,* on the basis of chemical and X-ray analyses, contained no chitin. Foucart et al. [1965] tested tubes of a pogonophoran (which phylum is considered a close relative of the Hemichordata [Marcus, 1958, and others]) and found 33% chitin, but application of the same tests [Foucart et al., 1965] to tubes of *Rhabdopleura* sp. and *Cephalodiscus inaequatus* (<*C. hodgsoni*) showed neither chitin nor cellulose. Though these workers found a great deal of glycine, they reached no conclusions about the material's protein structure.

Although the particular form of the coenecium does vary somewhat within a given species, it appears to be the single most reliable feature for the characterization and identification of species. Various characters of the individual animals inhabiting the colony, the zooids, have also been used, such as the length of the body and arms, the number of arms, the number of buds, and the presence and patterns of color [Harmer and Ridewood, 1913; Van der Horst, 1939]. Unfortunately, there are several objections to putting excess taxonomic weight on such characters, since they are very similar in most species, and within single colonies the animals may show marked variability; furthermore, poor preservation, which is all too common, may lead to the loss or alteration of zooidal characters.

Primarily on the basis of coenecial structure, Ridewood [1907] established 2 subgenera, *Idiothecia* and *Demiothecia,* and Andersson [1907] erected a third, *Orthoecus.* In discussing the makeup of the genus, Ridewood [1918b] presents an extensive account of these 3 subgenera and remarks that, although zooidal characters had been used previously to help distinguish them, it is less confusing to use coenecial characters alone. His key [Ridewood, 1920] is based almost entirely on the latter characters to the subgeneric level. John [1931], in describing *C. kempi,* could not fit it into any of these and so erected a fourth subgenus, *Acoelothecia.* (For descriptions of these subgenera see the key in the present paper.) Since *Demiothecia* includes the type species, *C. dodecalophus,* Johnston and Muirhead [1951] point out that this name must be suppressed in favor of *Cephalodiscus.* Bayer [1962], who placed his new species *C. atlanticus* in this subgenus, tacitly made this same alteration. Hyman [1959], however, disregards this point and continues to use the name *Demiothecia* (which she also spells '*Desmiotheca*' [p. 156] and '*Desmiothecia*' [pp. 156, 158]). Van der Horst [1939] lists the same 4 subgenera, while mentioning that possibly they should be considered genera. Van der Horst and Helmcke [1956] observed that, since the subgenera are based on coenecial characters, and since *Atubaria* differs from *Cephalodiscus* only in lacking a coenecium, either *Atubaria* should be considered a fifth subgenus or the others should be elevated to generic rank. Despite their observation that the zooids of all species are very similar, these authors have chosen the latter alternative of separate genera, a decision with which no one else seems to have agreed. In the present paper, the conventional subgeneric designation is employed.

Genus ***Cephalodiscus*** M'Intosh

Cephalodiscus M'Intosh, 1882, pp. 337–348. (Type species, *Cephalodiscus dodecalophus* M'Intosh, by monotypy.)

Definition. M'Intosh [1882, p. 348] gives a generic definition with the original description. Since that is based on a single species, however, it would be better to cite a broader definition which takes into account the whole range of variation of the genus as presently constituted, thus: 'Zooids small, living as a social community within a secreted coenecium, from the orifices or ostia of which they can emerge at will; coenecium with a common branching cavity or a separate tubular cavity for each zooid and its buds. Body of the zooid consisting of three parts, with separate divisions of coelom. Alimentary canal U-shaped, mouth ventral, behind the stalk of the shield; anus on the antero-dorsal surface of the trunk, near the bases of the arms. One pair of pharyngeal pores or gill-slits, near the collar-canals. A more or less tubular notochord projecting from the antero-dorsal wall of the pharynx; below it, in the shield, a pericardial sac with heart. Gonads simple, one pair, opening by short ducts near the anus. Trunk prolonged posteriorly or ventrally, according to the degree of extension of the zooid, into a stalk, with a terminal sucker, around the edge of which buds are produced' [Ridewood, 1918b, pp. 14–15].

Key to the Subgenera, Species, and Forms of *Cephalodiscus* M'Intosh

This diagnostic key includes 16 currently valid species, the forms of 2 of them, and a listing of their synonyms. It is based mainly on that of Ridewood [1920] with portions from his earlier keys [Ridewood, 1906, 1907] and additions from subsequent sources with observations on specimens in the present collection. The key is designed to make identification possible on the basis of coenecial characters alone. Although notes on the zooids are included, these should be considered only supplementary, since the zooids may be poorly preserved (in which event the arms are frequently uncountable and the color patterns lost) or occasionally absent; in the case of *C. atlanticus,* they have never been seen. In addition to its function in identification, the final step in each entry is designed as a diagnosis of the species involved.

1. Cavities in the coenecium in the form of tubes. Each tubular space with a single ostium and occupied by 1 zooid and its buds. Arms without end swellings or refractive beads 2
   Cavity of the coenecium continuous and occupied in common by the zooids and their buds....... 11

2(1). Coenecium in the form of a branching system, with the newest tubes at the apices of the branches (subgenus ***Idiothecia*** Ridewood 1907, p. 90) 3
   Coenecium unbranched, in the form of a hemisphere, cone, cake, or club, with the newest tubes at the edges; each tube extending from base to apex or side of colony, basal ends of tubes blind (subgenus ***Orthoecus*** Andersson 1907, p. 92) 7

3(2). Internal ends of the tubes communicating by a labyrinthic system. Branches massive (up to 115 mm long and 55 mm broad), coenecium friable, cream-colored, speckled, opaque, with abundant fragments of shell and other particles embedded; each ostium with a short, blunt lip or spine but no peristomial tube. Zooids 4.5 mm long, blackish; arms, 8–9 pairs; females unknown .... ***C. agglutinans*** Harmer & Ridewood 1913 (Falkland Islands)
   Internal ends of the tubes blind 4

4(3). Branches massive (generally 15–25 mm broad or more), each ostium with a short peristomial tube (2.5–4.5 mm from the surface of the branch). Zooidal arms usually 7 or 8 pairs.... 5
   Branches slender to medium (5–12 mm broad). Zooidal arms usually 6 pairs 6

5(4). Coenecium fragile, with abundant fragments of shell and other particles embedded. Zooids 3.5 mm long, white or pale green; arms usually 8 pairs ....
   C. evansi Ridewood 1918 (northern New Zealand)
   Coenecium grayish or brownish, translucent, without spines; peristomial tube blunt, more or less sessile, with single inflated triangular lip. Zooids 4.0–6.0 mm long, blackish; arms usually 7 pairs (may be 6 or 8), each with 2 black bands along the axis ....
   ..***C. nigrescens*** Lankester 1905 (Antarctic), p. 90

6(4). Branches fairly long, 5–10 mm broad, not fragile, with numerous long spines, brownish; ostia numerous except on main stem, with or without peristomial tubes. Zooids 1.6–1.8 mm long, blackish when alive, brown preserved, with blackish margin on anterior edge of shield; arms usually 6 pairs ....
   ......***C. gilchristi*** Ridewood 1906 (South Africa)
   Coenecium large (up to 132 mm long), subcylindrical, branches sparse, about 12 mm broad, including peristomes (3–5 mm without peristomes), some branches connected by solid bars; each ostium with a single-lipped peristome that extends beyond surface about 4 mm. Zooids 2.5 mm long, whitish; arms, 6 pairs ....
   .....***C. levinseni*** Harmer 1905 (southwest Japan)

7(2). Colony of long narrow tubes in sparse coenecial material; tubes only lightly embedded or free along most of length, mostly vertical and nearly parallel. Zooidal arms usually 8 pairs........ 8
   Colony of variable size and form, flattened, platelike or cakelike to bulky and massive; tubes firmly embedded in common coenecial substance, projecting little, closely crowded near base, gradually becoming more separated toward apex of colony. Zooidal arms 3 or 8 pairs........... 9

8(7). Colony up to 10 × 20 cm, common coenecial substance moderately firm; each ostium with a single thick lip, edge of ostium thick. Zooids 4–5 mm long, blackish, fading to pale brown when preserved ....*C. solidus* Andersson 1907 (Antarctic), p. 92
Colony to 7 cm high, tubes more or less separate and parallel to somewhat straggling, common coenecial subtance spongy, enclosing only basal ends of tubes; ostium without definite lip, almost always transverse, edge of ostium thin. Zooids 4–11 mm long, brownish or grayish............ ***C. densus*** Andersson 1907 (Antarctic) (>***rarus*** Andersson 1907; young form with small lax colony) (> ***anderssoni*** Gravier 1912), p. 94

9(7). Colony a small round plate (diameter 7–10 mm, thickness 3–4 mm); only very few (8–15) tubes; Colony orange alive, pale preserved; ostia broad and oval, without definite lip. Zooids 2.2 mm long, pale; arms 3 pairs....................... ..***C. indicus*** Schepotieff 1909 (Ceylon and India)
Colony flat (15 mm thick) and sandy to long (to 125 mm and more) and fleshy; tubes (may be very large number depending on colony's size) more or less vertical, with triangular peristomial lips. Zooidal arms usually 8 pairs..10

10(9). Colony very firm, may appear as ball of gray sand (most particles about 0.1 mm diameter) with tubes mostly inconspicuous to pale orange, fleshy mass with tubes extended; tubes without midribs, tapering progressively toward basal end, peristomial lips blunt. Zooids 3.0–9.4 mm long, with 2 black bands in axis of each arm, pinnules light tan with row of brown spots ....***C. fumosus*** John 1931 (Antarctic), p. 96
Colony very friable, thin-walled, yellowish, bearing particles up to 2 mm diameter; tubes average 10 mm long, with midrib along dorsal side, with constant diameter except slightly enlarged basal end, peristomial lips sharply pointed. Zooids 0.75–1.15 mm long, cream to brown in color, no pigment bands on arms..........***C. australiensis*** Johnston and Muirhead 1951 (southwest Australia)

11(1). Coenecium branching with numerous spines. Arms of zooids commonly with end swellings bearing refractive beads (subgenus ***Cephalodiscus*** s.s. (>***Demiothecia*** Ridewood 1907, p. 104).......12
Coenecium of meshwork of bars and spines; spaces between meshwork irregular and continuous (subgenus ***Acoelothecia*** John 1931)..............17

12(11). Colony up to 200 or 250 mm in height, coenecium amber-colored or pale; branches 3.5 mm or more wide..................................13
Colony small and delicate, coenecium orange, branches 2 mm or less wide..................15

13(12). Colony irregularly branched and straggling, some branches fused into network, crossbars usually solid, coenecium pale orange-brown or colorless, branches 3.5–6 mm broad; all ostia sessile, spines not obviously related to ostia, about 3–8 mm long; cavity of tube incompletely subdivided by ridges and partitions of irregular shape; distance between branches about 22 mm. Zooids whitish, arms usually 6 pairs (may be 5 pairs).. ....***C. dodecalophus*** M'Intosh 1882 (southern South America)
Colony irregularly branched, crossbars may be absent, older coenecium darker than younger, branches 4–6 mm broad; terminal ostia generally funnel-shaped, others sessile, spines on ostia, usually 2–5 mm long; cavity of tube with smooth inner surface; distance between branches about 10 mm. Zooids red with 12 arms or brown with 10 or 11 arms (***C. hodgsoni*** Ridewood 1907 (Antarctic) (>***aequatus*** Andersson 1907) (>***inaequatus*** Andersson 1907), p. 105)..........14

14(13). Older coenecium deep amber; internal diameter of tubes 1.5–5 mm (average 3 mm), ostia of fairly uniform size (1.5–2 mm × 1.0–1.5 mm), much more commonly lateral than terminal; terminal spines average 0.75 mm in diameter, much forked; crossbars commonly connect spines, sometimes in meshwork; commonly more than 10 spine tips on each branchlet............... ....***C. hodgsoni,*** form A, p. 105
Older coenecium dull brown; internal diameter of tubes 2.5–6 mm (average 4.5 mm), ostia funnel-shaped and large (2.5–3.5 mm × 1.5–2 mm), mostly terminal; terminal spines average 1.0 mm, little forked; crossbars between spines very rare or absent; rarely as many as 8 or 9 spine tips on each branchlet........................... ....***C. hodgsoni,*** form B, p. 105

15(12). Branches of coenecium biserial, alternate. Ostia at ends of very short side branches, funnel-shaped, bearing 2–3 spines 0.8–1.5 mm long. Zooids blackish, with 1 or 4 pairs of arms. Females unknown.............................. ....***C. sibogae*** Harmer 1905 (East Indies)
Branches of coenecium random..................16

16(15). Ostia at ends of main branches, funnel-shaped, bearing 3, 4, or 5 spines 2 mm long. Zooids 1.3 mm long, orange with black spots; arms 5 pairs with end swellings; numerous buds. Males unknown............................***C. gracilis***
Ostia on short lateral branchlets, flaring, bearing 1–5 spines 4–7 mm long. Zooids unknown...... ....***C. atlanticus*** Bayer 1962 (Caribbean)

17(11). Short tuftlike colonies with radiating spines on branches, no definite ostia, coenecium light brown; basal region of thick bars and spines forming meshwork from which spines radiate on all sides except basally..................... ***A. kempi*** John 1931, form A (Falkland Islands)
Large branching cylindrical colonies; close array of spines along sides of massive cylindrical branching stems............................. ***A. kempi*** John 1931, form B (Falkland Islands)

Subgenus *Idiothecia* Ridewood

Idiothecia Ridewood, 1906, p. 191. (Characterized in key to species of *Cephalodiscus.*)—Ridewood, 1907, p. 10.—Andersson, 1907, p. 10.—Schepotieff, 1909, p. 435.—Harmer and Ridewood, 1919, pp. 532, 536–540, 557, 558.—Ridewood, 1918a, pp. 15–25; 1918b, p. 8; 1920, pp. 407–409.—John, 1931, p. 256.—Johnston and Muirhead, 1951, p. 93.

Definition. 'Cavity of the tubarium multiple; each ostium leading into an unbranched, tubular cavity occupied by one polypide and its buds and having no connection with the other cavities of the tubarium' [Ridewood, 1907, p. 10].

Cephalodiscus (Idiothecia) nigrescens Lankester

Fig. 2

Cephalodiscus nigrescens Lankester, 1905, pp. 400–402; pl. 8 (preliminary note).—Ridewood, 1907, pp. 10–15, 20–49; text figs. 9–16; pl. 1; pl. 2, fig. 2; pl. 3; pl. 4, figs. 10–15; pl. 5, figs. 23–30, 37–40; pl. 7, figs. 60–75; 1912, pp. 540–554; 1 fig.—Gravier, 1913, pp. 75–77; fig. A.—Ridewood, 1918a, pp. 31–37, 73–74; pl. 1; pl. 2, figs. 3–5; 1918b, pp. 7, 13–15; pl. 7, fig. 2; 1920, pp. 408, 409.—John, 1931, pp. 233–235, 256–257; text fig. 6; pl. 37, fig. 5.—Johnston and Muirhead, 1951, pp. 95–97; pl. 1, fig. 1; pl. 2.

New records. A total of 30 specimens (some of which may be fragments of colonies) from 7 stations, as follows: No. 1: *Westwind* Sta. 7 (no. 91), Weddell Sea, 210 meters; 1 colony, USNM 11979. Nos. 2–5: *Glacier* Sta. BL-16, Ross Sea, 142 meters; 4 broken colonies, USNM 11980. No. 6: *Edisto* Sta. ED-20, TR-5, Weddell Sea, 384 meters; 1 colony, USNM 11981. No. 7: *Edisto* Sta. ED-21, TR-7, Weddell Sea, 412 meters; 1 colony, USNM 11982. Nos. 8–27: *Edisto* Sta. ED-28, TR-14, Antarctic Peninsula, 130 meters; 20 colonies, USNM 11983. Nos. 28–29: *Edisto* Sta. ED-35, TR-18, Antarctic Peninsula, 154 meters; 2 colonies, USNM 11984. No. 30: *Edisto* Sta. ED-36, TR-19, Antarctic Peninsula, 307 meters; 1 colony, USNM 11985.

Previous records.

Antarctic: Coulman I., near Victoria Land, Ross Sea, 100 fathoms [Lankester, 1905; Ridewood, 1907]; evidently in vicinity of Coulman I., about 300 fathoms [Ridewood, 1912]; South of Jenny I., Graham Land, 250 meters [Gravier, 1913]; McMurdo Sound, Ross Sea, 348–549 meters [Ridewood, 1918a]; Davis Sea, 120 fathoms [Ridewood, 1918b]; Bismarck Str., Palmer Archipelago, 93–126 meters and 315 meters [John, 1931]; Enderby Land, 300 meters, 209–180 meters, and 220 meters; Adélie Coast, 640 meters; Ingrid Christensen Coast, 437 meters; off Cape Bruce, Mac. Robertson Land, 219 meters [Johnston and Muirhead, 1951].

Diagnosis. Given in key, page 88.

Discussion. The specimens of *C. nigrescens* all were very readily identifiable, since they agree closely with the diagnosis [Ridewood, 1918a, p. 73]. Although there is some variability in the coenecial characters as described by Ridewood [1918a], his descriptions and photographs indicate these well. Coenecial characters alone were adequate for identification, although from each sample at least one zooid was dissected out to determine how well it conformed to the diagnosis. In all cases except one (no. 6 below) the paired black bands on the arms were clearly visible.

No. 2 (Figure 2A) is a large colony with a heavy cylindrical base from which 3 long, tapering arms arise perpendicularly. The colony is attached at the center to a flat white sheet of ectoproct colony, which is 1.5 mm thick and tapers from a width of 40 mm (due to branching) to only 3.5 mm basally, where it has been broken off. The coenecial material completely engulfs the ectoproct (and so probably killed it), and is devoid of tubes basally. This gives the ectoproct the appearance of a keel in the *Cephalodiscus* colony. There is no evidence that the colony was attached to anything but this thin ectoproct. The base is about 6 cm long, appearing broken at one end, and roughly cylindrical, its diameter varying between 18 and 24 mm. Although the base contains a large number of tubes, they are all empty and sealed off by septa roughly 2 mm below the tip of the lip. One branch is 140 mm long from its connection with the base to the point at its apical end, where it obviously has been broken off. It is nearly cylindrical, being 20 mm thick along most of its length, tapering slightly toward the apex. Most of the zooids of the first 3 cm of the base are missing, while their density progressively increases apically. The middle branch, which arises directly over the ectoproct keel, is 125 mm long and terete, tapering from 20 to 10 mm in diameter, and is broken off apically. The basal section, 2 cm long, lacks zooids. The third branch, which appears intact, is 150 mm long and tapers from 15 to 8 mm in diameter. There are no zooids in the first 3 cm.

No. 3 is only a fragment representing the trunk of

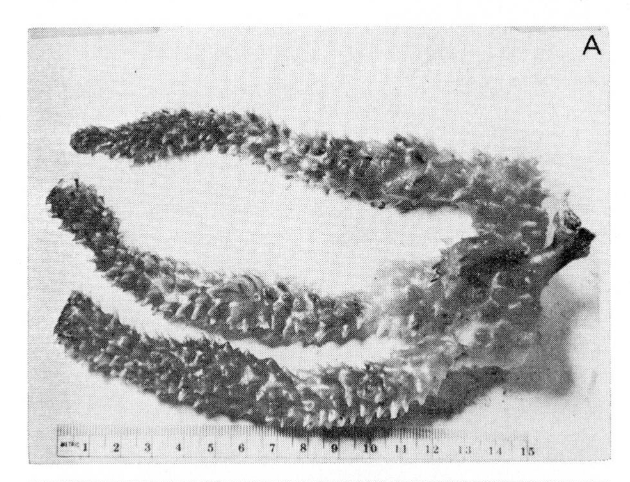

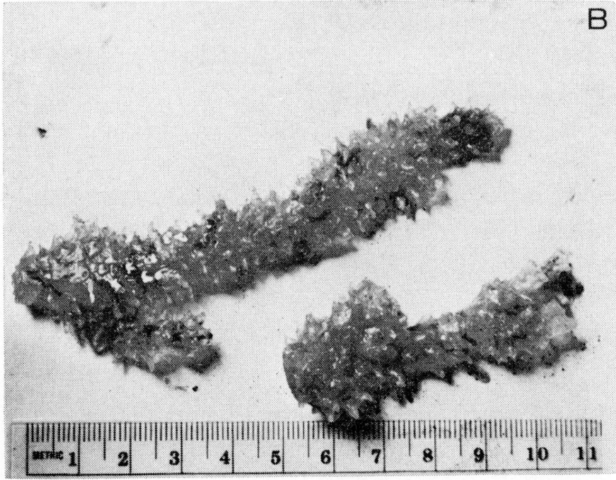

Fig. 2. *C. nigrescens:* A, no. 2, showing 3 long branches attached to large basal portion and thin ectoproct colony that supported the *Cephalodiscus* colony; B, nos. 28 and 29, fragmentary colonies showing zooids in branches as black strips and bases that lack zooids.

what was probably a rather large colony. Other branches in the jar are possibly part of the same colony, but none was found that was a good fit. The trunk is almost a perfect cylinder, 7 cm long and 22 mm in diameter. It bears the bases of 3 branches, each round in section and 13 mm in diameter. One of the branches is a 3-cm-long stub bearing a few zooids, while the others are broken flush with the base. The trunk is of particular interest because all of its zooidal tubes are closed over by coenecial material, some below the ends, as in no. 2, others completely covered over so that only bumps remain on the surface. Most of these tubes are of the same color as the rest of the coenecium, while 2 showed black beneath the surface. Dissection of one of the latter showed the covering to be completely independent of the contained tube, as reported by Ridewood [1907] for this same species. The black material proved to be unidentifiable dirtlike particles resting on an inner septum; there was no evidence that it represented a decomposed zooid, but that possibility is not excluded.

No. 4 is a 2-cm-long base that bears 2 broken branches, one 90 mm long, the other 95 mm long, both 15 mm in diameter. The former branch bears zooids above the first 3.5 cm, while the latter has them along its entire length.

No. 6 has been stained brown, possibly by being preserved with some other heavily pigmented organism. The zooids of this colony are also stained and so fail to show the well-defined armbands characteristic of this species. The form of the coenecium, however, is such as to leave no doubt about its identification.

Nos. 28, 29 (Figure 2B): These two basal portions show covered tubes similar to those already mentioned. In the photograph, note the absence of black strips (indicating zooids) in no. 29 (on the right).

Of particular interest is the evidence of the mode of growth of *C. nigrescens.* All the larger colonies lacked zooids in the base and progressively up the larger branches. This implies that as the colony grows the basal zooids die and are replaced at the apex. It is of course possible that the zooids, still alive, abandoned their tubes and moved up, but this does not seem likely. Because the empty tubes obviously were closed over from outside, it seems probable that zooids were absent from them while others in the colony were still very much alive; the latter zooids must have been able to range a considerable distance from their tubes to seal up the old ones. The progressive covering over of the tubes, by first inner then outer septa, implies that the process takes a considerable length of time. One possible reason for it is the strengthening of the colony as it grows upward.

This species, as evidenced by no. 2, can attain a considerable size, even when attached to very small objects. This would seem to indicate that it must ordinarily live in very quiet water, as would the observation (unfortunately made only on long-preserved material) that its branches are very fragile, readily breaking even when handled very carefully. A further characteristic of *C. nigrescens* that deserves mention is its extremely low density. A specimen was placed in sea water for 6 hours, during which time it remained floating at the surface. After this period, its specific gravity was roughly determined by diluting the sea water of known density with fresh water until the colony sank. This crude determination indicated a specific gravity of 1.01, considerably below that of the water in which it lives. Studies in antarctic waters [Bunt and Lee, 1969], which are of very low temperature and usually of high salinity (to over 37‰), have shown specific gravities of 1.0266 to 1.0300. It would be of great interest to perform such a determination on fresh material, but even the present test may partially explain how this species can stand on such a thin support and how its branches are able to grow up so straight and perpendicular to the base. It is of interest that individuals of all the other species at hand sank in the test sea water of T = 25.0°C, S = 33.1‰, sp. gr. = 1.0219.

On the basis of the foregoing observations, certain additions may be made to the characterization of *C. nigrescens:* The tubes are nearly sessile, with the lips thickened and nearly closing off the ostia. The colony, apart from the base, is of nearly uniform thickness along its whole length and frequently branching, with most of the resultant branches as large as the main stem. The tubes are not free in the colony, nor do they have any distinctive color of their own. The colony may seem flabby and flexible, but it generally breaks readily and cleanly right across the tubes, which are then revealed in a more or less symmetrical cluster of 5 or more at a given point, depending on the diameter of the branch. In the intact preserved colony, the zooids are distinct as short, black cylinders, especially conspicuous near the apices of branches.

Range. Specimens of *C. nigrescens* in the present collection come from the Ross and Weddell seas, from both of which regions the species has previously been recorded. It has also been found at several scattered locations around Antarctica, except on the Pacific side [Johnston and Muirhead, 1951]. Its known latitudinal range is rather restricted, from 65°46′S [Johnston and Muirhead, 1951] to 77°46′S [Ridewood, 1918a]. The present collection, obtained between 65°58′S and 77°40′S, does not extend this range. The previously recorded bathymetric range of *C. nigrescens* is 93–126 meters [John, 1931] to 640 meters [Johnston and Muirhead, 1951]; the material in the present collection, from 130 to 412 meters, lies well within this range.

Subgenus ***Orthoecus*** Andersson

Orthoecus Andersson, 1907, p. 11.—Gravier, 1912a, p. 146.—Ridewood, 1918a, pp. 16–26; 1920, pp. 407, 409.—John, 1931, p. 255.—Johnston and Muirhead, 1951, pp. 93, 101, 107, 109, 113.

Orthoëcus; Harmer and Ridewood, 1913, pp. 532, 538–540.

Orthœcus; Ridewood, 1918b, p. 8.

Definition. 'Colony not branching, but in the form of a cake, or cone, or mass of irregular shape. Each ostium of the coenecium leading into a tube which is occupied by one zooid and its buds. The tubes embedded in common coenecial substance, either for their whole length or towards their blind ends only; either closely set and parallel, more or less vertical, or irregularly bent and straggling' [Ridewood, 1918a, p. 74].

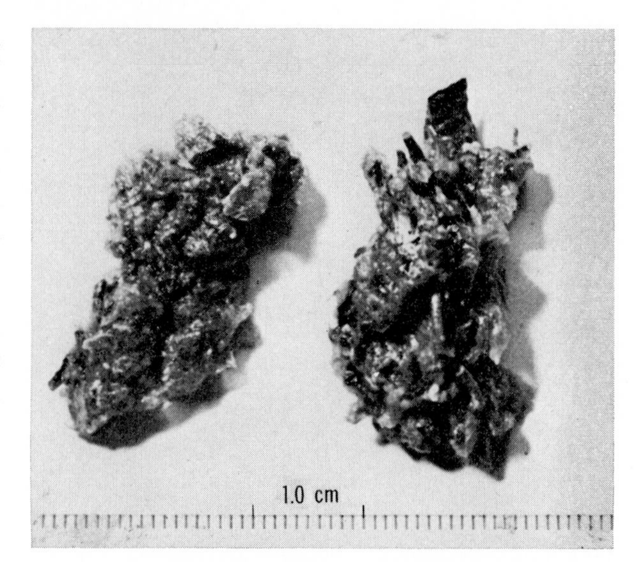

Fig. 3. *C. solidus*, 2 fragments, evidently of same colony, containing many types of inclusions.

Cephalodiscus (Orthoecus) solidus Andersson

Figs. 3, 4

Cephalodiscus solidus Andersson, 1907, pp. 11, 21–102 passim; pl. 2, fig. 4; pl. 3, fig. 10; pl. 7, figs. 59, 64; pl. 8, figs. 68, 86, 87.—Ridewood, 1918b, pp. 15–21; text figs. 2, 3; pl. 7, figs. 3–5.—Johnston and Muirhead, 1951, pp. 98–99.

New records. Two small fragments, evidently representing portions of a single colony from *Burton Island* Sta. 3, Ross Sea, 434 meters; USNM 11986.

Previous records.

Antarctic: North of Joinville I., 104 meters [Andersson, 1907]; off Mertz Glacier, 318 fathoms [Ridewood, 1918b]; off Enderby land, 2 locations, 209–

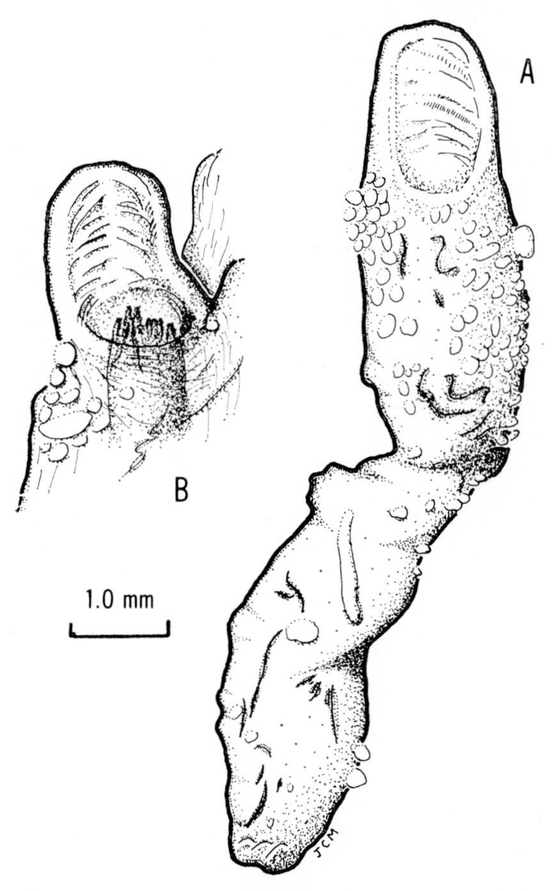

Fig. 4. *C. solidus:* A, strongly wrinkled entire tube dissected out of colony; some of sand grains and other particles have been eliminated for clarity; B, peristome of tube in colony with partly decomposed zooid at ostium; most sand grains have been eliminated for the sake of clarity.

180 meters and 220 meters; off Adélie Coast, 640 meters [Johnston and Muirhead, 1951].

Diagnosis. Given in key, page 89.

Discussion. The colony (Figures 3 and 4) is a flabby, friable, pale brown mass of tubes, rather similar in general form to *C. nigrescens,* despite being in a different subgenus. Ridewood [1918a, b, 1920] has discussed this resemblance at length. The coenecium contains a wide variety of debris, including siliceous spicules, foraminiferan tests, portions of ectoproct skeletons, shell fragments. Apically, each tube is sessile except for a slightly inflated lip reminiscent of *C. nigrescens.* It is thus unlike *C. densus,* in which the tubes lack a lip. Unlike *C. nigrescens,* its tubes are distinctly separate, blind-ended, and more highly colored than the common coenecial material; they also are rather easily freed from the coenecium. In these respects the tubes resemble those of *C. fumosus,* but they differ from those of the latter species in being much more flexible, generally not tapering, and not having a conspicuous basal region; their lips are also more inflated and squarish. In the original description, Andersson [1907] gave very little information about the coenecial structure and only one figure (pl. 2, fig. 4), which may be atypical in depicting such a regular arrangement of tubes. Since the zooids of the present material are so poorly preserved, they cannot be compared with Andersson's descriptions. Ridewood [1918b], in studying specimens found by the Australasian Antarctic Expedition, also reexamined the original material of *C. solidus* from the Swedish South Polar Expedition. Most of his descriptive remarks and all of his figures concern the coenecial structure. The material in the present collection, while not exactly as Ridewood [1918b] described *C. solidus,* agrees sufficiently well to justify the identification. In addition to the points cited above, the following characters conform to Ridewood's description: the tubes (Figure 4A) are of nearly uniform diameter except for frequent wrinkles and contractions; the walls of the tubes consist of numerous dark, very thin layers; further, the lips are oriented in many different directions with reference to the colony.

One tube dissected out is 12 mm long, its diameter ranging from 0.9 mm at the basal (blind) end to 2.0 mm at the ostial end, because of an abrupt flaring, which appears abnormal. The lip, the only portion projecting above the surface of the colony, is 1.5 mm long. The zooid within the tube is retracted so the tips of its arms lie 3.3 mm below the surface of the colony, or 4.8 mm from the end of the lip. In another case, however (Figure 4B), the zooid extends slightly from the end of the tube. Two zooids were removed, both in very poor condition and entirely unsuitable for illustrating or dissecting. Their color is a uniform drab brown except for a red proboscis line on one. One of the zooids had 12 arms 2.0 mm long, though some may have been missing; its body length, counting the arms, was 6.0 mm. No buds were discernible. On the second zooid, the decomposed arms were uncountable and unmeasurable, but at least 5 buds were present. Previous reports of *C. solidus* contain no illustrations of complete zooids, so it is especially regrettable that the present material is not suitable for drawing.

Range. *C. solidus* has been taken only 5 times previously. Its recorded range of latitude is from 62°55′S

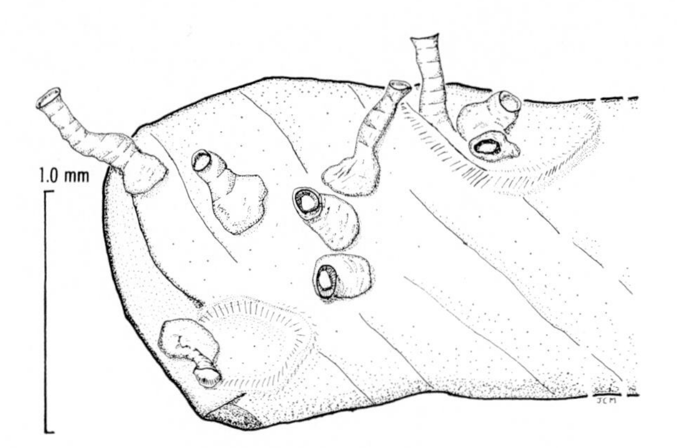

Fig. 5. *C. densus*, basal end of unoccupied tube bearing numerous tiny, dark tubes, which probably are the tests of folliculinid ciliates.

[Andersson, 1907] to 66°55′S [Ridewood, 1918b], so the present record of 72°08′S extends its range considerably southward. It has been taken in widely scattered areas around Antarctica, but neither in the Ross Sea nor on the Pacific side. The depth of the present specimen, 433 meters, lies well within the previously recorded bathymetric range.

Cephalodiscus (Orthoecus) densus Andersson

Figs. 5, 6

Cephalodiscus densus Andersson, 1907, pp. 12, 30–101 passim; pl. 3, figs. 7, 8; pl. 5, fig. 37.—Ridewood, 1918a, pp. 37–48, 75–76; text figs. 4–8; pl. 3; pl. 5, figs. 6–8; 1918b, pp. 22–23; pl. 7, fig. 1; 1920, pp. 408, 410; 1921, pp. 433–439; text figs. a–f; pl. 12.—John, 1931, pp. 235–238, 256; text fig. 7; pl. 33, fig. 6; pl. 35, fig. 1; pl. 36, figs. 2–6; pl. 37, figs. 1–4; pl. 38.—Johnston and Muirhead, 1951, pp. 97–98; pl. 1, fig. 2; pl. 3, fig. 4.

Cephalodiscus rarus Andersson, 1907, pp. 12–13, 20–86 passim; pl. 2, figs. 5, 6; pl. 3, figs. 9, 11, 15; pl. 4, fig. 22; pl. 5, figs. 28, 29, 32, 33; pl. 6, figs. 46, 47; pl. 7, figs. 53–55, 61, 62.—Ridewood, 1918a, pp. 39–40 (synonymized with *C. densus*).

Cephalodiscus varus; Andersson, 1907, p. 86 (erroneous spelling).

Cephalodiscus anderssoni Gravier, 1912a, pp. 146–150 (preliminary account); 1913, pp. 77–85; fig. B, figs. 1–16.—Ridewood, 1920, pp. 408, 410 (synonymized with *C. densus*).

New records. A total of 24 specimens (some possibly only fragments of colonies) from 12 stations, as follows: Nos. 1–4: *Edisto* Sta. ED-8, Ross Sea, 321 meters; 4 colonies, USNM 11615. Nos. 5–6: *Atka* Sta. 1A, Ross Sea, 549 meters; 2 colonies, USNM 11987. No. 7: *Westwind* Sta. 3, Weddell Sea, 210 meters; 1 colony, USNM 11988. No. 8: *Atka* Sta. 29, Wilkes Land, 135 meters; 1 colony, USNM 11989. Nos. 9–11: *Burton Island* Sta. 5, Davis Sea, 80 meters; 3 colonies, USNM 11990. Nos. 12–14: *Edisto* Sta. ED-14, TD-2, Weddell Sea, 1000–1100 meters; 3 colonies, USNM 11991. No. 15: *Edisto* Sta. ED-15, TD-3, Weddell Sea, 1280 meters; 1 colony, USNM 11992. No. 16: *Edisto* Sta. ED-20, TR-5, Weddell Sea, 384 meters; 1 colony, USNM 11993. Nos. 17–18: *Edisto* Sta. ED-21, TR-6, Weddell Sea, 394 meters; 2 colonies, USNM 11994. No. 19: *Edisto* Sta. ED-21, TR-7, Weddell Sea, 412 meters; 1 colony, USNM 11995. No. 20: *Edisto* Sta. ED-21, TR-8, Weddell Sea, 412 meters; 1 colony, USNM 11996. Nos. 21–24: *Edisto* Sta. ED-28, TD-4, Antarctic Peninsula, 130 meters; 4 colonies, USNM 11997.

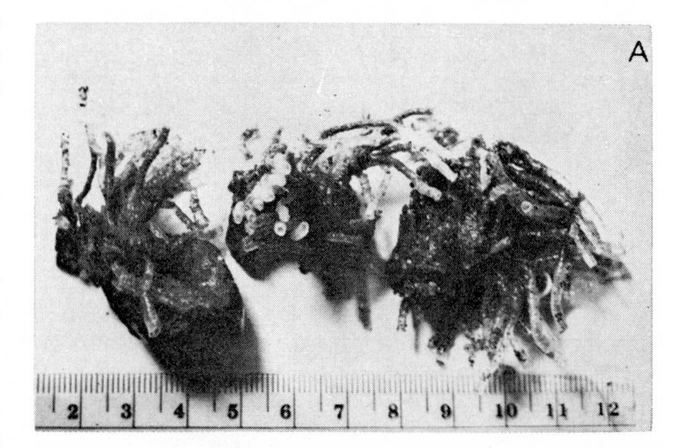

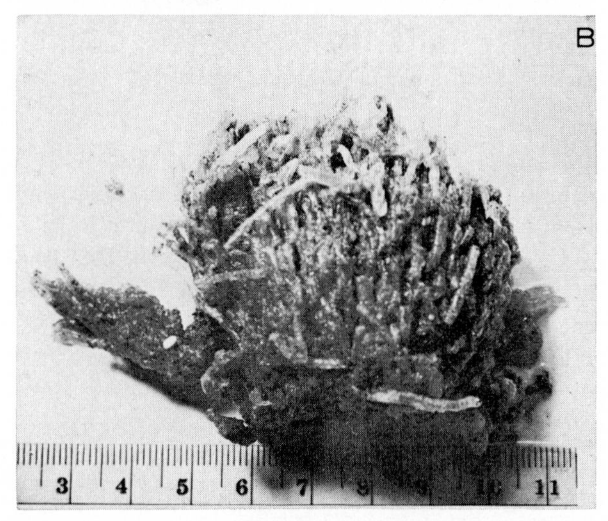

Fig. 6. *C. densus:* A, nos. 12, 13, 14, three small, straggling colonies attached to rocks; B, no. 21, large, fragile colony containing large amount of fine sand basally.

Previous records.

Antarctic: Southeast of Seymour I., Graham Land, 150 meters [Andersson, 1907]; McMurdo Sound, Ross Sea, at 5 stations, 50–300 fathoms; Coulman I., Ross Sea, 190 fathoms; Davis Sea, 120 fathoms [Ridewood, 1918a]; Bismarck Str., Palmer Archipelago, 126 meters [John, 1931]; outside MacKenzie Bay, off Mac. Robertson Land, 540 meters; off Enderby Land, 180 meters, 209–180 meters and 300 meters; off Adélie Coast, 640 meters; Knox Coast, 502–695 meters; Ingrid Christensen Coast, 437 meters; Cape Bruce, Mac. Robertson Land, 219 meters [Johnston and Muirhead, 1951]; southwest of Snow Hill I., 125 meters; west of Joinville I., Graham Land, 104 meters [Andersson, 1907]; south of Jenny I., 215 meters [Gravier, 1912a, 1913].

Subantarctic: Kerguelen I., 25 fathoms and 20–60 fathoms and 91 meters [Ridewood, 1921; Johnston and Muirhead, 1951]; 54°06′00″S, 57°46′00″W, 140–144 meters [John, 1931].

Elsewhere: South Atlantic, 32°42′00″S, 2°05′00″W, 75 meters (probably an erroneous record) [John, 1931].

Diagnosis. Given in key, page 89.

Discussion. All specimens of *C. densus* in the present collection were readily identified, since this species is quite distinctive in form. It is easily recognized by its tubes, which are separate except at the base, are of constant diameter, and nearly always have transverse ostia showing no enlargement of lips. The zooids examined agree with previously published descriptions. The general color of the tubes ranges from pale gray to faintly pink. The colonies break rather easily, always parallel to the tubes, which remain intact.

Nos. 1 and 2 may be part of a single colony, since both have the same appearance. Basally they are nearly black, mainly from the inclusion of coarse sand and small pebbles, which evidently were their substrate. The tubes are crooked and spaghettilike, roughly 1.5 mm outside diameter except for occasional knobby enlargements. All ostia are distinctly transverse. Colony no. 1, which appears to be essentially complete, flares widely from its basal end, being 40 mm tall, 30 mm wide at the base, and 45 mm wide shortly below the apex. One tube is 32 mm long, containing a retracted pale brown zooid lying between 6 and 12 mm from the blind basal end: Colony no. 2 is much more fragmentary; it is 45 mm long and nearly 20 mm in diameter along its whole length.

No. 5 consists of a few apparently empty tubes and a considerable mass of small black pebbles bound together by basal coenecial material.

No. 8 (Figure 5) is a collection of fragments, evidently representing only a small portion of what appears to be a single colony. There are only 3 fragments of basal material and 4 tubes, only one of which is unbroken and contains an intact zooid. Neither the coenecial material nor the tubes bear any sand. Several of the tubes bear large numbers of tiny, dark-brown, empty tubes, which are probably the tests of heterotrich ciliates of the family Folliculinidae. These tests are very similar in appearance to tubes of *C. densus* but for their minute size. At first they were considered newly formed tubes of buds recently released from the parent *Cephalodiscus*, but they are too small for this. Unfortunately, since all are empty, identification cannot be positive; but they correspond to illustrations of folliculinid tests [Andrews, 1955], and a *Cephalodiscus* tube, though not previously recorded as a substrate for them, would be similar to the types of substrates from which they have been gathered elsewhere.

Nos. 12, 13, 14 (Figure 6A): These 3 straggling colonies preserve their original attachment to large, black rocks.

In no. 17 the tubes are faintly pinkish and covered with coarse sand. One zooid dissected out is light brown and bears 7 pairs of arms and a single large, attached bud.

No. 20: This colony is represented by only a few small fragments in the same container with a specimen of *C. fumosus.* Upon first examination it was difficult to distinguish them from each other, since they were both coated with the same types of sand in the same pattern. This is a nearly uniform covering of fine, transparent sand grains about 0.1 mm in diameter, with occasional red and black grains about twice this diameter. There was no evidence that the two colonies had been joined together, but it is a distinct possibility. In any event, they must have been living in contact with the same sand deposit.

No. 21 (Figure 6B) is a very fragile colony, falling apart under even gentle handling. The basal half of the colony is only very loosely bound and appears gray because of adhering sand particles. At the base there are several black pebbles. Apically, the colony appears pinkish because of the color of its tubes, which are here conspicuous. Other inclusions in the

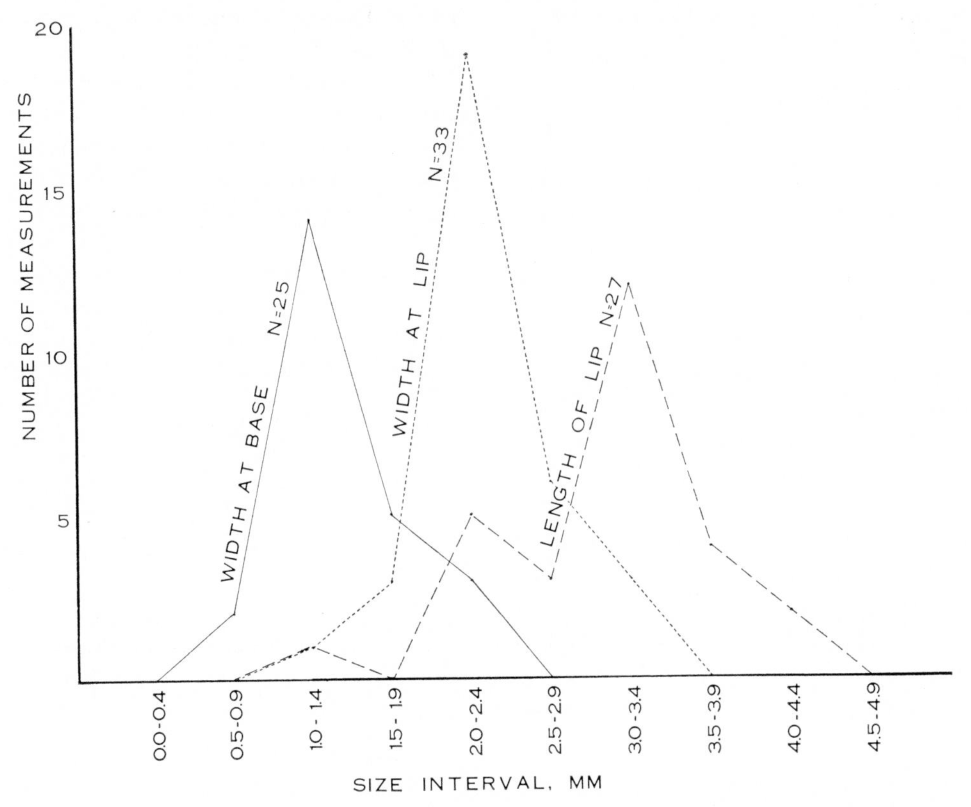

Fig. 7. Graphs of distributions of 3 indicated measurements of tubes from different colonies of *C. fumosus.*

colony are fragments of ectoproct colonies and a siliceous sponge bearing long, straight, glassy spicules.

Ridewood [1918a] comments that fairly long tubes of *C. densus* commonly contain a number of septa below the zooid, and he illustrates 2 such tubes in his text-figure 4. Examination of all specimens in the present collection failed to reveal such septa. In most colonies, the zooids were contracted completely into the basal ends of their tubes, even in tubes up to 35 mm in length and thus longer than those shown with septa by Ridewood.

Range. C. densus appears to be the most widely distributed species of *Cephalodiscus,* having been reported previously from a total of 20 stations on all but the Pacific side of Antarctica and over a latitudinal range from 49°28′S at Kerguelen [Johnston and Muirhead, 1951] to the most southerly stations at 77°45′S [Ridewood, 1918a]. John [1931] records it from as far north as 32°40′S, but Johnston and Muirhead [1951, p. 98] point out that this was probably due to an error in labeling; instead, they consider the station to have been south of the Falkland Islands. Its heretofore recorded bathymetric range, from 46 meters [Ridewood, 1921] to 640 meters [Johnston and Muirhead, 1951], is the greatest of any species. In the present collection, *C. densus* was found at the greatest number of stations, 12 of the 23, nearly all around Antarctica, including the 4 most southerly at 77°40′S and the second most northerly at 65°08′S, both of which lie within the range cited above. It was taken from depths ranging between 30 meters (the shallowest) and 1280 meters (the second deepest). Both the latter depth, at *Edisto* Sta. ED-12, TD-3, and a depth of 1000–1100 meters at *Edisto* Sta. ED-14, TD-4, far exceed the previously known maximum depth. The new record, 1280 meters is, indeed, just twice the 640 meters recorded previously.

Cephalodiscus (Orthoecus) fumosus John

Figs. 7–11

Cephalodiscus fumosus John, 1931, pp. 238–241, 256; pl. 34, fig. 1; pl. 35, fig. 3; pl. 36, fig. 1.

New records. A total of 82 colonies from 4 *Edisto* stations, all at the same loaction in the Weddell Sea, as follows: Nos. 1–14: Sta. ED-20, TR-5, 384 meters;

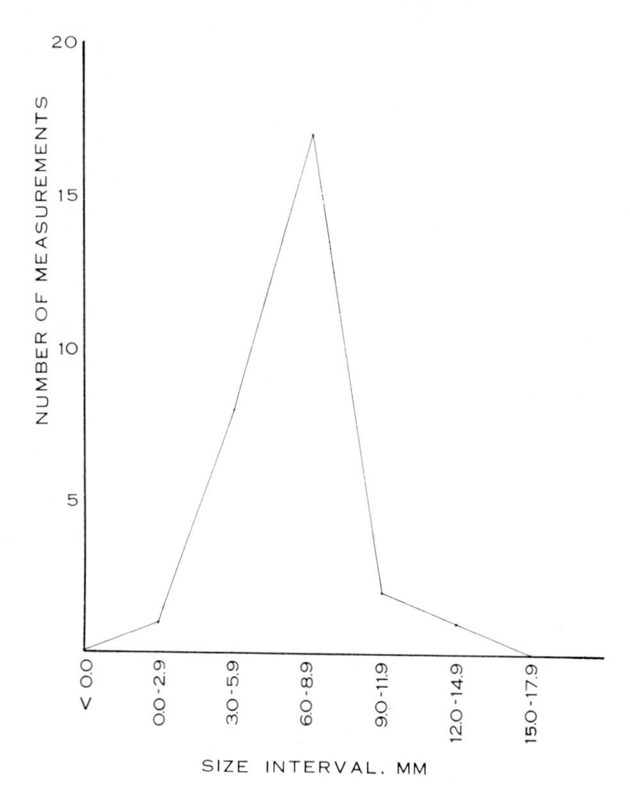

Fig. 8. Graph of distribution of lengths of peristomes of 29 tubes from different colonies of *C. fumosus.*

14 colonies, USNM 11998. Nos. 15–60: Sta. ED-20, TR-6, 394 meters; 46 colonies, USNM 11999. Nos. 61–79: Sta. ED-21, TR-7, 412 meters; 19 colonies, USNM 12000. Nos. 80–82: Sta. ED-21, TR-8, 412 meters; 3 colonies, USNM 12001.

Previous records.

Antarctic: Clarence I., near Antarctic Peninsula, 342 meters [John, 1931].

Diagnosis. 'Colony unbranched, in the form of a brittle arenaceous cake, measuring 90 mm. across and 16 mm. high and consisting of evenly distributed (5 to 7 mm. apart) vertically disposed tubes of uniform diameter. Tubes pale brown, buried in the common mass, which is formed of a dense agglutination of minute sand grains. Each ostium produced on one side into a short thick triangular lip, which can fold over like an operculum. Those in the middle of the colony 14 mm. long, those near the edge shorter. Width of the tube near the ostium 1.4 mm. Length of fairly extended zooid from tip of arms to base of body 3.2 mm., length of arms 1.2 mm., width of body 1 mm. Colour of preserved zooids—dorsal side of body and stolon light brown, ventral side brownish white. Proboscis with dark brown upper and lower edges. Usually eight pairs of arms, each with two black longitudinal bands running along the entire length as in *C. nigrescens.* No end-swellings with refractive beads. Female and hermaphrodite zooids found in the same colony. Up to ten buds on each zooid. The colony has a smoky grey colour which suggested the specific name.' [John, 1931, pp. 238–239.]

Discussion. C. fumosus previously was known only from the original description [John, 1931], which was based on 1 whole colony and a fragment of a second. The material in the present collection, although all from the same region, shows considerable variation in both the coenecial and zooidal characters. For this reason, it has been examined in detail. Measurements and other characters of 42 colonies were determined (Table 1). Tubes from 33 of these colonies were measured, the results being graphed in Figures 7 and 8. In addition, 13 zooids were dissected out, all from different colonies among the 42 mentioned above. Table 2 shows the results of this study.

As seen in Table 1 and Figures 9, 10, 11, there is great variation in the shape of the colonies of *C. fumosus* as well as in the amount of contained sand and in general appearance. In general, the colonies tend to be sandier when smaller, then gradually outgrow this condition, possibly because the zooids farther from the base are less able to obtain sand from the substrate for incorporation into the colony. However, 1 small colony (no. 72, Figure 10F) was almost devoid of sand. As the peristomes protrude farther, especially in the longer specimens, the appearance of the colonies becomes less sandy and more fleshy. In his original description John [1931] commented that the lips frequently act as opercula over the ostia; but in the present material, in which the lips are shaped essentially as he described them, this was observed only where the lips were obviously broken and unnaturally bent, probably as a result of collection and subsequent handling. The type specimen was a very flat cake, which some of the present specimens approximate.

The tubes, which are nearly circular in section (see Figure 11A), lie quite close together and open along the sides and apex of the colony in nearly uniform distribution, as John [1931] remarked (Figure 11C–E). Since all the tubes originate at the base, where their ends are blind (Figure 11B), their length is a function of their ostial positions in the colony,

TABLE 1. Measurements* and Observations in 42 Colonies of *C. fumosus*

Colony	Thickness, mm	Height, mm	Width, mm	Shape	Color	Inclusions
1	20	44	22	Long chunk	Tannish	Sparse sand;† few spicules
2	28	103	52	Long, thick sheet	Grayish tan	Thick sand; sponge
3	13	50	6–48	Triangular	Grayish	Fine sand
4	30–17	32	42	Hemisphere	Gray	Thick sand; spicules
5	38	. . .	68	Triangular	Grayish	Thick sand
6	23	85	51	Stack of flat sheets	Tannish	Sparse, fine sand; pebbles; spicules
7	18	28	32	Wedge	Gray	Thick sand
8	12	100	30–45	Flat sheet	Gray-brown	Sparse, fine sand
9	40	50	60	Chunk; ridge on top	Grayish	Sparse, fine sand; ectoprocts
10	10	112	53	Long, flat sheet	Grayish	Same as 9
11	49–25	20	70	Large, flat cake	Gray	Thick, fine sand
12	28	19	55	Rectangular bar	Grayish to glassy	Thick spicules above; thick sand below
13	18–26	20	25	Tapered	Gray	Thick, fine sand
14	20	72	25	Malletlike	Tan-gray	Sparse sand; spicules
15	14	84	31–59	Flat cake	Gray-tan	Sparse sand
16	5–7	40	13–18	Tapered	Pale gray	Same as 14
17	4–17	37	37	Cake	Grayish	Same as 14
18	40	50	35–60	Solid, rounded	Drab gray	Thick sand
20	8	44	22	Flat strip	Tan	Sparse, fine sand
21	20–30	125	30–85	Flat vase	Brownish	Sparse sand
48	25–30	100	45–77	Bulging chunk	Grayish	Sparse, fine sand
49	18–22	102	28–47	Clublike	Apex red-brown, base gray-brown	Fine sand; spicules; ectoprocts
55	35	60	10–45	Pyramidal	Gray	Thick fine sand
56	26	57	53	Solid chunk	Tan	Sand; spicules
57	13	45	30–40	Flat fan	Gray	Sand; ectoprocts; spicules; pebbles
58	10–25	85	25–40	Cake	Amber	Sand; spicules
59	20	15	20	Cake	. . .	Sand
60	10	30	20	Flat	Tan	Sand; spicules
61	22	78	45	Flat rectangle	Tannish	Sparse, fine sand
62	15	130	35	Flat strip	Tan-gray	Sparse sand
63	15–30	93	35–50	Flat, tapered	Tan-gray	Thick sand
64	17	95	20–35	Dumbbell	Tannish	Sparse sand
65	9	35	32	Pyramidal	Gray	Thick sand
66	9	46	16–57	Flat fan	Grayish	Fine sand
67	10–32	90	60	Long taper	Grayish	Sparse sand; ectoprocts
68	11	34	16	Subcylinder	Gray-tan	Sparse sand
69	17	124	5–80	Flat cactus	Orange-brown	Very sparse sand
70	20–30	22	20–30	Trapezoidal	Tan	Sparse sand
71	96	25–40	to 55	Flat cake	Gray	Thick sand
72	25	44	32	Nearly ovoid	Orange-tan	Very sparse sand
80	21	39	32	Rounded hump	Brownish red	Very sparse sand
81	40	30	40	Hemisphere	Gray-brown	Thick sand
Minimum	5	15	5			
Maximum	49	130	80			

* 'Height' is the distance along the basal-apical axis, which was determined by the presence of rocks, pebbles, and/or especially thick sand basally, or by the orientation of the peristomes, most of which always opened apically. Of the remaining two dimensions, 'width' was taken as the larger one, 'thickness' as the smaller. In cases of broad, flat colonies, it often happened that 'height' was the smallest dimension, while at other times it could be the greatest.

† The designation 'sand' usually refers to transparent grains approximately 0.1 mm in diameter uniformly distributed in the coenecium. Frequently, there were also scattered red and black grains about 0.2 mm in diameter. 'Spicules' indicates straight, fine siliceous sponge spicules reaching a length of several centimeters.

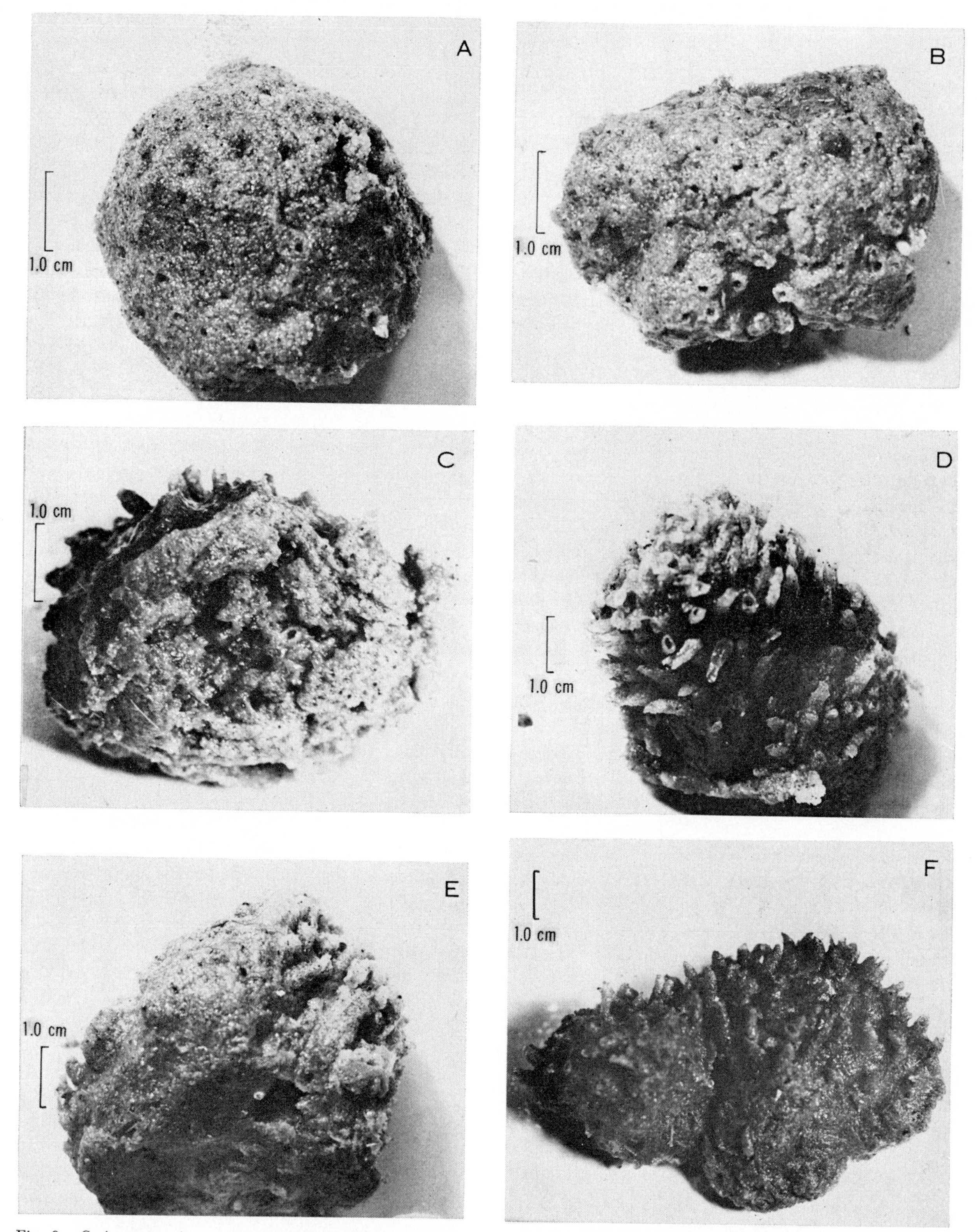

Fig. 9. *C. fumosus:* A, no. 81, spheroidal colony completely covered with fine sand; B, no. 21, slightly irregularly shaped colony almost completely covered with fine sand except for a few short peristomes; C, no. 51, side view; D, no. 18, moderately sandy colony, seen from above; E, same colony from basal end; F, no. 50, irregular colony with apically projecting peristomes.

extending from a minimum of 1 cm to the total height of the colony, which may be well over 100 cm (Table 1). The description by John [1931] of the color of the tubes is confirmed, except they might be a little better described as 'pale orange-brown' than as 'pale brown.' One colony (no. 57) was unusual in having completely colorless tubes, which otherwise were like those of the other colonies.

From observations on basal attachments, it is possible to theorize on the mode of growth of *C. fumosus* colonies. It appears that the colony spreads out, forming the 'cake' described by John [1931], until it has covered its basal substrate. It then begins growing upward, frequently enlarging above the base in order to accommodate more zooidal tubes.

In some of the larger colonies, a few tubes were found to be empty, and their ostia had been closed over by coenecial material that was built up on the outside several layers thick (Figure 12A), much in the manner described in the basal portions of *C. nigrescens.* However, unlike the case of the latter species, this seems to be a rather rare phenomenon in *C. fumosus.* No colony examined had more than about 10 or 12 out of several hundred tubes closed in this manner. The closed tubes tend to be in the basal half of the colony but are by no means clustered at the base, being usually of individual occurrence. In specimen 69 (Figure 10E) certain regions of the colony had grown out in such a way as to cover the ostia of tubes in adjacent regions. Several of the tubes so covered were closed over, their contained zooids probably having been killed by lack of food. One closed tube in no. 58 contains a mass of black material similar to that observed in some of the closed tubes of *C. nigrescens,* which may or may not be remnants of a decomposed zooid.

In all of the colonies, including the longest, the zooids were found to lie less than 1 cm from the ostial openings, usually less than one-half this distance. The reason for this, in all tubes more than a few centimeters long that were examined, is that the interior is completely closed off by a series of septa (Figure 12B). Although these septa are extremely difficult to see and are easily confused with wrinkles in the surface of the tubes (in the drawing they are exaggerated for purposes of illustration), one can demonstrate their presence by dissecting the tubes or by using a fine hypodermic needle to force bubbles or colored fluid such as India ink into the compartments separated by the septa. Because of the firmness of the common coenecial material, it is extremely difficult to dissect out a whole tube of *C. fumosus* of any great length. Fragments of tubes examined, however, showed the presence of septa occurring from a few millimeters above the blind basal ends up to just below the zooids, perhaps an average of roughly 4 mm apart. The resultant chambers are filled with liquid, probably sea water. Thus as the tube is lengthened, it is clear that the zooid lays down successive septa to cut off unused portions of the tube in a manner reminiscent of (though not nearly so regularly as) the chambered shell of the cephalopod *Nautilus.* This would be of obvious benefit to the animal by eliminating long open sections of tube that would otherwise cause stagnation of the water and possibly harbor unwanted commensals.

Similar septa have also been recorded from *C. densus* [Ridewood, 1918a], though they seem to be lacking in the present material of that species. Among the *Idiothecia* species, Lankester [1905] records septa from near the basal ends of the tubes of *C. nigrescens* as does Ridewood [1918a] from *C. evansi,* but in both cases the septa do not continue nearly so far apically, and they are much closer together than in *C. fumosus.*

When sand and siliceous sponge spicules (the most commonly encountered inclusions) are abundant, they lend a distinct gray color to the colony. Otherwise, the color of the tubes predominates, and the colony appears pinkish to orange-brown. A complete range of intermediate colors also occurs. In some cases, the colony may be very sandy basally and free of sand near the apex (Figures 10A and B), the color and general appearance varying accordingly.

Although specimens of *C. fumosus* exhibit two widely different appearances, there is considerable gradation between them, sometimes observable in a single colony. Furthermore, an entirely fleshy colony may represent a more mature form or one that lived where sand was scarcer. For these reasons, the forms have not been given special designations, as given by John [1931] for *C. kempi* and by Ridewood [1918b] for *C. hodgsoni.*

Although in his diagnosis, John [1931, p. 238] said that the tubes are of 'uniform diameter,' he commented later in the work [p. 240] that the tube 'becomes slightly narrower at the blind end.' In the present study, a decrease in the diameter of a single tube of as much as one-half has been seen. Although the data presented in Figure 7 are mainly of different tubes at each end, they clearly show reduction basally. The dimensions of the tubes and lips (Figure 7) are nearly all greater than those recorded by John (see

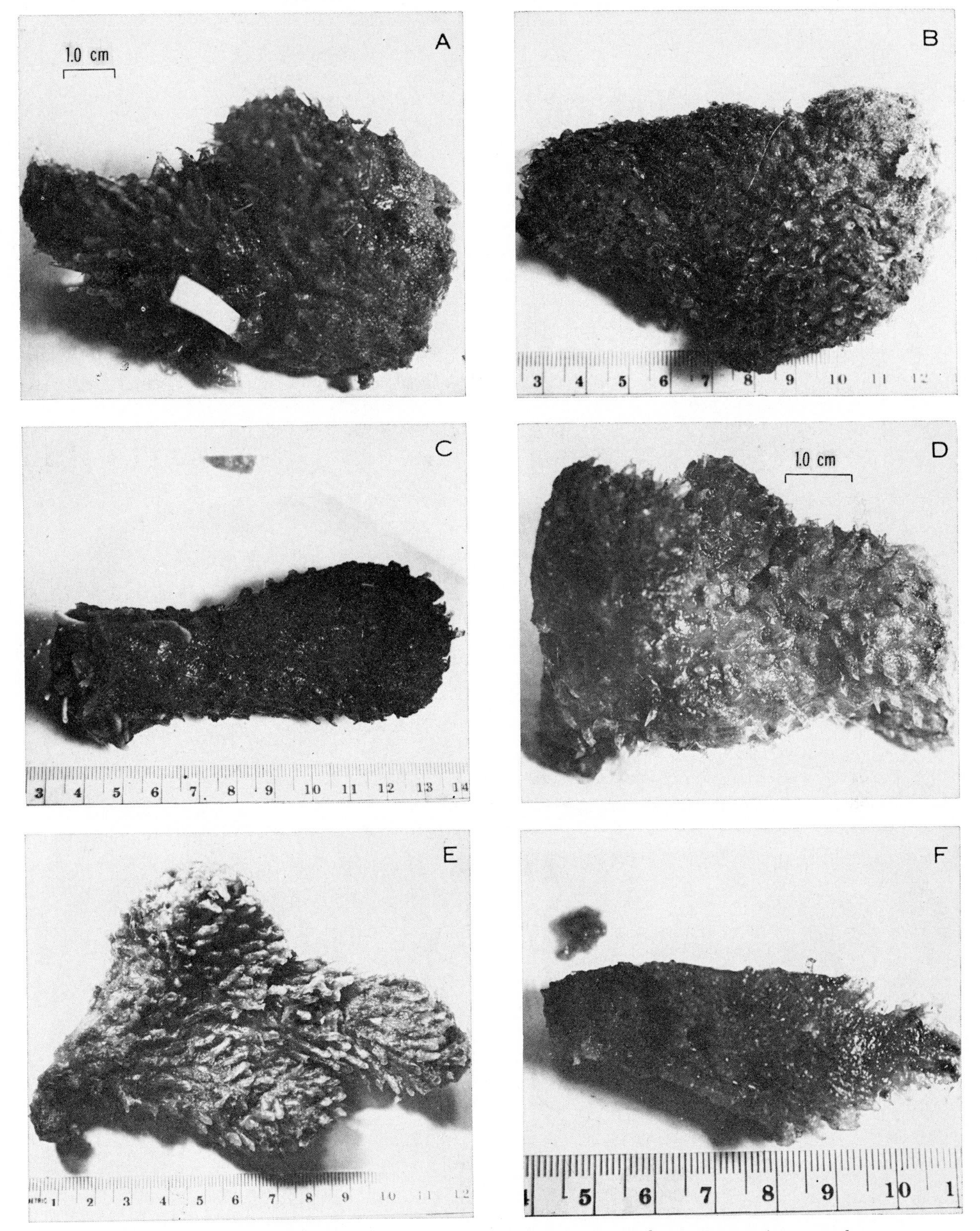

Fig. 10. *C. fumosus:* A, no. 52; B, no. 53; C, no. 49, long, club-shaped colony with prominent peristomes and numerous ectoproct colonies embedded in base; D, no. 48, large, fleshy colony; E, no. 69, large, flattened, fleshy colony with very long peristomes; F, no. 54, cut edge of portion of dissected colony showing arrangement of tubes.

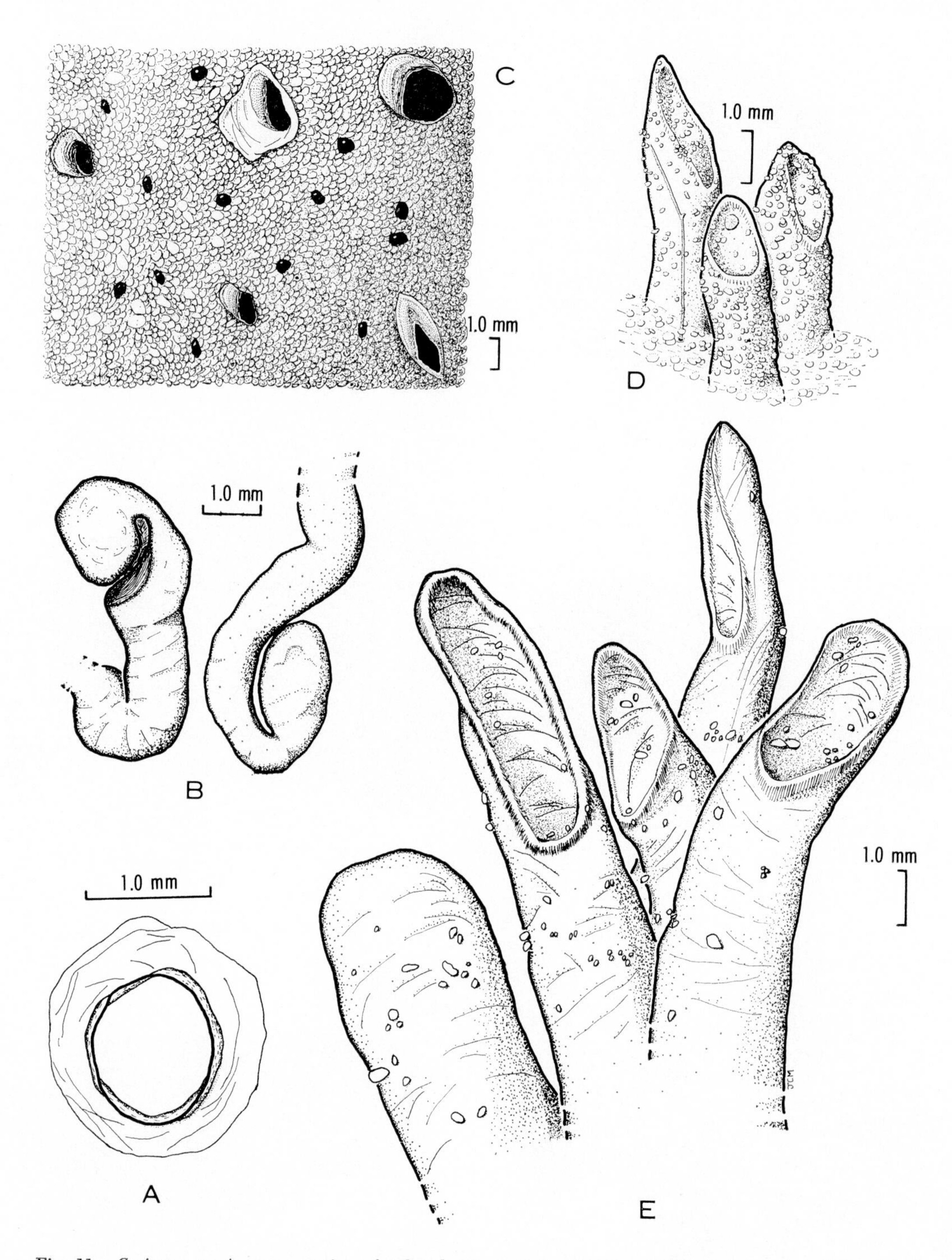

Fig. 11. *C. fumosus:* A, cross section of tube showing nearly circular outline and complete lack of midrib; B, basal ends of 2 small tubes showing how they are closed off; C, small portion of surface of very sandy colony, showing fine, colorless sand grains and occasional larger, black sand grains as well as peristomes nearly flush with surface; D, three peristomes projecting slightly above surface of moderately sandy medium-size colony; note long sponge spicule and foraminiferan test attached to peristomes; E, five peristomes from apical end of large, fleshy, nearly sandless colony.

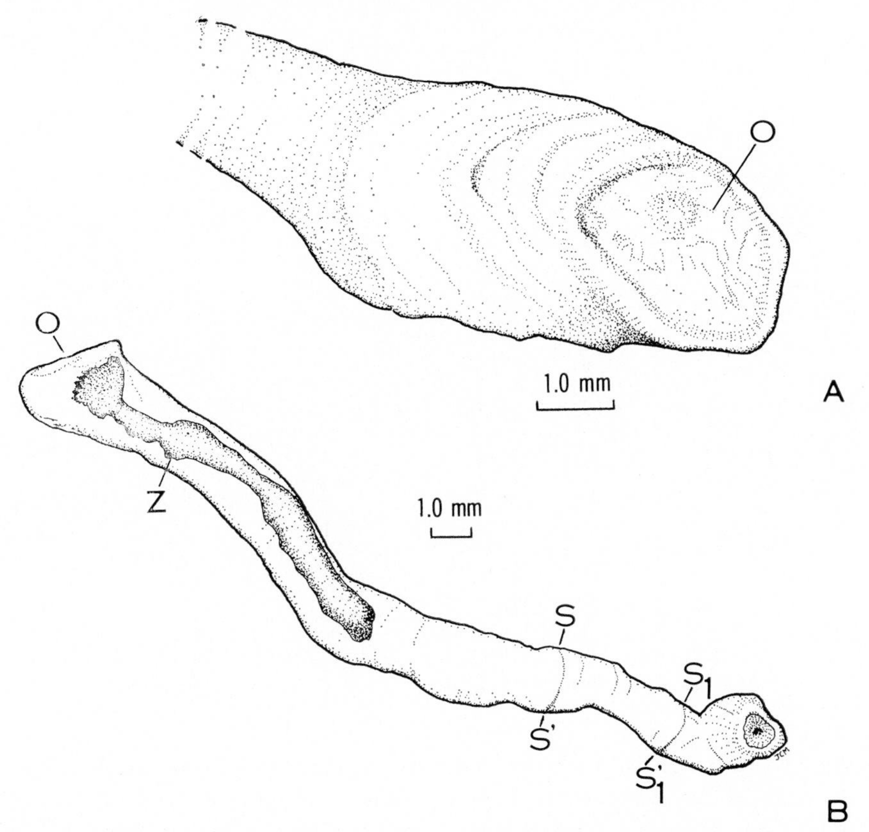

Fig. 12. *C. fumosus:* A, closed end of tube from specimen no. 58, a fairly large, fleshy colony showing several layers of coenecial material covering ostium (O, ostium); B, apical end of occupied tube from same colony showing 2 septa, whose distinctness has been exaggerated for the sake of illustration (O, ostium; Z, zooid in tube; S-S′, septum at the base of zooidal compartment; S_1-S_1', next septum).

diagnosis), probably because of the smaller size of his colonies, which were only 16 mm thick.

As other authors have noted [Johnston and Muirhead, 1951], the length of a zooid can vary considerably, depending upon its degree of contraction. It is still worth noting that several of the lengths here determined (Table 2) in the manner described by John [1931] are considerably greater than the 3.2 mm which he recorded for a 'fairly extended zooid' [p. 238], while the width and arm length are also somewhat greater than he records (see diagnosis). The reason for these greater dimensions is once again probably that John's material was a smaller colony, so the zooids may have been less mature. This is also reflected in the number of buds, which John records as being up to 10, while here a maximum of 21 was found. In the case of the zooid with 21 buds, some of these had the form of nearly complete zooids, which in turn bore less well-developed buds. Such a second generation of budding has been reported previously in *C. densus* [Ridewood, 1921].

TABLE 2. Dimensions and Variable Characters of 13 Zooids of Colonies Listed in Table 1

Colony	Length, mm	Width, mm	Color*	Arms, number	Arm Length, mm	Buds, number
1	4.5	1.2	B	14	1.7	4
17	4.7	1.2	T	16	1.7	5
18	7.0	1.0	T	16	3.0	8
20	7.4	1.0	T	16	1.7	8
55	5.5	1.0	T	18	2.0	3
56	7.0	1.2	B	18	3.0	13
57	9.4	1.2	T	20	3.0	10
58	8.6	1.0	T	16	3.3	21
59	4.8	1.0	B	..	1.8	5
70	6.0	1.1	B	14	2.0	8
71	7.9	1.3	T	16	1.7	3
80	6.8	1.3	T	16	2.3	12
81	6.3	1.2	T	12	1.3	12
Minimum	4.5	1.0		12	1.3	3
Maximum	9.4	1.3		20	3.3	21
Mean	6.6	1.1		16	2.2	

* Those zooids with the color pattern typical of the species (see diagnosis) are denoted 'T'; those in poor condition, which were a uniform drab brown, are denoted 'B.'

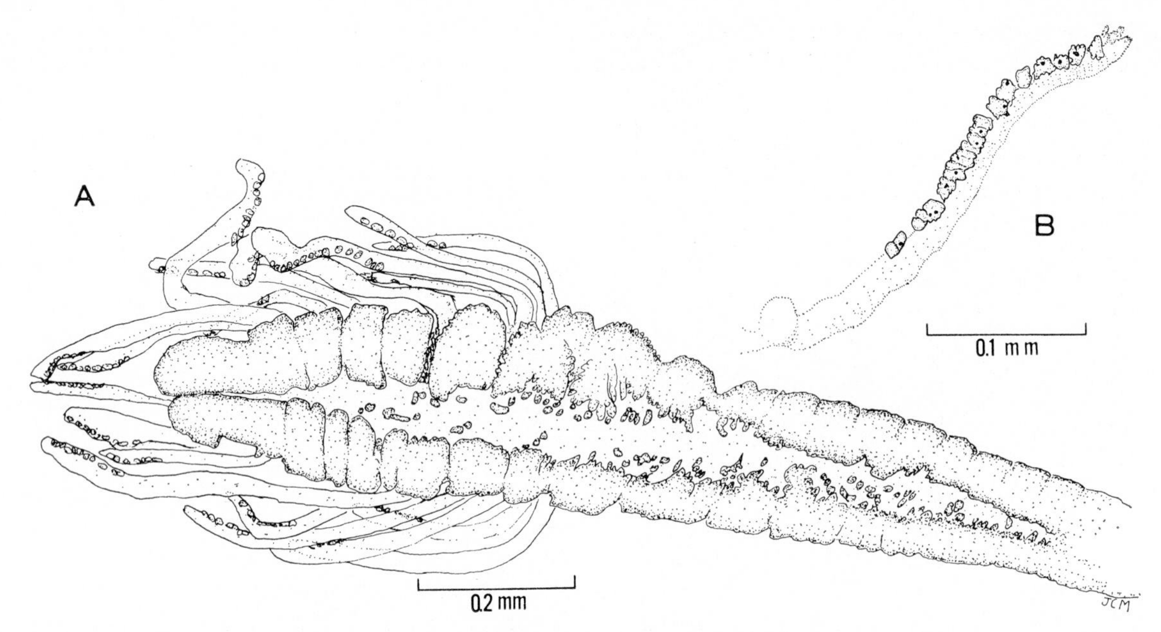

Fig. 13. *C. fumosus:* A, enlargement of arm of zooid from colony 72 showing detail of lateral black stripes and some of the distal pinnules; B, distal end of pinnule of arm of zooid from colony 57.

One fact of great importance is the discovery that, in poorly preserved material (as evidenced by partial decomposition and extreme fragility of the zooids), the characteristic coloration, including the arm stripes, becomes obliterated and the entire animal turns a uniform drab brown. Thus, while the description of the color in John [1931] seems to hold good (it was seen in material displaying many different coenecial characters), it must be remembered that it may be lost if preservation is faulty.

A few remarks can be made regarding the development of buds. When the buds are immature, they are attached to the parent zooid by a relatively thick, stubby stolon and are clustered behind and around the parent. As they develop further and eventually become detached, they move apically, often between the anterior end of the parent and the ostium. Their stolons also lengthen and become thinner until they are only fine filaments entwined around the parent.

John (see diagnosis) states that the zooids usually bear 16 arms. In the present material, 6 of the 12 zooids with countable arms examined bore 16 arms, while two each had 14 and 18, and one each bore 12 and 20 arms (see Table 2). Thus this is essentially in agreement with John's statement. Since John [1931] fails to present figures showing details of the arms and pinnules of *C. fumosus,* I have included them here (Figure 13). Otherwise, I was unable to find any zooidal characters that John has not already described and illustrated well.

All of the foregoing statements, where applicable, have been incorporated into the key presented on pp. 88–89.

Range. The original, heretofore unique, material of *C. fumosus* was from a single station [John, 1931] at 61°25′30″S, 53°46′00″W, off the Antarctic Peninsula. Although the present specimens came from 4 different stations, these were made at the same location, 77°40′S, 35°30′W, over a period of only 3 days. This is much farther south than previously recorded. The original discovery of *C. fumosus* was from 342 meters, somewhat shallower than the present samples, which range from 384 to 412 meters. *Cephalodiscus densus* was also taken at these 4 stations and *C. nigrescens* at 2 of them.

Subgenus *Cephalodiscus* M'Intosh

Cephalodiscus M'Intosh, 1882, p. 348; 1887, pp. 1–37. —Johnston and Muirhead, 1951, pp. 92, 93.—Bayer, 1962, p. 308.

Demiothecia Ridewood, 1906, p. 191 (defined in key to species of *Cephalodiscus*); pp. 7, 8.—Andersson, 1907, pp. 8–10.—Gravier, 1913, p. 73.—Harmer and Ridewood, 1913, pp. 532, 538–540.—Ridewood, 1918a, pp. 16, 18, 19, 24, 25; 1918b, p. 8; 1920, p. 410.—John, 1931, p. 257.

Fig. 14. *C. hodgsoni*, form A, no. 1, large colony with typical coenecial structure.

Desmiotheca; Hyman, 1959, p. 156 (erroneous spelling).

Desmiothecia; Hyman, 1959, pp. 156, 158 (erroneous spelling).

Definition. 'Colony branching. Each ostium of the coenecium leading into a cavity which is continuous throughout the colony, and is occupied in common by the zooids and their buds. Transverse sections of the branches showing the central cavity surrounded by a wall of coenecial substance, usually of irregular thickness, and sometimes with inwardly projecting bars and ridges' [Ridewood, 1918a, p. 66].

Cephalodiscus (Cephalodiscus) hodgsoni Ridewood

Fig. 14

Cephalodiscus hodgsoni Ridewood, 1907, pp. 3, 49–62; text fig. 17; pl. 2, fig. 1; pl. 4, figs. 21, 22; pl. 5, figs. 31–36, 41–47; pl. 6, figs. 48–59; pl. 7, figs. 76–84.—Harmer and Ridewood, 1913, p. 559. —Ridewood, 1918a, p. 48; pl. 4; pl. 5, figs. 1–5; 1918b, pp. 9–13; pl. 6.—Krumbach, 1927, pp. 465–467.—John, 1931, pp. 228–233, 258; text figs. 1–3; pl. 33, figs. 1–5.—Johnston and Muirhead, 1951, pp. 93–95, 113; pl. 3, figs. 2, 3.

Cephalodiscus aequatus Andersson, 1907, pp. 9, 10, 25–85 passim; pl. 5, fig. 38; pl. 6, fig. 49; pl. 7, fig. 52; pl. 8, fig. 66.—Johnston and Muirhead, 1951, pp. 93, 113 (synonymized with *C. hodgsoni*).

Cephalodiscus inaequatus Andersson, 1907, pp. 10, 19–102 passim; pl. 1; pl. 2, figs. 1, 2; pl. 3, figs. 12–14, 16; pl. 4, figs. 17–21, 24, 26, 27; pl. 5, figs. 31, 35, 36, 40; pl. 6, figs. 42–45, 50; pl. 7, figs. 51, 60, 63; pl. 8, figs. 65, 67, 69, 70, 84, 85.—Harmer and Ridewood, 1913, p. 559 (synonymized with *C. hodgsoni*).

New records. A total of 37 colonies, of which 30 are form A and 7 are form B, from 9 stations, as follows: Form A: No. 1: *Edisto* Sta. ED-8, Ross Sea, 321 meters; 1 colony, USNM 11607. Nos. 2–4: same station, 3 colonies, USNM 11606. Nos. 5–8: *Northwind* Sta. 8, Ross Sea, 135 meters; 4 colonies, USNM 12002. Nos. 9–18: *Staten Island* Sta. SI-1, 10, Victoria Land, 439 meters; 10 colonies, USNM 12003. No. 19: *Staten Island* Sta. SI-2, 10, Victoria Land, 128 meters; 1 colony, USNM 12004. No. 20: *Edisto* Sta. ED-18, TR-4, Weddell Sea, 1554 meters; 1 colony, USNM 12005. Nos. 21–30: *Edisto* Sta. ED-38, TR-20, Antarctic Peninsula, 154 meters; 10 colonies, USNM 12006. Form B: No. 1: *Edisto* Sta. ED-8, Ross Sea, 321 meters; 1 colony, USNM 11610. No. 2: same station; 1 colony, USNM 11611. No. 3: *Atka* Sta. 1A, Ross Sea, 549 meters; 1 colony, USNM 12007. No. 4: *Atka* Sta. 29, Wilkes Land, 135 meters; 1 colony, USNM 12008. No. 5: *Glacier* Sta. 16, Ross Sea, 142 meters; 1 colony, USNM 12009. No. 6: *Staten Island* Sta. SI-1, 10, Victoria Land, 439 meters; 1 colony, USNM 12010. No. 7: *Staten Island* Sta. SI-2, 10, Victoria Land, 128 meters; 1 colony, USNM 12011.

Previous records.

Antarctic: East end of Barrier, 78°16′14″S, 197° 41′47″E [*sic*], 100 fathoms, and at 3 other localities, 130–300 fathoms [Ridewood, 1907]; McMurdo Sound, Ross Sea, at 7 stations, 140–300 fathoms [Ridewood, 1918a]; Commonwealth Bay, Adélie Coast, 354 fathoms; Davis Sea, 120 fathoms; Shackleton Glacier, 110 fathoms [Ridewood, 1918b]; Gauss Station, Wilhelm II Coast, 350 meters [Krumbach, 1927]; Signy I., South Orkney I., 244–344 meters; Neumayer Channel, Palmer Archipelago, 259 meters; Bismarck Str., Palmer Archipelago, 93–126 meters [John, 1931]; Mac. Robertson Land, 163 and 219 meters [Johnston and Muirhead, 1951]; north of Joinville I., Graham Land, 104 meters; southeast of Seymour I., Graham Land, 150 meters [Andersson, 1907].

Subantarctic: South Atlantic, 54°07′00″S, 36°23′ 00″W, 160–330 meters [John, 1931].

Diagnosis. Given in key, page 89.

Variation. Ridewood [1918a] found sufficient consistent coenecial variations in this species to justify the designation of 2 growth forms, A and B. (See key, this paper, for distinctions between these.) Subsequent workers have followed this procedure, and the 2 forms are here listed separately.

Discussion. Most of the specimens of *Cephalodiscus hodgsoni* were readily identifiable, but in a few cases it was difficult to distinguish examples of form A from *C. dodecalophus* M'Intosh. Examination of the characters employed in the key (p. 89) resolved the question in every case.

Form A, no. 1 (Figure 14) is a fairly large colony with 2 branches arising from a central one. The base has evidently been broken off, but there are no other pieces in the jar with it. One branch has two further side branches. Each branch has a central dark portion roughly 5 mm in diameter, which bears the sessile ostia. The rest of the colony consists of the long, forked spines typical of form A. These spines are amber in color and increase the diameter of the branches to about 30 mm. The branches, which appear intact, are, respectively, 82 and 86 mm long from their mutual base. The side branches (on the longer branch) measure 40 and 29 mm long. The basal stem is 22 mm long from the branches to its point of breakage.

Form B, no. 3 is a colony more or less typical of form B. It is attached to a mass of small, dark pebbles by means of its basal coenecial material.

Range. C. hodgsoni is another very frequently taken species, fully as common as *C. densus* in certain locations, though not as widespread. It has been reported previously from 23 stations by 7 different expeditions around Antarctica except the Pacific and mid-Atlantic sides. In the present collection, *C. hodgsoni* is represented from 9 stations, some of them rather widely scattered. The specimen from Sta. ED-18, TR-2 establishes a new record in the mid-Atlantic gap. Its recorded latitudinal range is from 54°07′00″S [John, 1931] to 78°20′30″S [Ridewood, 1918a], which the present collection does not extend. Previous bathymetric records for this species are from a range of 93–126 meters [John, 1931] to 655 meters [Ridewood, 1918a]. In the present collection, the depth ranges from the second shallowest station, SI2, 10, at 128 meters, to the deepest, ED-18, TR-4, at 1554 meters. The latter depth thus establishes a new record over twice as deep as the previous maximum. Of all the published bathymetric records for the genus, the deepest is 900 meters, recorded by Ushakov [1962] off East Antarctica for some *Cephalodiscus* colonies of unspecified species. The new record of 1554 meters for *C. hodgsoni* is thus also a new record for the genus.

SUMMARY

Zoogeographical Considerations

Although *Cephalodiscus* has been known less than a century, it is evidently quite common at moderate depths throughout the Antarctic. As early as 1907, Andersson [1907, p. 16] remarked, 'Innerhalb des antarktischen Gebietes in der Gegend von Grahamland kann man ohne Übertreibung sagen, dass die Gattung *Cephalodiscus* einen stark hervortretenden Charakterzug der Meeresfauna bildet.' Gravier [1912b, p. 153] soon amplified this observation thus: 'Le *Cephalodiscus*, au moins en certains points, paraît être un des types les plus caractéristiques de la faune antarctique des fonds de moyenne profondeur.' These observations have been well borne out as virtually every subsequent major antarctic biological expedition has procured some specimens of the genus. The material on which this report is based, a respectable collection from an expedition that was only incidentally biological, tends to confirm the earlier belief that *Cephalodiscus* is indeed a characteristic part of the antarctic fauna.

Ekman [1953, p. 228], in discussing the great number of species and high degree of endemism in antarctic waters as compared with those of the arctic, remarks that 'the Antarctic has six species of the Pterobranchia genus *Cephalodiscus* which is missing altogether north of the tropics.' Although synonymization of *C. aequatus* with *C. hodgsoni* reduces the number of antarctic species to 5, and the single report of *C. levinseni* from near Japan makes this statement not strictly true, it still reflects the situation quite well. Table 3 shows the number of records of species of *Cephalodiscus* by general region, following the format used by Johnston and Muirhead [1951]. It is evident that *Cephalodiscus* is almost predictably taken in antarctic waters, while it is progressively rarer farther north until in the tropics and north temperate regions only 1 species has been collected at 2 stations, and 4 of the 5 such species are known only from single fragmentary colonies. Further, only 2 of the 5 antarctic species have been reported from the subantarctic and none from farther north, but, aside from *C. fumosus,* all of them have come from several different locations around Antarctica. This, then, indicates a need for the extremely low and constant temperatures of antarctic waters for optimal development of *Cephalodiscus.* It is of interest that the present report includes all 5 species previously found in

TABLE 3. Frequency of Collection of Species of *Cephalodiscus* in Five Zones

The table is based on all earlier expeditions and Operation Deep Freeze; in the latter all annual activities are considered as one. The number of stations producing *C. gilchristi* is uncertain; Ridewood [1906] reports 9, but Gilchrist [1915] gives no number. Those species denoted 'fr' are known only from single fragmentary colonies.

Zones	Species	Number of Expeditions	Number of Stations
Antarctic	*C. nigrescens*	8	23
	C. solidus	4	6
	C. densus	7	32
	C. fumosus	2	5
	C. hodgsoni	8	31
Subantarctic	*C. agglutinans*	1	1
	C. densus	3	4
	C. dodecalophus	2	3
	C. hodgsoni	1	1
	C. kempi	1	10
South temperate	*C. evansi*	1	1
	C. gilchristi	2	9 + ?
	C. australiensis	1	1
Tropical	*C. indicus*	1	2
	C. sibogae (fr)	1	1
	C. gracilis (fr)	1	1
	C. atlanticus (fr)	1	1
North temperate	*C. levinseni* (fr)	1	1

the Antarctic but none not already known from that region.

Remarks on Substrate and Growth

Apparently all species of *Cephalodiscus* need a solid substrate. In the case of *C. nigrescens*, the basal end sometimes bears an attached small rock or, more often, an indentation indicating that a rock has broken off. Several specimens of *C. densus* were still attached to cobbles, while others contained aggregations of pebbles in their bases. Both *C. fumosus* and *C. nigrescens* were frequently attached to ectoproct colonies basally. Thus I agree with Gilchrist [1915] that specimens of *Cephalodiscus* found on soft bottoms have probably been dislodged and carried there by water movements.

A rather curious feature of the bathymetric distribution of *Cephalodiscus*, which Johnston and Muirhead [1951, p. 116] have reviewed briefly, is that the shallowest finds, even from the intertidal zone, are those recorded from tropical regions, while the deepest finds, which now extend to the still relatively shallow depth of 1554 meters, are from antarctic areas. Species of *Cephalodiscus* from southern temperate and subtropical regions have generally been found at intermediate depths.

Van der Horst [1939] explicitly asserts that the colony of *Cephalodiscus* is enlarged by asexually produced buds that remain near the tubes of their parents and secrete new coenecial material, but he does not indicate the source of this observation, nor have I been able to find records of any species which had been kept alive long enough to prove this. However, there is good circumstantial evidence for such a conclusion. By studying a series of buds in different states of development, Masterman [1898] early established that they do indeed develop directly into the adult zooidal form. Andersson [1907] observed living *C. dodecalophus* and *C. inaequatus* (<*C. hodgsoni*) and recorded that they were able to crawl a considerable distance from their tubes. Gilchrist [1915] found the same ability in *C. gilchristi*, the only other species observed alive, and Harmer [1915] pointed out the probability of such a method of colonial enlargement on the basis of these observations. As far as the original formation of the colony is concerned, Gilchrist [1917], who described the ovulation, cleavage, and tissue formation of *C. gilchristi*, and Andersson [1907], Harmer [1905], and John [1932], who described several 'planula' larvae of *Cephalodiscus*, are undoubtedly correct in assuming that these sexually produced larvae are responsible. Further, Schepotieff [1909] found not only these 'planula' larvae but also more advanced ones that he considered to be pelagic in *C. indicus*, a tropical species.

Another important point is the occasional extreme variability of coenecial form among some of the species of *Cephalodiscus*. As mentioned earlier, the nature of the coenecium is one of the most reliable diagnostic features in all of the species, but in nearly all of them it can vary greatly without seriously hampering identification. *C. fumosus* presents an extreme example of this variability in that it ranges from a chunk resembling a ball of sand to an elongated fleshy-appearing colony with peristomes extending prominently.

That the coenecia do serve so well for identification is very fortunate, because zooidal characters may often be obscured or lost in preservation. Ridewood [1918a] reviews some of the color changes reported to occur with time in well-preserved specimens of several species. In addition to these problems, these animals are especially subject to poor preservation because they are likely to be misidentified in casual shipboard identification (some of the labels in the jars in the present collection bear the tentative identification 'sponge,' while no station record mentions pterobranchs) and because they quite probably contain a great deal of water, which would overdilute the preservative. In the present case, the single *C. solidus* and several of the *C. fumosus* have completely lost their color and were unsuitable for dissection because of poor preservation. That they had been preserved 10 years was clearly not the cause of the trouble because other specimens from the same stations but in different containers were in excellent condition. John [1931, p. 226], in commenting on the loss of many zooids in this manner, observes that 'when the zooids had been liberated from the coenecium and fixed in Bouin's fluid, the results were good.' In the present collection, 2 entire small colonies of *C. fumosus* (nos. 81 and 82) that were fixed and kept in Bouin's solution are in the best condition of all the specimens of this species. All the other specimens in the collection were preserved in 70% ethanol. Some of the station records indicated the use of 5% formalin for the original fixing, but it is not clear whether this applied to the *Cephalodiscus* colonies. Other fixing solutions that reportedly have given good results are 5% chromic acid followed by transfer to alcohol [Ridewood, 1918b] and a sequence of mixtures of absolute alcohol and xylol followed by preservation in 10% formalin sea water [Gilchrist, 1917]. In addition to the problems of poor preservation, there is also a strong possibility that the zooids may contract severely if they are not first relaxed. The zooids are commonly pictured with the stalk arising ventrally, but Ridewood [1918b] conjectured that this may be an artifact of preservation. Andersson [1903] reported that use of $MgSO_4$ produces well-extended zooids, and he later [Andersson, 1907] figured living specimens of *C. inaequatus* (<*C. hodgsoni*) in which the stalk clearly arises terminally, and which thus look very similar to *Atubaria heterolopha* as depicted by Komai [1949]. The various steps necessary to relax and carefully fix the animals would hardly be easy to perform on any significant scale while sorting through a large mass of freezing sediment in a howling gale aboard a rolling ship in the Antarctic, but it is to be hoped that someone on future collecting trips will try to make the effort.

GLOSSARY

Apex: the end or side of a colony opposite the point of attachment.
Arm: one of the appendages (usually paired) extending from the anterior end of a zooid toward the open end of the tube.
Base: the end or side of a colony attached to the substrate.
Bud: an asexually reproduced individual attached by a stalk to its parent zooid.
Coenecium: the common secreted investment that encloses the zooids.
Lip: a projection of one side of the ostium beyond the general ostial end.
Ostium: the opening of a tube to the outside at that tube's apical end.
Peristome: the free portion of a tube, i.e., that extending beyond the general surface of the colony.
Pinnule: a filamentous branchlet of an arm.
Polypide: term formerly used for 'zooid' (see below).
Polyparium: term formerly used for 'coenecium' (see above).
Spine: a thin, tapered projection of the coenecium, usually associated with an ostium.
Tubarium: term formerly used for 'coenecium' (see above).
Tube: the more or less cylindrical hollow portion of the coenecium (often set off by color and/or consistency from the rest of the coenecial material) that houses a zooid and its buds.
Zooid: an individual animal inhabiting a tube in the colony.

Acknowledgments. I am indebted to Dr. Frederick M. Bayer of the School of Marine and Atmospheric Science, University of Miami, for encouraging me to undertake this problem as part of a course in invertebrate systematics and for turning over to me the material on which he had earlier begun work at the Smithsonian Institution. Dr. Bayer made available to me his manuscript and notes and also provided much of the pertinent literature as well as helpful advice during the preparation of this paper. The specimens were obtained during Operations Deep Freeze I–IV by the U.S. Navy and are preserved in the U.S. National Museum of Natural History, Smithsonian Institution. Excluded from consideration are those specimens from Operations Deep Freeze I and II that were not personally examined by me. I am also grateful to Henry B. Roberts of the Smithsonian Institution for providing copies of the original station data sheets pertaining to the material reported. During this investigation I was supported at the University of Miami by a Robert E. Maytag Graduate Fellowship, for which I also wish to express my appreciation. Contribution 1340 of the Rosenstiel School of Marine and Atmospheric Science, University of Miami.

REFERENCES

Andersson, K. A.
1903 Eine Wiederentdeckung von *Cephalodiscus.* Zool. Anz., *26:* 368–369.
1907 Die Pterobranchier der schwedischen Südpolarexpedition 1901–1903 nebst Bemerkungen über *Rhabdopleura normani* Allman. *In* Wiss. Ergebn. Schwed. Südpolarexped., *5*(10): 1–122, 8 pls. Kungl. Boktryekeriet, Stockholm.
Andrews, E. A.
1955 More folliculinids (Ciliata Heterotricha) from British Columbia. J. Fish. Res. Bd Can., *12*(1): 143–145, 6 figs.
Barrington, E. J. W.
1965 The biology of Hemichordata and Protochordata. vi + 176 pp., 82 figs. Oliver and Boyd, Edinburgh and London.

Bayer, F. M.
1962 A new species of *Cephalodiscus* (Hemichordata: Pterobranchia), the first record from the tropical Western Atlantic. Bull. Mar. Sci. Gulf Caribb., *12*(2): 306–312, 6 figs., 1 table.

Bulman, O. M. B.
1955 Graptolithina with sections on Enteropneusta and Pterobranchia. *In* R. C. Moore (ed.), Treatise on invertebrate paleontology. Part V. 101 pp., 72 figs. Geological Society of America and University of Kansas Press, Lawrence, Kansas.

Bunt, J. S., and C. C. Lee
1969 Observations within and beneath antarctic sea ice in McMurdo Sound and the Weddell Sea, 1967–1968: Methods and data. Inst. Mar. Sci. Univ. Miami, Publ. 69-1: 1–12, 2 maps, 4 pls., 8 tables. Institute of Marine Sciences of the University of Miami, Miami, Florida.

Dawydoff, C.
1948 Classe des Ptérobranches. *In* P. Grassé (ed.), Traité de zoologie, *11:* 454–489, figs. 75–113. Masson et Cie, Paris.

Ekman, S.
1953 Zoogeography of the sea. xiv + 417 pp., 121 figs., 49 tables. Sidgwick and Jackson, London.

Foucart, M., F. S. Bricteux-Grégoire, and C. Jeuniaux
1965 Composition chimique du tube d'un pogonophore (*Siboglinum* sp.) et des formations squelettiques de deux pterobranches. Sarsia, *20:* 35–41, 2 tables.

Gilchrist, J. D. F.
1915 Observations on the Cape *Cephalodiscus (C. gilchristi,* Ridewood) and some of its early stages. Ann. Mag. Nat. Hist., (8)*16*(Art. 30): 233–243, 1 pl.
1917 On the development of the Cape *Cephalodiscus (C. gilchristi,* Ridewood). Q. Jl. Microsc. Sci., *62:* 189–211, pls. 13, 14.

Gravier, C.
1912a Sur une espèce nouvelle de *Cephalodiscus (C. anderssoni,* nov. sp.) provenant de la seconde expédition antarctique française. Bull. Mus. Hist. Nat., Paris, *18*(3): 146–150.
1912b Sur la répartition géographique des espèces actuellement connues du genre *Cephalodiscus* MacIntosh. Bull. Mus. Hist. Nat., Paris, *18*(3): 151–153.
1913 Ptérobranches. *In* Deuxième Expédition Antarctique Française (1908–1910). Pt. 4: 71–86, 17 figs. Masson et Cie, Paris.

Harmer, S. F.
1887 Appendix to Report on *Cephalodiscus dodecalophus. In* C. Wyville Thompson and J. Murray (eds.), Rep. scient. results voy. H.M.S. *Challenger* during the years 1873–76. Zool. *20*(52): 39–47, 3 figs.
1903 On new localities for *Cephalodiscus.* Zool. Anz., *26:* 593–594.
1905 The Pterobranchia of the Siboga-Expedition with an account on other species. *In* M. Weber (ed.), Résultats des Explorations à bord du *Siboga,* Monogr. *26bis:* 1–132, 2 figs., 14 pls. E. J. Brill, Leiden.
1915 Appendix to Observations on the Cape *Cephalodiscus* (*C. gilchristi,* Ridewood) and some of its early stages. Ann. Mag. Nat. Hist., (8)*16*(Art. 30): 243–246.

Harmer, S. F., and W. G. Ridewood
1913 The Pterobranchia of the Scottish National Antarctic Expedition (1902–1904). Trans. R. Soc. Edinb., *49,* part 3 (7): 531–565, 5 text-figs., 2 pls.

Hyman, L. H.
1959 The invertebrates: Smaller coelomate groups. Vol. 5. vii + 783 pp., 240 figs. McGraw-Hill, New York.

John, C. C.
1931 *Cephalodiscus.* 'Discovery' Rep., *3:* 223-260, text-figs. 1–7, pls. 33–38. Discovery Committee, London.
1932 On the development of *Cephalodiscus.* 'Discovery' Rep., *6:* 191–204, pls. 43–44.

Johnston, T. H., and N. G. Muirhead
1951 *Cephalodiscus.* Rep. BANZ Antarct. Res. Exped., (B)*1*(3): 91–120, 30 text-figs., 3 pls.

Komai, T.
1949 Internal structure of the pterobranch *Atubaria heterolopha* Sato, with an appendix on the homology of the 'notochord.' Proc. Japan Acad., *25*(7): 19–24, 4 figs.

Kozlowski, R.
1966 On the structure and relationships of graptolites. *J. Paleontol., 40:* 489–501, 12 text figs.

Krumbach, T.
1927 *Cephalodiscus. In* Von Drygalski, E. (ed.), Dt. Südpol.-Exped., 1901–1903, *19* (Zool., *11*): 465–467. Georg Reimer, Berlin.

Lankester, E. R.
1905 On a new species of *Cephalodiscus (C. nigrescens)* from the Antarctic Ocean. Proc. R. Soc. (London), ser. B, *76:* 400–402, pl. 8.

Marcus, E.
1958 On the evolution of the animal phyla. Q. Rev. Biol., *33:* 24–58.

Masterman, A. T.
1898 On the further anatomy and the budding process of *Cephalodiscus dodecalophus* (M'Intosh). Trans. R. Soc. Edinb., *39*(3): 507–527, 5 pls.

M'Intosh, W. C.
1882 Preliminary notice of *Cephalodiscus,* a new type allied to Prof. Allman's Rhabdopleura, dredged in H.M.S. 'Challenger.' Ann. Mag. Nat. Hist., (5) *10* (Art. 59): 337–348.
1887 Report on *Cephalodiscus dodecalophus,* M'Intosh, a new type of the Polyzoa, procured on the voyage of H.M.S. Challenger during the years 1873–76. *In* C. Wyville Thompson and J. Murray (eds.), Rep. scient. results voy. H.M.S. *Challenger* during the years 1873–76. Zool., *20*(52): 1–37, 2 figs., 7 pls. Neill, Edinburgh.

Ridewood, W. G.
1906 A new species of *Cephalodiscus (C. gilchristi)* from the Cape Seas. Mar. Invest. S. Afr., *4:* 173–192, 5 figs., 3 pls.
[It is evident from the text that the following paper was completed before the present one, and that the author considered it published, even though it actually did not appear until 1907. However, the British Museum catalogue cites it under 1907, as does Ridewood himself in later papers.]

1907 Pterobranchia. *Cebhalodiscus*. *In* E. R. Lankester (ed.), National Antarctic Exped. 1901–1904. Nat. Hist., *2* (Zool.): 67 pp., 17 text figs., 7 pls. Trustees of the British Museum, London.

1912 On specimens of *Cephalodiscus nigrescens* supposed to have been dredged in 1841 or 1842. Ann. Mag. Nat. Hist., (8) *10* (Art. 45): 550–555, 1 fig.

1918a *Cephalodiscus*. *In* Nat. Hist. Rep. Br. Antarct. 'Terra Nova' Exped., 1910, Zool., *4* (2): 11–82, 5 pls., 1 map. Trustees of the British Museum, London.

1918b Pterobranchia. *In* Scient. Rep. Australasian Antarctic Exped., 1911–14. Ser. C, *3* (2): 1–25, 3 text figs., pls. 6–7. Government Printer, Sydney.

1920 A key for the ready identification of the species of *Cephalodiscus*. Ann. Mag. Nat. Hist., (9) *5* (Art. 54): 407–410.

1921 On specimens of *Cephalodiscus densus* dredged by the 'Challenger' in 1874 at Kerguelen Island. Ann. Mag. Nat. Hist., (9) *8* (Art. 42): 433–440, pl. 12.

Rothschild, N. M. V.

1965 A classification of living animals. vi + 134 pp. Longmans, Green and Co., London.

Rudall, K. M.

1955 The distribution of collagen and chatin. *In* R. Brown and J. F. Danielli (eds.), Symp. Soc. Exp. Biol., *9*: 49–71.

Schepotieff, A.

1909 Die Pterobranchier des Indischen Ozeans. Zool. Jb., Abt. Systematik, *28* (4): 429–448, pls. 7–8.

Ushakov, P. V.

1962 Some characters of the distribution of bottom fauna off the coast of East Antarctica. (Transl. 1964.) Inf. Bull. Sov. Antarct. Exped., *4* (5) (Art. 40): 287–292.

Van der Horst, C. J.

1939 Hemichordata. *In* Bronns Klassen und Ordnungen des Tierreichs, *4* (Abt. 4) (B. 2) (T. 2): i–xiii + 1–737, 733 figs. Akademische Verlagsgesellschaft m. b. H., Leipzig.

Van der Horst, C. J., and J. G. Helmcke

1956 Cephalodiscidae. *In* W. Kükenthal and T. Krumbach (eds.), Handb. Zool., *3* (2. Hälfte) (8): 33–66, 50 figs. Walter de Gruyter & Co., Berlin.
[The volume in which this paper appears bears the date 1932, but it is evident from the text and references that this paper was written in 1939 or shortly thereafter. However, the publisher informs me that it was published June 29, 1956, the delay presumably having been due to wartime conditions.]

ANTARCTIC AND SUBANTARCTIC CAPRELLIDAE (CRUSTACEA: AMPHIPODA)

JOHN C. MCCAIN

Hazleton Laboratories, Inc., Falls Church, Virginia 22046

W. SCOTT GRAY, JR.

Division of Crustacea, Smithsonian Institution, Washington, D. C. 20560

Abstract. This paper reviews the Caprellidae of the Antarctic and Subantarctic. Six species are described as new and a distribution record added to expand the known caprellid fauna of this region to 21 species.

INTRODUCTION

Our knowledge of antarctic and subantarctic Caprellidae is due primarily to the works of Barnard [1930, 1931, 1932], Mayer [1903], Schellenberg [1926, 1931], and Stebbing [1883, 1888]. All together, less than 20 papers are concerned with antarctic-subantarctic caprellids. In them 14 valid species are enumerated. This paper reviews the known species, adds a new genus, describes six new species, and includes a new distribution record to expand the known regional fauna to 21 species.

SYSTEMATICS DISCUSSION

Method of Illustration

Illustrations of the whole mounts were made with a microprojector and those of dissected appendages with a camera lucida. The first illustrations were pencil sketches, which were later copied on Ethulon tracing film.

KEY TO SPECIES OF CAPRELLIDAE OF THE ANTARCTIC-SUBANTARCTIC REGION

1. Gills on pereonites 2–4 2
   Gills on pereonites 3–4 7
2 (1). Pereopod 5 6-segmented 3
   Pereopod 5 4-segmented 4
3 (2). Flagellum of antenna 2 of 2 or 3 articles, male abdomen with small anterior papillalike appendages .... ***Pseudoprotomima hedgpethi,*** new species, p. 133
   Flagellum of antenna 2 of 4 articles, male abdomen without small anterior papillalike appendages ....***Pseudoprotomima hurleyi,*** p. 135
4 (2). Pereopod 3 absent ..***Caprellina longicollis,*** p. 116
   Pereopod 3 6-segmented 5
5 (4). Gill on pereonite 2 subequal to gill on pereonite 3 6
   Gill on pereonite 2 less than ½ size of gill on pereonite 3 ....***Dodecas reducta,*** p. 121
6 (5). Gill on pereonite 4 subequal to gill on pereonite 3 ....***Dodecas elongata,*** p. 119
   Gill on pereonite 4 less than ½ size of gill on pereonite 3 ....***Dodecas eltaninae,*** new species, p. 119
7 (1). Mandibular palp absent 8
   Mandibular palp present 10
8 (7). Ventral spine present between insertions of basis of gnathopods 2 ......***Caprella equilibra,*** p. 113
   Ventral spine not present between insertions of basis of gnathopods 2 9
9 (8). Cephalon with anteriorly directed projection, posterior margin of propodus of pereopods 5–7 concave ....***Caprella penantis,*** p. 114
   Cephalon without anteriorly directed projection, posterior margin of propodus of pereopods 5–7 convex ....***Caprella ungulina,*** p. 115
10 (7). Pereopod 5 2- to 4-segmented 11
   Pereopod 5 6-segmented 13
11 (10). Pereopod 3 absent ....***Caprellinoides mayeri,*** p. 116
   Pereopod 3 1-segmented ....***Aeginoides gaussi,*** p. 112
   Pereopod 3 2-segmented ....***Mayerella magellanica,*** new species, p. 124
   Pereopod 3 5-segmented ....***Pseudododecas bowmani,*** new genus, new species, p. 131
   Pereopod 3 6-segmented 12
12 (11). Gill on pereonite 4 shorter than gill on pereonite 3 ....***Dodecasella georgiana,*** p. 121
   Gill on pereonite 4 longer than gill on pereonite 3 ....***Dodecasella elegans,*** p. 121
13 (10). Pereopod 3 absent ....***Pseudaeginella tristanensis,*** p. 131
   Pereopod 3 1-segmented 14

Pereopod 3 2-segmented 15
Pereopod 3 6-segmented ....
..................*Paraproto condylata*, p. 127

14 (13). Flagellum of antenna 2 biarticulate ....
..........*Protella trilobata*, new species, p. 128
Flagellum of antenna 2 of 3 articles ....
....................*Triantella solitaria*, p. 135

15 (13). Male abdomen with pair of true appendages, female without true appendages ....
..........*Luconacia vemae*, new species, p. 123
Male abdomen with 2 or more pairs of true appendages, female with pair of true appendages
................*Protellopsis kergueleni*, p. 131

Genus *Aeginoides* Schellenberg, 1926

Type species: *Aeginoides gaussi* Schellenberg, 1926 (by monotypy)

Diagnosis. Flagellum of antenna 2 of 4–8 articles, swimming setae absent; mandibular palp 3-segmented, setal formula for terminal article 1-x-1, molar absent; outer lobe of maxilliped subequal to inner lobe; gills on pereonites 3 and 4; pereopod 3 1-segmented, pereopod 4 absent or small knob, pereopod 5 4-segmented; abdomen of male and female with 2 pairs of biarticulate appendages and raised anterior projection.

Aeginoides gaussi Schellenberg, 1926

Figs. 1, 2

Aeginoides gaussi Schellenberg, 1926, pp. 465–467, fig. 1.—Barnard, 1930, pp. 442–443, fig. 63; 1932, pp. 305–306, fig. 169c, d.—Stephensen, 1947, p. 79, fig. 26.

Distribution. Type locality: 65°59′S, 89°33′E. Other records: Scotia Ridge, Peter I Island, Cape Adare, Oates Coast, Adélie Coast, and Davis Sea. Depth range: 37–1501 meters.

Diagnosis. Since this genus is monotypic, the characters of the genus are diagnostic for the species.

Material examined.

Vema, cruise 17: Sta. 45, 62°33′S, 59°26′W, April 24, 1961, 600–604 meters, 5 ♂♂, 1 ♀. Sta. 65, 50°18′S, 54°11′W, May 14, 1961, 1498–1501 meters, 1 ovigerous ♀.

Eltanin, cruise 6: Sta. 410, 61°18′S, 56°09′W, December 31, 1962, 220–240 meters, 1 ♂.

Eltanin, cruise 9: Sta. 671, 54°41′S, 38°38′W, August 23, 1963, 220–320 meters, 1 ♂.

Eltanin, cruise 12: Sta. 1003, 62°41′S, 54°43′W, March 15, 1964, 210–220 meters, 1 ♂, 2 juvenile ♀♀.

Eastwind: Sta. 66-010, 62°43.5′S, 61°51.8′W, February 1, 1966, 183 meters, 4 ♀♀. Sta. 66-012, 62°23′S, 60°51′W, February 3, 1966, 393–417 meters, 1 ♀. Sta. 66-035, 62°12′S, 54°25′W, February 16, 1966, 402–407 meters, 1 ♂, 2 ♀♀.

Glacier: NZOI Sta. E187, 72°18′S, 170°13′E, January 20, 1965, 37–42 meters, 1 ♂.

Shore collections (Arnaud): Many stations in Géologie Archipelago, 66°39′S, 139°55′E, and Cape Géodésie, 66°40′S, 139°51′E, Adélie Coast, 1961–1965, 50–170 meters, 50+ specimens.

Description. Body spinose to smooth, in most spinose, cephalon and pereonite 1 smooth, dorsal surface of pereonite 2 with pair of spines slightly anterior to midlength and pair of spines at posterior, insertion of gnathopod 2 with spine, anterolateral margin produced into triangular projection; dorsal surface of pereonites 3 and 4 with pair of spines slightly posterior to midlength, dorsal surface of pereonite 5 with pair of small humps at midlength, pereonites 6 and 7 smooth. Length of largest male 34 mm, largest female 27 mm, smallest ovigerous female 19.7 mm. Antenna 1 approximately ½ body length. Antenna 2 not extending beyond articles 1 and 2 of antenna 1, flagellum of 4–8 articles. Mandibular palp 3-segmented, setal formula for terminal article from 1-8-1 to 1-20-1. Left mandible with toothed incisor, toothed lacinia mobilis, 2 accessory plates, and up to 8 spines in setal row. Right mandible similar to left except lacinia mobilis serrate. Palp of maxilla 1 with 3–7 terminal spines and 3–6 subterminal setae, outer lobe with 6 serrate spines. Outer lobe of maxilla 1 with 7–10 setae, inner lobe with 5–8 setae. Outer lobe and palp of maxilliped not distinctive, inner lobe with 3–4 large apical spines and 2–4 setae. Propodus of gnathopod 1 with 3–4 proximal grasping spines, palm with numerous paired spines; grasping margin of propodus and dactylus not distinctly serrate. Propodus of gnathopod 2 with proximal tooth, palmar margin variously spinate and inflated in large males. Gills elongate, elliptical. Pereopod 3 of single small article; pereopod 4 absent; pereopod 5 of 4 articles, terminal article dactyliform; pereopods 6 and 7 6-segmented, propodus with up to 2 pairs of grasping spines. Abdomen of male and female with 2 pairs of biarticulate appendages and anterior raised projection, penes small and lateral.

Remarks. As we have indicated in the description, the body spination is quite variable, and there seemed to be little pattern in the reduction of one set of spines over another. Figure 1d illustrates the most spinose

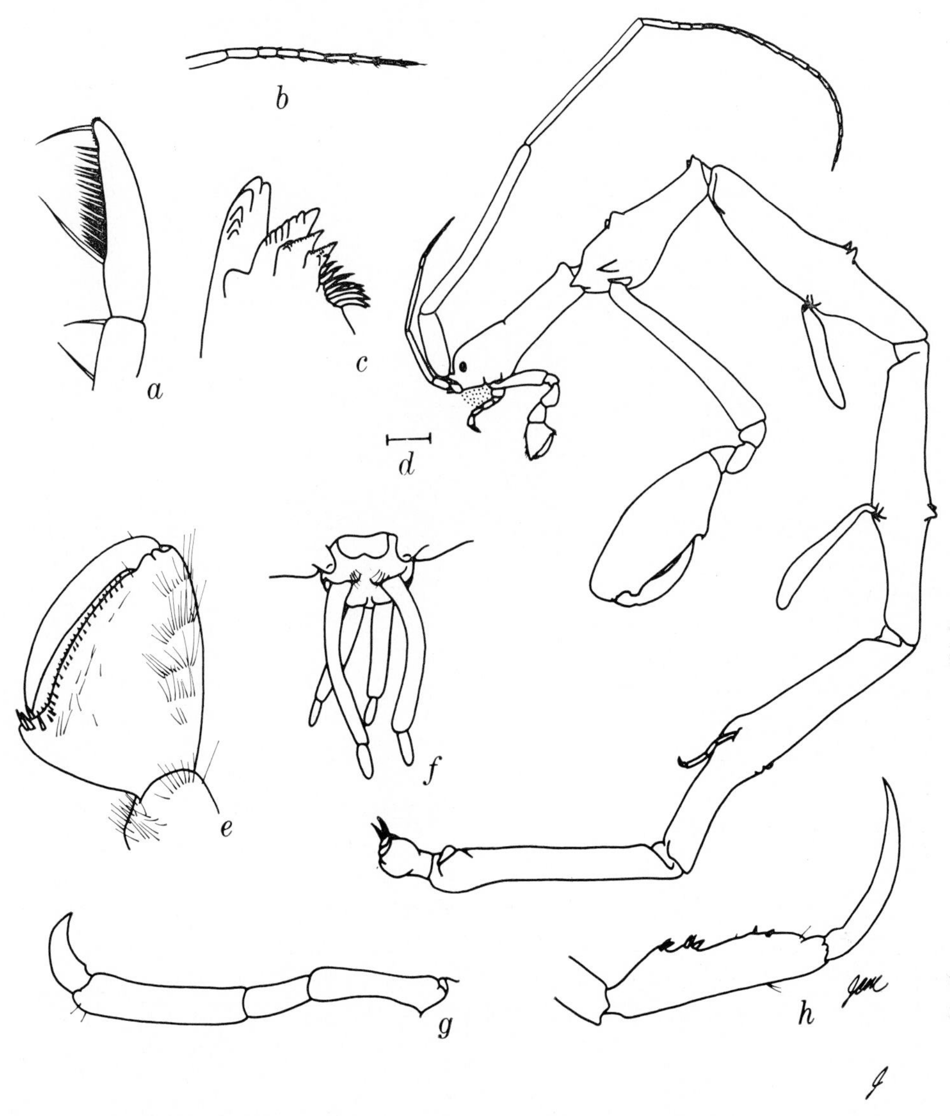

Fig. 1. *Aeginoides gaussi*, large male: *a*, terminal article of mandibular palp; *b*, flagellum of antenna 2; *c*, right mandible; *d*, lateral view; *e*, gnathopod 1; *f*, abdomen; *g*, pereopod 5; *h*, pereopod 7. Scale shown for *d* is 1 mm.

specimen in our collection. In many of the large males the palm of the propodus of gnathopod 2 was inflated and in several specimens thin-walled, much as in large males of *Phtisica marina.*

Genus *Caprella* Lamarck, 1801

Type species: *Cancer linearis* Linnaeus, 1767 (subsequent designation by Dougherty and Steinberg, 1953)

Diagnosis. Flagellum of antenna 2 biarticulate, swimming setae usually present; mandibular palp absent, molar present; outer lobe of maxilliped larger or equal to inner lobe; gills on pereonites 3 and 4; pereopods 3 and 4 absent, pereopod 5 6-segmented; abdomen of male with pair of appendages and pair of lobes, female with pair of lobes.

Remarks. We have used a key instead of figures to separate species of *Caprella,* since they are all quite similar in over-all appearance.

Caprella equilibra Say, 1818

Fig. 3

Distribution. Type locality: South Carolina. Other

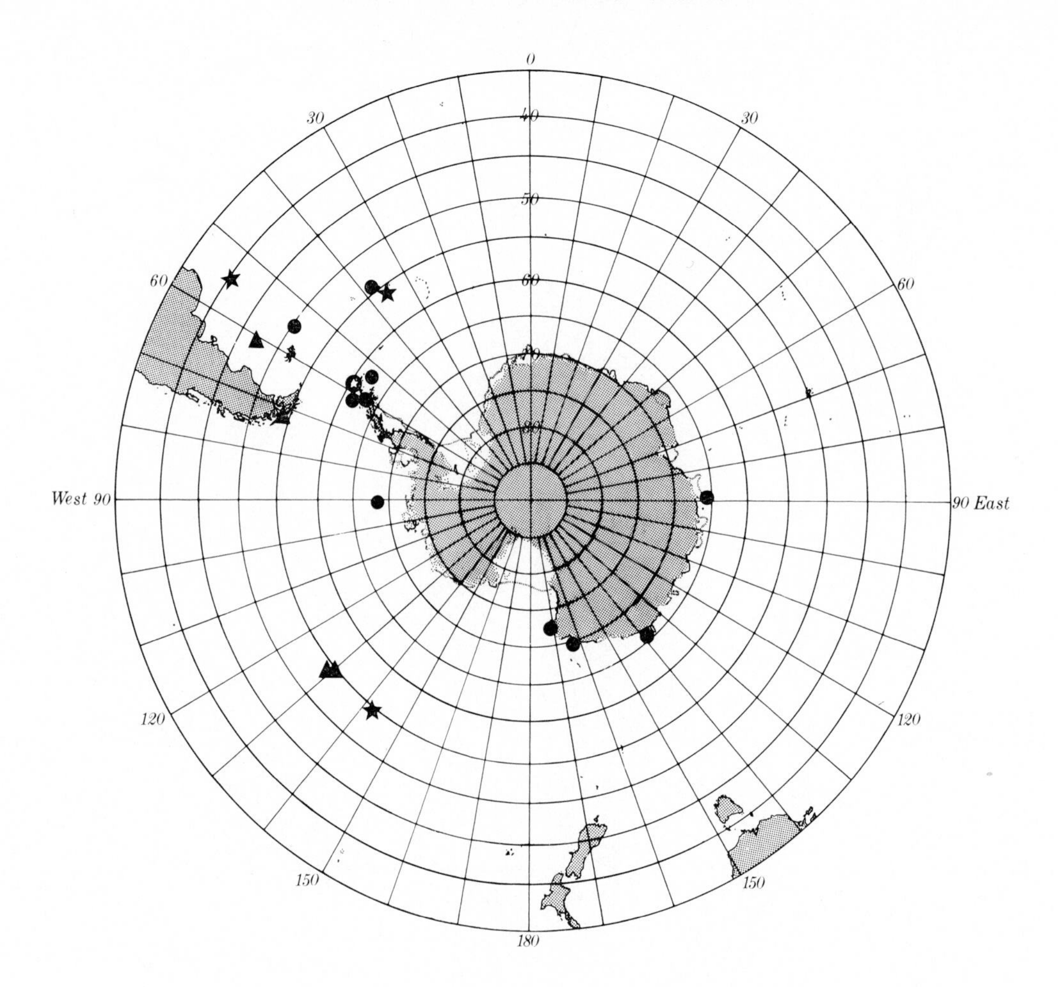

Fig. 2. Distribution records of *Aeginoides gaussi* (circle), *Dodecas eltaninae* (triangle), *Dodecasella georgiana* (star), *Pseudododecus bowmani* (star in circle).

records: Almost cosmopolitan. Depth range: Surface to ±3000 meters.

Material examined.

Vema, cruise 16: Sta. 40, 42°48′S, 63°11′W, May 27, 1960, 70 meters, 1 ♂.

Vema, cruise 17: Sta. 12, 43°30′S, 74°55′W, March 23, 1961, 112 meters, 18 ♂♂, 7 ♀♀. Sta. 71, 40°11′S, 60°27′W, May 18, 1961, 44 meters, 1 juvenile ♀.

Vema, cruise 18: Sta. 42, 34°15′S, 52°22′W, April 25, 1962, 40 meters, 1 ovigerous ♀.

Remarks. This species was recently described and illustrated fully by McCain [1968] in his monograph of the Caprellidae of the western North Atlantic. It is an almost cosmopolitan species and its occurrence in the South Atlantic and South Pacific has previously been recorded.

Caprella penantis Leach, 1814

Fig. 3

Distribution. Type locality: Devonshire coast, England. Other records: Almost cosmopolitan

Material examined.

Vema, cruise 18: Sta. 27, 52°41′S, 75°21′W, February 24, 1962, 470–562 meters, 2 ♂♂.

Eltanin, cruise 11: Sta. 958, 52°56′S, 75°00′W, February 5, 1964, 92–101 meters, 100+ specimens. Sta. 960, 52°40′S, 74°58′W, February 6, 1964, 64 meters, 1 ♂, 3 ♀♀, 4 juveniles. Sta. 966, 53°40′S, 66°20′W, February 10, 1964, 81 meters, 2 ♂♂, 1 ovigerous ♀. Sta. 967, 53°42′S, 66°19′W, February 10, 1964, 81 meters, 25+ specimens. Sta. 969, 54°56′S, 65°03′W, February 10–11, 1964, 229–265 meters, 4 juveniles.

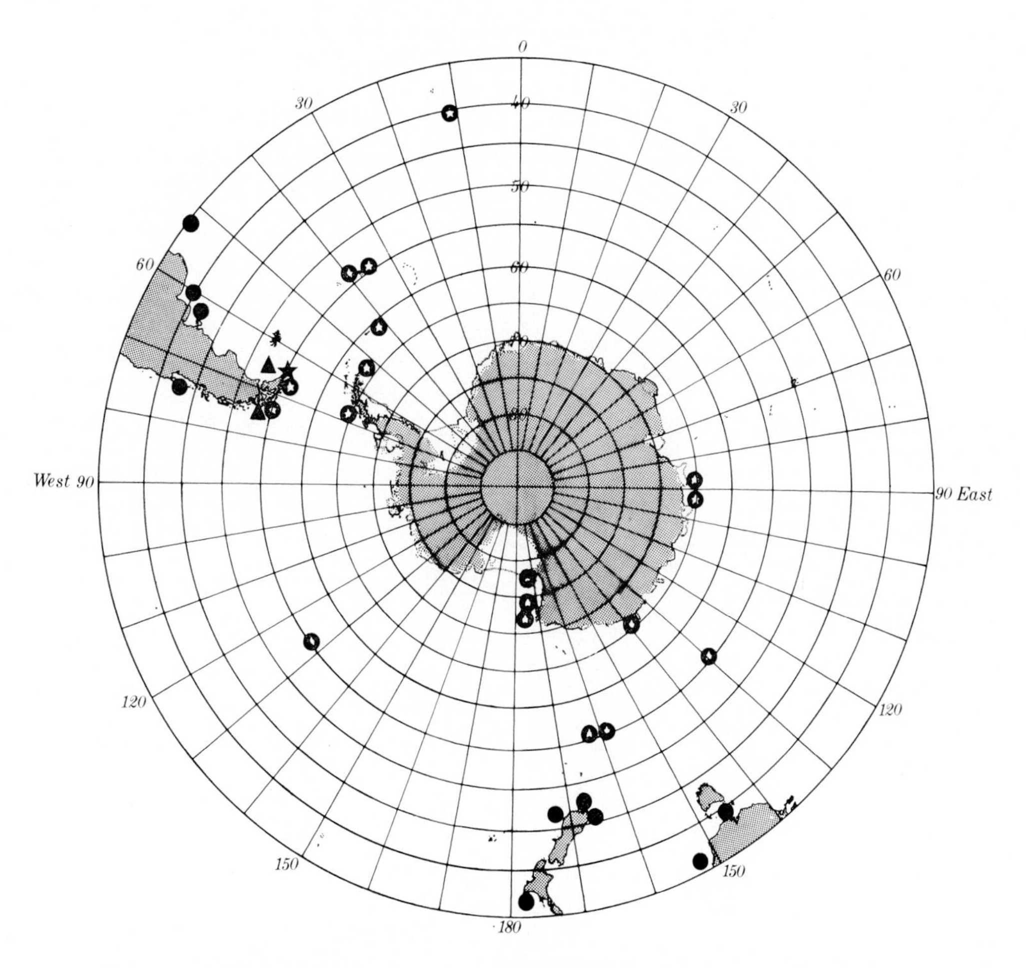

Fig. 3. Distribution records of *Caprella equilibra* (circle), *Caprella penantis* (triangle), *Caprella ungulina* (star), *Caprellinoides mayeri* (star in circle).

Remarks. McCain [1968] discusses this species and suggests that it should perhaps be divided into several species on the basis of the setation of gnathopod 2 and the pereopods. The *Vema* and *Eltanin* specimens belong to that portion of *C. penantis* that bear only a moderate number of setae on these appendages. Consult *McCain* [1968] for the synonymy of this species.

Caprella ungulina Mayer, 1903

Fig. 3

Caprella ungulina Mayer, 1903, p. 127, pl. 5, fig. 36, pl. 8, figs. 30, 31.—Schellenberg, 1931, pp. 266, 272. —McCain, 1966, p. 92.

Distribution. Type localities: Bahía York, Isla de los Estados, off Tierra del Fuego; 51°23′N, 130°34′W; and Galapagos Islands. Other records: Puerto Pantalon, Falkland Islands, and off Tierra del Fuego. Depth range: 7.3–1602 meters.

Material examined.

Vema, cruise 17: Sta. 47, 55°07.2′S, 66°29.3′W, May 4, 1961, 71 meters, 1 ♂.

Remarks. Our specimen was a 10-mm male that agreed closely with Mayer's description and figures. This species is recognized by the convex posterior margin and the unguliform appearance of the propodus of pereopods 5–7.

We have been unable to examine the type series and have, therefore, refrained from designating a lectotype.

Caprella sp.

Material examined.

Vema, cruise 17: Sta. 13, 46°59.5′S, 75°54′W, March 24, 1961, 2657 meters, 1 ♀.

Eltanin, cruise 11: Sta. 959, 52°55′S, 75°00′W, February 6, 1964, 92–101 meters, 3 juveniles.

Remarks. We were unable to identify these specimens with any degree of certainty except to the generic level. The *Vema* female had a dorsally directed cephalic spine but lacked any other diagnostic feature. The *Eltanin* specimens were extremely immature, and we were able to make generic determination only because of the presence of swimming setae on antenna 2.

Genus ***Caprellina*** Thomson, 1879a

Type species: *Caprellina novae-zealandiae* Thomson, 1879 (by monotypy)

Diagnosis. Flagellum of antenna 2 of 2–5 articles, swimming setae absent; mandibular palp 3-segmented, setal formula for terminal article 1-*x*-1, molar absent; outer lobe of maxilliped slightly longer than inner lobe; gills on pereonites 2–4; pereopods 3 and 4 absent, pereopod 5 3- or 4-segmented, inserted slightly posterior to midlength of pereonite 5; abdomen of male and female with pair of biarticulate appendages and pair of uniarticulate appendages.

Caprellina longicollis (Nicolet, 1849)

Caprella longicollis Nicolet, 1849, pp. 251–252, pl. 4, fig. 3.—Bate, 1862, p. 362, pl. 57, fig. 4.—Carus, 1885, p. 389.

Caprella brevicollis Nicolet, 1849, pp. 252–253, pl. 4, fig. 4.—Reed, 1897, pl. 11 (4).

Caprellina novae-zealandiae Thomson, 1879a, p. 247, pl. 10, fig. D-6; 1879b, p. 330.

Caprellina longicollis.—Mayer, 1882, pp. 27–28, figs. 4–5; 1890, pp. 15–16, pl. 6, fig. 4; 1903, p. 30.—Thomson and Chilton, 1886, p. 141.—Stebbing, 1910a, pp. 470–471.—Barnard, 1930, p. 440.—McCain, 1969, pp. 289–290, fig. 2.

Caprella nicoleti Reed, 1897, p. 11 (4).

Caprellinopsis longicollet.—Hutton, 1904, p. 261.

Caprellinopsis longicollis.—Chilton, 1909, pp. 605, 648.—Chevreux, 1913, p. 85.—Thomson, 1913, p. 245.—Thomson and Anderton, 1921, p. 113.

Distribution. Type locality: Chile. Other records: South Africa and New Zealand. Depth range: Surface to 123 meters.

Remarks. This species was recently described and illustrated by McCain [1969]. It is by far the most common New Zealand species. It is quite possible that it occurs in the subantarctic islands south of New Zealand and Chile, but its position as a true subantarctic species is questionable.

Genus ***Caprellinoides*** Stebbing, 1888

Type species: *Caprellinoides tristanensis* Stebbing, 1888 (by monotypy)

Diagnosis. Flagellum of antenna 2 of 2–5 articles, swimming setae absent; mandibular palp 3-segmented, setal formula for terminal article 1-*x*-1, molar absent; outer lobe of maxilliped larger than inner lobe; gills on pereonites 3 and 4; pereopods 3 and 4 absent; pereopod 5 usually 3-segmented, occasionally reduced; abdomen of male and female with pair of setose lobes.

Caprellinoides mayeri (Pfeffer, 1888)

Figs. 3–5

Caprellina mayeri Pfeffer, 1888, pp. 137–139, pl. 3, fig. 4.

Caprellinoides tristanensis Stebbing, 1888, pp. 1238–1240, pl. 141.—Barnard, 1932, p. 301.—Stephensen, 1949, p. 56.

Caprellinoides Mayeri.—Mayer, 1890, p. 88, pl. 5, figs. 57–58, pl. 6, figs. 15, 26, pl. 7, fig. 48.—Chevreux, 1913, p. 86.

Piperella grata Mayer, 1903, p. 59, pl. 2, fig. 29, pl. 7, figs. 40–45, pl. 9, figs. 24–25, 62.

Caprellinoides mayeri.—Chilton, 1913, pp. 54, 61–62.—Schellenberg, 1931, pp. 265, 272.—Barnard, 1932, pp. 302–303, fig. 167.

Caprellinoides antarctica Schellenberg, 1926, pp. 467–470, fig. 2.

Caprellinoides spinosa Barnard, 1930, p. 441, fig. 62.

Distribution. Type locality: South Georgia. Other records: Antarctica, Scotia Ridge, Tristan da Cunha, Tierra del Fuego, mid-South Pacific, Macquarie Island. Depth range: 22–1153 meters.

Diagnosis. Since this genus is monotypic, the characters of the genus are diagnostic for the species.

Material examined.

Vema, cruise 17: Sta. 23, 53°47′S, 70°17.5′W, March 29, 1961, 269–280 meters, 1 ♂, 1 ♀. Sta. 51, 55°17.5′S, 66°00′W, May 4, 1961, 205–207 meters, 2 ♂♂, 2♀♀.

Eltanin, cruise 12: Sta. 1003, 62°41′S, 54°43′W, March 15, 1964, 210–220 meters, 25+ specimens. Sta. 1081, 60°35′S, 40°44′W, April 13, 1964, 631–641 meters, 2 ♀♀.

Eltanin, cruise 15: Sta. 1343, 54°50′S, 129°50′W, November 7, 1964, 567–604 meters, 5 ♂♂, 1 ♀, 4

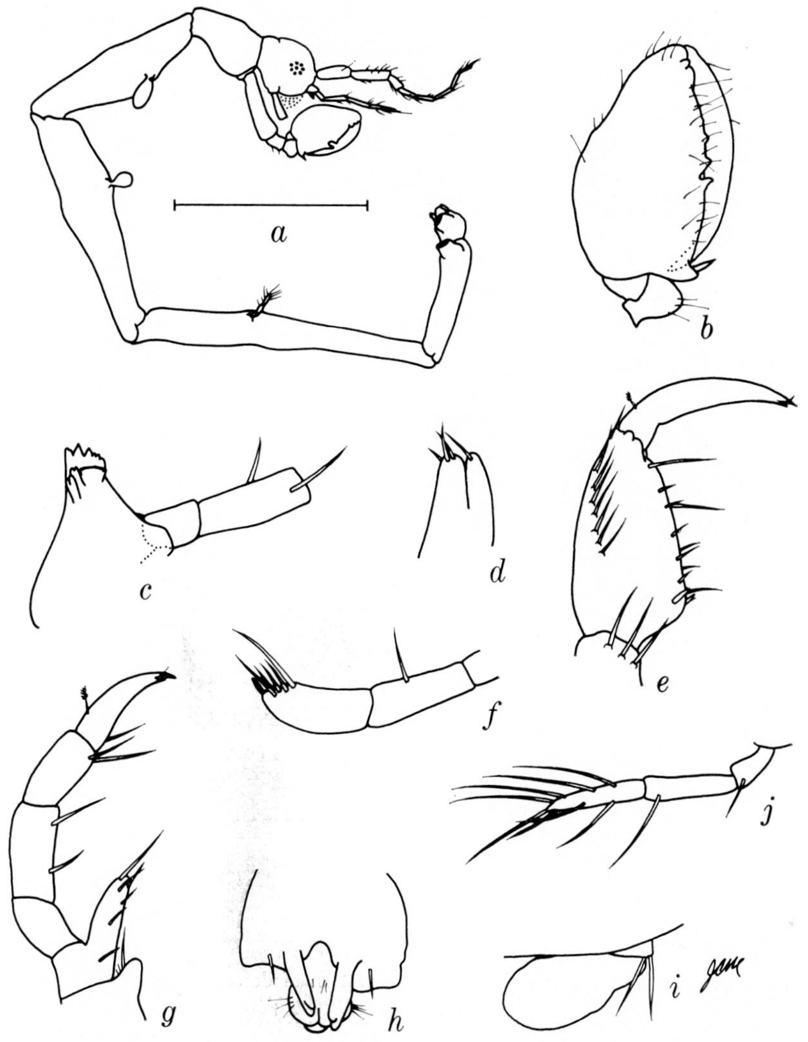

Fig. 4. *Caprellinoides mayeri: a, b, d, f–j*, male; *c* and *e*, female; *a*, lateral view; *b*, gnathopod 2; *c*, right mandible; *d*, maxilla 2; *e*, gnathopod 1; *f*, mandibular palp; *g*, maxilliped; *h*, abdomen; *i*, gill on pereonite 3; *j*, pereopod 5. Scale shown for *a* is 1 mm.

juveniles. Sta. 1345, 54°50′S, 129°48′W, November 7, 1964, 915–1153 meters, 3 ♂♂. Sta. 1346, 54°49′S, 129°48′W, November 7, 1964, 549 meters, 25+ specimens.

Eltanin, cruise 16: Sta. 1417, 54°24′S, 159°01′E, February 10, 1965, 79–93 meters, 1 ♂. Sta. 1418, 54°32′S, 159°02′E, February 10, 1965, 86–101 meters, 1 ♂, 1 ♀.

Atka: Sta. 22A, 72°17.2′S, 170°19.3′E, January 11, 1958, 36 meters, 25+ specimens. Sta. 23, 72°05.8′S, 172°15.2′E, January 12, 1958, 392 meters, 1 ♂.

Burton Island: Sta. 3, 72°08′S, 172°10′E, January 13, 1958, 499 meters, 2 ♂♂, 1 ♀. Sta. 5, 66°32.9′S, 93°00.9′E, January 29, 1958, 80 meters, 10 ♂♂, 4 ♀♀.

Glacier: Sta. 6, 73°40′S, 175°17′E, November 9, 1956, 521 meters, 1 ♂.

Staten Island: Sta. 2, 71°21.5′S, 170°05′E, January 23, 1959, 2 ♂♂.

Westwind: Sta. 9, no. 108, 62°24′S, 59°45′W, January 26, 1958, 167 meters, 1 ♂.

Rotoiti: NZOI Sta. C732a, 54°29.5′S, 158°58.5′E, November 25, 1961, 22 meters, 1 ♂, 3 ♀♀.

Shore collections (Arnaud): Several stations in Géologie Archipelago, 66°39′S, 139°55′E, 1962–1965, 31–85 meters, 20 ♂♂, 8 ♀♀.

Description. Body spination variable (see remarks). Length of largest male 18 mm, female 11 mm, smallest ovigerous female 5.2 mm. Antenna 2 approximately

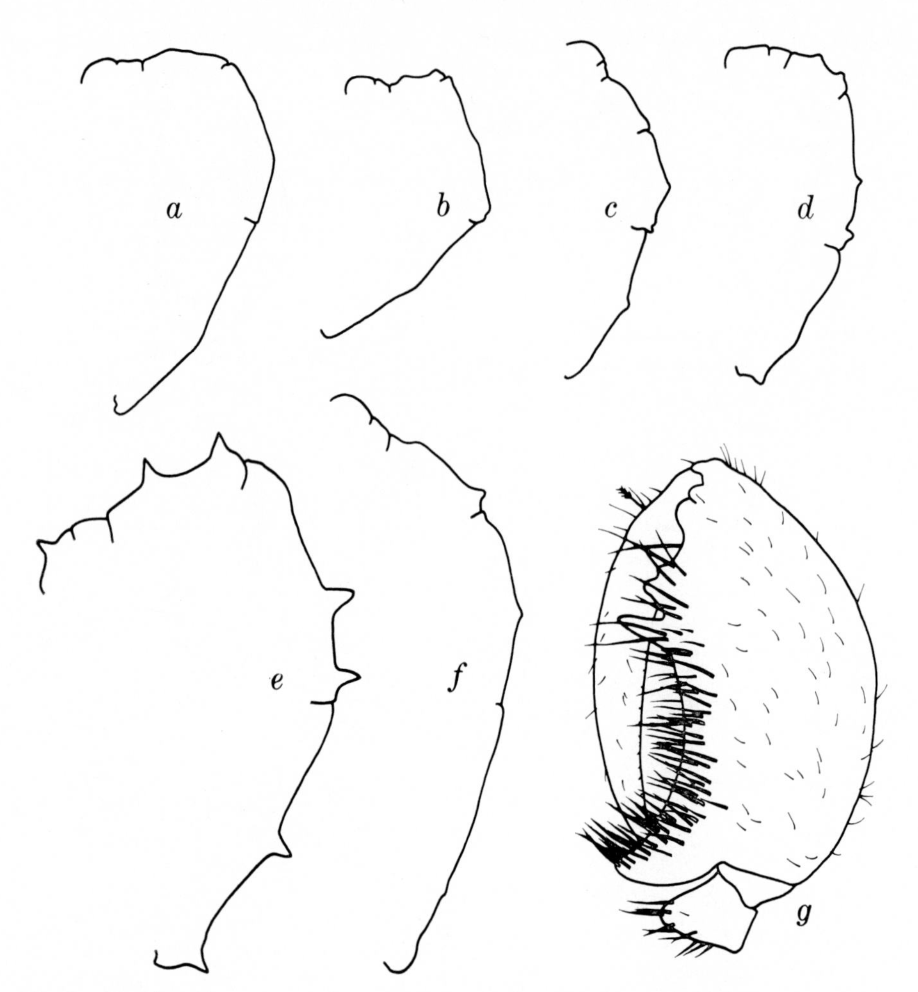

Fig. 5. *Caprellinoides mayeri: a–f,* variation in pattern of dorsal ornamentation; *g,* variant gnathopod 2.

equal in length to length of peduncle of antenna 1, flagellum of 2–5 articles. Mandible with 3-segmented palp, setal formula for terminal article varies from 1-3-1 to 1-7-1 with short terminal spinelike process; incisor toothed; right lacinia mobilis serrate, left toothed; setal row of several stout spines; molar absent. Outer lobe of maxilla 1 with 5 or 6 serrate spines. Outer lobe of maxilla 2 small with 1–3 setae, inner lobe with 2–4 setae. Inner lobe of maxilliped small with few setae, outer lobe much larger with several setae, terminal article of palp slightly serrate on grasping margin. Propodus of gnathopod 1 with 2 proximal grasping spines, grasping margin of propodus and dactylus slightly serrate, dactylus bifid at tip. Propodus of male gnathopod 2 with single proximal grasping spine, palm with notch at variable position from midlength to distal; female gnathopod 2 similar to male except palmar notch less distinct. Gills small, rounded. Pereopods 3 and 4 absent, vestige represented by several setae on raised knob; pereopod 5 usually 3-segmented, occasionally reduced; propodus of pereopods 6 and 7 with single proximal grasping spines, palmar margin distinctly concave with single spines set on knobs and evenly spaced along margin. Abdomen of male with pair of setose lobes, penes large and lateral; female with pair of setose lobes.

Remarks. We have been unable to find differences that adequately separate the four species of *Caprellinoides.* We are, therefore, placing all four species under Pfeffer's name, *C. mayeri,* the oldest available name. *C. mayeri* and *C. spinosa* were described as spiny forms, while *C. antarctica* and *C. tristanensis* were described as being smooth. The material available to us consisted of a much larger collection than was available to previous authors, and we observed all degrees of variation in spination, from entirely smooth to spiny, as illustrated in Figure 5. Even the smoothest specimens showed slight elevations on the dorsal surface that corresponded with the pattern of spination found in the spiny forms. Barnard [1930] remarks that his species, *C. spinosa,* differs from *C. antarctica* only in the presence of spines, which he found even on the embryos. McCain [1968] discusses the value of body spines as a specific character and cautions against their use.

Schellenberg [1926] based his separation of *C. antarctica* from *tristanensis* primarily on the basis of the relative lengths of the penultimate and ultimate articles of pereopod 5. In the material available to us this pereopod was reduced to a single article in one specimen, and in another specimen the ratio of the ultimate article to the penultimate article of the right and left sides varied within the limits set by Schellenberg to separate the two species. We feel, therefore, that the use of this character to separate species is of little value.

Figure 5g illustrates an unusual variant of gnathopod 2 that was relatively rare but present on both smooth and spiny forms. This variant gnathopod was found only on large male specimens.

Genus **Dodecas** Stebbing, 1883

Type species: *Dodecas elongata* Stebbing, 1883 (by monotypy)

Diagnosis. Flagellum of antenna 2 of 2–5 articles, swimming setae absent; mandibular palp 3-segmented, setal formula for terminal article 1-*x*-1 or 1-*x*-2, molar absent; outer lobe of maxilliped subequal to inner lobe; gills on pereonites 2–4; pereopod 3 6-segmented, pereopod 4 absent, pereopod 5 4-segmented; abdomen of male with 2 pairs of biarticulate appendages and anterior pair of papillalike appendages, female with 2 pairs of biarticulate appendages and raised anterior projection.

Dodecas elongata Stebbing, 1883

Dodecas elongata Stebbing, 1883, p. 207; 1888, pp. 1233–1237, pls. 139–140.—Mayer, 1890, p. 15, pl. 5, figs. 7–9, pl. 6, fig. 2.

Distribution. Type locality: Baie Rhodes and off Baie de Londres, Kerguelen Island, 173.6–201.2 meters.

Remarks. This species has not been collected since the *Challenger* Expedition, and the type material was not available to us. Large males of this species have an extremely elongated carpus on gnathopod 2, nearly twice the length of the propodus. This character, in combination with the three pairs of gills being subequal, easily separates this species from the following new species, *D. eltaninae.*

Dodecas eltaninae new species

Figs. 2, 6

Distribution. Type locality: 54°49′S, 129°48′W. Other records: Southern South America and mid-South Pacific. Depth range: 248–604 meters.

Diagnosis. Gill on pereonite 4 much smaller than that on pereonite 3, carpus of gnathopod 2 not longer than propodus in large males.

Material examined.

Vema, cruise 17: Sta. 18, 53°55′S, 71°16.8′W, March 28, 1961, 248–262 meters, 2 ♂ paratypes AMNH 13126.

Vema, cruise 18: Sta. 12, 47°09′S, 60°38′W, no date, 424–428 meters, 1 ♂ paratype AMNH 13127.

Eltanin, cruise 15: Sta. 1343, 54°50′S, 129°50′W, November 7, 1964, 567–604 meters, 1 juvenile ♀ paratype USNM 123746. Sta. 1346, 54°49′S, 129°48′W, November 7, 1964, 549 meters, 1 ♂ holotype USNM 123744, 7 ♂♂, 6 ♀♀ paratypes USNM 123745.

Description. Male holotype: Body smooth. Length 9.4 mm. Antenna 2 approximately equal in length to articles 1 and 2 of antenna 1, flagellum 2-segmented. Setal formula for terminal article of mandibular palp 1-4-1. Mandibles with toothed incisor, left lacinia mobilis toothed, right serrate, 2 accessory plates, setal row with numerous spines. Palp of maxilla 1 with 6 terminal spines and 4 subterminal setae, outer lobe with 6 spines. Outer lobe of maxilla 2 with 8 setae, inner lobe with 9 setae. Outer lobe of maxilliped small with several terminal setae, inner lobe with 3 large serrate spines and single smooth spine, penultimate article of palp with large distal projection. Propodus of gnathopod 1 with 3 proximal grasping spines, grasping margins of propodus and dactylus not serrate. Carpus of gnathopod 2 slightly elongate, propodus with proximal grasping spine, grasping margin with paired spines. Gill on pereonite 4 rounded and much smaller than gill on pereonite 3, gill on pereonite 3 elliptical. Pereopod 3 6-segmented, terminal article dactyliform; pereopod 4 absent; pereopod 5 4-segmented, terminal article dactyliform. Pereopods 6 and 7 6-segmented, propodus with paired proximal grasping spines, grasping margin with several spines. Abdomen with 2 pairs

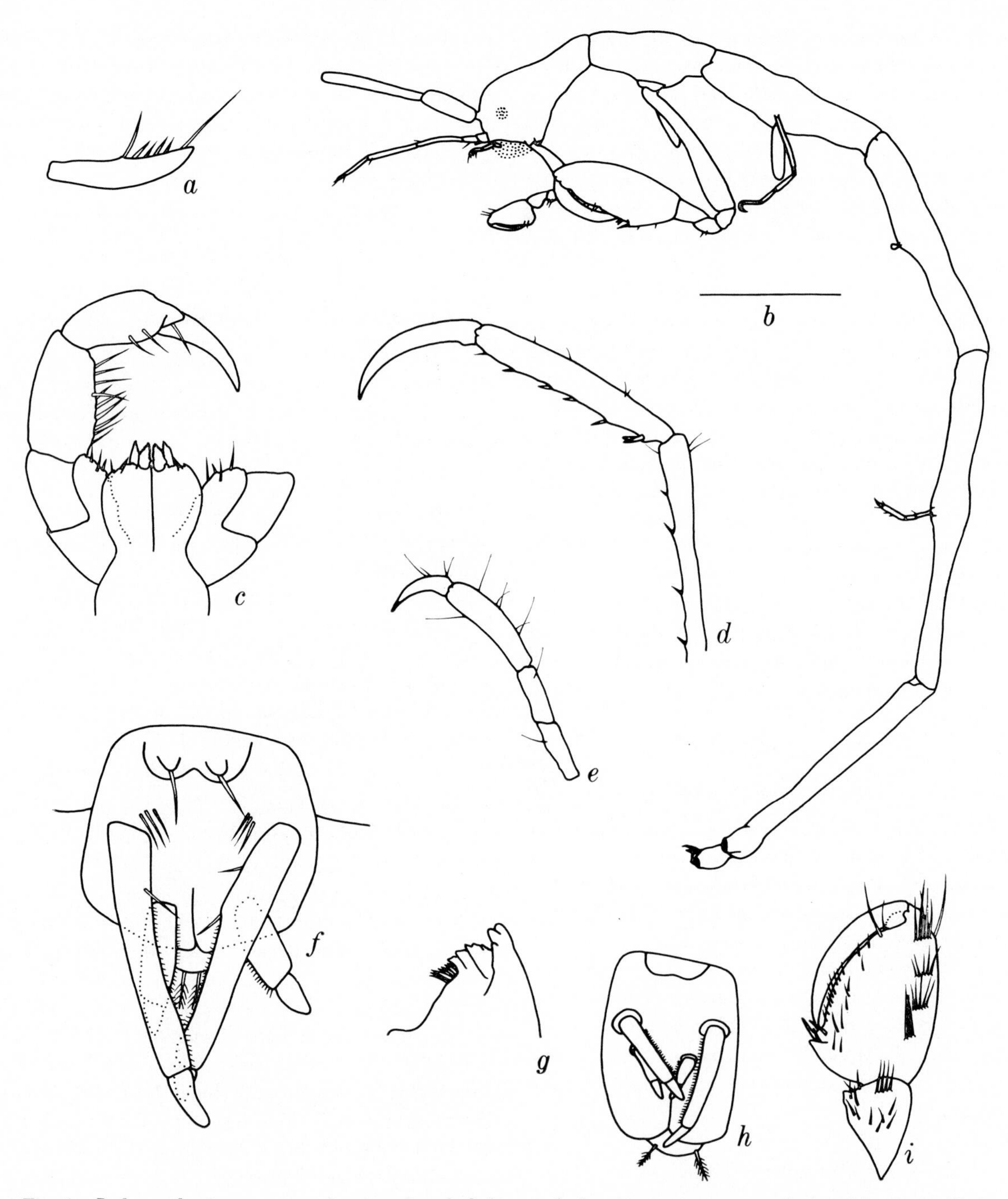

Fig. 6. *Dodecas eltaninae*, new species: *a–g*, *i*, male holotype; *h*, female paratype; *a*, terminal article of mandibular palp; *b*, lateral view; *c*, maxilliped; *d*, pereopod 7; *e*, pereopod 5; *f*, abdomen; *g*, left mandible; *h*, abdomen; *i*, gnathopod 1. Scale shown for *b* is 1 mm.

of biarticulate appendages and anterior papillalike appendages with long seta at tip.

Remarks. The holotype was the largest male in our collection. The largest female measured 10.9 mm and differed from the holotype in that the setal formula for the terminal article of the mandibular palp was 1-9-1-1 and its tip was more pointed than in the holotype. The flagellum of antenna 2 of the largest female had 3 articles on one side and 2 on the other. Antenna 2 of the largest female was approximately equal in length to articles 1 and 2 of antenna 1, as in the holotype; however, in other specimens antenna 2 was as long as the peduncle of antenna 1. In females and males smaller than the holotype the carpus was not elongate, as it was in the holotype.

Two of the three large serrate spines on the inner lobe of the maxilliped hide the more anterior spine, and at first it appears to be only 2 serrate spines, as illustrated. This type of maxilliped with unusually large inner lobes and huge serrate spines is typical of those genera that lack a molar on the mandible and demonstrates the close affinity between the *Dodecas* group of caprellids and the *Proto* group.

The genus *Dodecas* is presently composed of 7 species, including *Dodecas eltaninae*. *D. eltaninae* differs from all these species in that the gill on pereonite 4 is much smaller than that on pereonite 3. In *D. reducta* Barnard, 1932, the gill on pereonite 4 is slightly smaller than the gill on pereonite 3; however, the gill on pereonite 2 is very small. The gill on pereonite 2 of *D. eltaninae* is almost the size of that on pereonite 3.

The specific name is in honor of the USNS *Eltanin*, which collected the holotype.

Dodecas reducta Barnard, 1932

Dodecas reducta Barnard, 1932, p. 303, fig. 169b.

Distribution. Type locality: South Georgia.

Remarks. This species is known from a single male collected at South Georgia by the *Discovery* Expedition. The gills on pereonite 2 are quite reduced in size relative to those of pereonites 3 and 4, and the dorsal surface of pereonite 3 bears a pair of spines. These characters separate *D. reducta* from the other species of the genus.

Genus *Dodecasella* Barnard, 1931

Type species: *Dodecasella elegans* Barnard, 1931 (by original designation)

Diagnosis. Flagellum of antenna 2 of 3–7 articles, swimming setae absent; mandible with 3-segmented palp, setal formula for terminal article 1-*x*-1, molar absent; outer lobe of maxilliped longer than inner lobe, inner lobe wider than outer; gills on pereonites 3 and 4; pereopod 3 6-segmented, pereopod 4 absent, pereopod 5 4-segmented; abdomen of male and female with 2 pairs of biarticulate appendages.

Dodecasella elegans Barnard, 1931

Dodecasella elegans Barnard, 1931, p. 430; 1932, pp. 304–305, figs. 168, 169b.

Distribution. Type locality: South Georgia.

Remarks. This species is discussed under the remarks on the other species of the genus, *D. georgiana*. This species is known from over 50 specimens taken by the *Discovery* Expedition. In males the gills on pereonite 4 are extremely elongated, as long as pereonites 3 and 4 combined.

Dodecasella georgiana (Schellenberg, 1931)

Figs. 2, 7

Dodecas georgiana Schellenberg, 1931, pp. 262–264, 272, fig. 136.

Distribution. Type locality: South Georgia, Cumberland Bay, and Maiviken, 75–310 meters. Other records: 40°14.6′S, 55°24.7′W; 53°56′S, 140°19′W. Depth range: 75–1475 meters.

Diagnosis. Gills on pereonite 4 smaller than those on pereonite 3, basis of pereopod 3 in males longer than remaining articles.

Material examined.

Vema, cruise 15: Sta. 131, 40°14.6′S, 55°24.6′W, April 3, 1959, 1475 meters, 1 ♀.

Eltanin, cruise 23: Sta. 1691, 53°56′S, 140°19′W, May 14, 1966, 362–567 meters, 1 juvenile ♀.

Remarks. We are not able to identify either of these small female specimens (10 and 12 mm) with certainty. They resemble Schellenberg's description and figures of *Dodecasella georgiana*. Both specimens were slightly smaller than his ovigerous 16-mm female and agree in the configuration of gnathopod 2. The *Vema* female agrees with Schellenberg's female in that the gill on pereonite 4 is smaller than that on pereonite 3; however, in the *Eltanin* female the gill on pereonite 4 was not notably smaller than that on pereonite 4. It is unfortunate that only females were available to us. Even though our specimens agree fairly well with Schellenberg's female, we are still hesitant to identify ours as being conspecific, since our specimens were collected at a considerable distance from South Georgia.

Since Schellenberg did not provide figures of the mouth parts of his specimens, we have included them here. In general, the mouth parts resemble those of *Dodecas*, as does the over-all appearance of the species. The setal formula for the terminal article of the mandibular palp of the *Vema* specimens was 1-5-1 for the right and 1-3-1 for the left. This variation, while not unreported, is unusual.

We are herein transferring *D. georgiana* from

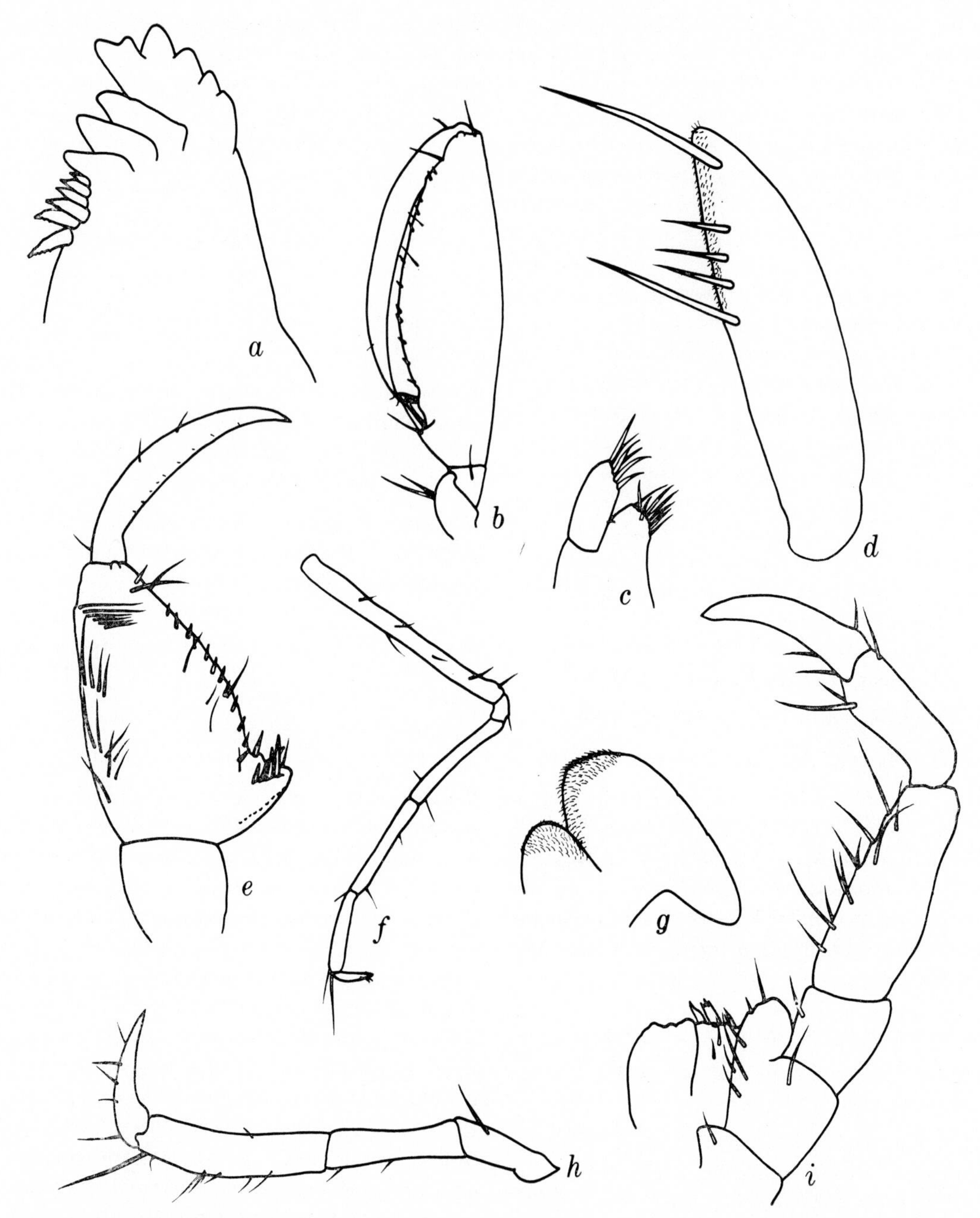

Fig. 7. *Dodecasella georgiana,* small female: *a,* left mandible; *b,* gnathopod 2; *c,* maxilla 2; *d,* terminal article of mandibular palp; *e,* gnathopod 1; *f,* pereopod 3; *g,* lower lip; *h,* pereopod 5; *i,* maxilliped.

Dodecas to *Dodecasella.* Schellenberg clearly states that *D. georgiana* differs from the other species of *Dodecas* in that it lacks gills on pereonite 2, the primary character separating *Dodecas* and *Dodecasella. Dodecasella georgiana* differs from *D. elegans* Barnard, 1931, the only other species of the genus, in that the gills on pereonite 4 are shorter than those on pereonite 3 in *D. georgiana* and the gills on pereonite 4 are longer than those on pereonite 3 in *D. elegans.* The basis of pereopod 3 of the males of *D. georgiana* is longer than the remaining articles combined, whereas in *D. elegans* this is not true.

Genus ***Luconacia*** Mayer, 1903

Type species: *Luconacia incerta* Mayer, 1903 (by monotypy)

Diagnosis. Flagellum of antenna 2 biarticulate, swimming setae absent; mandibular palp 3-segmented, setal formula for terminal article 1-*x*-1, molar present; outer lobe of maxilliped larger than inner lobe; gills on pereonites 3 and 4; pereopods 3 and 4 2-segmented, pereopod 5 6-segmented and inserted near midlength on pereonite 5; abdomen of male with pair of appendages and pair of lobes, female without lobes or appendages.

Remarks. We have altered the generic diagnosis from that of McCain [1968] by omitting the presence of knobs on the terminal article of the mandibular palp. The knobs that are present on the palp of *L. incerta* are not present on the palp of the following new species; hence this character is considered only of specific value.

Luconacia vemae new species

Figs. 8, 9

Distribution. Type locality: 41°57′S, 59°03′W. Other records: Off coast of Argentina and southern Chile. Depth range: 24–101 meters.

Diagnosis. Cephalon with paired dorsal spines, dorsal surface of pereonites without spines; terminal article of mandibular palp without knobs.

Material examined.

Vema, cruise 16: Sta. 37, 51°52′S, 67°01′W, May 16, 1960, 101 meters, 1 ♀ paratype AMNH 13128.

Vema, cruise 17: Sta. 25, 53°20.5′S, 69°32.8′W, March 29, 1961, 44 meters, 1 juvenile paratype AMNH 13129. Sta. 29, 52°43.7′S, 69°53.7′W, April 1, 1961, 24 meters, 2 ♂♂, 2 ♀♀ paratypes AMNH 13130. Sta. 47, 55°07.2′S, 66°29.3′W, May 4, 1961, 71 meters, 2 ♂ paratypes AMNH 13131. Sta. 76, 41°57′S, 59°03′W, May 23, 1961, 81 meters, 1 ♂ holotype AMNH 13132, 2 ♂♂, 1 ♀ paratypes AMNH 13133.

Eltanin, cruise 11: Sta. 958, 52°56′S, 75°00′W, February 5, 1964, 92–101 meters, 1 ♀ paratype USNM 123749. Sta. 981, 52°44′S, 67°42′W, February 14, 1964, 40–49 meters, 1 juvenile ♀ paratype USNM 123748.

Description. Male holotype: Cephalon with paired dorsal spines, anterolateral margins of pereonites 2–4 with large projections. Length 13.5 mm. Antenna 2 slightly longer than peduncle of antenna 1. Flagellum of antenna 1 with 11 articles. Mandibular palp 3-segmented, setal formula for terminal article 1-11-1. Left mandible with 5-toothed incisor, 5-toothed lacinia mobilis, setal row of 3 serrate spines. Right mandible with 5-toothed incisor, serrate lacinia mobilis, setal row of 2 serrate spines. Palp of maxilla 1 with 4 terminal spines, outer lobe with 6 spines. Inner and outer lobe of maxilla 2 with 5 terminal setae. Outer lobe of maxilliped with 1 terminal seta and small spine set in notch on medial margin, inner lobe with 1 plumose and 2 nonplumose setae, penultimate article of palp with medial fringe of setae distally. Propodus of gnathopod 1 with 2 grasping spines, grasping margins of dactylus and propodus not serrate. Propodus of gnathopod 2 with proximal tooth and notch at midlength, dactylus with setae along distal portion of grasping margin. Pereopods 3 and 4 2-segmented, both articles with several setae; pereopods 5–7 missing. Abdomen with pair of appendages and pair of setose lobes, appendage with small apical papillae and short fringe of small teeth.

Remarks. The pair of dorsal cephalic spines was present on most specimens, but on the smaller specimens they were quite small and in one juvenile male absent. The largest female measured 10.2 mm and differed from the holotype in that the abdomen lacked the pair of appendages and lobes and the anterolateral margin of pereonite 4 did not bear a projection. The female gnathopod 2 was not notched as deeply and the notch was more distal than in the male. The setal formula for the terminal article of the mandibular palp varied from 1-2-1 to 1-11-1.

Pereopod 5 was missing on all specimens except the single juvenile male, which lacked dorsal cephalic spines. It was fully segmented, and the terminal article was dactyliform; however, it was much smaller than pereopods 6 and 7. Judging from the position of the insertion of pereopod 5 at the midlength of pereonite 5, it is probable that pereopod 5 of the holotype was similar to that of the juvenile male.

The genus *Luconacia* was heretofore monotypic. This new species differs from *L. incerta* in that it lacks the dorsal spines on pereonites 2–4. The abdominal appendage of *L. vemae* does not bear as complete a fringe of small teeth around the papillae as in *L. incerta.* The knobs on the terminal article of the mandibular palp of *L. incerta* are not present in *L. vemae.*

The specific name is in honor of the R/V *Vema,* which collected the holotype.

See also remarks under *Triantella solitaria.*

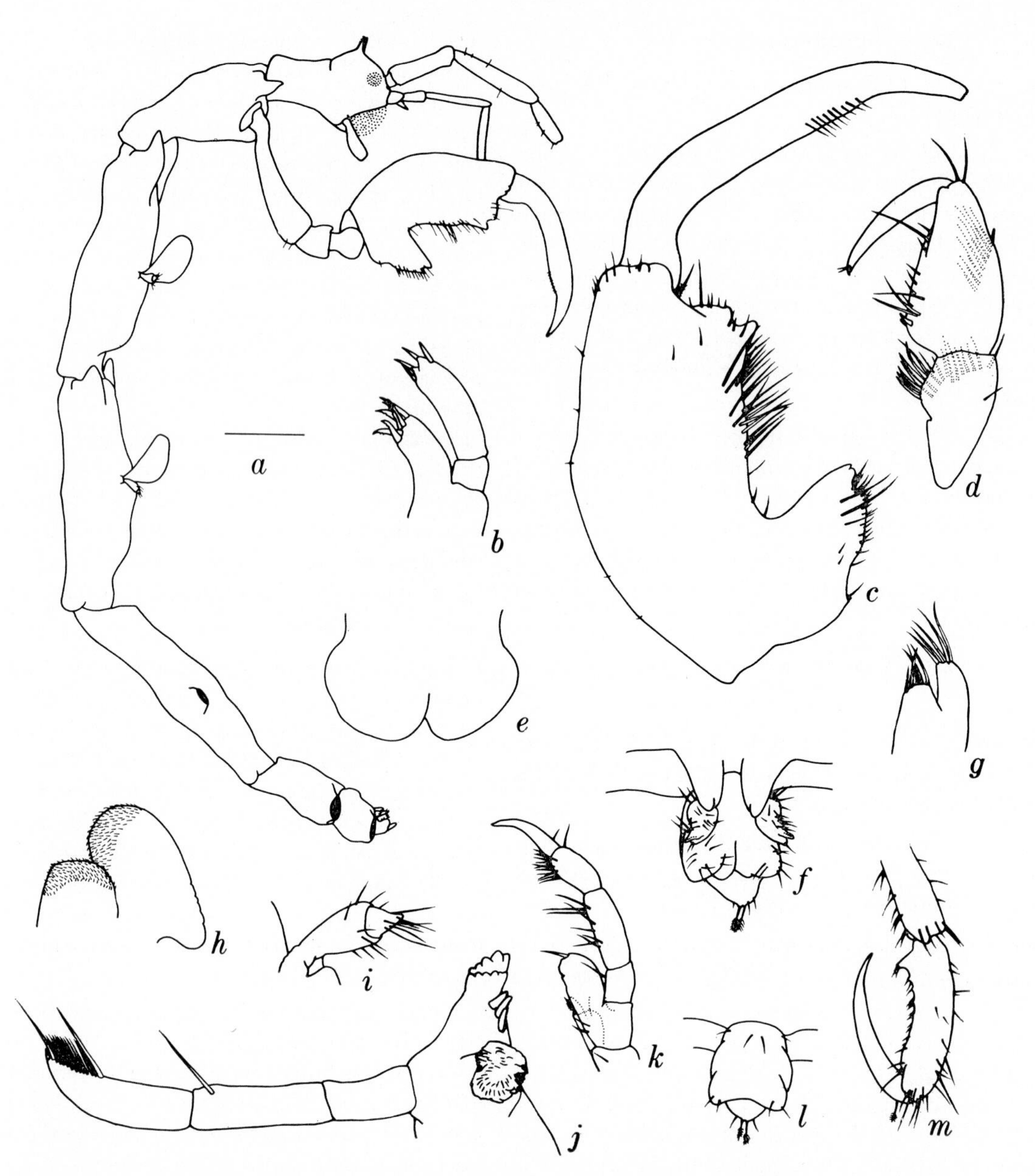

Fig. 8. *Luconacia vemae*, new species: *a–k*, male holotype; *l–m*, female paratype; *a*, lateral view; *b*, maxilla 1; *c*, gnathopod 2; *d*, gnathopod 1; *e*, upper lip; *f*, abdomen; *g*, maxilla 2; *h*, lower lip; *i*, pereopod 4; *j*, left mandible; *k*, maxilliped; *l*, abdomen; *m*, pereopod 6. Scale shown for *a* is 1 mm.

Genus ***Mayerella*** Huntsman, 1915

Type species: *Mayerella limicola* Huntsman, 1915 (by monotypy)

Diagnosis. Flagellum of antenna 2 biarticulate, swimming setae absent; mandibular palp 3-segmented, setal formula for terminal article 1 or 2, molar present; outer lobe of maxilliped larger than inner lobe; gills on pereonites 3 and 4; pereopods 3 and 4 2-segmented, pereopod 5 2- to 4-segmented; abdomen of male with pair of uniarticulate appendages and pair of lobes, female with pair of lobes.

Mayerella magellanica new species

Figs. 9, 10

Distribution. Type locality: 43°25′S, 75°05′W. Other records: Coasts of Chile and Argentina. Depth range: 70–676 meters.

Diagnosis. Male abdomen with penes in contact

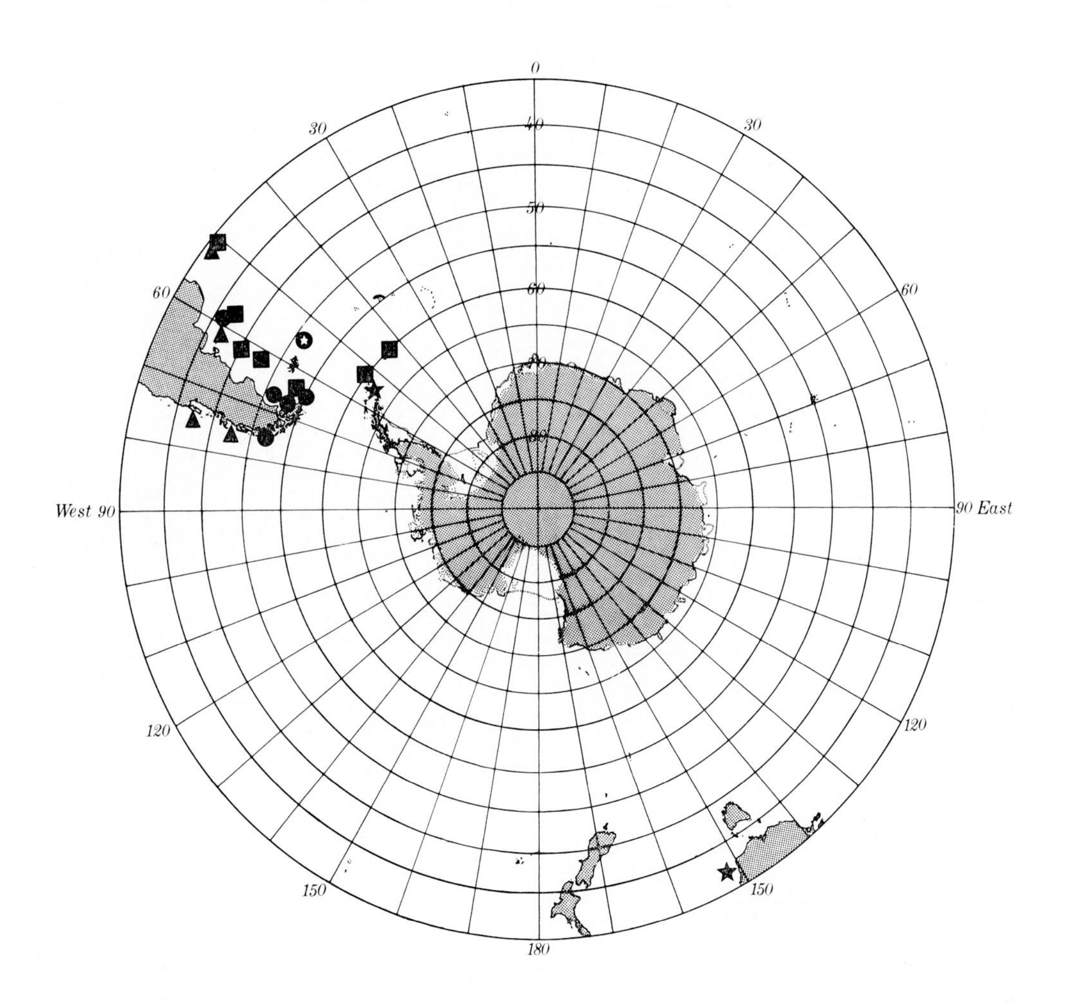

Fig. 9. Distribution records of *Luconacia vemae* (circle), *Mayerella magellanica* (triangle), *Paraproto condylata* (star), *Protella trilobata* (star in circle), and *Pseudoprotomima hedgpethi* (square).

medially, appendages medial; pereopod 5 4-segmented.

Material examined.

Vema, cruise 17: Sta. 11, 43°25′S, 75°05′W, March 23, 1961, 152 meters, 1 ♂ holotype AMNH 13134, 100+ paratypes AMNH 13190. Sta. 12, 43°30′S, 74°55′W, March 23, 1961, 112 meters, 8 ♂♂, 7 ♀♀ paratypes AMNH 13135. Sta. 15, 47°02′S, 75°36′W, March 24, 1961, 642 meters, 1 ♀ paratype AMNH 13136. Sta. 68, 41°16′S, 60°03′W, May 18, 1961, 70 meters, 14 ♂♂, 12 ♀♀ paratypes AMNH 13137. Sta. 74, 41°27′S, 59°33′W, May 23, 1961, 71 meters, 1 ♀ paratype AMNH 13138.

Vema, cruise 18: Sta. 9, 36°17′S, 53°21′W, February 4, 1962, 547–676 meters, 1 ♂ paratype AMNH 13139.

Description. Male holotype: Body smooth. Length 7.3 mm. Antenna 2 subequal in length to articles 1 and 2 of antenna 1. Setal formula for terminal article of mandibular palp 2, apex pointed. Left mandible with 5-toothed incisor, 5-toothed lacinia mobilis, setal row of 3 serrate spines. Right mandible with 5-toothed incisor, lacinia mobilis deeply serrate, setal row of 2 serrate spines. Palp of maxilla 1 with 4 terminal spines and 2 subterminal setae, outer lobe with 7 spines. Outer lobe of maxilla 2 with 9 setae, inner with 8 setae. Outer lobe of maxilliped with 1 terminal seta and several setae on anterior surface, medial margin serrate; inner lobe with 3 nonplumose setae; dactylus of palp with grasping margin serrate and with subterminal seta. Propodus of gnathopod 1 without true grasping spine, grasping margin of dactylus and propodus serrate. Palm of propodus of gnathopod 2 with proximal tooth and spine, notch distal, and row of well-spaced stout spines. Gills elliptical, third pair

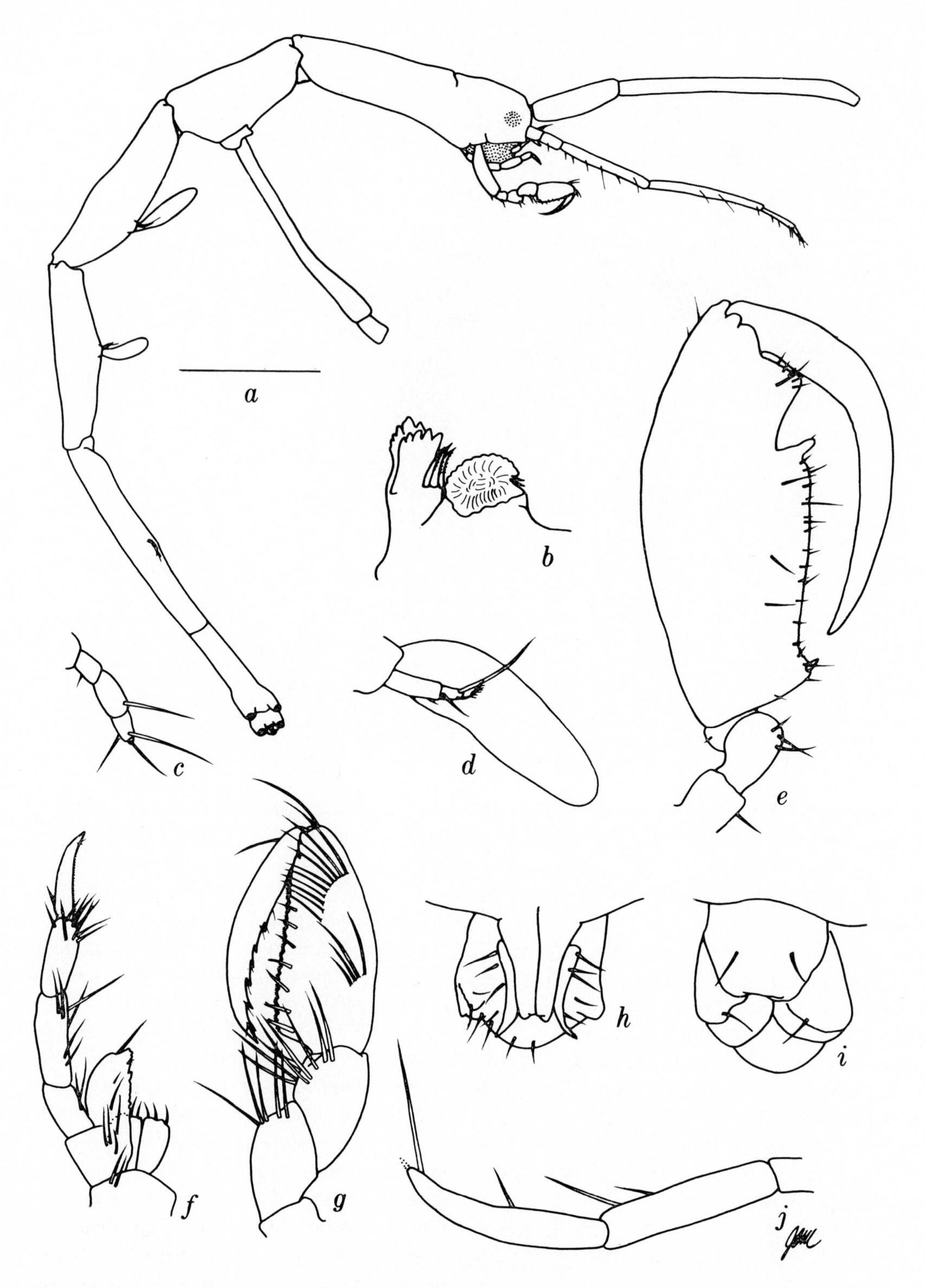

Fig. 10. *Mayerella magellanica*, new species: *a–h*, *j*, male holotype; *i*, female; *a*, lateral view; *b*, left mandible; *c*, pereopod 5; *d*, pereopod 4; *e*, gnathopod 2; *f*, maxilliped; *g*, gnathopod 1; *h*, abdomen; *j*, mandibular palp. Scale shown for *a* is 1 mm.

usually larger than fourth. Pereopods 3 and 4 2-segmented, terminal article with several setae. Pereopod 5 4-segmented, terminal article with single seta. Pereopods 6 and 7 missing. Abdomen with pair of long articulate appendages originating near medial portion of abdomen, penes long and in contact medially.

Remarks. The largest male in our collection measured 9 mm in length and the largest female 4.2 mm. The largest female was ovigerous, and we were unable to find any smaller ovigerous specimens. The terminal article of the mandibular palp bore either 1 or 2 setae. In smaller specimens antenna 2 was slightly smaller than the peduncle of antenna 1. The propodus of gnathopod 2 of females had poorly developed notches, and the female abdomen lacked appendages but had well-developed lobes.

Mayerella magellanica differs from the other species of *Mayerella, M. limicola* Huntsman, 1915, and *M. redunca* McCain, 1968, primarily by the shape and position of the abdominal appendages of the male. In *M. redunca* these appendages are very long and slender and recurved at their tips, whereas those of *M. limicola* and *M. magellanica* are much shoiter and the penes are much more medial than in *M. redunca.* The abdominal appendages of *M. limicola* are appreciably more lateral and anterior than those of *M. magellanica.* Pereopod 5 was 4-segmented in all the specimens of *M. magellanica* that we examined. Pereopod 5 of *M. limicola* is consistently 3-segmented, and that of *M. redunca* is either 2- or 3-segmented.

Specific name refers to zoogeographic province from which this species was collected.

Genus **Paraproto** Mayer, 1903

Type species: *Proto condylata* Haswell, 1885 (by present selection)

Diagnosis. Flagellum of antenna 2 of 3–14 articles, swimming setae absent; mandible with 3-segmented palp, setal formula for terminal article 1-*x*-1, molar absent; outer lobe of maxilliped larger than inner lobe; gills on pereonites 3 and 4; pereopods 3–5 6-segmented; abdomen of male with 2 pairs of biarticulate appendages and anterior pair of papillalike appendages, female with 2 pairs of biarticulate appendages and raised anterior projection.

Remarks. The generic diagnosis remains unaltered from that of Mayer, 1903, with the exception of the anterior pair of papillalike appendages on the male abdomen and the raised anterior projection of the female abdomen.

Paraproto condylata (Haswell, 1885)

Figs. 9, 11

Proto condylata Haswell, 1884 (1885), pp. 993–995, pl. 48, figs. 1–4.—Mayer, 1890, p. 14.

Paraproto condylata.—Mayer, 1903, p. 25, pl. 1, fig. 10, pl. 6, fig. 20.—Stebbing, 1910b, p. 651.

Distribution. Type locality: Australia. Other records: Scotia Ridge. Depth range: To 220 meters.

Diagnosis. Body smooth, without spines; merus, carpus, and propodus of gnathopod 2 subequal in large males.

Material examined.

Eltanin, cruise 12: Sta. 1003, 62°41′S, 54°43′W, March 15, 1964, 210–220 meters, 5 ♂♂, 8 ♀♀, 12 juveniles.

Description. Body smooth. Length of largest male 27.5 mm, of largest female 19.9 mm, of smallest ovigerous female 15.1 mm. Antenna 2 reaching approximately midlength of peduncle article 2 of antenna 1, flagellum of 3–9 articles. Mandible with 3-segmented palp, setal formula for terminal article 1-6-1 to 1-16-1; incisor and lacinia mobilis present with several accessory plates; setal row of 5 or 6 stout spines; molar absent. Outer lobe of maxilla 1 with 6 serrate spines. Inner lobe of maxilla 2 with more than 2 setae, outer lobe with more than 7 setae. Maxilliped usual, terminal article of palp with comb setae. Propodus of gnathopod 1 with more than 2 grasping spines, grasping margin of dactylus and propodus not serrate. Propodus of gnathopod 2 of large males with palmar projection at midlength, palm concave; dactylus short, reaching midlength of propodus; carpus approximately length of propodus; smaller males and females with less concave palm of propodus, carpus length reduced relative to propodus length. Gills elongate. Penultimate article of pereopods 3 and 4 with palmar margin bearing 1 or 2 spines, ultimate article dactyliform. Pereopod 5 6-segmented, ultimate article dactyliform, penultimate article without distinctive palmar margin. Pereopods 6 and 7 apparently broken from all available specimens. Abdomen of male with 2 pairs of long biarticulate appendages and small pair of anterior papillalike appendages, female abdomen similar to male except lacking papillalike appendages and with anterior raised projection.

Remarks. The genus *Paraproto* is presently composed of 3 species, *P. condylata* (Haswell, 1885), *P. gabrieli* Stebbing, 1914, and *P. spinosa* (Haswell,

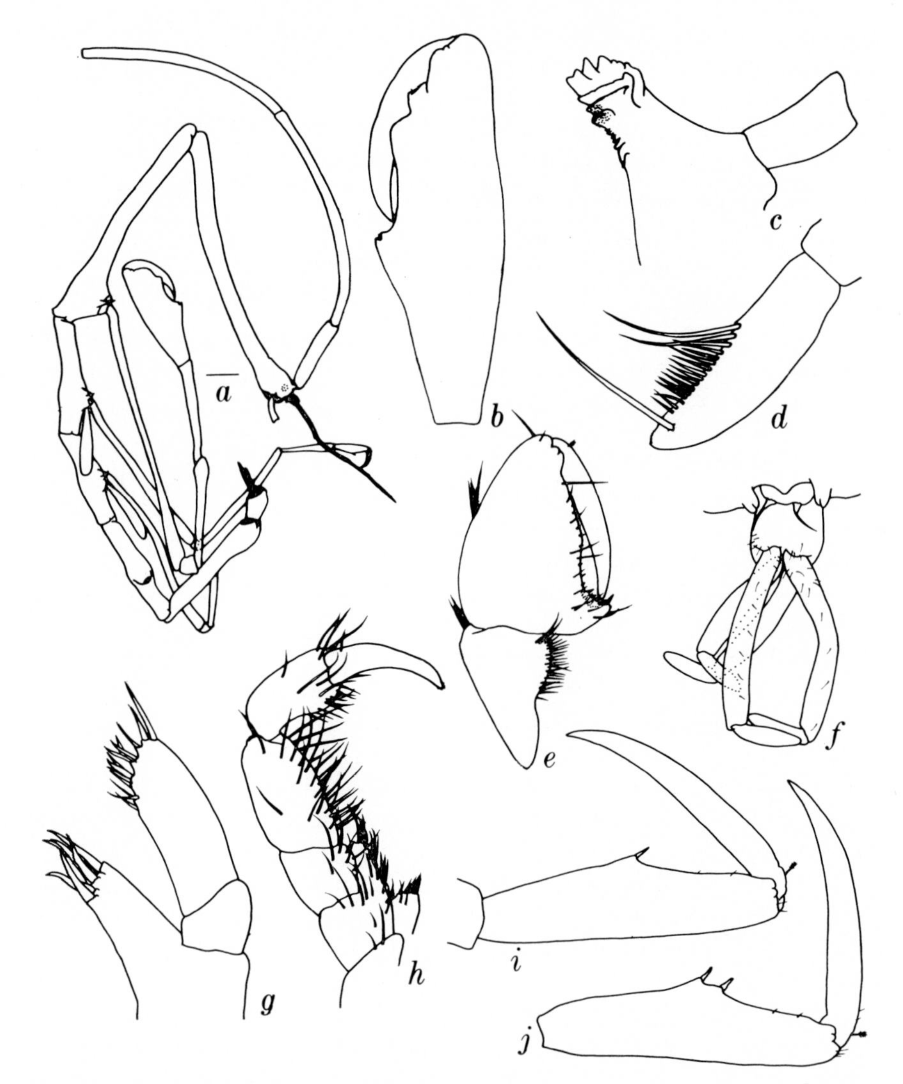

Fig. 11. *Paraproto condylata*, large male: *a*, lateral view; *b*, gnathopod 2; *c*, right mandible; *d*, terminal article of mandibular palp; *e*, gnathopod 1; *f*, abdomen; *g*, maxilla 1; *h*, maxilliped; *i*, pereopod 4; *j*, pereopod 3. Scale shown for *a* is 1 mm.

1885). The first species lacks the body spination characteristic of the last 2 species.

Genus ***Protella*** Dana, 1852

Type species: *Protella gracilis* Dana, 1853 (by monotypy)

Diagnosis. Flagellum of antenna 2 biarticulate, swimming setae absent; mandibular palp 3-segmented, setal formula for terminal article 1-x-1 or 1-x-y-1, molar present; outer lobe of maxilliped larger than inner lobe; gills on pereonites 3 and 4; pereopods 3 and 4 1-segmented, pereopod 5 6-segmented; abdomen of male with pair of appendages and pair of lobes, female with pair of lobes.

Protella trilobata new species

Figs. 9, 12–13

Distribution. Type locality: 51°58′S, 56°38′W, 646–845 meters.

Diagnosis. Pereopods 3 and 4 terminally constricted, length less than ¾ length of gill, peduncular article 3 of antenna 1 much shorter than article 2.

Material examined.

Eltanin, cruise 8: Sta. 558, 51°58′S, 56°38′W, March 14, 1963, 646–845 meters, 1 ♂ holotype USNM 123750, 8 ♂♂, 11 ♀♀ paratypes USNM 123751.

Description. Male holotype: Body smooth except

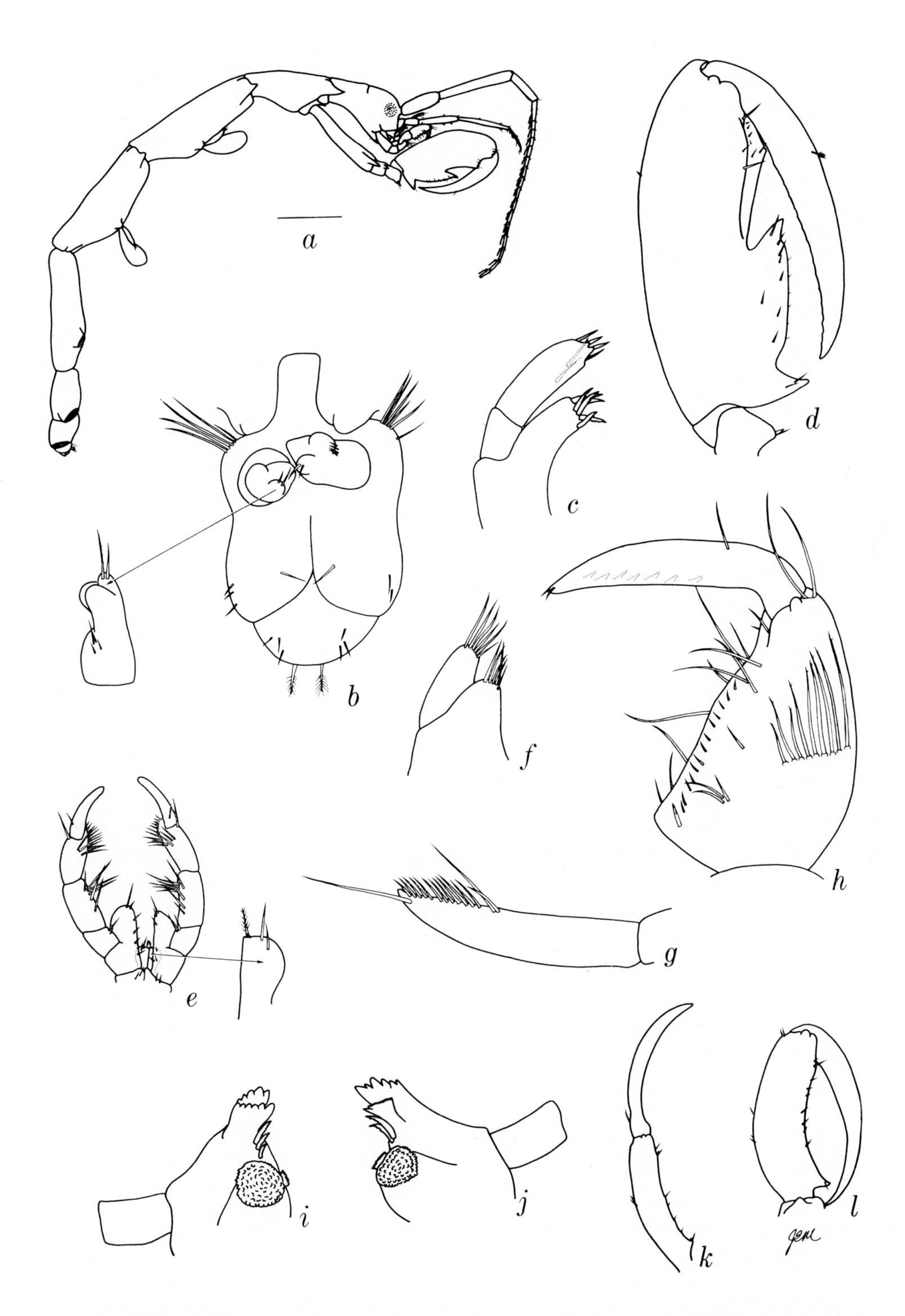

Fig. 12. *Protella trilobata*, new species, male holotype: *a*, lateral view; *b*, abdomen; *c*, maxilla 1; *d*, gnathopod 2; *e*, maxilliped; *f*, maxilla 2; *g*, terminal article of mandibular palp; *h*, gnathopod 1; *i*, left mandible, *j*, right mandible; *k*, pereopod 5; *l*, pereopod 7. Scale shown for *a* is 1 mm.

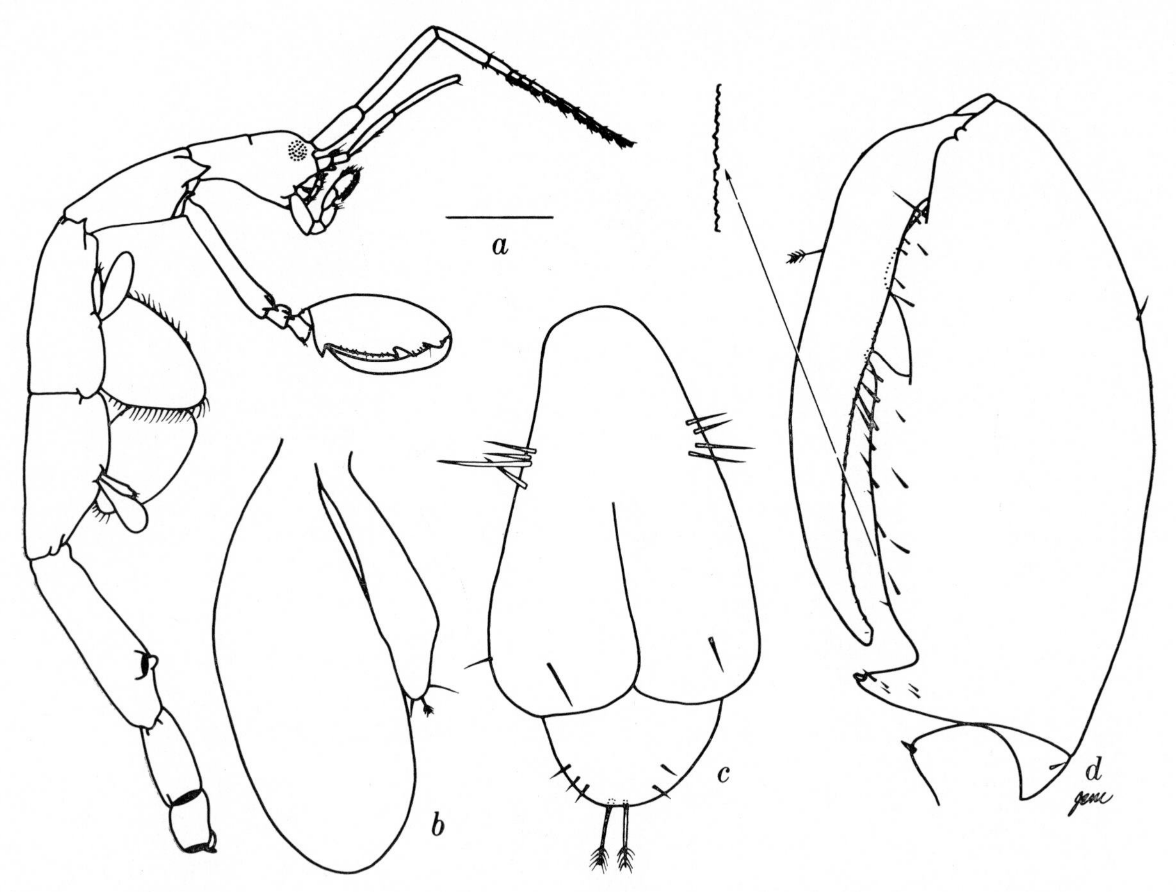

Fig. 13. *Protella trilobata*, new species, female paratype: *a*, lateral view; *b*, gill and appendage of pereonite 3; *c*, abdomen; *d*, gnathopod 2. Scale shown for *a* is 1 mm.

for anterolateral processes on pereonites 2–4. Length 9.2 mm. Antenna 1 shorter than body length, article 3 much shorter than article 2. Antenna 2 approximately equal in length to the peduncle of antenna 1. Setal formula for terminal article of mandibular palp 1-13-1. Left mandible with 5-toothed incisor, 6-toothed lacinia mobilis, setal row of 3 serrate spines, molar with flake. Right mandible with 6-toothed incisor, lacinia mobilis serrate, setal row of 2 serrate spines, molar with flake. Palp of maxilla 1 with 4 terminal spines and 2 subterminal setae, outer lobe with 6 serrate spines. Inner and outer lobes of maxilla 2 with 6 terminal setae. Outer lobe of maxilliped not serrate with 5 or 6 marginal setae; inner lobe with 1 plumose and 2 nonplumose setae; penultimate article of palp produced distally with distal fringe of setae on medial margin, ultimate article dactyliform and rounded terminally. Propodus of gnathopod 1 subtriangular without distinct grasping spines, proximal portion of grasping margin minutely serrate; grasping margin of dactylus not distinctly serrate. Propodus of gnathopod 2 with proximal tooth and notch at midlength, dactylus minutely serrate. Gills elliptical to club-shaped. Pereopods 3 and 4 1-segmented, distal end constricted but not segmented. Pereopod 5 6-segmented, penultimate article with weakly developed grasping margin, ultimate article dactyliform. Pereopods 6 and 7 usually with well-developed propodus. Abdomen with pair of distally trilobate appendages and pair of lobes.

Remarks. The largest male in our collection measured 11.9 mm, and the largest female 9.3 mm. The setal formula for the terminal article of the mandibular palp varied from 1-13-1 to 1-18-1. No other significant variation was noted.

Protella trilobata differs from the other species of *Protella*, *P. gracilis* Dana, 1853, and *P. similis* Mayer, 1903, in that peduncular article 3 of antenna 1 is much shorter than article 2 in *P. trilobata*, whereas in the other species the third article is as long or longer than the second article. Pereopods 3 and 4 of *P. trilo-*

bata are less than ¾ of the length of the gills, whereas in the other species these pereopods are longer than ¾ of the gill length. In both *P. gracilis* and *P. similis* pereopods 3 and 4 are somewhat rod-shaped and lack the terminal constriction found in those pereopods of *P. trilobata.*

The specific name refers to the 3-lobed appearance of the terminal portion of the male abdominal appendage.

Genus *Protellopsis* Stebbing, 1888

Type species: *Protellopsis kergueleni* Stebbing, 1888 (by monotypy)

Diagnosis. Flagellum of antenna 2 biarticulate, swimming setae absent; mandibular palp 3-segmented, setal formula for terminal article 1-*x*-2, molar present; outer lobe of maxilliped larger than inner lobe; gills on pereonites 3 and 4; pereopods 3 and 4 2-segmented, pereopod 5 6-segmented; abdomen of male with 2 pairs of appendages and questionable pair of lobes.

Protellopsis kergueleni Stebbing, 1888

Protellopsis kergueleni Stebbing, 1888, pp. 1241–1244, pl. 142.—Mayer, 1890, p. 17, pl. 5, figs. 12–13; 1903, p. 32.—Chevreux, 1913, p. 86.

Distribution. Type locality: Off Baie Greenland, Kerguelen Island, 5.5 meters.

Remarks. This species was based on 2 male specimens taken by the *Challenger* Expedition near Kerguelen Island. To our knowledge, no other specimens of this species have been collected. The type material was not available to us.

Genus *Pseudaeginella* Mayer, 1890

Type species: *Aeginella tristanensis* Stebbing, 1888 (by monotypy, subsequent designation by McCain, 1968)

Diagnosis. Flagellum of antenna 2 biarticulate, swimming setae absent; mandibular palp 3-segmented, setal formula for terminal article 1-*x*-1, molar questionable; outer lobe of maxilliped larger than inner lobe; gills on pereonites 3 and 4; pereopods 3 and 4 absent, pereopod 5 6-segmented; abdomen of male and female with pair of lobes.

Pseudaeginella tristanensis (Stebbing, 1888)

Aeginella tristanensis Stebbing, 1888, pp. 1249–1251, pl. 143.

Pseudaeginella tristanensis.—Mayer, 1890, pp. 37–38, pl. 5, fig. 51, pl. 6, fig. 14.—Barnard, 1932, pp. 300–301, fig. 166; 1940, p. 486.—Stephensen, 1949, pp. 52, 56, fig. 23. [Not Stebbing, 1895, p. 402, or Mayer, 1903, p. 59.]

Distribution. Type locality: Off Nightingale Island, Tristan da Cunha, 201.2 meters. Other records: Tristan da Cunha and East London, South Africa. Depth range: 3–201.2 meters.

Remarks. This species was not present in our collection. McCain [1968, p. 100] remarks on the remarkable similarity between *P. tristanensis* and *Fallotritella biscayensis.* The latter genus differs from *Pseudaeginella* primarily in the presence of small uniarticulate appendages on pereonites 3 and 4. It is, therefore, quite possible that the two genera are synonymous; and extreme care should be taken to avoid overlooking these minute appendages.

Genus *Pseudododecas* new genus

Type species: *Pseudododecas bowmani,* new species (by present designation)

Diagnosis. Flagellum of antenna 2 of 9 articles, swimming setae absent; mandibular palp 3-segmented, setal formula for terminal article 1-25-1, molar absent; outer lobe of maxilliped subequal to inner lobe; gills on pereonites 3 and 4; pereopod 3 5-segmented, pereopod 4 absent, pereopod 5 4-segmented and inserted near midlength of pereonite 5; abdomen of female with 2 pairs of biarticulate appendages.

Remarks. This new genus is closely related to *Dodecas* and its allied genera, but the small 5-segmented appendage on pereonite 3 separates it from *Dodecas* and other caprellid genera.

The generic name is derived from the Greek term 'pseudo,' meaning false, in combination with *Dodecas,* the name of a closely allied genus.

Pseudododecas bowmani new species

Figs. 2, 14

Distribution. Type locality: South Shetland Islands, 61°44′S, 55°56′W, 769 meters.

Diagnosis. Since this genus is monotypic, the characters of the genus are diagnostic for the species.

Material examined.

Eltanin, cruise 12: Sta. 997, 61°44′S, 55°56′W, March 14, 1964, 769 meters, 1 ♀♀ holotype USNM 123747.

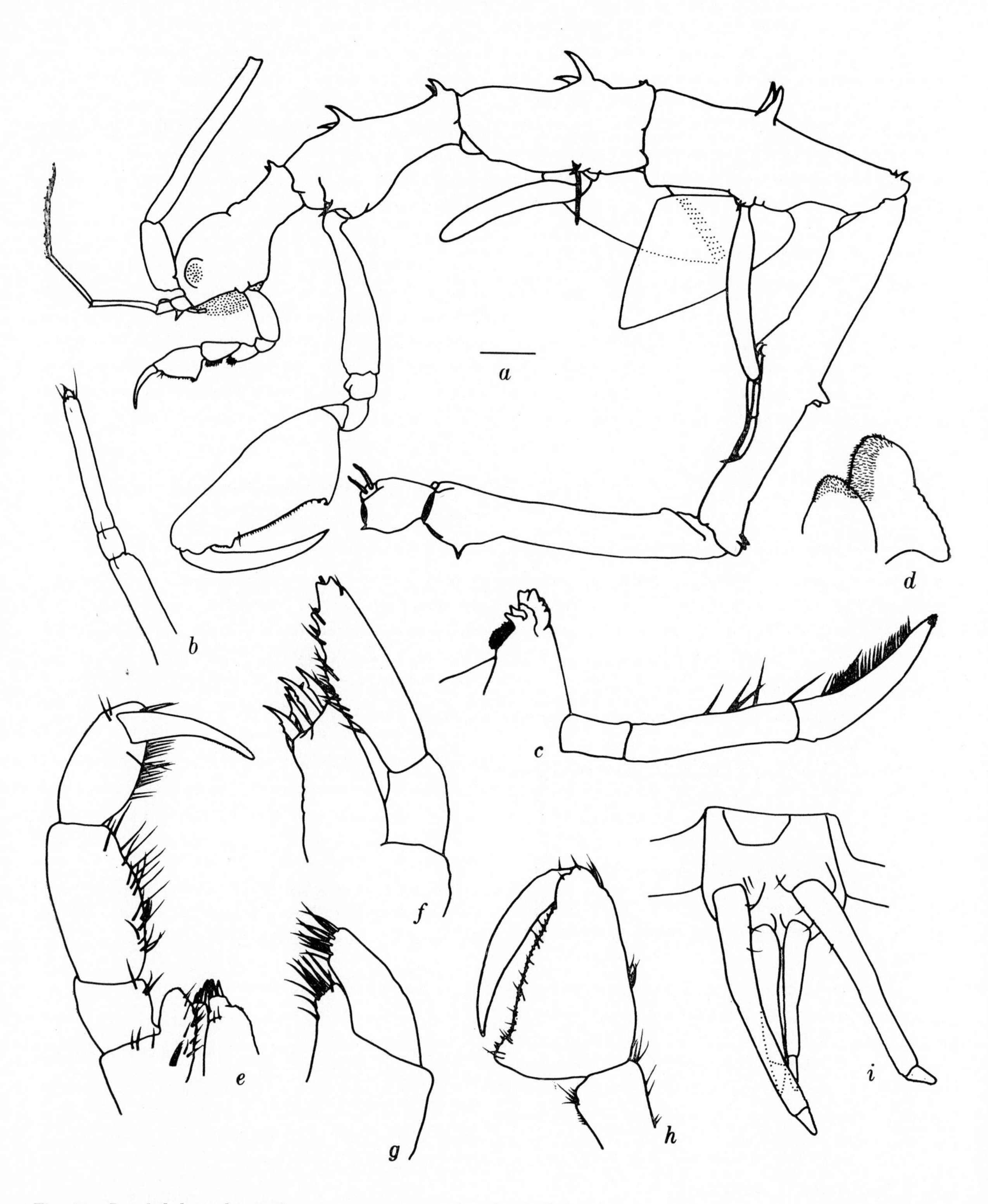

Fig. 14. *Pseudododecas bowmani*, new genus, new species, female holotype: *a*, lateral view; *b*, pereopod 3, terminal articles; *c*, right mandible; *d*, lower lip; *e*, maxilliped; *f*, maxilla 1; *g*, maxilla 2; *h*, gnathopod 1; *i*, abdomen. Scale shown for *a* is 1 mm.

Description. Female holotype: Body spinose, pereonite 1 with pair of posterodorsal spines; pereonites 2, 4, and 5 with pair of dorsal spines at midlength and pair of posterodorsal spines; pereonite 3 with pair of dorsal spines slightly posterior to midlength and pair of posterodorsal spines; pereonite 6 with single posterodorsal spine; pereonite 7 with dorsal spines at base of pereopod 7; pereonite 2 with lateral spine above insertion of gnathopod 2; pereonites 3 and 4 with lateral spine above insertion of gills. Length 25 mm. Setal formula for terminal article of mandibular palp 1-25-1. Mandibles with incisor, lacinia mobilis, 2 accessory plates, and setal row of 14 spines, molar absent. Palp of maxilla 1 with 4 spines and numerous setae, outer lobe with 6 large apical spines. Outer lobe of maxilla 2 with 9 setae, inner lobe with 10. Outer lobe of maxilliped with numerous setae, apically serrate; inner lobe with 3 large spines and several setae; distal end of penultimate article of palp produced slightly beyond insertion of ultimate article. Propodus of gnathopod 1 triangular with 3 grasping spines, grasping margin not serrate with double row of short spines; grasping margin of dactylus not serrate. Palm of propodus of gnathopod 2 with 3 proximal grasping spines set in notch, double series of small spines on margin, and distal projection; dactylus not serrate. Gills elliptical and elongate. Pereopod 3 of 5 articles, terminal article minute; pereopod 4 absent; pereopod 5 of 4 articles, terminal article dactyliform. Pereopods 6 and 7 missing from holotype. Abdomen with 2 pairs of biarticulate appendages and raised proximal projection.

Remarks. Specific name is in honor of Dr. Thomas E. Bowman of the Division of Crustacea at the United States National Museum.

Genus ***Pseudoprotomima*** McCain, 1969

Type species: *Pseudoprotomima hurleyi* McCain, 1969 (by original designation)

Diagnosis. Flagellum of antenna 2 of 2–4 articles, swimming setae absent; mandibular palp 3-segmented, setal formula for terminal article 1-*x*-1, molar absent; outer lobe of maxilliped subequal to inner lobe; gills on pereonites 2–4; pereopods 3–5 6-segmented; abdomen of male and female with 2 pairs of appendages, males with or without pair of papillalike appendages, females with or without small anterior raised projection.

Remarks. This genus was originally established with hesitation, since it differs from *Protomima* Mayer, 1903, only in that it bears a truly 6-segmented appendage on pereonite 5 instead of the 5-segmented appendage of *Protomima.* The following new species changes the generic diagnosis so that the number of articles in the flagellum of antenna 2 varies from 2 to 3, the males bear a pair of small papillae on the abdomen that may represent appendages, and the female abdomen bears a raised projection.

Pseudoprotomima hedgpethi, new species

Figs. 9, 15

Distribution. Type locality: 41°41′S, 59°19′W. Other records: Off east coast of South America and South Orkney Islands. Depth range: 71–403 meters.

Diagnosis. Flagellum of antenna 2 with 2 or 3 articles, male abdomen with anterior pair of papillalike appendages and 2 pairs of biarticulate appendages.

Material examined.

Vema, cruise 14: Sta. 5, 45°51′S, 61°52′W, February 3, 1958, 107 meters, 25+ paratypes AMNH 13140. Sta. 6, 46°47.7′S, 62°47′W, February 4, 1958, 105 meters, 25+ paratypes AMNH 13141. Sta. 14, 54°23′S, 65°35′W, February 19, 1958, 75 meters, 25+ paratypes AMNH 13142.

Vema, cruise 17: Sta. 74, 41°27′S, 59°33′W, May 23, 1961, 71 meters, 25+ paratypes AMNH 13143. Sta. 75, 41°41′S, 59°19′W, May 23, 1961, 82 meters, 1 ♂ holotype AMNH 13144, 92 ♂♂, 62 ♀♀ paratypes AMNH 13145. Sta. 76, 41°57′S, 59°03′W, May 23, 1961, 81 meters, 5 ♂♂, 1 ♀ paratype AMNH 13146. Sta. 88, 45°11′S, 60°55′W, June 11, 1961, 110 meters, 1 ♂ paratype AMNH 13147. Sta. 89, 45°02′S, 61°18′W, June 11, 1961, 102 meters, 50+ paratypes AMNH 13148. Sta. 90, 44°53′S, 61°43′W, June 11, 1961, 99 meters, 6 ♂♂, 3 ♀♀ paratypes AMNH 13149. Sta. 91, 44°45′S, 62°11′W, June 11, 1961, 98 meters, 1 ♂ paratype AMNH 13150. Sta. 102, 34°25′S, 52°19′W, June 27, 1961, 73 meters, 100+ paratypes AMNH 13151.

Eltanin, cruise 12: Sta. 1082, 60°50′S, 42°55′W, April 14, 1964, 293–311 meters, 1 ♀ paratypes USNM 123753. Sta. 1084, 60°22′S, 46°50′W, April 15, 1964, 293–403 meters, 1 juvenile ♀ paratype USNM 123752.

Description. Male holotype: Body smooth. Length 13.5 mm. Antenna 1 approximately ⅔ body length, flagellum with 12 articles. Antenna 2 sparsely setose, flagellum with 2 articles. Setal formula for terminal article of mandibular palp 1-7-1. Mandibles with incisor, lacinia mobilis, several accessory plates; setal

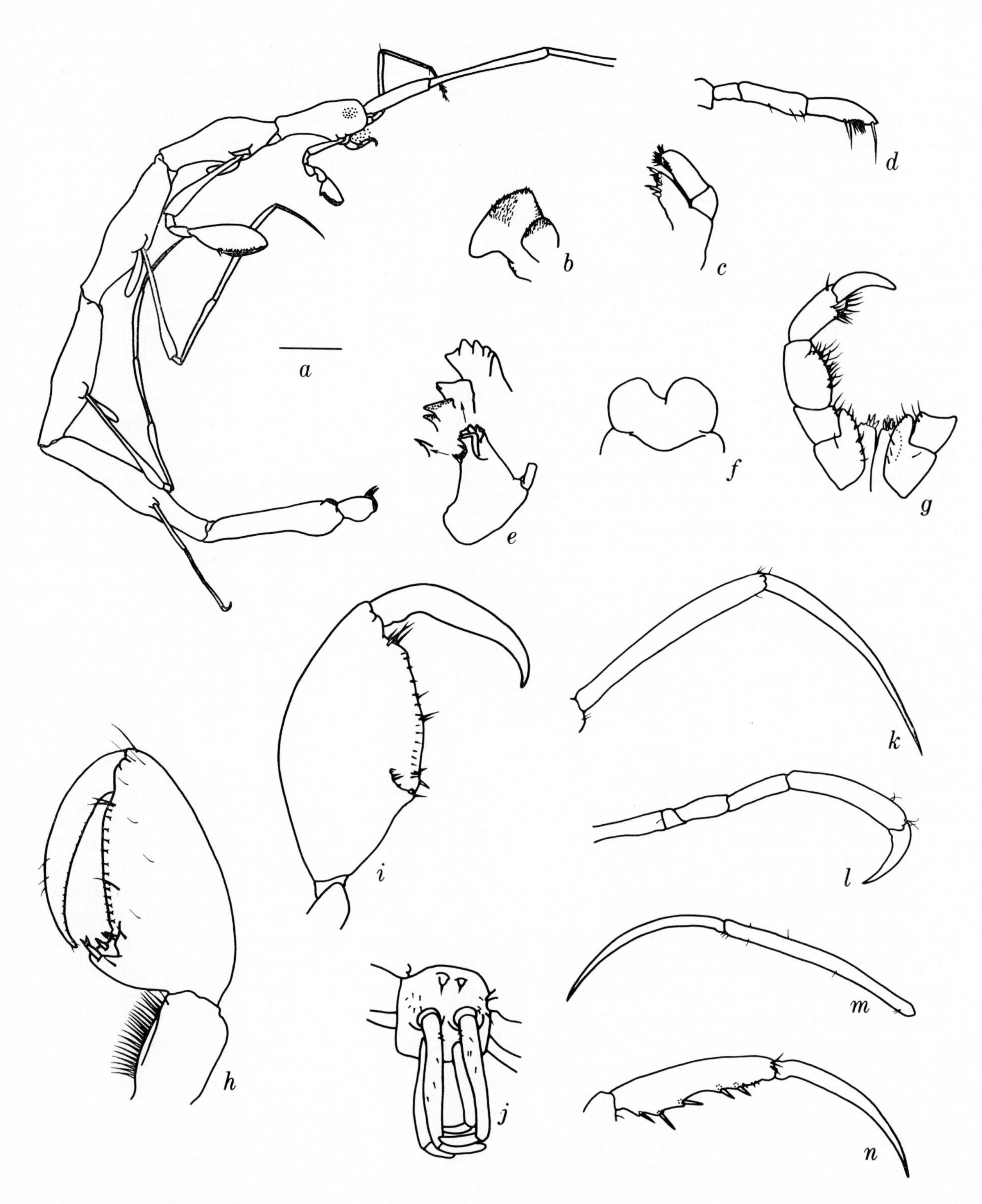

Fig. 15. *Pseudoprotomima hedgpethi*, new species: *a–m*, male holotype; *n*, male paratype; *a*, lateral view; *b*, lower lip; *c*, maxilla 1; *d*, mandibular palp; *e*, right mandible; *f*, upper lip; *g*, maxilliped; *h*, gnathopod 1; *i*, gnathopod 2; *j*, abdomen; *k*, pereopod 3; *l*, pereopod 5; *m*, pereopod 7; *n*, pereopod 3. Scale shown for *a* is 1 mm.

row of right mandible with 3 small and 2 large spines, that of left with 7 small and 2 large; molar absent. Palp of maxilla 1 with 9 apical and 3 subapical setae, outer lobe with 5 apical spines. Lobes of maxilla 2 with 5 or 6 apical setae. Outer lobe of maxilliped with 8 or 9 marginal setae, margin only slightly serrate; inner lobe with 3 large apical spines; palp not distinctive, inner distal margins of articles 2 and 3 setose. Propodus of gnathopod 1 subtriangular, grasping margin with 5 proximal grasping spines, palm not serrate but with distinct row of short paired spines; dactylus not serrate. Propodus of gnathopod 2 inflated at midlength with spine notch at approximately ⅓ length, grasping margin with row of short paired spines. Gills elongate, subelliptical. Pereopods 3 and 4 long, propodus with only few small distal setae, dactylus long and approximately equal in length to propodus. Pereopod 5 shorter than 3 and 4, terminal article dactyliform. Abdomen with 2 pairs of biarticulate appendages and anterior pair of small papillalike appendages, penes lateral.

Remarks. The holotype is a small adult male. Although not the largest male in our collection, it was chosen because some larger males show the same inflation and collapsing of the palm of gnathopod 2 as is common in *Phtisica marina.* The largest male in our collection measured 15.3 mm and agreed closely with the holotype except that the spine notch on the propodus of gnathopod 2 is more proximal and the spines on the outer lobe of maxilla 1 appeared to be degenerate. The largest female, 8.1 mm, resembles the holotype except that the flagellum of antenna 2 consists of 3 articles and lacks the typically male papillalike appendages on the abdomen but instead has a raised projection. The outer lobe of maxilla 1 of the largest female showed the same degenerate condition as the largest male. The smallest ovigerous female measured 7 mm and agreed with the largest female except that the outer lobe of maxilla 1 bore 6 well-developed spines.

There was considerable variation in the spination of the grasping margin of the propodus of pereopods 3 and 4. Some large males had up to 6 large spines along with several setae, while smaller males and females had nearly smooth grasping margins. The setal formula for the terminal article of the mandibular palp varied from 1-1-1 to 1-7-1. The mandible of the larger males bore an incisor, lacinia mobilis, up to 2 accessory plates, and up to 9 spines in the setal row. In smaller specimens the incisor, lacinia mobilis, and the accessory plates were developed; but the setal row varied and in one case appeared to be absent.

Pseudoprotomima hedgpethi differs from *P. hurleyi* McCain, 1969, the only other species of this genus in that in the former the flagellum of antenna 2 consists of 2 or 3 articles and in the latter, 4 articles. The other main difference between the species is that the males of *P. hedgpethi* bear small papillalike appendages in the anterior portion of the abdomen that are lacking in *P. hurleyi.* These papillalike appendages, reminiscent of the third pair of appendages of *Phtisica marina,* emphasize the close relationship of *Phtisica* and *Pseudoprotomima.*

The specific name is in honor of Dr. Joel W. Hedgpeth, Director of the Yaquina Marine Biological Laboratory of Oregon State University, in gratitude for his enthusiastic support of this project.

Pseudoprotomima hurleyi McCain, 1969

Pseudoprotomima hurleyi McCain, 1969, pp. 292–293, fig. 3.

Distribution. Type locality: 44°45′S, 174°30′E. Other records: 41°10′S, 176°58′E. Depth range: 788–1609 meters.

Remarks. This species appears to be limited to the deeper water near New Zealand. For differences between this species and *P. hedgpethi* see the remarks under that species.

Genus ***Triantella*** Mayer, 1903

Type species: *Triantella solitaria* Mayer, 1903 (by monotypy)

Diagnosis. Flagellum of antenna 2 3-segmented, swimming setae absent; mandibular palp 3-segmented, setal formula for terminal article 1-*x*-1, molar not described; outer lobe of maxilliped larger than inner lobe; gills on pereonites 3 and 4; pereopods 3 and 4 1-segmented, questionably with terminal constriction representing partially fused article, pereopod 5 6-segmented; abdomen of male with pair of appendages and pair of lobes, female questionably without lobes or appendages.

Triantella solitaria Mayer, 1903

Triantella solitaria Mayer, 1903, p. 32, pl. 1, fig. 18, pl. 6, figs. 38–40, pl. 9, figs. 9, 36, 59.—Schellenberg, 1931, pp. 264–265, 272.

Distribution. Type locality: Atlantic Ocean south of La Plata, 52 fathoms. Other records: Northern Argen-

TABLE 1. Distribution of Antarctic-Subantarctic Caprellidae by Biogeographic Provinces*

	Antipodean	Kerguelenian	Magellanic	Tristan	Georgian	Scotian	Rossian	Antarctic
Number of species at each station								
Total	3	3	12	3	5	5	2	2
Endemic species	1	2	2	1	2	1	0	0
Species from warmer water	2	0	9	2	3	4	2	2
Species from colder water	0	1	4	1	2	2	2	2
Occurrence of each species								
Aeginoides gaussi			X		X	X	X	X
Caprella equilibra	X†		X†					
Caprella penantis			X†	X†				
Caprella ungulina			X†					
Caprellina longicollis	X†		X†					
Caprellinoides mayeri		X	X	X	X	X	X	X
Dodecas elongata		X						
Dodecas eltaninae			X					
Dodecas reducta					X			
Dodecasella elegans					X			
Dodecasella georgiana					X†			
Luconacia vemae			X†					
Mayerella magellanica			X†					
Paraproto condylata						X†		
Protella trilobata			X					
Protellopsis kergueleni		X						
Pseudaeginella tristanensis				X				
Pseudododecas bowmani						X		
Pseudoprotomima hedgpethi			X†			X†		
Pseudoprotomima hurleyi	X							
Triantella solitaria			X†					

* Provinces are arranged from left to right in approximate order of decreasing water temperature.

† Indicates species also found in warmer water outside subantarctic-antarctic region.

tina, 37°50′S, 56°11′W; Port William, Falkland Islands.

Remarks. Mayer described this species from a single male specimen of 7 mm in length. Later Schellenberg reported on another male and a female of the same length. Schellenberg's specimens differed from Mayer's in that the flagellum of antenna 2 was composed of 2 articles instead of 3 articles. The morphological similarity between *T. solitaria* and *Luconacia vemae* is striking, and the primary feature separating the two genera seems to be the number of articles in the flagellum of antenna 2. All of our specimens of *L. vemae* had biarticulate flagella; other differences from *T. solitaria* included fully 2-segmented pereopods 3 and 4, well-developed lateral projections on pereonite 2, and the ratio of the lengths of pereopods 3 and 4 to the gill (much less than 1:2, whereas in *T. solitaria* it exceeds 1:2).

Mayer states that he obtained his material of *T. solitaria* from the Stockholm Museum; however, the types apparently were not deposited there, and a search of most of the larger museums in the United States and Europe has not revealed their location. It is quite possible that the type of *T. solitaria* may be a juvenile of *L. vemae* and that the two genera and species should be combined. Since Schellenberg's specimens of *T. solitaria* had biarticulate flagella on antenna 2, they may prove to belong to *L. vemae.*

DISTRIBUTION SECTION

Zoogeography

The distribution of several species of antarctic-subantarctic caprellids conforms roughly to the biogeographical provinces as proposed by Knox [1960] and modified by Powell [1965] and Kusakin [1967]. Some of the species overlap these provinces; however, their distributions roughly coincide with the subantarctic-antarctic water masses. Table 1 presents the distribution of caprellids in the biogeographic provinces of the antarctic-subantarctic water masses as proposed by Knox [1960].

It is evident from the analysis at the bottom of Table 1 that endemism in the Caprellidae is low, less than 50%, in most of Knox's biogeographic provinces. On the other hand, we find an almost 50% endemism if we consider the Scotia Ridge, Tierra del Fuego, Falkland Islands, and Antarctic Peninsula regions as a faunistic unit. This faunistic unit is composed of 3 distinct groups of caprellids, the endemic group, a more northern, warmer-water group, and a southern, colder-water group. The northern, warmer-water group has affinities with Knox's Magellanic–Central-Chilean–Argentinian provinces and the southern, colder-water group with his Antarctic-Rossian provinces. We feel that the faunistic unit as delineated above is a true biogeographic province, at least for the Caprellidae. The family Caprellidae is known for its wide-ranging species [McCain, 1968], and therefore the 5 endemic species out of 11 found in this province are unusual. Admittedly, we are dealing with a small number of species and our samples are probably biased, owing to the large number of samples we had from this province as compared to the other portions of the antarctic-subantarctic region. This province, far from being a bridge, as suggested by other authors, is an area of mixing, since no warmer-water species cross this arc into the Antarctic proper and no colder-water species cross this arc into the more northern regions.

It is regrettable that we do not have samples from Tristan da Cunha and the Kerguelen Islands, since these areas are at the northern limit of the subantarctic water mass. Stebbing [1888] cites 2 species from the Kerguelen Islands, both of which are endemic. Therefore, from this tenuous evidence we can only speculate that the Kerguelenian province of Knox, Powell, and Kusakin is valid for the Caprellidae.

Stebbing also includes 1 species from Tristan da Cunha that is endemic; however, 2 other species known from this island are wide-ranging species with warmer-water affinities. Knox tentatively groups Tristan da Cunha and Gough Island together as a single biogeographic province. We cannot believe in the uniqueness of the Tristan province, at least as a subantarctic province, because of the presence of the 2 wide-ranging warmer-water species.

McCain [1969] reviews the Caprellidae of New Zealand and finds only 3 species occurring there, one endemic, one undoubtedly subantarctic, and one a wide-ranging warmer-water species. As with the Tristan province, we do not have sufficient evidence to comment on the distinctiveness of the Antipodean province.

Bathymetry

McCain [1966] reviews the Caprellidae reported from depths greater than 400 meters. He lists 18 species that occur at these depths and describes a new species, *Abyssicaprella galatheae,* which is the deepest reported caprellid, with a depth range of 3501–4004 meters. The following species may be added to McCain's list of caprellids occurring below 400 meters:

Aeginoides gaussi	37–1501 meters
Caprellinoides mayeri	22–1153 meters
Dodecas eltaninae	248–604 meters
Dodecasella georgiana	75–1475 meters
Mayerella magellanica	70–676 meters
Protella trilobata	646–845 meters
Pseudododecas bowmani	769 meters
Pseudoprotomima hedgpethi	71–403 meters
Pseudoprotomima hurleyi	788–1609 meters

The bathymetric range of *Aeginoides gaussi* increases directly with the distance from the antarctic continent, and it is found at greater depths to the north. The relationship has not been found in the other species of caprellids from the antarctic-subantarctic region.

Acknowledgments. We are indebted to Dr. Gerald W. Thurman of the American Museum of Natural History (AMNH) for specimens collected by the R/V *Vema* and to the Smithsonian Oceanographic Sorting Center for specimens collected by the U. S. Coast Guard Cutters *Burton Island, Glacier, Staten Island, Atka, Westwind,* and the USNS *Eltanin* of the U. S. Antarctic Research Program. We would also like to thank Dr. Patrick M. Arnaud of the Station Marine d'Endoume et Centre d'Océanographie, Marseille, for his shore collection of caprellids from the Adélie Coast. Dr. Arnaud made these collections from a small boat, using a Charcot rectangular dredge. Dr. D. E. Hurley of the New Zealand Oceanographic Institute (NZOI) kindly supplied material collected by the USCGC *Glacier* during the joint U. S.–New Zealand Balleny Islands Expedition and by the HMNZS *Rotoiti* at Macquarie Island. Drs. David L. Pawson and Donald F. Squires of the United States National Museum (USNM) contributed caprellids they collected while aboard the USCGC *Eastwind* in the Antarctic Peninsula area.

This work was supported by National Science Foundation grants GA-1217 (Office of Antarctic Programs) and GW-2911 (Research Participation for High School Teachers).

REFERENCES

Barnard, K. H.

1930 Crustacea. Part XI. Amphipoda. Br. Antarct. 'Terra Nova' Exped., Nat. Hist. Rep., Zool., *8* (4): 307–454, 63 figs.

1931 Diagnosis of new genera and species of amphipod Crustacea collected during the *Discovery* investigations, 1925–1927. Ann. Mag. Nat. Hist., ser. 10, 7 (40): 425–430.

1932 Amphipoda. 'Discovery' Rep., *5:* 1–326, figs. 1–174, 1 pl.

1940 Contributions to the crustacean fauna of South Africa. XII. Further additions to the Tanaidacea, Isopoda, and Amphipoda, together with keys for the identification of the hitherto recorded marine and freshwater species. Ann. S. Afr. Mus., *32* (5): 381–543, figs. 1–35.

Bate, C. Spence

1862 Catalogue of the specimens of amphipodous Crustacea in the collection of the British Museum. iv + 399 pp., 58 pls. British Museum, London.

Carus, J. V.

1885 Coelenterata, Echinodermata, Vermes, Arthropoda. *In* Prodromus faunae mediterraneae sive descriptio animalium maris Mediterranei incolarum quam comparata silva rerum quatenus innotuit adiectis locis et nominibus vargaribus eorumque auctoribus in commodum zoologorum. *1:* xi + 525 pp. E. Schweizerbart, Stuttgart.

Chevreux, E.

1913 Amphipodes. *In* Deuxième Expédition Antarctique Française (1908–1910) commandée par le Dr. Jean Charcot. Sciences Naturelles: Documents Scientifiques. Pp. 79–186, 62 figs. Masson, Paris.

Dana, J. D.
1852 On the classification of the Crustacea Choristopoda or Tetradecapoda. Am. J. Sci. Arts, ser. 2, *14:* 297–316.
1853 Crustacea. United States Explor. Exped., *14* (part 2) : 689–1618.

Chilton, C.
1909 The Crustacea of the subantarctic islands of New Zealand. *In* C. Chilton (ed.), The Subantarctic Islands of New Zealand. Phil. Inst. Canterbury, *2:* 601–671, figs. 1–19.
1913 Revision of the Amphipoda from South Georgia in the Hamburg Museum. Mitteil. Naturh. Mus. Hamb., *30:* 53–63.

Haswell, W. A.
1884 (1885) Revision of Australian Laemodipoda. Proc. Linn. Soc. N.S.W., *9:* 993–1000, pls. 48–49.

Huntsman, A. G.
1915 A new caprellid from the Bay of Fundy. Contrib. Can. Biol., sessional pap. 39b. Pp. 39–42, pls. 5–6.

Hutton, F. W.
1904 Index Faunae Novae Zealandiae. viii + 372 pp. Dulau, London.

Knox, G. A.
1960 Littoral ecology and biogeography of the southern oceans. Proc. R. Soc. London, ser. B, *152* (949) : 577–624, figs. 54–73.

Kusakin, O. G.
1967 K faune Isopoda i Tanaidacea shel"fovykh zon antarkticheskikh i subantarkticheskikh vod. *In* Issledovniya Fauny Morey, *4* (12) : 220–380, 64 figs. Akad. Nauk SSSR, Moscow.

Lamarck, J. B. P. A. de
1801 Système des animaux sans vertèbres, ou, tableau général des classes, des ordres et des genres de ces animaux . . . précède du discours d'ouverture du cours de zoologie, donné dans le muséum national d'histoire naturelle l'an 8 de la République. viii + 432 pp. Paris.

Leach, W. E.
1814 Crustaceology. *In* Edinburg Encyclopaedia; conducted by David Brewster . . . with the assistance of gentlemen eminent in science and literature. Edinburg. Vol. 7, part 2, pp. 385–437.

McCain, J. C.
1966 *Abyssicaprella galatheae,* a new genus and species of abyssal caprellid (Amphipoda: Caprellidae). Galathea Rep., *8:* 91–95, figs. 1–3.
1968 The Caprellidae (Crustacea: Amphipoda) of the Western North Atlantic. Bull. U.S. Natn. Mus., *278:* vi + 147, figs. 1–56.
1969 New Zealand Caprellidae (Crustacea: Amphipoda). N. Z. J. Mar. Freshwat. Res, *3* (2) : 286–295, 3 figs.

Mayer, P.
1882 Die Caprelliden des Golfes von Neapel und der angrenzenden Meersabschnitte. Eine Monographie. Fauna Flora Golf. Neapel, *6:* x + 201, figs. 1–39, pls. 1–10.
1890 Die Caprelliden des Golfes von Neapel und der angrenzenden Meeresabschnitte. Nachtrag zur Monographie derselben. Fauna Flora Golf. Neapel, *17:* vii + 157, pls. 1–7.
1903 Die Caprellidae der *Siboga*-Expedition. Siboga Exped., *34:* 1–160, pls. 1–10.

Nicolet, H.
1849 Crustáceos. *In* C. Gay (ed.), História física y política de Chile segun documentos adquiridos en ésta república durante doce años de residencia en ella y publicada bajo los suspicios del supremo gobierno. *3:* 115–318, pls. 1–4. Paris.

Pfeffer, G.
1888 Die Krebse von Süd-Georgien nach der Ausbeute der Deutschen Station 1882–83. Teil 2. Die Amphipoden. Jb. Wiss. Anst. Hamburg, *5:* 77–142, pls. 1–3.

Powell, A. W. B.
1965 Mollusca of antarctic and subantarctic seas. *In* Biogeography and ecology in Antarctica. Monogr. Biol., *15:* 333–380, figs. 1–6.

Reed, E. C.
1897 Catálogo de los Crustáceos Amfípodos y Lemodípodes de Chile. Revta Chil. Hist. Nat., *1:* 9–11.

Say, T.
1818 An account of the Crustacea of the United States. J. Acad. Nat. Sci. Philadelphia, *1:* 374–401.

Schellenberg, A.
1926 Die Caprelliden und *Neoxenodice caprellinoides* n. g., n. sp. der Deutschen Südpolar-Expedition, 1901–1903. Dt. Südpol.-Exped., *18* (zool., 10) : 465–473, figs. 1–3.
1931 Gammariden und Caprelliden des Magellangebietes, Südgeorgiens und der Westantarktis (in English). *In* Further Zool. Results Swed. Antarct. Exped., *2* (6) : 1–290, figs. 1–136, 1 pl.

Stebbing, T. R. R.
1876 Description of a new species of sessile-eyed crustacean, and other notices. Ann. Mag. Nat. Hist., ser. 4, *17:* 73–80, pls. 4–5.
1883 The *Challenger* Amphipoda. Ann. Mag. Nat. Hist., ser. 5, *11:* 203–207.
1888 Report on the Amphipoda collected by H.M.S. *Challenger* during the years 1873–76. Rep. *Challenger* (zool.), *29* (67) : xxiv + 1737, i–xii, pls. 1–210.
1895 Two new amphipods from the West Indies. Ann. Mag. Nat. Hist., ser. 6, *15:* 397–403, pls. 14–15.
1910a General catalogue of South African Crustacea (Part V. of S. A. Crustacea, for the marine investigations of South Africa). Ann. S. Afr. Mus., *6* (4) : 281–593, pls. 41–48.
1910b Scientific results of the trawling expedition of H.M.C.S. *Thetis.* V. Crustacea Amphipoda. Mem. Aust. Mus., *4* (12) : 567–658, pls. 47–60.
1914 A new Australian caprellid. Australian Zool., *1:* 27–28, pl. 3.

Stephensen, K.
1947 Tanaidacea, Isopoda, Amphipoda, and Pycnogonida. Scient. Results Norw. Antarct. Exped. Norske Vidensk. Akad. Oslo, *27:* 1–90, figs. 1–26.
1949 The Amphipoda of Tristan de Cunha. Results Norw. Scient. Exped. Tristan da Cunha, Norske Vidensk. Akad. Oslo, *19:* 1–61, figs. 1–23.

Thomson, G. M.
1879a New Zealand Crustacea, with descriptions of new species. Trans. Proc. N. Z. Inst., *11:* 230–248, pl. 10.
1879b Additions to the amphipodous Crustacea of New Zealand. Ann. Mag. Nat. Hist., ser. 5, *4:* 329–333, pl. 16.
1913 The natural history of Otago Harbour and the adjacent sea, together with a record of the researches carried on at the Portobello Marine Fish-Hatchery. Part I. Trans. Proc. N. Z. Inst. (new issue), *45:* 225–251, pl. 10.

Thomson, G. M., and T. Anderton
1921 History of the Portobello Marine Fish-Hatchery and Biological Station. Bull. Bd. Sci. Art N. Z., *2:* 1–131.

Thomson, G. M., and C. Chilton
1886 Critical list of the Crustacea Malacostraca of New Zealand. Trans. Proc. N. Z. Inst., *18:* 141–159.

POSTNAUPLIAR DEVELOPMENTAL STAGES OF THE COPEPOD CRUSTACEANS *CLAUSOCALANUS LATICEPS*, *C. BREVIPES*, AND *CTENOCALANUS CITER* (CALANOIDA: PSEUDOCALANIDAE)

Gayle A. Heron

Department of Oceanography, University of Washington, Seattle, Washington 98195

Thomas E. Bowman

Division of Crustacea, Smithsonian Institution, Washington, D. C. 20560

Abstract. A new species of *Ctenocalanus, C. citer,* is established. Its 6 copepodid stages and the 6 copepodid stages of *Clausocalanus laticeps* and *C. brevipes* are described from specimens collected in the Pacific sector of the antarctic and subantarctic regions by the USNS *Eltanin.*

INTRODUCTION

The zoologist working up a plankton collection is often confronted with large numbers of immature copepods in his samples. Realizing that copepodid stages have not been described in the vast majority of copepod species, he is forced to assign the immature forms to genera only or even to lump them as 'copepodid stages.' But life cycles and patterns of distribution are characteristics of species, not of higher taxonomic categories; and much information is lost when the species of copepods are not identified.

As a contribution to the solution of this vexing problem we present herein descriptions and illustrations of all copepodid stages of 3 of the smaller species of calanoids, all in the family Pseudocalanidae: *Clausocalanus laticeps* Farran (1929), *Clausocalanus brevipes* Frost and Fleminger (1968), and *Ctenocalanus citer* new species. The last is very similar to *Ctenocalanus vanus* Giesbrecht (1888), until now the only species in the genus. The detailed study of the copepodid stages of *C. citer,* however, revealed small but constant differences from *C. vanus.*

MATERIALS AND METHODS

Copepodid stages of the 3 species were selected from samples collected by the USNS *Eltanin* in the Pacific sector of the antarctic and subantarctic regions under the auspices of the U.S. Antarctic Research Program during cruises 15 (October–November 1964), 17 (March 1965), and 18 (June 1965). Because of the small size of the species studied (body length as little as 0.35 mm in copepodid I), we used samples collected with a half-meter 'microplankton net' having a mesh aperture of 76 μ. The net was towed vertically (sample 737) or obliquely (the remaining samples) from depth to the surface. Data for the 13 samples used are given in Table 1.

About 300 specimens were dissected and mounted. Because of the generally poor preservation of the samples it was not usually possible to obtain satisfactory mounts of all appendages from a single specimen. In the legend for each figure we have indicated the body length of the specimen used; the letter refers to the scale (in millimeters) at which the figure was drawn. The length was measured from the anterior margin of the rostrum to the posterior margin of the caudal rami. In the descriptions we give the average length followed by the range and number of specimens measured, both in parentheses. Where an appendage of a particular copepodid stage appeared identical in all 3 species, the appendage of only one species is illustrated and described. Illustrations of all legs are in posterior view, unless noted otherwise. Table 3 summarizes the arrangement and numbers of setae on the mouth parts.

All of the illustrations are based on specimens collected from the antarctic area except Figure 2, which is based on a specimen collected off Daytona Beach, Florida, by the U.S. Fish and Wildlife Service.

Ctenocalanus citer VI

1. Female, outer spine of exopod 2 and proximal outer spines of exopod 3 of fourth leg, 1.19 mm.

Ctenocalanus vanus VI

2. Female, outer spine of exopod 2 and proximal outer spines of exopod 3 of fourth leg, 1.18 mm (Fish and Wildlife Service, 29° N, 80°32.7′W, SE off Daytona Beach, Florida, April 24, 1953, 9 fm, USNM 203789).

Although outside of the aims of the study, note was taken of the relative abundance of the different copepodid stages in the samples examined (Table 2).

DIAGNOSIS OF THE NEW *CTENOCALANUS*

Ctenocalanus citer new species

Figs. 1, 16–18, 31–36, 54–58, 71–77, 94–99, 130–150

Diagnosis. Closely resembling *C. vanus* but differing in several respects: (1) First antenna shorter, reaching posterior margin of caudal ramus (first antenna of *C. vanus* exceeds caudal ramus by several segments). (2) Outer spines of exopod of fourth leg directed distally (Figure 1), with axis of spine at only a slight angle to axis of exopod (spines of *C. vanus* robust, expanded at base [Figure 2], with greater angle between axis of spine and axis of exopod). (3) Male with 1-segmented right fifth leg (male of *C. vanus* without right fifth leg).

Detailed descriptions of adults (copepodid VI) and younger stages (copepodids I–V) are given elsewhere in this paper.

Types. Holotype, female, USNM 123670; allotype, male, USNM 123671; paratypes, 15 females, 2 males, USNM 123672; all from sample 853.

Etymology. The specific name, from the Latin citer ('lying near'), refers to the orientation of the outer spines on the exopod of the fourth leg.

Distribution. According to published reports, *Ctenocalanus vanus* has a nearly worldwide distribution. The lack of detail in most of these reports makes it impossible to determine which species, *C. vanus* or *C. citer,* the authors actually had. Statements in 2 reports, however, suggest that the authors had *C. citer.* Björnberg [1963] noted 2 different forms of *C. vanus* along the coast of Brazil, 'one with longer and one with shorter antennae.' It seems likely that these forms were *C. vanus* and *C. citer,* respectively. Unterüberbacher [1964] observed male specimens of *Ctenocalanus* with a 1-segmented right fifth leg along the coast of South-West Africa; these may have been *C. citer.*

DESCRIPTIONS OF DEVELOPMENTAL STAGES

The admirable monograph of Frost and Fleminger [1968] makes it possible now to identify with confidence the species of *Clausocalanus.* Heretofore attempts to identify members of this genus have been fraught with uncertainty and frustration. Frost and Fleminger distinguished the species of *Clausocalanus* by characters that they found more useful than the mouth parts, stating that the latter are 'essentially similar among all species.' They provide illustrations of the mouth parts of only the type species, *C. mastigophorus* (Claus). The mouth parts of *C. laticeps* and *C. brevipes* agree with the illustrations of *C. mastigophorus* by Frost and Fleminger in most respects. We have found only 2 minor differences, 1 of which is an apparent, not a real, difference.

TABLE 1. Station Data on USNS *Eltanin* Samples Examined for Study

Sample Number	Cruise	Location	Date	Local Time	Sample Depth, m	Bottom Depth, m	Surface Temp., °C
737	15	50°04'S, 94°50'W	Oct. 13, 1964	1525–1545	0–250	4816	4.02
771	15	59°07'S, 105°00'W	Oct. 24, 1964	0508–0550	0–485	4627	4.16
787	15	57°51'S, 108°38'W	Oct. 28, 1964	1553–1613	0–250	4687	4.40
809	15	56°00'S, 134°28'W	Nov. 9, 1964	0057–0140	0–500	3182	3.12
845	15	56°02'S, 143°18'W	Nov. 15, 1964	0708–0730	0–250	878	2.60
851	15	56°02'S, 144°40'W	Nov. 16, 1964	0455–0531	0–500	2608	1.35
853	15	55°17'S, 145°02'W	Nov. 17, 1964	0323–0345	0–250	3100	2.60
866	15	56°05'S, 149°46'W	Nov. 19, 1964	1927–1950	0–250	3438	3.05
940	17	56°00'S, 135°03'W	March 26, 1965	0725–0745	0–250	3219	5.10
947	17	56°56'S, 135°02'W	March 27, 1965	0534–0556	0–240	3292	5.35
954	17	59°00'S, 135°32'W	March 28, 1965	1300–1320	0–250	3840	2.42
955	17	61°07'S, 134°20'W	March 30, 1965	0141–0207	0–250	4435	1.80
1137	18	54°46'S, 99°11'W	June 5, 1965	0727–0753	0–220	4389	5.78

The exopod of the second antenna of *C. mastigophorus* is shown in plate 5, figure b, of Frost and Fleminger as having 4 short segments between the second and distal segments. Our specimens of *C. laticeps* and *C. brevipes* have 3 short segments and an incomplete suture near the distal end of the second segment. Frost and Fleminger show 2 setae on the first short segment of *C. mastigophorus*, whereas we can find but 1 in the corresponding position in the other species. Giesbrecht's figure of *C. arcuicornis* [1892, pl. 10, fig. 2] agrees with our specimens.

Frost and Fleminger's plate 5, figure c, shows an 8-segmented maxilliped, whereas Giesbrecht [1892, pl. 10, fig. 10] shows only 7 segments. The extra segment, bearing 2 setae, is distal to the second long segment and is the one that Gurney [1931] considered the first endopod segment. Gurney states that in different calanoids this segment is sometimes separate from and sometimes united with the long segment. In all specimens of *Clausocalanus*, including *C. mastigophorus*, and in several other genera of calanoids that we have examined, a line that could be interpreted as a suture is present on the medial surface at the distal end of the second long segment. There is only a faint, incomplete line or no line on the lateral surface. As a result the maxilliped appears to have 8 segments when viewed medially (as by Frost and Fleminger) and 7 segments when viewed laterally (as by Giesbrecht). In this paper we consider the maxilliped to be 7-segmented, and we suggest that the apparent

TABLE 2. Dominance of Stages by Sample (Specimens Dissected for Study)

	Clausocalanus laticeps									***Clausocalanus brevipes***									***Ctenocalanus citer***								
	I	II	III	IV ♀	IV ♂	V ♀	V ♂	VI ♀	VI ♂	I	II	III	IV ♀	IV ♂	V ♀	V ♂	VI ♀	VI ♂	I	II	III	IV ♀	IV ♂	V ♀	V ♂	VI ♀	VI ♂
737								+	+																		
771									+						+	+											
787																	+	+								+	
809			+									+														+	+
845		+			+	+	+	+	+										+							+	+
851																									+	+	+
853	+		+		+		+		+						+	+			+	+	+	+	+	+	+		+
866										+									+							+	+
940	+									+	+								+	+		+				+	
947	+	+	+	+						+	+	+	+	+						+	+	+	+			+	
954			+					+	+										+	+					+	+	+
955									+								+			+						+	+
1137																	+		+	+						+	

TABLE 3. Armatures of Second Antenna to Maxilliped, Stages I–VI, of ***Clausocalanus laticeps*** (a), ***C. brevipes*** (b), and ***Ctenocalanus citer*** (c)

	Stage I			Stage II			Stage III			Stage IV						Stage V						Stage VI					
Species	a	b	c	a	b	c	a	b	c	a♀	a♂	b♀	b♂	c♀	c♂	a♀	a♂	b♀	b♂	c♀	c♂	a♀	a♂	b♀	b♂	c♀	c♂
Second antenna																											
Coxa	1	1	1	1	1	1	1	1	1	1	1	1	1	1	1	1	1	1	1	1	1	1		1		1	1
Basis	2	2	2	2	2	2	2	2	2	2	2	2	2	2	2	2	2	2	2	2	2	2	1	2	1	2	2
Endo. 1	2	2	2	2	2	2	2	2	2	2	2	2	2	2	2	2	2	2	2	2	2	2		2		2	
Endo. 2	4/6	4/6	4/6	5/6	5/6	5/6	6/6	6/6	6/6	7/7	7/7	7/7	7/7	7/7	7/7	8/7	8/7	8/7	8/7	8/7	8/7	9/7	7/6	9/7	7/6	9/7	7/7
Exo. 1	2	2	2	2	2	2	2	2	2	2	2	2	2	2	2	2	2	2	2	2	2	2		2		2	2
Exo. 2	3	3	3	3	3	3	3	3	3	3	3	3	3	3	3	3	3	3	3	3	3	3	1	3	1	3	3
Exo. 3	1	1	1	1	1	1	1	1	1	1	1	1	1	1	1	1	1	1	1	1	1	1	1	1	1	1	1
Exo. 4	1	1	1	1	1	1	1	1	1	1	1	1	1	1	1	1	1	1	1	1	1	1	1	1	1	1	1
Exo. 5	1	1	1	1	1	1	1	1	1	1	1	1	1	1	1	1	1	1	1	1	1	1	1	1	1	1	1
Exo. 6	4	4	4	4	4	4	4	4	4	4	4	4	4	4	4	4	4	4	4	4	4	4	4	4	4	4	4
Mandible																											
Basis	4	4	4	4	4	4	4	4	4	4	4	4	4	4	4	4	4	4	4	4	4	4		4		4	1
Endo. 1	4	4	4	4	4	4	4	4	4	4	4	4	4	4	4	4	4	4	4	4	4	4	1	4	1	4	1
Endo. 2	6	6	6	7	7	7	8	8	8	9	9	9	9	9	9	10	10	10	10	10	10	11	9	11	9	11	9
Exo. 1	1	1	1	1	1	1	1	1	1	1	1	1	1	1	1	1	1	1	1	1	1	1	1	1	1	1	1
Exo. 2	1	1	1	1	1	1	1	1	1	1	1	1	1	1	1	1	1	1	1	1	1	1	1	1	1	1	1
Exo. 3	1	1	1	1	1	1	1	1	1	1	1	1	1	1	1	1	1	1	1	1	1	1	1	1	1	1	1
Exo. 4	1	1	1	1	1	1	1	1	1	1	1	1	1	1	1	1	1	1	1	1	1	1	1	1	1	1	1
Exo. 5	2	2	2	2	2	2	2	2	2	2	2	2	2	2	2	2	2	2	2	2	2	2	2	2	2	2	2
First maxilla																											
Gnath.	9	9	9	10	10	10	11	11	11	14	14	14	14	14	14	14	14	14	14	14	14	14		14		14	
Prox. lobe	4	4	4	4	4	4	5	5	5	5	5	5	5	5	5	5	5	5	5	5	5	5		5		5	
Dist. lobe	4	4	4	4	4	4	4	4	4	4	4	4	4	4	4	4	4	4	4	4	4	4	(2)*	4	(2)	4	(3)
Epipod	4	4	4	6	6	6	8	8	8	9	9	9	9	9	9	9	9	9	9	9	9	9	6	9	6	9	6
Exo.	7	7	7	7	7	7	8	8	8	9	9	9	9	9	9	10	10	10	10	10	10	11	11	11	11	11	11
Basis	3	3	3	3	3	3	4	4	4	4	4	4	4	4	4	5	5	5	5	5	5	5	(4)	5	(4)	5	(4)
Endo. 1																											
(mesial)	2	2	2	2	2	2	3	3	3	3	3	3	3	3	3	4	4	4	4	4	4	4	(3)	4	(3)	4	(2)
(distal)	2	2	2	3	3	3	3	3	3	4	4	4	4	4	4	4	4	4	4	4	4	5	(3)	5	(3)	5	(4)
Endo. 2	5	5	5	5	5	5	6	6	6	6	6	6	6	6	6	7	7	7	7	7	7	7	6	7	6	7	6
Second maxilla																											
Endite 1	3	3	3	5	5	5	5	5	5	5	5	5	5	5	5	5	5	5	5	5	5	5	4	5	4	5	4
2	3	3	3	3	3	3	3	3	3	3	3	3	3	3	3	3	3	3	3	3	3	3	(2)	3	(2)	3	(2)
3	3	3	3	3	3	3	3	3	3	3	3	3	3	3	3	3	3	3	3	3	3	3	(3)	3	(3)	3	(3)
4	3	3	3	3	3	3	3	3	3	3	3	3	3	3	3	3	3	3	3	3	3	3	(3)	3	(3)	3	(3)
Endo. 1	4	4	4	4	4	4	4	4	4	4	4	4	4	4	4	4	4	4	4	4	4	4	(3)	4	(3)	4	(3)
Endo. 2	1	1	1	1	1	1	1	1	1	1	1	1	1	1	1	1	1	1	1	1	1	1	1	1	1	1	1
Endo. 3	1	1	1	1	1	1	1	1	1	1	1	1	1	1	1	1	1	1	1	1	1	1	1	1	1	1	1
Endo. 4	1	1	1	1	1	1	1	1	1	1	1	1	1	1	1	1	1	1	1	1	1	1	1	1	1	1	1
Endo. 5	3	3	3	3	3	3	3	3	3	3	3	3	3	3	3	3	3	3	3	3	3	3	3	3	3	3	3
Maxilliped																											
Coxa	6	6	6	7	7	7	9	9	9	9	9	9	9	9	9	9	9	9	9	9	9	9	1	9	1	9	1
Basis	3	3	3	4	4	4	5	5	5	5	5	5	5	5	5	5	5	5	5	5	5	5	3	5	3	5	3
Endo. 1	1	1	1	1	1	1	1	1	1	2	2	2	2	2	2	3	3	3	3	3	3	4	4	4	4	4	4
Endo. 2							1	1	1	2	2	2	2	2	2	3	3	3	3	3	3	4	4	4	4	4	4
Endo. 3										1	1	1	1	1	1	2	2	2	2	2	2	3	2	3	2	3	3
Endo. 4				1	1	1	2	2	2	2	2	2	2	2	2	3	3	3	3	3	3	4	2	4	2	4	3
Endo. 5	4	4	4	4	4	4	4	4	4	4	4	4	4	4	4	4	4	4	4	4	4	4	3	4	3	4	4

* Numbers in parentheses indicate a variable count.

extra segment results from the pattern of sclerotization at the major flexure of the appendage.

Copepodid I

Clausocalanus laticeps

Figs. 3–7

Body 0.42 mm long (0.38–0.45 mm, based on 13 specimens). Prosome slender, 2-segmented, with incipient suture on distal segment, and distinctly separated from 2-segmented urosome in lateral view (Figure 3). Rostrum delicate and difficult to discern. Caudal rami usually extended parallel to one another in dorsal view (Figure 4).

First antenna 13-segmented, with most proximal segments faintly defined. Second antenna with segmentation and setation resembling those of that appendage of *C. brevipes* (Figure 10). Width of terminal segment of exopod measures less than half the length (Figure 5). Mandible with segmentation, setation, and shape of masticatory blade resembling those of that appendage of *C. brevipes* (Figure 11). First maxilla, second maxilla, and maxilliped with segmentation and setation resembling those of those appendages of *C. brevipes* (Figures 12 ,13, 14).

First leg (Figure 6), endopod 1-segmented, with 5 setae. Exopod 1-segmented, with incipient suture, 3 short outer setae, 1 terminal seta, and 3 inner setae. Second leg (Figure 7), endopod 1-segmented, with 6 setae. Exopod 1-segmented, with 3 outer spines, 1 long terminal spine having a toothed hyaline flange on outer margin, and 3 inner setae.

Clausocalanus brevipes

Figs. 8–15

Body 0.42 mm long (0.40–0.46 mm, based on 9 specimens). Prosome 2-segmented, with incipient suture on distal segment, distinctly separated from 2-segmented urosome in lateral view (Figure 8). Rostrum delicate and difficult to discern. Caudal rami extended at a slightly acute angle to one another (Figure 9).

First antenna 13-segmented, with proximal segments faintly defined. Second antenna (Figure 10), coxa with 1 seta on distal inner margin. Basis with 2 setae. Endopod 2-segmented, with 2 setae on first segment, 4 setae on inner lobe, and 6 setae on outer lobe. Exopod with 6 distinct segments; rudimentary articulative line across distal third of second segment. Segmentation, including obscure line, similar to that of stage VI. First segment with 2 setae on inner margin; second segment with 3 setae; third, fourth, and fifth segments each with 1 seta, and sixth segment with 1 lateral and 3 terminal setae. Width of terminal segment of exopod approximately two-thirds the length. Mandible (Figure 11), masticatory blade armed with 1 seta and 7 teeth, bearing fragile hyaline cusps. Segmentation of this appendage similar to that of stage VI. Basis with 4 setae. Endopod 2-segmented; first segment with 4 setae, second segment with 6 terminal setae and several setules on posterior surface. Exopod 5-segmented; each of first 4 segments with 1 seta; fifth segment with 2 terminal setae. First maxilla (Figure 12) similar to that of stage VI female in general shape. Gnathobase bears 9 spines. Proximal and distal small inner lobes each with 4 setae. Epipod with 4 long setae, exopod with 7 long setae, and basipod with 3 setae. Endopod 2-segmented, with faint suggestion of suture on inner margin of first segment; 2 setae on each lobe of first segment; distal segment with 5 terminal setae, outermost seta with row of setules on outer margin. Second maxilla (Figure 13): first 4 endites each with 3 setae. Endopod 5-segmented; first segment produced anteriorly into an endite with 4 setae; 3 succeeding segments each with 1 seta; terminal segment with 3 setae. Maxilliped (Figure 14), 4-segmented; first segment with 1 short proximal seta, 3 median setae, and 2 setae on distal prominence; second segment with 2 median setae and 1 distally; third segment with 1 seta; terminal segment with 4 setae.

First leg with segmentation and setation resembling those of that appendage of *C. laticeps* (Figure 6). Second leg (Figure 15), endopod 1-segmented with 6 setae. Exopod 1-segmented, with 3 outer spines, 1 long terminal spine having a toothed hyaline flange on outer margin, and 3 inner setae.

Ctenocalanus citer new species

Figs. 16–18

Body 0.41 mm long (0.35–0.45 mm, based on 20 specimens). Prosome slender and 2-segmented, with incipient suture on distal segment, distinctly separated from 2-segmented urosome in lateral view (Figure 16). Rostral filaments delicate and difficult to discern.

First antenna 13-segmented, with most proximal segments faintly defined; length of terminal segment approximately that of penultimate segment. Second antenna, first maxilla, second maxilla, and maxilliped with segmentation and setation resembling that of those appendages of *Clausocalanus brevipes* (Figures 10, 12, 13, 14). Mandible with segmentation, setation,

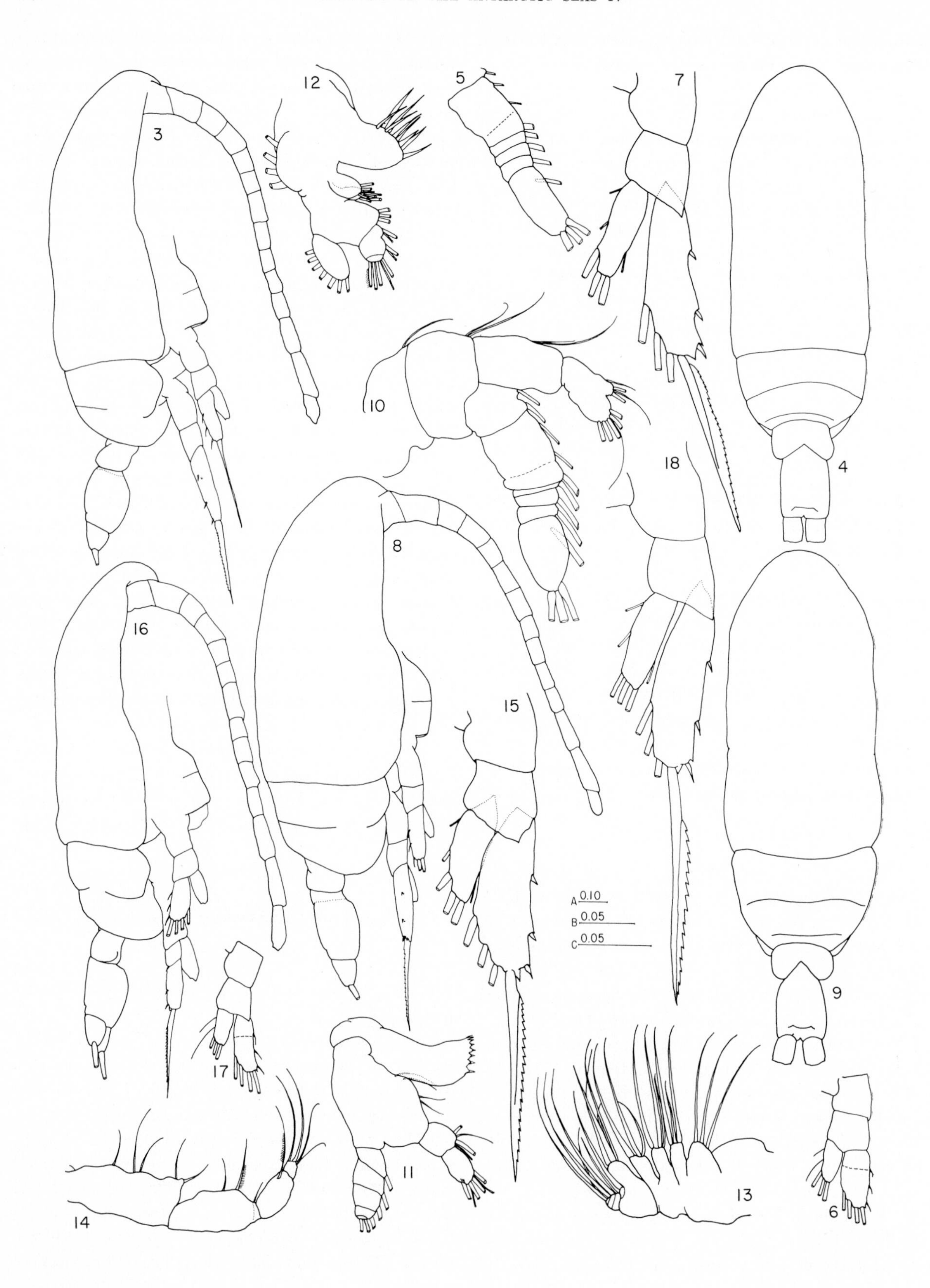
3
4
5
6
7
8
9
10
11
12
13
14
15
16
17
18
A 0.10
B 0.05
C 0.05

and shape of masticatory blade resembling those of that appendage of *C. brevipes* (Figure 11).

First leg (Figure 17), endopod 1-segmented, with 5 setae. Exopod 1-segmented with incipient suture, 3 short outer setae, 1 terminal seta, and 3 inner setae. Second leg (Figure 18), endopod 1-segmented, with 6 setae. Exopod 1-segmented, with 3 outer spines, 1 long terminal spine having a toothed hyaline flange on outer margin, and 3 inner setae.

Comparison of the Three Species, Stage I

The last 2 segments of the first antenna of *Ctenocalanus citer* are approximately equal in length; this was the characteristic most readily observable for distinguishing this species from the 2 species of *Clausocalanus*, when the first antenna was not damaged.

The segmentation and setation of the second antenna of *C. brevipes* are similar to those of the second antenna of *C. laticeps*, but the difference in the length–width ratio proved a reliable characteristic for distinguishing between the 2 species in stages I–V.

The caudal rami of *C. laticeps* usually extend parallel to one another in dorsal view, whereas the caudal rami of *C. brevipes* extend at a slightly acute angle to one another. This characteristic is discernible only in stages I and II.

Copepodid II

Clausocalanus laticeps

Figs. 19–25

Body 0.56 mm long (0.51–0.61 mm, based on 12 specimens). Prosome slender, 3-segmented with incipient suture on posterior segment; urosome 2-segmented (Figure 19). Bifurcate rostrum present. Caudal rami usually extend parallel to one another in dorsal view (Figure 20).

First antenna 18-segmented, with most proximal segments faintly defined. Second antenna (Figure 21) differs from that of stage I by addition of 1 seta on inner lobe of second segment of endopod. Width of terminal segment of exopod approximately half the length. Mandible with segmentation, setation, and shape of masticatory blade resembling those of that appendage of *Ctenocalanus citer* (Figure 32). First maxilla (Figure 22) differing from that of stage I by additions of 1 spine on gnathobase, 2 long setae on epipod, and 1 seta on first segment of endopod. Second maxilla with segmentation and setation resembling those of second maxilla of *C. citer* (Figure 33). Maxilliped (Figure 23), 5-segmented; differing from that of stage I by additions of 1 seta each on proximal segment, second segment, and additional segment (which corresponds to fourth segment of endopod of stage VI). Slight pattern of sclerotization surrounding bases of 2 distal setae on inner side of second segment, becoming more conspicuous in succeeding stages.

First leg with segmentation and setation resembling those of first leg of *C. citer* (Figure 33). Second leg (Figure 24) coxa with 1 inner seta. Endopod 1-segmented, with 6 setae. Exopod 2-segmented; first segment with 1 outer spine; second segment with 2 outer spines, 1 long terminal spine having a toothed hyaline flange on outer margin, and 4 inner setae. Third leg (Figure 25), endopod 1-segmented, with 6 setae. Exopod 1-segmented, with 3 outer spines, 1 long terminal spine having a toothed hyaline flange on outer margin, and 3 inner setae.

Clausocalanus brevipes

Figs. 26–30

Body 0.58 mm long (0.51–0.65 mm, based on 13 specimens). Prosome 3-segmented, with incipient suture on posterior segment; urosome 2-segmented (Figure 26). Bifurcate rostrum present. Caudal rami usually extended at a slightly acute angle to one another in dorsal view (Figure 27).

First antenna 18-segmented, with most proximal segments faintly defined. Second antenna with seg-

---

***Clausocalanus laticeps* I**
3. Lateral view, 0.40 mm, B.
4. Dorsal view, 0.41 mm, B.
5. Terminal exopod segment of second antenna, 0.43 mm, A.
6. First leg, 0.43 mm, C.
7. Second leg, 0.41 mm, A.

***Clausocalanus brevipes* I**
8. Lateral view, 0.42 mm, B.
9. Dorsal view, 0.42 mm, B.
10. Right second antenna, 0.44 mm, A.
11. Right mandible, 0.44 mm, A.
12. Left first maxilla, 0.44 mm, A.
13. Right second maxilla, 0.44 mm, A.
14. Left maxilliped, 0.42 mm, A.
15. Second leg, 0.44 mm, A.

***Ctenocalanus citer* I**
16. Lateral view, 0.40 mm, B.
17. First leg, 0.41 mm, C.
18. Second leg, 0.40 mm, A.

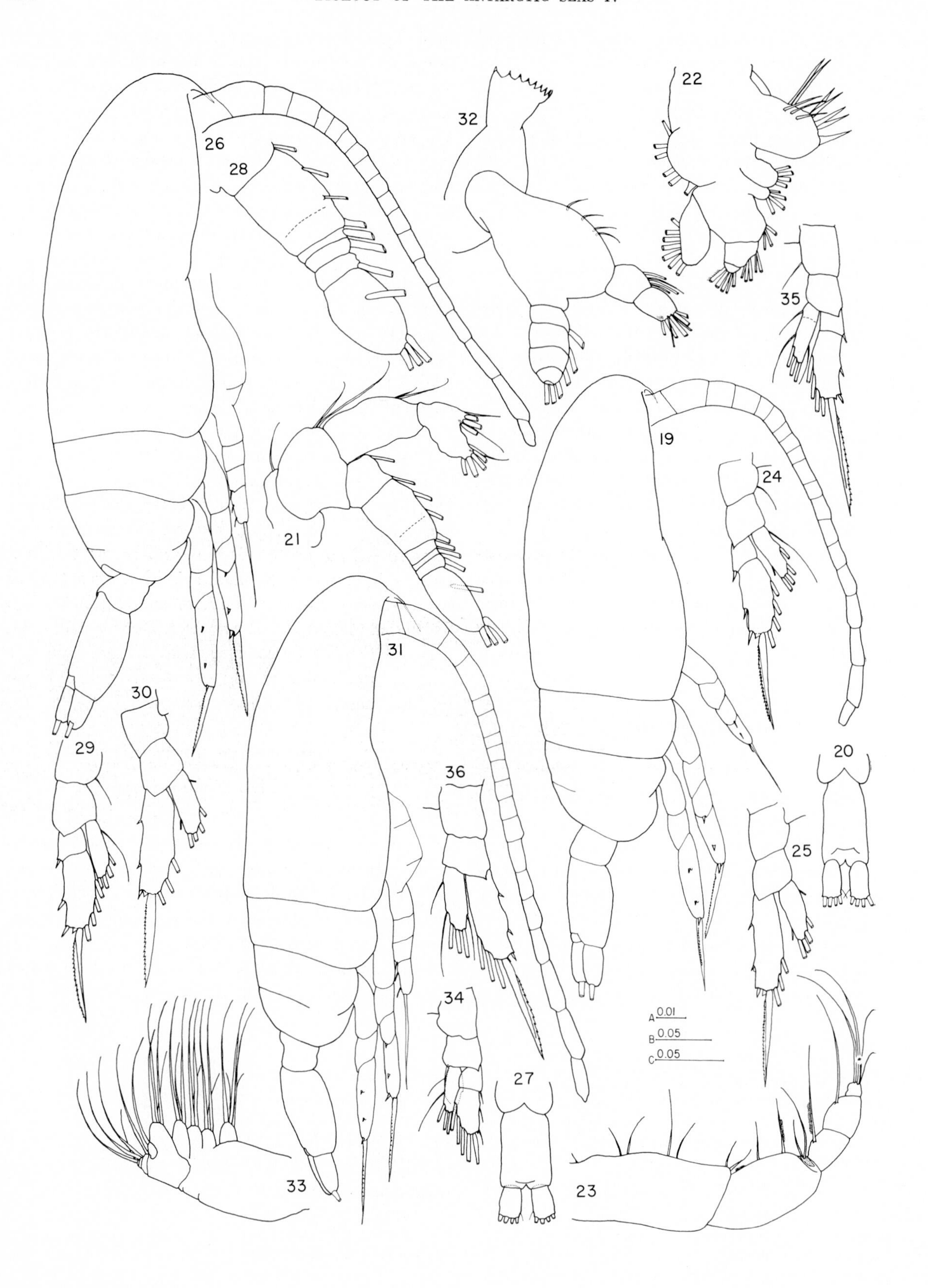
26
28
32
22
35
19
24
21
31
30
29
36
20
25
34
A 0.01
B 0.05
C 0.05
27
33
23

mentation and setation resembling those of that appendage of *C. laticeps* (Figure 21). Width of terminal segment of exopod approximately three-quarters the length (Figure 28). Mandible with segmentation, setation, and shape of masticatory blade resembling those of that appendage of *Ctenocalanus citer* (Figure 32). First maxilla and maxilliped with segmentation and setation resembling those of those appendages of *Clausocalanus laticeps* (Figures 22, 23). Second maxilla with segmentation and setation resembling those of that appendage of *Ctenocalanus citer* (Figure 33).

First leg with segmentation and setation resembling those of that appendage of *Ctenocalanus citer* (Figure 33). Second leg (Figure 29), coxa with 1 inner seta. Endopod 1-segmented, with 6 setae. Exopod 2-segmented; first segment with 1 outer spine; second segment with 2 outer spines, 1 long terminal spine having a toothed hyaline flange on outer margin, and 4 inner setae. Third leg (Figure 30), endopod 1-segmented, with 6 setae. Exopod 1-segmented, with 3 outer spines, 1 long terminal spine having a toothed hyaline flange on outer margin, and 3 inner setae.

Ctenocalanus citer

Figs. 31–36

Body 0.55 mm long (0.50–0.63 mm, based on 18 specimens). Prosome 3-segmented with incipient suture on posterior segment; urosome 2-segmented (Figure 31). Rostral filaments long, filiform.

First antenna 18-segmented, with most proximal segments faintly defined. Second antenna with segmentation and setation resembling that appendage of *Clausocalanus laticeps* (Figure 21). Mandible (Figure 32), masticatory blade armed with 1 seta and 8 teeth, each tooth with fragile hyaline cap. Setation of palp differing from that of stage I by addition of 1 seta on second segment of endopod. First maxilla with segmentation and setation resembling that appendage of *C. laticeps* (Figure 22). Second maxilla (Figure 33) with segmentation and setation similar to that of stage VI. Proximal endite with 5 setae; succeeding 3 endites each with 3 setae. Endopod 5-segmented; first segment with 4 setae; second to fourth segments, each with 1 seta; terminal segment with 3 setae. Maxilliped with segmentation and setation resembling that appendage of *C. laticeps* (Figure 23).

First leg (Figure 34), endopod 1-segmented, with 5 setae. Exopod 2-segmented; first segment with 1 short outer seta; second segment with 2 short outer setae, 1 terminal seta, and 4 inner setae. Second leg (Figure 35), coxa with 1 inner seta. Endopod 1-segmented, with incipient suture and 6 setae. Exopod 2-segmented; first segment with 1 outer spine, second segment with 2 outer spines, 1 long terminal spine having a toothed hyaline flange on outer margin, and 4 inner setae. Third leg (Figure 36), endopod 1-segmented, with 6 setae. Exopod 1-segmented, with 3 outer spines, 1 long terminal spine having a toothed hyaline flange on outer margin, and 3 inner setae.

Comparison of the Three Species, Stage II

The prosome of *Clausocalanus brevipes* appears more robust than those of *C. laticeps* and *Ctenocalanus citer* when compared dorsally or laterally.

Characteristics that serve to distinguish between *Clausocalanus laticeps* and *C. brevipes*, usually apparent at stage II and in some cases more pronounced in succeeding stages, are the following:

The first antenna of *C. brevipes* is slightly longer.

In stages II and III, the length of the terminal flanged spine of the exopod of the second leg of *C. laticeps* is approximately the same or greater than the length of the 2 exopod segments combined; in *C. brevipes* the spine is shorter than the 2 exopod segments combined.

***Clausocalanus laticeps* II**
19. Lateral view, 0.58 mm, B.
20. Urosome, dorsal view, 0.51 mm, B.
21. Right second antenna, 0.53 mm, A.
22. Left first maxilla, 0.55 mm, A.
23. Left maxilliped, 0.60 mm, A.
24. Second leg, 0.51 mm, C.
25. Third leg, 0.51 mm, C.

***Clausocalanus brevipes* II**
26. Lateral view, 0.58 mm, B.
27. Urosome, dorsal view, 0.58 mm, B.
28. Terminal exopod segment of second antenna, 0.61 mm, A.
29. Second leg, 0.61 mm, C.
30. Third leg, 0.62 mm, C.

***Ctenocalanus citer* II**
31. Lateral view, 0.53 mm, B.
32. Left mandible, 0.60 mm, A.
33. Left second maxilla, 0.50 mm, A.
34. First leg, 0.53 mm, C.
35. Second leg, 0.53 mm, C.
36. Third leg, 0.53 mm, C.

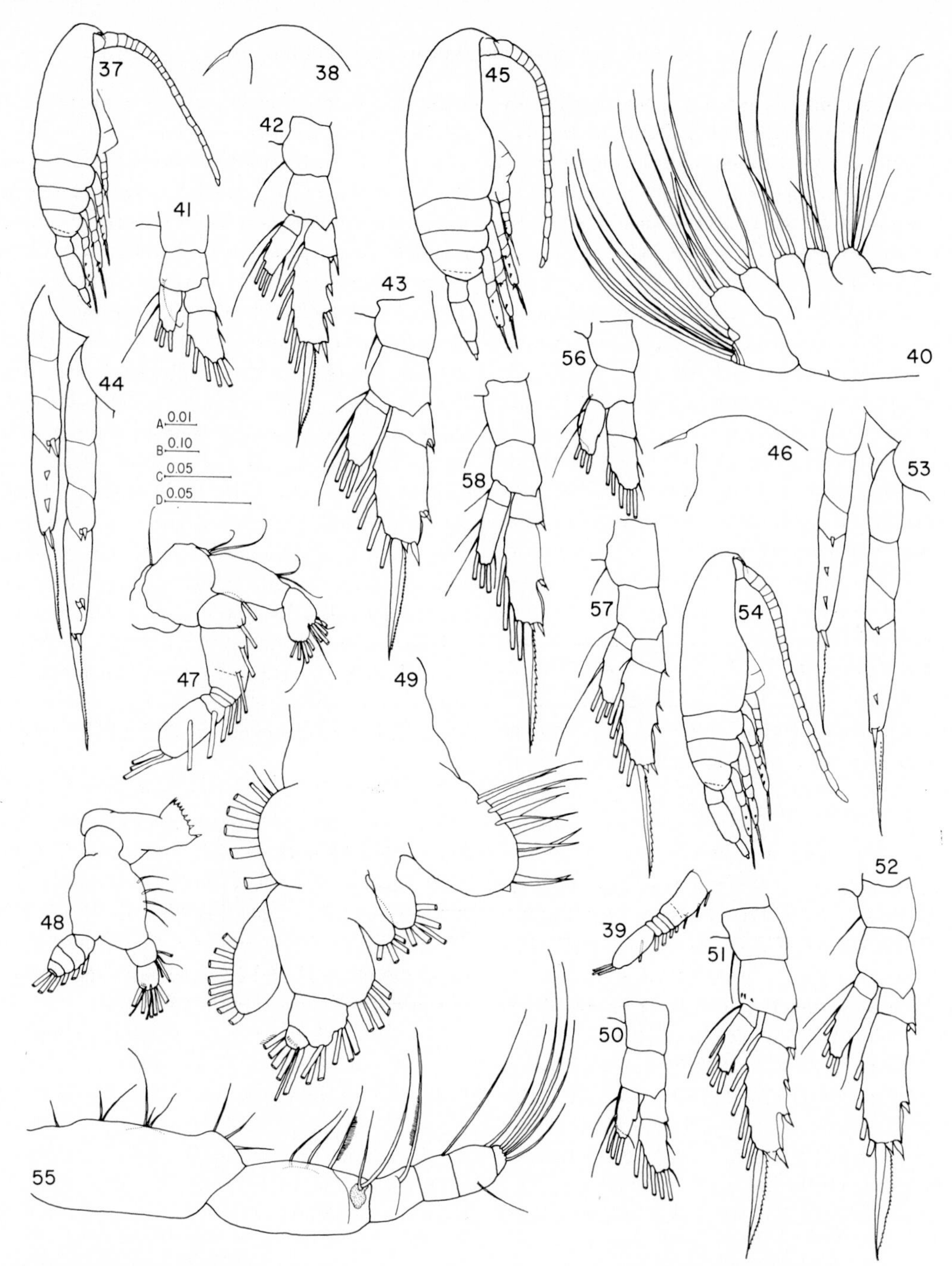

***Clausocalanus laticeps* III**

37. Lateral view, 0.72 mm, B.
38. Rostral region, lateral view, 0.72 mm, D.
39. Terminal exopod segment of second antenna, 0.71 mm, C.
40. Right second maxilla, 0.62 mm, A.
41. First leg, 0.72 mm, C.
42. Second leg, 0.71 mm, C.
43. Third leg, 0.73 mm, C.
44. Legs 2, 3, lateral view, 0.72 mm, C.

***Clausocalanus brevipes* III**

45. Lateral view, 0.80 mm, B.
46. Rostral region, lateral view, 0.80 mm, D.
47. Left second antenna, 0.75 mm, C.
48. Left mandible, 0.81 mm, C.
49. Left first maxilla, 0.74 mm, A.
50. First leg, 0.78 mm, C.
51. Second leg, 0.72 mm, C.
52. Third leg, 0.72 mm, C.
53. Legs 2, 3, lateral view, 0.72 mm, C.

***Ctenocalanus citer* III**

54. Lateral view, 0.80 mm, B.
55. Left maxilliped, 0.67 mm, A.
56. First leg, 0.79 mm, C.
57. Second leg, 0.79 mm, C.
58. Third leg, 0.79 mm, C.

In *C. laticeps* the outer spine of the first segment of the exopod of the second leg is slightly longer than the medial outer spine of the second segment; in *C. brevipes* the outer spine of the first segment is slightly shorter or the same length as the medial outer spine of the second segment.

In *C. brevipes* the hyaline flanges of the terminal spines of the exopods are slightly more expanded, causing the teeth to appear farther apart than in *C. laticeps.*

In *C. laticeps* the medial teeth of the flanged terminal spines are smaller in size than the distal teeth, and all are directed distally; in *C. brevipes* all the teeth are approximately constant in size, and the more medial teeth are directed outward, whereas the posterior teeth are directed distally.

Copepodid III

Clausocalanus laticeps

Figs. 37–44

Body 0.71 mm long (0.62–0.78 mm, based on 13 specimens). Prosome slender, 4-segmented, with incipient suture on posterior segment; urosome 2-segmented (Figure 37). Bifurcate rostrum usually directed posteriorly (Figure 38).

First antenna 21-segmented, extending approximately to posterior margin of second prosomal segment. Second antenna with segmentation and setation resembling those of that appendage of *C. brevipes* (Figure 47). Width of terminal segment of exopod approximately one-third the length (Figure 39). Mandible with segmentation, setation, and shape of masticatory blade resembling those of that appendage of *C. brevipes* (Figure 48). First maxilla with segmentation and setation resembling those of that appendage of *C. brevipes* (Figure 49). Second maxilla (Figure 40) with segmentation and setation similar throughout stages II–VI. Maxilliped with segmentation and setation resembling those of that appendage of *Ctenocalanus citer* (Figure 55).

First leg (Figure 41), basis with 1 distal seta on anterior inner corner. Endopod 1-segmented, with 5 setae, row of setules on convex process of inner margin. Exopod 2-segmented; first segment with 1 short outer seta; second segment with 2 short outer setae, 1 terminal seta, and 4 inner setae. Second leg (Figure 42), coxa with 1 inner seta. Endopod 2-segmented; first segment with 1 seta; second segment with 5 setae. Exopod 2-segmented; first segment with 1 outer spine and 1 inner seta; second segment with 3 outer spines, 1 long terminal spine having a toothed hyaline flange on outer margin, and 5 inner setae. Third leg (Figure 43), coxa with 1 inner seta. Endopod 2-segmented; first segment with 1 inner seta; second segment with 6 setae. Exopod 2-segmented; first segment with 1 outer spine; second segment with 2 outer spines, 1 long terminal spine having a toothed hyaline flange on outer margin, and 4 inner setae. Lappets, which expand on anterior side of bases of 2 medial outer spines of exopod, observable in lateral view (Figure 44). Fourth leg endopod, 1-segmented, with 6 setae. Exopod 1-segmented, with 3 outer spines, 1 long terminal spine having a toothed hyaline flange on outer margin, and 3 inner setae.

Clausocalanus brevipes

Figs. 45–53

Body 0.75 mm long (0.71–0.81 mm, based on 11 specimens). Prosome 4-segmented, with incipient suture on posterior segment; urosome 2-segmented (Figure 45). Bifurcate rostrum directed ventrally (Figure 46).

First antenna 21-segmented, extending beyond posterior margin of third prosomal segment. Second antenna (Figure 47) differing from that of stage II by addition of 1 seta on inner lobe of second segment of endopod and by having small setules on outer lobe. Width of terminal segment of exopod measures slightly more than half the length. Mandible (Figure 48) differing from that of stage II by addition of 1 seta on second segment of endopod and by having small setules on outer surface. First maxilla (Figure 49) differing from that of stage II by addition of 1 spine on gnathobase, 1 seta on proximal small inner lobe, 2 setae on epipod, and 1 seta each on exopod, basis, and first and second segments of endopod. Second maxilla with segmentation and setation resembling those of that appendage of *C. laticeps* (Figure 40). Maxilliped with segmentation and setation resembling those of that appendage of *Ctenocalanus citer* (Figure 55).

First leg (Figure 50), basis with 1 distal seta on anterior inner corner. Endopod 1-segmented, with 5 setae and row of setules on convex process of inner margin. Exopod 2-segmented; first segment with 1 short outer seta; second segment with 2 short outer setae, 1 terminal seta, and 4 inner setae. Second leg (Figure 51), coxa with 1 inner seta. Basis with several small spines on posterior surface. Endopod 2-segmented; first segment with 1 seta; second segment with 5 setae. Exopod 2-segmented; first segment with 1 outer spine and 1 inner seta; second segment with 3 outer spines, 1 long terminal spine having a toothed hyaline flange on outer margin, and 5 inner setae. Third leg (Figure 52), coxa with 1 inner seta. Endopod 2-segmented; first segment with 1 inner seta; second segment with 6 setae. Exopod 2-segmented; first segment with 1 outer spine; second segment with 2 outer spines, 1 long terminal spine having a toothed hyaline flange on outer margin, and 4 inner setae. Bases of 2 medial, outer spines of exopod without lappets in lateral view (Figure 53). Fourth leg endopod 1-segmented, with 6 setae. Exopod 1-segmented, with 3 outer spines, 1 long terminal spine having a toothed hyaline flange on outer margin, and 3 inner setae.

Ctenocalanus citer

Figs. 54–58

Body 0.72 mm long (0.60–0.82 mm, based on 8 specimens). Prosome slender, 4-segmented, with incipient suture on posterior segment; urosome 2-segmented (Figure 54).

First antenna 21-segmented, extending to posterior margin of urosome. Second antenna and first maxilla with segmentation and setation resembling those of those appendages of *Clausocalanus brevipes* (Figures 47, 49). Mandible with segmentation, setation, and shape of masticatory blade resembling those of that appendage of *C. brevipes* (Figure 48). Second maxilla with segmentation and setation resembling those of that appendage of *C. laticeps* (Figure 40). Maxilliped (Figure 55), 6-segmented, differing from that of stage II by addition of 2 setae on proximal segment and 1 seta each on second segment, additional segment (which corresponds to second segment of endopod of stage VI), and inner surface of penultimate segment.

First leg (Figure 56), basis with 1 distal seta on anterior inner corner. Endopod 1-segmented, with 5 setae; row of setules on convex process of inner margin. Exopod 2-segmented; first segment with 1 short outer seta; second segment with 2 short outer setae, 1 terminal seta, and 4 inner setae. Second leg (Figure 57), coxa with 1 inner seta. Endopod 2-segmented; first segment with 1 seta; second segment with 5 setae. Exopod 2-segmented; first segment with 1 outer spine and 1 inner seta; second segment with 3 outer spines, 1 long terminal spine having a toothed hyaline flange on outer margin, and 5 inner setae. Third leg (Figure 58), coxa with 1 inner seta. Endopod 2-segmented; first segment with 1 inner seta; second segment with 6 setae. Exopod 2-segmented; first segment with 1 outer spine; second segment with 2 outer spines (the medial with toothed inner margin), 1 long terminal spine having a toothed hyaline flange on outer margin, and 4 inner setae. Fourth leg, endopod 1-segmented, with 6 setae. Exopod 1-segmented, with 3 outer spines, 1 long terminal spine having a toothed hyaline flange on outer margin, and 3 inner setae.

Comparison of the Three Species, Stage III

The rostrums of the 3 species have become a distinguishing characteristic by this stage: the bifurcate rostrum of *Clausocalanus laticeps* is directed posteriorly; the bifurcate rostrum of *C. brevipes* is directed ventrally; the rostral filaments of *Ctenocalanus citer* are filiform and long.

The prosome of *Clausocalanus brevipes* appears more robust in lateral view than that of *C. laticeps* and *Ctenocalanus citer*.

The size difference of the outer spines of the second leg and the lappets expanding at bases of outer spines of exopods of the third and fourth legs distinguish *Clausocalanus laticeps* from *C. brevipes.*

The small teeth on the inner margin of the medial outer spine of the second segment of the exopod of the third leg of *Ctenocalanus citer* are a useful aid in distinguishing it from the 2 species of *Clausocalanus* in this and all succeeding stages.

The number of teeth on the flanged spines of the second, third, and fourth legs proved useful as an aid for distinguishing between the 3 species in succeeding stages; differences could be noted between second legs of the species in this stage (Table 4).

Copepodid IV

Clausocalanus laticeps

Figs. 59–65

Female body 0.91 mm long (0.85–0.95 mm, based on 5 specimens). Male body 0.91 mm long (0.90–0.93 mm, based on 5 specimens). Prosome 4-segmented; urosome 3-segmented (Figure 59, female; Figure 60, male).

First antenna 24-segmented, extending approximately to midlength of third prosomal segment. Second antenna with segmentation and setation resembling those of that appendage of *C. brevipes* and *Ctenocalanus citer* (Figures 68, 73). Width of terminal seg-

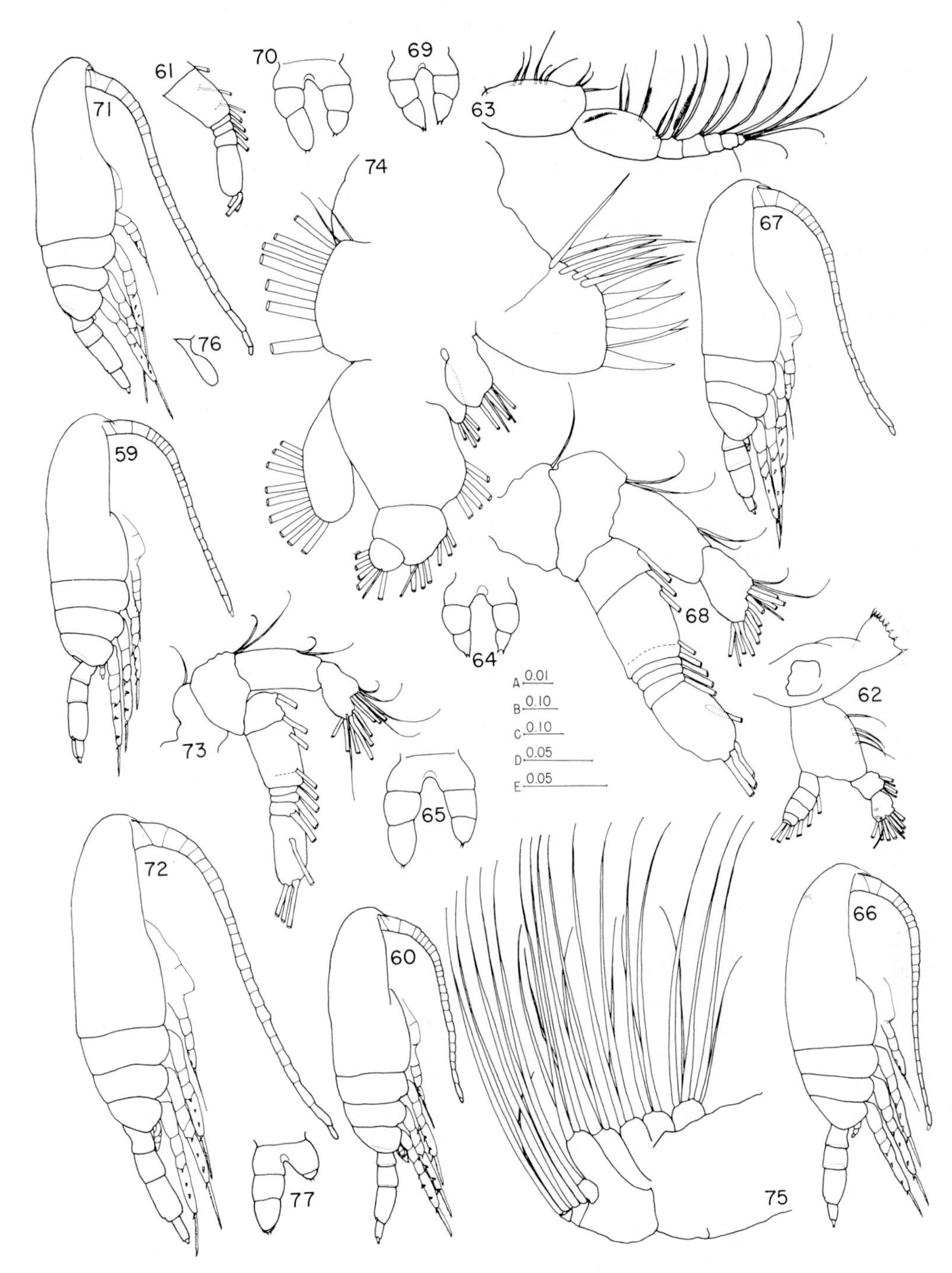

***Clausocalanus laticeps* IV**

59. Female, lateral view, 0.95 mm, B.
60. Male, lateral view, 0.91 mm, B.
61. Male, terminal exopod segment of second antenna, 0.93 mm, D.
62. Male, left mandible, 0.91 mm, D.
63. Female, right maxilliped, 0.91 mm, D.
64. Female, fifth legs, 0.90 mm, E.
65. Male, fifth legs, 0.91 mm, E.

***Clausocalanus brevipes* IV**

66. Female, lateral view, 0.98 mm, B.
67. Male, lateral view, 0.93 mm, B.
68. Male, right second antenna, 1.01 mm, D.
69. Female, fifth legs, 0.91 mm, E.
70. Male, fifth legs, 0.93 mm, E.

***Ctenocalanus citer* IV**

71. Female, lateral view, 0.92 mm, B.
72. Male, lateral view, 1.08 mm, C.
73. Male, second antenna, 0.95 mm, D.
74. Male, first maxilla, 1.10 mm, A.
75. Male, left second maxilla, 0.95 mm, A.
76. Female, fifth leg, 0.80 mm, E.
77. Male, fifth legs, 0.95 mm, E.

TABLE 4. Approximate Range of Teeth on Flanged Spines of Exopods of Legs in ***Clausocalanus laticeps, C. brevipes,*** and ***Ctenocalanus citer***

	Stage I	Stage II		Stage III		Stage IV			Stage V			Stage VI		
	Leg 2	Leg 2	Leg 3	Leg 2	Leg 3	Leg 2	Leg 3	Leg 4	Leg 2	Leg 3	Leg 4	Leg 2	Leg 3	Leg 4
C. laticeps	19–23	19–23	19–24	27–29	25–30									
Female						26–31	32–34	29–34	32–38	38–41	36–40	40–44	45–49	47–51
Male						29–31	29–33	30–31	30–33	32–38	35–40	46–55	50–66	51
C. brevipes	19–22	19–23	19–20	19–23	21–26									
Female						17–24	20–26	21–28	25–28	32–34	34–36	31–33	32–37	39–41
Male						19–24	22–28	22–30	28–30	21–28	32–36	67–70	73–75	
C. citer	15–18	16–18	12–13	14–17	15–19									
Female						16–18	20–22	18–19	21–23	21–23	22–23	19–23	22–25	22–26
Male						18–21	19–20	20–21	17–22	19–26	22–24	39–51	35–58	45–51

ment of exopod is approximately half the length (Figure 61). Mandible (Figure 62) differing from that of stage III by the addition of 1 seta on second segment of endopod. First maxilla and second maxilla with segmentation and setation that resemble those of first and second maxillae of *C. citer* (Figures 74, 75). Maxilliped (Figure 63) 7-segmented; differing from that of stage III by the addition of 1 seta on each of first 2 segments of endopod and additional segment (which corresponds to third segment of endopod of stage VI); row of setules arising near proximal outer margin of second segment.

First leg with segmentation and setation similar to that of stage III. Second leg with segmentation and setation similar to that of stage III, with an increase in size of spines and setae. Third leg differing from that of stage III by addition of 1 inner seta on first segment of exopod and 1 outer spine on second segment of exopod. Fourth leg, endopod 2-segmented; first segment with 1 seta; second segment with 6 setae. Exopod 2-segmented; first segment with 1 outer spine; second segment with 3 outer spines, 1 long terminal spine having a toothed hyaline flange on outer margin, and 5 inner setae. Fifth legs of female (Figure 64), with 2-segmented left and right rami; terminal segments apically bifurcate. Fifth legs of male (Figure 65) with 2-segmented left and right rami, left slightly longer; terminal segments each with 2 apical setae.

Clausocalanus brevipes

Figs. 66–70

Female body 0.95 mm long (0.91–1.02 mm, based on 5 specimens). Male body 0.98 mm long (0.93–1.02 mm, based on 5 specimens). Prosome 4-segmented, urosome 3-segmented (Figure 66, female; Figure 67, male).

First antenna 24-segmented, extending approximately to midlength of fourth prosomal segment. Second antenna (Figure 68) differing from that of stage III by addition of 1 seta on each of inner and outer lobes of second segment of endopod. Width of terminal segment of exopod approximately three-fifths the length.

***Clausocalanus laticeps* V**

78. Female, lateral view, 1.25 mm, B.
79. Male, lateral view, 1.14 mm, B.
80. Male, rostral region, lateral view, 1.14 mm, C.
81. Male, left second antenna, 1.14 mm, C.
82. Male, right second maxilla, 1.14 mm, A.
83. Female, fifth legs, 1.35 mm, D.
84. Male, fifth legs, 1.31 mm, D.

***Clausocalanus brevipes* V**

85. Female, lateral view, 1.19 mm, B.
86. Male, 1.30 mm, lateral view, B.
87. Male, rostral region, lateral view, 1.30 mm, C.
88. Terminal exopod segment of second antenna, 1.22 mm, C.
89. Male, left mandible, 1.32 mm, C.
90. Female, right first maxilla, 1.36 mm, C.
91. Male, left maxilliped, 1.31 mm, C.
92. Female, fifth legs, 1.36 mm, D.
93. Male, fifth legs, 1.31 mm, D.

***Ctenocalanus citer* V**

94. Female, lateral view, 1.00 mm, B.
95. Male, lateral view, 1.21 mm, B.
96. Male, rostral region, lateral view, 1.21 mm, C.
97. Male, right second antenna, 1.35 mm, C.
98. Female, fifth leg, 0.90 mm, D.
99. Male, fifth leg, 1.15 mm, D.

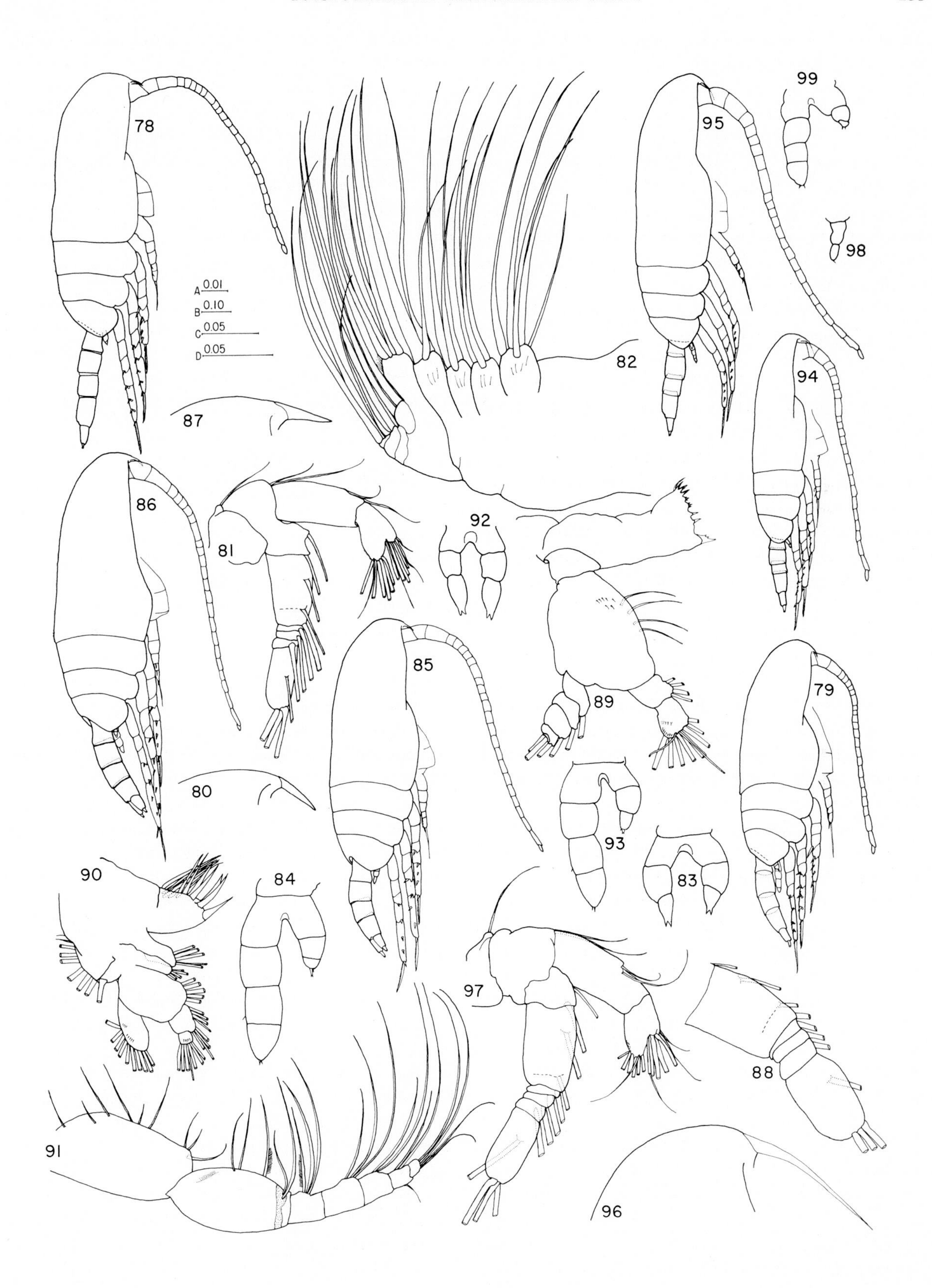
78
95
99
98
A 0.01
B 0.10
C 0.05
D 0.05
82
94
87
86
81
92
85
89
79
80
93
83
90
84
97
88
91
96

Mandible with segmentation, setation, and shape of masticatory blade resembling those of that appendage of *C. laticeps* (Figure 62). First maxilla and second maxilla with segmentation and setation resembling those of first and second maxillae of *Ctenocalanus citer* (Figures 74, 75). Maxilliped with segmentation and setation resembling those of that appendage of *Clausocalanus laticeps* (Figure 63). First and second legs with segmentation and setation similar to that of stage III; second leg with an increase in size of spines and setae. Third leg differing from that of stage III by the addition of 1 inner seta on first segment of exopod and 1 outer spine on second segment of exopod. Fourth leg, endopod 2-segmented; first segment with 1 seta; second segment with 6 setae. Exopod 2-segmented; first segment with 1 outer spine; second segment with 3 outer spines, 1 long terminal spine having a toothed hyaline flange on outer margin, and 5 inner setae. Fifth legs of female (Figure 69), with 2-segmented left and right rami; terminal segments apically bifurcate. Fifth legs of male (Figure 70), with 2-segmented left and right rami, left slightly longer; terminal segments each with 2 apical setae.

Ctenocalanus citer

Figs. 71–77

Female body 0.85 mm long (0.75–0.95 mm, based on 8 specimens). Male body 1.00 mm long (0.93–1.10 mm, based on 6 specimens). Prosome 4-segmented, urosome 3-segmented (Figure 71, female; Figure 72, male).

First antenna 24-segmented, extending approximately to posterior margin of urosome. Second antenna (Figure 73) differing from that of stage III by the addition of 1 seta each on inner and outer lobes of second segment of endopod. Mandible and maxilliped with segmentation, setation, and shape of masticatory blade that resemble those of mandible and maxilliped of *Clausocalanus laticeps* (Figures 62, 63). First maxilla (Figure 74) differing from that of stage III by the addition of 3 spines on gnathobase and 1 seta each on epipod, exopod, and first segment of endopod. Second maxilla (Figure 75) with segmentation and setation similar throughout stages II–VI.

First and second legs with segmentation and setation similar to that of stage III; second leg with an increase in size of spines and setae. Third leg differs from that of stage III by addition of 1 inner seta on first segment of exopod and 1 outer spine on second segment of exopod. Fourth leg, endopod 2-segmented; first segment with 1 seta; second segment with 6 setae. Exopod 2-segmented; first segment with 1 outer spine; second segment with 3 outer spines, 1 long terminal spine having a toothed hyaline flange on outer margin, and 5 inner setae. Fifth leg of female (Figure 76) with 1-segmented left ramus, variable in shape or absent. Fifth leg of male (Figure 77) with 2-segmented left ramus, 1-segmented right ramus; terminal segments each with 2 apical setae.

Comparison of the Three Species, Stage IV

The differences noted as being apparent in the comparison of the 3 species in stage III increase in conspicuousness with each succeeding stage. Anatomical features of the fifth legs, female and male, of the 3 species cannot be assumed as specifically diagnostic at this stage.

Copepodid V

Clausocalanus laticeps

Figs. 78–84

Female body 1.26 mm long (1.20–1.35 mm, based on 3 specimens). Male body 1.14 mm long (1.05–1.21 mm, based on 5 specimens). Prosome and urosome 4-segmented (Figure 78, female; Figure 79, male). Bifurcate rostrum directed posteriorly (Figure 80).

First antenna 24-segmented, extending approximately to posterior margin of third prosomal segment. Second antenna (Figure 81) differing from that of stage IV by addition of 1 seta on inner lobe of second segment of endopod. Width of terminal segment of exopod more than one-third the length. Mandible with segmentation and setation resembling those of that appendage of *C. brevipes* (Figure 89). First maxilla with segmentation and setation resembling those of that appendage of *C. brevipes* (Figure 90). Second maxilla (Figure 82) with segmentation and setation similar throughout stages II–VI. Few setules on interior margins of endites. Maxilliped with segmentation and setation resembling those of that appendage of *C. brevipes* (Figure 91).

First to fourth legs with segmentation and setation similar to that of stage VI, female, except spines and setae not fully developed. Fifth legs of female (Figure 83) with 2-segmented left and right rami, terminal segments apically bifurcate. Fifth leg of male (Figure 84) with left ramus 3-segmented, penultimate segment with 1 short seta on outer margin; right ramus 2-segmented, terminal segments each with rudimentary spinous elements.

Clausocalanus brevipes

Figs. 85–93

Female body 1.27 mm long (1.19–1.36 mm, based on 3 specimens). Male body 1.29 mm long (1.22–1.35 mm, based on 5 specimens). Prosome and urosome 4-segmented (Figure 85, female; Figure 86, male). Bifurcate rostrum directed ventrally (Figure 87).

First antenna 24-segmented, extending approximately to posterior margin of fourth prosomal segment. Second antenna with segmentation and setation resembling those of that appendage of *C. laticeps* and *Ctenocalanus citer* (Figures 81, 97). Width of terminal segment of exopod more than one-half the length (Figure 88). Mandible (Figure 89) differing from that of stage IV by addition of 1 seta on second segment of endopod; setules on inner surface of basis. Caps complex on some teeth of masticatory blade, when found intact. First maxilla (Figure 90) differing from that of stage IV by addition of 1 seta each on exopod, basipod, and first and second segments of endopod; few small setules on outer surface of second segment of endopod and exopod. Second maxilla with segmentation and setation resembling those of that appendage of *Clausocalanus laticeps* (Figure 82). Maxilliped (Figure 91) differing from that of stage IV by addition of 1 seta each on first, second, third, and fourth segments of endopod.

First to fourth legs with segmentation and setation similar to that of stage VI, female, except spines and setae not fully developed. Fifth legs of female (Figure 92) with 2-segmented left and right rami, terminal segments apically bifurcate. Fifth leg of male (Figure 93) with 3-segmented left ramus, penultimate segment with 1 short seta on outer margin; right leg 2-segmented; terminal segments each with rudimentary spinous elements.

Ctenocalanus citer

Figs. 94–99

Female body 0.98 mm long (0.80–1.11 mm, based on 5 specimens). Male body 1.18 mm long (1.01–1.35 mm, based on 13 specimens). Prosome and urosome 4-segmented (Figure 94, female; Figure 95, male). Rostral filaments long, filiform (Figure 96).

First antenna 24-segmented, extending approximately to posterior margin of urosome. Second antenna (Figure 97) differing from that of stage IV by the addition of 1 seta on inner lobe of second segment of endopod. Mandible with segmentation, setation, and shape of masticatory blade resembling those of mandible of *Clausocalanus brevipes* (Figure 89). First maxilla and maxilliped with segmentation and setation resembling those of those appendages of *C. brevipes* (Figures 90, 91). Second maxilla with segmentation and setation resembling those of that appendage of *C. laticeps* (Figure 82).

First to fourth legs with segmentation and setation similar to that of stage VI female, except spines and setae not fully developed. Fifth leg of female (Figure 98) with 1- or 2-segmented left ramus, sometimes absent. Fifth leg of male (Figure 99) with 3-segmented left ramus, 2-segmented right ramus; terminal segments each with rudimentary spinous elements.

Comparisons of the Three Species, Stage V

Anatomical features of the fifth legs, female and male, of the 3 species cannot be assumed as specifically diagnostic at this stage without also considering the other distinguishing characteristics, which were noted as being apparent in stage III. The differences increase in conspicuousness with each succeeding stage.

Copepodid VI

Clausocalanus laticeps

Figs. 100–115

Female. Body 1.47 mm long (1.35–1.55 mm, based on 10 specimens). Prosome and urosome 4-segmented (Figure 100). Head broad in lateral view. Genital field with operculum extending laterally over more than two-thirds the width of genital segment (Figure 101).

First antenna 24-segmented, extending to midlength of fourth prosomal segment; faint suture dividing segments 23 and 24. Second antenna and second maxilla with segmentation and setation resembling those of those appendages of *C. brevipes* (Figures 118, 119). Mandible (Figure 102) differing from that of stage V by the addition of 1 seta on second segment of endopod. First maxilla (Figure 103) differs from that of stage V by the addition of 1 seta each on exopod and first segment of endopod. Maxilliped (Figure 104) differing from that of stage V by addition of 1 seta each on first 4 segments of endopod.

First leg (Figure 105), coxa with row of setules on distal margin of anterior surface. Basis with 1 distal seta on anterior inner corner, row of hairs on inner margin, and setules on anterior surface. Endopod 1-segmented, with 5 setae; row of setules on convex process of inner margin. Exopod 3-segmented; first segment with 1 short outer seta; second segment with

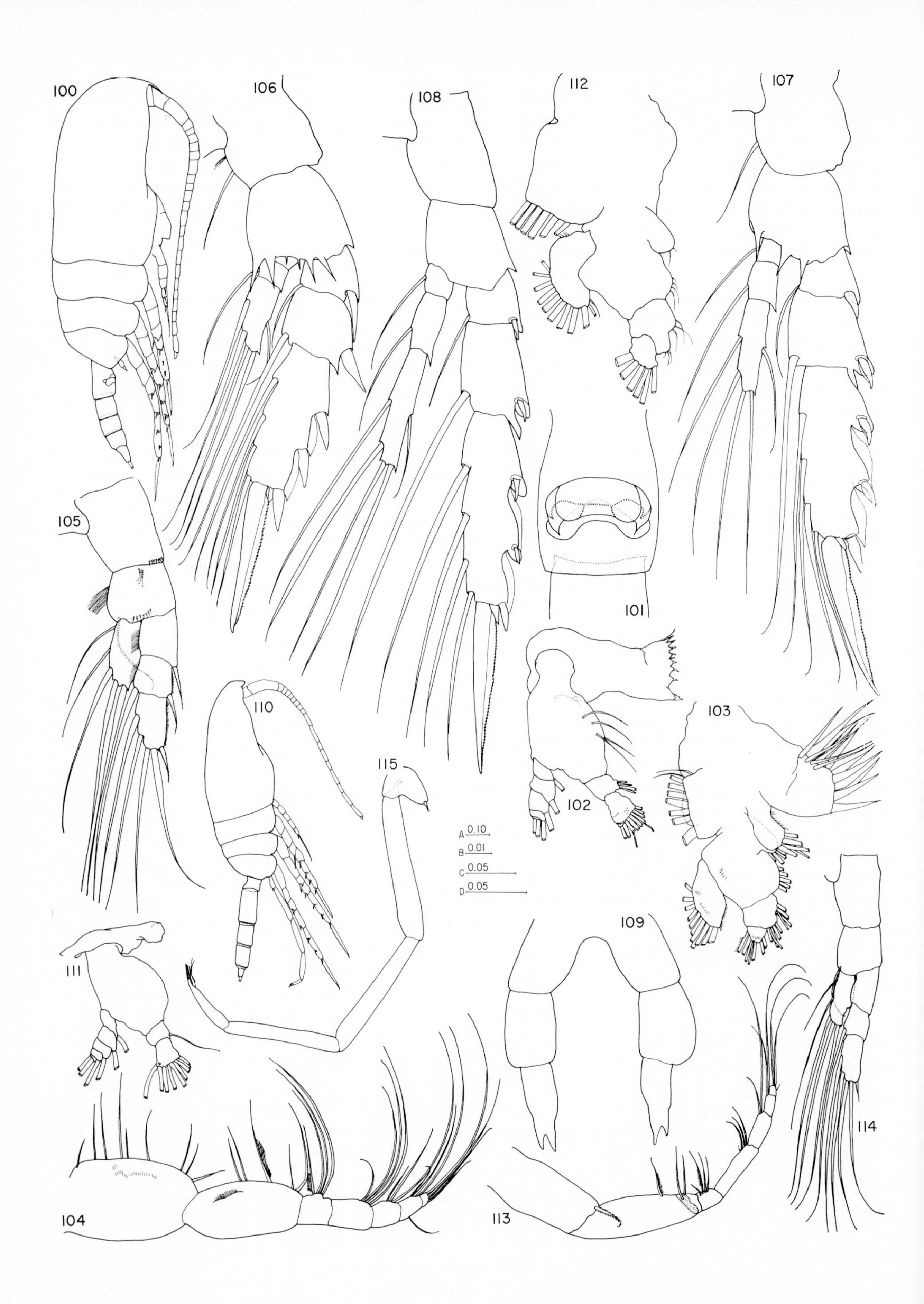

100
106
108
112
107
105
101
110
115
103
102
A 0.10
B 0.01
C 0.05
D 0.05
109
111
104
113
114

1 short outer seta and 1 inner seta; third segment with 1 short outer seta, 1 terminal seta, and 3 inner setae; each of 3 outer setae lined with short hairs; hairs on outer margins of second and third segments. Second leg (Figure 106), coxa with 1 inner seta; hairs on inner margin. Pattern of spines on distal margin of posterior surface of basis, slightly variable in shape and number, shape of 4 larger spines being most constant. Exopod 2-segmented; first segment with 1 seta; second segment with 5 setae. Exopod 3-segmented; first and second segments each with 1 outer spine and 1 inner seta; third segment with 3 outer spines, 1 long terminal spine, and 4 inner setae. Third leg (Figure 107), coxa with 1 inner seta. Basis with spines on distal margin of posterior surface, variable in shape and number, absent in 1 specimen. Endopod 3-segmented; first and second segments each with 1 inner seta; third segment with 5 setae. Exopod 3-segmented; first and second segments each with 1 outer spine and 1 inner seta; third segment with 3 outer spines, 1 long terminal spine, and 4 inner setae. Fourth leg (Figure 108) endopod 3-segmented; first and second segments each with 1 inner seta; third segment with 5 setae. Exopod 3-segmented; first and second segments each with 1 outer spine and 1 inner seta; third segment with 3 outer spines, 1 long terminal spine, and 4 inner setae; lappets, expanding on anterior side of bases of first 4 outer spines, extending almost to apices of spines. Fifth legs (Figure 109) with 2-segmented left and right rami, segments approximately equal in length; terminal segments apically bifurcate.

Male. Body 1.12 mm long (1.00–1.30 mm, based on 12 specimens). Prosome 4-segmented, urosome 5-segmented (Figure 110). Rostrum modified into a rounded protuberance.

First antenna 19-segmented, extending beyond posterior margin of second prosomal segment; segments 1-2-3, 10-11, 12-13, and 20-21 fused. Second antenna with segmentation and setation resembling those of that appendage of *C. brevipes* (Figure 126). Mandible (Figure 111), masticatory blade reduced to rounded process. Palp differs from that of female by lack of 4 setae from basis and 3 setae from first endopod. First maxilla (Figure 112) without gnathobase, remaining proximal segments markedly reduced; number and extent of reduction of setae on these segments variable between specimens as well as between right and left appendages of each specimen. Basis heavily sclerotized. Second maxilla resembling that appendage of *Ctenocalanus citer* (Figure 144) and variable in extent of reduction. Maxilliped (Figure 113) differing from that of female by lack of 8 setae from coxa, 2 setae from basis, and 1 seta each from third, fourth, and fifth segments of endopod.

First leg (Figure 114) differing from that of female by lack of outer short seta from each of first and second segments of exopod. Second to fourth legs similar to those of female, except for an increase in length of outer spines of exopod. Fifth leg (Figure 115) with 5-segmented left ramus extending to posterior margin of caudal rami, each segment shorter than the one preceding; complex sclerotization on anterior surface of basis surrounding articulation of ramus; penultimate segment with 1 seta and 2 spines, extending from distal margin and row of short hairs on outer margin; terminal segment inserted subapically, with row of hairs and distally 2 setae and a spine. Right ramus with 1 or 2 small segments supported by elongation of basis, with 3 setae on distal segment.

Clausocalanus brevipes

Figs. 116–129

Female. Body 1.48 mm long (1.41–1.54 mm, based on 5 specimens). Prosome and urosome 4-segmented (Figure 116). Genital field situated in approximate center of segment (Figure 117).

First antenna 24-segmented, extending beyond posterior margin of fourth prosomal segment; suture faint between segments 23 and 24. Second antenna (Figure 118) differing from that of stage V by addition of 1 seta on inner lobe of second segment of endopod. Mandible with segmentation, setation, and

---

***Clausocalanus laticeps* VI**

100. Female, lateral view, 1.55 mm, A.
101. Female, genital field, 1.37 mm, D.
102. Female, right mandible, 1.40 mm, C.
103. Female, left first maxilla, 1.50 mm, D.
104. Female, right maxilliped, 1.55 mm, C.
105. Female, first leg, 1.55 mm, C.
106. Female, second leg, 1.52 mm, C.
107. Female, third leg, 1.52 mm, C.
108. Female, fourth leg, 1.52 mm, C.
109. Female, fifth legs, 1.52 mm, B.
110. Male, lateral view, 1.18 mm, A.
111. Male, right mandible, 1.18 mm, C.
112. Male, right first maxilla, 1.12 mm, B.
113. Male, left maxilliped, 1.18 mm, C.
114. Male, first leg, 1.15 mm, C.
115. Male, fifth legs, 1.15 mm, C.

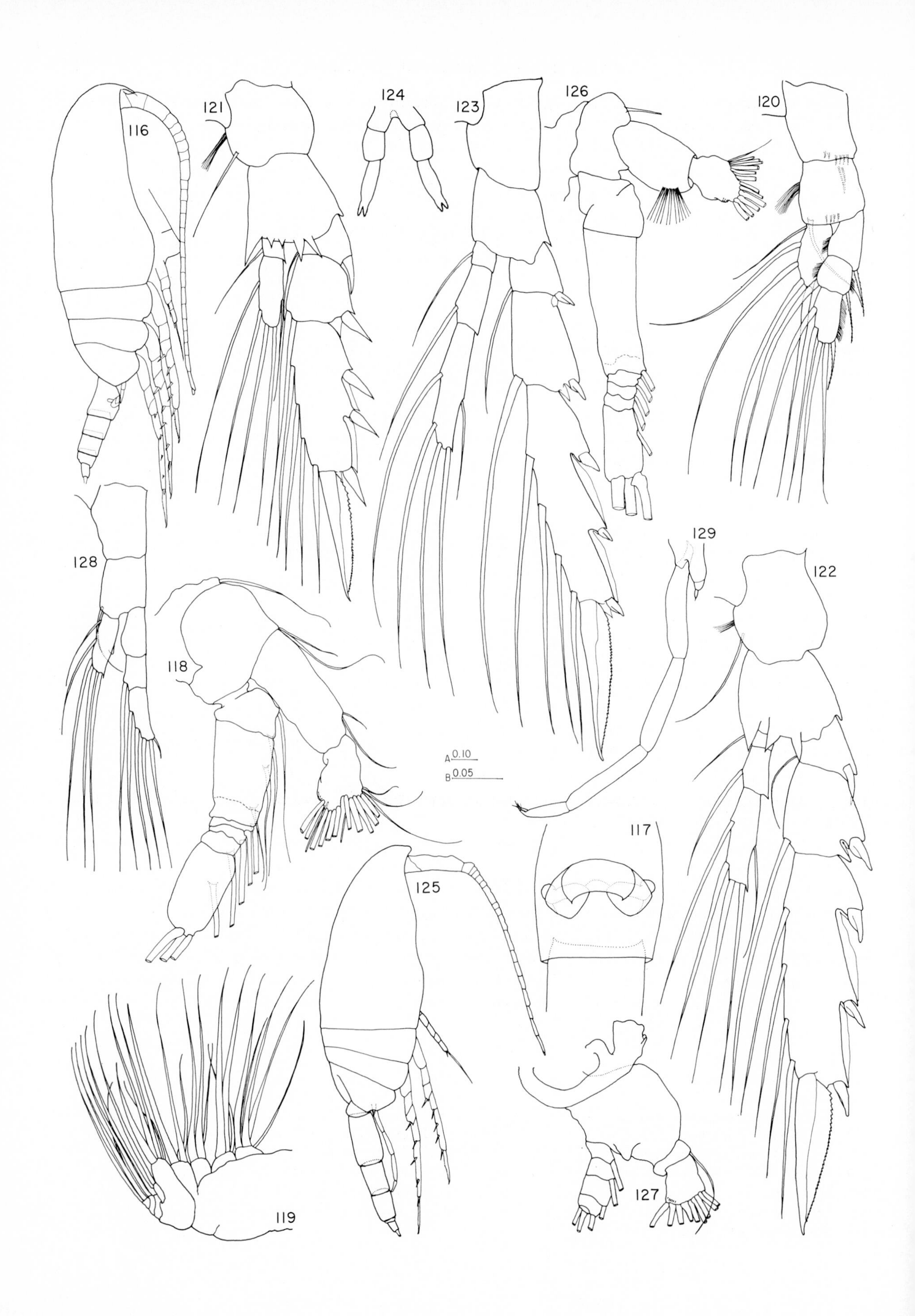

116
117
118
119
120
121
122
123
124
125
126
127
128
129
A 0.10
B 0.05

shape of masticatory blade resembling those of that appendage of *C. laticeps* (Figure 102). First maxilla and maxilliped with segmentation and setation resembling those of the first maxilla and maxilliped of *C. laticeps* (Figures 103, 104). Second maxilla (Figure 119) with segmentation and setation similar to all stages succeeding stage I, each stage exhibiting gradual increase in length of seta and size of appendage.

First leg (Figure 120), coxa with row of setules on distal margin of anterior surface. Basis with 1 distal seta on anterior inner corner, row of hairs on inner margin, and setules on anterior surface. Endopod 1-segmented, with 5 setae; row of setules on convex process of inner margin. Exopod 3-segmented; first segment with 1 short outer seta; second segment with 1 short outer seta and 1 inner seta; third segment with 1 short outer seta, 1 terminal seta, and 3 inner setae; each of 3 outer setae lined with short hairs; hairs on inner margin of first and second segments and outer margins of second and third segments. Second leg (Figure 121), coxa with 1 inner seta; hairs on inner margin. Basis with pattern of spines on distal margin of posterior surface. Endopod 2-segmented; first segment with 1 seta; second segment with 5 setae. Exopod 3-segmented; first and second segments each with 1 outer spine and 1 inner seta; third segment with 3 outer spines, 1 long terminal spine, and 4 inner setae. Third leg (Figure 122), coxa with 1 inner seta. Basis with pattern of spines on distal margin of posterior surface. Endopod 3-segmented; first and second segments each with 1 inner seta, third segment with 5 setae. Exopod 3-segmented; first and second segments each with 1 outer spine and 1 inner seta; third segment with 3 outer spines, 1 long terminal spine, and 4 inner setae. Fourth leg (Figure 123), endopod 3-segmented; first and second segments each with 1 inner seta; third segment with 5 setae. Exopod 3-segmented; first and second segments each with 1 outer spine and 1 inner seta; third segment with 3 outer spines, 1 long terminal spine, and 4 inner setae. Fifth legs (Figure 124), with 2-segmented left and right rami; terminal segments longer than penultimate segment; apical bifurcation lined with short hairs.

Male. Body 1.43 mm long (1.34–1.50 mm, based on 3 specimens). Prosome 4-segmented, urosome 5-segmented (Figure 125). Rostrum modified to a rounded protuberance.

First antenna 19-segmented, extending beyond posterior margin of third prosomal segment; segments 1-2-3, 10-11, 12-13, and 20-21 fused. Second antenna (Figure 126) differing from that of female by lack of the following setae: 1 each from coxa and from basis, which now bears a blunt-tipped seta; 2 from first segment of endopod; 2 from inner lobe and 1 from outer lobe of second segment of endopod; and 2 each from first 2 segments of exopod. Mandible (Figure 127), masticatory blade reduced to rounded process. Palp differs from that of female by lack of 4 setae from basis and 3 setae from first endopod segment. Basis with round process near lateral margin on posterior surface. First maxilla with segmentation and setation resembling those of that appendage of *C. laticeps* (Figure 112); appendage variable in extent of reduction. Second maxilla resembling that appendage of *Ctenocalanus citer* (Figure 144); appendage variable in extent of reduction. Maxilliped resembling maxilliped of *Clausocalanus laticeps* (Figure 113).

First leg (Figure 128) differing from that of female by lack of outer short seta from each of first and second segments of exopod. Second and third legs similar to those of female, except for greater length of outer spines on exopod. Fourth leg, protopod and endopod similar to female; exopod missing. Fifth leg (Figure 129), with 5-segmented left ramus extending approximately to posterior margin of third prosomal segment; complex sclerotization on anterior surface of basis surrounding articulation of ramus; penultimate segment with 1 seta and 2 spines extending from distal margin; terminal segment developed subapically, with row of short hairs and 1 distal spine. Right ramus is 2-segmented, with distal seta on terminal segment.

***Clausocalanus brevipes* VI**
116. Female, lateral view, 1.54 mm, A.
117. Female, genital field, 1.54 mm, B.
118. Female, right second antenna, 1.45 mm, B.
119. Female, left second maxilla, 1.45 mm, B.
120. Female, first leg, 1.51 mm, B.
121. Female, second leg, 1.45 mm, B.
122. Female, third leg, 1.54 mm, B.
123. Female, fourth leg, 1.54 mm, B.
124. Female, fifth legs, 1.45 mm, B.
125. Male, lateral view (fourth leg missing), 1.50 mm, A.
126. Male, left second antenna, 1.50 mm, B.
127. Male, right mandible, 1.50 mm, B.
128. Male, first leg, 1.50 mm, B.
129. Male, fifth legs, 1.50 mm, B.

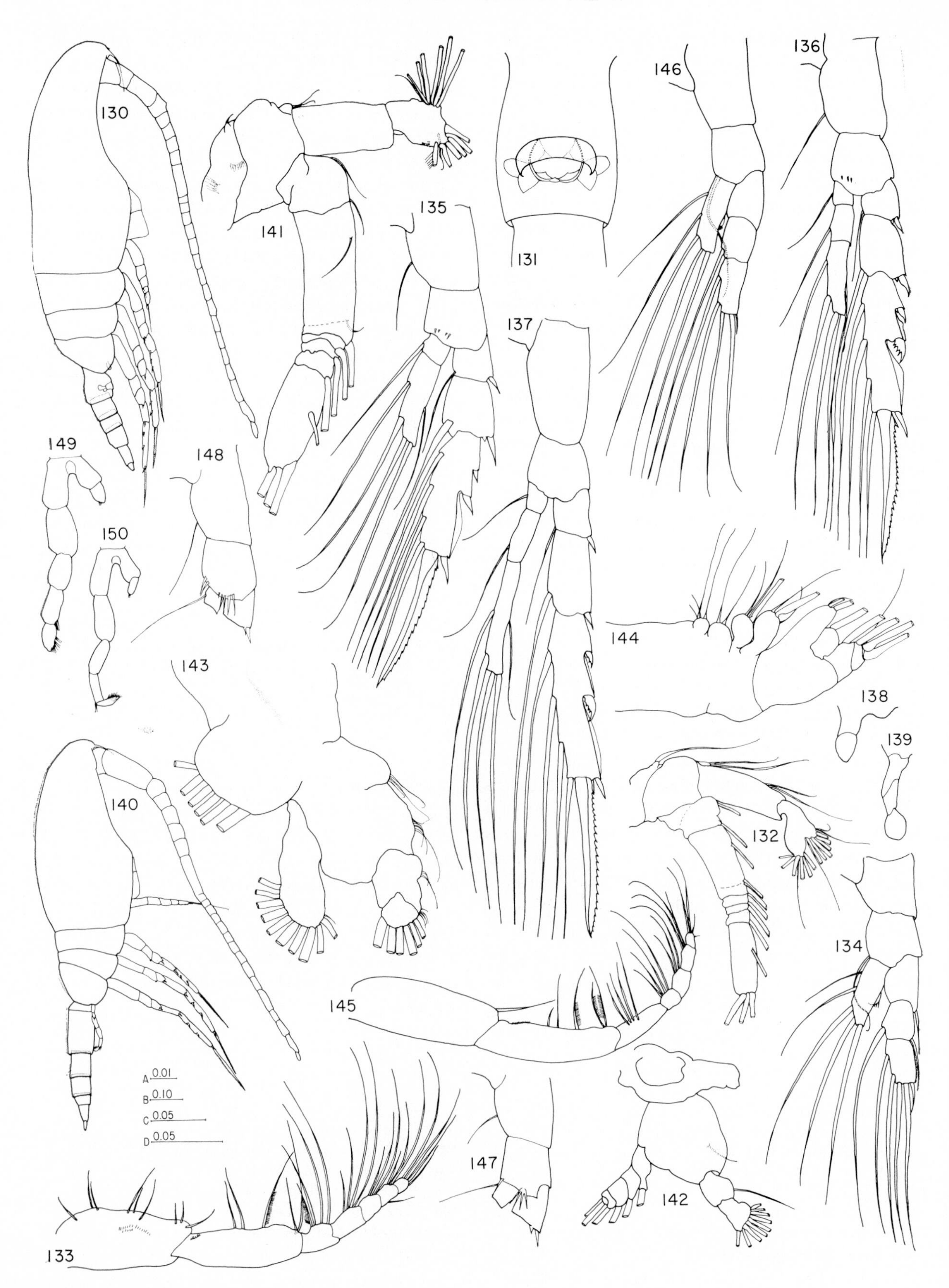
130
141
135
131
146
136
137
149
148
150
143
144
138
139
140
132
134
145
A 0.01
B 0.10
C 0.05
D 0.05
147
142
133

Ctenocalanus citer

Figs. 130–150

Female. Body 1.17 mm long (0.80–1.41 mm, based on 18 specimens). Prosome and urosome 4-segmented (Figure 130). Genital field situated in posterior half of segment (Figure 131).

First antenna 24-segmented, extending approximately to posterior margin of caudal rami; suture faint between segments 1 and 2. Second antenna (Figure 132) differing from that of stage V by addition of 1 seta on inner lobe of second segment of endopod. Mandible with segmentation, setation, and shape of masticatory blade resembling those of mandible of *Clausocalanus laticeps* (Figure 102). First maxilla with segmentation and setation resembling those of that appendage of *C. laticeps* (Figure 103). Second maxilla with segmentation and setation resembling those of that appendage of *C. brevipes* (Figure 119). Maxilliped (Figure 133) differing from that of stage V by addition of 1 seta each on first 4 segments of endopod.

First leg (Figure 134), basis with 1 distal seta on anterior inner corner. Endopod 1-segmented, with 5 setae; row of setules on convex process of inner margin. Exopod 3-segmented; first segment with 1 short outer seta; second segment with 1 short outer seta and 1 inner seta; third segment with 1 short outer seta, 1 terminal seta, and 3 inner setae; hairs on outer margin of third segment. Second leg (Figure 135), coxa with 1 inner seta. Basis with several spines near distal margin of posterior surface, variable in size and number. Endopod 2-segmented; first segment with 1 seta, second segment with 5 setae. Exopod 3-segmented; first and second segments each with 1 outer spine and 1 inner seta; third segment with 3 outer spines, 1 long terminal spine, and 4 inner setae. Third leg (Figure 136), coxa with 1 inner seta. Basis with several spines, variable in number, near distal margin of posterior surface. Endopod 3-segmented; first and second segments each with 1 inner seta; third segment with 5 setae. Exopod 3-segmented; first and second segments each with 1 outer spine and 1 inner seta; third segment with 3 outer spines, 1 long terminal spine, and 4 inner setae; second outer spine of exopod with strong teeth on inner margin; other outer spines with or without small teeth; if present, variable in size. Fourth leg (Figure 137), 3-segmented endopod; first and second segments each with 1 inner seta, third segment with 5 setae. Exopod 3-segmented; first and second segments each with 1 outer spine and 1 inner seta; third segment with 3 outer spines, 1 long terminal spine, and 4 inner setae; first 4 outer spines with teeth on medial margin, often obscure, depending on position of mount. Fifth leg variable in shape, most frequently resembling Figure 138; legs of several specimens resembling Figure 139.

Male. Body 1.22 mm long (0.99–1.45 mm, based on 15 specimens). Prosome 4-segmented, urosome 5-segmented (Figure 140).

First antenna 22-segmented, extending approximately to distal margin of caudal rami; segments 1-2, 8-9 fused, the latter separated from succeeding segment by incomplete suture. Second antenna (Figure 141) differing from that of female by lack of 2 setae each from first segment of endopod, inner lobe of second segment of endopod, and first 2 segments of exopod. Mandible (Figure 142), masticatory blade reduced to rounded process. Palp differs from that of female by lack of 3 setae each from basis and first segment of endopod and 2 setae from second segment of endopod. First maxilla (Figure 143) without gnathobase and with marked reduction of remaining proximal segments; number and extent of reduction of setae on these segments variable among specimens, as well as between right and left maxillae of each specimen. Second maxilla (Figure 144) reduced by alteration of setae; setae of proximal endites diminished in size and

***Ctenocalanus citer* VI**

130. Female, lateral view, 1.25 mm, E.
131. Female, genital field, 1.24 mm, D.
132. Female, right second antenna, 1.28 mm, A.
133. Female, right maxilliped, 1.30 mm, A.
134. Female, first leg, 1.25 mm, A.
135. Female, second leg, 1.25 mm, A.
136. Female, third leg, 1.25 mm, A.
137. Female, fourth leg, 1.25 mm, A.
138. Female, fifth leg, 1.30 mm, A.
139. Female, fifth leg, 1.26 mm, A.
140. Male, lateral view, 1.45 mm, D.
141. Male, left second antenna, 1.39 mm, A.
142. Male, right mandible, 1.30 mm, A.
143. Male, right first maxilla, 1.39 mm, C.
144. Male, left second maxilla, 1.41 mm, B.
145. Male, right maxilliped, 1.45 mm, A.
146. Male, first leg, 1.31 mm, A.
147. Male, second leg, protopod, 1.10 mm, A.
148. Male, third leg, protopod, 1.10 mm, A.
149. Male, fifth legs, 1.10 mm, A.
150. Male, fifth legs, 1.19 mm, A.

varying in number among specimens; 4 setae on fourth and fifth segments of endopod, excluding 1 fine seta of terminal segment, with structure altered from stiff, spinous setae into elongate, flexible, slightly plumose setae. Maxilliped (Figure 145), differing from that of female by lack of 8 setae from coxa, 2 setae from basis, and 1 seta from fourth segment of endopod.

First leg (Figure 146) differing from that of female by lack of outer short seta from each of first and second segments of exopod. Second leg differing from that of female in pattern of spines on distal margin of posterior surface of basis (Figure 147). Third leg differing from that of female in size of spines on distal margin of posterior surface of basis (Figure 148). Fourth leg resembling that of female. Fifth legs (2 forms were noted: Figures 149, 150) with 4-segmented left ramus; penultimate segment with several short setae extending from distal margin; terminal segment with group of short setae near inner margin. Right ramus 1-segmented, with 2 small distal spines.

Esterly [1924] described 2 forms of the male fifth leg for *C. vanus,* collected off the coast of California. The illustrations resemble those noted for *C. citer* but lack the right ramus.

Comparison of the Three Species, Stage VI

The differences noted in all of the preceding stages between *Clausocalanus laticeps* and *C. brevipes* in the shape of the terminal segment of the exopod of the second antenna could not be discerned in this stage.

The genital fields of the 3 species differ most conspicuously in their location on the genital segment. *C. laticeps* differs from the other 2 species by the greater expanse of the operculum; *C. brevipes* has the narrowest aperture of the free margin of the operculum; *Ctenocalanus citer* has the most complex pattern of sclerotization in the area of the seminal receptacles.

Clausocalanus laticeps and *C. brevipes* have complex patterns of sclerotization on the anterior surface of the basis of the fifth leg, surrounding the articulations of the rami. This feature evidently induces the appendage to become oriented in lateral view in a mounted preparation; and this position is the one in which the appendage has usually been illustrated, basis omitted, for species of the genus. When the fifth leg is mounted in posterior view, the basis can be distinctly defined.

DISCUSSION

In all species of calanoids in which the copepodid stages have been studied, the basic pattern is similar. The regular increase in body segments and in the number of swimming legs during development can be predicted with confidence for most calanoid species. In general, copepodid I has 2 pairs of swimming legs, copepodid II has 3 pairs, and copepodid III has 4 pairs. The division of the first 4 legs into segments is somewhat variable, but a few generalizations can be made: (1) Copepodid I always has 2 pairs of legs with unisegmental rami. (2) When a leg first appears, its rami are unisegmental. (3) Adult segmentation is developed in copepodid V.

Development of the first antenna varies among calanoid species, especially in copepodids I–III. This apparent variation may be attributed partly to difficulty in determining the exact numbers of antennal segments in early copepodids. Careful examination of many specimens was necessary before it was concluded that there are 13 segments in copepodid I and 18 in copepodid II. This is the highest number of segments yet reported for copepodid I (most calanoids are stated to have 9 or 10 segments), and we must admit that it is sometimes a rather arbitrary decision whether a suture is present or not. In copepodid III and older stages the segmentation is identical with that of *Pseudocalanus* as reported by Oberg [1905], which is a member of the same family (Pseudocalanidae) as *Clausocalanus* and *Ctenocalanus.*

Acknowledgments. The study was supported by National Science Foundation grant GA773 to Dr. I. Eugene Wallen, Office of Oceanography and Limnology, Smithsonian Institution. We are grateful to Dr. Wallen for making funds available from his grant.

Small wooden forceps with a hog's eyelash attached to each tip, designed by Dr. Martin W. Johnson, greatly facilitated the manipulation of the minute copepodids and their appendages. Our special thanks go to Dr. Johnson, who with his usual generosity donated a handmade pair of these forceps.

REFERENCES

Björnberg, T. K. S.
1963 On the marine free-living copepods off Brazil. Bolm. Inst. Oceanogr., *13*(1): 3–142, figs. 1–51.

Esterly, C. O.
1924 The free-swimming Copepoda of San Francisco Bay. Univ. Calif. Publs. Zool., *26*(5): 81–129, figs. A–P, tables 1–4.

Farran, G. P.
1929 Crustacea. Part 10. Copepoda. Br. Antarct. 'Terra Nova' Exped., 1910, Nat. Hist. Rep., Zool., *8*(3): 203–306, figs. 1–37, pls. 1–4.

Frost, B., and A. Fleminger
1968 A revision of the genus *Clausocalanus* (Copepoda: Calanoida) with remarks on distributional patterns in diagnostic characters. Bull. Scripps Instn Oceanogr., *12:* 1–99, figs. 1–11, pls. 1–67, tables 1–4, charts 1–16.

Giesbrecht, W.
1888 Elenco dei Copepodi pelagici raccolti dal tenente di vascello Gaetano Chierchia durante il viaggio della R. Corvetta 'Vettor Pisani' negli anni 1882–1885, e dal tenente di vascello Francesco Orsini nel Mar Rosso, nel 1884. Atti Accad. Lincei, Rendiconti, Rome, ser. 4, *4* (2): 330–338.
1892 Systematik und Faunistik der pelagischen Copepoden des Golfes von Neapel und der angrenzenden Meeres-abschnitte. Fauna Flora Golf. Neapel, monogr. 19. ix + 1–831, pls. 1–54.

Gurney, R.
1931 British fresh-water Copepoda. Vol. 1. lii + 238 pp., figs. 1–344. Ray Society, London.

Oberg, Max
1905 Die Metamorphose der Plankton-Copepoden der Kieler Bucht. Wiss. Meeresunters., Abt. Kiel. Neue Folge, *9:* 39–103, pls. 1–7.

Tanaka, Otohiko
1960 Pelagic Copepoda. Spec. Publ. Seto Mar. Biol. Lab., Biol. Results Jap. Antarctic Res. Exped. *10:* 1–95, pls. 1–40.

Unterüberbacher, H. K.
1964 Zooplankton studies in the waters off Walvis Bay with special reference to the Copepoda. *In* The Pilchard of South West Africa. Admin. South West Africa, Mar. Res. Lab., Investigational Rep. *11:* 1–42, pls. 1–36.

Vervoort, W.
1963 Pelagic Copepoda. Part I. Copepoda Calanoida of the families Calanidae up to and including Euchaetidae. Atlantide Rep., *7:* 77–194, figs. 1–23.

BENTHIC OSTRACODA (MYODOCOPINA: CYPRIDINACEA) FROM THE SOUTH SHETLAND ISLANDS AND THE PALMER ARCHIPELAGO, ANTARCTICA

LOUIS S. KORNICKER

Division of Crustacea, Smithsonian Institution, Washington, D. C. 20560

Abstract. Benthic Ostracoda (Myodocopina: Cypridinacea) from the South Shetland Islands and the Palmer Archipelago, Antarctica, are described, including 5 new species. Cypridinacea described from within the Antarctic Convergence by earlier authors were reexamined; supplementary descriptions of some of these are presented, and some are referred to different species. The data suggest depth zonation of Ostracoda in antarctic waters.

INTRODUCTION

The South Shetland Islands and the Palmer Archipelago are two island groups situated off the northwest coast of the Antarctic Peninsula (Figure 1). Anvers Island, the largest and southernmost member of the Palmer Archipelago, is the location of a United States base for antarctic exploration. Greenwich Island, which lies about 280 km northeast of Anvers Island, is a member of the South Shetland Islands and the location of one of the Chilean bases. The present study is founded on collections by Mr. James K. Lowry in 1967 from Arthur Harbor, Anvers Island, and by Dr. V. A. Gallardo in 1967–1968 from the vicinity of Greenwich Island. These collections were supplemented with 3 samples obtained in the area by Dr. Waldo L. Schmitt in 1963 while aboard the USS *Staten Island.* Collecting localities are shown in Figure 1. Station data are presented in Tables 1–3.

Myodocopid ostracoda from the South Shetland Islands and the Palmer Archipelago have not previously been reported. Daday [1908], however, described 2 new species collected by the French Antarctic Expedition (1903–1905) from nearby Booth Island (Wandel Island), which lies off the Graham Coast, and Kornicker [1969*b*] described a new species collected in 1964 during cruise 12 of the USNS *Eltanin* from the Bransfield Strait at a locality just northeast of the present study area. An additional species was reported by Skogsberg [1920] from near Snow Hill Island, which lies off the eastern coast of the Antarctic Peninsula. These localities are also shown in Figure 1.

Only 2 species (*Philomedes trithrix, Parasterope lowryi*) were collected in Arthur Harbor, Anvers Island, Palmer Archipelago, where the depth was 39 meters. One of these, *P. trithrix,* was also collected in Melchior Harbor, Gamma Island, Palmer Archipelago, where water depth was 45.7–49.4 meters. Both species were also present in the South Shetland Islands, where *P. trithrix* was collected at depths of 33 to 355 meters and *P. lowryi* at depths of 47 and 58 meters. At depths comparable to those sampled on the Palmer Archipelago, the South Shetland Islands provided 3 additional species, *Philomedes charcoti, Scleroconcha gallardoi,* and *Doloria levis.* Specimens of *S. gallardoi* were also found at greater depth. Three additional species, all members of the genus *Philippiella*—*P. skogsbergi, P. spinifera, P. pentathrix* — were collected in the vicinity of the South Shetland Islands at depths of 188 to 355 meters. Table 4 shows the distribution of myodocopids in the South Shetland Islands with samples arranged according to increasing depth. The data suggest depth zonation of Ostracoda in antarctic waters.

Prior studies of Cypridinacea collected from within the boundary marked by the Antarctic Convergence, but somewhat distant from the Antarctic Peninsula, include those by Brady [1907] and Barney [1921] from McMurdo Sound, Müller [1908] from Gauss Station, which was on the coast of Antarctica opposite the Antarctic Peninsula, and Scott [1912] from the South Orkney Islands. Of particular interest is the work of Skogsberg [1920] in the vicinity of South Georgia, because some of the species described from there are also found in the South Shetland Islands. I described one new species from South Georgia [Kornicker, 1969*a*] and one from the Amundsen Sea

TABLE 1. Stations at Which Ostracoda Were Collected by James K. Lowry, in Arthur Harbor, Anvers Island, Palmer Archipelago

For both, the depth was 39 meters on a soft bottom consisting mainly of silts; temperatures varied from —1.7°C in winter to 0°C in summer; salinity normally was 34 parts per thousand.

Station No.	Lat. S.	Long. W.	Collecting Dates (1967)
II	64°45′51″	64°05′51″	May 17, June 20, July 15
III	64°45′46″	64°06′17″	Feb. 20, July 14

[1969b]. Localities within the Antarctic Convergence from which Cypridinacea have been collected are shown in Figure 2. Localities of the widespread pelagic *Gigantocypris muelleri* [Tibbs, 1965] are omitted from the map. Species reported within the Antarctic Convergence are listed in Table 5.

I have taken the opportunity herein to give supplementary descriptions of *Philomedes charcoti* and *Philomedes laevipes* based on specimens obtained on loan from the National Museum of Natural History, Paris, and *Philippiella quinquesetae, Parasterope ohlini,* and *Scleroconcha appelloefi* based on specimens obtained on loan from the Swedish State Museum (Riksmuseum), Stockholm. The first two species were described from Booth Island by Daday [1908]; the remaining species were described from South Georgia by Skogsberg [1920].

Supplementary descriptions are also presented of 3 species, *Philomedes orbicularis, Philomedes assimilis, Philomedes antarctica,* described by Brady [1907] from McMurdo Sound. These are based on specimens in the type series of each species obtained on loan from the British Museum (Natural History) and the Hancock Museum, Newcastle-on-Tyne. Specimens identified from the South Orkney Islands as *Philomedes assimilis* and *Asterope australis* by Scott [1912] were obtained on loan from the Royal Scottish Museum, Edinburgh, and these species are discussed herein.

TABLE 2. Stations at Which Ostracoda Were Collected by Dr. V. A. Gallardo (with Peterson 0.1-m² Grab)

Station No.	Lat. S.	Long. W.	Depth, m	Bottom	Date
1	62°59.6′	59°26.2′	37	Sand	Dec. 21, 1967
2	62°59.4′	60°33.5′	99	Mud	Dec. 21, 1967
21	62°28.8′	59°39.6′	146	Mud	Jan. 11, 1968
22	62°28.5′	59°39.4′	196	Mud	Jan. 11, 1968
23	62°28.1′	59°39.0′	225	Mud	Jan. 11, 1968
24	62°27.7′	59°39.1′	228	Mud	Jan. 11, 1968
25	62°28.0′	59°38.3′	200	Mud	Jan. 11, 1968
26	62°28.2′	59°38.2′	90	Mud	Jan. 11, 1968
27	62°28.5′	59°38.4′	90	Sandy mud	Jan. 11, 1968
29	62°29.5′	59°40.1′	49	Sandy mud	Jan. 12, 1968
31	62°29.1′	59°41.0′	39	Sandy mud	Jan. 12, 1968
32	62°29.2′	59°41.5′	71	Sandy mud	Jan. 12, 1968
34	62°29.0′	59°42.1′	38	Sandy mud	Jan. 12, 1968
35	62°29.0′	59°42.6′	48	Sandy mud	Jan. 12, 1968
37	62°28.4′	59°41.4′	33	Sandy mud	Jan. 12, 1968
41	62°27.2′	59°37.6′	220	Sandy mud	Jan. 13, 1968
42	62°28.0′	59°36.6′	82	Sandy mud	Jan. 13, 1968
43	62°27.8′	59°36.4′	270	Sandy mud	Jan. 13, 1968
44	62°27.6′	59°37.5′	249	Mud	Jan. 13, 1968
45	62°27.4′	59°38.5′	139	Mud	Jan. 13, 1968
47	62°27.7′	59°40.5′	66	Sandy mud	Jan. 13, 1968
48	62°28.2′	59°40.9′	73	Sandy mud	Jan. 13, 1968
51	62°28.8′	59°40.6′	79	Sandy mud	Jan. 13, 1968
52	62°27.2′	59°37.8′	252	Mud	Jan. 17, 1968
54	62°26.1′	59°38.3′	347	Mud	Jan. 17, 1968
55	62°26.1′	59°37.5′	355	Mud	Jan. 17, 1968
56	62°25.8′	59°37.0′	274	Fine sand	Jan. 17, 1968
58	62°26.7′	59°40.1′	90	Mud	Jan. 17, 1968
60	62°27.9′	59°41.2′	54	Sandy mud	Jan. 17, 1968
61	62°28.1′	59°39.8′	188	Mud	Jan. 17, 1968

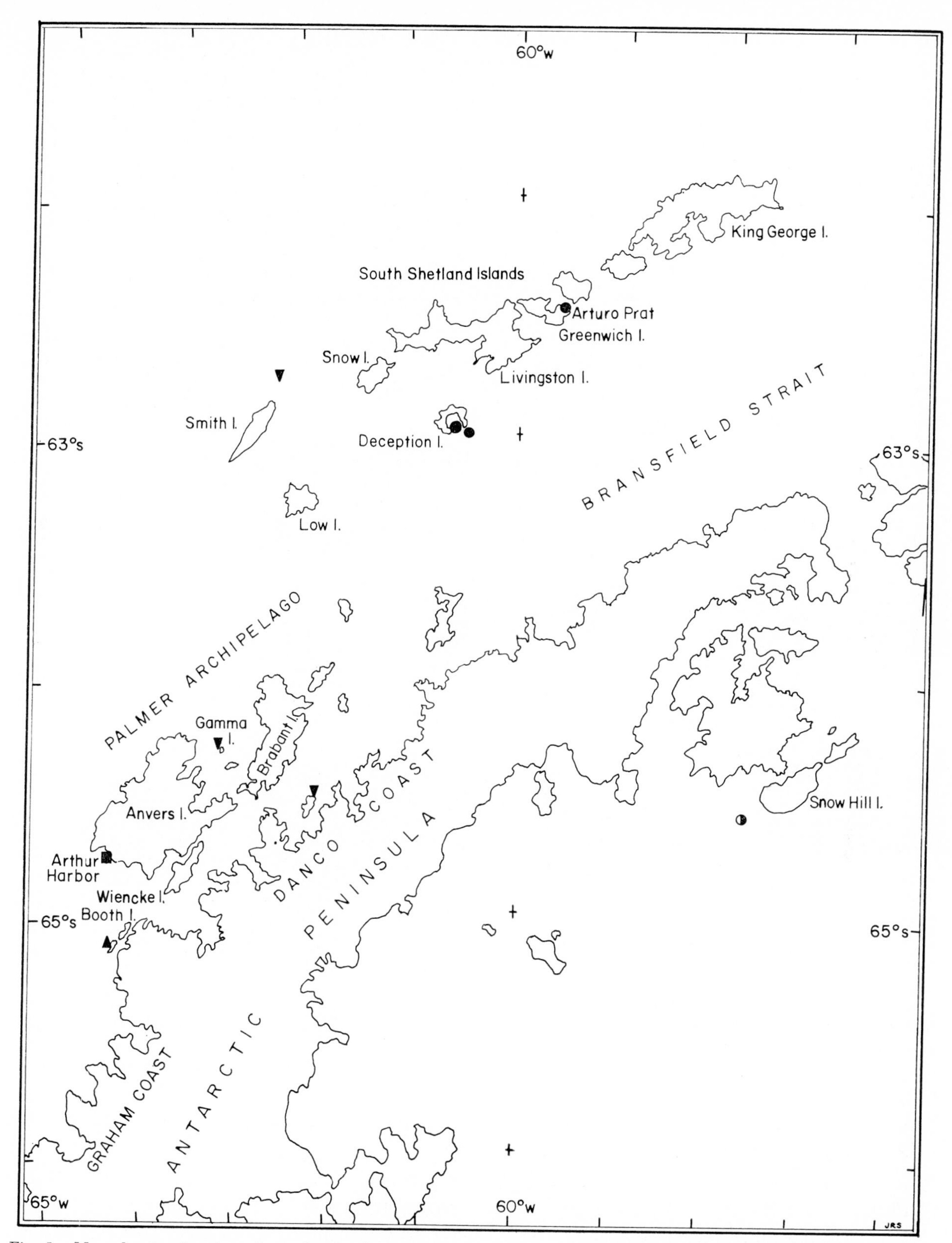

Fig. 1. Map showing location of samples in the vicinity of the Antarctic Peninsula from which Cypridinacea have been reported. Legend: filled-in circle, collections by Dr. V. A. Gallardo; inverted triangle, collections by Dr. Waldo L. Schmitt; square, collections by James K. Lowry; triangle, collections by the French Antarctic Expedition (1903–1905) described by Daday [1908]; half-filled circle, Station 6 of Swedish Antarctic Expedition described by Skogsberg [1920]. The base map is adapted from the Antarctic Strip Chart 16384-5, Naval Oceanographic Office.

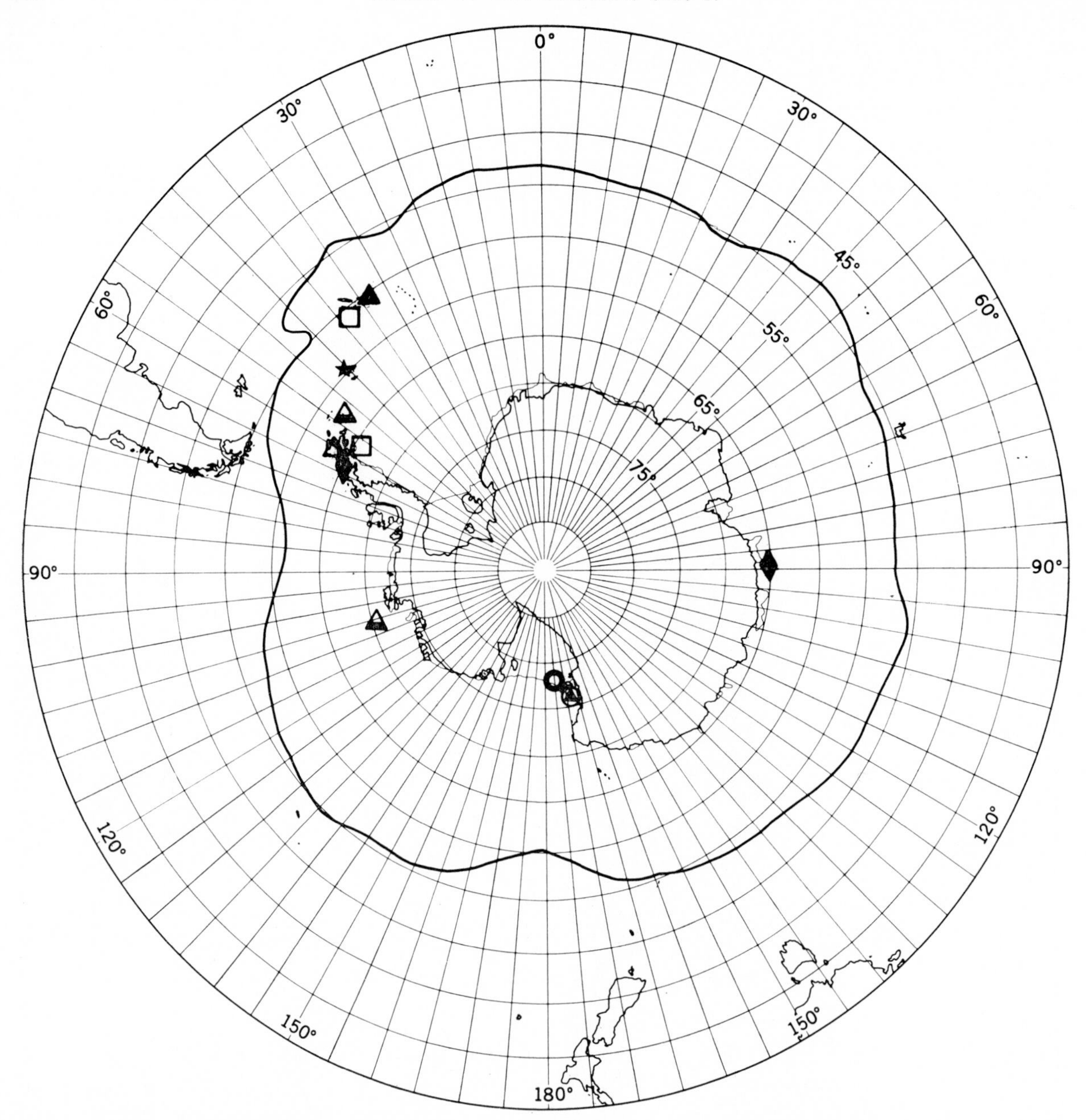

Fig. 2. Localities within the Antarctic Convergence (heavy line on map) from which Ostracoda of the suborder Cypridinacea have been collected. Legend: open triangle, study area of present report; open circle, Brady [1907]; half-filled diamond, Daday [1908]; filled diamond, Müller [1908]; star, Scott [1912]; open square, Skogsberg [1920]; triangle within circle, Barney [1921]; filled triangle, Kornicker [1969*a*]; half-filled triangle, Kornicker [1969*b*].

Family CYPRIDINIDAE

Genus ***Doloria*** Skogsberg, 1920

Type species: *Cypridina (Doloria) levis* Skogsberg, 1920, p. 225

Doloria levis (Skogsberg, 1920)

Fig. 3

Cypridina (Doloria) levis Skogsberg, 1920, pp. 225–237, figs. 31–34.

Doloria levis (Skogsberg); Poulsen, 1962, p. 146.

Material. USNM 125492, 1 juvenile female, USS *Staten Island,* Station 61-63, anchorage off Livingston Island, False Bay, South Shetland Islands, 62°42′S, 60°22′W, bottom depth 31.1 meters, bottom consisting of muddy coarse sand, gravel, and rocks.

Distribution. South Georgia [Skogsberg, 1920, p. 237] and South Shetland Islands. Collected at depths

TABLE 3. Stations at Which Ostracoda Were Collected by Dr. Waldo L. Schmitt

USS *Staten Island* Stations	Date, 1963	Lat. S	Long. W	Depth, m	Bottom	Locality
32–63	Feb. 6	64°19′14″	62°59′11″	45.7	Mud and sand	Melchior Harbor at anchorage off Gamma Island, Palmer Archipelago
37–63	Feb. 9	64°34′	62°00′	45.7 or 49.4	Gravel and sand	Danco Coast, Wilhelmina Bay, Foyn Harbor
61–63	Feb. 25	62°42′	62°22′	31.1	Mud, muddy coarse sand, gravel, rocks	South Shetland Islands, anchorage off Livingston Island

ranging from 24 to 310 meters. In the vicinity of South Georgia, bottom temperatures are reported by Skogsberg [1920, p. 237] as 1.2°C and 1.45°C.

Description of juvenile ♀ (? N-1). Carapace distorted; measurements of distorted specimen: length 2.19 mm, height 1.64 mm. Small eggs present within body.

First antenna: Sensory bristle of fifth joint with 8 long proximal filaments and 3 short, slender distal filaments; in remaining characteristics similar to that described by Skogsberg [1920, p. 226].

Second antenna: Similar to that described by Skogsberg [1920, p. 226].

Mandible: Proximal side of dorsal margin of second endopodite joint with 6 medium to long slender bristles and 7 short, pectinate bristles; surface spines not observed on first and second joints of endopodite; limb otherwise similar to that described by Skogsberg [1920, p. 227].

Maxilla (Figures 3a, b): Similar to that described by Skogsberg [1920, p. 229], except for a few minute spines near the middle of the largest d-bristle.

Fifth limb (Figure 3c): Similar to that described by Skogsberg [1920, p. 232].

Sixth limb (Figure 3d): First endite with 4 bristles, 2 long, 2 short; second endite same; third endite with 4 bristles, 3 long, 1 short; fourth endite with 3 or 4 bristles; end joint with 12 bristles; all long bristles of end joint with long marginal spines; 3 bristles present in place of epipodial appendage.

Seventh limb: 38 bristles present, each bristle tapering distally and with 1–5 bells (proximal ventral bristle without bell); otherwise similar to that described by Skogsberg [1920, p. 234].

Furca (Figure 3e): Each lamella with 8 claws separated from lamella by suture; all claws with teeth along concave margin.

Lateral eye (Figure 3j): Well developed with about 25 ommatophores in lateral view.

Medial eye and rod-shaped organ (Figures 3f, g): Medial eye fairly small. Rod-shaped organ thumblike.

Upper lip (Figures 3h, i): Lip consisting of single lobe with shallow ventral groove in posterior part; hairs present on smooth posterior surface and on small single lower lip.

Remarks. The specimen from the South Shetland Islands differs from that described by Skogsberg [1920] in having a few minute spines or teeth near the middle of the concave margin of the largest d-bristle of the maxilla. According to Skogsberg, all d-bristles are smooth. In this respect the South Shetland specimen seems to be intermediate between *D. levis* from South Georgia and *Doloria pectinata* [Skogsberg, 1920] from the Falkland Islands and Tierra del Fuego, on which all d-bristles are strongly pectinate.

Family Philomedidae

Genus ***Philomedes*** Liljeborg, 1853

Type species: *Philomedes longicornis* Liljeborg, 1853

Philomedes charcoti Daday, 1908

Figs. 4, 5

Philomedes charcoti Daday, 1908, p. 9, figs. 11, 12.

This species was described originally by Daday [1908] from collections made by the French Antarctic Expedition (1903–1905) near Booth Island. Daday described only the adult male. A supplementary description of the adult male is presented herein, based on a specimen obtained on loan from the National Museum of Natural History, Paris. The adult female is described for the first time from specimens in the Chilean collection from Bransfield Strait.

Material. The number, source, and collection data for specimens examined during this study are listed below. Four additional adult males from Booth Island in the Paris collection and 9 additional specimens from Station 1 in the Chilean collection were not studied in detail.

USNM	Number of Specimens	Locality	Date Collected	Remarks
—	1 adult ♂	Near Booth Island	Aug. 17, 1904	Paris museum
125442	1 adult ♀	Station 1	Dec. 21, 1967	Chilean collection
125443	1 adult ♀	Station 1	Dec. 21, 1967	Chilean collection
125444	1 adult ♀	Station 1	Dec. 21, 1967	Chilean collection

(Daday [1908, p. 12] states that specimens were collected from off the coast of the islands of Booth-Wandel on August 17, 1904. The label in the vial of specimens I examined states, '*Philomedes charcoti* Dad. Auct. det. 1908,' and to one side, 'Ins. Booth Wandel, Charcot, 1905.')

Supplementary description of male (Figure 4). Carapace with reticulate ornamentation appearing scalelike on anterior part of shell (Figure 4*a*); infold behind rostrum with 13 bristles (Figure 4*b*); infold anterior to caudal process with distribution of bristles similar to that on female.

Second antenna: Endopodite 3-jointed: first joint with 5 short bristles; second joint with 3 bristles; third joint reflexed with 1 long proximal bristle and 2 short subterminal bristles (Figure 4*c*); ridges present at tip of third joint. Exopodite: bristle of second joint with about 24 ventral spines, distal spines weak; bristles of joints 3–8 with natatory hairs; ninth joint with 6 bristles, 4 long, 1 medium, 1 short.

Mandible: Coxale endite small, strongly bifurcate, with short bristle at base (Figure 4*i*). Basale: medial surface with clusters of spines and 6 bristles, 5 proximal and 1 near middle; ventral margin with 6 bristles; dorsal margin with 3 bristles, 1 near middle and 2 terminal. Exopodite about 60% length of first endopodite joint and bearing 2 long bristles near tip. Endopodite: first joint with clusters of spines on medial surface and 4 ventral spines; second joint with clusters of spines on medial surface; ventral margin with 2 bristles in proximal group and 3 in distal group; dorsal margin with bristles roughly in 2 groups, 3 bristles in proximal group, 5 in distal group. End joint with 3 claws and 4 bristles, all claws with medial spines along middle section.

TABLE 4. Distribution of Myodocopida in the Vicinity of the Shetland Islands

Numerals in species columns represent the number of specimens of each species in sample.

Station*	Depth, m	*Doloria levis*	*Philomedes trithrix*	*Philomedes charcoti*	*Scleroconcha gallardoi*	*Parasterope lowryi*	*Philippiella skogsbergi*	*Philippiella spinifera*	*Philippiella pentathrix*
61–63	31.1	1							
37	33		8		2				
1	37			12					
34	38		4						
31	39		5		1				
35	48		2						
29	49		1						
60	54		1						
47	66		7			2			
32	71		1						
48	73		3						
51	79		2						
42	82		11						
26	90		6						
27	90		6						
58	90		36			3			
2	99				1				
45	139		1						
21	146		4						
61	188		3						1
22	196		5						
25	200		9						
41	220		6						
23	225		3		1				
24	228		3						
44	249		4						
52	252		5						
43	270		2						
56	274		26		12		3		11
54	347		3		1				
55	355		22		2			3	3

* Specimens from stations 61–63 were collected by Dr. Waldo L. Schmitt, the remainder by Dr. V. A. Gallardo.

Seventh limb (Figure 4d): Right limb with total of 15 bristles, 10 on one side and 5 on the other, 2 + 3 in distal group and 3 + 7 in proximal group, each bristle with 3 or 4 bells; terminal comb with about 13 teeth; 2 pegs present opposite comb. Left limb with only 12 bristles. (Specimen described by

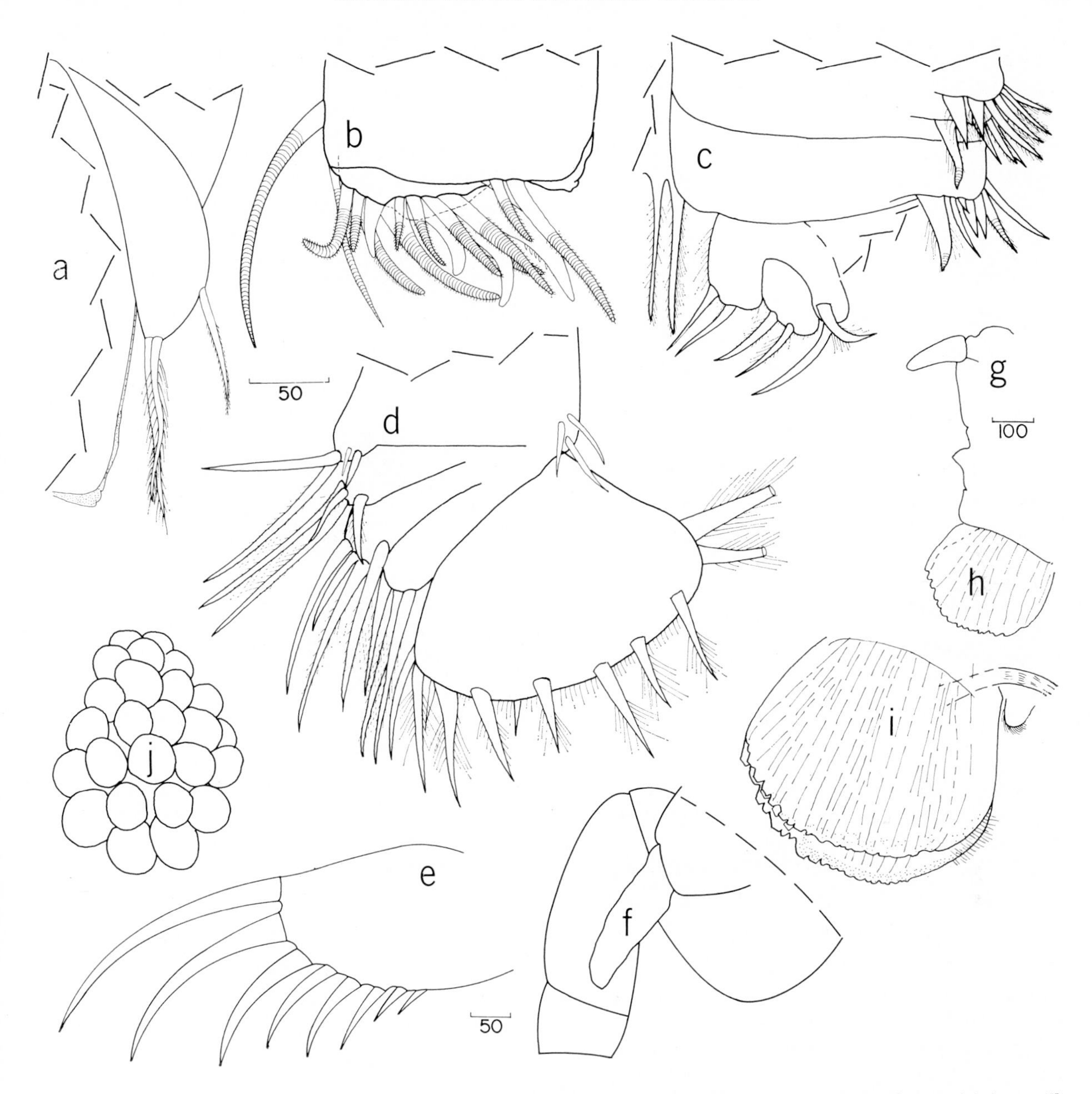

Fig. 3. *Doloria levis*, juvenile ♀: *a*, exopodite and cutting edge of first endopodite joint of maxilla; *b*, tip of left maxilla, outside view; *c*, part of fifth limb, anterior view; *d*, sixth limb; *e*, left lamella of furca; *f*, rod-shaped organ and joints 1–3 of first antenna, without bristles; *g*, rod-shaped organ and medial eye; *h*, *i*, upper lip; *j*, left lateral eye. (Same scale in microns: *a–d*; *e, f, i, j*; *g, h*.)

Daday [1908, p. 10] has 13 bristles, 6 on one side, 7 on the other.)

Furca (Figures 4f, g): Each lamella with 9 claws; claws 1–4 stouter than claws 5–9, all claws with marginal teeth; hairs at base of some claws.

Rod-shaped organ (Figure 4e): Organ elongate, 1-jointed, tapering to acute tip.

Eyes: Lateral eye with numerous (at least 20) ommatophores (Figure 4*h*). Medial eye similar in size to lateral eye (Figure 4*e*).

Description of female (Figure 5). Carapace with large rostrum and incisur, and truncate posterior (Figure 5*a*).

Ornamentation (Figures 5b, e): Surface with numerous pits within hexagons; hexagons seen more easily with transmitted light; surface appearing scale-

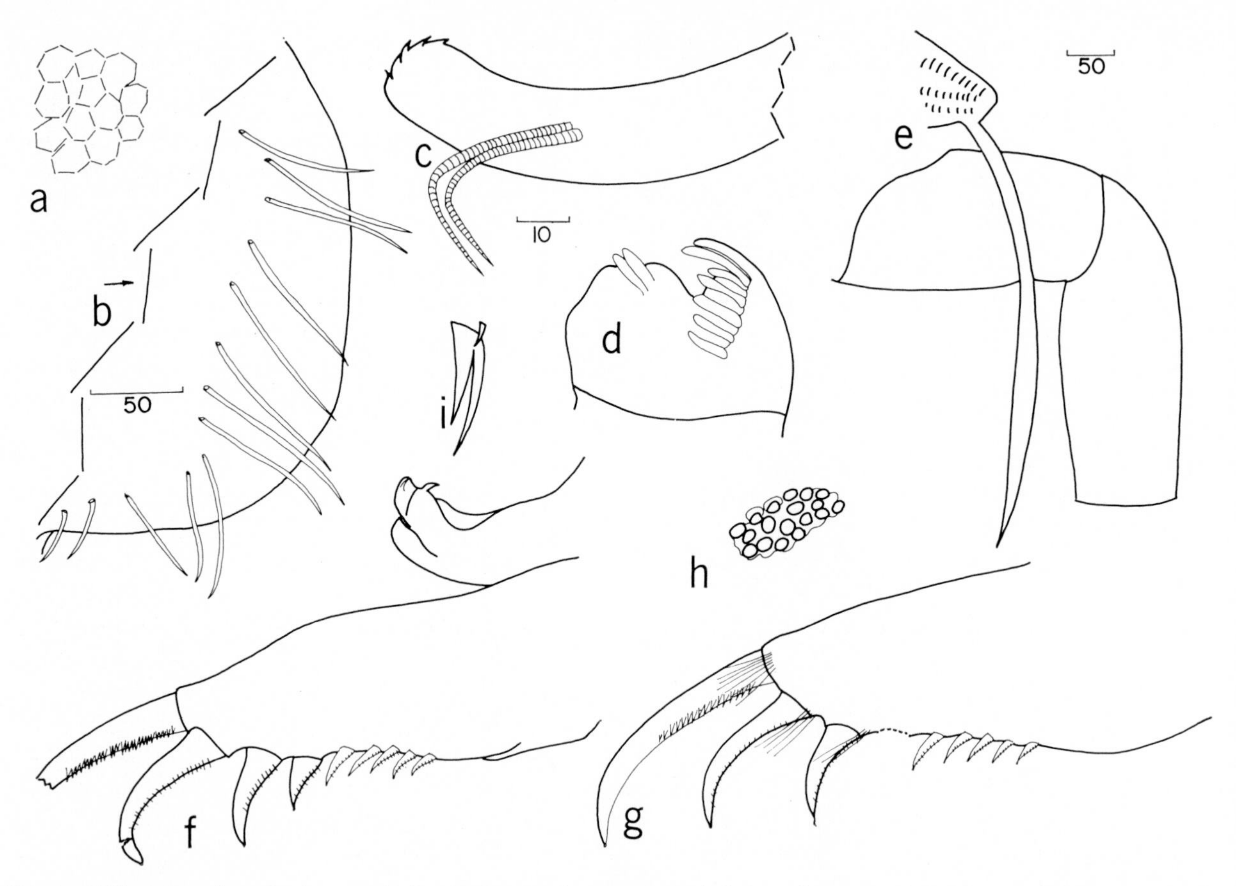

Fig. 4. *Philomedes charcoti*, ♂: *a*, reticulate structure of carapace, medial view; *b*, rostrum, left valve, medial view; *c*, tip of third joint endopodite, right second antenna, medial view; *d*, tip of seventh limb, bristles not shown; *e*, rod-shaped organ and medial eye; *f*, left lamella of furca and copulatory organ; *g*, right lamella, medial view; *h*, right lateral eye; *i*, coxale endite of right mandible, medial view. (Same scale in microns: *a*, *b*; *c*, *d*, *i*; *e–h*.)

like on anterior part of shell; few long hairs present being more numerous along free margin.

Infold (Figures 5c, d): Area behind rostrum with row of about 7 bristles; 3 bristles present along ventral margin of rostrum; 1 short bristle present below incisur; 7 bristles present on infold along anteroventral margin; infold along posterior margin with numerous groups of 1–4 short bristles near inner margin; middle of posterior infold with 5 bristles; infold along anteroventral margin with striations parallel to margin.

Selvage: Consists of narrow striated ridge with wide lamellar prolongation with fringed margin; lamellar prolongation in area of rostrum and incisur with segmented appearance.

Size: USNM 125442, length 1.87 mm, height 1.29 mm; USNM 125443, length 1.75 mm, height 1.24 mm; USNM 125444, length about 1.9 mm (shell distorted).

First antenna: Joints 1–5 with clusters of spines along dorsal margin; joints 1–2 with clusters of long spines on medial surface; second joint with 2 bristles, 1 ventral, 1 dorsal (some limbs also with 1 lateral bristle); third joint with 3 bristles, 1 ventral, 2 dorsal; fourth joint with 3–4 bristles, 2–3 ventral, 1 dorsal; sensory bristle of fifth limb with 4 proximal and 5 terminal filaments; sixth joint with spinous medial bristle. Seventh joint: a-bristle spinous; b-bristle with 1 proximal and 3 terminal filaments; c-bristle with 3 proximal and 5 terminal filaments. Eighth joint: d- and e-bristles bare and about equal in length to c-bristle; f-bristle with 3 proximal and 4 or 5 terminal filaments; g-bristle with 2 or 3 proximal and 5 terminal filaments.

Second antenna: Endopodite 2-jointed (Figure 5g): first joint with 6 short, bare bristles; second joint with 1 long, spinous ventral bristle and 1 long, bare subterminal bristle. Exopodite: first joint with small

TABLE 5. Cypridinacea within the Antarctic Convergence (excluding *Gigantocypris*)

Species	Palmer Archipelago	South Shetland Islands	McMurdo Sound	Antarctic Peninsula	Gauss Station	South Orkney Islands	South Georgia	Amundsen Sea
Cypridinidae								
Doloria levis		x					x	
Cypridina glacialis			x					
Vargula antarctica					x			
Philomedidae								
Philomedes orbicularis			x					
P. rotunda							x	
P. assimilis			x		x	x		
Philomedes species A			x					
P. trithrix	x	x						
P. charcoti		x		x				
Scleroconcha appelloefi				?x			x	
S. gallardoi		x						
Cylindroleberididae								
Parasterope ohlini							x	
P. lowryi	x	x						
P. curta							x	
P. glacialis					x			
Philippiella skogsbergi		x						
P. quinquesetae							x	
P. spinifera		x				?x	x	
P. pentathrix		x						
Sarsiellidae								
Spinacopia antarctica				x				
S. octo								x
S. menziesi							x	

medial spine on distal margin; joints 2–8 with comb of short spines and basal spine on distal margins; ninth joint with 7 bristles, 4 long, 3 shorter; all long bristles broken off near base, stumps bare.

Maxilla (Figures 5h–j): Coxale bristle short and spinous; precoxale and coxale with fringe of long hairs; basale with 3 distal bristles; exopodite with 3 bristles, 2 long, 1 short; first endopodite joint with 1 alpha- and 4 beta-bristles; end joint typical for genus; first endite with 9 bristles, second endite with 6 bristles, third endite with 9 bristles.

Mandible: Coxale: endite spinous, bifurcate at tip; small bristle present at base of endite. Basale: ventral margin with 5 spinous bristles; dorsal margin with 1 bristle near middle and 2 terminally; medial side with 5 bristles, 3 stout and pectinate, 2 slender and spinous. Exopodite reaching about two-thirds distance up first endopodite joint, and bearing 2 long subterminal bristles. Endopodite: ventral margin of first joint with 4 terminal bristles, 3 long, 1 short; dorsal margin of second joint with 2 bristles in proximal group and 5 in distal group; ventral margin of second joint with 2 bristles in proximal group and 3 in distal group; end joint with 3 claws, 2 long, 1 short, and several bristles.

Fifth limb (Figures 5k, l): Epipodial appendage with 54 bristles; main tooth with 4 teeth typical of genus; tip of large triangular tooth of second joint rounded (appears worn); bristles of second and third joint typical for genus; fourth + fifth joints with 5 spinous bristles.

Sixth limb (Figure 5m): 3 bristles in place of epipodite; first endite with 3 bristles, 1 long, 2 short; second endite with 1 proximal and 3 distal bristles; third endite with 1 proximal and 8 distal bristles; fourth endite with 1 proximal and 6 or 7 distal bristles; end joint with 17–18 bristles, all bristles on limb spinous; end joint hirsute.

Seventh limb (Figure 5n): Each limb with 15–16 bristles, 5 distal, 10–11 proximal; each bristle with 3–7 bells and marginal spines distally; terminal comb with 12 teeth; most teeth with secondary proximal teeth and lateral flange; 2 spinous pegs present opposite comb.

Rod-shaped organ (Figure 5p): Elongate, 1- or 2-jointed with short spines at tip.

Eyes: Lateral eyes elongate with 2 ommatophores (Figure 5*q*); medial eye large and pigmented (Figure 5*p*).

Anterior and upper lip (Figure 5r): Anterior with rounded process near middle; upper lip projecting anteriorly, hirsute.

Furca (Figure 5o): Each lamella with 10 claws: claws 1–5 stout, remaining claws more slender, all with marginal teeth; tips of claws 1–3 or 1–5 appear shortened by wear on all specimens, tips being rounded on some and broken on others; long hairs present at base of some claws.

Comparisons. *P. charcoti* differs from *P. eugeniae* Skogsberg, 1920, in having 2 instead of 3 bristles on the second joint of the endopodite of the second antenna of the female. *P. charcoti* differs from *P. rotunda* Skogsberg, 1920, in shape of carapace and in having 2 instead of 3 pegs on the seventh limb and in having fewer bristles on the limb. The female of *P. charcoti* differs from *P. laevipes* Daday in shape of carapace.

Brady [1907] described 2 new species of *Philomedes* each having a carapace with a shape similar to that of *P. charcoti: P. assimilis* and *P. antarctica.* I have examined type material of these species presently at the British Museum (Natural History) and find they both differ from *P. charcoti* in having 5 bristles on the second joint of the endopodites of the second antennae of the female and in having only 9 bristles on the seventh limb.

Remarks. The female specimens described above did not have eggs in their brood chambers; therefore, that criterion could not be used to identify them as adults. The presence of unextruded eggs within the body places them as not being younger than the N-1 stage. Bristles on the exopodite of the second antenna were broken on all specimens. This is characteristic of adults of many species and may indicate that the present specimens are adult. However, only bristles of the distal joints are usually broken in previously described species, whereas all bristles were broken on the present specimens; so two different phenomena may be involved.

The main claws on the furcae of all specimens were rounded or broken or showed other signs of wear. This suggests that the stoutness of these claws is in part due to removal of the more slender distal part of each claw.

Philomedes trithrix, new species

Figs. 6–12

Holotype. USNM 125200, gravid ♀, length 2.61 mm, height 1.97 mm. Valves and some appendages in alcohol, remaining appendages on slides.

Paratypes from Arthur Harbor, Anvers Island. USNM 125201, gravid ♀, Station 3, July 14, 1967; USNM 125202, ♀ without eggs, Station 2, June 20, 1967; ♀ without eggs (returned to Mr. James K. Lowry), Station 2, June 20, 1967; USNM 125203, USNM 125204, adult males, Station 2, May 17, 1967; USNM 125208, ♀ without eggs, Station 2, July 15, 1967; USNM 125209–125212, 4 juveniles from Station 2, July 15, 1967; USNM 125213–125215, 3 juveniles from Station 3, February 20, 1967.

Paratypes from off Greenwich Island. USNM 125390, adult ♀, Station 45; USNM 125391, adult female, Station 47; USNM 125392, adult ♀, Station 37; USNM 125393, adult ♀, 10 eggs in brood chamber, Station 55; USNM 125394, juvenile ♂, Station 47; USNM 123395, adult ♀, 5 eggs in brood chamber, Station 42; USNM 125396, adult ♀ with eggs in brood chamber (not dissected), Station 58. In addition, 182 specimens were returned to Tomás Cekalovic K., Curador del Museo, Departamento de Zoología, Instituto Central de Biología, Universidad de Concepción, Concepción, Chile.

Paratype from Melchior Harbor. USNM 125493, adult ♀, Stations 32–63.

Type locality. Arthur Harbor, Anvers Island, Station 3.

Other localities. Arthur Harbor, Anvers Island, Station 2; Greenwich Island vicinity, Stations 21, 22, 23, 24, 25, 26, 27, 29, 31, 32, 34, 35, 37, 41, 42, 43, 44, 45, 47, 48, 51, 52, 54, 55, 56, 58, 60, 61; USS *Staten Island*, Stations 32–63, Melchior Harbor, off Gamma Island.

Etymology. The specific name 'trithrix,' from the Greek 'tri,' three, and 'thrix,' hair, refers to the three bristles on the second joint of the endopodite of the female second antenna.

Distribution. The number of specimens collected at each station is as follows: Arthur Harbor, Station 2 (numeral following date is number of specimens): May 17, 1967—2; June 20, 1967—2; July 15, 1967—5. Arthur Harbor Station 3: February 20, 1967—3; July 14, 1967—2. Greenwich Island vicinity (numbers

Fig. 5. *Philomedes charcoti*, ♀ carapace: *a*, complete specimen, length 1.87 mm; *b*, detail of punctations on shell; *c*, posterior right valve, medial view; *d*, anterior left valve, medial view; *e*, detail of reticulations from posterior of right valve, medial view; *f*, dorsal muscles of right valve, medial view. Appendages: *g*, endopodite, right second antenna; *h*, left maxilla, medial view; *i*, right maxilla, marginal spines not shown, lateral view; *j*, bristles on end joint of right maxilla, not all bristles shown, lateral view; *k*, part of right fifth limb, anterior view; *l*, part left fifth limb; posterior view; *m*, sixth limb, marginal spines of bristles not shown; *n*, tip of seventh limb; *o*, furca; *p*, rod-shaped organ and medial eye; *q*, lateral eye; *r*, anterior of animal from medial eye to upper lip. (Same scale in microns: *c, d, f, h, i, m, o, p, r; e, g, j–l, q*. Figures *o, q*, from USNM 125443, figure *p* from USNM 125444, remaining figures from USNM 125442.)

a
b
c
d
50
e
f
g
h
i
j
k
50
l
m
n
10
o
p
q
r

are station and number of specimens): 21—4, 22—5, 23—3, 24—3, 25—9, 26—6, 27—6, 29—1, 31—5, 32—1, 34—4, 35—2, 37—8, 41—6, 42—11, 43—2, 44—4, 45—1, 47—7, 48—3, 51—2, 52—5, 54—3, 55—22, 56—26, 58—36, 60—1, 61—3. One specimen was collected in Melchior Harbor, off Gamma Island, USS *Staten Island* Stations 32–63. Specimens were collected at depths ranging from 33 meters to 355 meters and on bottoms consisting of sand, fine sand, sandy mud, silts, and mud. In Arthur Harbor, the temperature ranged from −1.7°C to 0°C, and the salinity was normally 34 parts per thousand.

Description of female (Figures 6, 7a, b, h). Carapace oval in lateral view with greatest height near middle (Figures 6*a*, *b*); broad rostrum overlapping large wide incisur; each valve with small caudal process not always showing in lateral view (Figures 6*c*, *d*).

Ornamentation (Figure 6e): Surface with faint pits, numerous small spinelike hairs, and few scattered very long hairs.

Infold: Infold of rostrum with 14–18 long, spinous bristles paralleling anterior margin and 3–4 bristles paralleling ventral margin (Figure 7*a*); infold posterior to incisur with short bristle; anteroventral part of infold striate and with about 16 spinous bristles; caudal process with 2–4 bristles along outer edge and 3–5 bristles at middle of infold (Figures 6*c*, *d*); infold along posterior and ventral margins with numerous minute bristles along inner margin.

Selvage: Striate lamella prolongation with fringed edge consisting of long and short spines present on all except hinge margins of each valve.

Muscle scars: Central muscle scars not clearly visible laterally.

Size: USNM 125200, gravid ♀, length 2.61 mm, height 1.97 mm; USNM 125201, gravid ♀, length 2.59 mm; height 1.99 mm; USNM 125202, adult ♀, length 2.65 mm, height 2.00 mm; adult ♀, length 2.58 mm, height 1.90 mm; USNM 125208, ♀ without eggs, length 2.50 mm, height 1.91 mm; USNM 125390, adult ♀, eggs within body, bristles of second antennae broken, length 2.33 mm, height 1.77 mm; USNM 125391, adult ♀, eggs within body, bristles of second antennae broken, length 2.46 mm, height 1.81 mm; USNM 125393, gravid ♀, 10 eggs in brood chamber, length 2.59 mm, height 1.91 mm; USNM 125493, ♀ without eggs, length 2.48 mm, height 1.79 mm. Range: length 2.33–2.65 mm; height 1.77–2.00 mm.

First antenna: First and second joint with spines in clusters on medial and lateral surfaces; second joint with ventral, dorsal, and lateral bristle, all with wreaths of long spines; third joint with 1 ventral and 2 dorsal spinous bristles; fourth joint with 4 spinous bristles, 3 ventral, 1 dorsal; bristle on fifth joint with 4 proximal and 4 terminal filaments; bristle on sixth joint with wreath of long spines and short spines distally; a-bristle spinous; b-bristle with 1 proximal filament and 4 terminal filaments; c-bristle with 5 proximal and 4 terminal filaments; d- and e-bristles bare and of equal length; f- and g-bristles with 3 proximal and 4 terminal bristles.

Second antenna: Endopodite 2-jointed (Figure 6*f*): First joint with 5 bristles, 4 in proximal group, 1 distal; second joint elongate with proximal, distal, and terminal bristle: proximal bristle 2 to 2½ times length of joint, with numerous long spines along middle part and short spines distally; distal bristle about half length of proximal bristle, with few long spines near middle and short spines distally; terminal spine usually recurved, bare; short spines forming few small clusters on joint surface. Exopodite: First joint with small terminal medial spine; joints 2 to 8 with minute spines forming clusters along distal margin; bristle of second joint extends well past ninth joint; ninth joint with 6 bristles, 4 long, 1 medium and 1 short; bristles on joints 6 to 8 and 4 long bristles on ninth joint broken; medium and short bristle of ninth joint with spines, all remaining bristles without spines and natatory hairs.

Two gravid females (USNM 125200, USNM 125201) and 1 adult female without eggs (USNM 125202) from Anvers Island and others from Greenwich Island

Fig. 6. *Philomedes trithrix* Kornicker, ♀: *a*, complete carapace, length 2.65 mm, height 2.0 mm; *b*, left valve, medial view; *c*, posterior right valve, medial view; *d*, posterior left valve, medial view; *e*, detail of surface, right valve, lateral view; *f*, right second antenna, medial view; *g*, exopodite of left mandible, lateral view; *h*, posterior margin of basale of right mandible, medial view; *i*, main tooth and distal part of second joint of right fifth limb, posterior view; *j*, right sixth limb, medial view, spines and hairs on bristles not shown; *k*, tip of seventh limb; *l*, lateral eye, medial eye, rod-shaped organ, anterior of body, upper lip; *m*, upper lip and mouth, ventral view, anterior toward bottom of page; *n*, protozoon on protopodite of second antenna. (Same scale in microns: *c–h*, *j*, *m*; *i*, *n*. Figure *a* from USNM 125202, figures *e*, *m*, *n* from USNM 125201, remaining figures from USNM 125200.)

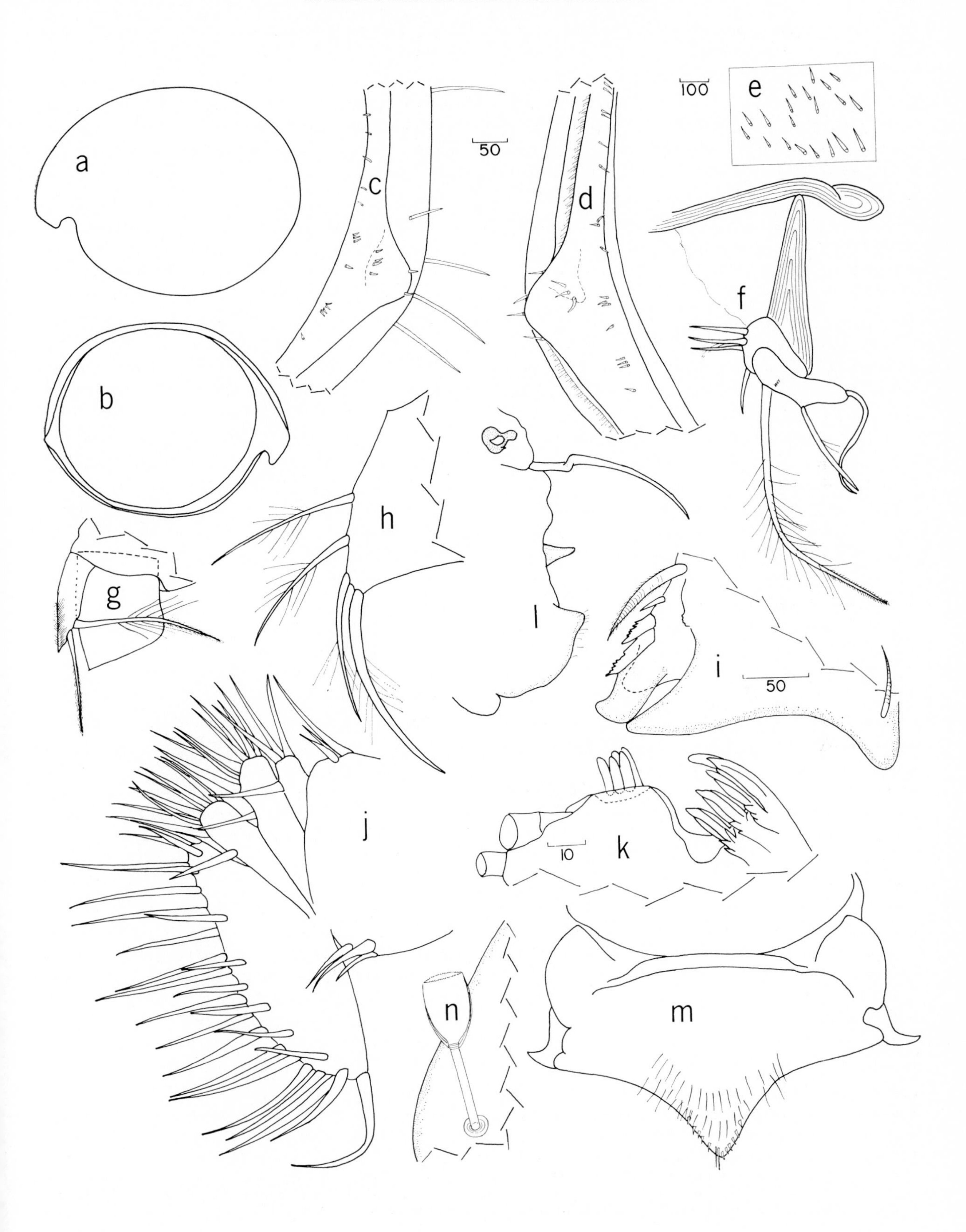
a
b
c
50
d
100
e
f
g
h
l
i
50
j
k
10
m
n

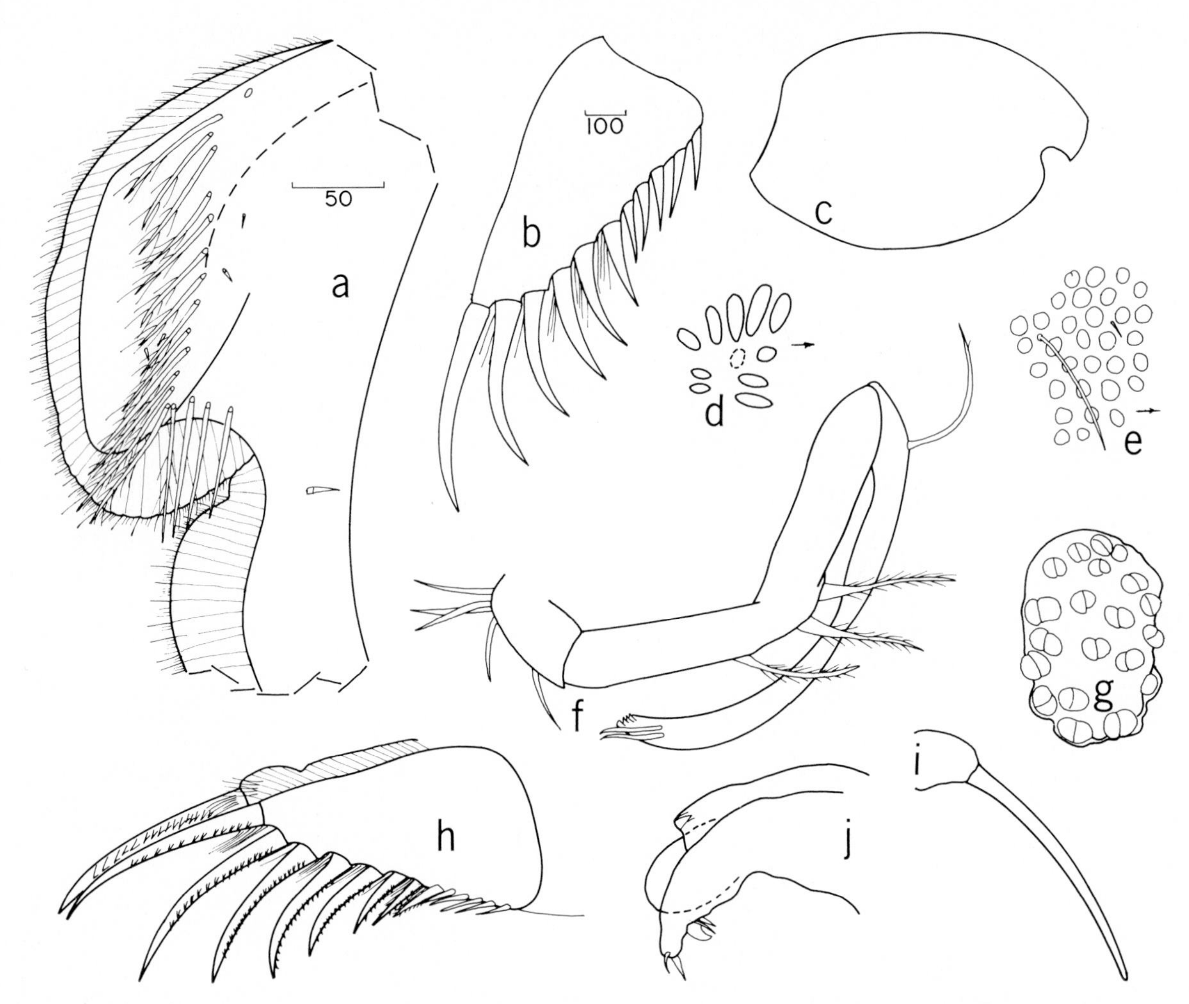

Fig. 7. *Philomedes trithrix*, ♀: *a*, anterior right valve, medial view; *b*, left lamella of furca. ♂: *c*, complete carapace, length 2.53 mm; *d*, central muscle scars of right valve (not all scars visible), lateral view; *e*, detail of punctations on right valve near middle, lateral view; *f*, endopodite, second antenna; *g*, lateral eye. ♀: *h*, furca. ♂: *i*, medial eye and rod-shaped organ; *j*, copulatory organ. (Same scale in microns: *b, h, i; a, e–g, j*. Figures *a, b* from USNM 125200; figure *h* from USNM 125208; remaining figures from USNM 125203.)

stations are as described above. One adult female without eggs (USNM 125208) differs from the above description in having the long bristles of the exopodite unbroken. The long bristles on joints 6 to 9 are about 3 times the length of bristles on joints 2 to 5 and have natatory hairs.

Mandible (Figures 6g, h): Coxale endite large, bifurcate, with short, triangular teeth and long, thin spines; small bristle present at basis of endite; basale with 6 medial-proximal bristles: 3 with wreaths of long hairs, 3 with stout marginal teeth; 4 spinous bristles forming row on lateral surface paralleling ventral margin (proximal bristle of the 4 is on ventral margin on some specimens); 3 distal bristles present on ventral margin; dorsal margin of basale with 2 terminal bristles and 2 bristles distal to middle, all bristles with long spines; long spines forming clusters on medial and lateral surface of coxale and basale; exopodite about 80% length of first endopodite joint and bearing 2 bristles longer than exopodite; first endopodite joint with 5 ventral spines; dorsal margin of second endopodite joint with 3 groups of bristles: proximal group with 2–4 bristles, middle group with 1 bristle, distal group with 6 bristles; ventral margin with 2 groups of bristles: proximal group with 4 bristles, distal group with 3 bristles; medial surface of joint with spines forming clusters; end joint with 3 bare claws and 4 bristles.

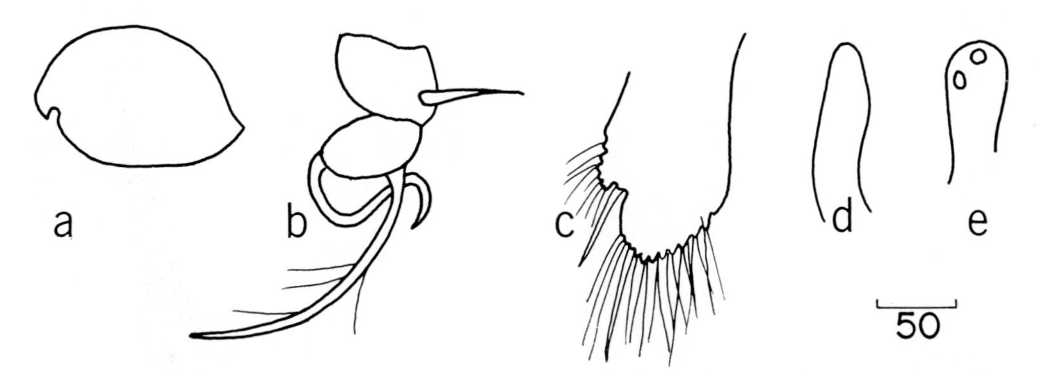

Figure 8. *Philomedes trithrix*, ? ♂ instar II, USNM 125213: *a*, complete carapace, length 0.83 mm; *b*, endopodite of second antenna; *c*, sixth limb; *d*, seventh limb; *e*, lateral eye. (Same scale in microns: *b–e*.)

Maxilla: Normal for genus; first endopodite joint with 1 alpha- and 4 beta-bristles, all with marginal spines.

Fifth limb (Figure 6i): Epipodial appendage with 58 bristles; first endite with 6 bristles; main tooth of first exopodite joint with large, square tooth followed by 2 teeth with serrate edge and small spinelike tooth; triangular, smooth protuberance present anterior to main tooth; small bristle present on posterior side near outer corner of large tooth of second endopodite joint; posterior side of second joint with usual 4 bristles; third endopodite joint with usual 3 inner bristles and 2 outer bristles, both outer bristles with long marginal spines; fourth and fifth joints hirsute, with 7 spinous bristles.

Sixth limb (Figure 6j): Epipodial appendage with 3 spinous bristles; first endite with 3 spinous bristles, 2 short, 1 long; second endite with 4 spinous bristles, 1 proximal, 3 terminal; third and fourth endite each with 9 spinous bristles, 1 proximal, 8 terminal; end joint with 27–28 spinous bristles; lateral and medial surfaces of end joint hirsute.

Seventh limb (Figure 6k): Each limb with total of 36–43 bristles, each with 3–8 bells and a few short marginal spines distally; terminal comb with 9–15 curved teeth having lateral wings and 1 or 2 small teeth on each side at base; 7–12 pegs about 3½ times long as wide present opposite comb.

Furca (Figures 7b, h): Distribution of claws normal for genus; 11 to 13 claws present; claws 7 to 10 or 13 with fine spines along anterior and posterior margins and could be considered secondary; claws 1–6 with teeth along concave margin; long spines forming medial clusters at base of claws.

Rod-shaped organ and medial eye (Figure 6l): Medial eye well developed. Rod-shaped organ elongate, widening in middle, weakly 2-jointed on some specimens.

Lateral eye (Figure 6l): Small with only 2 or 3 ommatophores.

Upper lip and anterior (Figures 6l, m): Upper lip simple, hirsute, projecting anteriorly. Single pointed projection present in the middle of anterior margin between upper lip and rod-shaped organ.

Eggs: 13 oval eggs in brood chamber of USNM 125200; 7 eggs in chamber of USNM 125201; 10 eggs in chamber of USNM 125393; 5 eggs in chamber of USNM 123395.

Epibiota: Many appendages with filamentous and stalked protistans (Figure 6n).

Description of adult male (Figures 7c–g, i, j): Carapace more elongate than female and with prominent caudal process (Figure 7c); surface with few scattered long and short hairs; small pits, usually faint or absent on female, abundant on male (figure 7e); central muscle scars consisting of about 11 individual scars (Figure 7d). Size: USNM 125203, length 2.53 mm, height 1.55 mm; USNM 125204, length 2.61 mm, height 1.55 mm.

First antenna: Second, third, fourth, sixth joints with clusters of spines on medial surface; second joint with 3 bristles: 1 ventral, 1 dorsal, 1 lateral; third joint with 1 ventral and 2 dorsal bristles; fourth joint with 2 spinous bristles: 3 ventral, 1 dorsal; sensory bristle of small fifth joint with many filaments on the broad proximal part and 4 terminally; sixth joint with short medial bristle; bristles of joints 7–8 typical for genus.

Second antenna: Endopodite 3-jointed (Figure 7f): First joint with 5 bristles, 4 proximal and 1 distal;

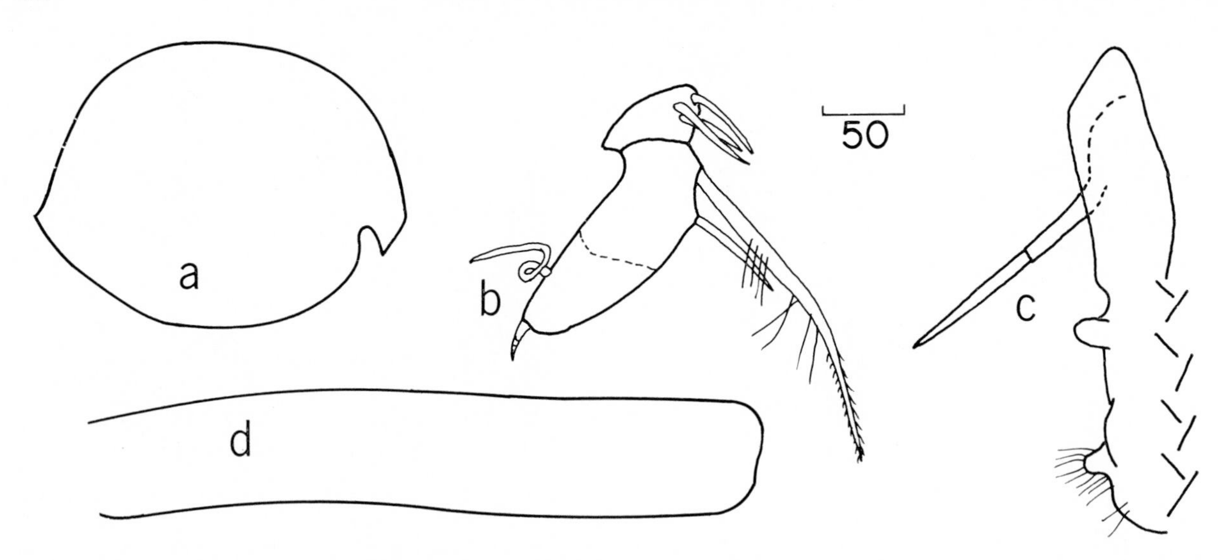

Fig. 9. *Philomedes trithrix*, ? ♀ instar III, USNM 125209: *a*, complete carapace, length 1.52 mm; *b*, endopodite of second antenna; *c*, anterior of animal including rod-shaped organ, medial eye, upper lip; *d*, seventh limb. (Same scale in microns: *b–d*.)

distal bristle about same length as proximal bristles; second joint with 3 spinous bristles; third joint with 1 proximal and 2 short subterminal bristles; ridges present at tip of third joint. Exopodite: bristle of second joint without marginal spines or natatory hairs; bristles of joints 3–8 with natatory hairs; ninth joint with 4 long and 1 short bristles, short bristle with few hairs, long bristles with many long hairs.

Mandible: Coxale endite small, bifurcate at tip, with short bristle at base. Basale: medial surface with clusters of spines and 6 bristles, 5 proximal and 1 near middle; ventral margin with 5 spinous bristles; dorsal margin with 4 bristles as on female. Exopodite about 60% length of first endopodite joint and bearing 2 bristles similar to female. Endopodite: First joint with clusters of spines on medial surface and 5 ventral spines; second joint with clusters of spines on medial surface; ventral margin of second joint with 3 bristles in proximal and distal groups; dorsal margin as in female; end joint as in female.

Maxilla, fifth limb: Reduced, typical for genus.

Sixth limb: Epipodial appendage with 3 plumose bristles; first endite with 3 bristles; second endite with 1 proximal and 3 distal bristles; third endite with 1 proximal and 8 terminal bristles; fourth endite with 1 proximal and 6 terminal bristles; end joint with 18 bristles; surface of endites 2–4 and end joint hirsute.

Seventh limb: Each limb with total of 29 to 31 bristles (1 specimen), each bristle with 2–4 bells; terminal comb with about 10 curved teeth; 3 pegs present opposite comb.

Furca: Each lamella with 9 to 10 claws.

Rod-shaped organ and medial eye (Figure 7i): Similar to female.

Lateral eye (Figure 7g): Large with about 20 divided ommatophores in lateral view.

Copulatory organ (Figure 7j): Elongate but small, about two-thirds the length of claw 1 of furca.

Upper lip: Similar to that of female.

Description of ? male instar II (Figure 8).

Carapace: Similar in shape to adult male (Figure 8*a*); size: USNM 125213, length 0.83 mm, height 0.58 mm.

First antenna: Third joint right limb with 2 bristles, 1 ventral, 1 dorsal; left limb with 3 bristles, 1 ventral, 2 dorsal; fourth joint with 1 dorsal and no ventral bristles.

Second antenna: Ninth joint of exopodite with 3 bristles; all exopodite bristles with short marginal teeth or spines.

Endopodite (2-jointed) (Figure 8b): First joint with 1 bristle; second joint with 1 spinous ventral bristle and 1 recurved terminal bristle.

Mandible: Dorsal margin of basale with 1 bristle near middle and 2 terminal.

Sixth limb: Bilobed with 1 bristle between lobes; both lobes with long marginal spines (Figure 8*c*).

Seventh limb: Short, bare (Figure 8*d*).

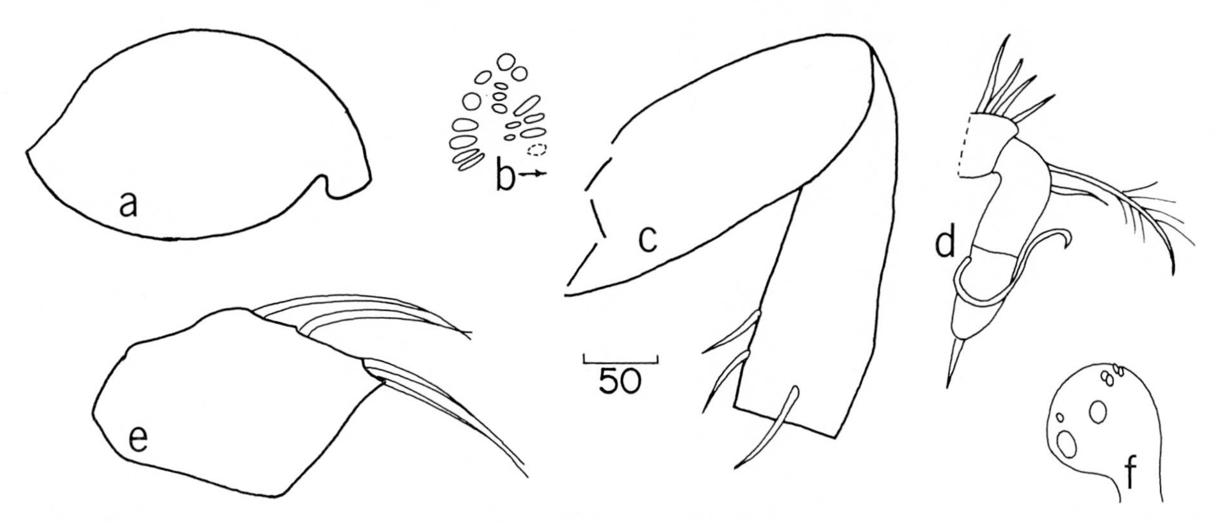

Fig. 10. *Philomedes trithrix*, ? ♂ instar ? IV, USNM 125214: *a*, complete carapace, length 1.52 mm; *b*, muscle scars, right valve, lateral view; *c*, joint 1, 2, of first antenna; *d*, endopodite, second antenna; *e*, mandibular basale with bristles on dorsal margin; *f*, lateral eye. (Same scale in microns: *c–f*.)

Furca: Each limb with 6 claws. Lateral eye small with 2 or 3 ommatophores (Figure 8*e*).

Description of ? female instar III (Figure 9).

Carapace: Similar in shape and ornamentation to adult female, but with more prominent caudal process (Figure 9*a*).

Size: USNM 125209, length 1.52 mm, height 1.22 mm; USNM 125210, length 1.65 mm, height 1.22 mm.

First antenna: Fourth joint with 1 ventral and 1 dorsal bristle, number of bristles on remaining joints similar to adult female.

Second antenna: Ninth joint with 3–4 bristles, shortest of these with short spines, others bare; bristles of joints 2–8 bare; first joint with short medial spine.

Endopodite (weakly 3-jointed) (Figure 9b): First joint with 3 bristles; second joint with 2 ventral bristles; third joint with 1 coiled or recurved dorsal bristle and 1 short terminal bristle.

Fifth limb: Epipodial appendage with 39 bristles.

Sixth limb: Has numerous bristles.

Seventh limb: Long, bare (Figure 9*d*).

Furca: Each lamella with 8 claws.

Medial eye: Similar to adult.

Rod-shaped organ: 1- or 2-jointed with 1 or 2 hairs at tip (Figure 9*c*).

Description of ? male instar ? IV (Figure 10).

Carapace: Similar in shape to adult male (Figure 10*a*).

Size: USNM 125214, length 1.52 mm, height 1.01 mm.

First antenna: Ventral margin of second joint with 2 bristles (Figure 10*c*); fourth joint with 1 dorsal and 2 ventral bristles.

Second antenna: Endopodite (Figure 10*d*): Similar to ? female instar III.

Mandible: Dorsal margin of basale with 4 bristles (Figure 10*e*).

Sixth limb: Numerous bristles.

Seventh limb: 4 proximal and 4 distal tapering bristles.

Lateral eye: Small with about 3 ommatophores (Figure 10*f*).

Furca: Each lamella with 8 claws.

Description of male instar ? V (Figure 11).

Carapace: Similar in shape to adult male (Figure 11*a*).

Size: USNM 125215, length 1.84 mm, height 1.22 mm; USNM 125394, length 1.82 mm, height 1.15 mm.

First antenna: Second joint with 2 ventral bristles and no dorsal bristles; fourth joint with 2 ventral and 1 dorsal bristles; number of bristles on other joint same as on adult female.

Second antenna: Endopodite (3-jointed) (Figure 11*d*): First joint with 4 short bristles; second joint with 1 long and 2 short ventral bristles; third joint with 1 long, recurved dorsal bristle and 2 short terminal bristles.

Exopodite: Bristles of joints 6–9 short and without natatory hairs.

Mandible: Dorsal margin of basale with 4 bristles as in adult.

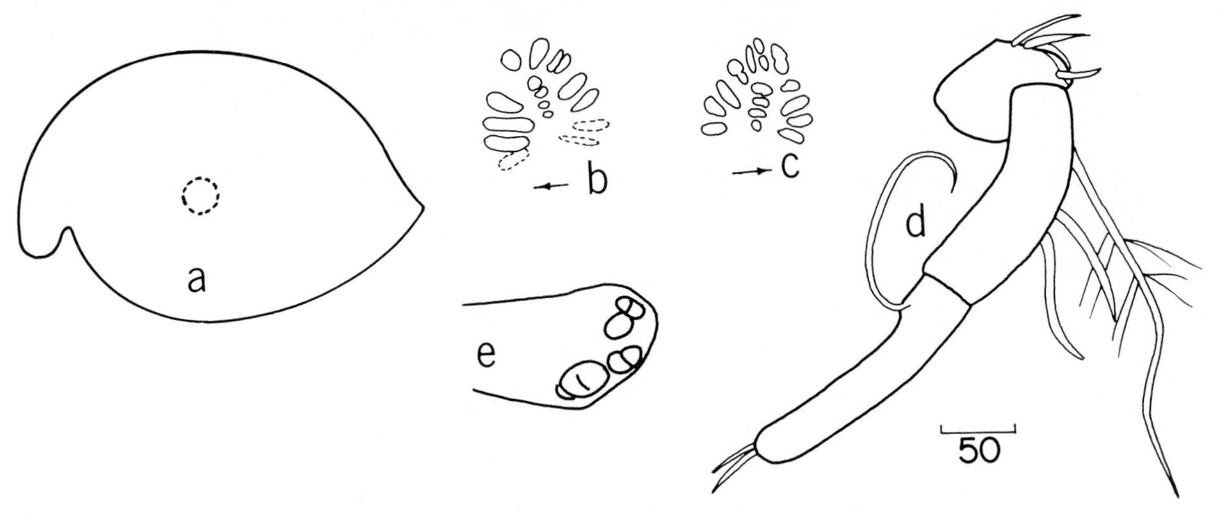

Fig. 11. *Philomedes trithrix*, ♂ instar ? V, USNM 125215: *a*, complete carapace, length 1.84 mm; *b*, muscle scars, left valve, lateral view; *c*, muscle scars, right valve, lateral view; *d*, endopodite, second antenna; *e*, lateral eye. (Same scale in microns: *d, e.*)

Seventh limb: 9 tapering bristles, 5 in distal group and 4 in proximal group; terminus with comb of about 11 teeth opposing 2 pegs.

Furca: Each lamella with 9 claws.

Lateral eye (Figure 11e): USNM 12515, similar to adult female; USNM 125394, similar to adult male but smaller.

Description of female instar ? V (Figure 12).

Carapace: Similar in shape and ornamentation to adult female (Figure 12*a, b*).

Size: USNM 125211, length 1.87 mm, height 1.65 mm; USNM 125212, length 2.16 mm, height 1.63 mm.

First antenna: Fourth joint with 3 bristles, 2 ventral and 1 dorsal; number of bristles on remaining joints similar to adult female.

Second antenna: Ninth joint of exopodite with 5 bristles; bristles of joints 2–8 and long bristles of ninth joint bare; bristles of joints 6–9 relatively short. Endopodite (2-jointed) (Figure 12*c*): First joint with 4 bristles; second joint as on adult female.

Seventh limb: Each limb with 37–39 tapering bristles, each with 2–3 bells; terminal comb with about 13 teeth; 5–8 pegs present opposite comb.

Furca: Each lamella with 9–10 claws.

Development. An insufficient number of instars (8) were examined to do more than sketch the development of the species. The reproductive organs are not well developed in early instars, so they are not useful for determining the sex of specimens. In late stages the morphology of the endopodite of the second antenna is useful for determining sex, but in early stages it is not sufficiently known to be used with certainty. It was observed that in the N-1 stage the male and female could easily be distinguished by the lateral outline of the carapace, which resembles that of the adult; i.e., the female has a minute caudal process, whereas the male has a prolonged one. It was also observed that similar differences were present in the shape of carapaces of earlier instars whose sexes were unknown. The shape of the carapace was then used to identify the sex of the early instars. I am uncertain about the reliability of the method and therefore must consider the sex determinations of the earlier instars to be tentative. This is indicated by a '?' before the sex indication. I presented a key [Kornicker, 1969*a*] for identifying instars I, II, and III; that key is used herein for identifying those stages.

Sexual dimorphism. The carapace of the male is more elongate than the female and has a more prominent caudal process, more distinct surface pits, and fewer short surface hairs. Sexual differences between the first and second antennae, mandible, maxilla, and fifth limb are of the usual type for the genus. The sixth limb of the male has fewer terminal bristles on the third and fourth endites and only 18 bristles on the end joint compared to 27–28 on the end joint of the female. The seventh limb of the male bears only 29–31 bristles and 3 terminal pegs, compared to 39–43 bristles and 7–12 terminal pegs in the female. The furca of the male bears 9–10 claws, compared to 11–13 in the female.

Remarks. Possibly the males described herein as *P. trithrix* are not conspecific with that species. The

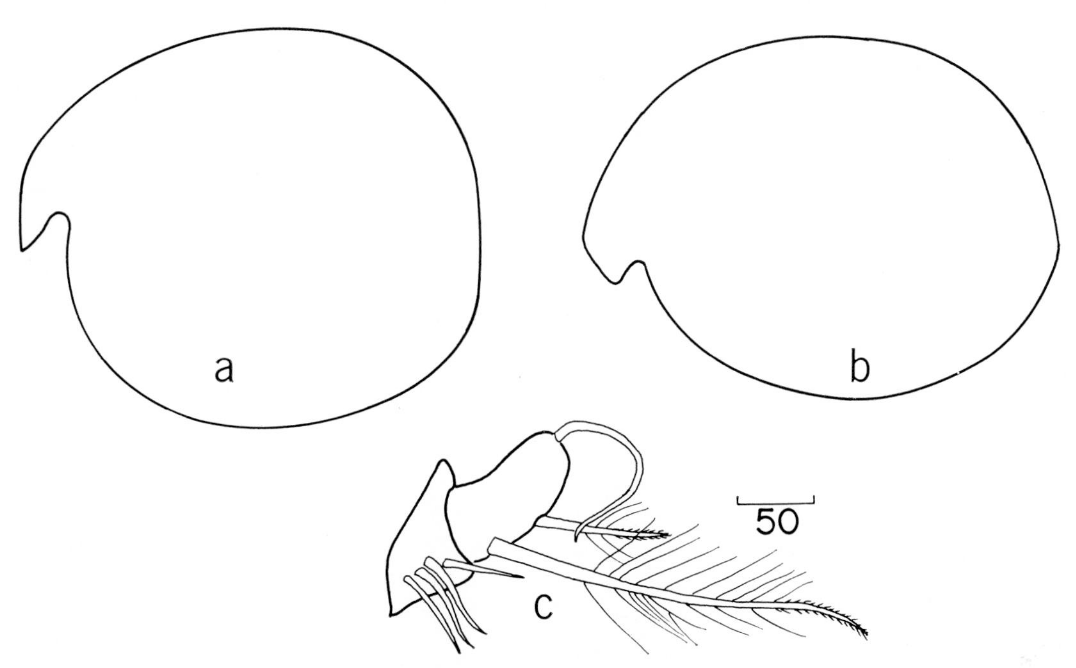

Fig. 12. *Philomedes trithrix*, ♀ instar ? V: *a*, complete carapace, length 1.87 mm; *b*, complete carapace, length 2.16 mm; *c*, endopodite, second antenna. (Figure *a* from USNM 125211; figures *b*, *c*, from USNM 125212.)

distinct pitting on the carapace of the male and the presence of only 3 pegs on the male seventh limb compared to 7–12 on the female are differences that suggest nonconspecificity. Males and females have been considered conspecific because of the following reasons: (1) The number of bristles on the seventh limbs of both sexes (29–31 on male, 39–43 on female) is higher than previously reported for species of the genus. (2) The dorsal margins of the basales of the mandibles of both the males and females have 4 bristles having similar positions on the margin. Not many species in the genus have precisely this number.

Comparisons. P. trithrix differs from *P. charcoti* in having more bristles on the seventh limb and in having more than 2 bristles on the second joint of the endopodite of the second antenna.

P. trithrix differs from *P. rotunda* Skogsberg, 1920, in having more bristles on the female seventh limb (37–43 bristles compared to 23–26). The terminus of the seventh limb on the female of *P. rotunda* has only 3 pegs compared to 7–12 for *P. trithrix*. The dorsal margin of the basale of the mandible of *P. rotunda* bears only 3 bristles compared to 4 on *P. trithrix*. The length of the female *P. rotunda* is 1.9–2.1 mm compared to 2.33–2.65 mm for *P. trithrix*. The carapace of *P. trithrix* has a small caudal process that is not present on *P. rotunda*.

In the key to species of *Philomedes* given by Poulsen [1962, p. 345] the present species would key out to *Philomedes curvata* Poulsen, 1962. *P. trithrix* differs from that species in having more bristles on the end joint of the sixth limb and also on the seventh limb. The female of *P. trithrix* has a lateral eye that is absent on *P. curvata*.

The carapace of the adult female of *P. trithrix* is similar to the carapace of *P. orbicularis* illustrated by Brady [1907, pl. 1, figs. 1, 2]; however, the appendages illustrated by Brady [pl. 1, figs. 8, 9, 11] do not resemble the appendages of *P. trithrix*. Possibly the appendages illustrated by Brady are not from *P. orbicularis* (see discussion herein under *P. orbicularis*). When *P. orbicularis* becomes better known, it may be necessary to place *P. trithrix* in its synonymy.

Philomedes laevipes Daday, 1908

Fig. 13

Philomedes laevipes Daday, 1908, p. 12, figs. 13, 14.

This species was described by Daday [1908] from specimens collected near Booth Island by the French Antarctic Expedition (1903–1905). Müller [1912] thought the species may have been based on juveniles of *Philomedes orbicularis* Brady, 1907, and placed it in the synonymy of that species, preceding the reference by a question mark. Skogsberg [1920, p. 418] suggested the possibility that *P. laevipes* is a larva of *P. charcoti* Daday.

I applied to the National Museum of Natural History, Paris, for the types of *P. laevipes* and received 4 disarticulated valves in preservative, but no appendages. The label with the vial states, '*Philomedes laevipes* Dad. Auct. det. 1908, No. 822—drague 25^{m} —(2-1-05)' and 'Museum Paris, I. Wiencke, Mis. Charcot, 1905.' One of the valves (Figure 13) resembles in outline a specimen illustrated by Daday [1908, p. 13, fig. 13a], except for having an incisur placed more dorsally on the anterior margin. My measurements of the valve in its present state are length 0.71 mm, height 0.6 mm.

Wiencke Island, recorded on the label as the locality of the 4 specimens, is situated off the southeastern coast of Anvers Island and is a member of the Palmer Archipelago. Booth Island, which is recorded by Daday [1908, p. 15] as the locality of the specimens he described, lies southward, close to the coast of the Antarctic Peninsula. Coordinates for Wiencke Island are 64°30′S, 63°15′W; those for Booth Island, which is about 30 km from Wiencke Island, are 65°03′S, 64°00′W. Probably, the specimens I received from Paris are those described by Daday, but whether they were collected closer to Wiencke Island than to Booth Island is unknown.

Daday [1908, figs. 13, 14] illustrated the carapaces of 2 specimens and also some appendages, presumably from one of the specimens. The absence of bristles on the seventh limb and the well-developed sixth limb of the appendages (Figures 14*d, f*) indicate that the specimen is a stage III instar. The small size of the specimens (greatest length 0.9 mm) suggests that all specimens he included in the type series are juveniles. The rounded posterior of the carapace of the largest specimen he illustrated (Figure 13*a*) indicates that it is female; the smaller specimen with an angular posterior (Figure 13*b*) may be male.

The possibility of at least the larger specimen being a larva of *P. charcoti* as suggested by Skogsberg can be eliminated, because neither the male nor the female of *P. charcoti* (the female is described for the first time herein) has a carapace with a rounded posterior.

Because it seems unlikely that this species can be known with certainty, I consider it to be a species dubia.

Philomedes assimilis Brady, 1907

Fig. 14

Philomedes assimilis Brady, 1907, p. 5, pl. I, figs. 16–21, pl. II, figs. 1–6; Müller, 1908, p. 87, pl. VI, figs. 9–17, pl. VII, figs. 14–16. — Scott, 1912, p. 586. — [?] Barney, 1921, p. 178.

Philomedes antarctica Brady, 1907, p. 5, pl. 3, figs. 1–10.

This species was described by Brady [1907] from collections of the National Antarctic Expedition (1901–1904) obtained at their winter quarters in the McMurdo Sound area. Müller [1908] reported it from the Gauss Station in collections made by the Deutsche Südpolar-Expedition (1901–1903). Scott [1912] listed the species in collections of the Scottish National Antarctic Expedition (1902–1904) from the South Orkney Islands, but the identification needs verification. Barney [1921] again reported the species from McMurdo Sound in collections made by the British Antarctic 'Terra Nova' Expedition in 1911 and 1912.

I received from the British Museum (Natural History) 37 specimens in alcohol labeled '*Philomedes assimilis* Brady, 1907. 7.2 45–64. (part) Types.' The description below is based on a specimen from the vial. The specimen contains unextruded eggs in the ovaries and a second antenna with broken bristles on the exopodite, indicating it to be a mature female. I also received 5 slides from the Hancock Museum, Newcastle-on-Tyne. Particulars concerning these slides are tabulated below:

Slide Number	Contents of Label	Description of Slide
1	'*Philomedes assimilis* G. S. Brady, Mawson Expedition, 5–50 fathoms, 1912'	5 whole dry specimens
2	'*Philomedes assimilis*, n.sp. types, *Discovery*.'	5 whole dry specimens
3	'*Philomedes assimilis* n.sp. ♀, W. 6, 30.IX .03, No. 12 Hole D. net (246).' [W. 6 = WQ]	appendages and fragmented carapace on slide
4	'*Philomedes assimilis* n. sp., W. 6, 30.IX .03, No. 12 Hole D. net (246).' [W. 6 = WQ]	appendages and fragmented carapace on slide
5	'*Philomedes assimilis* n. sp. ♂, *Discovery* 13. II. 04, WQ. hist. D. net'	appendages and fragmented carapace on slide

Slide 2 contains many air inclusions, making details on most appendages difficult to observe. The specimen is a juvenile ♀. Comparison of the second antenna and mandible on this slide with illustrations of these appendages given by Brady [1907, pl. 1, figs. 18, 19] indicates that Brady's illustrations are based on the specimen on slide 2.

Although air inclusions are present in slide 3, appendages are easily visible. A few have escaped from under the cover slip. The specimen on this slide is an adult ♀. In the description below, appendages on slides 2 and 3 are compared with those of the adult ♀ from the British Museum. The adult ♂ on slide 5 is poorly preserved.

Brady [1907, p. 5] states that all specimens of *P. assimilis* were collected at the winter quarters and lists dates and localities. The specimens on slides 3 and 4 from the Hancock Museum are undoubtedly from the station listed by Brady as 'September 30, 1903.—No. 12 hole, D. net 246.' The specimen on slide 5 is probably from the station listed by Brady as 'Hut Point. — February 13, 1904, D. net 264.' Locality data are not given on slide 2, but because the specimens were collected by the *Discovery* and it is noted on the slide that the specimens represent 'new species' and 'types,' it may be assumed that they are

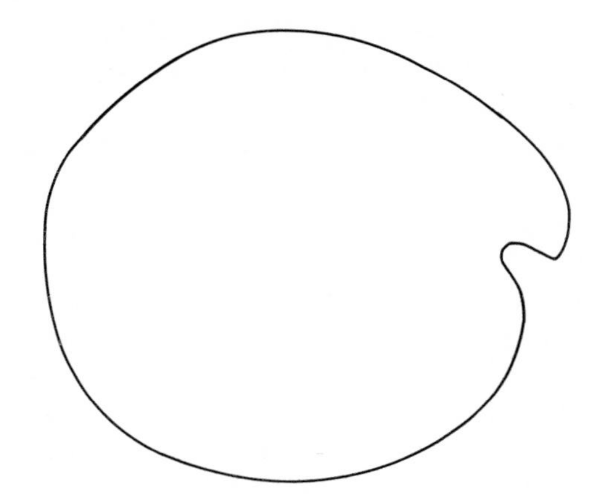

Fig. 13. *Philomedes laevipes*, left valve, medial view, length 0.71 mm.

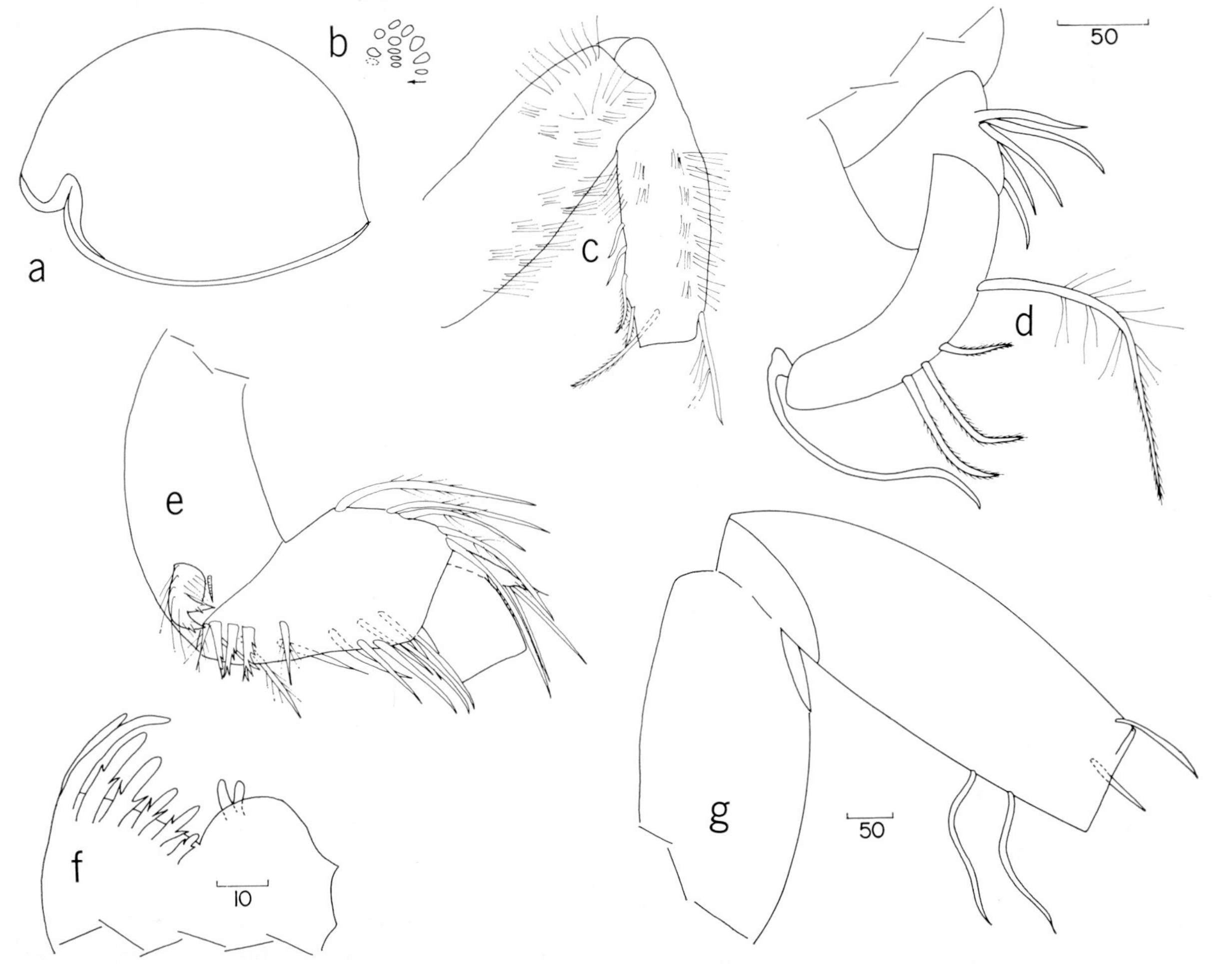

Fig. 14. *Philomedes assimilis*, ♀: *a*, complete carapace, length 1.91 mm; *b*, muscle scars, left valve, lateral view; *c*, joints 1 and 2 of left first antenna, medial view; *d*, endopodite, left second antenna, medial view; *e*, proximal part, left mandible, medial view; *f*, tip of seventh limb, bristles not shown. Adult ♂: *g*, joints 1 and 2 of left first antenna, medial view. (Same scale in microns: *c*, *e*, *g*.)

from the winter quarters. Slide 1 containing specimens collected in 1912 by the Mawson expedition is not pertinent to the present investigation.

Supplementary description of adult female (Figures 14a–f).

Carapace: Length 1.91 mm, height 1.35 mm (Figure 14*a*).

Left first antenna: Medial surface, ventral and dorsal margins of first and second joints extremely spinous; few spines present on ventral and dorsal margins of joints 3–5; ventral margin of second joint with 4 short bristles, dorsal margin with 1 subterminal bristle, lateral surface with 1 slender spinous-bristle (Figure 14*c*); third joint with 3 bristles, 1 ventral, 2 dorsal; fourth joint with 5 bristles, 4 ventral, 1 dorsal; sensory bristle of fifth limb with 5 proximal and 5 distal filaments; medial bristle of sixth joint spinous. Seventh joint: a-bristle spinous; b-bristle with 1 proximal and 4 terminal filaments; c-bristle with 5 proximal and 5 terminal filaments. Eighth joint: d- and e-bristles bare and about equal in length to c-bristle; f-bristle with about 4 proximal and 5 distal filaments; g-bristle with 3 proximal and 5 distal filaments.

The left first antenna on slide 3 from the Hancock Museum is similar to the above except for having only 1 dorsal bristle on the third joint (number of filaments on bristles of seventh and eighth joints not determined). The right first antenna on slide 3 differs in having 3 bristles on the ventral margin and 2 subterminal bristles on the dorsal margin of the second joint.

The right first antenna on slide 2, which belongs to a juvenile ♀, differs from that described above in having 3 ventral bristles and 2 subterminal dorsal bristles on the second joint and only 3 ventral bristles on the fourth joint. (Filaments of bristles of seventh and eighth joints were not counted; some of these bristles are broken.)

Left second antenna: Protopodite with long spines along anterodorsal margin and clusters of short spines along ventral margin. Endopodite 2-jointed (Figure 14*d*): First joint with 5 short bare bristles; second joint elongate with 1 long and 3 short, spinous ventral bristles and 1 long, recurved subterminal bristle. Exopodite: ninth joint with 7 bristles, 4 long, 3 shorter; bristles of joints 6–8 and 4 long bristles of joint 9 broken; joints 2–8 with comb of short spines and basal spine on distal margins and clusters of spines on medial surface; medial spine present on first joint.

The second antennae on the specimen on slide 3 from the Hancock Museum are similar to the above with the exception that the first joint of the endopodite of both limbs bears 6 small, bare bristles.

The left second antenna of the juvenile ♀ on slide 2, which is probably the one illustrated by Brady [1907, pl. 1, fig. 18], differs from that described above in having an endopodite with 6 bare bristles on the first joint and 3 ventral and 1 subterminal recurved bristles on the second joint. All bristles of the exopodite are unbroken. The number of bristles on the ninth joint of the exopodite could not be discerned on the slide.

Left mandible (Figure 14e): Coxale: Endite spinous, bifurcate at tip; small bristle present near base of endite. Basale: Dorsal margin with 1 proximal bristle, 2 near middle and 2 terminal; medial surface with 5 proximal bristles near ventral margin and with 3 stout, pectinate, 2 slender, spinous, and 1 short, slender bristle nearer to middle; ventral margin with 4 distal bristles; lateral surface with 5 slender bristles. Exopodite with hirsute tip and 2 subterminal bristles. Endopodite: Ventral margin of first joint with 4 terminal bristles; dorsal margin of second joint with 2 bristles in proximal group and 6 in distal group; ventral margin of second joint with 3 bristles in proximal and distal groups; end joint with 3 claws, 2 long, 1 short, each with minute teeth along middle of concave or ventral margin, and 3 or 4 bristles; spines present on medial surfaces of coxale, basale, and second joint of endopodite.

The left mandible on slide 3 from the Hancock Museum is similar to above. The right mandible differs in having 5 terminal bristles on the ventral margin of the first endopodite joint and 3 bristles in the proximal group on the dorsal margin of the second endopodite joint.

Details on the mandibles of the juvenile ♀ on slide 2 cannot be observed. The dorsal margin of the basale has 1 fairly long mid-bristle as shown by Brady [1907, pl. 1, fig. 19] and 2 terminal bristles. The exopodite has 2 subterminal bristles, not 3 as indicated on Brady's illustration. The dorsal margin of the basale of the right mandible has 2 bristles near the middle and 2 terminal bristles.

Left seventh limb (Figure 14f): Limb with 9 bristles, 5 distal, 4 proximal; each bristle with 4–5 bells and marginal spines distally; terminal comb with 11 teeth; most teeth with lateral flange; 2 pegs present opposite comb.

The seventh limbs could not be found on slide 3 from the Hancock Museum. On slide 2, which contains a juvenile ♀, the seventh limbs are similar to that described above with the exceptions that the bristles are more strongly tapered (a juvenile characteristic) and bear only 3 or 4 bells; the terminal comb and pegs seem similar to those described above but could not be clearly observed.

Lateral eyes (from slide 3, Hancock Museum): Eyes small, each with 2 divided ommatophores.

Supplementary description of male (from slide 5). Left first antenna, second joint (Figure 14g): Ventral margin with 2 bristles; dorsal margin with short terminal bristle, lateral surface with short distal bristle.

Remarks. Scott [1912, p. 586] referred specimens collected in Scotia Bay, South Orkney Islands, to *Philomedes assimilis* Brady. His total discussion of the species is as follows: 'One or two specimens of a *Philomedes,* which I ascribe to the species mentioned [*P. assimilis*], occurred in a small sample of dredged material from Scotia Bay, collected in April 1903; Station 325, 60°43′42″S., 44°38′33″W. The length of the specimen represented by the drawing [Figure 16] is 1.8 mm'.

In order to verify Scott's identification, I wrote to the Royal Scottish Museum, Edinburgh, asking for the opportunity to examine the specimens of *Philomedes assimilis* Brady listed by Scott and collected by the Scottish Antarctic Expedition (1902–1904) at the South Orkney Islands. Mr. David Heppell kindly forwarded 3 vials containing *Philomedes.* The contents of the labels and kinds of specimens in each vial are as follows:

Vial Number	Contents of Label	Contents of Vial
1	'*Philomedes assimilis* Brady' '*Philomedes* Stat A "Scotia" coll. 1904, Dredge & Trap, 2 April' '1921 . 143 . 1144'	5 specimens of *Philomedes,* 2 gravid females, 3 juveniles; condition of specimens indicates that they had once been dry
2	'*P. assimilis* Brady' '*Philomedes* "Scotia" S. Orkneys, 1903' '1921 . 143 . 1144'	27 specimens of *Philomedes,* 1 specimen of Cylindroleberidinae, 1 podocopid and 1 amphipod
3	'*Philomedes ?australis,* Rendhe Stat A "Scotia" Antarctic, April 1903' '1921 . 143 . 1145'	1 specimen of *Philomedes*

The vials contain a total of 33 specimens, many more than the 'one or two specimens' mentioned by Scott [1912, p. 586]. All carapaces of specimens of *Philomedes* in the vials have a truncate posterior and resemble the carapace of *P. assimilis.*

An adult female with unextruded eggs within the body and with bristles of the exopodite of the second antenna broken was selected from vial 2 and dissected. Carapace dimensions of the preserved specimen are length 1.91 mm, height 1.32 mm. The appendages show that it is *Philomedes assimilis,* thereby corroborating Scott's finding of the species in the vicinity of the South Orkney Islands. In brief, characteristics of the appendages used in the identification are presence of 4 bristles on the ventral margin of the second joint of the first antenna, 5 bristles on the second joint of the endopodite of the second antenna, 5 bristles on the dorsal margin of the basale of the mandible, 9 bristles on the seventh limb, and 2 pegs opposite the terminal comb.

Philomedes antarctica Brady, 1907

Fig. 15

Philomedes antarctica Brady, 1907, p. 5, pl. III, figs. 1–10.

This species was correctly placed in synonymy with *Philomedes assimilis* Brady by Müller [1908]. The present supplementary description is to supply additional evidence for the referral. Brady [1907] described the species from specimens collected presumably at the winter quarters by the National Antarctic Expedition (1901–1904). According to Brady (p. 5), '*P. antarctica* was found rather sparingly in four of the gatherings taken at No. 4 hole in a depth of five fathoms.'

I received from the Hancock Museum 2 slides labeled as follows: slide 1: '*Philomedes antarctica* n.sp., ♂, type, *Discovery,* 10. V. 03, 5 fathoms'; slide 2: '*Philomedes antarctica* n.sp., ♀, type, *Discovery,* 10. V. 03, 5 fathoms.' From the British Museum (Natural History) I received a vial containing 2 specimens in alcohol and the label '*Philomedes antarctica* Brady, 1907. 7. 2. 65–67, (Part) Types.' When the shells of the 2 specimens were removed, it was observed that one is an adult male and the other a female, probably also an adult. The brief supplementary descriptions that follow are based on the British Museum specimens, but the descriptions are followed by comparisons with the Hancock Museum slides. The slides contain air inclusions, making details of the

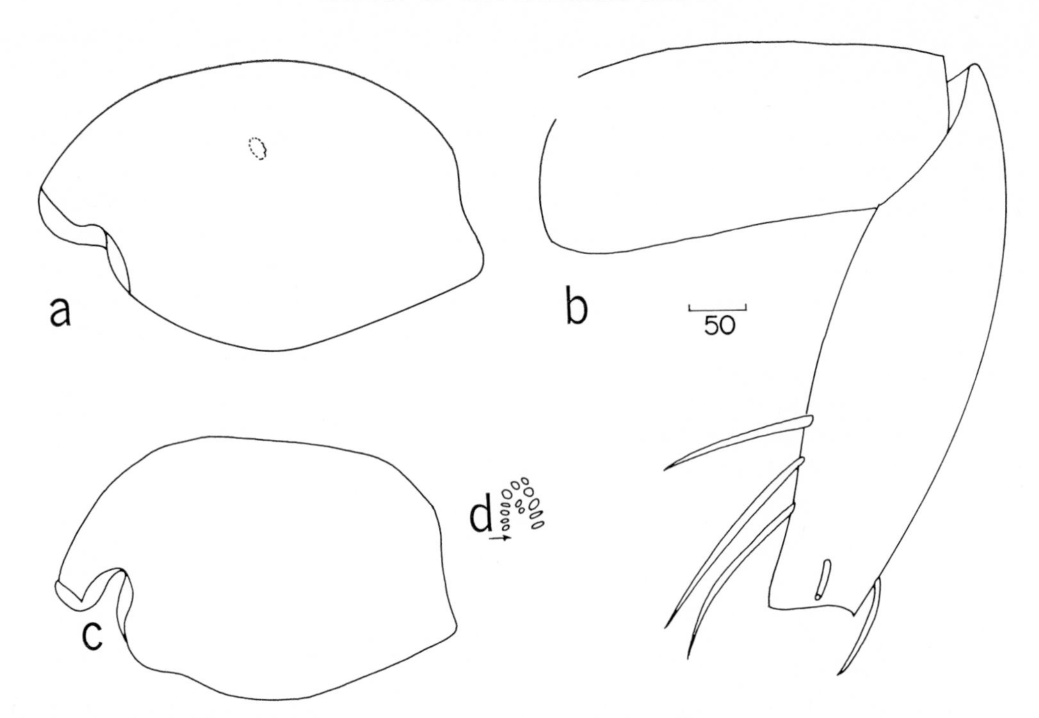

Fig. 15. *Philomedes antarctica*, adult ♂: *a*, complete carapace, length 2.12 mm; *b*, joints 1 and 2 of right first antenna, medial view. ♀: *c*, complete carapace, length 1.11 mm; *d*, muscle scars, right valve, lateral view.

appendages difficult to discern. The male on the Hancock Museum slide is adult. On the slide marked ♀, the endopodite of the second antenna is obscure, but the bristles on the exopodite are very long, indicating that the specimen is adult; the parallel sides of the bristles on the seventh limbs support the probability of the specimen's being adult. If the specimen is adult, it is unquestionably female; if it is not an adult it could be either male or female. I have assumed herein that is is female.

Supplementary description of female.

Carapace: Length 1.82 mm, height 1.11 mm (Figure 15*c*).

First antenna: First and second joints extremely spinous; few spines present along margins of third and fourth joints; second joint with 4 ventral bristles, 1 subterminal dorsal bristle, and 1 distal lateral bristle; third joint with 1 ventral and 2 dorsal bristles; fourth joint with 1 dorsal and 4 ventral bristles; remaining joints with usual accompaniment of bristles.

The first antennae on the slide from the Hancock Museum are not well preserved; however, it is possible to determine that the ventral margin of the second joint on each appendage bears 4 bristles.

Second antenna: Protopodite with long spines on dorsal margin and shorter spines on ventral margin. Endopodite 2-jointed; first joint with 6 bare bristles; second joint with 4 ventral bristles and 1 subterminal recurved bristle. All bristles of exopodite unbroken, and some quite long.

The endopodite of each second antenna on the slide from the Hancock Museum is obscure. The bristles on the exopodites are unbroken, and some are extremely long like those on the British Museum specimen.

Mandible: Dorsal margin of basale with 3 bristles near middle and 2 subterminal. (The mandibles could not be found on the slide from the Hancock Museum.)

Seventh limb: Each limb with 9 bristles, 5 distal and 4 proximal; 2 pegs present opposite comb.

The seventh limbs of the specimens on the slide from the Hancock Museum are fairly obscure but appear to be similar to those of the British Museum specimen.

Supplementary description of male.

Carapace: Length 2.12 mm, height 1.22 mm (Figure 15*a*).

First antenna: Second joint, right limb: Ventral margin with 3 bristles distributed along margin; dorsal margin with 1 subterminal bristle; lateral surface with 1 distal bristle (Figure 15*b*). Left limb with only

2 bristles on ventral margin, otherwise same as right limb.

The first antennae on the slide from the Hancock Museum are poorly preserved, but the ventral margin of one bears 3 bristles as on the right limb of the British Museum specimen.

Philomedes orbicularis Brady, 1907

Fig. 16

Philomedes orbicularis Brady, 1907, p. 4, pl. 1, figs. 1–3, 6–15.

Brady [1907] described this species from two collections made on May 23, 1902, and June 15, 1902, presumably at the winter quarters. He did not select a holotype. At the end of the description, Brady states (p. 5), 'But one or two small specimens which occurred along with *P. orbicularis*, and which I at first took to be the young of that species, were very distinctly angulated posteriorly, and I now think that they belong probably to the following species, *P. assimilis*, especially as they have not the villous covering of *P. orbicularis*.'

In answer to my request for material, I received from the British Museum (Natural History) a vial with 5 specimens and the following label: '*Philomedes orbicularis* Brady, 1907-7-2. 35–44, (part) Types.' The jar in the British Museum from which the specimens were obtained contains the additional notation '*Discovery* Exp.' (McKenzie, in litt., 1969). The 5 specimens are from a lot of 25 specimens in a tube that contains the label '15-VI-02 WQ' and 3 undecipherable letters, which could stand for *Discovery*, National Antarctic, or D. J. Scourfield (McKenzie, in litt., 1969). The notation '15-VI-02 WQ' no doubt refers to June 15, 1902, winter quarters.

The carapaces of the 5 specimens do not resemble those of the globose hirsute female carapace with a rounded posterior illustrated by Brady [1907, pl. 1, figs. 1, 2]. All carapaces are much smaller (maximum length 1.85 mm) than the 2.5 mm given by Brady as the length of the species, and all possess a posteroventral corner. Upon further inquiry to the British Museum, I was kindly informed by Dr. K. A. McKenzie that none of the specimens in the remaining part of the type series approaches a length of 2.5 mm.

The largest of the 5 specimens, a gravid female having a length of 1.85 mm and height of 1.31 mm, has a carapace with a posterior less truncate than *P. assimilis*, but not nearly as rounded as *P. orbicularis*. The endopodite of the second antenna is similar to that of *P. assimilis*, but the specimen differs from that species in having only 1 ventral bristle on the second joint of the first antenna, 7 bristles on the dorsal margin of the mandibular basale, and 11 bristles on the seventh limb. I designate this specimen '*Philomedes* sp. A' to aid in later references to it.

At this time, I received from the Hancock Museum, Newcastle-on-Tyne, additional specimens of *P. orbicularis* from the type series. These consist of 2 whole dry mounts in a cardboard slide and 6 glass slides with appendages and fragmented, squashed carapaces. The cardboard slide contains a juvenile specimen and a large specimen (Figure 16; length 2.45 mm; height 2.08 mm) similar in appearance to the female illustrated by Brady [pl. 1, figs. 1, 2]. This specimen has a minute caudal process similar to *P. trithrix*, but it is hardly visible in lateral view. The label on the slide is as follows: '*Philomedes orbicularis* n. sp. (old & young), *Discovery*, 15. VI. 02., W. b. D. net.' The 'W. b.' on the label and the 'W6' on the glass slides noted below probably should be W. Q., for winter quarters. At first Brady considered this species a new genus, but then he crossed out the new genus and referred the species to *Philomedes;* I have omitted the crossed-out name from the label here and on the slides listed below.

The contents of the labels on the glass slides are as follows:

Slide Number	Label
1	'*Philomedes orbicularis* n. sp. ♀, tumid form, *Discovery*, 15. VI. 02, W. 6 D. net'
2	'*Philomedes orbicularis* n. sp. ♀, *Discovery*, 15. VI. 02, W. 6 D. net'
3	'*Philomedes orbicularis* n. sp. ♀, *Discovery*, D. 23, V. 02'
4	'*Philomedes orbicularis* n. sp. ♀, *Discovery*, 15. VI. 02, W. 6 D. net'
5	'*Philomedes orbicularis* n. sp. ♂, *Discovery*, 15. VI. 02, W. 6 D. net'
6	'*Philomedes orbicularis* n. sp. ♂, *Discovery*, 15. VI. 02, W. 6 D. net'

The contents of the slides are described briefly below:

Slide 1. Most appendages are obscure. A first antenna bears only 2 bristles on the fourth joint, indicating that the specimen is an early instar. Carapace hirsute.

Slide 2. This specimen is a female with unbroken bristles on the exopodite of the second antenna. The ventral margin of the second joint of a first antenna

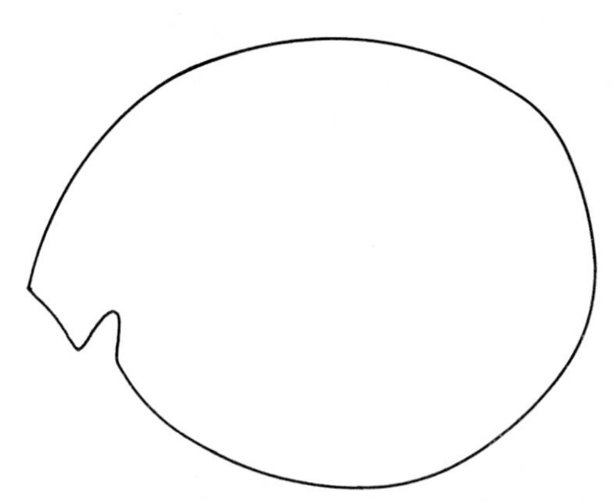

Fig. 16. *Philomedes orbicularis*, complete specimen, length 2.45 mm.

bears only 1 bristle. The second joint of an endopodite of a second antenna bears 5 bristles. A seventh limb bears about 11 bristles (this limb has escaped from beneath the cover slip). The dorsal margin of the basale of the left mandible bears 4 bristles, whereas that of the right mandible bears at least 5 and possibly 7.

Slide 3. The ventral margin of the second joint of each first antenna bears 4 bristles. The endopodite of the second antenna bears 6 bristles on the first joint and 5 on the second joint. The dorsal margin of the basale of the left mandible bears 1 bristle near the middle and 2 terminally; the terminal bristles are obscure on the basale of the right mandible, but at least 1 and possibly 2 bristles are near the middle. The posteroventral corner of one of the valves is prominent.

Slide 4. The dorsal margin of the basale of the mandible bears 7 bristles. The seventh limbs are obscure but bear about 10 bristles.

Slide 5. The seventh limb of this adult male bears 9 bristles. The second joint of the first antenna bears only 1 ventral bristle.

Slide 6. The ventral margin of the second joint of this adult male bears only 1 bristle. The dorsal margin of the basale of one mandible bears 5 bristles, the other mandible bears 6 bristles. The seventh limb bears about 9 bristles.

The fragmented and squashed condition of the carapaces of females on the above slides do not permit relating them with certainty to the carapace illustrated by Brady [pl. 1, figs. 1, 2]. Probably the female on slide 3 is not *P. orbicularis* because of the prominent posteroventral corner on one of the valves. This specimen, however, is the only one on the slides that has a mandible with only 3 bristles on the dorsal margin of the basale, similar to the one illustrated by Brady [pl. 1, fig. 11]. The presence of 4 bristles on the ventral margin of the second joint of the first antenna on slide 3 suggests that it is *P. assimilis.* If the mandible illustrated by Brady is drawn correctly and was obtained from a female similar to that illustrated by him [pl. 1, figs. 1, 2], the females on slides 2 and 4 cannot be *P. orbicularis* because the mandibular basales bear more than 3 dorsal bristles. The specimens on those two slides are probably *P.* sp. A.

All specimens on the slides, with the exception of that on slide 1, which is an early instar and indeterminate, are not *P. trithrix.* The presence of only 9 to 11 bristles on their seventh limbs makes this quite clear.

The type series of *P. orbicularis* contains at least 3 species: *P. orbicularis, P. assimilis, P.* sp. A. Possibly the female carapace illustrated by Brady [pl. 1, figs. 1, 2] is conspecific with *P. trithrix,* but not conspecific either with the female appendages he illustrated or with the male. Unless additional specimens in the type series become available, it may be necessary to examine the appendages of the dried adult female in the collection of the Hancock Museum, Newcastle-on-Tyne, to resolve this problem.

Philomedes species A

Fig. 17

Philomedes orbicularis Brady [part], 1907, p. 4.

Material. One specimen, a gravid female, from a vial identified as *Philomedes orbicularis* by Brady, and part of the type series. The specimen was collected at the winter quarters at McMurdo Sound, June 15, 1902, by members of the National Antarctic *(Discovery)* Expedition. The specimen is in the collection of the British Museum (Natural History).

Description of female. Carapace with broad rostrum and incisur and truncate posterior (Figure 17*a*); surface with scattered hairs; reticulations visible in transmitted light; length 1.85 mm, height 1.31 mm.

First antenna: First and second joints extremely spinous; fewer spines present on joints 3–5; second joint with 3 bristles, 1 ventral, 1 lateral, 1 dorsal; third joint with 3 bristles, 1 ventral, 2 dorsal; fourth joint with 5 bristles, 1 dorsal, 4 ventral; remaining joints normal for genus.

Second antenna: Protopodite spinous along ventral and dorsal margins. Endopodite 2-jointed (Figure 17*c*): First joint with 5 proximal and 2 distal bristles;

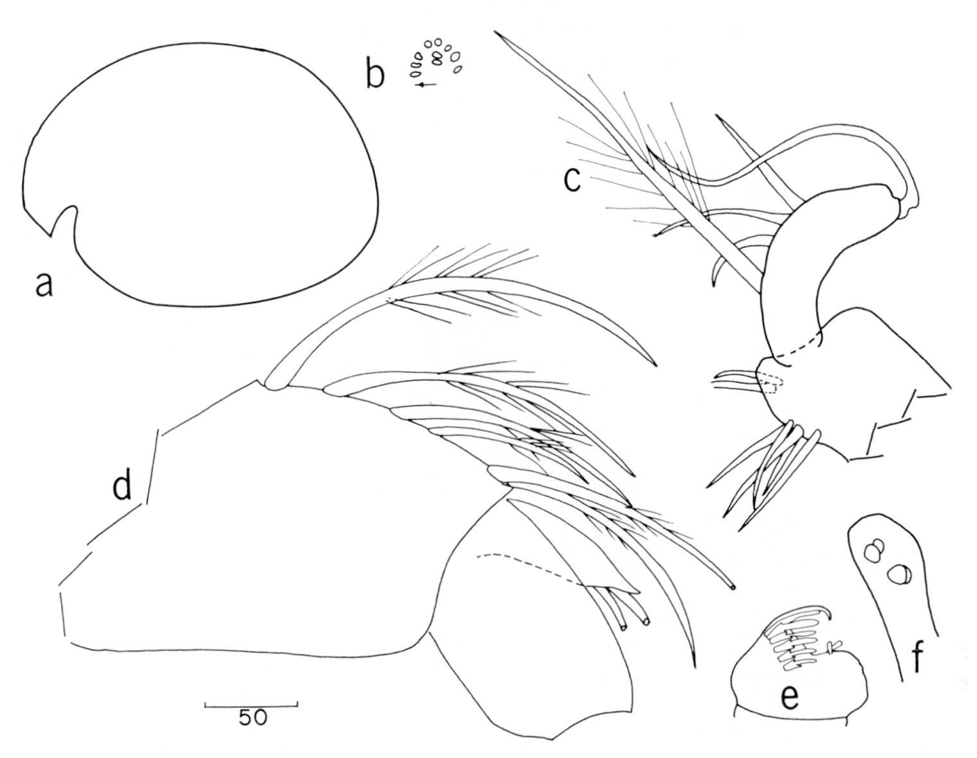

Fig. 17. *Philomedes* species A, gravid ♀: *a*, complete carapace, length 1.85 mm; *b*, central muscle scars, left valve, lateral view; *c*, endopodite, second antenna, medial view; *d*, basale, exopodite, and first endopodite joint of left mandible, medial view; *e*, tip of seventh limb, bristles not shown; *f*, right lateral eye. (Same scale in microns: *c–f*.)

second joint with 4 ventral bristles, 1 long spinous, 3 short bare, and 1 recurved terminal bristle. Exopodite: Bristles of joints 6 to 8 and 4 long bristles of joint 9 broken; bristles of joints 2 to 5 with spines forming row along middle of ventral margin and without natatory hairs; ninth joint with total of 7 bristles; small basal spines present on joints 2 to 8; medial spine present on distal margin of first joint; a cluster of long spines present on medial surface of first joint near distal end; joints 2 to 8 with clusters of short spines forming up to 6 rows on medial surface.

Left mandible: Coxale endite spinous with bifurcate tip and small bristle near base. Basale: Dorsal margin with 6 bristles; medial surface with 5 proximal bristles near ventral margin, 3 stout and pectinate, 2 slender and spinous; ventral margin with 4 distal bristles; lateral surface with 5 slender bristles. Exopodite about 75% length of dorsal margin of first endopodite joint; outer bristle with a few short spines, inner bristle with wreaths of long spines. Endopodite: Ventral margin of first joint with 4 terminal bristles; dorsal margin of second joint with 3 bristles in proximal group and 6 in distal group; ventral margin of second joint with 3 bristles in proximal and distal groups; end joint with 4 bristles and 3 claws, 2 long, 1 short, with minute spines near middle of ventral margin; spines forming clusters present on medial surfaces of all joints (Figure 17*d*).

Right mandible: Basale differs from that of left mandible in having an additional short, slender, spinous proximal medial bristle near ventral margin and 5 bristles, instead of 4, in addition to the 2 terminal bristles on dorsal margin (proximal bristle on dorsal margin is missing, but socket is clearly visible).

Seventh limb (Figure 17e): One limb with 11 bristles (5 distal, 6 proximal), other limb with 12 bristles (5 distal, 7 proximal); each bristle with 3 to 4 bells. Terminal comb with 13 teeth, middle tooth longer than others and recurved, lateral teeth with flare at base; 2 pegs present opposite comb.

Furca: Each lamella with 10 claws; claws decrease in length proximally along each lamella.

Lateral eye (Figure 17f): Eye small with about 2 ommatophores.

Eggs: 14 eggs present in marsupium.

Comparisons. The carapace of *P.* sp. A resembles those of *P. assimilis* and *P. charcoti.* Some appendages of these species showing differences between each are compared below:

Appendage	*P. charcoti*	*P. assimilis*	*P.* species A
First antenna, number of ventral bristles on second joint of female	1	4	1
Second antenna, number of bristles on second joint of endopodite of female	2	5	5
Mandible, number of bristles on dorsal margin of basale	3	5	6–7
Seventh limb, number of bristles	12–16	9	11–12

Genus ***Scleroconcha*** Skogsberg, 1920

Type species: *Philomedes (Scleroconcha) Appellöfi* Skogsberg, 1920 = *Scleroconcha appelloefi* (Skogsberg).

Scleroconcha gallardoi new species

Fig. 18

Holotype. USNM 125397, valves and some appendages in alcohol, remaining appendages on slides, ♀ with 27 eggs in brood chamber, length 4.07 mm, height 2.73 mm.

Paratypes. USNM 125398, adult ♀; USNM 125399, gravid ♀; USNM 125440, adult ♀; USNM 125441, N-1 ♂. In addition, 15 juveniles were returned to Universidad Concepción, Instituto Central de Biología, Departamento de Zoología, Concepción, Chile.

Type locality. Off Greenwich Island, Station 31, 62° 29.1′S, 59°41.0′W, Jan. 12, 1968, bottom depth 39 meters, sandy mud bottom.

Other localities. Off Greenwich Island: Stations 23, 37, 54, 55, 56. In Bransfield Strait: Station 2.

Etymology. This species is named for its collector, Dr. V. A. Gallardo.

Distribution. The distribution of specimens in the collection is as follows (numerals represent Station—number of specimens): 2—1, 23—1, 31—1, 37—2, 54—1, 55—2, 56—12.

Description of adult female. Carapace (Figure 18*a*) identical except for size to *Scleroconcha appelloefi* (Skogsberg, 1920).

Size: USNM 125397, ♀ with 27 eggs in brood chamber, length 4.07 mm, height 2.73 mm; USNM 125398, ♀ without eggs, length 3.96 mm, height 2.70 mm; USNM 125340, ♀ with 24 eggs in brood chamber, length 4.13 mm, height 2.62 mm; USNM 125341, ♀ with unextruded eggs, length 4.04 mm, height 2.43 mm. Range of lengths 3.96 mm to 4.13 mm; range of heights 2.43 mm to 2.73 mm.

First antenna: First joint bare; second joint with 3 bristles of which ventral bristle is longest and dorsal bristle shortest, lateral bristle almost reaching middle of fifth joint; third joint with 3 bristles, 1 ventral, 2 dorsal; fourth joint with 4 ventral and 2 dorsal bristles; sensory bristle of fifth joint with 7 proximal and 5 terminal filaments; sixth joint with long, spinous medial bristle. Seventh joint: a-bristle spinous, slightly longer than bristle on sixth joint; b-bristle with 2 proximal and 4 terminal filaments; c-bristle with about 5 proximal and 5 terminal filaments. Eighth joint: d- and e-bristles bare, subequal, slightly longer than b-bristle; f-bristle with about 6 proximal and 4 terminal filaments. Joints 2 to 5 with short spines forming clusters along ventral and dorsal margins.

Second antenna: Exopodite: First joint with short medial spines on distal margin; distal margins of joints 2–8 with short spines in row, no basal spine present; bristles of joints 2–4 shorter than length of exopodite and without marginal spines or hairs; bristles of joints 5–8 bare or with natatory hairs, distal part broken off; ninth joint with 7 bristles consisting of 4 long and 3 short bristles; long bristles with natatory hairs distally on proximal part and distal part broken off, short bristles with long marginal hairs. Endopodite 2-jointed (Figure 18*c*): First joint with 6 subequal ventral bristles consisting of 5 proximal and 1 distal; medial bristle in proximal group longer than others; second joint with extremely long, spinous ventral bristle reaching past ninth joint of exopodite and shorter, bare terminal bristle.

Mandible: Coxale: Short bristle at base of endite; endite spinous with bifurcate tip. Basale: Dorsal margin with 1 long, spinous bristle anterior to middle and 2 terminal; ventral margin with 9 spinous bristles; medial side with 3 pectinate bristles and 2–3 with slender spines. Exopodite ⅔ length of first endopodite joint and has 2 long, spinous subterminal bristles. Endopodite: Ventral margin of first joint with 3–4 spinous bristles; dorsal margin of second joint with 4–5 bristles in proximal group and 6 spinous bristles in distal group; medial side between groups with short pectinate bristle; ventral margin of second joint with 3–4 spinous bristles in subterminal group and 3 in terminal group; end joint with 3 claws and about 4

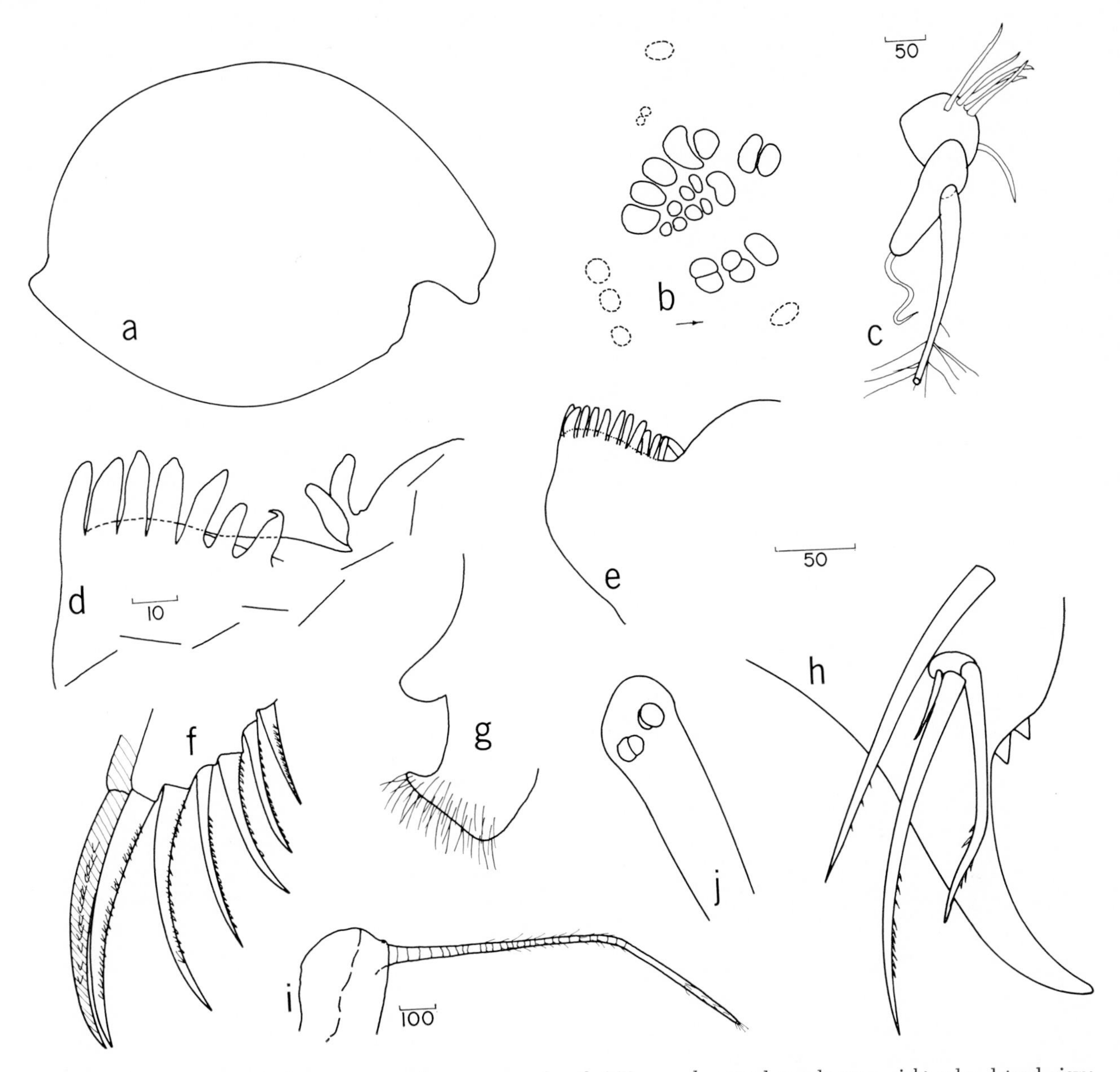

Fig. 18. *Scleroconcha gallardoi*, ♀: *a*, complete carapace, length 2.73 mm; *b*, central muscle scars, right valve, lateral view; *c*, endopodite, left second antenna; *d*, tip of left seventh limb; *e*, tip of right seventh limb; *f*, distal portion of furca; *g*, anterior and upper lip; *h*, part of second joint of exopodite of left fifth limb, posterior view; *i*, medial eye and rod-shaped organ; *j*, left lateral eye. (Same scale in microns: *b*, *f*, *i*; *c*, *j*; *e*, *h*. Figure *b* from USNM 125399, remaining figures from USNM 125397.)

bristles; dorsal claw reaching slightly past midpoint of longest claw; all claws with a few proximal teeth along concave margin. Basale and first and second joints of endopodite with marginal spines.

Maxilla: Anterior bristle of basale short and with short marginal spines; first endopodite joint with 5 beta-bristles. End joint: 1 stout b-bristle present with 4 teeth on concave margin and 3 on convex margin; 2 stout d-bristles present, 1 bare, other with 1 tooth on concave margin.

Fifth limb (Figure 18h): Similar to *S. appelloefi*. Epipodial appendage with 65 bristles.

Sixth limb: First endite with 1 terminal bristle and 2 medial bristles, one medial bristle about 1/3 length of other; second endite with 4 bristles, 3 terminal and 1 medial; third endite with 10 bristles, 9 terminal and

1 medial; fourth endite with 9 bristles, 8 terminal and 1 medial; end joint with 35 bristles, 19 marginal and 16 medial (submarginal); medial surface of third and fourth endite with few spines forming clusters; medial surface of end joint hirsute. Number of bristles in place of epipodial appendage: USNM 125397 with 4 bristles on one limb and 6 on other; USNM 125440 with 3 bristles on each limb.

Seventh limb (Figures 18d, e): Each limb with 10 bristles; 6 in terminal group, 3 on each side; 4 in proximal group, 2 on each side; each bristle with 3 to 9 bells and marginal spines distally. Comb teeth and pegs with smooth margins; number and attitude of comb teeth and number of pegs on each limb of 4 adult females are shown in the following table (opposing limbs are indicated by 'A' and 'B'):

USNM	Number of Comb Teeth		Number of Pegs		Attitude of Comb
	Limb A	*Limb B*	*Limb A*	*Limb B*	
125397	16	17	2	1	perpendicular
125398	13	12	1	1	perpendicular
125399	9	9	1	1	parallel
125440	11	11	1	1	parallel

('Parallel' indicates that long axes of comb teeth are oriented parallel to sides of limb; 'perpendicular' indicates that comb teeth are oriented perpendicular to sides of limb.)

Furca (Figure 18f): Each lamella with 16 claws; claw 3 more slender than claw 4; claw 3 longer than claw 4 on some specimens, shorter on others.

Eyes: Medial eye large, pigmented (Figure 18*i*). Lateral eye elongate with 2 divided ommatophores (Figure 18*j*).

Upper lip and anterior (Figure 18g): Anterior with projection near middle; upper lip hirsute.

Eggs: USNM 125397 with 27 eggs in brood chamber; USNM 125398 with 24 eggs in brood chamber.

Rod-shaped organ: Elongate, jointed proximally (Figure 18*i*).

Description (part) of juvenile (N-1 stage) male.

Ornamentation of carapace: Similar to adult female.

Size (USNM 125441): Length 3.02 mm, height 2.03 mm.

First antenna: Fourth joint with 3 ventral and 2 dorsal bristles; remaining joints with same number of bristles as on adult female.

Second antenna: Endopodite (3-jointed): First joint with 2 short bristles with few spines; second joint with 1 long proximal ventral bristle with wreaths of long marginal spines, and 6 short distal ventral bristles with short marginal spines; third joint with 1 long, bare dorsal bristle and 2 short, subequal terminal bristles, longest with short marginal spines, other bare. Exopodite: Bristle on second joint with 5 proximal spines; some bristles on ninth joint with short spines, remaining bristles of exopodite bare.

Seventh limb: Each limb with 4 terminal and 4 proximal bristles, each with 2 to 6 bells and marginal spines distally; terminal comb with 3 or 4 smooth teeth oriented perpendicular to margin of limb; 1 smooth peg present opposite comb.

Comparisons. The carapace of *S. gallardoi* appears identical to that of *S. appelloefi* except for its larger size. The species may be identified by the number of comb teeth on the seventh limb and by the structure of the teeth margins. Some characteristics of adult females of the two species are compared below:

Characteristic	*S. appelloefi*	*S. gallardoi*
Carapace length	3.3–3.6	3.96–4.13
Seventh limb, number of comb teeth	7–8	9–17
Seventh limb, margins of comb teeth	denticulate	smooth
Furca, number of claws	13–15	15–16

Discussion of variability of the comb teeth on the seventh limb. Four adult females are in the collection: USNM 125397 from Station 31, USNM 125398 from Station 2, and USNM 125399 and USNM 125440 from Station 56. The first 2 specimens have 16–17 and 12–13 comb teeth, respectively, and the teeth are oriented with their long axes almost perpendicular to the sides of the limbs. The second 2 specimens have 9 and 11 comb teeth, respectively, and the teeth are oriented with their long axes almost parallel to the sides of the limbs. The difference in orientation between the comb teeth of the first and second pairs of specimens could be interpreted as indicating the presence of 2 species in the collection. The sample from Station 56, which contained 2 specimens with comb teeth oriented parallel to the sides of the limb, also contains several juveniles, presumably members of the same species as the adults. One of these (USNM 125441), a juvenile male at the N-1 stage of development, was dissected. Unlike the adults in the sample, the comb teeth of its seventh limbs are oriented perpendicular to the sides of the limbs. Because of this, differences in orientation of the comb teeth have been interpreted herein as being the result of individual variability.

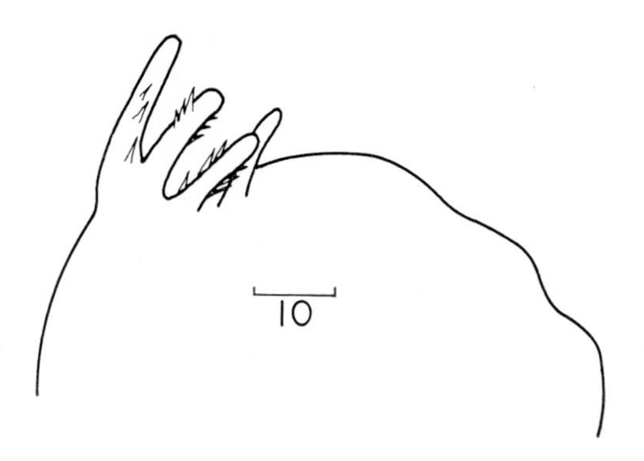

Fig. 19. *Scleroconcha appelloefi*, juvenile, tip of seventh limb.

Scleroconcha appelloefi (Skogsberg, 1920)

Fig. 19

Philomedes (Scleroconcha) Appellöfi Skogsberg, 1920, pp. 419–428, figs. LXXVII–LXXXIII.

Streptoleberis appellöfi (Skogsberg); Sylvester-Bradley, 1961, p. 400, fig. 422–1.

Scleroconcha appelófi (Skogsberg); Poulsen, 1962, p. 395 [key].

Skogsberg [1920] described this species from specimens collected by the Swedish Antarctic Expedition in 1902 in the vicinity of South Georgia (Stations, 20, 22, 24, and 30). One specimen was collected southwest of Snow Hill Island (Station 6), an island east of the Antarctic Peninsula (see Figure 1). According to Skogsberg [1920, p. 428], that specimen is a larva, too young for certain identification.

I applied to the Swedish State Museum (Riksmuseum), Stockholm, for types of this species and received 28 glass slides and a vial within a vial, both containing alcohol. One of the slides contains a specimen from Station 20, 18 slides contain specimens from Station 22, 3 slides contain specimens from Station 24, and 6 slides contain specimens from Station 30. All stations are located in the vicinity of South Georgia. The slides with mounted appendages have many inclusions of air, making it difficult to study appendage morphology.

The label in the larger of the two vials reads as follows: 'Svenska Südpolar exp. 1901–03. No. 20. 6/5 1902. 250 m. Kleine Steine. 54°12′s. Br.—36°50′w. L. Süd-Georgien. Antarctic Bai.' A second label indicates that the museum sample number is 139.

The small vial within the larger vial contains only the right valve of a single specimen. Because no appendages could be observed within the vial, I did not open it. The larger vial contained a left and right valve, a body without furca, and a third mandible. The body and valves are from a juvenile specimen about to molt. The left seventh limb was removed from the body for study after which it was placed in a small vial with the remainder of the body, the valves, and the third mandible (the size of the third mandible indicates it is from a specimen much smaller than the other specimen in the vial). This vial and the vial containing a right valve were then put inside the larger vial received from the Swedish State Museum and returned to the museum.

Description of seventh limb of juvenile. Limb with 6 bristles in distal group, 3 on each side; 4 in proximal group, 2 on each side; bristles taper distally; terminal comb with 6 curved teeth with marginal spines; toothlike peg present opposite comb (Figure 19).

Family Cylindroleberididae

Genus ***Parasterope*** Poulsen, 1965

Parasterope lowryi new species

Figs. 20–22, 23*a*, *b*

Holotype. USNM 125130, valves and some appendages in alcohol, remaining appendages on slide; ♀ with 17 eggs in brood chamber; length 2.33 mm, height 1.69 mm.

Paratypes. USNM 125131, juvenile ♀, valves and some appendages in alcohol, first antenna on slide; length 1.60 mm, height 1.16 mm. USNM 125132, undissected juvenile in alcohol; length 1.55 mm, height 1.08 mm. One undissected juvenile in alcohol returned to James K. Lowry, Virginia Institute of Marine Science, Gloucester Point, Virginia; length 1.43 mm, height 1.08 mm.

Additional material (all juveniles and not identifiable with certainty as P. lowryi): USNM 125489, ♂, length 1.76 mm, height 1.22 mm, Station 58; 4 specimens, 2 from Station 58 and 2 from Station 47, were returned to Universidad de Concepción, Instituto Central de Biología, Departamento de Zoología, Concepción, Chile.

Type locality. Arthur Harbor, Antarctic Peninsula, February 20, 1967, Station III, 64°45′46″S, 64°06′17″W; bottom depth 39 meters. Bottom temperature varied at Station III between −1.7°C in winter and 0°C in summer; salinity was normally around 34 parts per thousand; soft bottom composed mainly of

Fig. 20. *Parasterope lowryi*, ♀ USNM 125130: *a*, complete carapace, length 2.33 mm; *b*, posterior right valve, medial view; *c*, anterior right valve, lateral view; *d*, anterior right valve, medial view; *e*, detail of surface pores and reticulations. Central muscle scars, lateral view: *f*, right valve; *g*, left valve. Appendages: *h*, left first antenna, lateral view, dorsal bristles not shown; *i*, endopodite and part of protopodite of right second antenna, medial view; *j*, basale endite, right mandible, medial view; *k*, exopodite, right mandible, lateral view; *l*, left maxilla, medial view (reconstructed from folded appendage). (Same scale in microns: *b–d*; *h, i, l*; *j, k*.)

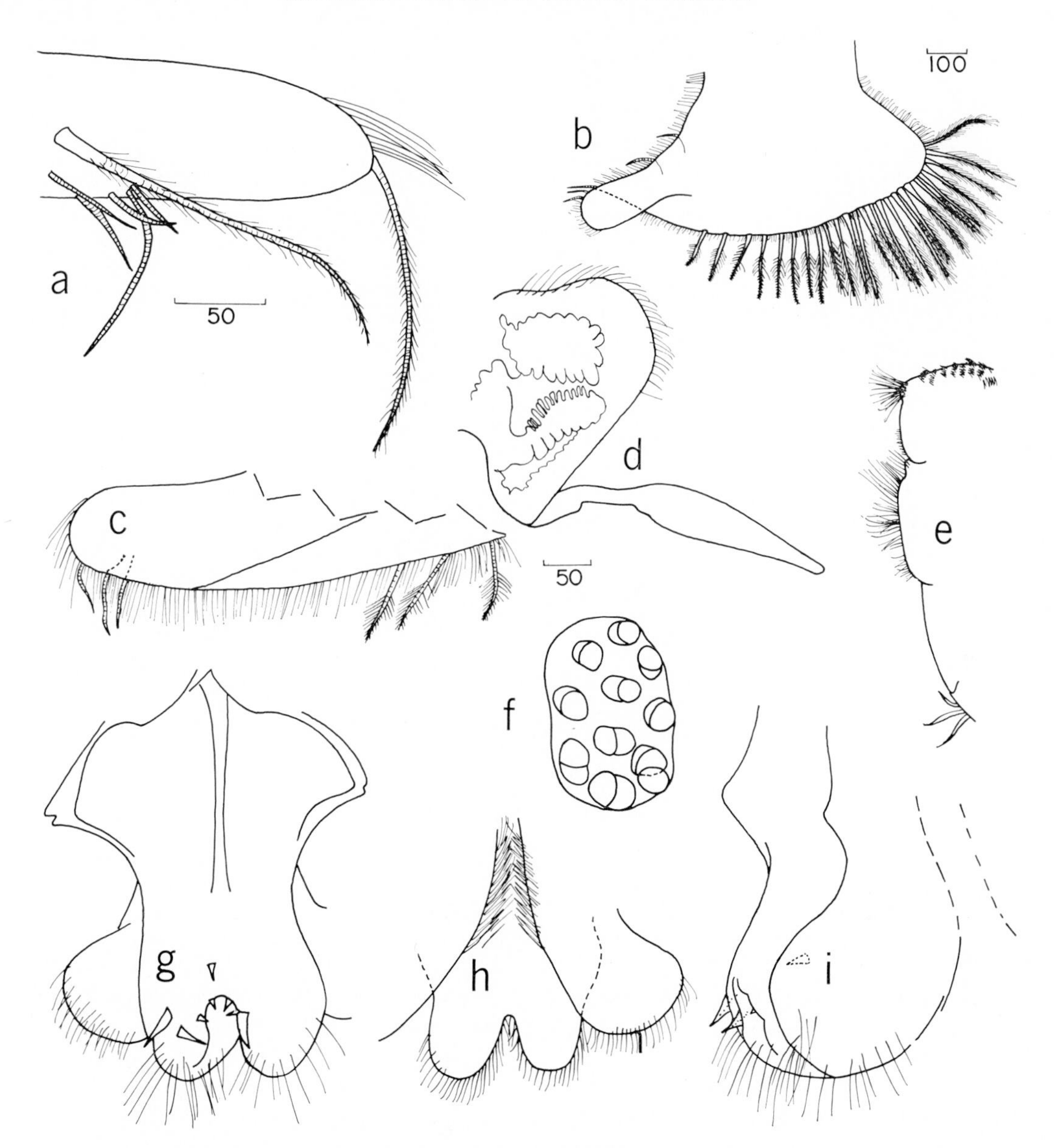

Fig. 21. *Parasterope lowryi*, ♀ USNM 125130; *a*, comb of fifth limb with epipodial bristles, lateral view; *b*, left sixth limb, lateral view; *c*, tip of right sixth limb, medial view; *d*, medial eye and rod-shaped organ; *e*, posterior and proximal 3 bristles of furca; *f*, right lateral eye. Upper lip: *g*, anterior view; *h*, posterior view; *i*, lateral view. (Same scale in microns: *b*, *e*; *c*, *d*, *f–i*.)

silts (Lowry, written communication, 1968). Paratypes are from same sample as holotype.

Other localities. In vicinity of Greenwich Island, Stations 47 and 58.

Etymology. This species is named for its collector, James K. Lowry.

Distribution. Arthur Harbor, Anvers Island, Station III (4 specimens) and Stations 47 and 58 in vicinity of Greenwich Island (4 juveniles, identification questionable). Specimens were collected at depths ranging from 39 to 90 meters in mud, silts, and sandy mud.

Description of female (Figures 20, 21; 23a, b).

Carapace: Oval in lateral view with greatest height posterior to middle (Figure 20*a*); rostrum and incisur of usual type for genus (Figure 20*c*).

Ornamentation (Figure 20e): Surface with numerous minute pores and scattered short hairs; large

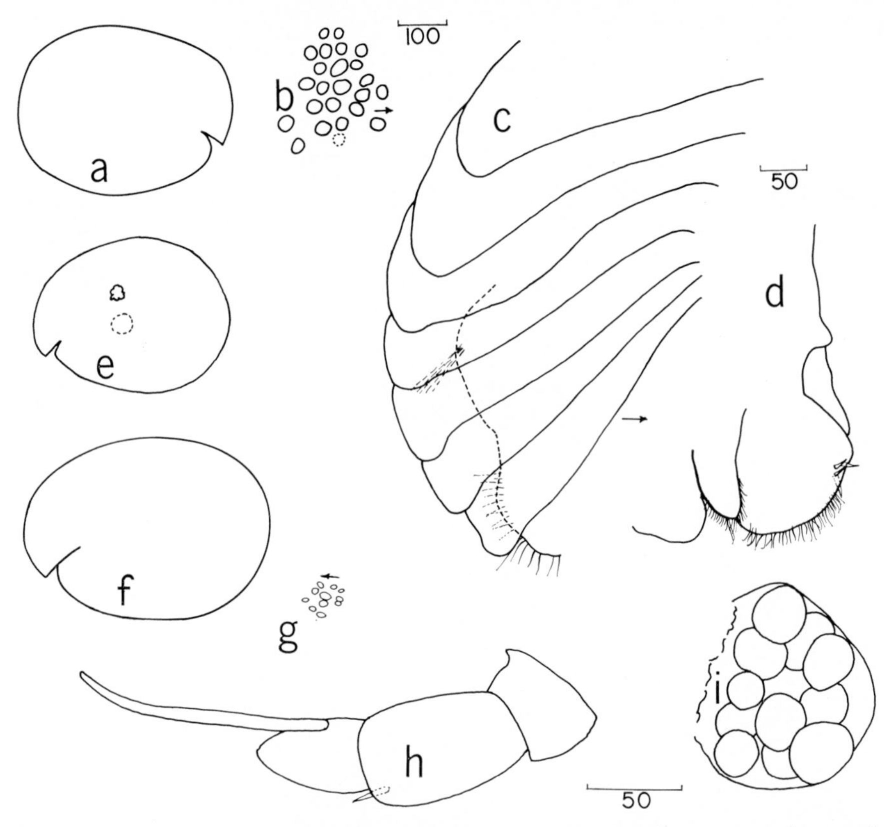

Fig. 22. *Parasterope lowryi*, juvenile: *a*, complete carapace, length 1.60 mm; *b*, central muscle scars, right valve, lateral view; *c*, posterior and 'gills'; *d*, anterior and upper lip; *e*, complete carapace, length 1.43 mm; *f*, complete carapace, length 1.76 mm; *g*, sketch of visible central muscle scars, left valve, lateral view; *h*, endopodite, second antenna; *i*, lateral eye. (Same scale in microns: *c, d; h, i*. Figures *a–d*, from USNM 125131; figure *e*, from USNM 125132; figures *f–i*, juvenile male from USNM 125489.)

but faint reticulations present but visible only under high magnification (more easily seen in some parts of shell than others).

Infold: Area above rostrum with about 70 bare bristles (Figure 20*d*); area below incisur with about 37 bare bristles; area along ventral margin with about 27 bristles; ridge along inner margin of posterior infold with about 47 broad, hyaline, triangular bristles varying in size; some hyaline bristles adajacent to each other, others with 1 or 2 short, slender bristles separating them; area posterior to ridge with only 2 bristles in dorsal half and 11 or 12 in ventral half (Figure 20*b*).

Selvage: Consists of narrow striated ridge with narrow lamellar prolongation with smooth margin; prolongation broadening slightly along lower margin of incisur and bearing marginal hairs.

Muscle scars (Figures 20f, g): Central muscle scars obscure but consisting of about 22 or 23 individual scars, 2 middle scars larger than others.

Pore canals: Radial canals abundant with some ending in minute bristle on infold. Numerous normal pores or punctae visible under high magnification.

Size: Holotype length 2.33 mm, height 1.69 mm; height as percent of length 72.5.

First antenna (Figure 20h): First joint with numerous short spines; second joint with surface spines, 1 long, spinous dorsal bristle, and 1 short lateral bristle with faint marginal spines; third joint with short ventral bristle and 6 long, spinous dorsal bristles, next-to-last bristle without long marginal spines; fourth joint with long, spinous dorsal bristle and 2 short, spinous ventral bristles; sensory bristles of fifth joint with 6 long terminal filaments; sixth joint with long medial bristle with short marginal spines; seventh joint with fairly long a-claw pectinate at tip, b-bristle with 4 marginal filaments, c-bristle with 5 marginal filaments; eighth joint with bare e-bristle about ⅔ length of c-bristle, no d-bristle, f-bristle with

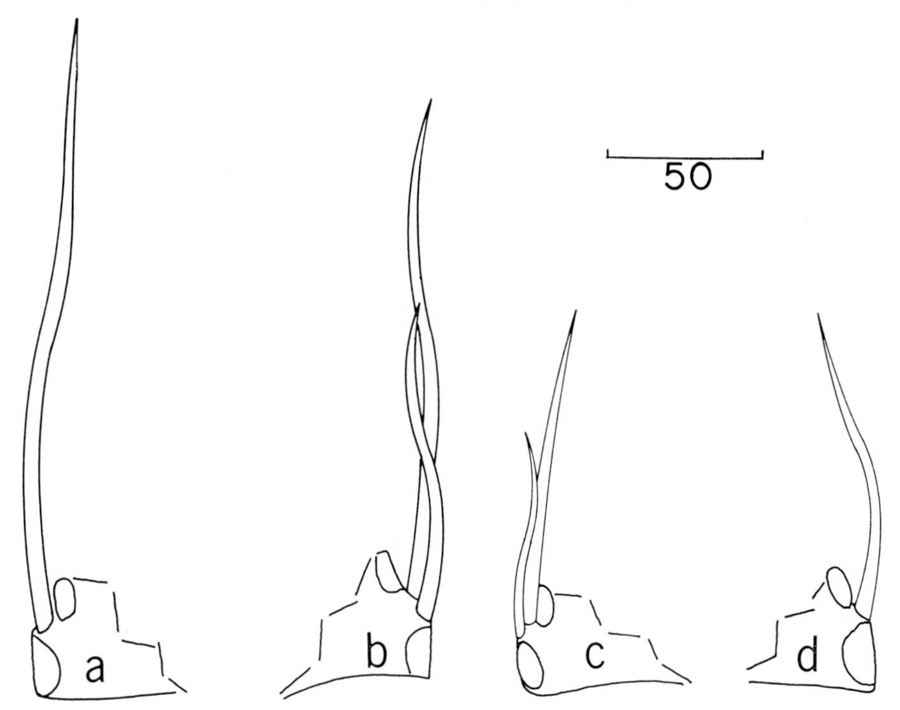

Fig. 23. Comparison of proximal bristles on second joint of mandibular endopodite, medial view, *Parasterope lowryi*, ♀ USNM 125130: *a*, left limb; *b*, right limb. *Parasterope ohlini*, ♀: *c*, left limb; *d*, right limb.

4 or 5 marginal filaments, and g-bristle with 6 marginal filaments.

Second antenna (Figure 20i): Protopodite with small medial bristle and spines along anterior margin. Endopodite distinctly 3-jointed with terminal bristle considerably longer than stem (stem length 70% of bristle length). Exopodite: 2 small basal spines present on joint 3, 3 on joint 4; joints 5 to 9 with 1 stout basal spine, spine on joint 9 larger than others (about half length of joint); joints 2 to 8 with small spines in row along distal margin; bristle of second joint with short marginal hairs and reaching past ninth joint; bristles of joints 3–8 with natatory hairs; ninth joint with 4 bristles, 2 long with natatory hairs, 1 medium with few marginal hairs, 1 short and bare; inner margin of first joint with long distal hairs.

Mandible (Figures 20j, k, 23a, b): Small bare bristle present at basis of ventral branch of coxale endite; endite broken off on both limbs of specimen; basale endite with glandular peg, 1 slender dwarf bristle, 4 spinous terminal bristles, and 3 triaenid bristles; 1 triaenid bristle present on ventral margin of basale near endite; triaenid bristles with 4 to 7 pairs of marginal spines below terminal pair; dorsal margin of basale with 2 long terminal bristles, both with short marginal spines; exopodite reaching about ⅔ length of first endopodite joint, hirsute and with 2 short subterminal bristles. Endopodite: First joint with 3 spinous bristles, most slender of 3 with only short spines; ventral margin of second joint with 3 long, spinous terminal bristles; dorsal margin with 1 or 2 proximal bristles (1 on left limb, 2 on right), and stout a-, b-, c-, and d-bristles with short marginal spines; longest proximal bristle about 36% length of a-bristle; long, slender bristle with short marginal spines between b- and c-bristles. Medial bristles of second endopodite joint: 1 near base of b-bristle; 2 oblique rows of about 7 bristles in each row present between b- and c-bristles. One long, stout, spinous lateral bristle between c- and d-bristles. One slender, spinous terminal bristle distal to d-bristle. End joint with 5 bristles and 1 stout claw; claw about same length as longer margin of second endopodite joint and with small teeth along middle of concave margin.

Maxilla (Figure 20l): Proximal endite with 4 bristles, 3 long and 1 short; distal endite with 3 long bristles; epipodite acuminate, hirsute, about half length of basale; dorsal and ventral margins of basale each with fairly long proximal and distal bristle; ventral margin also with bristle anterior to middle; first joint of endopodite with short dorsal and long, bare ventral bristle; end joint with long, bare terminal bristle.

Fig. 24. *Parasterope ohlini*, ♀: *a*, complete carapace, length 2.14 mm; *b*, anterior right valve, medial view; *c*, central muscle scars, reticulate ornamentation and pores, left valve, lateral view; *d*, posterior right valve, medial view; *e*, posteroventral portion of infold right valve, medial view; *f*, part of left first antenna, lateral view, not all bristles shown; *g*, endopodite and part of protopodite, right second antenna, medial view; *h*, proximal part of coxal endite of left mandible, medial view; *i*, epipodite and part of basale of maxilla; *j*, epipodial bristles of right fifth limb, lateral view. (Same scale in microns: *b*, *d*; *c*, *e–g*; *h*, *i*.)

Fifth limb (Figure 21a): Dorsal margin of comb bare; distal margin with long hairs; comb-bristles consisting of 2 rows of long and short bristles with marginal hairs; distal comb-bristles longer than proximal bristles; lateral side of comb with 1 long, spinous bristle extending past end of comb and 5 slender bristles ventral to long bristle.

Sixth limb (Figures 21b, c): Lateral sole with 3 slender bristles; ventral-posterior margin with 25 spinous bristles; anterior margin with 2 slender bristles; margins and medial surface hirsute.

Seventh limb: Distal group with 6 bristles, 3 on each side; proximal group with 6 or 7 bristles, 3 to 4 on each side; each bristle with 3 or 4 bells; terminus with opposing combs, each having about 15 teeth with marginal spines.

Furca: Each lamella with 9 pectinate claws, the posterior 2 being secondary.

Lateral eye (Figure 21f): Well developed, with 11 bifid ommatophores in lateral view.

Medial eye and rod-shaped organ (Figure 21*d*): Medial eye large, with hairs along dorsal margin. Rod-shaped organ broad near middle.

Upper lip (Figures 21g–i): Consisting of paired hirsute lobes with 2 anterior spines on right lobe, 1 on left lobe, and 4 smaller spines between lobes; lateral hirsute flap present.

Eggs: 17 oval eggs in brood chamber of holotype.

Posterior (Figure 21e): Dorsum forming almost right angle with tuft of hairs on corner and spines forming rows along dorsal surface; posterior margin hirsute. Gills present, not reaching posterior margin of dorsum.

Description of juvenile female, USNM 125131 (Figures 22a–d). Carapace similar to adult but somewhat less tumid (Figure 22*a*); length 1.60 mm, height 1.16 mm, height as percent of length 72.5. First antenna: ventral margin of third joint with only 4 bristles; medial bristle of distal pair with only short spines, others also with wreaths of long hairs; sensory bristle of fifth joint same as adult. Second antenna: Protopodite and endopodite similar to adult; first joint of exopodite with tuft of long spines subterminal on inner margin; ninth joint with only 3 bristles. Mandible: Coxale endite broken off distal to small bristle at basis; basale endite and basale with same number of bristles as adult, but triaenid bristles with only 2–5 pairs of marginal spines below terminal pair; dorsal margin of second endopodite joint without proximal bristle; medial surface with only about 9 bristles. Fifth limb with 66 bristles on epipodite. Sixth limb similar to female but with only 12 ventral-posterior bristles on right limb and 16 on left; lateral sole less pronounced. Seventh limb with 4 bristles in distal group, 2 on each side, and 6 in proximal group, 2 on each side; each bristle tapering distally and with 1 or 2 bells; terminus with opposing combs, each with about 12 teeth. Lateral eye similar to adult, but with only about 9 bifid ommatophores in lateral view. Frontal organ similar to adult but somewhat more crinkled in proximal half. Furca: Each lamella with 6 pectinate claws followed by 1 spinous, annulate secondary claw. Gills: 7 on each side, overlapping posterior margin of dorsum (Figure 22*c*). Maxilla, dorsum, and upper lip similar to those of adult (Figure 22*d*).

Comparisons. P. lowryi is closely related to *Parasterope ohlini* (Skogsberg, 1920, p. 493). Some of the differences between adult females of the two species are listed below:

Characteristic	*P. ohlini*	*P. lowryi*
Number of bristles on infold between posterior ridge and shell margin	21	13
Marginal spines of terminal bristles on first and second joints of endopodite of maxilla	present	lacking

Proximal bristles on the ventral margin of the second endopodite joint of the mandible of *P. lowryi* are longer than those on the mandible of *P. ohlini* (Figure 23), but this characteristic may not be useful without knowing more about individual variability.

Parasterope ohlini (Skogsberg, 1920)

Figs. 23*c*, *d*, 24, 25

Asterope ohlini Skogsberg, 1920, pp. 493–497, figs. XCI, XCII.

Parasterope ohlini (Skogsberg); Poulsen, 1965, p. 363 [key].

This species was described by Skogsberg [1920] from specimens collected in the vicinity of South Georgia by the Swedish Antarctic Expedition, 1902. I applied to the Swedish National Museum for types and received 12 glass slides with mounted specimens and 3 vials with whole specimens in alcohol. One slide contains specimens from Station 22; 1 slide contains specimens from Station 25; 6 slides contain specimens from Station 33; 3 slides contain specimens from Station 34. Vial number 168 contains 1 specimen from Station 22; vial number 169 contains 10 specimens from Station 33; vial number 170 contains 4 specimens from Station 34. All specimens had been identified by Skogsberg. Some of the slides are in good condition, others have air inclusions. The supplementary description below is based on a female with eggs in the marsupium obtained from vial number 169 from Station 33, South Georgia.

Description of female.

Carapace: Oval in lateral view with greatest height posterior to middle (Figure 24*a*); rostrum and incisur of usual type for genus.

Ornamentation: Surface with numerous minute pores and scattered short hairs; large but faint reticulations present, but visible only under high magnification (Figure 24*c*).

Infold: Area above rostrum with about 82 bare bristles; area below incisur with about 46 bare bristles (Figure 24*b*); area along ventral margin with about

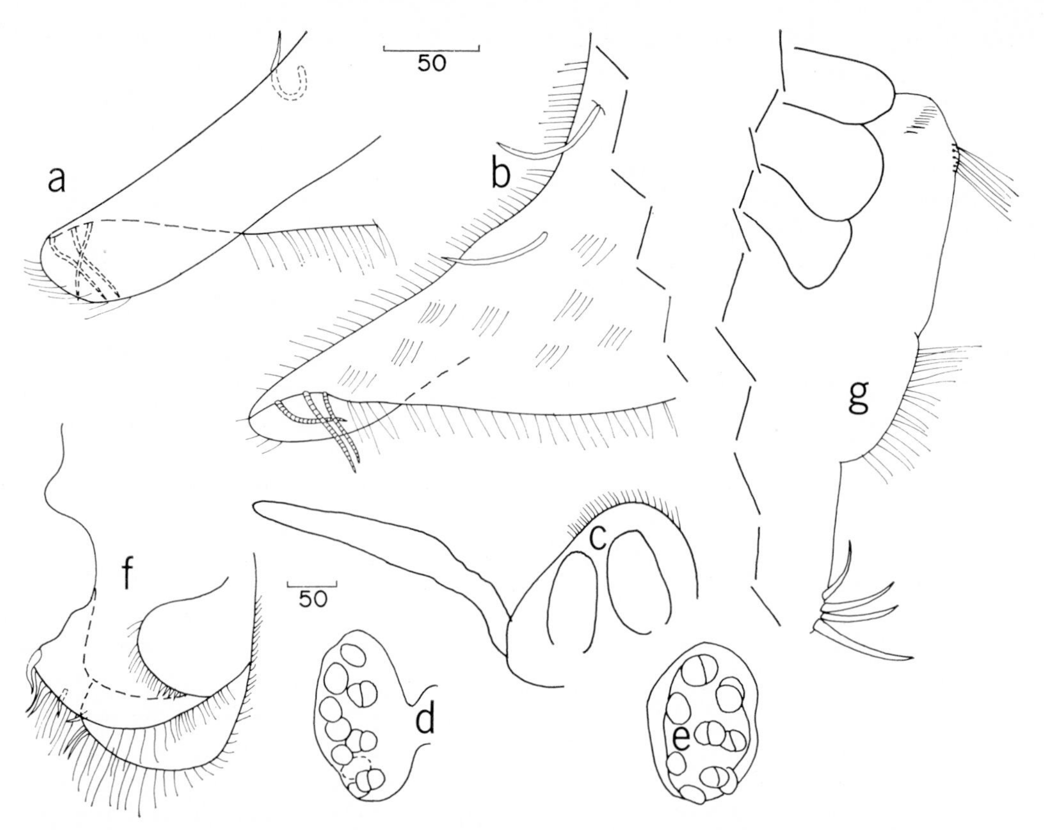

Fig. 25. *Parasterope ohlini*, ♀: *a*, anterior end left sixth limb, lateral view; *b*, anterior right sixth limb, medial view; *c*, medial eye and rod-shaped organ; *d*, left lateral eye; *e*, right lateral eye; *f*, upper lip, oblique view from left; *g*, posterior showing ends of 3 'gills' and proximal 4 bristles of furca. (Same scale in microns: *a*, *b*; *c–f*.)

28 bristles; ridge along inner margin of posterior infold with about 52 broad, hyaline, triangular bristles varying in size; some hyaline bristles adjacent to each other (especially in ventral part), others with 1 or 2 (rarely 3) short, slender bristles separating them; area posterior to ridge with about 6 bristles in dorsal half and 16 in ventral half (Figures 24*d*, *e*).

Muscle scars: Central muscle scars consisting of about 13 individual ovoid scars (Figure 24*c*).

Pore canals: Radial and normal pore canals numerous, similar to those of *P. lowryi*.

Size: Carapace length 2.14 mm, height 1.55 mm.

First antenna (Figure 24f): First joint spinous; fourth joint with spines forming clusters along dorsal margin; ventral tip of a-claw of seventh joint with minute teeth forming medial and lateral row.

Second antenna (Figure 24g): Protopodite with small medial bristle and spines along anterior margin; stem of 3-jointed endopodite about 75% length of terminal bristle (measured on 3 specimens). Exopodite: Inner margin of first joint with long distal hairs; joint 3 with minute basal spine; joint 4 with 2 minute basal spines; joints 5 to 9 with 1 stout basal spine, spine on joint 9 larger than others (almost half length of joint); joints 2 to 8 with small spines in row along distal margin; bristle of second joint with short marginal hairs and reaching past ninth joint; bristles of joints 3 to 8 with natatory hairs; ninth joint with 4 bristles, 2 long with natatory hairs, 1 medium with few marginal hairs, and 1 short and bare.

Mandible: Slender bristle present at base of ventral branch of coxale endite (Figure 24*h*). Basale: Endite with glandular peg, 1 slender dwarf bristle, 4 spinous terminal bristles, and 3 triaenid bristles; 1 triaenid bristle present on ventral margin of basale near endite; triaenid bristles with 4 to 7 pairs of marginal spines below terminal pair. Second joint of endopodite with 1 or 2 proximal bristles (1 on right limb, 2 on left), longest bristle 24% length of a-bristle.

Maxilla: Epipodite and dorsal margin of basale hirsute (Figure 24*i*). Proximal endite with 4 bristles, 3 long, 1 short.

Fifth limb (Figure 24j): Dorsal margin of comb bare; comb-bristles consisting of 2 rows of long and short bristles with marginal hairs; distal comb-bristles longer than proximal bristles; lateral side of comb with 1 long, spinous bristle extending past end of comb and 6 slender bristles ventral to long bristle.

Sixth limb (Figures 25a, b): Each limb with 3 anteroventral bristles and upper and lower anterior bristle; right limb with 21 posteroventral bristles.

Lateral eye (Figures 25d, e): Each eye with 9 or 10 ommatophores.

Upper lip (Figure 25f): Consisting of paired hirsute lobes with several spines on each lobe; lateral hirsute flap present.

Medial eye and rod-shaped organ (Figure 25c): Medial eye hirsute; rod-shaped organ elongate, 1-jointed.

Remarks. Differences between the specimen I examined and those described by Skogsberg [1920, p. 493] are listed below:

Part	Herein	Skogsberg [1920]
First antenna		
Lateral surface of 1st joint	spinous	spines not shown
a-claw of 7th joint	pectinate at tip	smooth
Second antenna		
Anterior margin of protopodite	with spines	not described
Inner margin of 1st exopodite joint	with spines	not described
Mandible		
Coxale endite	bristle present at base of ventral branch	not shown
Number of pairs of spines below terminal pair on triaenid bristles	4–7	2–5
Number of proximal bristles on dorsal margin of 2nd endopodite joint	1 or 2	1
Maxilla		
Epipodite and dorsal margin of basale	hirsute	no hairs shown
Sixth limb		
Number of anteroventral bristles	3	2

Genus ***Philippiella*** Poulsen, 1965

The name *Philippiella* is preoccupied by a mollusk. The name is retained herein, but will have to be changed when a new name is proposed by Dr. Poulsen, who has been notified.

Philippiella pentathrix new species

Figs. 26–28

Holotype. USNM 125483, valves and some appendages in alcohol, remaining appendages on slide; ♀ with 11 eggs in brood chamber; length 2.47 mm, height 1.42 mm.

Paratypes. USNM 125484, ♀ with 12 eggs in brood chamber, from Station 55; USNM 125485, undissected ♀ with eggs in brood chamber, from Station 56; USNM 125486, undissected ♀ with eggs in brood chamber, from Station 56. In addition, the following specimens were returned to the Universidad de Concepción, Instituto Central de Biología, Departamento de Zoología, Concepción, Brazil: 1 juvenile from Station 55; 9 gravid ♀♀ from Station 56; 1 specimen from Station 61.

Type locality. Station 55, off Greenwich Island, 62°26′06″S, 59°37′30″W, depth 355 meters, mud bottom.

Other localities. Off Greenwich Island: Stations 56, 61.

Etymology. The specific name 'pentathrix' from the Greek 'pente,' five, and 'thrix,' hair, refers to the 5 bristles on the dorsal margin of the basale of the mandible.

Distribution. The number of specimens collected at each station in the vicinity of Greenwich Island is as follows (Station — number of specimens): 55 — 3, 56 — 13, 61 — 1. Specimens were collected at depths of 188–355 meters, in mud and fine sand.

Description of female.

Carapace: Oval in lateral view with rostrum and incisur of usual type for genus (Figures 26a, d; 28a).

Ornamentation: Surface smooth; few bristles present along posterior margin.

Infold: About 44 bristles forming row just within and parallel to anterodorsal margin above incisur; about 36 bristles, mostly short, present between row of 44 bristles and list; numerous bristles present on list above incisur and between list and incisur; numerous bristles also present below incisur and on infold along anteroventral margin; list paralleling posterior margin with about 29 hyaline, broad, transparent spines posterior to numerous long and short slender spines; 5 short processes forming row between list

Fig. 26. *Philippiella pentathrix*, ♀, USNM 125483: *a*, complete carapace, length 2.47 mm; *b*, central muscle scars, left valve, lateral view; *c*, posteroventral part of infold; *d*, incisur, medial view, bristles on infold not shown; *e*, maxilla; *f*, comb of fifth limb showing epipodial bristles, lateral view; *g*, right sixth limb, medial view; *h*, anterior tip left sixth limb, lateral view; *i*, right lateral eye; *j*, medial eye; *k*, rod-shaped organ; *l*, upper lip; *m*, posterior process. USNM 125496: *n*, coxale endite, tip of dorsal process not shown. (Same scale in microns: *c–e; f, h, n.*)

and posterior shell margin; numerous long and short bristles present between list and posterior margin of valve (Figure 26*c*).

Selvage: Narrow lamellar prolongation, with smooth margin present except near middle of lower margin of incisur, where prolongation bears long marginal hairs.

Muscle scars: Central muscle scars obscure but consisting of 18 or more individual scars (Figure 26*b*).

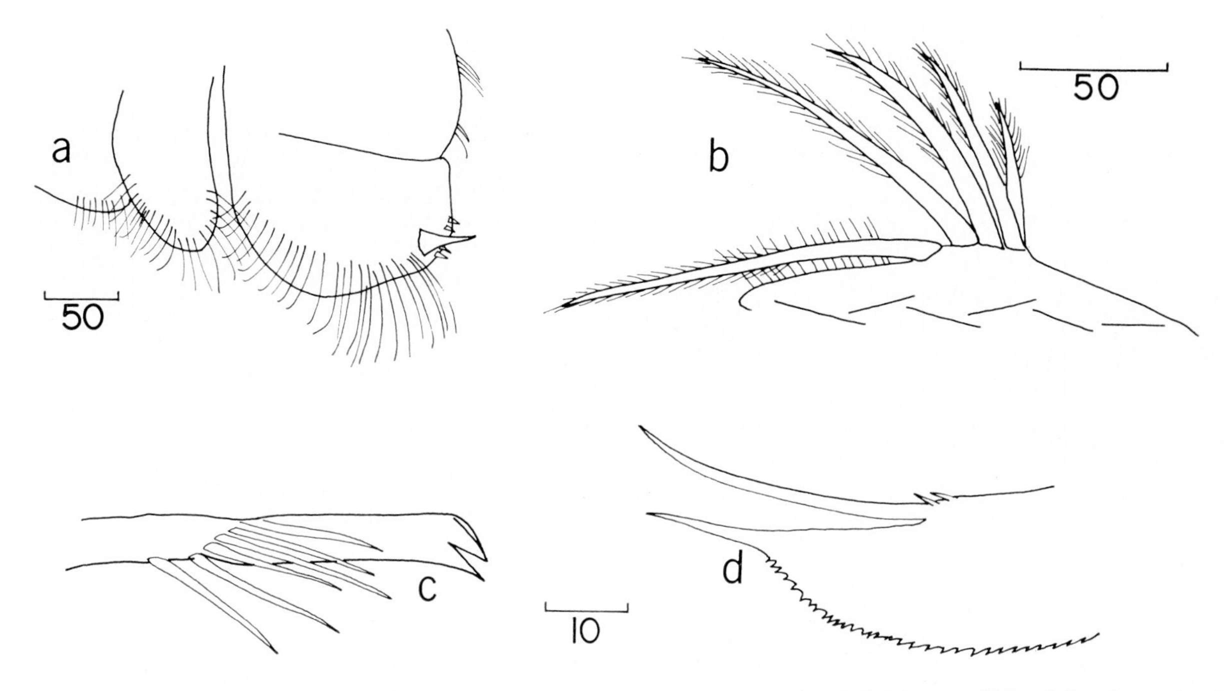

Fig. 27. *Philippiella pentathrix*, ♀, USNM 125484: *a*, upper lip; anterior to right; *b*, bristles at middle of dorsal margin of mandibular basale, anterior to right; *c*, tip of ventral branch of coxale endite; *d*, tip of dorsal branch of coxale endite. (Same scale in microns: *c*, *d*.)

Pore canals: Radial pore canals numerous, ending in minute spines along posterior margin.

Size: Holotype (USNM 125483) length 2.47 mm, height 1.42 mm; USNM 125484, length 2.29 mm, height 1.29 mm; USNM 125485, length 2.25 mm, height 1.27 mm; USNM 125486, length 2.28 mm, height 1.24 mm.

First antenna (Figure 28c): Similar to *P. quinquesetae.*

Second antenna (Figure 28b): Exopodite: Bristle of second joint with long, slender ventral spines; bristles on joints 3–6 with small marginal spines and natatory hairs; basal spine of ninth joint not bifurcate. In other respects limb similar to that of *P. quinquesetae.*

Mandible (Figures 26n, 27b–d, 28d): Similar to that of *P. quinquesetae* except in having fewer spines at tip of ventral branch of coxale endite.

Maxilla (Figures 26e, 28e): Epipodite reaching past middle of basale; tip of epipodite elongate, hirsute; hairs also present along dorsal and ventral margins of broad proximal part of epipodite; basale and proximal dorsal margin of first endopodite joint hirsute; ventral margin of basale with 6–7 bristles near middle and 1 long, plumose terminal bristle; dorsal margin of basale with fairly long proximal bristle and 3 to 4 shorter distal bristles; proximal endite with 4 bristles, 3 long and 1 short; distal endite with 3 bristles, middle bristle shorter than outer two.

Fifth limb: Exopodite bristle longer than end of comb; 2 bristles present below base of exopodite bristle and 4 additional bristles present near ventral margin (Figure 26*f*). Epipodial appendage with 70 plumose bristles.

Sixth limb (Figures 26g, h): Anterior margin with 1 upper and 1 lower bristle; anteroventral corner with 2 small bristles; anterior margin of lateral sole with 3 bristles; posteroventral margin with 20–21 spinose bristles; limb extremely hirsute.

Seventh limb (Figure 28f): End combs each with 14–17 spinose teeth; each limb with 18 bristles with 3–4 or 3–6 bells.

Furca: Each lamella with 8 claws, including 1 posterior backward-pointing bristle; claws curved with teeth along concave margins.

Upper lip (Figures 26l, 27a): Hirsute with anterior spines.

Eyes: Lateral eye small with 3 ommatophores (Figure 26*i*); medial eye about twice size of lateral eye and pigmented (Figure 26*j*).

Rod-shaped organ (Figure 26k): Elongate with rounded tip and restriction near middle on USNM

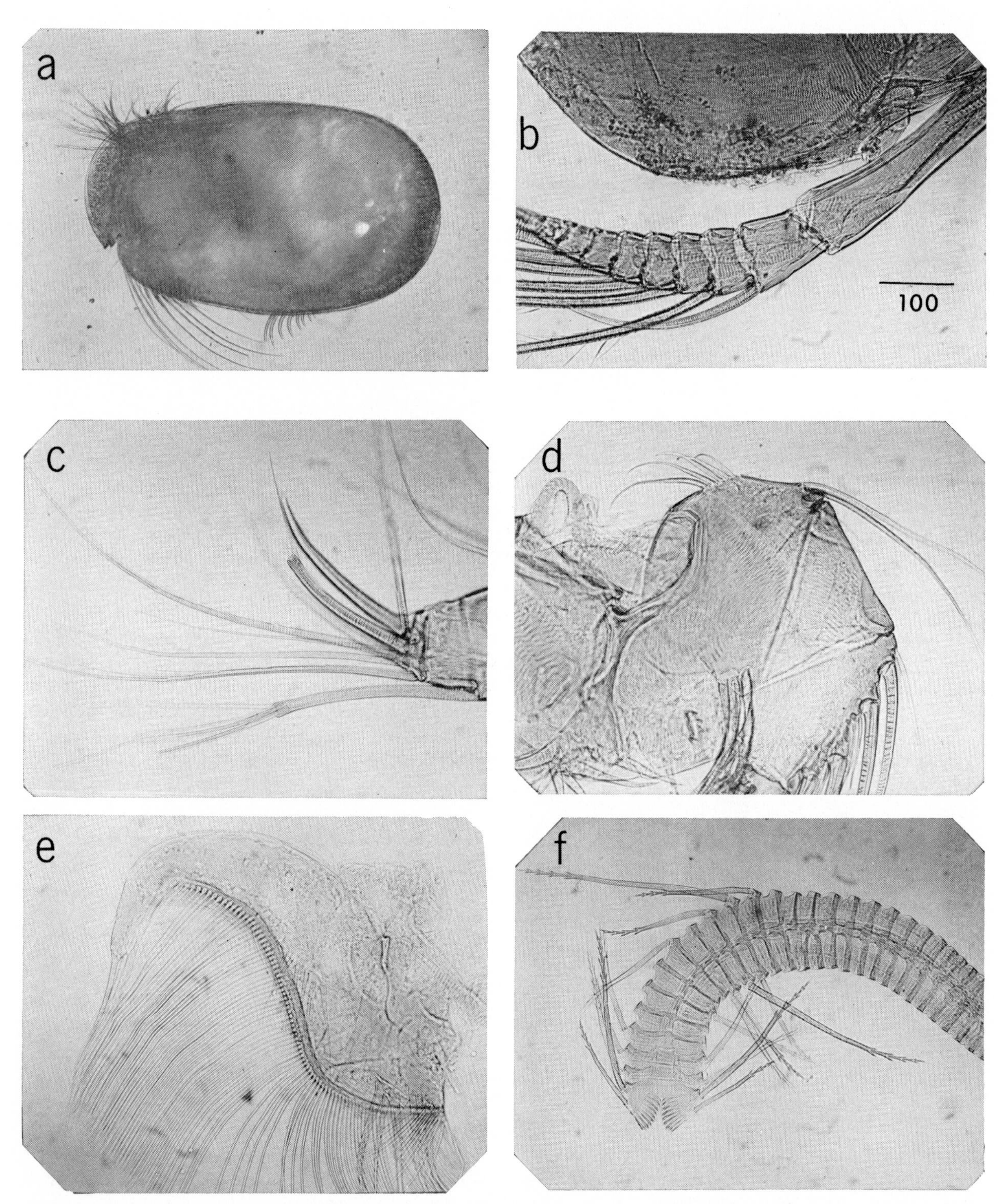

Fig. 28. *Philippiella pentathrix*, ♀: *a*, complete specimen, length 2.25 mm; *b*, second antenna; *c*, tip of first antenna; *d*, part of mandible; *e*, maxilla; *f*, seventh limb. (Same scale in microns: *b–f*. Figure *a* from USNM 125485; remaining figures from USNM 125483.)

125483; distal end of proximal half inserted into proximal end of distal half of organ on USNM 125484.

Posterior (Figure 26m): Posterior margin with spines; thumblike, spinose dorsal process present.

Eggs: 11 eggs in brood chamber of USNM 125483; 12 eggs in brood chamber of USNM 125484.

Comparisons. P. pentathrix is closely related to *P. quinquesetae* (Skogsberg, 1920). Differences are listed below:

Characteristic	*P. quinquesetae*	*P. pentathrix*
Carapace length, mm	2.88–3.1	2.25–2.47
Lower bristle of anterior margin of 6th limb	lacking	present
Number of posteroventral bristles on 6th limb	22–24	20–21
Number of bristles on 7th limb	22–26	18
Number of ommatophores in lateral eye	9–10	3

Philippiella spinifera (Skogsberg, 1920)

Figs. 29, 30

Asterope spinifera Skogsberg, 1920, p. 476, figs. LXXXVI–LXXXVIII.

Philippiella spinifera (Skogsberg); Poulsen, 1965, p. 344.

Holotype. Asterope spinifera Skogsberg, 1920, specimen on slide of Swedish State Museum (Riksmuseum), Stockholm.

Material. USNM 125487, ♀ with eggs in brood chamber, Station 55 from near Greenwich Island. One gravid ♀ returned to Universidad de Concepción, Instituto Central de Biología, Departamento de Zoología, Concepción, Brazil. One juvenile lost.

Distribution. The 3 specimens in the collection are all from Station 55 in the vicinity of Greenwich Island. Skogsberg [1920, p. 483] describes specimens collected from the vicinities of South Georgia and Tierra del Fuego. Specimens have been collected at depths ranging from 70 m to 355 meters, in mud and in grayish clay with scattered stones; bottom temperature at South Georgia was 1.45°C [Skogsberg, 1920, p. 483]. Hartmann-Schröder and Hartmann [1965, p. 324] listed *Cylindroleberis* cf. *spinifera* from a depth of 260 meters off Chile.

Supplementary description of female.

Carapace (Figures 29a–c): As described by Skogsberg [1920, p. 476] and with characteristic transverse medial ridge on posterior of right valve [Skogsberg, p. 477, fig. 3].

Muscle scars (Figure 29b): Central muscle scars consisting of numerous (about 40) individual ovoid scars.

Size: USNM 125487, length 2.63 mm, height 1.74 mm.

First antenna (Figure 30a): Posterior or ventral bristles of fourth joint reaching only to about middle of fifth joint; medial surface of third joint with short spines forming clusters near ventral margin; longer spines forming clusters on medial surface of fourth joint near ventral margin; in remaining characteristics similar to first antennae described by Skogsberg [1920].

Skogsberg [1920, p. 478] described the sensory bristle of the fifth joint as having 7 filaments, but illustrated [fig. 4, p. 477] a sensory bristle having 8 filaments. The specimen I studied had 7 filaments on the sensory bristle.

Second antenna (Figure 30b): Medial surface of protopodite with slender spines forming clusters on anterior and posterior parts; spines also along anterior margins, those along posterior margin stouter than those along anterior margin; ventral margin of bristle of second joint of exopodite with long, slender spines; bristles on joints 3 and 4 with small marginal spines and natatory hairs; bristles on joints 5–8 and 2 long bristles on joint 9 with natatory hairs; 2 shorter bristles on ninth joint spinous; joints 3–9 with basale spines increasing in size on distal joints; joints 3–8 with short spines along distal margins; endopodite distinctly 3-jointed, with long terminal bristle.

Mandible (Figure 30c): 4–5 triaenid bristles on basale endite with 9–17 pairs of spines; 1 triaenid bristle on basale near endite with 5–9 pairs of spines; proximal dwarf bristle shorter than distal bristle; glandular peg large, about same length as short dwarf bristle; dorsal margin of basale with 14–15 short bristles and 2 long terminal bristles; slender spines forming clusters present in vicinity of dorsal margin; exopodite slightly longer than half length of dorsal margin of first endopodite joint. Endopodite: dorsal margin of first joint with 4 stout terminal spines; ventral margin of first joint with 3 long, spinous bristles; stout dorsal claw of end joint with about 8 short, slender spines near middle of concave margin; in remaining characteristics, limb similar to that described for *P. spinifera* by Skogsberg [1920, p. 478].

Maxilla (Figure 29d): Tip of epipodite elongate, hirsute; dorsal margin of basale with hairs forming

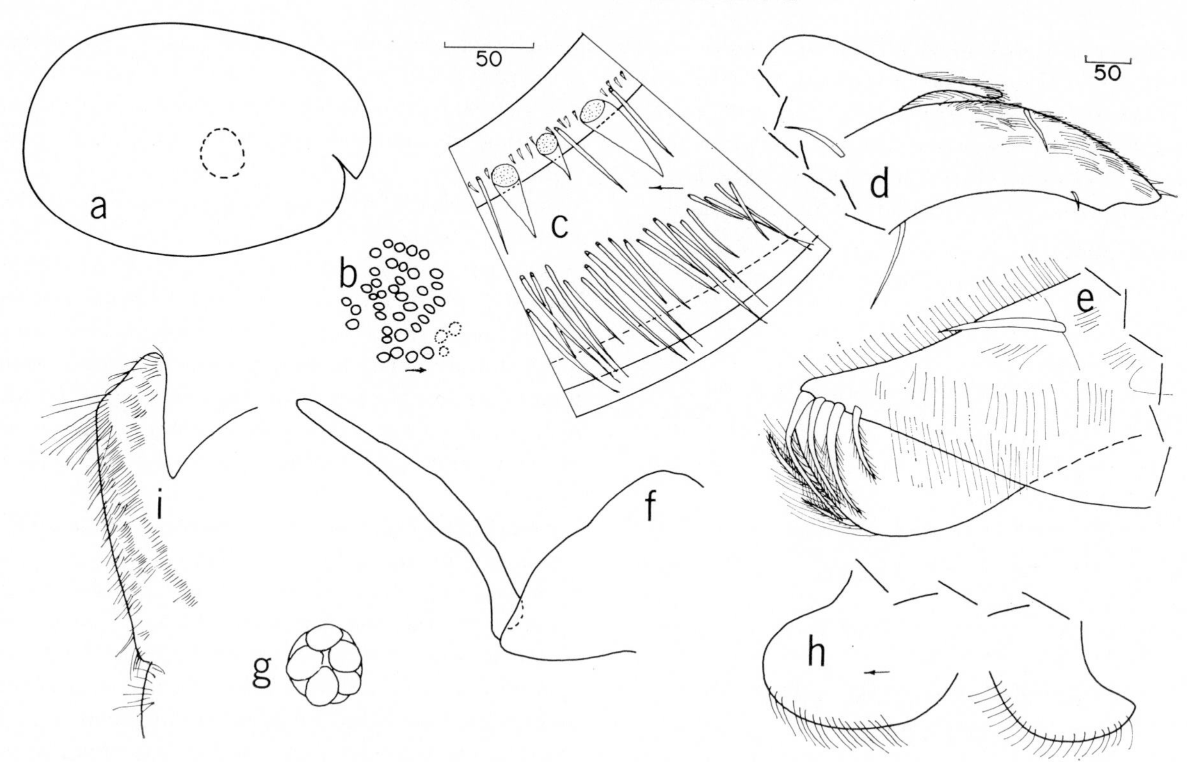

Fig. 29. *Philippiella spinifera,* ♀ USNM 125487: *a,* complete specimen, length 2.63 mm; *b,* central muscle scars, right valve, lateral view; *c,* posteroventral part, infold, right valve, medial view; *d,* epipodite and part of basale of maxilla; *e,* anterior tip, sixth limb, medial view; *f,* medial eye and rod-shaped organ; *g,* lateral eye; *h,* upper lip; *i,* posterior process. (Same scale in microns: *d, f–i; c, e.*)

clusters; in remaining characteristics similar to maxilla of *P. spinifera* described by Skogsberg [1920, p. 481].

Fifth limb (Figure 30d): Exopodite bristle slightly longer than end of comb; 2 small bristles present below base of exopodite bristle and 4 additional bristles present near ventral margin; epipodial appendage with about 84 bristles.

Sixth limb (Figures 29e, 30e): Anteroventral corner with 3–4 spinous bristles; lateral flap with 2 short, slender spinous bristles; anterior margin with upper and lower bare bristles; posteroventral margin with 26 spinose bristles; anterior and posterior margins and ventral surface hirsute.

Seventh limb (Figure 30f): End combs each with about 18 spinose teeth; each limb with 18 bristles with 3–5 bells.

Furca: Each lamella with total of 10 claws as in *P. spinifera* described by Skogsberg [1920, p. 483].

Rod-shaped organ (Figure 29f): Elongate, somewhat widened near middle.

Posterior (Figure 29i): Posterior margin with spines and thumblike, spinose dorsal process.

Eyes: Medial eye fairly large, without hairs (Figure 29f); lateral eye with 5 or 6 ommatophores (Figure 29g).

Upper lip (Figure 29h): Hirsute with a hirsute flap on each side of mouth.

Eggs: USNM 125487 with 9 eggs in brood chamber.

Remarks. Specimens from near Greenwich Island differ slightly from those described by Skogsberg [1920] from South Georgia and Tierra del Fuego; differences are compared in the following table:

Characteristic	Specimens Described Herein	Specimens Described by Skogsberg [1920]
Carapace length, mm	2.63	2.55–2.60
End claw of mandible	spines near middle of concave margin	smooth
Dorsal margin of basale of mandible, number of bristles	14–15	10–12

In addition, Skogsberg described the lateral eye as 'well developed,' whereas the eyes on the specimen from Greenwich Island have only about 6 ommatophores. I examined a mounted specimen from the type series of Skogsberg labeled '*Asterope spinifera* n. sp. ♀, Ant. Sept. -01-03, st. 34 (71).' A lateral eye on that specimen is larger than that on the specimen from Greenwich Island and has about 9 scattered ommatophores.

Philippiella skogsbergi new species

Fig. 31

Holotype. USNM 125488, valves and some appendages in alcohol, remaining appendages on slides, ♀ with 16 eggs in brood chamber, length 3.55 mm, height 2.19 mm.

Paratypes. USNM 125490, undissected juvenile preserved in alcohol. In addition, 1 undissected juvenile returned to Universidad de Concepción, Instituto Central de Biología, Departamento de Zoología, Concepción, Brazil.

Type locality. Vicinity of Greenwich Island, Station 56, 62°25.8′S, 59°37.0′W. Paratypes collected in same sample as holotype.

Etymology. The species is named in honor of Dr. Tage Skogsberg.

Distribution. The 3 specimens in the collection were all collected at Station 56 at a depth of 274 meters in fine sand.

Description of female. Shape similar to *P. spinifera* (Figure 31*a*); complement of medial bristles on infold similar to *P. spinifera* except for bristles forming outer row along posteroventral margin, which are less numerous by almost half than those on *P. spinifera* (compare Figures 29*c* and 31*b*).

Size: USNM 125488, length 3.55 mm, height 2.19 mm.

First antenna: Longer of 2 ventral bristles on fourth joint as long as or slightly longer than ventral margin of fifth joint (Figure 31*d*); a-claw of seventh joint with smooth concave margin, not weakly pectinated as on a-claw of *P. spinifera.* In other characters, limb similar to that of *P. spinifera* described herein.

Second antenna: Protopodite with spines forming clusters along anterior margin and on anterior medial surface; posterior margin of protopodite smooth, without spines as on protopodite of *P. spinifera;* ninth joint of exopodite with 4–5 bristles; in remaining characteristics, limb similar to *P. spinifera* described herein.

Mandible (Figures 31e, f): Dorsal margin of basale with 9–10 short bristles and 2 long terminal bristles; 4–5 terminal spines on dorsal margin of basale, slightly shorter and more acute than those on mandible of *P. spinifera;* basale endite with 6 triaenid bristles with 7–11 pairs of spines and 4 spinous bristles; 1 triaenid bristle on basale near endite with 5–6 pairs of spines; coxale with bristle near base of endite; spine proximal to main spine of dorsal branch of coxale endite about same length as main spine; many spines of ventral branch with secondary crust on specimen examined; stout dorsal claw of end joint of endopodite without spines on concave margin; exopodite reaching about halfway up dorsal margin of first endopodite joint; in remaining characteristics, limb similar to that of *P. spinifera* described herein.

Maxilla and fifth limb: Similar to those on *P. spinifera* described herein.

Seventh limb: Each comb with approximately 17 spinous teeth; 25 bristles with 3–5 bells present on right limb, left limb with distal part missing.

Furca: Each lamella with 9 claws, otherwise similar to *P. spinifera* described herein.

Posterior: Similar in size and outline to those of *P. spinifera* described herein.

Rod-shaped organ (Figure 31h): Elongate with rounded tip, wrinkled in proximal part.

Eyes: Medial eye smooth, pigmented (Figure 31*g*); lateral eye larger than that on *P. spinifera* described herein and with more (about 16) ommatophores; black pigment surrounding ommatophores (Figure 31*i*).

Eggs: USNM 125488, with 16 eggs in brood chamber.

Comparisons. This species is closely related to *P. spinifera.* Some differences are listed below:

Characteristic	*P. skogsbergi*	*P. spinifera*
Carapace length, mm	3.55	2.55–2.63
First antenna, length of longer of 2 ventral bristles on 4th joint as % of length of ventral margin of 5th joint	100	50
Second antenna, posterior margin of protopodite, smooth (S), with stout spines (W)	S	W
Seventh limb, number of bristles	25	17–20
Number of furcal claws	9	10
Distribution of bristles in outer row on posteroventral infold (see Figures 29*c* and 31*b*)	sparse	dense

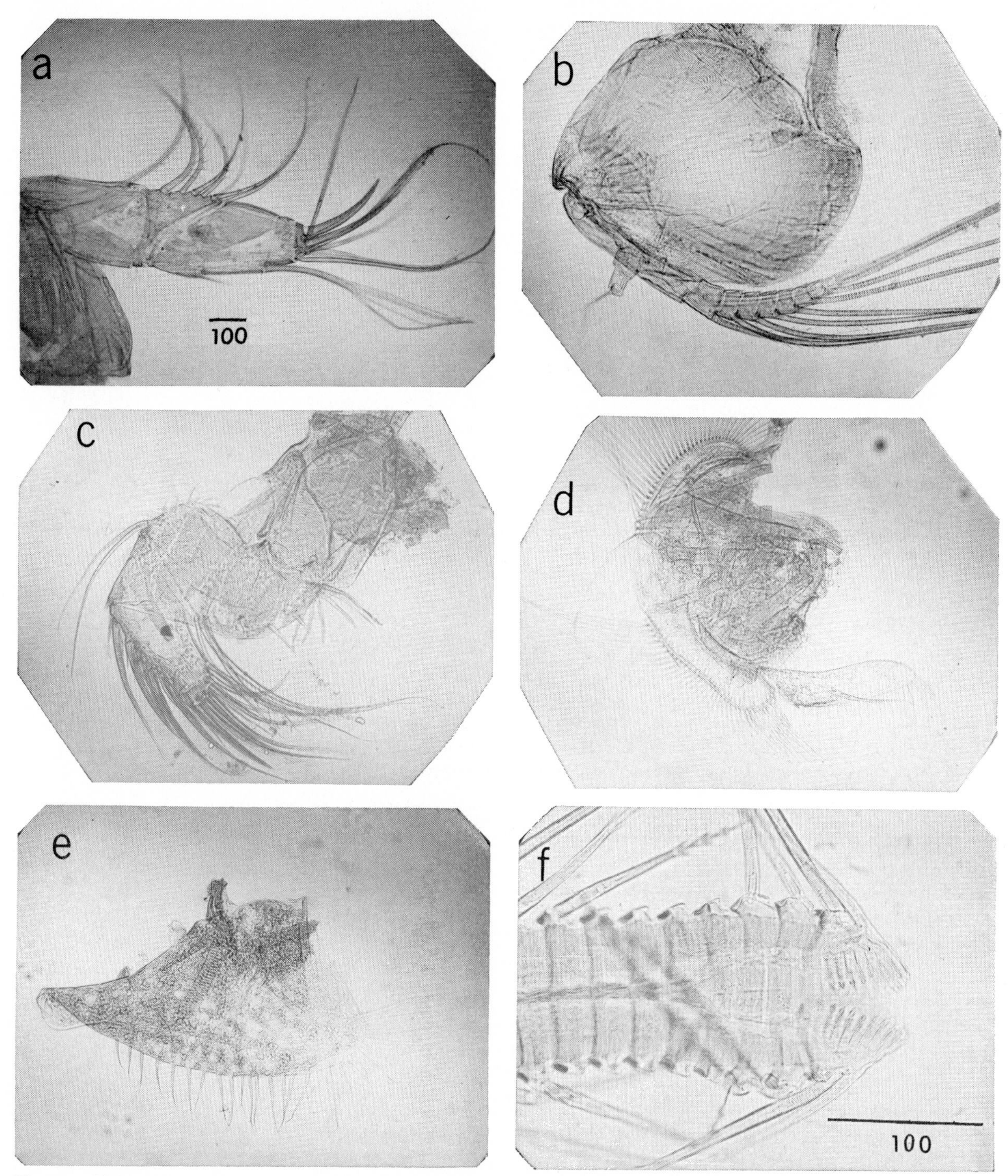

Fig. 30. *Philippiella spinifera*, ♀ USNM 125487; *a*, first antenna; *b*, second antenna; *c*, mandible; *d*, fifth limb; *e*, sixth limb; *f*, tip of seventh limb. (Same scale in microns: *a–e*.)

Considerable variability in size and number of ommatophores was observed in the lateral eye of *P. spinifera*. Until the reason for this is understood, it is probably best not to use the relative development of the lateral eyes to differentiate *P. spinifera* and *P. skogsbergi*.

Remarks. Scott [1912, p. 586] referred a specimen collected in Scotia Bay, South Orkney Islands, to

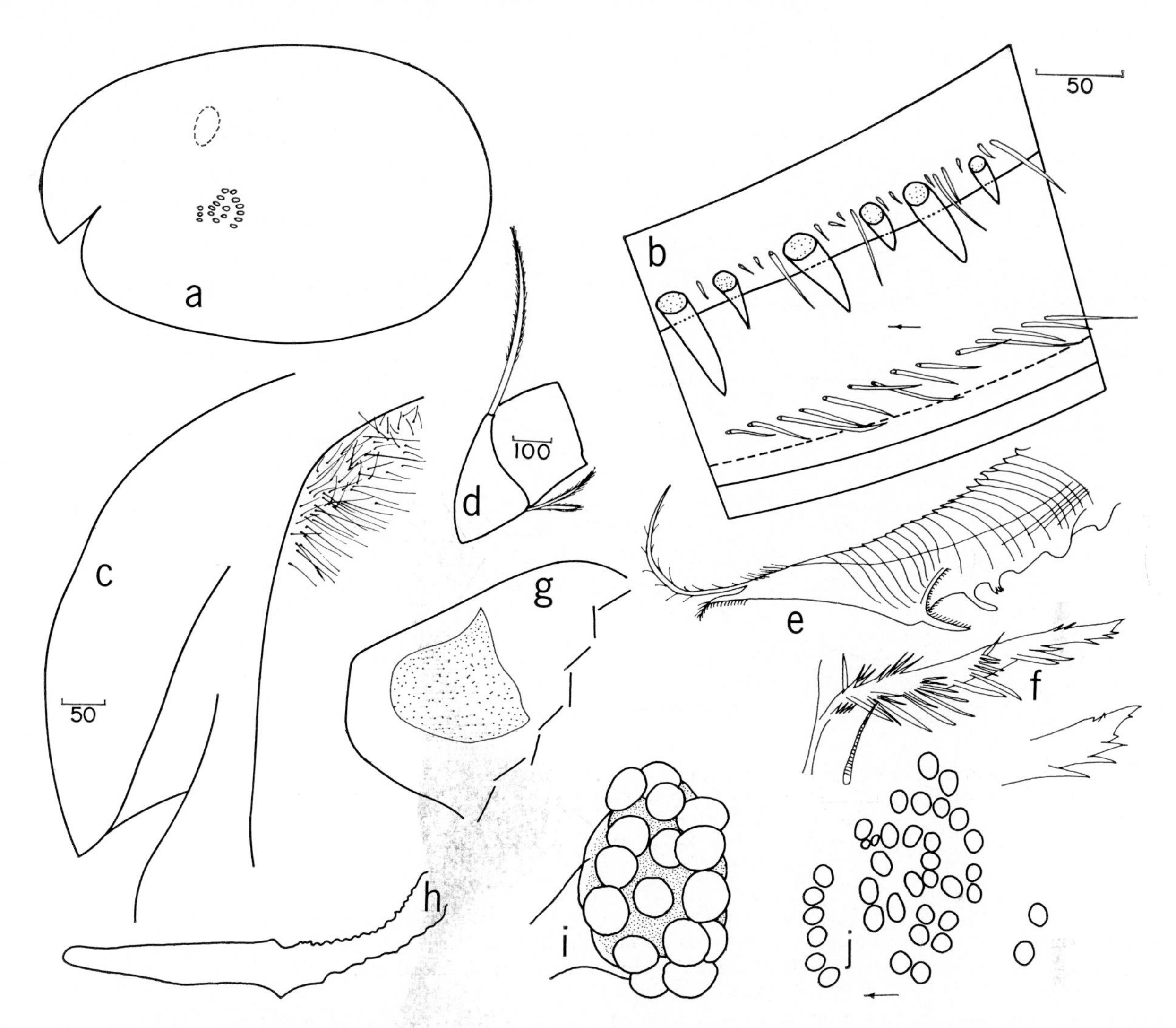

Fig. 31. *Philippiella skogsbergi*, ♀ USNM 125488: *a*, complete carapace, length 3.55 mm; *b*, posteroventral part of infold, right valve, medial view; *c*, anterodorsal part, right valve, showing hairs on vestment proximal to infold, medial view; *d*, fourth and fifth joints of first antenna; *e*, distal part of dorsal branch of mandibular coxale endite; *f*, ventral branch of coxale endite and bristles near its base; *g*, medial eye; *h*, rod-shaped organ; *i*, lateral eye; *j*, central muscle scars, left valve with animal removed, lateral view (more individual scars are visible on valve after animal is removed). (Same scale in microns: *d, j*; *c, g–i*; *b, e, f*.)

Asterope australis Brady. His total statement concerning the specimen is as follows: 'This species was obtained in a small gathering of dredged material collected in Scotia Bay, South Orkneys, on 3rd June 1903; Station 325, 60°43′42″S, 44°38′33″W. The length of the specimen—a female—represented by drawing (fig. 18) is 2.75 mm.' *Asterope australis* was described by Brady [1890] from specimens from the South Pacific; the description is based on exterior characteristics of the carapace, which are of little discriminatory importance in this group. Brady [1898] described a few appendages of *A. australis* using specimens collected in New Zealand; but, as already noted by Skogsberg [1920, p. 483], the New Zealand specimens belong to a much smaller species. I wrote to the Royal Scottish National Museum, Edinburgh, for the specimen identified by Scott and received from Mr. David Heppell a vial containing the undissected preserved specimen and 3 labels with the following information: '*Asterope australis*, Stat. 325 fms Bwy Pt, S. Orkneys, June 1903,' 'Renche Dredge, Stat 325 fms, June 3, Bay Pt, S. Orkneys,' '1921 . 143 . 1145.' My measurements of the preserved specimen, a gravid female with 11 eggs in the brood chamber,

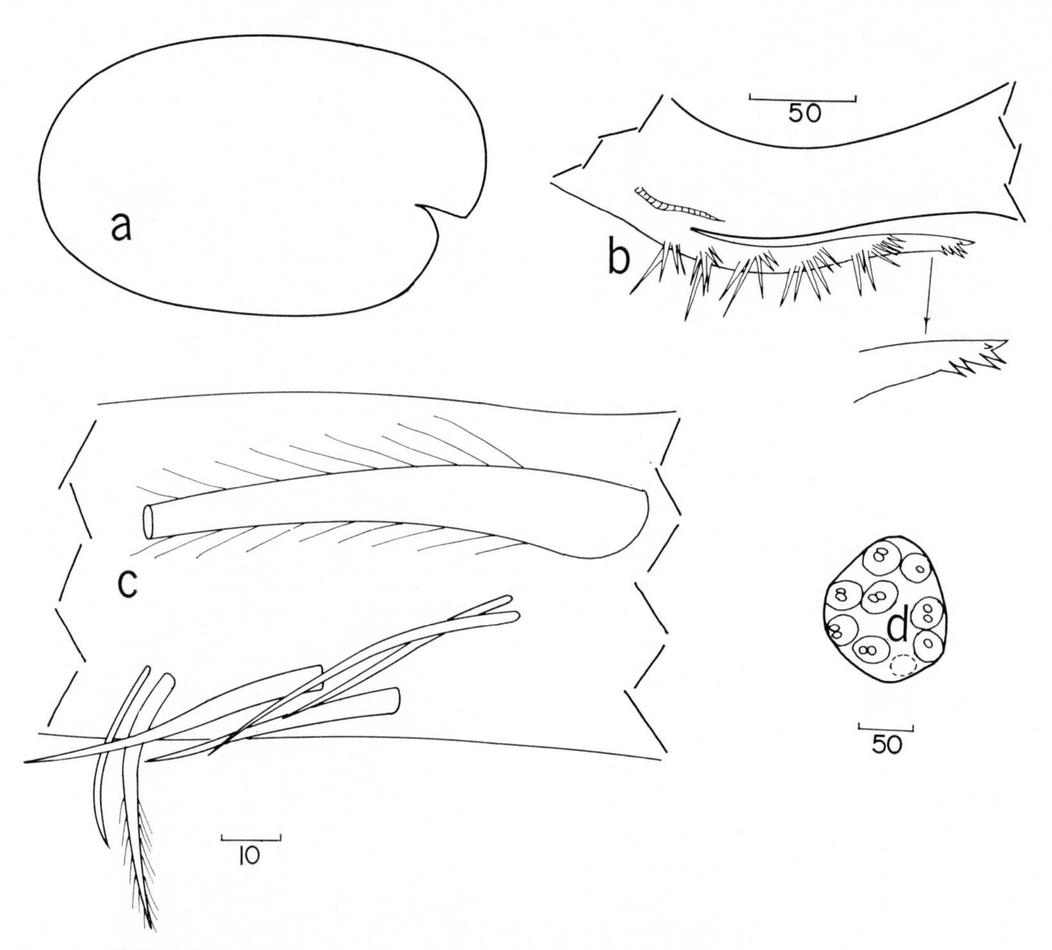

Fig. 32. *Philippiella quinquesetae*, ♀: *a*, complete carapace, length 2.88 mm; *b*, proximal part of coxale endite showing ventral branch; *c*, part of comb of fifth limb showing epipodial bristles, lateral view, bristles of ventral margin not shown; *d*, lateral eye.

are length 2.73 mm, height 1.78 mm. The central muscle scars are similar to those of *Philippiella spinifera.*

Permission was obtained from Dr. Rodger Waterston, Royal Scottish Museum, to dissect the specimen. The well-developed d- and e-bristles of the first antenna place the specimen in the genus *Philippiella.* The distribution of spines and bristles on the dorsal margin of the mandibular basale indicates that the specimen is closely related to *P. spinifera* and *P. skogsbergi.* Other characters of the specimen are intermediate between *P. spinifera* and *P. skogsbergi:* The carapace is 0.10 mm longer than previously reported specimens of *P. spinifera* and 0.82 mm shorter than *P. skogsbergi.* The length of the longer of the 2 ventral bristles on the fourth joint of the first antenna is as long as the ventral margin of the fifth joint; the posterior margin of the protopodite of the second antenna bears a few stout spines; the seventh limb bears 21 bristles; the furca bears 9 claws on one lamella and 10 on the other; the lateral eye is well developed, with about 20 ommatophores. Primarily because of its size, I have considered the specimen to be *P. ? spinifera,* as recorded in Table 5. The valves (unseparated) and appendages were put in a vial with alcohol and returned to the Royal Scottish Museum.

Philippiella quinquesetae (Skogsberg, 1920)

Fig. 32

Asterope quinquesetae Skogsberg, 1920, p. 470, figs. LXXXIV, LXXXV.

Philippiella quinquesetae (Skogsberg); Poulsen, 1965, p. 344.

Skogsberg [1920] described this species from the Swedish Antarctic Expedition's Station 34 off South

Georgia at a depth of 252–310 meters. I received from the Swedish State Museum 8 glass slides with mounted valves and appendages and a vial containing 5 specimens in alcohol, R. M. S. 162. Some of the slides are in good condition; others have air inclusions. The brief supplementary description that follows is based on a gravid female that had been preserved in alcohol.

Supplementary description.

Carapace: Length 2.88 mm, height 1.64 mm (Figure 32*a*).

Second antenna: Anterior margin and anterior half of medial surface of protopodite with slender spines forming clusters, posterior margin with short, stout spines; small medial bristle present on protopodite; bristles of second joint of exopodite with long, slender ventral spines; bristles of joints 2–8 with long natatory hairs; ninth joint with 4 bristles, 1 short with short marginal spines, 1 medium with short marginal spines, 2 long with natatory hairs; fourth to sixth joints with 3 small basal spines; seventh and eighth joints with single large spines; ninth joint with single bifurcate spine; second to eighth joints with short spines along distal margins.

Mandible: Small bristle at base of ventral branch of coxale endite (Figure 32*b*); tip of ventral branch with 6 spines, 5 stout and 1 minute.

Maxilla: Proximal endite with 4 bristles, 3 long, 1 short; epipodial appendage and basale hirsute; marginal spines on long terminal bristles of basale and on first and second joints of exopodite. In other respects as described by Skogsberg [1920].

Fifth limb: Exopodite bristle longer than end of comb; 2 bristles below base of exopodite bristle and 4 additional bristles near ventral margin (Figure 32*c*); dorsal margin of comb bare; long hairs at distal end of comb; epipodial appendage with 73 bristles.

Sixth limb: Anterior margin with 1 upper and no lower bristle; anteroventral end with about 5 bristles (obscure on specimen examined); posteroventral margin with 23 spinose bristles; limb extremely hirsute and similar in shape to that of *P. pentathrix.*

Seventh limb (1 limb examined): Limb with 22 bristles, 13 on one side, 9 on other; each bristle with 3 to 4 bells; each comb with 17 to 19 spinose teeth.

Eggs: 16 eggs present in marsupium.

Furca: Each lamella with 7 claws and 2 small bristles.

Lateral eye: Each eye with 9 or 10 ommatophores (Figure 32*d*).

Acknowledgments. I wish to thank Mr. Roy Olerod of the Naturhistoriska Rikmuseet, Stockholm, for the opportunity to examine specimens of *Philomedes rotunda, Scleroconcha appelloefi, Philippiella quinquesetae, Philippiella spinifera,* and *Parasterope ohlini* from South Georgia described by Skogsberg [1920]. Also, J. Forest of the Muséum National d'Histoire Naturelle, Paris, for the opportunity to examine specimens of *Philomedes charcoti* and *Philomedes laevipes* from Booth Island, described by Daday [1908], and Dr. K. G. McKenzie of the British Museum (Natural History) and Dr. John R. A. Gray of the Hancock Museum, Newcastle-on-Tyne, for the opportunity to examine specimens described by Brady [1907] of *Philomedes orbicularis, Philomedes assimilis,* and *Philomedes antarctica* from McMurdo Sound, and David Heppell and Dr. A. Rodger Waterston of the Royal Scottish Museum, Edinburgh, Scotland, for the opportunity to examine specimens of *Philomedes assimilis* and *Asterope australis* described by Scott [1912] from the South Orkney Islands. I also wish to thank Dr. J. Laurens Barnard, Dr. Waldo L. Schmitt, and Miss Lucile McCain for reviewing and editing the manuscript, and Jack R. Schroeder for final preparation of figures from my camera lucida drawings.

REFERENCES

Barney, R. W.
1921 Crustacea, Part V. Ostracoda. *In* Nat. Hist. Rep. Br. Antarct. 'Terra Nova Exped.', 1910. Zool., *3* (7): 175–190, 6 figs.

Brady, G. S.
1890 On Ostracoda collected by H. B. Brady, Esq., L.L.D., F.R.S., in the South Sea Islands. Trans. R. Soc. Edinb., *35* (2): 489–525, 4 pls.
1898 On new or imperfectly-known species of Ostracod chiefly from New Zealand. Trans. Zool. Soc. London, *14* (8): 429–452, 5 pls.
1907 Crustacea, Ostracoda. *In* Nat. Antarct. Exped., 1901–1904. *3:* 1–9, pls. 1–3.

Daday, E. de
1908 Ostracodes marins. *In* Expédition Antarctique Française, 1903–1905, *16:* 1–15, figs. 1–14.

Hartmann-Schröder, Gesa, and Gerd Hartmann
1965 Zur Kenntnis des Sublitorals der chilenischen Küste unter besonderer Berucksichtigung der Polychaeten und Ostracoden. Mitteilungen Hamb. Zool. Mus. Inst. Erganzungsband, *62:* 1–384.

Kornicker, L. S.
1969*a* Morphology, ontogeny, intraspecific variation of *Spinacopia,* a new genus of myodocopid ostracod (Sarsiellidae). Smithson. Contr. Zool., *8:* 1–50, 26 figs., 6 pls.
1969*b* Ostracoda (Myodocopina) from the Peru-Chile trench and the Antarctic Ocean. Smithson. Contr. Zool., *32:* 1–42, 25 figs.

Liljeborg, W.
1853 De Crustaces ex ordinibus tribus: Cladocera, Ostracoda, et Copepoda, in Scania occurrentibus. Om de inom Skane Forekommande Crustaceer af ordiningere Cladocera, Ostracoda och Copepoda. (In Swedish.) xvi + 22 pp., 27 pls. Lund, Sweden.

Müller, G. W.
1908 Die Ostracoden der Deutschen Südpolar-Expedition, 1901–1903. Dt. Südpol.-Exped., 1901–1903. X, Zoologie, *2* (2): 51–181, pls. 4–19, text-fig. 1.
1912 Ostracoda. Tierreich, *31:* i-xxxiii, 1–434.

Poulsen, E. M.
1962 Ostracoda-Myodocopa. 1. Cypridiformes-Cypridinidae. (In English.) Dana Rep., no. 57, pp. 1–414, 181 figs.
1965 Ostracoda-Myodocopa. 2. Cypridiniformes-Rutidermatidae, Sarsiellidae and Asteropidae. (In English.) Dana Rep., no. 65, pp. 1–484, 156 figs.

Scott, Thomas
1912 The Entomostraca of the Scottish National Antarctic Expedition, 1902–1904. Trans. R. Soc. Edinb., *48:* 521–600, pls. 1–14.

Skogsberg, T.
1920 Studies on marine ostracods. 1. Cypridinids, Halocyprids and Polycopids. Zool. Bidr. Fran Upps., supplement 1, pp. 1–884, 153 figs.

Sylvester-Bradley, P. C.
1961 Myodocopida, pp. 387–406, figs. 311–330. *In* R. C. Moore (ed.), Treatise on invertebrate paleontology. *3* (Q): 1–442, 334 figs. Geological Society of America and University of Kansas Press, Lawrence, Kansas.

Tibbs, J. F.
1965 Observations on *Gigantocypris* (Crustacea: Ostracoda) in the Antarctic Ocean. Limnol. Oceanogr., *10:* 480–482.

A REVIEW OF THE PYCNOGONID GENUS *PANTOPIPETTA* (FAMILY AUSTRODECIDAE, EMENDED) WITH THE DESCRIPTION OF A NEW SPECIES

JOEL W. HEDGPETH

Marine Science Center, Oregon State University, Newport, Oregon 97365

JOHN C. MCCAIN

Hazleton Laboratories, Inc., Falls Church, Virginia 22046

Abstract. This paper reviews the taxonomic status of the pycnogonid genus *Pantopipetta* Stock, 1963, placing it in the family Austrodecidae. The species of *Pantopipetta* collected by the *Eltanin*, *Vema*, and *Acona* are described including a new species found off the coast of Oregon. A key to the known species of *Pantopipetta* is presented.

INTRODUCTION

This report is based on the collections made by the USNS *Eltanin* of the United States Antarctic Research Program and the R/V *Vema* of Lamont-Doherty Geological Observatory, Columbia University, in the deep waters south of 16°S between 18°E and 75°W and by R/V *Acona* of Oregon State University off the coast of Oregon. These collections yielded 28 specimens belonging to 4 species of the relatively rare genus *Pantopipetta*.

Family AUSTRODECIDAE Stock, 1954, emend.

Definition. Eye tubercle well developed, reaching over origin of proboscis or erect, situated at anterior margin of cephalic somite. Proboscis slender, tube- or pipette-shaped, often annulated distally. Chelifores absent. Palpi 5- to 9-jointed. Ovigers reduced to 4 to 7 joints or developed to 10 joints, with special spines on terminal joints and a terminal claw. Legs moderately long to attenuated, with prominent coxal spurs and, in males of many species, with femoral cement gland processes. Tarsus usually short; auxiliary claws present or absent. Genera: *Austrodecus* (littoral to 920 meters); *Pantopipetta* (66 to 5024 meters).

Discussion. In 1904 Loman published a brief paper on the nature of the pycnogonid proboscis, appended to a description of a new genus and species, *Pipetta weberi* from *Siboga* Station 127 in the Banda Sea at 2081 meters [Loman, 1904]. In this paper he did not suggest a family affiliation, but in the *Siboga* Report [Loman, 1908] he ascribed *Pipetta* to the Colossendeidae. Eight, or possibly nine, species are now recognized for this genus, now known as *Pantopipetta* [Stock, 1963] because *Pipetta* is a preoccupied name (Protozoa). The genus was excluded from the Colossendeidae by Calman and Gordon [1933], who defined the Colossendeidae in the following terms:

'Cephalon short, neck absent; proboscis large, usually exceeding trunk length. Trunk of 4–6 somites, fused or free. Chelophores with two-jointed scape in larva, sometimes persisting in adult. Palp with 8–10, usually 9 segments, inserted on a ventral process similar to and usually contiguous with that which bears the oviger. Oviger large in both sexes, with 10 segments and a terminal claw which is often much reduced; several rows of special spines on segments 7–10. Legs 8–12; coxal segments relatively short, a genital pore on each second coxa in both sexes; auxiliary claws absent. Species often of large size.'

Stock [1963] included *Pantopipetta* in the Colossendeidae on the basis that 'no ovigerous males are known. This, together with the structure of the oviger, in which segments 2, 4, and 6 are elongated, is a strong indication for attributing the genus to the family Colossendeidae.' The lack of ovigerous males hardly seems a good systematic character, especially of familial rank [Hedgpeth, 1948]. The ovigers of most specimens of *Pantopipetta* are indeed 10-jointed; but the terminal segments bear a single row of triangular denticulate spines more suggestive of the type found in ammotheids, and further, they occur only in a single row, rather than in the fields or patches characteristic of *Colossendeis*. In *Pantopipetta aconae* new species, the ovigers are reduced to 4 or 5 segments, or perhaps the specimens are not completely mature. However, the males are recognizable by the presence of well-developed cement gland tubes on the femora (another character not found in the colossendeids). Furthermore, the palps of *Pantopipetta* are 7- or 8-jointed, as opposed to the usual 9 of the colossendeids (including *Decolopoda* and *Dodecolopoda*). All of these characters (the different spination of the ovigers and their reduction in one species, the presence of femoral cement gland tubes, and the differently proportioned palpi) would require altering the diagnosis of the Colossendeidae as presently understood.

All specimens of *Pantopipetta* have very long, attenuated proboscides, and in several of the species the proboscis is annulated distally. This trait is shared with species of the genus *Austrodecus*, for which Stock [1954] established the family Austrodecidae, which he diagnosed as follows:

'Ocular tubercle strong, reaching far over the implantation of the proboscis. Ocular tubercle forms a gradual transition into the cephalic segment. Proboscis slender, tube-shaped, annulated. No chelifores. Palpi well developed, 5- to 6-jointed. Ovigers strongly reduced in both sexes. 4- to 7-jointed. Terminal joint far the longest, without indication of a dorsal or a ventral margin, without special spines. Legs with intermediate distal joints. Auxiliary claws either present or absent.'

Certain features of *Pantopipetta aconae* agree with this diagnosis. The eye tubercle, while erect, is at the forward margin of the cephalic somite, and it does indeed form a gradual transition into the somite. (This feature is also present in *P. brevicauda.*) The ovigers are strikingly reduced, and evidently do not serve any cleaning function. There is a prominent, slender spinelike tube for the femoral cement gland of the male, similar to that encountered in some species of *Austrodecus* (for example, *A. tubiferum* Stock); in other species of *Austrodecus* the cement gland opening is on a conical process (for example, *A. glaciale* Hodgson, *A. profundum* Stock). In all species of *Austrodecus* so far described the terminal joint of the palp is set at an angle on the penultimate joint; this feature is not found in the species of *Pantopipetta.*

It is obvious that the genera *Austrodecus* and *Pantopipetta* are in some ways more closely related to each other than *Pantopipetta* is to colossendeids. The most conspicuous differences between *Pantopipetta* and the colossendeids are the presence of prominent cement-gland cones or tubelike processes and the presence of conspicuous long tubercles on the lateral processes and the occasional long spines on the legs. The long, slender, and essentially unadorned appendages of colossendeids are certainly a more objective taxonomic criterion than the absence of ovigerous males. This trait is also shared, to a lesser degree, by the Nymphonidae, and it is interesting to note that it is in these two groups of pycnogonids that the ovigers are equally developed in both sexes, with the terminal segments arranged to form a prehensile cleaning structure for grooming the long, smooth legs. It should also be noted that this structure is so arranged that it functions as a cleaner of the dorsal surfaces [Hedgpeth, 1964], and hence strong development of spines and tubercles on the dorsal surface is incompatible with use of the ovigers for grooming. In the Colossendeidae the ovigers evidently no longer function as egg-mass carriers.

In most of the described species of *Pantopipetta* the ovigers appear to be long enough to function as groomers, but the development of long dorsal spurs on the coxae and femora and the reduction of the special spines on the terminal segments of the ovigers suggest that this grooming function is not as significant in *Pantopipetta* as it is in such genera as *Colossendeis* and *Nymphon.*

The structure of the proboscis in *Austrodecus* and *Pantopipetta* indicates that species of these two genera are very similar ecologically and that they feed, as Fry [1965] suggested for antarctic species of *Austrodecus,* on individual polypides of bryozoans or similar prey, and that *Pantopipetta* is therefore the deep-water analog of *Austrodecus.*

In any event, it seems inadvisable to compromise the comparatively discrete morphological diagnosis of the Colossendeidae by including *Pantopipetta,* in view of the obvious relationship of these forms to *Austrodecus;* and we therefore suggest that the concept of the Austrodecidae be amended to include *Pantopipetta.*

Genus ***Pantopipetta*** Stock, 1963

Definition. Eye tubercle erect, situated at anterior margin of cephalic somite. Proboscis slender, pipette-shaped, annulated distally in several species. Ovigers

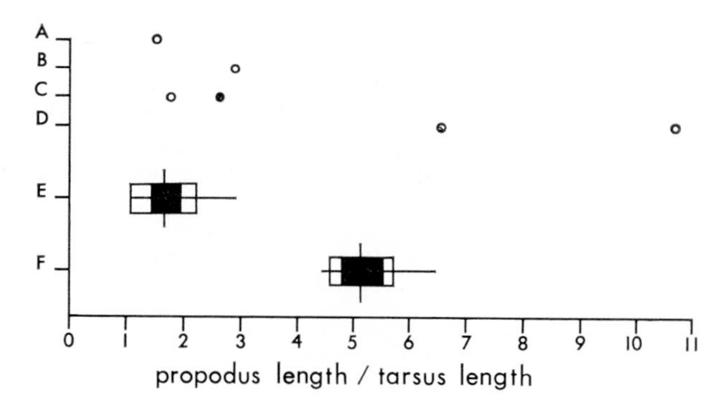

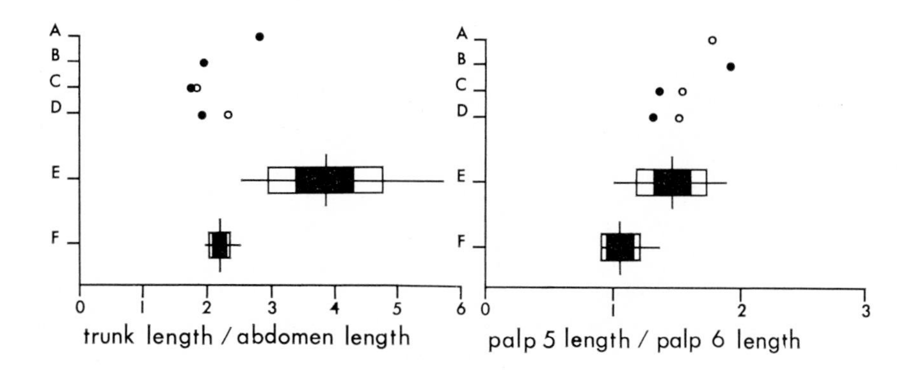

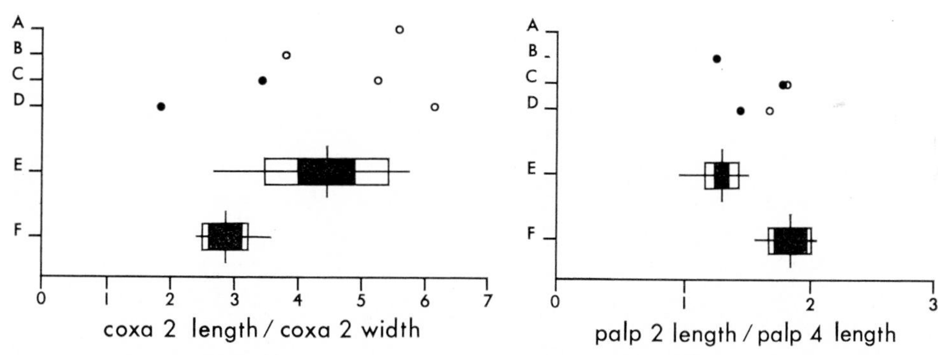

Fig. 1. Mensural ratios of 6 species of *Pantopipetta:* A, *P. weberi;* B, *P. longituberculata;* C, *P. capensis;* D, *P. australis;* E, *P. brevicauda;* F, *P. aconae.* Solid circles represent measured specimens, open circles published figures; the horizontal line represents range, the vertical line the mean; the black rectangle represents the value of 2 standard errors either side of the mean, and the white rectangle the value of 1 standard deviation either side of the mean.

of 10 joints, reduced to 4 or 5 joints in one species, terminal segments of 10 jointed ovigers with single row of triangular denticulate spines. Palpi 7- or 8-jointed in most species. Legs long, auxiliary claws present or absent. Species: *Pantopipetta aconae* new species; *P. australis* (Hodgson, 1914); *P. auxiliata* Stock, 1968; *P. brevicauda* Stock, 1963; *P. capensis* (Barnard, 1946); *P. longituberculata* (Turpaeva, 1955); *P. oculata* Stock, 1968; *P. weberi* (Loman, 1904).

Discussion. The genus *Pantopipetta* is presently composed of 8 species, including *P. aconae,* which is herein described as new. Table 1 presents the characters used to separate these species.

Figure 1 graphically illustrates the major mensural differences between 6 of these species. In all cases *Pantopipetta brevicauda* Stock, 1963, is significantly different from *P. aconae,* new species. Regrettably, we do not have sufficient data on the remaining species to comment on their statistical differences, yet several

TABLE 1. Discrimination of the Species of ***Pantopipetta***

Many of the ratio values in this table were computed from published figures, and their accuracy is therefore questionable. The high ratio values for *P. australis* and *P. capensis* were taken from figures, whereas the low values were taken from the *Vema* specimens. The ratio values for *P. brevicauda* and *P. aconae* were taken from our specimens, the remainder from figures.

Species	Eye tubercle	Coxa 1, dorsal	Lateral processes	Abdomen length	Ovigeral spines, segments 7:8:9:10	Propodus length: tarsus length	Palp segment 5 length: palp segment 6 length	Palp segment 2 length: palp segment 4 length	Coxa 2 length: coxa 2 greatest width	Trunk length: abdomen length	Distribution
P. aconae, new species	low–high	2 spurs on legs 1–4	1–4 with low tubercles	to end of coxa 2 of leg 4	none	4.4–6.4	0.9–1.4	1.6–2.1	2.6–3.6	4.4–6.4	Oregon, 1400–2000 meters
P. australis (Hodgson, 1914)	high	smooth	smooth	to end of coxa 2 of leg 4	3:2:2:3	6.5–10.7	1.3–1.5	1.4–1.7	1.8–6.1	1.9–2.3	Antarctica and South Georgia, 2450–3725 meters
P. auxiliata Stock, 1968	high	2 spurs on coxa 1 of legs 1–4	1–4 with 1 spur	to end of coxa 2 of leg 4	3:2:2:3	4.5	1.7	1.3	1.7	1.9	Natal, South Africa, 68–69 meters
P. brevicauda Stock, 1963	high	smooth	smooth	to end of coxa 1 of leg 4	2:2:2:3 3:1:2:3 3:2:2:3 3:2:2:4 3:2:2:5 4:3:2:6 3:3:2:4 4:4:3:6 5:2:3:4	1.1–2.9	1.0–1.9	1.0–1.5	2.7–5.8	2.5–5.7	Quadrant south of 16°S between 13°E and 75°W, 1806–5024 meters
P. capensis (Barnard, 1946)	high	smooth	2–4 with dorsal spur, 1 with low tubercle	to end of coxa 2 of leg 4	4:2:2:3 3:2:2:4	1.8–2.7	1.3-1.5	1.4-1.8	3.4–5.3	1.8	South Africa, 841–2890 meters

P. longituberculata (Turpaeva, 1955)	high	smooth	smooth	to end of coxa 2 of leg 4	4:4:4:4	2.88	1.9	1.2	3.8	1.9	Kurile Islands, 4850 meters
P. oculata Stock, 1968	high	1–3 with 4 spurs, 4 with 1 spur	smooth	to end of coxa 2 of leg 4	3:2:2:3	4.3	1.5	2.1	1.7	1.6	Andaman Islands, 66 meters
P. weberi (Loman, 1904)	low?	smooth	smooth	to end of coxa 2 of leg 4	?	1.50	1.8	?	5.6	2.1	Banda Sea, 2081 meters
P. sp. Stock, 1963	high	smooth	1–4 with low tubercles	to end of coxa 2 of leg 4	?	4 or less	?	?	about 4	?	South Africa, 1326–3256 meters

points are obvious. It can be seen that the propodus length divided by the tarsus length of *P. australis* (Hodgson, 1914) is much higher than that of the other species and that *P. brevicauda*, *P. longituberculata*, and *P. weberi* have in common a very low ratio, with *P. aconae* falling as an intermediate between the latter species and *P. australis*.

Pantopipetta brevicauda, as the name implies, is characterized by its short abdomen. In *P. brevicauda* the ratio of trunk length divided by abdomen length is relatively high, with a mean around 4. The other species have a relatively low ratio, around 2. The low-ratio species would need quite a large degree of variation not to differ significantly from *P. brevicauda*. It should be mentioned that although *P. brevicauda* is characterized by the abdomen reaching the distal end of coxa 1 of leg 4, in some cases it extends beyond this to approximately midlength of coxa 2.

Turpaeva [1955] mentions that palp segment 2 is almost twice the length of palp segment 4 in *Pantopipetta australis*, whereas in *P. longituberculata* and *P. weberi* these segments are approximately equal in length. *P. longituberculata* does have a smaller ratio than *P. australis;* however, in those species where a large number of specimens were examined, the ratio varied considerably. The use of this ratio as a taxonomic character seems, therefore, less useful than propodus length divided by tarsus length and trunk length divided by abdomen length; and this ratio should probably be best used after it is calculated for a larger number of specimens than are known at this time. A similar situation exists for the length of palp segment 5 divided by palp segment 6.

It is apparent from the graph of the ratios of the length of coxa 2 divided by its greatest width that this ratio is subject to considerable variation. We put little faith in the ratios obtained from published figures and therefore, except in the case of *P. brevicauda* and *P. aconae*, this ratio seems to be of little value until many more specimens of the other species are examined.

Stock's descriptions [1968] of *Pantopipetta auxiliata* and *P. oculata* were published after this manuscript was submitted for publication; therefore, the ratios are not included in Figure 1. Both of these species fall as intermediates along with *P. aconae* in the ratio of propodus length divided by tarsus length. They differ from all other species so far assigned to the genus in having auxiliary claws. Stock associates this character with the relatively shallow-water habitat in this genus, although he attaches no significance

TABLE 2. Measurements (mm) of Species of ***Pantopipetta*** from Published Records and Specimens Cited in This Report

Character	*P. australis**	*P. capensis†*		*P. longi-tuberculata††*	*P. weberi§*	*P. auxiliata‖*	*P. oculata¶*
Proboscis, length	1.90	3.38	3.50	3.10	3.70	1.32	1.03
Proboscis, width	0.18	0.20					
Somite 1, length	0.61	0.58		0.42			
Somite 2, length	0.39	0.60		0.52			
Somite 3, length	0.43	0.75		0.56			
Somite 4, length	0.55	0.97		0.72			
Trunk, length	1.98	2.90		2.22	2.70		
Abdomen, length	1.03	1.62	1.75	1.14	1.30	0.55	0.48
Second lateral process, width	1.17	2.40				0.72	0.58
Coxa 1, length	0.28	0.30		0.30		0.21	0.19
Coxa 2, length	0.55	0.99		0.94		0.27	0.23
Coxa 3, length	0.21	0.30		0.28		0.13	0.13
Femur, length	1.49	2.15	2.00	1.75		0.56	0.45
Tibia 1, length	1.41	2.00	1.75	1.85		0.56	0.45
Tibia 2, length	0.95	1.50	1.50	1.54		0.55	0.45
Tarsus, length	0.07	0.30		0.25		0.06	0.06
Propodus, length	0.75	0.80		0.72		0.27	0.26
Claw, length	0.20	0.27		0.27		0.10	0.10
Coxa 2, distal diameter	0.30	0.29					
Palp segment 1, length		0.24		0.13			
Palp segment 2, length	1.00	1.67		1.41			
Palp segment 3, length	0.15	0.25		0.28			
Palp segment 4, length	0.70	1.16		1.14			
Palp segment 5, length	1.30	0.23		0.21			
Palp segment 6, length	0.10	0.17		0.11			
Palp segment 7, length	0.07	0.14		0.09			
Palp segment 8, length	0.08	0.15		0.17			
Oviger segment 1, length		0.15		0.05			
Oviger segment 2, length	0.13	0.25		0.10			
Oviger segment 3, length	0.07	0.13		0.19			
Oviger segment 4, length	0.27	0.50		0.38			
Oviger segment 5, length	0.11	0.17		0.20			
Oviger segment 6, length	0.32	0.58		0.52			
Oviger segment 7, length	0.13	0.23		0.20			
Oviger segment 8, length	0.11	0.15		0.16			
Oviger segment 9, length	0.11	0.13		0.10			
Oviger segment 10, length	0.12	0.15		0.07			
Oviger segment 11, length	0.14	0.04		0.03			

* ♀ from *Vema*, cruise 14, Sta. 22.
† Left column, ♀ from *Vema*, cruise 14, Sta. 32; right column, ♀ described by Barnard [1954].
†† Specimen described by Turpaeva [1955].
§ ♀ described by Loman [1904].
‖ ♀ described by Stock [1968].
¶ ♀ described by Stock [1968].

to the presence or absence of auxiliary claws in *Rhynchothorax*. The function of auxiliary claws is not known, and their significance in classification is uncertain.

Pantopipetta oculata has a high ratio, over 2, for the length of palp segment 2: the length of palp segment 4; while *P. auxiliata* is close to the ratios of the other species. In the remaining ratios both species are close to those of the other species considered.

KEY TO THE SPECIES OF ***Pantopipetta***

1. Abdomen short, extending to distal end of coxa 1 of leg 4 . . . . ***P. brevicauda,*** p. 223
   Abdomen long, extending to distal end of coxa 2 of leg 4 2

2(1). All lateral processes smooth 3
   Any or all lateral processes with dorsal spurs or tubercles 6

3(2). Propodus more than 5 times length of tarsus . . . . ***P. australis,*** p. 223

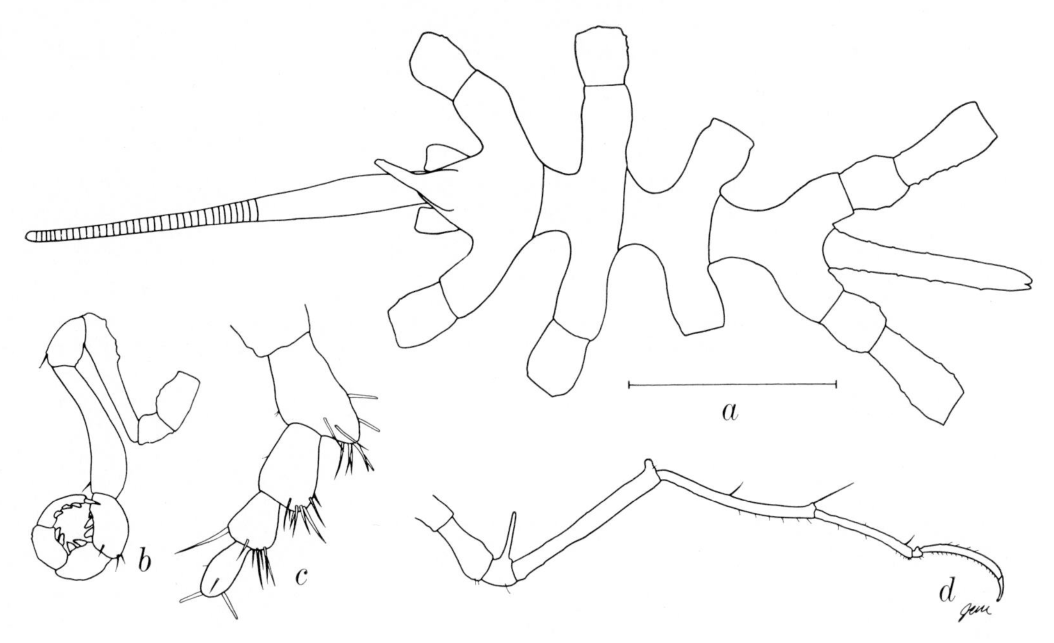

Fig. 2. *Pantopipetta australis: a,* dorsal view; *b,* oviger; *c,* terminal segments of palp; *d,* right leg 3. Scale above letter *a* represents 1 mm.

Propodus less than 5 times length of tarsus 4
4(3). Eye tubercle short .... ***P. weberi***
Eye tubercle long 5
5(4). Coxa 1 with dorsal spurs; with auxiliary claws ... . ***P. oculata***
Coxa 1 without dorsal spurs; no auxiliary claws ... . ***P. longituberculata***
6(2). Coxa 1 dorsal spurs 7
Coxa 1 without dorsal spurs 8
7(6). Lateral processes with low dorsal tubercles; no auxiliary claws .... ***P. aconae*** new species, p. 227
Lateral processes with dorsal spur; with auxiliary claws .... ***P. auxiliata***
8(6). Lateral processes 2–4 with dorsal spur .... ***P. capensis,*** p. 226
Lateral processes 2–4 with low dorsal tubercles ... . ***P.*** sp. Stock, 1963

Pantopipetta australis (Hodgson, 1914)

Figs. 1–2; Table 2

Pipetta australis Hodgson, 1914, p. 159; 1915, pp. 141–142; 1927; pp. 314–315, fig. 2.—Turpaeva, 1955, p. 327.

Material.

Vema, cruise 14: Sta. 22, 55°19′S, 37°57′W, 3725 meters, March 6, 1958, 1 ♀.

Diagnosis. Coxa 1 and lateral processes smooth, palp 8-segmented, propodus more than 4 times as long as tarsus, abdomen reaching end of coxa 2 of leg 4.

Remarks. This record from south of South Georgia is the second for this species, which was originally collected off the *Gauss* winter quarters at a depth of 2450 meters. The single specimen agrees well with Hodgson's description of *P. australis;* however, the propodus is only a little over 6 times the tarsus length, instead of over 10 times the length, as Hodgson's figure shows. In his text, however, Hodgson states that the tarsus is very short, less than $1/5$ the length of the propodus. The specimen also differs from Hodgson's figure in that the ratio of the length and width of coxa 2 is much less than that figured by Hodgson.

P. australis differs from the other species of *Pantopipetta* primarily in the extreme elongation of the propodus relative to the tarsus and in the smoothness of the lateral processes and of the first coxa.

Pantopipetta brevicauda Stock, 1963

Figs. 1, 3–4; Table 3

Pantopipetta brevicauda Stock, 1963, pp. 336–338, figs. 9, 10a.

Material. (Numbers in parentheses at end of station data below correspond to column headings in Table 3.)

Vema, cruise 12:

Sta. 1, 38°58.5′S, 41°45′W, 5024 meters, April 6, 1957, 1 ♀ (1).

TABLE 3. ***Pantopipetta brevicauda,*** Measurements (mm)

Character	Stock, 1953		1	2	3	4	5	6	7	8	9	10	11	12	13	14	15	16	17	18
	♂	♀	(♀)	(♀)	(♀)	(♂)	(♂)	(♀)	(♀)	(♂)	(♀)	(♀)	(♂)	(♂)	(♀)	(♀)	(♂)	(♂)	(♀)	(♀)
Proboscis, length	3.16	3.07	3.80	3.22	3.90+	3.18	3.94	4.16	3.08		2.30	2.30	2.50	3.30	3.40	3.60	3.40	3.40	2.9+	3.6
Proboscis, width			0.23	0.18	0.22	0.15	0.20	0.20	0.18	0.13	0.12	0.21	0.16	0.23	0.25	0.25	0.21	0.20	0.23	0.22
Somite 1, length			0.43	0.40	0.60	0.40	0.34	0.43	0.33	0.15	0.32	0.27	0.35	0.26	0.35	0.32	0.28	0.33	0.28	0.25
Somite 2, length			0.45	0.48	0.63	0.68	0.59	0.58	0.46	0.33	0.45	0.53	0.43	0.65	0.68	0.60	0.62	0.60	0.67	0.68
Somite 3, length			0.58	0.61	0.75	0.78	0.72	0.77	0.59	0.46	0.65	0.66	0.52	0.82	0.80	0.70	0.78	0.78	0.82	0.80
Somite 4, length			0.68	0.70	0.93	0.86	0.84	0.88	0.74	0.40	0.65	0.70	0.59	0.92	0.98	0.92	0.86	0.90	0.95	0.85
Trunk, length	2.19	2.19	2.14	2.19	2.91	2.72	2.49	2.66	2.12	1.34	2.07	2.16	1.89	2.65	2.81	2.54	2.54	2.61	2.72	2.58
Abdomen, length	0.65	0.53	0.70	0.75	0.95	1.08	0.80	0.77	0.63	0.41		0.70	0.55	0.50	0.70	0.55	0.50	0.62	0.78	0.56
Second lateral process, width	1.67	1.74	1.60	1.65	2.20	1.80	1.95	1.80	1.70	1.15	1.40	1.60	1.20	2.13			1.95	2.06	2.20	2.10
Coxa 1, length	0.28	0.28	0.25	0.27	0.43	0.23	0.35	0.27	0.34	0.17	0.23	0.33	0.26	0.34	0.30	0.33	0.33	0.42	0.35	0.40
Coxa 2, length	1.02	0.98	0.95	1.08	1.40	0.84	1.18	1.30	1.10	0.72	0.75	1.16	0.80	1.10	1.30	0.90	1.26	1.32	0.97	1.05
Coxa 3, length	0.19	0.23	0.28	0.32	0.45	0.20	0.22	0.27	0.30	0.15	0.18	0.33	0.25	0.43	0.35	0.25	0.32	0.35	0.37	0.40
Femur, length	1.91	1.91	1.80	2.10	2.74	1.45	3.65	2.64	1.80	1.65	1.66	2.33	1.40	2.15	2.10	2.20	2.35	2.30	2.20	1.95
Tibia 1, length	1.86	1.86	1.75	2.30	2.80	1.16	2.55	2.20	1.95	1.65	1.35	2.30	1.60	2.15	2.00	2.40	2.26	2.25	2.10	1.98
Tibia 2, length	1.33	1.35	1.21	1.65	2.02	1.05	1.80		1.30	1.40	0.88	1.77	1.10	1.30	1.50	1.60	1.50	1.65	1.50	1.25
Tarsus, length	0.37	0.40	0.27	0.55	0.60	0.24	0.52		0.43	0.20	0.26	0.49	0.25	0.43	0.33	0.48	0.47	0.48	0.40	0.38
Propodus, length	0.56	0.60	0.60	0.65	0.67	0.70	0.65		0.60	0.58	0.37	0.65	0.29	0.73	0.53	0.80	0.57	0.75	0.72	0.65
Claw, length	0.28	0.28	0.26	0.50	0.30	0.26	0.32		0.26	0.23	0.17	0.26	0.24	0.23	0.27	0.30	0.23	0.30	0.22	0.23
Coxa 2, distal diameter			0.26	0.21	0.35	0.17	0.22	0.24	0.21	0.13	0.13	0.25	0.23	0.24	0.49	0.32	0.26	0.28	0.29	0.30
Cement gland from proximal femur							3.20			1.35			1.34	1.85			1.88	1.80		
Palp segment 1, length			0.17	0.12	0.20	0.12	0.13	0.15	0.12	0.10	0.03	0.11	0.14	0.14	0.16	0.18	0.14	0.18	0.16	0.18
Palp segment 2, length			1.68	1.55	1.96	1.46	1.92	2.00	1.48	1.25	1.10	1.93	1.18	1.40	1.50	1.70	1.60	1.50	1.60	1.65
Palp segment 3, length			0.28	0.20	0.27	0.21	0.25	0.25	0.24	0.17	0.14	0.28	0.20	0.24	0.25	0.25	0.27	0.33	0.22	0.18
Palp segment 4, length			1.40	1.15	1.46	1.30	1.46		1.55	0.90	0.90	1.55	0.85	1.00	1.28	1.30	1.20	1.10	1.20	1.10
Palp segment 5, length			0.22	0.19	0.23	0.15	0.22		0.15	0.14	0.13	0.23	0.14	0.17	0.22	0.20	0.15	0.20	0.21	0.23
Palp segment 6, length			0.13	0.12	0.14	0.08	0.19		0.10	0.13	0.10	0.13	0.08	0.15	0.15	0.18	0.15	0.15	0.14	0.13
Palp segment 7, length			0.13	0.10	0.15	0.06	0.08		0.11	0.11	0.07	0.12	0.08	0.10	0.12	0.12	0.10	0.10	0.12	0.11
Palp segment 8, length			0.20	0.15	0.21	0.10	0.18		0.13	0.12	0.03	0.20	0.18	0.17	0.17	0.10	0.15		0.13	0.14
Oviger segment 1, length					0.13	0.08	0.12	0.11	0.10	0.07		0.11	0.07	0.11	0.12	0.10	0.13	0.13	0.15	0.10
Oviger segment 2, length					0.20	0.15	0.29	0.31	0.16	0.25		0.24	0.15	0.39	0.41	0.30	0.31	0.33	0.30	0.38
Oviger segment 3, length					0.12	0.07	0.13	0.10	0.08	0.03		0.13	0.05	0.09	0.14	0.12	0.11	0.11	0.11	0.19
Oviger segment 4, length					0.26	0.27	0.43	0.45	0.13	0.39		0.45	0.24	0.54	0.45	0.53	0.23	0.57	0.48	0.39
Oviger segment 5, length					0.20	0.11	0.18	0.19	0.13	0.07		0.17	0.11	0.18	0.16	0.18	0.17	0.18	0.12	0.18
Oviger segment 6, length					0.15	0.22	0.53	0.48	0.30	0.10		0.45	0.35	0.50	0.50	0.55	0.47	0.50	0.33	0.42
Oviger segment 7, length					0.15		0.18	0.20	0.13	0.13		0.16	0.11	0.16	0.15	0.20	0.17	0.22	0.12	0.19
Oviger segment 8, length					0.12		0.18	0.12	0.09	0.10		0.11	0.09	0.18	0.08	0.13	0.11	0.12	0.11	0.14
Oviger segment 9, length					0.13		0.15	0.08	0.09	0.07		0.10	0.11	0.08	0.11	0.10	0.10	0.13	0.10	0.12
Oviger segment 10, length					0.16		0.10	0.12	0.13	0.07		0.13	0.07	0.12	0.12	0.17	0.11	0.12	0.13	0.14
Oviger segment 11, length					0.04		0.04	0.03	0.03	0.03		0.06	0.01	0.02	0.01	0.03	0.04	0.02	0.02	0.03

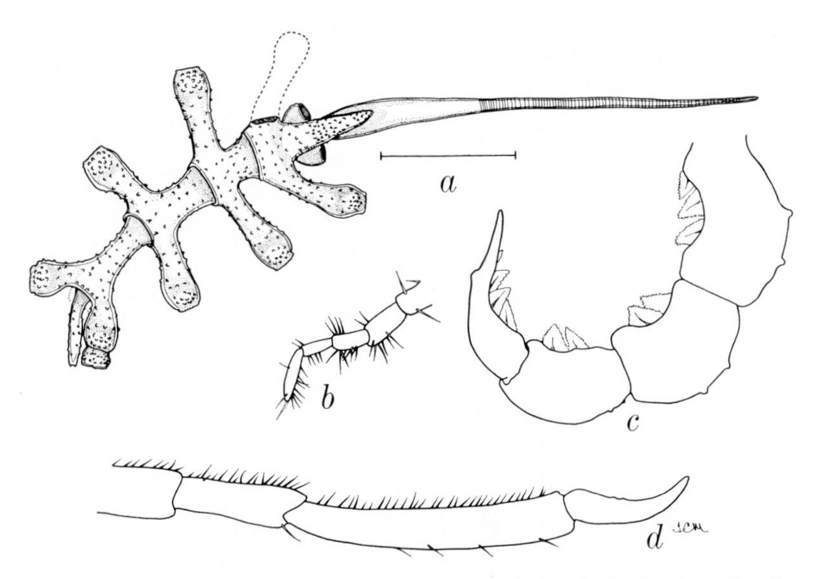

Fig. 3. *Pantopipetta brevicauda: a,* dorsal view; *b,* terminal segments of palp; *c,* terminal segments of oviger; *d,* right leg 3. Scale above letter *a* represents 1 mm.

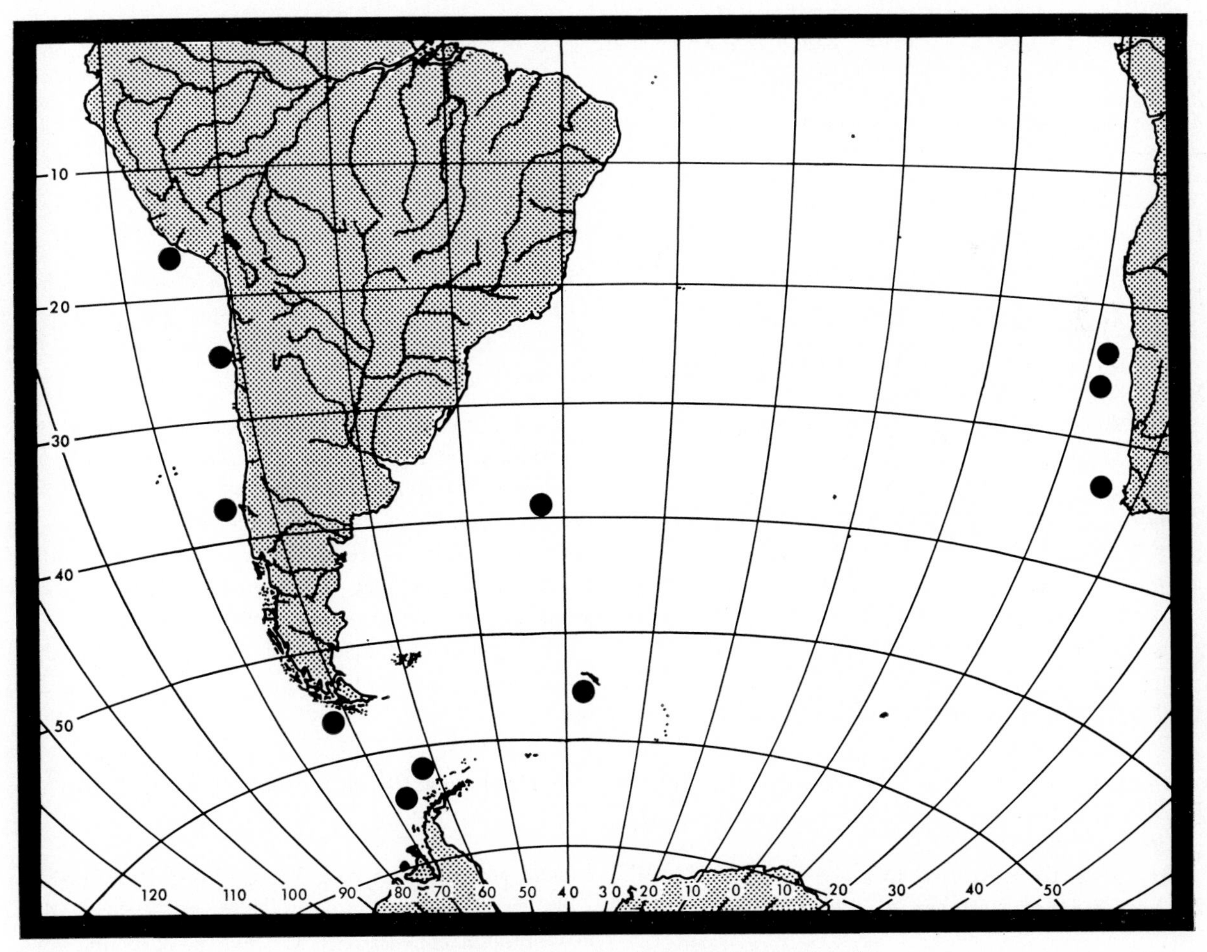

Fig. 4. Distribution records of *Pantopipetta brevicauda.*

TABLE 4. ***Pantopipetta aconae*****,** Measurements (mm)

Character	Holotype (♂)	1 (♂)	2 (♀)	3 (♂)	4 (♂)	5 (♀)	6 (♀)	7 (♂)	8 (♀)
Proboscis, length	1.32	1.40	1.55	1.30	1.30	1.70	1.50	1.40	1.50
Proboscis, width	0.13	0.13	0.13	0.13	0.15	0.15	0.13	0.12	0.10
Somite 1, length	0.30	0.30	0.30	0.24	0.25	0.27	0.25	0.20	0.25
Somite 2, length	0.40	0.45	0.46	0.42	0.45	0.53	0.50	0.42	0.45
Somite 3, length	0.40	0.48	0.47	0.42	0.42	0.54	0.48	0.42	0.45
Somite 4, length	0.35	0.45	0.42	0.37	0.37	0.56	0.43	0.36	0.36
Trunk, length	1.45	1.68	1.65	1.45	1.49	1.80	1.66	1.42	1.51
Abdomen, length	0.60	0.75	0.75	0.60	0.73	0.83	0.80	0.72	0.71
Second lateral process, width	0.90	1.10	0.92	0.90	0.95	1.18	0.93	0.95	0.82
Coxa 1, length	0.15	0.17	0.17	0.15	0.18	0.20	0.20	0.15	0.12
Coxa 2, length	0.35	0.40	0.43	0.36	0.35	0.40	0.40	0.42	0.33
Coxa 3, length	0.12	0.13	0.18	0.13	0.15	0.20	0.15	0.13	0.16
Femur, length	0.75	0.85	0.95	0.75	0.80	0.85	0.70	0.75	0.82
Tibia 1, length	0.71	0.80	0.87	0.70	0.75	0.93	0.92	0.70	0.82
Tibia 2, length	0.70	0.75	0.73	0.56	0.75	0.70	0.88	0.63	0.83
Tarsus, length	0.07	0.10	0.10	0.07	0.10	0.08	0.10	0.06	0.08
Propodus, length	0.45	0.50	0.44	0.35	0.48	0.40	0.50	0.32	0.44
Claw, length	0.15	0.20	0.23	0.15	0.20	0.25	0.25	0.20	0.23
Coxa 2, distal diameter	0.13	0.14	0.12	0.14	0.12	0.15	0.15	0.13	0.14
Cement gland from femur proximal	0.45	0.55		0.45	0.48			0.44	
Palp segment 1, length	0.12	0.10	0.10	0.10	0.08	0.08	0.09	0.10	0.12
Palp segment 2, length	0.62	0.67	0.67	0.60	0.60	0.82	0.75	0.68	0.65
Palp segment 3, length	0.09	0.10	0.11	0.10	0.10	0.13	0.11	0.10	0.10
Palp segment 4, length	0.32	0.38	0.40	0.30	0.30	0.40	0.48	0.35	0.39
Palp segment 5, length	0.09	0.10	0.10	0.15	0.11	0.14	0.12	0.10	0.10
Palp segment 6, length	0.09	0.11	0.11	0.11	0.11	0.12	0.11	0.11	0.09
Palp segment 7, length	0.05	0.05	0.06	0.04	0.05	0.05	0.04	0.06	0.05

Sta. 4, 25°33′S, 12°27′E, 2970 meters, May 3, 1957, 1 ♀ (2).

Sta. 6, 23°00′S, 08°11′E, 4047 meters, May 7, 1957, 1 ♀ (9).

Vema, cruise 14:

Sta. 22, 55°19′S, 37°57′W, 3725 meters, March 6, 1958, 1 ♂ (11).

Sta. 35, 29°44′S, 37°15′E, 4987 meters, April 26, 1958, 1 ♂ (8).

Vema, cruise 17:

Sta. 6, 37°57′S, 75°08′W, 4303 meters, March 21, 1961, 1 ♀ (10).

Eltanin, cruise 3:

Sta. 50, 16°12′S, 74°41′W–16°10′S, 74°41′W, 2599–2858 meters, June 15, 1962, 1 ♀ (3).

Sta. 65, 25°43′S, 71°07′W–25°42′S, 71°07′W, 3149–3257 meters, June 21, 1962, 1 ♂, 1 ♀ (5–6).

Eltanin, cruise 4:

Sta. 127, 61°45′S, 61°14′W, 4758 meters, August 1, 1962, 1 ♂ (4).

Eltanin, cruise 5:

Sta. 268, 64°01′S, 67°45′W–64°08′S, 67°44′W, 2763–2818 meters, October 20, 1962, 2 ♂, 4 ♀ (12–18).

Sta. 322, 56°04′S, 71°13′W–56°05′S, 71°09′W, 1806–2013 meters, November 7, 1962, 1 ♀ (7).

Diagnosis. Coxa 1 and lateral processes smooth or with low tubercles, palp 8-segmented, propodus more than 4 times as long as tarsus, abdomen reaching end of coxa 1 of leg 4.

Remarks. Stock [1963] described this species from several specimens collected with *Pantopipetta capensis* off Cape Point, South Africa, at a depth of 2744–2890 meters. The material collected by the *Eltanin* and *Vema* south of 16°S in the quadrant from 13°E to 75°W shows that *P. brevicauda* is a relatively abundant and wide-ranging deep-sea species.

This is the only species of *Pantopipetta* in which the abdomen does not extend to the end of coxa 2 of leg 4.

Pantopipetta capensis (Barnard, 1946)

Figs. 1, 5; Table 2

Pipetta capensis Barnard, 1946, p. 60; 1954, pp. 86–88, fig. 1.

Pantopipetta capensis Stock, 1963, p. 336, figs. 8a–e.

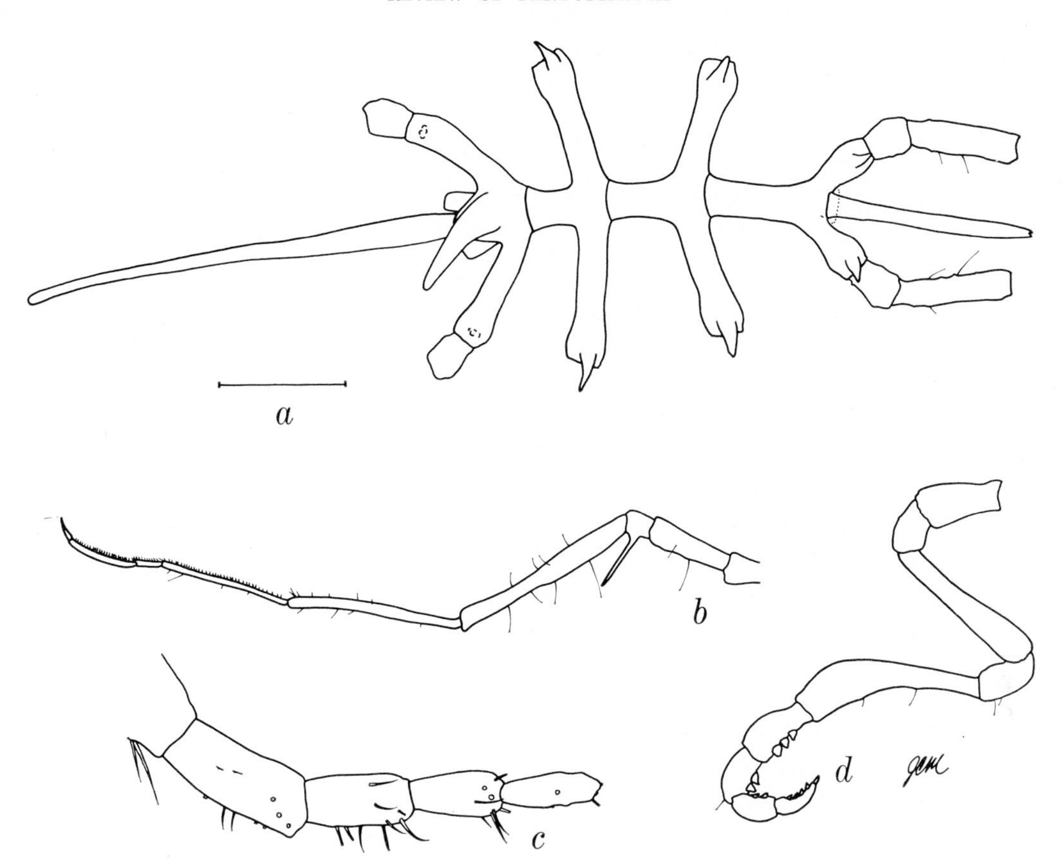

Fig. 5. *Pantopipetta capensis: a,* dorsal view; *b,* right leg 3; *c,* terminal segments of palp; *d,* oviger. Scale above letter *a* represents 1 mm.

Material.

Vema, cruise 14: Sta. 32, 34°35′S, 17°31′E, 1816 meters, April 6, 1958, 1 ♀.

Diagnosis. Coxa 1 smooth; lateral processes of legs 2–4 with dorsal spur, lateral processes of leg 1 with low tubercles; palp 8-segmented; propodus less than 4 times as long as tarsus; abdomen reaching end of coxa 2 of leg 4.

Remarks. Barnard described this species from a single female specimen collected off Cape Point, South Africa, at a depth of 841 meters. Stock's specimens and the *Vema* specimen were collected in the same general area, the former from depths of 2744 to 2890 meters.

This species can be distinguished from the other species of *Pantopipetta* by the spurs on the lateral processes of legs 2–4.

Pantopipetta aconae, new species

Figs. 1, 6; Table 4

Material. (Numbers in parentheses at end of station data below correspond to column headings in Table 4.)

Acona (anchor dredge):

Sta. 6, 44°33.5′N, 125°14.6′W, 2000 meters, June 6, 1963, 2 ♂, 3 ♀♀ paratypes USNM 123743 (2–6).

Sta. 31, 44°39.2′N, 125°11.0′W, 1400 meters, January 25, 1963, 1 ♂ holotype USNM 123740, 1 ♂, 2 ♀ paratypes USNM 123741 (7–8).

Sta. 39, 44°39.1′N, 125°11.0′W, 1420 meters, April 27, 1963, 1 ♂ paratype USNM 123742 (1).

Diagnosis. Coxa 1, legs 1–4 with pair of dorsal spurs, lateral processes with small spur, palp of 7 segments, propodus more than 4 times as long as tarsus, abdomen reaching end of coxa 2, leg 4.

Remarks. This species can easily be distinguished from the other species of *Pantopipetta* by the presence of a pair of dorsal spurs on coxa 1 of all the legs and the low tubercles on the lateral processes. None of the 9 specimens examined had fully developed ovigers or 8-segmented palps; however, several of the specimens bore ovigers of 4 or 5 segments. There is,

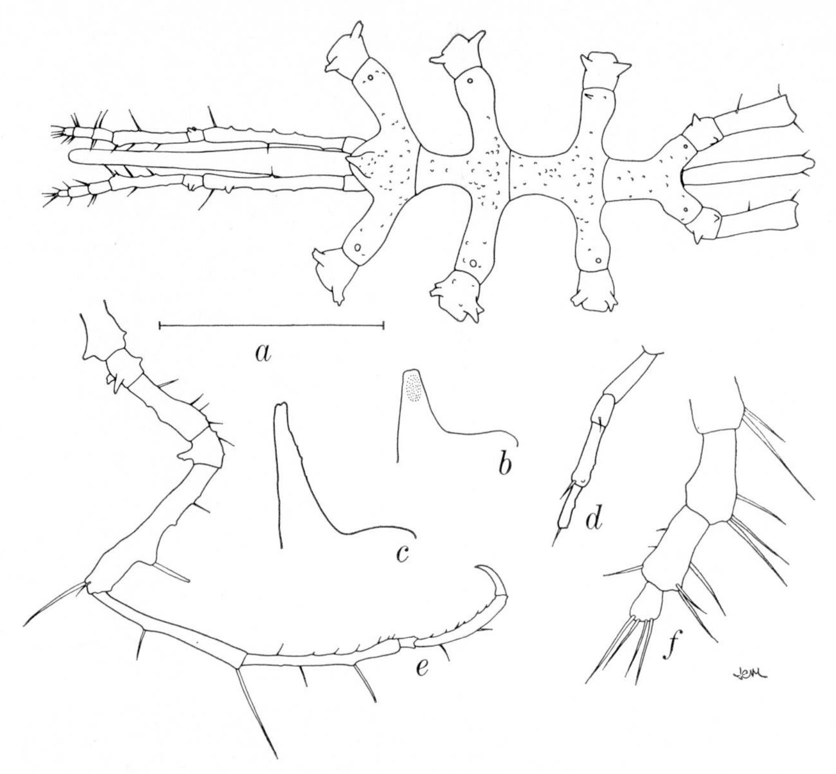

Fig. 6. *Pantopipetta aconae* new species: *a*, dorsal view, male holotype; *b*, eye tubercle; *c*, eye tubercle, male paratype; *d*, oviger, male paratype; *e*, left leg 3, male holotype; *f*, terminal segments of palp, male holotype. Scale above letter *a* represents 1 mm.

therefore, some doubt that we have adult individuals in our collection; yet we have no hesitation in describing this species as new, since the specimens are consistent in their differences from the other described species and are the first specimens of *Pantopipetta* collected from the northeastern Pacific.

The species name is in honor of the Oregon State University's former research vessel *Acona,* which collected the material upon which this species is based.

Acknowledgments. We would like to thank Dr. Robert J. Menzies, Department of Oceanography, Florida State University, for making available to us collections he made on the R/V *Vema* and for the use of his sketches of *Pantopipetta brevicauda,* which are included as Figure 3 in this report. This work was supported by grant GA-1217 from the National Science Foundation, Office of Antarctic Programs.

REFERENCES

Barnard, K. H.

1946 Diagnoses of new species and a new genus of Pycnogonida in the South African Museum. Ann. Mag. Nat. Hist., Ser. 11, *13:* 60–63.

1954 South African Pycnogonida. Ann. S. Afr. Mus., *41* (3) : 81–158, 34 figs.

Calman, W. T., and I. Gordon

1933 A dodecapodous pycnogonid. Proc. R. Soc. London, *113:* 107–115, 3 figs.

Fry, W. G.

1965 The feeding mechanisms and preferred foods of three species of Pycnogonida. Bull. Br. Mus. Nat. Hist., *12*(6) : 195–223, 8 figs., 5 pls.

Hedgpeth, J. W.

1948 The Pycnogonida of the Western North Atlantic and the Caribbean. Proc. U.S. Nat. Mus., *97:* 157–342, figs. 4–53, charts 1–3.

1964 Notes on the peculiar egg laying habit of an Antarctic prosobranch (Mollusca: Gastropoda). Veliger, *7*(1) : 45–46, fig. 1.

Hodgson, T. V.

1914 Preliminary report on the Pycnogonida of the German Southpolar Expedition 1901–1903 (in English). Zool. Anz., *45:* 158–165.

1915 The Pycnogonida collected by the *Gauss* in the antarctic regions, 1901–1903; preliminary report. Ann. Mag. Nat. Hist., ser. 8, *15:* 141–149.

1927 Die Pycnogoniden der Deutschen Südpolar-Expedition 1901–03. Dt. Südpol.-Exped., 1901–1903, *19:* 303–358, 17 figs.

Loman, J. C. C.
1904 *Pipetta weberi* n. g., et n. sp., with notes about the proboscis of the Pycnogonida. Tijdschr. Ned. Dierk. Vereen., ser. 2, *8:* 259–266, 7 figs.
1908 Die Pantopoden der Siboga-Expedition. Siboga Exped. Monogr. 40. Pp. 1–88, 15 pls.

Stock, J. H.
1954 Pycnogonida from Indo-West Pacific, Australian, and New Zealand waters. Vidensk. Meddr Dansk Naturh. Foren., *116:* 1–168, 81 figs.
1963 South African deep-sea Pycnogonida, with descriptions of five new species. Ann. S. Afr. Mus., *46*(12): 321–340, 10 figs.
1968 Pycnogonida collected by the *Galathea* and *Anton Brunn* in the Indian and Pacific Oceans. Vidensk. Meddr Dansk Naturh. Foren., *131:* 7–65, 22 figs.

Turpaeva, E. P.
1955 Novyye vidy mnogokolenchatykh (*Pantopoda*) Kurilo-Kamchatskoy Vpadiny [New species of Pantopoda from the Kurile-Kamchatka Trench]. [In Russian.] Trudy Instituta Okeanologii, *12:* 322–327, 2 figs.

THREE NEW SPECIES OF *DICYEMA* (MESOZOA: DICYEMIDAE) FROM NEW ZEALAND

ROBERT B. SHORT

Department of Biological Science, Florida State University, Tallahassee, Florida 32306

Abstract. Three new species of dicyemid Mesozoa are described from the renal organs of cephalopod hosts from New Zealand waters. Dicyemids and hosts are *Dicyema robsonellae* from *Robsonella australis*, and *D. knoxi* and *D. maorum* from *Octopus maorum*. This is the second report of dicyemids from New Zealand and the first for the genus *Dicyema*.

INTRODUCTION

During December 27–29, 1964, opportunity was provided for examination of octopuses for mesozoan parasites on Kaikoura Peninsula, at the Edward Percival Marine Laboratory of the Zoology Department of the University of Canterbury, New Zealand. The hosts were 10 of 11 specimens of *Octopus maorum* Hutton, 1880, and 12 of 16 specimens of *Robsonella australis* (Hoyle, 1885). Two species of *Dicyemennea* and 3 of *Dicyema* were found (Table 2); all were new. This paper deals with the 3 species of *Dicyema*. It, and one recently published on the 2 species of *Dicyemennea* [Short and Hochberg, 1969], to my knowledge, are the first on dicyemids from New Zealand waters and the first on dicyemids from these hosts.

The genera *Dicyema* and *Dicyemennea*, as Nouvel [1948, 1961] has indicated, are heterogeneous and poorly defined, being separated mainly by differences in the number of metapolar cells: 4 in *Dicyema*, 5 in *Dicyemennea*. Although the situation is not satisfactory, attempts at reclassification at the generic level would seem to be premature until additional information is available on a number of species, especially on structure of infusoriform larvae and stem nematogens. It seemed desirable, therefore, in the present paper to assign species to existing genera if they clearly fell within the generic boundaries as now recognized. This was done with the anticipation that revision would occur when more data were available. In this connection, rather detailed but not exhaustive reports are given on infusoriform larvae of the three species here described.

MATERIALS AND METHODS

The descriptions are based mainly on organisms in coverslip smears of infected octopus kidneys fixed in Sanfelice's or Bouin's fluid and stained with iron hematoxylin and eosin. Some observations were made also on infusoriform larvae and on cilia of vermiform stages in temporary preparations of material fixed and stored in a mixture of equal parts of sea water and 10% formalin.

In the descriptions, infusorigens are considered mature when they have apparently mature spermatozoa and oocytes.

Nematogens, unless stated otherwise, are considered to be primary nematogens; in some instances, of course, it is possible that secondary nematogens were present.

Measurements of nematogens and rhombogens, unless stated to the contrary, were made from medium-sized to large specimens. For the ratios of lengths of propolar cells to metapolars, and for the ratios of diameters of nuclei of propolar and metapolar cells, one pair of cells, or nuclei, per individual were measured.

All drawings were made with the aid of a camera lucida. Measurements of mesozoans are given in some instances as ranges, in others as averages with ranges in parentheses. Dorsal mantle lengths of hosts were measured.

Dicyema robsonellae sp. n.

Figs. 1–20, 69–72

Host. *Robsonella australis* (Hoyle, 1885).

Locality. Offshore about 5½ miles northeast of Kaikoura Peninsula, South Island, New Zealand, at about 20 fathoms.

TABLE 1. Peripheral Cell Numbers of ***Dicyema robsonellae***

Cell No.	Phase			Total
	Vermiform Embryo	Nematogen	Rhombogen	
18	1	1	3	5
19	...	...	1	1
20	48	10	15	73
21	1	...	1	2
Total	50	11	20	81

Type specimens. Syntypes on slides 882–1, 882–6 (USNM Helm. Coll. 24064, 24065) and other slides of 881, 882, 883, and 888 series (author's collection).

Description.

Dicyema: Peripheral cell number of vermiform stages (except for stem nematogen) typically 20 (Table 1): 4 propolars, 4 metapolars, 2 parapolars, 8 diapolars, 2 uropolars.

Nematogens: Body of rather uniform width except for cephalic enlargement, which is conspicuous in some larger specimens. Lengths of 10 longest individuals, 717 to 1402 μ; length of five smallest ones containing large vermiform embryos, 297 to 341 μ. Widest part of body usually at cephalic enlargement, at level of either parapolar or metapolar cells; widths of 10 widest nematogens 82 to 130 μ.

Trunk cells generally arranged in opposed pairs. Numerous small eosinophilic granules present in parapolar and trunk cells, sometimes also a few large eosinophilic masses. No verruciform cells.

Calotte orthotropal in some, plagiotropal in other individuals; variable in shape; sometimes rather smoothly rounded, or flattened to various degrees anteriorly; often of rather irregular shape, with anterior to lateral projections formed by propolar and metapolar cells. Calotte, in larger individuals especially, forming a thin cap over enlarged anterior end of the axial cell; propolar cells thinner than metapolars, tending to be transversely elongate so that the 4 metapolar cells form a band separating the propolar cap from the parapolar cells (Figure 2). Nuclei of propolar cells, in all but very young nematogens, usually slightly smaller than those of metapolars; ratios of diameters, 1:1.0 to 1:1.4 (20 individuals).

Parapolar cells typically longer than calotte, sometimes showing a slight tendency to partially surround the bases of the metapolars laterally.

Axial cell extending anteriorly to propolar cells, typically enlarged and rounded in calotte region, covered by thin propolar and metapolar cells.

Stem nematogen: Based on 3 specimens, 1 entire, the other 2 with intact axial cells but with 1 or more peripheral cells missing. All specimens immature. Length of entire individual, 366 μ. Axial cell number, 3 in all instances. Peripheral cell number of entire individual, 26: 4 propolar, 4 metapolar, 3 parapolar, 14 diapolar, 1 uropolar.

Vermiform embryo in middle axial cell of entire specimen with 19 peripheral cells, including 8 that comprise the calotte.

Vermiform embryos: At eclosion about 65 to 90 μ in length. Dimensions of 20 longest, apparently full-grown embryos, in axial cells of nematogens: total length, 71 μ (62 to 93 μ); width, 18 μ (15 to 20 μ); calotte length, 15 μ (12 to 20μ); ratios of calotte lengths to widths, 1:0.7 to 1:1.2; ratios of calotte lengths to total body lengths 1:3.2 to 1:6.5.

Trunk cells usually in opposed pairs.

Anterior end of axial cell ending bluntly between bases of propolar cells; axial cell nucleus usually in anterior half of cell, occasionally at middle; usually 1 axoblast present, sometimes 2, typically posterior to axial cell nucleus; data on 70 vermiform embryos: 58 with 1 axoblast each, 11 with 2, 1 with 3; axoblasts posterior to axial cell nucleus, except for 5 with single axoblast anterior to nucleus.

Rhombogens: Usually shorter and stockier than nematogens, otherwise generally similar in shape and proportions. Lengths of 10 longest specimens, 627 to 866 μ; lengths of 10 shortest rhombogens with mature infusoriforms, 214 to 301 μ. Widest part of body at level of parapolar or metapolar cells; widths of 10 widest rhombogens, 105 to 187 μ.

Accessory nuclei only occasionally observed, in diapolar and uropolar cells; when present, only 1 per cell and smaller than normal nucleus. Granules in parapolar and trunk cells similar to those in nematogens. No verruciform cells.

Calotte generally similar in shape to those of nematogens, more massive and wider in larger individuals. Nuclei of propolar and metapolar cells usually about equal in size, those of metapolars sometimes slightly larger; ratios of diameters (i.e., nuclei of propolars to nuclei of metapolars, 20 individuals), 1:1.0 to 1:1.2.

Axial cell rounded and typically enlarged at anterior end, covered anteriorly by propolar and metapolar cells that form a cap that is sometimes very thin in propolar region.

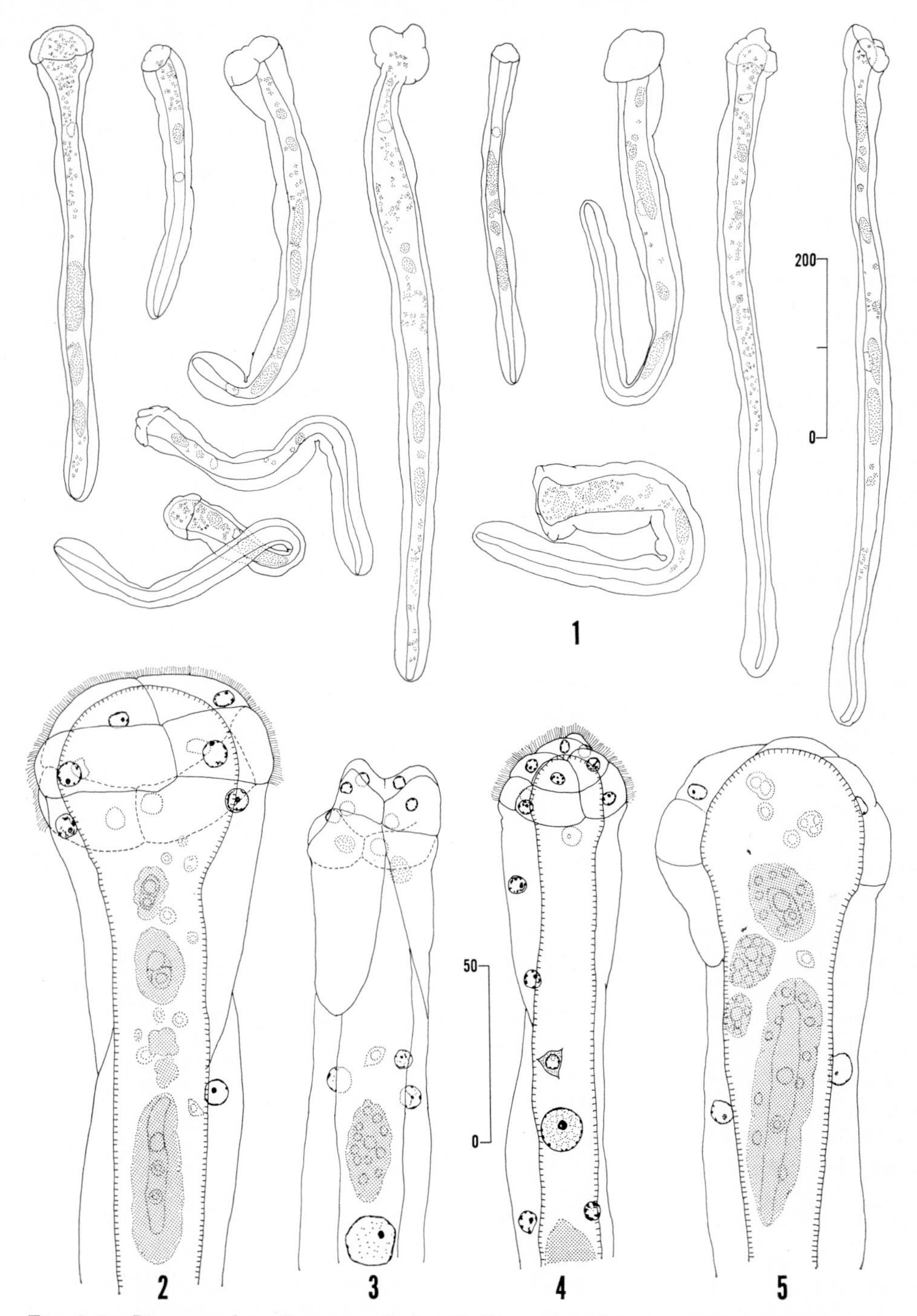

Figs. 1–5. *Dicyema robsonellae* sp. n. Scale with Figures 3 and 4 also applies to Figures 2 and 5. 1, entire nematogens of various sizes. 2–5, nematogens, anterior ends: 5, optical section.

Usually one infusorigen per rhombogen; of 50 medium to large rhombogens, 36 contained one infusorigen, 13 contained two, and one contained three. Infusorigens usually in anterior half of body.

Infusorigens: Not unusual, relatively small, with few cells, similar to Figure 59 of *D. maorum*. In 25 infusorigens: number of oogonia and oocytes, 6 to 18 (mode, 9); number of spermatogonia and spermatocytes, 2 to 4 (mode, 3).

Infusoriform larvae: Ovoid, rounded to bluntly pointed posteriorly. Dimensions of apparently mature larvae within axial cells of rhombogens: lengths of 25 larvae (excluding cilia) 37 μ (31 to 42 μ); widths of 9, 31 μ (24 to 35 μ); depths of 18, 31 μ (28 to 34 μ). Cilia at posterior end about 7 to 8 μ long.

Refringent bodies apparently solid; appearing in optical section from lateral and dorsal views usually about equal to or slightly smaller than the urn contents. Two cells (presumably the second ventrals) extending, one on each side, from mid-anteroventral region dorsolaterally; containing eosinophilic material in cytoplasm (Figure 69).

Capsule cells completely surrounding urn cells, occupying space anteroventral to urn cells, space which usually comprises the urn cavity; cytoplasm of capsule cells containing small eosinophilic granules. No cilia observed on cells (presumably ventral internals) forming the anterior border of the urn cavity.

Infusoriform composed typically of 37 cells: 33 somatic and 4 germinal. Results of cell counts on 22 infusoriforms: 19 with 37 cells, 1 with 38, and 2 with 36. Four urn cells, each containing 2 somatic nuclei and 1 germinal cell.

Remarks. Abundant material was available except for the stem nematogen and infusoriform stages. Only 25 infusoriform larvae were studied and, as indicated above, the description of the stem nematogen is based on 1 entire specimen and 2 individuals lacking 1 or more peripheral cells. The peripheral cell count of 26 was based on the 1 entire specimen and did not include a small problematical cell, which was judged to be an axoblast that had escaped from the axial cell. This cell number, then, should be verified on additional material.

Table 2 summarizes data on hosts and infections. Except for one stem nematogen (Host 888, slide 3), descriptions were based on material from hosts 881, 882, and 883. All hosts except 874 were collected by dredge at the locality given above. Host 874 was collected by a local fisherman, and its exact locality was not determined.

From Table 2 it is seen that 5 octopuses harbored *Dicyemennea rostrata* in addition to *D. robsonellae.* Hosts 889, 894, and those listed as parasitized with only *D. robsonellae* were immature specimens with light infections. The vermiform embryos within nematogens in this scanty material differed in certain respects from the above description: they were slightly smaller (60 to 81 μ for 10 longest), and a majority had 2 to 4 axoblasts instead of 1. However, it seems advisable for the present to consider these specimens as *D. robsonellae.* Nouvel [1947, p. 182] has pointed out that, in certain species, the number of germinal cells in vermiform embryos (at eclosion) may become reduced in the course of the cycle, and the present finding may be a manifestation of this tendency. Genetic variation between populations of the parasite or influence of different hosts may also be suspected.

As noted in the description and figures, there is a good deal of variability in shape of calottes of this species. In addition to projections frequently encountered, distinct grooves (Figure 6) were occasionally seen involving metapolar and propolar cells.

The refringent bodies of infusoriform larvae appear to be solid; but this point is not certain, since the

TABLE 2. Octopods and Dicyemid Parasites from Kaikoura, New Zealand

Hosts			
No.	Dorsal Mantle Length, cm	Sex	Dicyemids
Robsonella australis			
881	3.2	F	***Dicyema robsonellae*** ***Dicyemennea rostrata*** Short and Hochberg
882	3.5	F	
883	1.9	F	
889	1.6	?	
894	1.5	?	
884	1.9	F	***D. robsonellae***
885	1.8	F	
886	1.7	F	
888	1.5	?	
892	1.3	?	
887	1.3	F	Scanty infection, species of dicyemid not identified
893	1.4	?	
874	1.8	?	Negative
880	1.2	?	
890	1.2	?	
891	1.1	?	
Octopus maorum			
871	15.0	M	***Dicyema knoxi*** ***Dicyema maorum*** ***Dicyemennea kaikouriensis*** Short and Hochberg
872	13.0	M	
873	16.0	M	***D. knoxi*** ***D. maorum***
875	14.0	M	
876	14.0	M	
877	13.0	M	
878	14.0	M	
895	17.0	F	
896	13.0	M	
870	15.0	F	***D. knoxi***
879	16.5	M	Negative

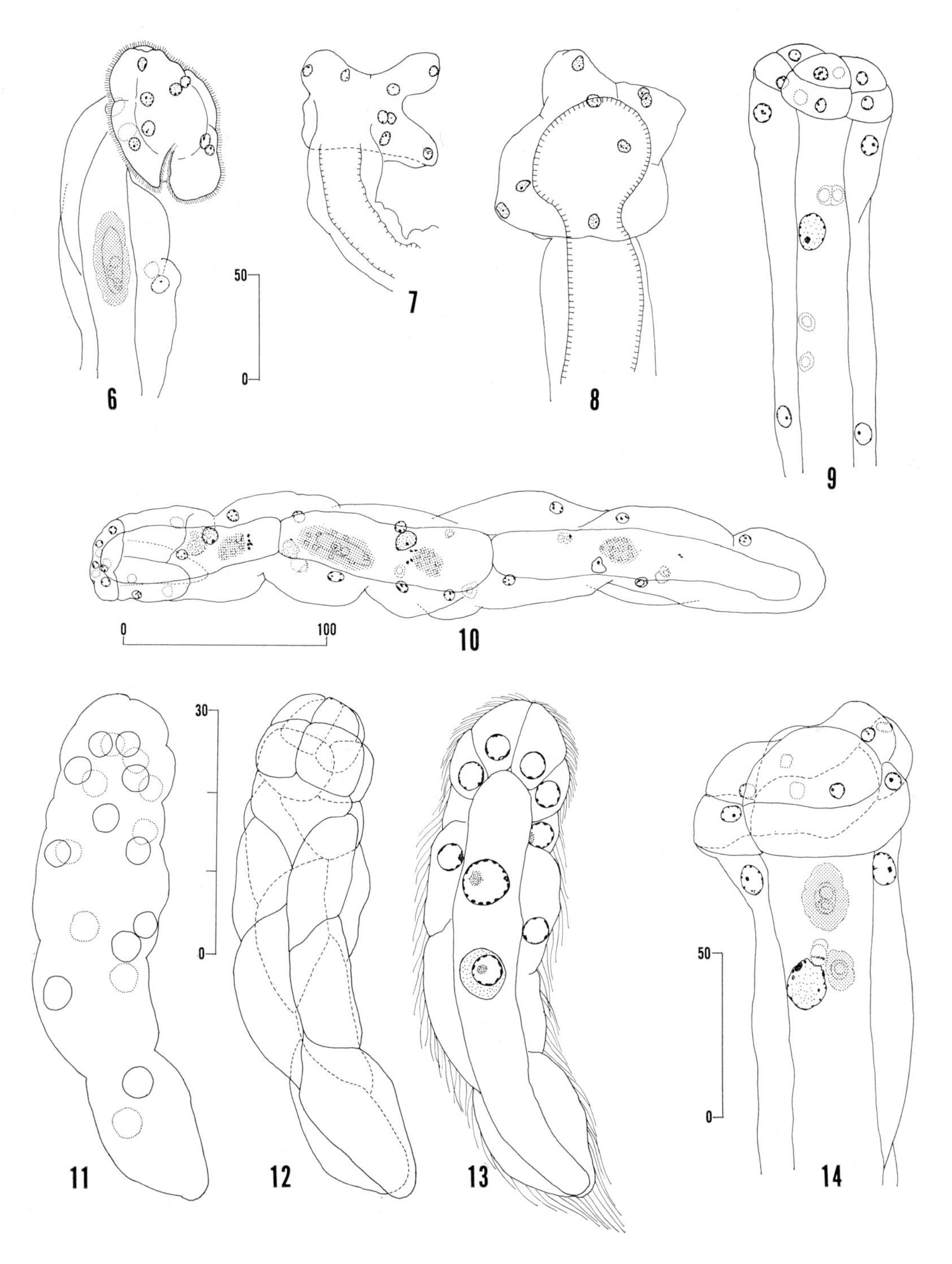

Figs. 6–14. *Dicyema robsonellae* (continued). Scale with Figure 6 also applies to Figures 7–9. Scale with Figures 11 and 12 also applies to Figure 13. 6–9 and 14, nematogens, anterior ends. 10, stem nematogen. 11–13, vermiform embryo about full grown, within axial cell of nematogen: 11, outline with peripheral cell nuclei; 12, peripheral cell outlines; 13, optical section.

only material studied was fixed, stained, and mounted. Further discussion of infusoriforms will occur after descriptions of the following 2 species.

Vermiform stages of *D. robsonellae* differ from all previously described species of *Dicyema* in typically having 20 peripheral cells. Occasionally a peripheral cell of a vermiform stage was small with a pycnotic nucleus, and it is suggested that degeneration of such cells may account, in part, for the infrequent occurrence of less than 20 peripheral cells.

Species have been reported with 20 peripheral cells in some individuals: *D. typus* van Beneden, *D. briarei* Short, and *D. typoides* Short; but in these species 20 is not the typical number. Besides in cell number, *D. robsonellae* differs from these 3 species with respect to the following 2 points: the extreme anterior extension of the axial cell, which in larger individuals leaves only a thin layer of propolars covering it anteriorly; and the shape of the calotte, which is often bulbous and flattened anteriorly, and which has projections and occasionally a ciliated groove.

The stem nematogen of *D. robsonellae* agrees with descriptions of this stage for other species of *Dicyema* in possessing (1) 3 axial cells, (2) 3 parapolar cells, (3) a single uropolar cell, and (4) a larger number of peripheral cells than do other vermiform stages of the respective species [Nouvel, 1947; Short, 1962]. The stem nematogen of *D. robsonellae* agrees also with that described for *Dicyemennea eledones* (Wagner) Whitman, in all the above respects except the number of axial cells, which is 2 in *D. eledones* [Nouvel, 1947].

Dicyema knoxi sp. n.

Figs. 21–36, 60–63

Host. *Octopus maorum* Hutton, 1880.

Incidence. In 10 of 11 large (apparently mature) hosts examined (Table 2).

Locality. Near shore of Kaikoura Peninsula, South Island, New Zealand.

Syntypes. On slides 870–2, 872–1 (USNM Helm. Coll. 24060, 24061) and other slides of the 870, 872, and 873 series (author's collection).

Description.

Dicyema: Peripheral cell number of vermiform stages typically 16 (Table 3): 4 propolars, 4 metapolars, 2 parapolars, 4 diapolars, 2 uropolars.

Nematogens: Body of rather uniform width throughout, rather elongate and narrow in older individuals. Lengths of 10 longest nematogens, 640 to 776 μ; lengths of 5 smallest nematogens with apparently full-grown vermiform embryos, 207 to 236 μ. Body sometimes widest at uropolar region, other times at parapolar or midbody; width of 10 widest individuals, 41 to 43 μ.

TABLE 3. Peripheral Cell Numbers of ***Dicyema knoxi***

Cell No.	Phase			Total
	Vermiform Embryo	Nematogen	Rhombogen	
15	2	...	...	2
16	65	10	20	95
Total	67	10	20	97

Trunk cells arranged generally in opposed pairs. Granular material in parapolar and trunk cells scanty and relatively fine. No verruciform cells.

Calotte usually orthotropal, occasionally plagiotropal; rounded to bluntly pointed anteriorly, tending to become more pointed in older individuals; usually slightly longer than wide, sometimes more elongate; lengths of 10 longest calottes, 31 to 44 μ; ratios of lengths to widths 1:0.7 to 1:1.1 (30 calottes). Propolar cells smaller than metapolars; ratios of lengths 1:1.2 to 1:2.3 (20 calottes). Nuclei of propolar cells smaller than those of metapolars; ratios of diameters, 1:1.1 to 1:2.0 (20 individuals).

Parapolar cells in younger nematogens about equal in length to calotte or slightly longer; in older individuals parapolar cells relatively longer, sometimes 2 to 3 times length of calotte. Ratios of calotte lengths to lengths of parapolars, 1:1.3 to 1:3.2 (16 nematogens, 31 parapolar cells).

Axial cell extending forward between bases of metapolar cells, usually bluntly pointed; axial cell nucleus often near anterior end of cell, especially in younger nematogens.

Vermiform embryos: At eclosion about 40 to 55 μ in length. Measurements of 20 longest, apparently full-grown embryos in axial cells of nematogens: total length, 49 μ (42 to 57 μ); width 11 μ (10 to 13 μ); calotte length, 14 μ (12 to 15 μ); ratios of calotte lengths to widths, 1:0.6 to 1:0.8; ratios of lengths of calottes to total body lengths, 1:2.8 to 1:4.3.

Peripheral cells usually arranged in opposed pairs.

Anterior end of axial cell ending bluntly between bases of metapolar cells; axial cell nucleus usually in anterior half of cell, with 2 axoblasts posterior to

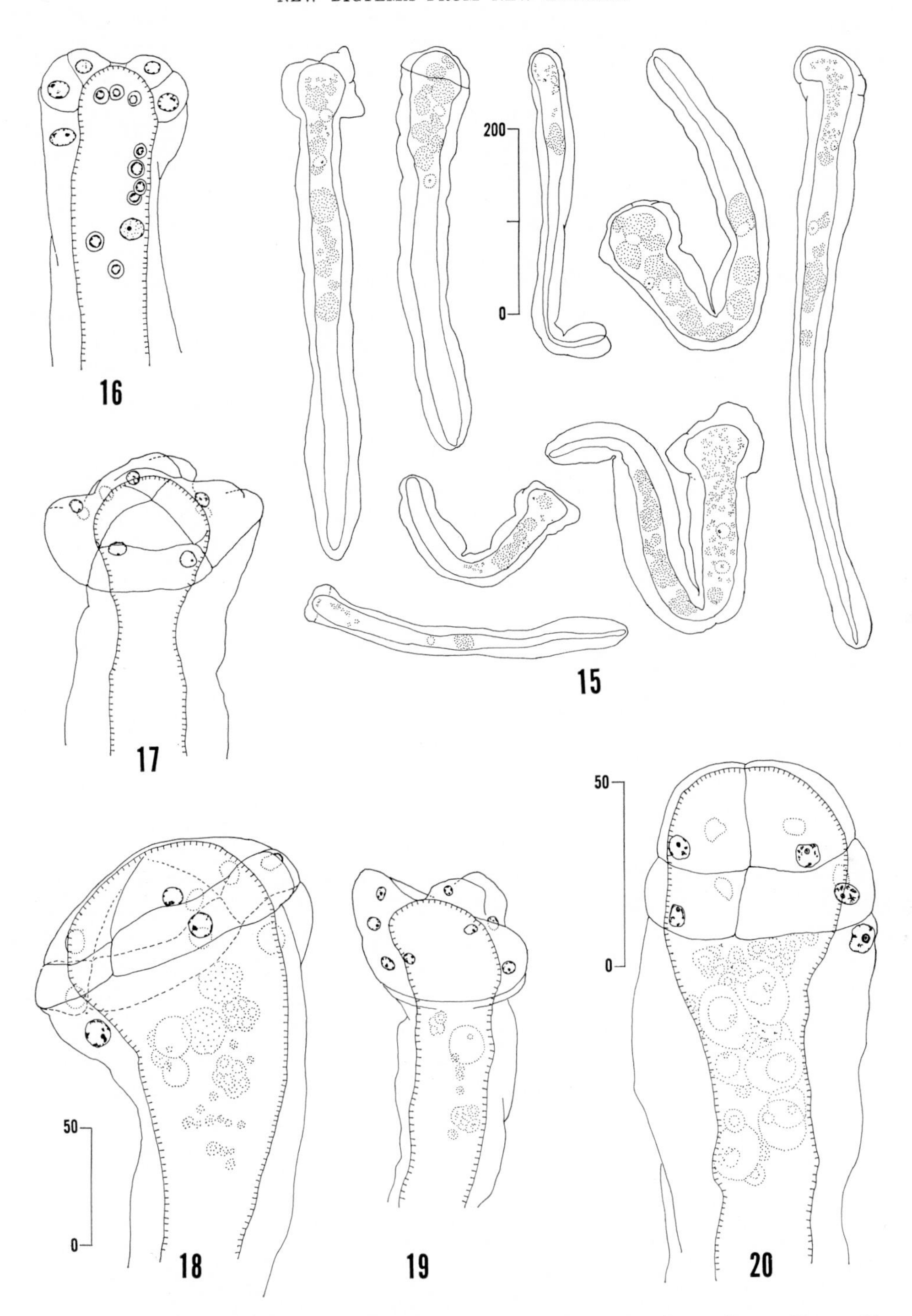

Figs. 15–20. *Dicyema robsonellae* (continued). Scale with Figure 18 also applies to Figures 16, 17, and 19. 15, entire rhombogens. 16–20, rhombogens, anterior ends.

nucleus; data on 66 vermiform embryos: 62 with axoblasts posterior to axial cell nucleus (8 with 1 axoblast, 52 with 2, 1 with 3, 1 with 4); 3 with axoblasts anterior to axial cell nucleus (2 with 1 axoblast, 1 with 2); 1 with 1 axoblast anterior and 1 posterior to axial cell nucleus. Axoblasts commonly elongate oval to fusiform, sometimes with sharply pointed ends.

Rhombogens: Body thickest anteriorly, usually in parapolar region, sometimes in metapolar region; of rather uniform thickness posterior to parapolars, sometimes tapering slightly posteriorly. Greatest body width of largest rhombogens seldom exceeding 88 μ. Width of 10 widest rhombogens, 90 μ (84 to 108 μ). Often a slight constriction between calotte and parapolar cells.

Rhombogens usually less than 800 μ in length; lengths of 10 longest individuals, 744 to 1288 μ; lengths of 10 smallest rhombogens with mature infusorigens, 180 to 240 μ.

Trunk cells arranged generally in opposed pairs. Granules in parapolar and trunk cells scanty, small, and irregular in shape. Well-defined accessory nuclei in peripheral cells not present. Frequently what appeared to be small axoblasts and faded remnants of axoblasts seen in peripheral cells and sometimes a few such cells per rhombogen were seen more or less flattened between the axial cell and a peripheral cell (Figure 36). No verruciform cells.

Calotte usually orthotropal, sometimes apparently plagiotropal; usually broadly pointed anteriorly, appearing more or less like an equilateral triangle in side view; sometimes more elongate and more pointed; longer than wide in 16 of 20 mature individuals, with ratios of lengths to widths varying from 1:1.14 to 1:0.61; average ratio, 1:0.89 (20 individuals). Propolar cells smaller than metapolars; ratios of lengths of propolar to metapolar cells 1:2.3 to 1:1.4, average ratio 1:1.8 (20 mature individuals). Nuclei of propolar cells smaller than those of metapolars; ratios of diameters of propolar to metapolar nuclei, 1:2 to 1:1.3; average ratio 1:1.6 (20 individuals).

Parapolar cells bluntly rounded to more attenuated posteriorly, usually about twice the length of calotte; ratios of calotte lengths to lengths of parapolars 1:1.3 to 1:3.5 (20 individuals, 39 parapolar cells).

Axial cell ending bluntly anteriorly, usually between bases of metapolar cells, sometimes extending to middle region of metapolar cells. Axial cell usually containing only 1 infusorigen (47 of 50 rhombogens), sometimes 2 (3 of 50 rhombogens). Infusoriform larvae generally in anterior ⅓ to ½ of axial cell.

Infusorigens: Relatively small, with few cells. In 25 infusorigens: number of oogonia and oocytes, 3 to 10 (mode, 6); number of spermatogonia and spermatocytes, 2 to 4 (mode, 3).

Infusoriform larvae: Ovoid, rounded to bluntly pointed posteriorly. Dimensions of apparently mature larvae within axial cells of rhombogens: lengths of 50 larvae (excluding cilia), 29 μ (25 to 33 μ); widths of 25, 20 μ (19 to 22 μ); depths of 25, 19 μ (17 to 24 μ). Cilia relatively short, at posterior end about 3 to 5 μ long.

Refringent bodies absent. One pair of cells (presumably the second ventrals: V2, Figures 60, 62, 63; see discussion below) extending, one on each side, from mid-anteroventral region dorsolaterally; containing eosinophilic granular material in cytoplasm, which is denser than that of other peripheral cells and more eosinophilic.

Capsule cells completely surrounding urn cells, occupying the space anteroventral to urn cells (space that usually comprises the urn cavity); cytoplasm of capsule cells containing small eosinophilic granules, which are most prominent anteroventral to urn cells and which are more conspicuous than material in presumed second ventral cells. No cilia observed on cells (presumed ventral internals: VI, Figure 63) forming the anterior border of the urn cavity.

Infusoriform typically composed of 37 cells: 33 somatic and 4 germinal. Results of cell counts of 16 infusoriforms: 13 with 37 cells, 1 with 36, and 2 with apparently 38. Each of the 4 urn cells containing 2 somatic nuclei and 1 germinal cell.

Remarks. Abundant material was available for study of this species, which is named in honor of Professor George A. Knox, University of Canterbury, Christ Church, N. Z. Vermiform stages of *D. knoxi* typically have 16 peripheral cells, and in this respect they differ from other described species, except the following: *D. bilobum* Couch and Short, *D. caudatum* Bogolepova, *D. monodi* Nouvel, and possibly *D. megalocephalum* Nouvel (reported to have 16 or fewer peripheral cells). Species that sometimes have 16 peripheral cells are *D. oligomerum* Bogolepova (usually 14 to 16) and *D. typus* van Beneden (typically 18 or 19, rarely 16, 17, or 20). In addition to other ways, *D. knoxi* differs from *D. bilobum, D. caudatum, D. megalocephalum,* and *D. oligomerum* in the anterior extent of the axial cell, which, in these species, reaches the propolar cells. *D. knoxi* differs from *D. monodi,* and also from *D. bilobum* and *D. megalocephalum,* in the shape of the calotte. Infusoriforms of *D. typus* are larger (40 μ long) than those of *D. knoxi* (29 μ) and possess conspicuous refringent bodies. Further discussion of infusoriform larvae will occur after description of the next species.

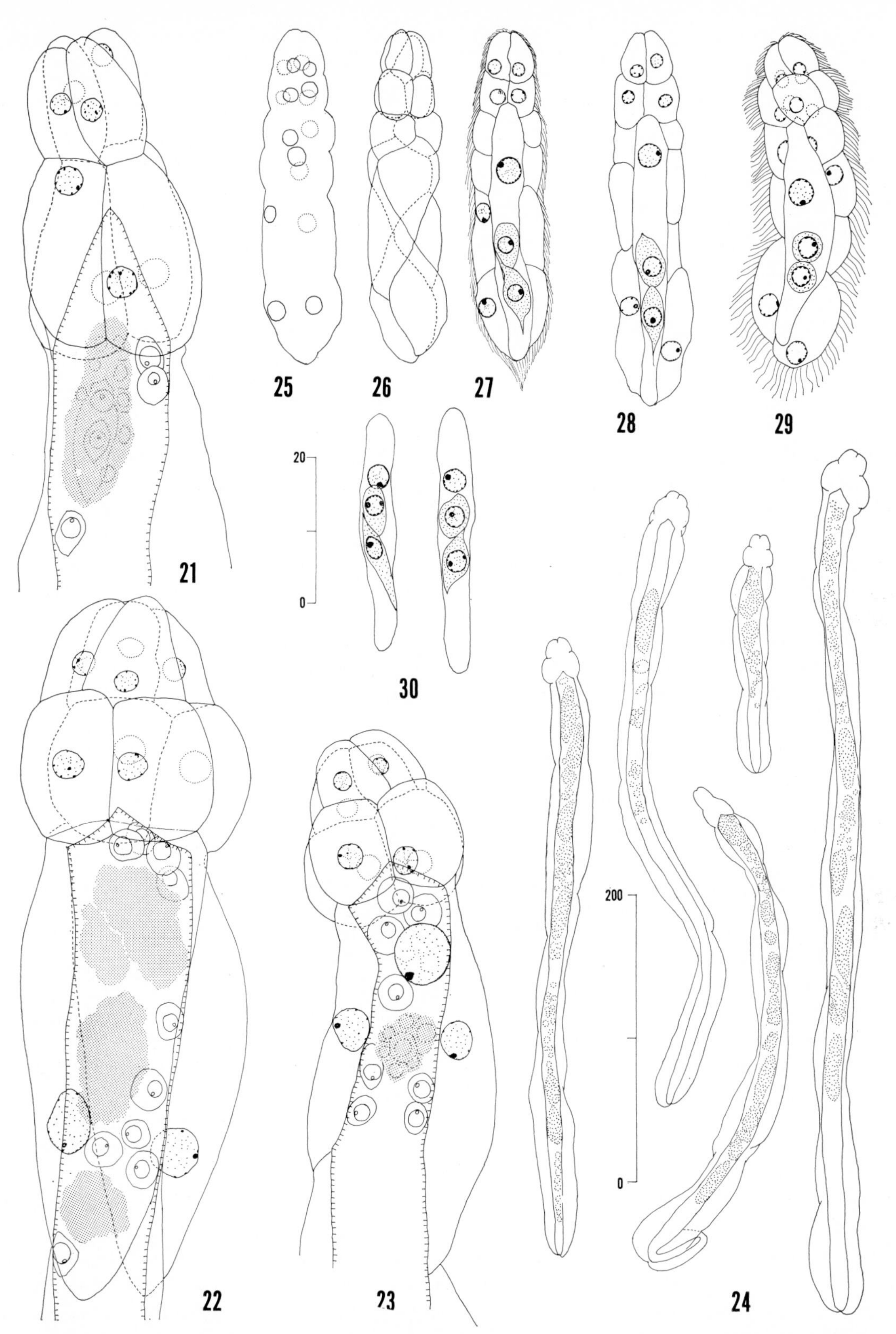

Figs. 21–30. *Dicyema knoxi* sp. n. Scale with Figure 30 also applies to Figures 21–23 and 25–29. 21–23, nematogens, anterior ends. 24, entire nematogens. 25–27, vermiform embryo about full grown, within axial cell of nematogen: 25, outline with peripheral cell nuclei; 26, peripheral cell outlines; 27, optical section. 28, vermiform embryo, about full grown within axial cell of nematogen, optical section. 29, young vermiform individual outside parent nematogen. 30, axial cells of vermiform embryos from within axial cells of nematogens.

Dicyema maorum sp. n.

Figs. 37–59, 64–68

Host. *Octopus maorum* Hutton, 1880.

Incidence. In 9 of 11 hosts examined (Table 2).

Locality. Near shore of Kaikoura Peninsula, South Island, New Zealand.

Syntypes. On slides 875–1, 876–6 (USNM Helm. Coll. 24062, 24063) and other slides of the 871, 872, 873, 875, and 876 series (author's collection).

Description.

Dicyema: Peripheral cell number of vermiform stages typically 16 (Table 4): 4 propolars, 4 metapolars, 2 parapolars, 4 diapolars, 2 uropolars.

Nematogens: Calotte and parapolar cells forming a conspicuous cephalic enlargement. Trunk of rather uniform width. Lengths of 10 longest individuals, 725 to 1350 μ; lengths of 5 smallest ones containing large vermiform embryos, 213 to 341 μ. Widest part of body at cephalic enlargement; parapolar region usually slightly wider than calotte, occasionally the reverse. Widths of 5 widest nematogens, 65 to 75 μ. Trunk cells arranged generally in opposed pairs, no conspicuous granules observed in peripheral cells; no verruciform cells.

Calotte of younger individuals orthotropal, more or less rounded; in older specimens flattened anteriorly, tending with age toward a disc shape with metapolar cells peripheral to propolars; propolar cells typically opposite to metapolars, with calotte sometimes appearing twisted from anterior view (Figure 37). Vacuoles often present in propolar cells near nuclei, sometimes also in metapolars, of larger individuals. Propolar cells of medium to large specimens widest anteriorly, narrowing posteriorly, more or less wedge-shaped with narrow part toward anterior end of axial cell (Figure 39). Propolar cells smaller than metapolars; ratios of diameters (widths) of propolar discs to diameters (widths) of calottes (in metapolar region) 1:1.4 to 1:2.1 (20 individuals). Nuclei of propolar cells smaller than those of metapolars; ratios of diameters, 1:1.1 to 1:1.7 (20 individuals).

Parapolar cells in older individuals longer than calotte, conspicuously enlarged immediately behind calotte, and partially surrounding metapolar cells laterally; usually slightly wider than calotte and forming widest part of body; parapolar cells sometimes indented, forming a constriction posterior to cephalic enlargement.

Axial cell extending forward between metapolar cells; in older individuals reaching to or almost to propolars, often separated from propolar cells by a thin region of metapolars (Figure 39).

TABLE 4. Peripheral Cell Numbers of ***Dicyema maorum***

Cell No.	Phase			Total
	Vermiform Embryo	Nematogen	Rhombogen	
15	...	...	4	4
16	55	10	22	87
Total	55	10	26	91

Vermiform embryos: At eclosion about 50 to 65 μ in length; measurements of 20 longest, apparently full-grown embryos, in axial cells of nematogens: total length, 51 μ (45 to 62 μ); width, 11 μ (9 to 12 μ); calotte length, 15 μ (12 to 18 μ); ratio of calotte lengths to widths, 1:0.4 to 1:0.7; ratio of calotte lengths to total body lengths, 1:3.1 to 1:4.1.

Trunk cells generally arranged in opposed pairs.

Anterior end of axial cell ending bluntly between bases of metapolar cells; axial cell nucleus in anterior half of cell (invariably in 60 specimens analyzed); usually single axoblast posterior to nucleus, occasionally 2; data on 65 vermiform embryos: 60 with 1 axoblast, 5 with 2; axoblasts with oval to fusiform shapes.

Rhombogens: Calotte and parapolar cells forming a conspicuous cephalic enlargement. Trunk of rather uniform width; neck region sometimes constricted. Lengths of 10 longest individuals 960 to 1568 μ; lengths of 10 shortest rhombogens with mature infusorigens, 224 to 325 μ.

Widest part of body at cephalic enlargement; parapolar region usually slightly wider than calotte, occasionally the reverse. Widths of 10 widest rhombogens 74 to 128 μ.

Peripheral cells with scanty, inconspicuous granules. No accessory nuclei seen in peripheral cells; no verruciform cells.

Calotte of younger individuals rounded anteriorly; in older specimens, usually flattened and expanded laterally into a disc shape with metapolar cells peripheral to propolars; propolar cells usually opposite to metapolars, sometimes apparently alternate (Figure 56). Propolar cells smaller than metapolars; ratios of diameters of propolar discs to widths of calottes, 1:2.1 to 1:1.7 (20 individuals). Nuclei of propolar cells smaller than those of metapolars; ratios of diameters, 1:2.0 to 1:1.3 (20 individuals). Calotte

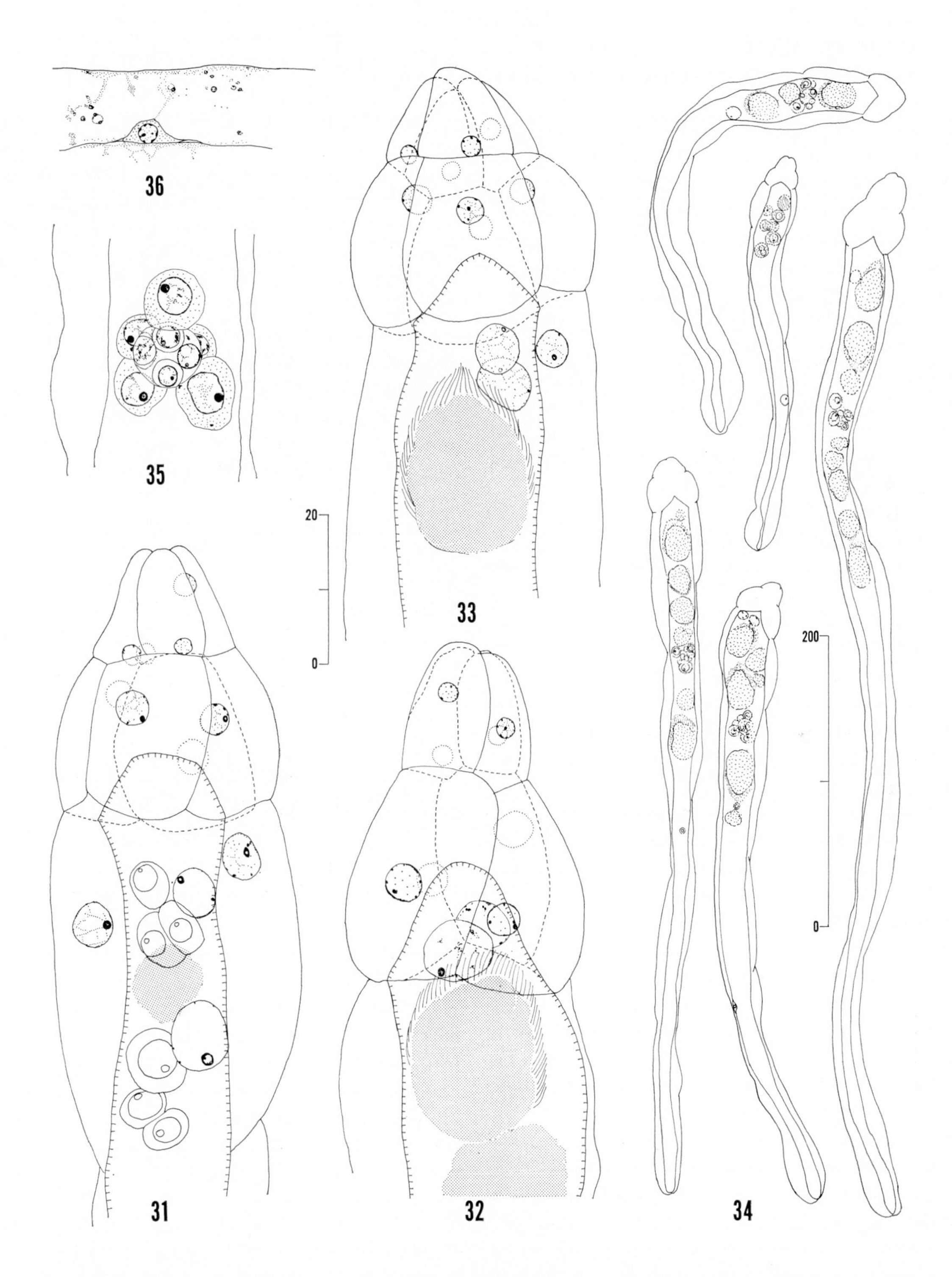

Figs. 31–36. *Dicyema knoxi* (continued). Scale with Figure 33 also applies to Figures 31, 32, 35, and 36. 31–33, rhombogens, anterior ends. 34, entire rhombogens. 35, infusorigen. 36, axoblast between axial cell and diapolar cell of rhombogen.

discs of larger individuals sometimes squarish, occasionally with border irregularities, including notches, which may be continuous with radial furrows in metapolar cells (Figure 49); propolar and metapolar cells of larger individuals typically vacuolated (Figure 55).

Parapolar cells enlarged immediately behind calotte; in older individuals partially surrounding metapolars posterolaterally; parapolars sometimes indented, forming a constriction posterior to cephalic enlargement.

Axial cell extending anteriorly, reaching (or almost reaching) the propolar cells; anterior end of axial cell in larger rhombogens enlarged and either rounded or with branches extending between metapolar cells and parapolars; branches sometimes reaching nearly to exterior (Figures 47, 48). Usually 1 infusorigen per rhombogen, occasionally 2 (all of first 50 rhombogens studied contained 1 infusorigen each, later occasional specimens were seen with 2 infusorigens); infusorigen usually near middle of body or in anterior half.

Infusorigens: Relatively small, with few cells. In 25 infusorigens: number of oogonia and oocytes, 4 to 10 (mode, 6); number of spermatogonia and spermatocytes, 1 to 4 (mode, 2).

Infusoriform larvae: Ovoid, rounded to bluntly pointed posteriorly. Dimensions of apparently mature larvae within axial cells of nematogens: lengths of 50 larvae (excluding cilia), 28 μ (23 to 33 μ); widths of 25, 19 μ (17 to 22 μ); depths of 25, 19 μ (17 to 21 μ). Cilia at posterior end about 6.0 to 7.5 μ long.

Refringent bodies absent. One pair of cells (presumably the second ventrals, Figures 64, 66–68; see discussion below) extending, one on each side, from near the mid-anteroventral region dorsolaterally, containing eosinophilic material in cytoplasm.

Capsule cells completely surrounding urn cells, occupying space anteroventral to urn cells (space that usually comprises urn cavity). Nuclei of capsule cells varying in position from usual posterodorsal one to lateral and occasionally anterolateral (Figures 64, 67); cytoplasm of capsule cells containing rather fine eosinophilic granules, which are most prominent anterolateral to urn cells and which are more conspicuous than material in presumed second ventral cells. No cilia observed on cells (presumably ventral internals) forming the anterior border of the urn cavity.

Infusoriform typically composed of 37 cells: 33 somatic and 4 germinal. Results of cell counts of 11 infusoriforms: 10 with 37 cells, 1 with 38. Each of the 4 urn cells containing 2 somatic nuclei and 1 germinal cell.

Remarks. The 9 infected hosts are the same as those that harbored *D. knoxi,* and 2 (871, 872) also were infected with *Dicyemennea kaikouriensis.*

Abundant material of this species was available, especially rhombogens, infusorigens, and infusoriforms. In the development of *D. maorum* there is a marked change in shape of the calotte from the vermiform embryos to large nematogens and rhombogens. There is also a concomitant change in the anterior extent of the axial cell. Both of these changes are especially pronounced in large rhombogens, which have disclike calottes, sometimes with distinct grooves, and axial cells with branches in the anterior region (Figures 47–49). Such anterior branching of the axial cell is unusual and, to my knowledge, has been reported hitherto only in *Conocyema adminicula* McConnaughey, 1949; this species has 5 metapolars, and 5 anterior branches of the axial cell are pictured [McConnaughey, 1949, Figures 40, 43–45].

Small cells of *D. maorum,* apparently axoblasts, were seen occasionally between cells of the calotte (Figure 38) of nematogens and rhombogens.

D. maorum is also unique among described species of *Dicyema* in having such an expanded cephalic end and a flattened calotte. In this respect it resembles certain species of *Dicyemennea* (e.g., *D. abbreviata* McConnaughey; *D. brevicephaloides* Bogolepova), *Dicyemodeca dogieli* Bogolepova, and *Conocyema deca* McConnaughey. Such comparisons, as well as those of the infusoriform larvae below, add weight to the contention [Nouvel, 1961] that generic criteria should consist of more than the number of cells of the calotte and that the 2 largest genera of dicyemids, *Dicyema* and *Dicyemennea,* are heterogeneous or polyphyletic.

Dicyema maorum differs from *D. knoxi,* described above, and all other described species of *Dicyema* with 16 peripheral cells (as well as those with other numbers) in the shape of the calotte. The infusoriform larva is also unique among those described, although it resembles very closely that of *D. knoxi.* Differences are discussed below.

Discussion of infusoriform larvae. Infusoriforms of *D. knoxi* and *D. maorum* are of about the average size of those 20 of the genus whose lengths have been reported; whereas infusoriforms of *D. robsonellae* are relatively large, only 3 species having been

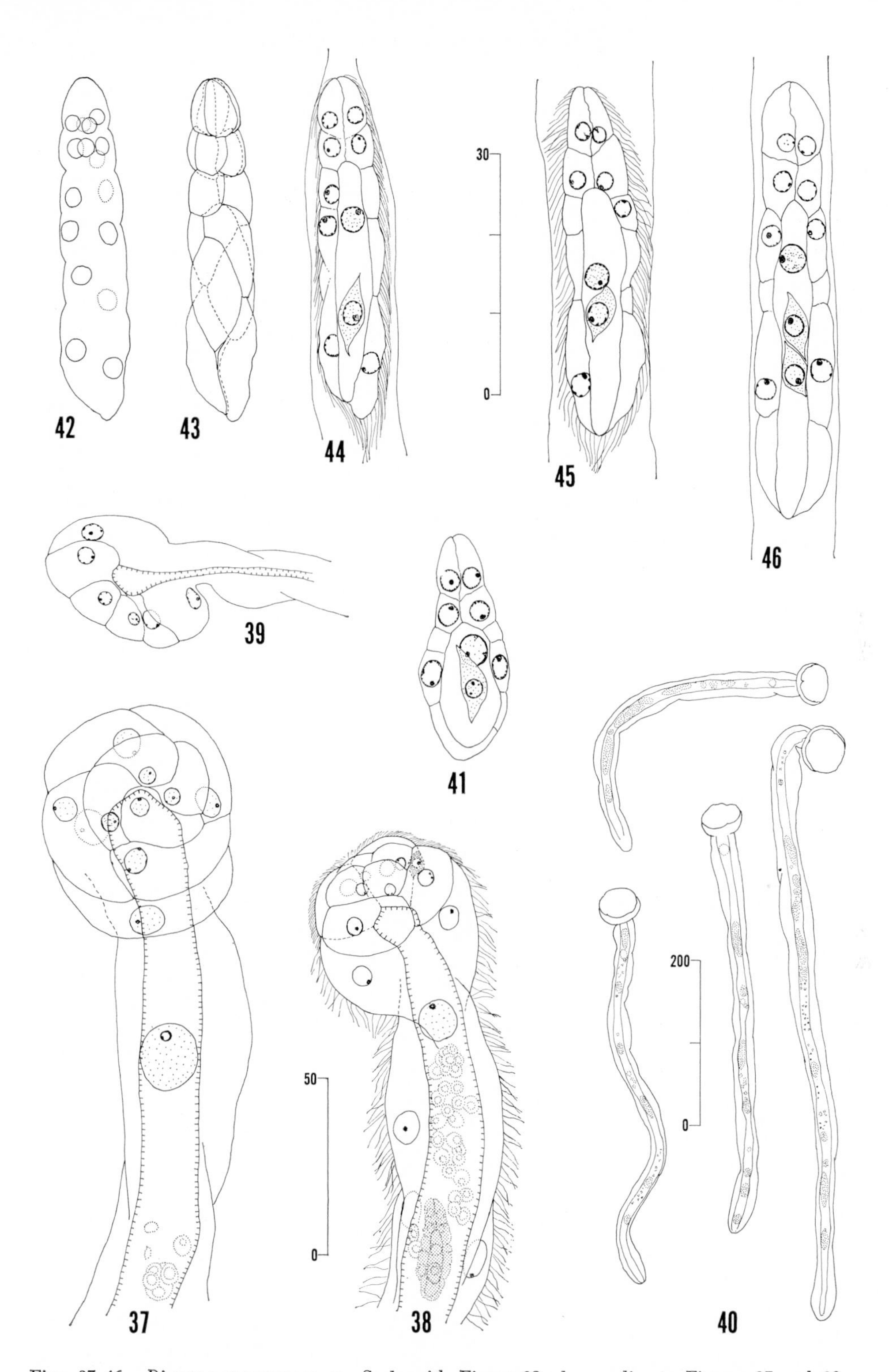

Figs. 37–46. *Dicyema maorum* sp. n. Scale with Figure 38 also applies to Figures 37 and 39. Scale with Figure 45 also applies to Figures 42–44 and 46. 37–39, nematogens, anterior ends: 38, cilia on parapolar and diapolar cells from material in mixture of formalin and sea water; 39, optical section. 40, entire nematogens. 41, vermiform embryo, immature, optical section. 42–44, vermiform embryo, about full grown, within axial cell of nematogen: 42, outline with peripheral cell nuclei; 43, peripheral cell outline; 44, optical section. 45 and 46, vermiform embryos, about full grown, within axial cells of nematogens, optical sections.

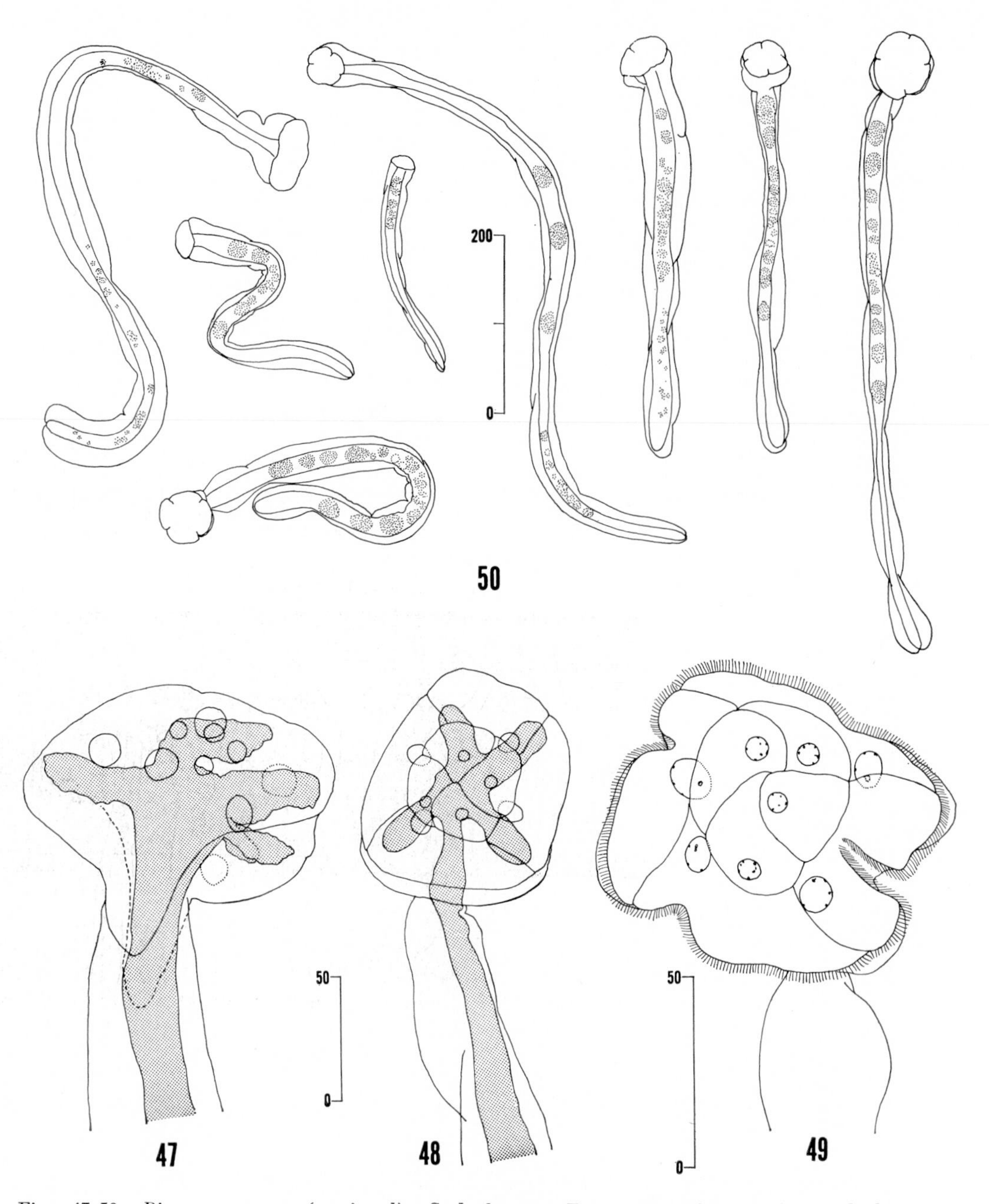

Figs. 47–50. *Dicyema maorum* (continued). Scale between Figures 47 and 48 applies to both. 47–49, rhombogens, anterior ends. 50, entire rhombogens.

reported with longer ones: *D. typus*, *D. macrocephalum*, and *D. briarei*.

The infusoriform larvae of *D. knoxi* and *D. maorum* are very similar to each other in size and morphology. In fact, the only readily discernible difference is in the length of cilia, which are longer on *D. maorum:* 6 to 7.5 μ at the posterior end, as compared to 3 to 5 μ in *D. knoxi*.

A fairly detailed analysis was made of cell numbers and positions of larvae of the 3 species; in all 3, the typical cell number is 37, and the cells are rather constant in position, as is to be expected. Deviations of 1 cell more or less from the typical 37 were noted in exceptional cases of each species; and in certain instances the missing cell, or the extra one, was identified: e.g., in one larva of *D. knoxi* 1 capsule cell (at least the nucleus) was missing, and in 1 larva of *D. maorum* there was an extra first ventral cell.

Of the 3 species described here only *D. robsonellae* has infusoriforms with refringent bodies, and the larvae of this species appear superficially typical. Infusoriforms of *D. knoxi* and *D. maorum* lack refringent bodies, and in this respect they are unusual for dicyemids; in fact, they are the only species of *Dicyema* with described infusoriforms that lack refringent bodies. Infusoriforms of some species of *Dicyemennea (D. eledones* and *D. lameeri)* have been reported with liquid or mucoid material in the apical cells instead of solid bodies [Nouvel, 1948], and recently Hoffman [1965] has described infusoriform larvae of *Dicyemennea brevicephaloides* and *D. parva* with no refringent bodies. Bogolepova [1962] had reported 2 small non–light-refracting bodies in *D. brevicephaloides.*

In larvae of all 3 species, cells were identified by comparing them with cells that appeared to be homologous in the infusoriform larva of *Dicyema aegira,* which is the most completely described [Short and Damian, 1966]. This study of *D. aegira* confirmed and extended earlier work by Nouvel [1948, 1961] on several species. In certain instances the homologies seem rather certain; in other cases they appear questionable. In such comparisons it must be realized that homologies can be judged best only on the basis of comparative developmental studies, and these have not been made. Nevertheless, it seems desirable to attach names to all the cells, and this has been done. The following cells were identified in all 3 dicyemids. Cell groups are the same as those given for *D. aegira* [Short and Damian, 1966]:

Peripheral cells with cilia: paired dorsals (2), median dorsal (1), dorsal caudals (2), lateral caudals (2), ventral caudal (1), laterals (2), and posteroventral laterals (2). The total for this group is 12, as it is for *D. aegira.*

Internal cells not ciliated: capsule cells (2), urn cells (4), germinal cells (4), dorsal internals (2), and ventral internals (2). In addition, 2 cells here designated the postcapsular cells (PC, Figures 60, 64, 68, 69) were seen lying between the capsule cells and the posterior ciliated cells. These cells have not been described hitherto. Another finding of interest in all 3 species is the absence of cilia on the ventral internal cells that form the anterior border of the urn. This lack of cilia is correlated with the obliteration of the urn cavity anteroventral to the urn cells by extension of the capsule cells into this region. In lacking cilia, the ventral internals are similar to the ventral internals of *Dicyema moschatum,* but in this latter species these cells are more posterior, squeezed between the capsule cells and peripheral cells, and the second ventrals (V2) bear the urn ciliature [Nouvel, 1948]. The two postcapsular cells bring the number of cells in this group of internal cells to 16 in *D. robsonellae, D. knoxi,* and *D. maorum,* in contrast to 14 in *D. aegira.*

Peripheral cells not ciliated: The couvercle (1) and enveloping cells (2) were easily identified, as were the apicals of *D. robsonellae.* The anterior laterals (reported in *D. aegira*) are missing, as they appear to be in *Dicyemennea gracile* [Nouvel, 1948] and *Pleodicyema delamarei* [Nouvel, 1961].

Identity of the cells listed above seems rather certain, but homologies of other unciliated peripheral cells presented difficulties, so here the picture is not entirely satisfactory. The cells in question are the anterior ventral cells (V1, V2, V3 ?) of all 3 species and the apicals (A) of *D. knoxi* and *D. maorum.* Infusoriforms of these last 2 species were studied before those of *D. robsonellae,* and at first it was thought that the large cells (here labeled V2, Figures 60, 63, 64, 68) were apical cells containing eosinophilic material instead of the usual refringent bodies, or mucoid material as described for a few species. The nuclei of these problematical cells are rounded, not collapsed as they usually appear when refringent bodies are present; and the cytoplasm, after Sanfelice's fixation, appeared rather uniformly dense and slightly eosinophilic, more so than the other peripheral cells. In some specimens, after Bouin's fixation, the cytoplasm contained a yellowish to dark reticulum throughout; in others after Bouin's, the cytoplasm appeared almost empty or contained a light, rather homogeneous eosinophilic cast with a barely visible reticulum that extended throughout all, or part of, the cell. Under certain circumstances, then, the appearance of these cells might suggest the interpretation that they are apical cells somewhat similar to those described with mucoid or liquid material. However, infusoriforms of *D. knoxi* and *D. maorum* were examined in a mixture of formalin and sea water, and nothing suggestive of vacuoles containing liquid or of refringent bodies was seen. The interpretation given below is proposed, pending a more exhaustive study.

During examination of infusoriform larvae of *D. robsonellae,* which have typical refringent bodies (vacuoles, at least) in the apical cells, a pair of cells somewhat similar in general appearance, shape, and

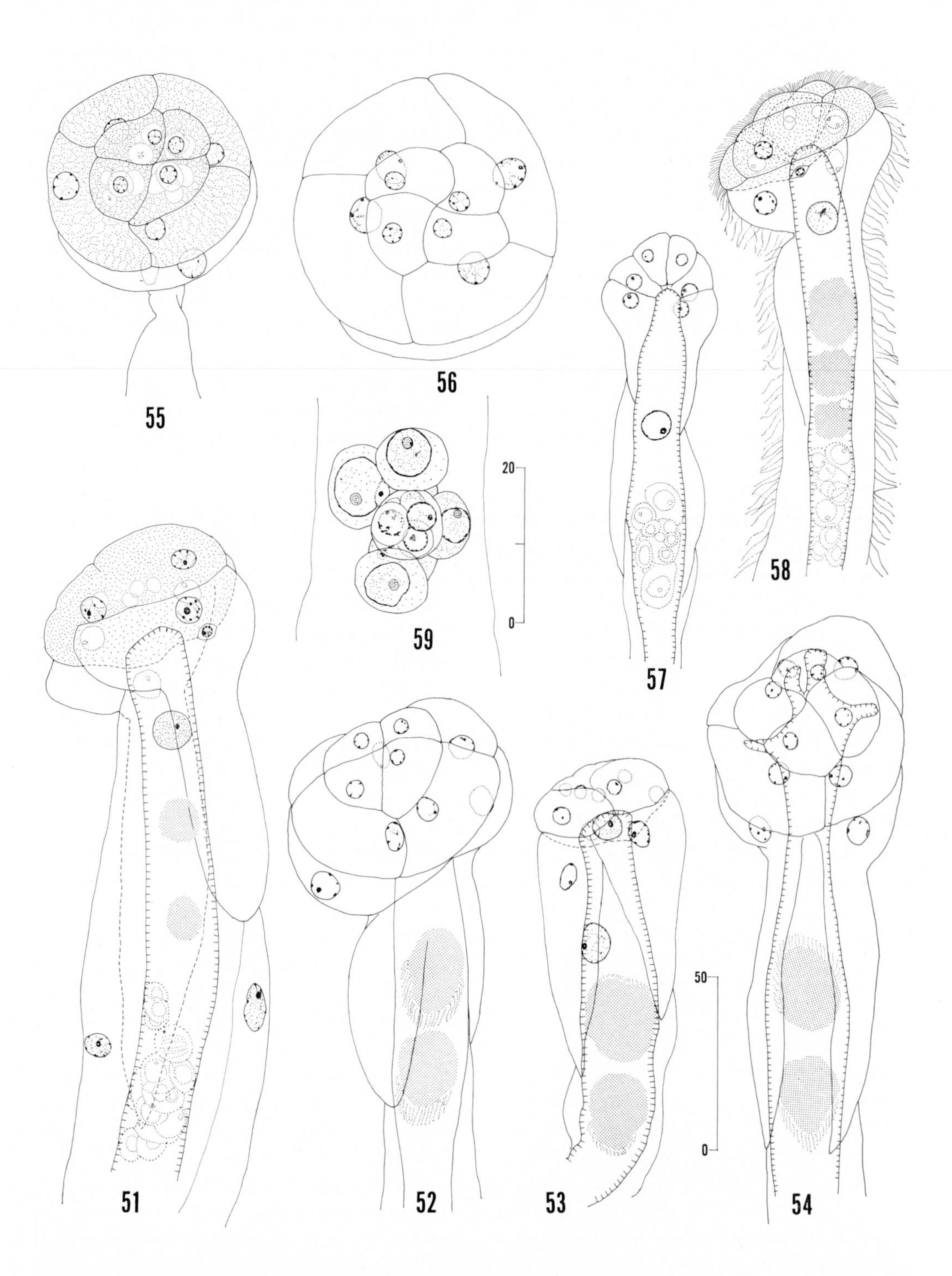

Figs. 51–59. *Dicyema maorum* (continued). Scale between Figures 53 and 54 applies to all figures except 59. 51–58, rhombogens, anterior ends: 51, 55, and 58, calottes stippled; 56, only calotte and edge of one parapolar cell (at bottom) shown, en face view; 55, vacuoles shown in propolar cells; 57, optical section of young individual; 58, cilia on parapolar and diapolar cells from material in mixture of formalin and sea water. 59, infusorigen.

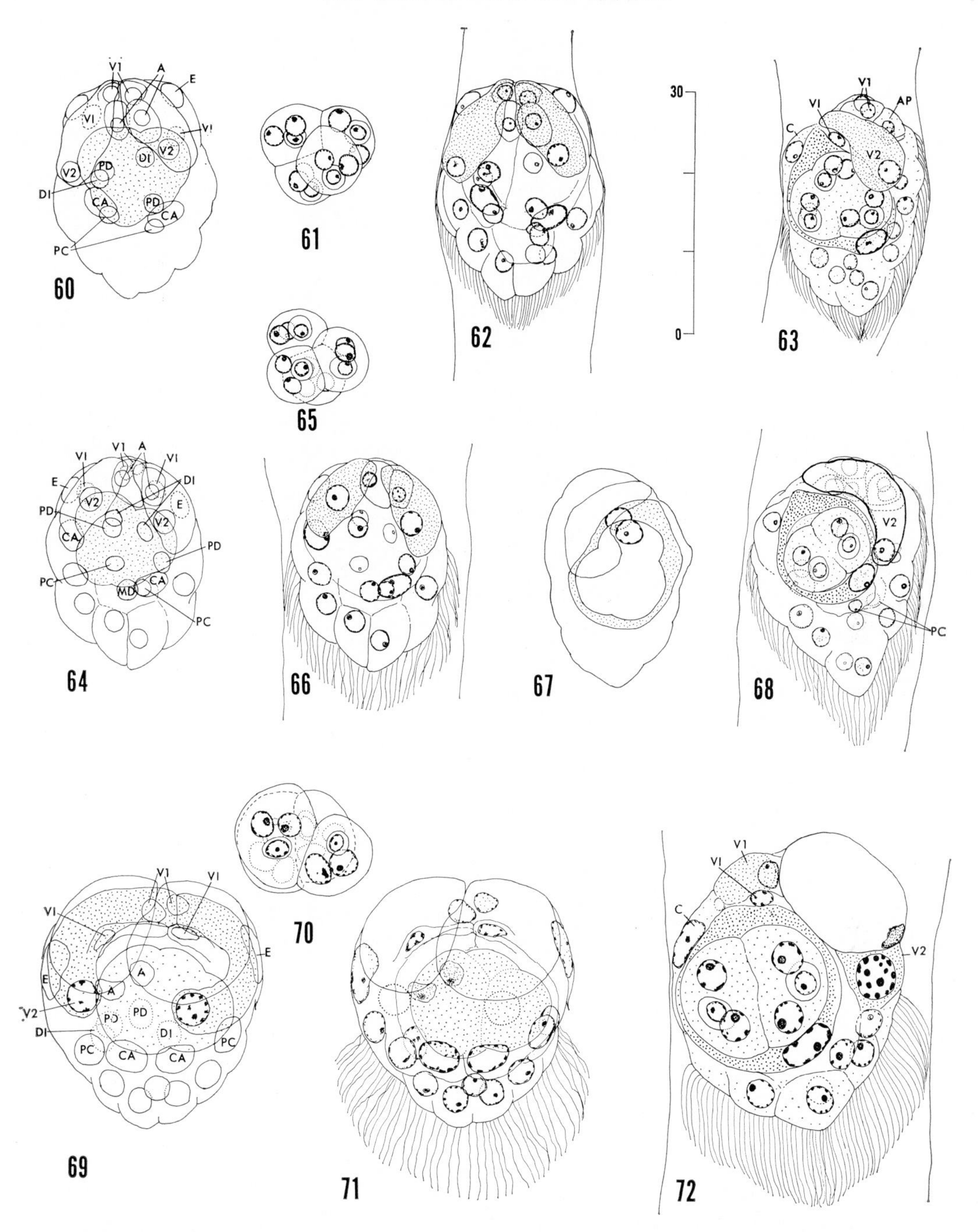

Figs. 60–72. Infusoriform larvae. Scale applies to all figures. Abbreviations denoting cells: A, apical; CA, capsule; C, couvercle; DI, dorsal internal; E, enveloping; MD, median dorsal; PC, postcaudal; PD, paired dorsal; VI, ventral internal; V1, first ventral; V2, second ventral. 60–63, *Dicyema knoxi*: 60, 61, and 62 are from same larva, dorsal view; 60, outlines showing positions of certain cells and nuclei, urn cells stippled; 61, urn cells; 62, second ventral cells stippled; 63, side view, capsule cells and second ventrals (V2) stippled more heavily than other cells. 64–68, *Dicyema maorum*: 64, 65, and 66, dorsal view; 64, outline showing positions of certain cells and nuclei, urn cells stippled; 65, urn cells; 66, second ventral cells stippled; 67, 68, side views; 67, capsule cells stippled showing their extension anterior to urn cells, one second ventral cell in outline; 68, optical section, capsule cells heavily stippled. 69–72, *Dicyema robsonellae*, ventral view: 69, outlines showing positions of certain cells and nuclei, second ventral cells stippled, urn stippled more lightly; 70, urn cells; 71, urn cells stippled; 72, side view, optical section.

position to the problematical cells of *D. knoxi* and *D. maorum* were noted. These cells of *D. robsonellae* seem to be most similar to the second ventrals (V2) of *D. aegira* and were designated as such (Figures 69, 72). Then on the basis of their similarity to the V2 cells of *D. robsonellae,* these cells of *D. knoxi* and *D. maorum* were also designated second ventrals (V2, Figures 60, 63, 64, 68). Following this, the small anterodorsal cells were labeled apicals (A) and the small cells anterior in position and ventral to the apicals were designated the first ventrals (V1). If this interpretation is correct, the third ventrals (V3) are missing. It is believed that this scheme is most probably the correct one, although other interpretations are possible; e.g., it could be argued that the apical cells are missing in these 2 species and that the cells here labeled A are the first ventrals, the cells labeled V1 are the second ventrals, and the large cells labeled V2 are the third ventrals. In any case, larvae of all 3 species have a total of 9 cells in this group of peripheral cells which are not ciliated, in contrast to a total of 13 for *D. aegira.* The difference lies in the absence in these species of the anterior laterals (2) and 1 pair of the anterior ventral group, probably the third ventrals.

Of over 50 described species of dicyemids, cell numbers have been reported for infusoriform larvae of only the following:

Dicyemennea gracile	37 cells [Nouvel, 1948]
D. adscita	39 cells [McConnaughey, 1951] (The author did not state the cell number, but according to his account the total appears to be 39.)
D. eltanini	37 cells [Short and Powell, 1969]
D. kaikouriensis	37 cells [Short and Hochberg 1969]
D. rostrata	39 cells [Short and Hochberg, 1969]
D. antarcticensis	39 cells [Short and Hochberg, 1970]
Pleodicyema delamarei	37 cells [Nouvel, 1961]
Dicyema moschatum	39 cells [Nouvel, 1948]
D. aegira	39 cells [Short and Damian, 1966]
D. robsonellae	37 cells [this paper]
D. knoxi	37 cells [this paper]
D. maorum	37 cells [this paper]

Examination of the cell numbers above, as well as a comparison of descriptions of infusoriform larvae in the present paper with the few earlier reports on other species, further supports the contention by Nouvel [1948] that specific differences exist in infusoriforms; and it seems evident that data on infusoriform larvae, after much more work, should eventually play an important role in reorganizing the classification of dicyemids at the generic level.

Acknowledgments. Collection and fixation of material were a joint effort of Mr. Edwin C. Powell and the author. I wish to thank Professor George A. Knox, University of Canterbury, for his kindness and for the privilege of working at the Edward Percival Marine Laboratory, and also Dr. and Mrs. Ian Mannering and Mr. Malcolm Mannering for aid in collecting octopuses as well as for other courtesies extended to us at the laboratory. Thanks are due Mr. Glenn M. Sponholtz for staining and mounting the material and Mr. F. G. Hochberg, Jr., for reading the manuscript, making several suggestions for improvements, and helping prepare the final copy. The work was supported by research grants GA-115, GA-265, and GA-1102 from the National Science Foundation through the Office of Antarctic Programs.

REFERENCES

Bogolepova, I. I.

1962 Dicyemidae of the far eastern seas. Part 2. New species of the genus *Dicyemennea* [in Russian, English summary]. Zool. Zh., *41:* 503–518, figs. 1–8.

Hoffman, E. G.

1965 Mesozoa of the sepiolid, *Rossia pacifica* (Berry). J. Parasit., *51:* 313–320, figs. 1–5.

McConnaughey, B. H.

1949 Mesozoa of the family Dicyemidae from California. Univ. Calif. Publs Zool., *55:* 1–34, figs. 1–89.

1951 The life cycle of the dicyemid Mesozoa. Univ. Calif. Publs Zool., *55:* 295–336, figs. 1–13.

Nouvel, H.

1947 Les Dicyémides. 1re partie: systématique, générations vermiformes, infusorigène et sexualité. Archs Biol., *58:* 59–220, figs. 1–59.

1948 Les Dicyémides. 2e partie: infusoriforme, tératologie, spécificité du parasitisme, affinités. Archs Biol., *59:* 147–223, figs. 1–34.

1961 Un dicyémide nouveau, *Pleodicyema delamarei* n. g., n. sp., parasite du céphalopode *Bathypolypus sponsalis.* Remarques sur la validité des genres *Dicyemodeca* Wheeler, *Pseudicyema* Nouvel et *Microcyema* v. Bened. (I) Vie Milieu, *12:* 565–574, figs. 1–16.

Short, R. B.

1962 Two new dicyemid mesozoans from the Gulf of Mexico. Tulane Stud. Zool., *9:* 101–111, figs. 1–66.

Short, R. B., and R. T. Damian
1966 Morphology of the infusoriform larva of *Dicyema aegira* (Mesozoa: Dicyemidae). J. Parasit., *52:* 746–751, figs. 1–12.

Short, R. B., and F. G. Hochberg, Jr.
1969 Two new species of Dicyemennea (Mesozoa: Dicyemidae) from Kaikoura, New Zealand. J. Parasit., *55:* 583–596, figs. 1–51.

1970 A new species of *Dicyemennea* (Mesozoa: Dicyemidae) from near the Antarctic Peninsula. J. Parasit., *56:* 517–522, figs. 1–21.

Short, R. B., and E. C. Powell
1969 *Dicyemennea eltanini* sp. n. (Mesozoa: Dicyemidae) from Antarctic waters. J. Parasit., *55:* 794–799, figs. 1–23.

DISTRIBUTION OF RECENT BENTHONIC FORAMINIFERA IN THE DRAKE PASSAGE

René Herb

Geological Institute, University of Bern, Bern, Switzerland

Abstract. Recent benthonic Foraminifera from 73 stations in the Drake Passage between South America and Antarctica have been analyzed qualitatively and quantitatively. A bathymetric zonation as well as an areal distinction of faunal provinces can be recognized. Thus, assemblages of the shelf and upper bathyal zone of Tierra del Fuego (Cape Horn province) differ from those of the antarctic shelf (South Shetland province). *Discanomalina vermiculata, Cibicides fletcheri, Eponides isabelleanus, Ehrenbergina pupa, Mississippina concentrica, Heronallenia kempii,* and *Polystomellina patagonica* are typical species of the Cape Horn province. *Reophax pilulifer, Cibicides refulgens, Cibicides grossepunctatus,* and *Cribrostomoides jeffreysi* are common in the South Shetland province. Slight differences are also seen between assemblages of the Cape Horn province and the area of the Falkland Islands and of the Burdwood Bank (Falkland subprovince), where *Elphidium crispum* and *Hoeglundina elegans* are common. Predominantly calcareous populations were recovered from shallow-water stations in all these areas. Generally, a progressive increase in the percentage of arenaceous Foraminifera is seen with greater depth. Calcareous perforate Foraminifera are rare in most of the southern deep-sea areas, whereas in the north the group is still dominant at many stations. Partly this distribution is probably controlled by a somewhat higher position of the $CaCO_3$ dissolution boundary in the cold Antarctic Bottom Water of the south. Significant arenaceous genera of the deep sea are *Rhabdammina, Hyperammina, Saccammina, Psammosphaera, Ammolagena, Reophax, Hormosina, Cribrostomoides, Recurvoides, Ammobaculites, Cyclammina, Martinottiella, Eggerella,* and *Karreriella.* Calcareous imperforate Foraminifera, such as different species of the genus *Pyrgo,* reach diameters up to 3.5 mm. A bathymetric zonation was established by plotting the upper depth limits of the most common and significant foraminiferal species. The significance of isobathyal and heterobathyal species relative to the bathymetric zonation or to other parameters is discussed. Displacement of shallow-water assemblages into deep water was recognized in several cases.

INTRODUCTION

As part of a long-term research project in the antarctic seas of the Drake Passage, the Scotia Sea, the Bellingshausen Sea, and the South Pacific between South America and New Zealand, the USNS *Eltanin* has been cruising in these areas since 1962. During three of these cruises, bottom samples and various oceanographic data were obtained from the Drake Passage, which separates South America from Antarctica. As a participant in this program, the author had the opportunity to study the Foraminifera from these bottom samples, which were obtained by various types of trawls, grabs, dredges, and corers, as will be discussed below.

The purpose of this work was to evaluate distributional patterns of benthonic Foraminifera rather than to attempt a taxonomic treatment, which has been done in earlier reports. Distributional patterns of the planktonic Foraminifera have been discussed elsewhere [Herb, 1968].

PREVIOUS WORK

A summary of earlier publications on antarctic Foraminifera has been given by McKnight [1962]. The works of D'Orbigny [1839] and Brady [1884], the well-illustrated volumes of Wiesner [1931] and Parr [1950], and the *Discovery* Reports of Heron-Allen and Earland [1932] and Earland [1933, 1934, 1936] are among the classic ones to be mentioned. Quantitative studies on benthonic Foraminifera of antarctic seas were published by Uchio [1960], Saidova [1961], McKnight [1962], Pflum [1966], and Kennett [1968]. Bandy and Echols [1964], reevaluating McKnight's data, proposed a bathymetric zonation of antarctic Foraminifera, and also plotted depth-temperature characteristics for selected species by comparing antarctic occurrences with those of temperate or tropical areas.

For the Drake Passage area Heron-Allen and Earland [1932] and Earland [1934] gave a detailed record of species encountered, as well as taxonomic discussions of many of them. These authors also compared their antarctic specimens with those of the D'Orbigny collection.

In a more recent investigation, Shishkevish [1964] studied the areal distribution of Foraminifera of the Scotia Ridge, including the southernmost South American shelf, the Burdwood Bank, and the Falkland Islands, as well as the northeastern part of the Drake Passage. Two ecological zones were distinguished, one corresponding to the shelf areas, and the other to the deep-sea floor near the Antarctic Convergence.

METHODS

A total of 73 surface sediment samples from the Drake Passage, the areas south and southeast of the Falkland Islands, and the Bransfield Strait were available for qualitative and quantitative investigation of benthonic Foraminifera. Of these, 37 were collected with a small biological trawl (Menzies trawl) with a mesh opening of 0.5 mm [Menzies, 1964]. Seven were taken with a Petersen grab, and 22 with other gear, such as a Blake trawl, a beam trawl, and a rock dredge. Detailed lists of the University of Southern California *Eltanin* stations, with position, depth, and gear used, are given in Savage and Geiger [1965]. In addition, 7 trawl samples, prefixed 'V,' taken from this area by the *Vema* (Lamont Geological Observatory), were used in this study.

Immediately after collection the samples were preserved in buffered formalin, and before being washed in the laboratory, about half of them were treated with a protoplasmic stain (rose bengal) to determine which specimens were living when collected. Quantitative analysis was performed on splits containing at least 300–500 specimens.

It is evident that the Menzies trawl, having a mesh opening of 0.5 mm, selectively concentrates larger-sized Foraminifera [Bandy and Rodolfo, 1964]. Of the planktonic Foraminifera only large specimens are collected, and therefore a distribution study of planktonic Foraminifera cannot be based on such material. In many cases, however, smaller Foraminifera, probably caught in lumps of mud, were also collected by trawls and dredges. Percentage figures based on such samples are not accurate. For obtaining more uniformly sized populations of benthonic Foraminifera, these samples could be washed through a 0.5-mm mesh screen in order to eliminate smaller Foraminifera. However, even then foraminiferal species having an average size around 0.5 mm would be represented by larger individuals only, and the respective percentage figures relative to the total population of benthonic Foraminifera would be too low. Therefore, an attempt was made to use specimens of all sizes within each sample in order to obtain the maximum possible information for each station. It must be pointed out, however, that the percentage figures in Tables 1 and 2 may not show the true proportions, especially among smaller species like *Ehrenbergina pupa*.

Of the 73 samples available, 5 furnished only a few individuals, insufficient for any quantitative interpretation. Another 5 samples contained a total of 20 to less than 100 specimens of benthonic Foraminifera. The quantitative relationship for these stations is given on Table 2 with special signatures or with figures in brackets. In Figures 3–6 they are marked by small circles.

Phleger cores were taken at many of the Menzies trawl stations, but the surface sediments in these cores contained relatively few benthonic Foraminifera, especially in the cores from areas south of the Antarctic Convergence. In the open sea north of this line the planktonic Foraminifera predominate to such an extent that a quantitative study of the rare benthonic forms becomes difficult. Quantitative data on benthonic Foraminifera of these cores are therefore incomplete and were not included in this report. The cores were used, however, for evaluating the distributional pattern of planktonic Foraminifera [Herb, 1968], and the ratio between benthonic and planktonic Foraminifera in the surface sediments (see Figure 11).

Because of the restricted ship time available, the sampling pattern is not always such as would be desirable, especially with regard to the determination of the bathymetric distribution. Thirty-five samples were taken in water deeper than 2800 meters and 25 in water shallower than 900 meters, but only 8 stations are located at depths between 900 and 2700 meters. Because of this irregular bathymetric control, the technique of plotting cumulative percentages of species as a function of depth, as used by Bandy and Echols [1964], cannot be applied. However, a reasonable areal sampling coverage was achieved, although gaps still exist, such as that for the area around the Burdwood Bank or the areas southwest

TABLE 1. Abundance of Benthonic Foraminifera Found at Specified Depths at *Eltanin* Stations in the Falkland Subprovince

	Sta. 344	Sta. 339	Sta. 340	Sta. 557
Percentages of types of Foraminifera				
Arenaceous	x	7	15	24
Calcareous imperforate	5	5	15	14
Calcareous perforate	95	88	70	62
Percentages of each genus and species				
1. *Rhizammina indivisa*	x		x	1
2. *Psammosphaera fusca*	x			2
3. *Triloculina elongata*	x			
4. *Miliolinella irregularis*	1			
5. *Quinqueloculina magellanica*	x			
6. *Quinqueloculina* cf. *vulgaris*	3		4	1
7. *Lagena* spp.	x			1
8. *Lenticulina* spp.	x	x	x	2
9. *Cassidula crassa*	4			
10. *Cassidulina subglobosa*	6			
11. *Cassidulina pulchella*	6			
12. *Ehrenbergina pupa*	6			
13. *Cibicides fletcheri*	42		x	x
14. *Eponides isabelleanus*	x	1		
15. *Bucella frigida*	7			
16. *Hoeglundina elegans*	x	1	23	38
17. *Pullenia subcarinata*	x		x	
18. *Elphidium crispum*	19			
19. *Elphidium lessonii*	x			
20. *Rhabdammina* spp., *Hyperammina* spp.		x	1	2
21. *Crithionina pisum*		x		
22. *Thurammina papillata*		3	x	
23. *Reophax pilulifer*		1	1	
24. *Reophax curtus*			x	4
25. *Recurvoides contortus*		3	5	1
26. *Cyclammina cancellata*			7	6
27. *Cyclammina pusilla*			1	1
28. *Quinqueloculina peruviana*			1	2
29. *Triloculina trigonula*		x	1	
30. *Pyrgo williamsoni*		x	x	2
31. *Pyrgo depressa*		2	3	2
32. *Pyrgo murrhina*		2	6	7
33. *Pyrgoella sphaera*		x		2
34. *Lenticulina occidentalis glabrata*			6	
35. *Vaginulina* sp. aff. *spinigera*			1	
36. *Dentalina* spp.			1	
37. *Rectoglandulina* sp.		1	3	
38. *Uvigerina bassensis*		1		
39. *Discanomalina vermiculata*		5		
40. *Cibicides wuellerstorfi*		76	38	4
41. *Rupertia stabilis*		x		
42. *Eponides tener tener*		1		
43. *Pelosinella bicaudata*				4
44. *Reophax scorpiurus*				3
45. *Karreriella* cf. *novangliae*				1

x indicates less than 1% occurrence of specimens.

and east-northeast of Station Tr-4-10, which is a trigger core sample only.

A rather conservative taxonomy for benthonic Foraminifera has been used here on the generic level. The proliferation of generic names, as observed in recent years and reflected in the 'Treatise of Invertebrate Paleontology,' has made, in my opinion, the use of such a modern classification difficult and, in many instances, impractical. A conservative attitude seems justified also in the light of the new possibilities for the study of ultrastructures of Foraminifera offered by the scanning electron microscope.

On the other hand, an accurate species determination is of special importance for ecological studies. In studying the Drake Passage material I noted that a considerable number of species found would require a thorough taxonomic revision, which could not be undertaken for the present report. For this reason, most of the important species have been figured on Plates 1–16.

OCEANOGRAPHY

An excellent picture of the submarine topography in the Drake Passage and surrounding areas is given by Heezen and Tharp [1961]. The bathymetric contours shown on the present map (Figure 1) at intervals of 1000 meters (plus a 200-meter line) were plotted by the author from sounding data given in Hydrographic Office Chart no. Misc. 15.254-11 and from precision depth recordings taken by the *Eltanin*. These contours are simplified, however, and show the over-all picture only. Since the deep-sea floor of the Drake Passage has an average depth of nearly 4000 meters over wide areas, the 4000-meter contour would show a much more complicated pattern if plotted in detail.

The outer edge of the shelf around the southern tip of South America is at depths between 180 and 250 meters, whereas the edge of the antarctic shelf is much deeper; in many places its depth is 650–700 meters.

Deep-sea troughs parallel the South American shelf to the southwest and the antarctic shelf to the north of the South Shetland Islands, but do not reach extreme depths (4385 meters south of Tierra del Fuego, 5250 meters north of the South Shetlands).

The structure of antarctic waters has recently been summarized by Gordon [1967]. For the study of benthonic Foraminifera the pattern of physical parameters of the Antarctic Bottom Water is of special interest. Gordon's data are here reported in Figure 2.

The Antarctic Convergence (Polar Front) is another important feature of the area. Its position and

TABLE 2. Abundance of Benthonic Foraminifera in Trawl and Grab Samples from the Drake Passage

X indicates less than 1% occurrence of specimens. O indicates species that are common in samples with more than 20 but less than 60 specimens in total. ● indicates species that are abundant in samples with more than 20 but less than 60 specimens in total. Parentheses indicate percentages for stations where only 60–100 specimens were recovered.

	STATIONS SOUTH OF THE ANTARCTIC CONVERGENCE																														
STATION NO.	127	303	257	390	129	135	260	394	298	283	265	140	400	145	276	268	138	412	432	430	428	416	415	272	418	1002	408	441	439	435	436
DEPTH IN METERS	4758	4077-4176	3867-4086	3825-4090	3678-3816	3715-3752	3761-3781	3724-3825	3777	3550-3693	3691-3693	3687	3537	3312-3532	3022-3043	2763-2818	1437	1180	935-884	681-1409	662-1120	494-507	406-465	412	426-311	265	225-223	156-253	128-165	73	73
ARENACEOUS FORAM. (IN % OF TOTAL BENTH. FORAM.)	99	97	94	99	85	98	95	100	99	99	99		100	95	100		99	92		93	77	85		82	8	5	67	2	3		12
CALC. IMPERF. "	1	x	2	-	1	x	2	x	1	x	1		x	2	x		1	6		3	2	4		8	1	1	13	1	x		1
CALC. PERF. "	x	2	3	x	14	1	2	-	-	x	-		-	3	-		x	2		4	21	12		10	91	94	21	97	96		87
> 73 m																															
1 PSAMMOSPHAERA PARVA		x	x															x													1
2 REOPHAX PILULIFER																	4	31	●		49	54	o	18	2	x	29		2	(20)	4
3 CRIBROSTOMOIDES JEFFREYSI			x																		x	10		4	1	4	2	1	x		4
4 QUINQUELOCULINA SEMINULUM																															
5 PYRGO DEPRESSA		x			x													1		2					x		6			(x)	
6 SIGMOILINA OBESA																															
7 ANGULOGERINA OCCIDENTALIS																									x	x		1			
8 CASSIDULINA PULCHELLA																															
9 CASSIDULINA CRASSA	x	x					1			x		2		x		o		1	●	1	2	x		9	x	2	x	x			
10 CASSIDULINA CRASSA ROSSENSIS																															5
11 CASSIDULINA SUBGLOBOSA			x									3							o											(x)	
12 ROSALINA VILARDEBOANA																									x	2		10	7	(10)	
13 DISCORBIS AFF. D. NITIDUS																															
14 DISCANOMALINA VERMICULATA																															
15 CIBICIDES REFULGENS						1												1			14	9	●		89	77	9	78	79	(50)	72
16 CIBICIDES GROSSEPUNCTATUS																o					2	1	o		2	9	x	7	10	(10)	8
17 CIBICIDES FLETCHERI																															
18 EPONIDES ISABELLEANUS																															
19 PULLENIA SUBCARINATA												x						x							x	1	x	x			
20 ROTALIA SP.																															
> 104 m																															
21 BATHYSIPHON FILIFORMIS		1			x	x		x				6		5											1	x					
22 RHIZAMMINA INDIVISA		x		!	x							1						1			2	x		1			1				
23 PILULINA SP.																									1			1	x		
24 SACCAMMINA SPHAERICA		1												1									o				x				1
25 REOPHAX NODULOSUS	1	x	x	x	2	x	x	x		1	x	1	1	3	1		1	x	●	3	5	1			x		x			(x)	
26 CYCLAMMINA PUSILLA	1	29	3	1	1	3	17	1	18	4	10	3	1	8	19		x					x									
27 QUINQUELOCULINA PERUVIANA																															
28 TRILOCULINA CF. ELONGATA	x													x						3											
29 PYRGO BULLOIDES																									x	x		x	x		
30 PYRGO WILLIAMSONI																						3		1			2				
31 MILIOLINELLA CRYPTELLA	x	x	x		x		x																								
32 EHRENBERGINA PUPA																															
33 PATELLINA CORRUGATA																															
34 HERONALLENIA KEMPII																															
35 BUCELLA FRIGIDA																															
36 HOEGLUNDINA ELEGANS																															
37 GYROIDINA NEOSOLDANII														x					o												
38 ELPHIDIUM SPP.																															
39 CIBICIDES LOBATULUS												1									x			x			x				
40 QUINQUELOCULINA MAGELLANICA	x													x				x													
> 205 m																															
41 BATHYSIPHON SPP.		2	30	36	2	6	21	4	15	x	14	x	11	9	1		1				2	2	o	3	x		22				
42 PSAMMOSPHAERA FUSCA																											1				
43 THURAMMINA PAPILLATA					x		x				x	x																			
44 PSAMMOPHAX CONSOCIATA			2	5		6		x	1		x	x	1	5									o		cf. 1	x					
45 REOPHAX CURTUS		x	4	x	x	x	1	x	2		x	x													x						
46 REOPHAX BREVIS				x				2									1								? 1						
47 REOPHAX DENTALINIFORMIS																				1	2			6			6				
48 RECURVOIDES CONTORTUS	2	3	x	1	2	1	1	2		1	1	x	x				4	54	●	77	14			12	x		6				
49 QUINQUELOCULINA VULGARIS																								x			1				
50 PYRGO ISABELLEANA																															
51 PYRGO PISUM																		x									2				
52 PYRGO MURRHINA	x	1					1		x		1	1				o	x		o			x					1				
53 MILIOLINELLA SUBROTUNDA					x	x	1															x				1	x	x			
54 CRUCILOCULINA TRIANGULARIS																															
55 GLANDULINA LAEVIGATA																		x									5				
56 BULIMINELLA ELEGANTISSIMA																															
57 PSEUDOBULIMINA CHAPMANI																									x	x					
58 EHRENBERGINA HYSTRIX GLABRA																															
59 RUPERTIA STABILIS																								x			x				
60 MISSISSIPPINA CONCENTRICA																															
61 PULLENIA BULLOIDES																				1	x								x		

TABLE 2. (continued)

	STATIONS SOUTH OF THE ANTARCTIC CONVERGENCE																													
STATION NO.	127	303	257	390	129	135	260	394	298	283	265	140	400	145	276	268	138	412	432	430	428	416	415	272	418	1002	408	441	439	435	436
DEPTH IN METERS	4758	4077-4176	3867-4086	3825-4090	3678-3816	3715-3752	3761-3781	3724-3825	3777	3550-3693	3691-3693	3687	3537	3312-3532	3022-3043	2763-2818	1437	1180	935-884	681-1409	662-1120	494-507	406-465	412	426-311	265	225-223	156-253	128-165	73	73
ARENACEOUS FORAM. } IN % OF TOTAL BENTH. FORAM.	99	97	94	99	85	98	95	100	99	99	99		100	95	100		99	92		93	77	85		82	8	5	67	2	3		12
CALC. IMPERF. "	1	x	2	-	1	x	2	x	1	x	1		x	2	x		1	6		3	2	4		8	1	1	13	1	x		1
CALC. PERF. "	x	2	3	x	14	1	2	-	-	x	-		-	3	-		x	2		4	21	12		10	91	94	21	97	96		87
> 411 m																															
62 RHABDAMMINA ABYSSORUM	9	x			x		x				x		1				23	1		3	3			4							
63 RHABDAMMINA IRREGULARIS		2	12		6	7	4	11	3	4			5			o	x					x									
64 PELOSINELLA BICAUDATA			1		x			x				x					5							6							
65 BATHYSIPHON ARENACEUS	1			1	6			1				x	1			o						14	o								
66 HYPERAMMINA ELONGATA		10					2			17			x	2	1									1							
67 CRIBROSTOMOIDES SUBGLOBOSUS	12	22	10•	1	21	17	7	20	19	41	17	18	36	8	38	•	5				x	1									x
68 CYCLAMMINA ORBICULARIS	1	x	1		1	5	4	1	5	9	10	4	x	3		o	6							1							
69 AMMOBACULITES AGGLUTINANS	1	2	1	1	9	9	2	4	2	3	5	5	4	2		o	45	2		2		1									
> 662 m																															
70 AMMOLAGENA CLAVATA	x	5	1	2	1	10	6	1	4	3	8	3	6	23	6																
71 HORMOSINA GLOBULIFERA	x	1		x	2	x		2	3	3	2	1	1	1	x		x			8											
72 AMMODISCUS SPP.										x																					
73 CRIBROSTOMOIDES CRASSIMARGO	2	1	4	48	7	24	7	21	5	4	7	2	14	7	23	•	cf.2				?x										
74 TEXTULARIA AGGLUTINANS																															
75 TEXTULARIA CF. PHILIPPINENSIS																															
76 KARRERIELLA NOVANGLIAE, WHITE SHELL																															
77 KARRERIELLA NOVANGLIAE, GRAY SHELL				x	x	x	1				x	x	x	x		o															
78 DOROTHIA PSEUDOTURRIS																															
79 TRILOCULINA TRIGONULA	x	x	x		x								x																		
80 MILIOLINELLA SUBVALVULARIS	x	x			x	x		x						x				2			x										
81 PYRGOELLA SPHAERA							x																								
82 ANGULOGERINA CARINATA BRADYANA					2	x																						x			
83 PYRULINA FUSIFORMIS							x							1																	
84 EHRENBERGINA PACIFICA																															
85 LATICARININA PAUPERATA																			o												
86 CIBICIDES WUELLERSTORFI																															
87 CIBICIDES BRADII												2																			
> 1805 m																															
88 EGGERELLA BRADYI BRADYI, WHITE SHELL																															
89 KARRERIELLA BRADYI, WHITE SHELL																															
90 MARTINOTTIELLA COMMUNIS, LARGE, TYPICAL						6	x	x				2																			
91 MARTINOTTIELLA COMMUNIS, SHORT																															
92 MARTINOTTIELLA NODULOSA	1	2		x	4	1	2	1	x	x	1	9	1	4																	
93 QUINQUELOCULINA SP.		x																													
94 UVIGERINA BRADYANA																															
95 ANGULOGERINA ANGULOSA																															
96 CIBICIDES CORPULENTUS																															
97 EPONIDES TENER STELLATUS				x								1																			
> 2762 m																															
98 REOPHAX INSECTUS		8	x		x						x	3			1																
99 NODOSINUM GAUSSICUM					x	x	1	1			1		1		10	o															
100 CIBICIDES SP. A												x																			
101 CIBICIDES PSEUDOUNGERIANUS	x					x						x																			
> 3137 m																															
102 ASTRORHIZA ANGULOSA								1																							
103 TOLYPAMMINA VAGANS		x		x	x			4			8	x	2	x																	
104 RHIZAMMINA ALGAEFORMIS																															
105 HYPERAMMINA CYLINDRICA	x		4		x	1	x	5		x	x	2																			
106 HORMOSINA NORMANI	56	x	1		11				2	2	x	x	2																		
107 HORMOSINA ROBUSTA		2	2				6	x	7	2	4		3	4																	
108 EGGERELLA BRADYI BRADYI, GRAY SHELL	x		1					x		x																					
109 EGGERELLA BRADYI NITENS	x	x	x	x	x		x	1	x					x																	
110 KARRERIELLA BRADYI, GRAY SHELL		x		x	x			x				x		x																	
111 TROCHAMMINA RUGOSA	x		cf.6		x	1	1	1		x	1																				
112 TROCHAMMINA SP. AFF. T. INFLATA	x	1	1																												
113 PYRGO WIESNERI	x	x			x																										
114 INVOLVOHAUERINA GLOBULARIS																															
115 KERAMOSPHAERA MURRAYI		x					x	x		x		x																			
116 MARGINULINA CF. PLANATA																															
117 MARGINULINA OBESA																															

TABLE 2. (continued)

	STATIONS NORTH OF THE ANTARCTIC CONVERGENCE																															
STATION NO.	152	230	249	112	155	311	120	356	126	250	315	307	362	244	234	148	384	115	305	V-17-61	372	322	V-17-54	161	970	369	V-17-51	162	219	370	217	V-17-48
DEPTH IN METERS	4172-4209	4185-4191	3989-4099	ca. 4008	3927	3911-4099	3825-3975	4136-3678	3733-3806	3678-3803	3806-3770	3642-3935	3590-3477	3457-3514	3349-3477	3294-3376	3138-3426	3074-3093	2782-2827	1814-1919	1953-1971	1806-2013	1274-1362	878	586-641	247-293	205	174	115	104-115	106-110	42
ARENACEOUS FORAM. } IN % OF TOTAL BENTH. FORAM.		95	5	25	20	69	71	14	31	46		25	61	40	21	90	88	83	2	5	33	41		36	4	x	x	-	1	1	1	4
CALC. IMPERF. "		2	1	8	2	4	2	8	3	8		5	19	19	18	5	4	3	3	1	3	7		11	11	15	11	3	8	2	7	4
CALC. PERF. "		4	94	68	78	27	27	77	66	46		69	20	41	60	5	9	14	95	94	64	52		52	85	85	88	97	91	97	92	91
> 73 m																																
1 PSAMMOSPHAERA PARVA			x		x		x																									
2 REOPHAX PILULIFER																							(1)									
3 CRIBROSTOMOIDES JEFFREYSI				4	1				x									x								x						1
4 QUINQUELOCULINA SEMINULUM								1												x					1	8			6	2	2	x
5 PYRGO DEPRESSA							1			2		x										1										
6 SIGMOILINA OBESA																										1	1					4
7 ANGULOGERINA OCCIDENTALIS																										1						1
8 CASSIDULINA PULCHELLA																																5
9 CASSIDULINA CRASSA			20	64	59	7	17	7	51	28		39	4	24	37		7	1	2	1				x		13	35	57	50	51	20	
10 CASSIDULINA CRASSA ROSSENSIS																									4							13
11 CASSIDULINA SUBGLOBOSA																1		6	5	1	40											
12 ROSALINA VILARDEBOANA																								x	1	x				x		x
13 DISCORBIS AFF. D. NITIDUS																									1	2	x					x
14 DISCANOMALINA VERMICULATA																								1	32	28	13	34	28	17	17	3
15 CIBICIDES REFULGENS																									x	16						
16 CIBICIDES GROSSEPUNCTATUS												cf.x																	x			
17 CIBICIDES FLETCHERI								?34														cf.→(10)			39	16	18	x		17	29	51
18 EPONIDES ISABELLEANUS																									1	2	5	3	8			12
19 PULLENIA SUBCARINATA																						x		x		3			1	2	1	1
20 ROTALIA SP.																										x	x					4
> 104 m																																
21 BATHYSIPHON FILIFORMIS		5	x	x	1		12	3	1	1		1		x	1	2	1				1									x		
22 RHIZAMMINA INDIVISA					x							2			x														1	x		
23 PILULINA SP.																																
24 SACCAMMINA SPHAERICA					12										1																	
25 REOPHAX NODULOSUS		x	x	x	x	1							1			2																
26 CYCLAMMINA PUSILLA		6	x	1	1	12	2	x	1	2	o	x	1			6	1	x			2	1	(1)	x							x	
27 QUINQUELOCULINA PERUVIANA																													1		3	
28 TRILOCULINA CF. ELONGATA																														x		
29 PYRGO BULLOIDES																										4						
30 PYRGO WILLIAMSONI																									1			x	2	x	2	
31 MILIOLINELLA CRYPTELLA		x		1					x	1		x	1				1		x										1		x	
32 EHRENBERGINA PUPA								x															(1)		1	8	6	x		7	10	
33 PATELLINA CORRUGATA																								x		x					2	
34 HERONALLENIA KEMPII																									1	x			x			
35 BUCELLA FRIGIDA								3																						x		
36 HOEGLUNDINA ELEGANS														5	7				1		3	1									x	
37 GYROIDINA NEOSOLDANII			2	1	15		9	12		9		11	5	7	7	x	1		3												cf.1	
38 ELPHIDIUM SPP.			11																	x					x	x			2	1	x	
39 CIBICIDES LOBATULUS			x			x						1		x					5	20	x	12	(3)	9		1	7		1		2	
40 QUINQUELOCULINA MAGELLANICA																									x		x	3				
> 205 m																																
41 BATHYSIPHON SPP.																							(6)				x					
42 PSAMMOSPHAERA FUSCA	o		1	14	x	9	4		17	2	o	2	3	2	2	8	3	35		x	10	1			4		x					
43 THURAMMINA PAPILLATA					x				x																							
44 PSAMMOPHAX CONSOCIATA						4	x		x	6		x	1	1	1	2	3	7	x			3										
45 REOPHAX CURTUS		1		1	x		x	1		x		x	1	1		1	1				x	x										
46 REOPHAX BREVIS				4	1				x									x														
47 REOPHAX DENTALINIFORMIS																																
48 RECURVOIDES CONTORTUS	o	x			x	x	2		x	1		x	1	x		x	x						(1)									
49 QUINQUELOCULINA VULGARIS																						x		x		x	cf.5					
50 PYRGO ISABELLEANA																										2						
51 PYRGO PISUM										x			3	x			x	x						x								
52 PYRGO MURRHINA			1	2	1	2	1	1	3	5		x	12	15	15	2	2	3	3	x	3	5	(1)	3								
53 MILIOLINELLA SUBROTUNDA																1								2	6							
54 CRUCILOCULINA TRIANGULARIS																							(6)				x					
55 GLANDULINA LAEVIGATA									x					x																		
56 BULIMINELLA ELEGANTISSIMA																										1						
57 PSEUDOBULIMINA CHAPMANI																																
58 EHRENBERGINA HYSTRIX GLABRA																								3		x						
59 RUPERTIA STABILIS																			40	2		3	(2)	9								
60 MISSISSIPPINA CONCENTRICA																								1	2	7	1					
61 PULLENIA BULLOIDES					5	x												x	x			x	4									

TABLE 2. (continued)

	STATIONS NORTH OF THE ANTARCTIC CONVERGENCE																															
STATION NO.	152	230	249	112	155	311	120	356	126	250	315	307	362	244	234	148	384	115	305	V-17-61	372	322	V-17-54	161	970	369	V-17-51	162	219	370	217	V-17-48
DEPTH IN METERS	4172-4209	4185-4191	3989-4099	ca. 4008	3927	3911-4099	3825-3975	4136-3678	3733-3806	3678-3803	3806-3770	3642-3935	3590-3477	3457-3514	3349-3477	3294-3376	3138-3426	3074-3093	2782-2827	1814-1919	1953-1971	1806-2013	1274-1362	878	586-641	247-293	205	174	115	104-115	106-110	42
ARENACEOUS FORAM. IN % OF TOTAL BENTH. FORAM.		95	5	25	20	69	71	14	31	46		25	61	40	21	90	88	83	2	5	33	41		36	4	x	x	-	1	1	1	4
CALC. IMPERF. "		2	1	8	2	4	2	8	3	8		5	19	19	18	5	4	3	3	1	3	7		11	11	15	11	3	8	2	7	4
CALC. PERF. "		4	94	68	78	27	27	77	66	46		69	20	41	60	5	9	14	95	94	64	52		52	85	85	88	97	91	97	92	91
> 411 m																																
62 RHABDAMMINA ABYSSORUM	●	3			x	x	x								1	2	6						(2)									
63 RHABDAMMINA IRREGULARIS	o	1	1		x	1	3			1	o	x	x	x		5	1						(1)									
64 PELOSINELLA BICAUDATA		x					x			x		1	x			x	x				2											
65 BATHYSIPHON ARENACEUS		1		x	x	x	x		x	x							x															
66 HYPERAMMINA ELONGATA		2							x			x				3	1															
67 CRIBROSTOMOIDES SUBGLOBOSUS	o		x	1	1	18	1		1	8	o	5	5	1	2	12	19	x		x	6	3									x	
68 CYCLAMMINA ORBICULARIS								x	x				x			2	x	x				1										
69 AMMOBACULITES AGGLUTINANS	●	2		x	x	1			1	1		x	5	1		4	10	x														
> 662 m																																
70 AMMOLAGENA CLAVATA		1	1			1	26		x	11	o		3	1	1	17	3	26		2	5			x								
71 HORMOSINA GLOBULIFERA																																
72 AMMODISCUS SPP.			x																			2	(1)									
73 CRIBROSTOMOIDES CRASSIMARGO				?x		2	2	2	x	x	o	x	10	1		5	8	2				cf.1										
74 TEXTULARIA AGGLUTINANS																								2		x						
75 TEXTULARIA CF. PHILIPPINENSIS																								16								
76 KARRERIELLA NOVANGLIAE, WHITE SHELL																			x					3								
77 KARRERIELLA NOVANGLIAE, GRAY SHELL			x			1	3	x	1	x		x	3	x	1	1	x	1				x		1								
78 DOROTHIA PSEUDOTURRIS																								17								
79 TRILOCULINA TRIGONULA				2				x				x	3				x							x		x						
80 MILIOLINELLA SUBVALVULARIS																																
81 PYRGOELLA SPHAERA						1						1			2								(3)	3								
82 ANGULOGERINA CARINATA BRADYANA																							(4)	1								
83 PYRULINA FUSIFORMIS					x	5			x	1		2	2	1		1		x						x								
84 EHRENBERGINA PACIFICA			10			x														19				3								
85 LATICARININA PAUPERATA			x		x			6																								
86 CIBICIDES WUELLERSTORFI			x			1			1		o	4		x					17	1	6	6	(1)	6								
87 CIBICIDES BRADII			7			2			x			1			4		x			1	x	x	(3)	1								
> 1805 m																																
88 EGGERELLA BRADYI BRADYI, WHITE SHELL			x																x	x	x											
89 KARRERIELLA BRADYI, WHITE SHELL																				x												
90 MARTINOTTIELLA COMMUNIS, LARGE, TYPICAL							2		x			3	13	23	9		5	8														
91 MARTINOTTIELLA COMMUNIS, SHORT												3		3			1					1										
92 MARTINOTTIELLA NODULOSA		1	x			2				x		1	1			3	x				x											
93 QUINQUELOCULINA SP.				x	x	1								x	1		x	x				2										
94 UVIGERINA BRADYANA												3		3			1					1										
95 ANGULOGERINA ANGULOSA			12	x				5												28												
96 CIBICIDES CORPULENTUS																			9	6	1	25	(1)									
97 EPONIDES TENER STELLATUS			25			5			x	4		7		5				2		1	2											
> 2762 m																																
98 REOPHAX INSECTUS			x		x				x																							
99 NODOSINUM GAUSSICUM		1		1					1				5				1															
100 CIBICIDES SP. A										x				x	1	x		1	9													
101 CIBICIDES PSEUDOUNGERIANUS						x						1		x		x																
> 3137 m																																
102 ASTRORHIZA ANGULOSA		3											x																			
103 TOLYPAMMINA VAGANS									x							x																
104 RHIZAMMINA ALGAEFORMIS		3						1		6							x															
105 HYPERAMMINA CYLINDRICA										x	o		1																			
106 HORMOSINA NORMANI		4								x																						
107 HORMOSINA ROBUSTA						x						x	x			4	x															
108 EGGERELLA BRADYI BRADYI, GRAY SHELL			x			2	x	x	x	x		1		x		x	x															
109 EGGERELLA BRADYI NITENS		2			x	6	x		x	x		x		x		x	x															
110 KARRERIELLA BRADYI, GRAY SHELL			x		x					x				x		x	x															
111 TROCHAMMINA RUGOSA																	x															
112 TROCHAMMINA SP. AFF. T. INFLATA						2	x			x																						
113 PYRGO WIESNERI								x						3																		
114 INVOLVOHAUERINA GLOBULARIS				2	x				x																							
115 KERAMOSPHAERA MURRAYI										x		x				x	x															
116 MARGINULINA CF. PLANATA																																
117 MARGINULINA OBESA									1	1		2		x			1															

significance were discussed in detail by Mackintosh [1946] and Gordon [1967]. The position given on Figure 2 is evaluated from *Eltanin* surface temperature data and may give the actual position at the surface. However, Mackintosh's and Gordon's maps show that the convergence may, within certain limits, change its position with time.

DISTRIBUTION OF BENTHONIC FORAMINIFERA: GENERAL REMARKS

For recognition of a bathymetric zonation of foraminiferal species in a given area, Bandy and Arnal [1960] introduced the technique of plotting quantitative data of foraminiferal samples by arranging the stations according to their depth. By working progressively from deep to shallow stations it is possible to eliminate the effect of faunal displacement, and a bathymetric zonation based on the upper depth limits of the different species can be achieved. For antarctic Foraminifera the technique has been applied by Bandy and Echols [1964] in a re-evaluation of McKnight's results of 1962. This technique has been used in the present study also, and a summary of the most important species showing their upper depth limits is given in Table 2. Since a comprehensive table listing all species encountered (over 250 for this report) would be impractical, rare or apparently insignificant species, including those which have been found at single stations only, have been omitted, and therefore are not listed in Table 2. An examination of this table shows that a bathymetric zonation based upon the upper depth limits of foraminiferal species is possible in this area.

In addition to the bathymetric zonation, a differentiation of a number of faunal provinces is possible, especially in the shallower waters of the shelf and the upper and middle bathyal zone. Comparing populations from similar depths, we find that those around Tierra del Fuego are in striking contrast to those found in the seas around the antarctic continent. The former area is here called the Cape Horn province, the latter one the South Shetland province. Foraminiferal populations in the shallow waters of the Burdwood bank and the Falkland Islands, including the Falkland trough in its shallower central and western part, differ somewhat from those collected in the Cape Horn area. Therefore this area is here tentatively called the Falkland subprovince. As will be shown in detail below, this distinction is preliminary because the number of samples available from there is still insufficient with respect to areal and bathymetrical distribution. Further investigations will have to be made to show whether this distinction is justified and can be defined more precisely. Foraminiferal populations of the Falkland subprovince were not included in Table 2, but are listed in a separate table (Table 1).

In the deep-sea basin of the Drake Passage between the northern and the southern shelf areas a distinction of northern and southern areas based upon the occurrence of selected foraminiferal species does not seem possible.

However, regarding the ratio of arenaceous to calcareous Foraminifera, populations south of the Antarctic Convergence often contrast with those found north of this line, as will be explained below. Therefore, the 2 areas were separated in Table 2 and listed in connection with the adjoining shelf and slope areas.

BATHYMETRIC FORAMINIFERAL ZONATION

Cape Horn Province

Bathymetric zone H 1, shelf. Five samples were available from depths between 42 and 174 meters. Four of these were collected in water depths between 104 and 174 meters; they are similar to each other with regard to their foraminiferal assemblage. The fifth was collected in shallower water (42 meters) and shows a less diversified composition: a number of species, which otherwise are typical for the shelf and the upper bathyal zone, were not found here. Therefore an attempt has been made to distinguish a zone H 1a (inner shelf) from a zone H 1b (outer shelf).

Zone H 1a, inner shelf (see Plate 1, left side): Common species (arranged in the order of their frequency) are

Cibicides fletcheri
Cassidulina crassa rossensis
Cassidulina pulchella
Eponides isabelleanus
Sigmoilina obesa
Trochammina aff. *squamata*
Rotalia sp.
Discanomalina vermiculata
Cribrostomoides jeffreysi
Pullenia subcarinata
Quinquelocuina seminulum
Angulogerina occidentalis
Discorbis aff. *D. nitidus*
Rosalina vilardeboana

All of these species occur in one or several of the following deeper zones also. However, except for *Eponides isabelleanus*, this population is mainly characterized by the predominance of small forms, such as *Cibicides fletcheri*, *Cassidulina pulchella*, and a small form of *Cassidulina crassa*, *C. crassa rossensis*. *Cribrostomoides jeffreysi* and *Trochammina squamata* are also represented by very small individuals. Their occurrence is restricted in the Cape Horn province to this one sample from the inner shelf. In the South Shetland province, however, larger individuals of *Cribrostomoides jeffreysi* are common in depths between 73 and 1437 meters, and *Trochammina squamata* occurs at a few stations in the deep sea south of the Antarctic Convergence. In the Falkland subprovince *Cassidulina pulchella* has been found at a depth of 119 meters (Station 344), but is missing at deeper stations. For identification of this characteristic species the reader is referred to Heron-Allen and Earland [1932, p. 357].

It is interesting to note that Shishkevish [1964] also noted relatively meager populations in the shallower parts of the shelf compared with its edge and suggested a possible subdivision of his 'Zone 1.'

Zone H 1b, upper depth limit 104 meters: In a depth range between 104 and 174 meters the following species are the most common:

Discanomalina vermiculata
Cassidulina crassa
Cibicides fletcheri
Ehrenbergina pupa
Pullenia subcarinata
Quinqueloculina seminulum
Pyrgo williamsoni
Heronallenia kempii
Discorbis aff. *D. nitidus*
Patellina corrugata
Rhizammina indivisa

The characteristic form *Discanomalina vermiculata* is abundant down to depths of 641 meters and in low percentages was still found in 878 meters. *Heronallenia kempii* is an infrequent but characteristic form of this province an is restricted in our samples to depths between 115 and 641 meters. Another well-known form is *Hoeglundina elegans* (D'Orbigny). It is rather rare here (less than 1% of the total benthonic Foraminifera), but it occurs in high percentages in depths between 570 and 860 meters in the Falkland subprovince.

Elphidium crispum (Linnaeus) is common at most stations in zones H 1b and H 2, but no living individuals have been observed in samples treated with rose bengal. This species, however, is very frequent at Station 344 on the Burdwood Bank (119 meters), where it occurs together with the less frequent *Elphidium lessonii*.

Cassidulina pulchella is restricted, apart from its occurrence in zone H 1a, to Stations 370 (115 meters) and 344 (119 meters) and was not found at deeper stations. *Bucella frigida* is also limited to the same two stations. The occurrence of these two species at Station 356 (3678–4136 meters) has to be explained by displacement (see p. 279). *Cibicides lobatulus* has its upper depth limit in this zone, but otherwise is more common in zones H 2–H 5. *Cassidulina crassa* is found in high frequencies on the shelf and in the upper bathyal zone, as well as in the lower bathyal zone between 3000 and 4000 meters.

Bathymetric zone H 2, upper depth limit 205 meters (see Plate 1, right side). Two samples (V-17-51 and 369) were available from depths between 200 and 300 meters, and another one (970) was taken between 586 and 641 meters.

Upper depth limits of 2 very significant species occur within this depth range: *Mississippina concentrica* (Parker and Jones) and *Buliminella elegantissima* (D'Orbigny), the former representing up to 7% of the total benthonic population of Foraminifera.

Sample 369 furnished an especially rich fauna. The following species are common or otherwise of special interest:

**Quinqueloculina inca*
Pyrgo isabelleana
**Spirillina tuberculata*
Ehrenbergina pupa
Cibicides fletcheri
**Cibicides dispars*
**Cibicides tenuimargo*
Cibicides refulgens
Eponides isabelleanus
Mississippina concentrica
**Anomalina umbilicatula*
Discanomalina vermiculata
**Polystomellina patagonica*
Cassidulina crassa

Asterisk indicates occurrence in frequencies of less than 1%.

One of the most common arenaceous species, *Psammosphaera fusca*, has its upper depth limit in this zone also. However, it occurs in higher percentages in the lower bathyal and the abyssal zone only.

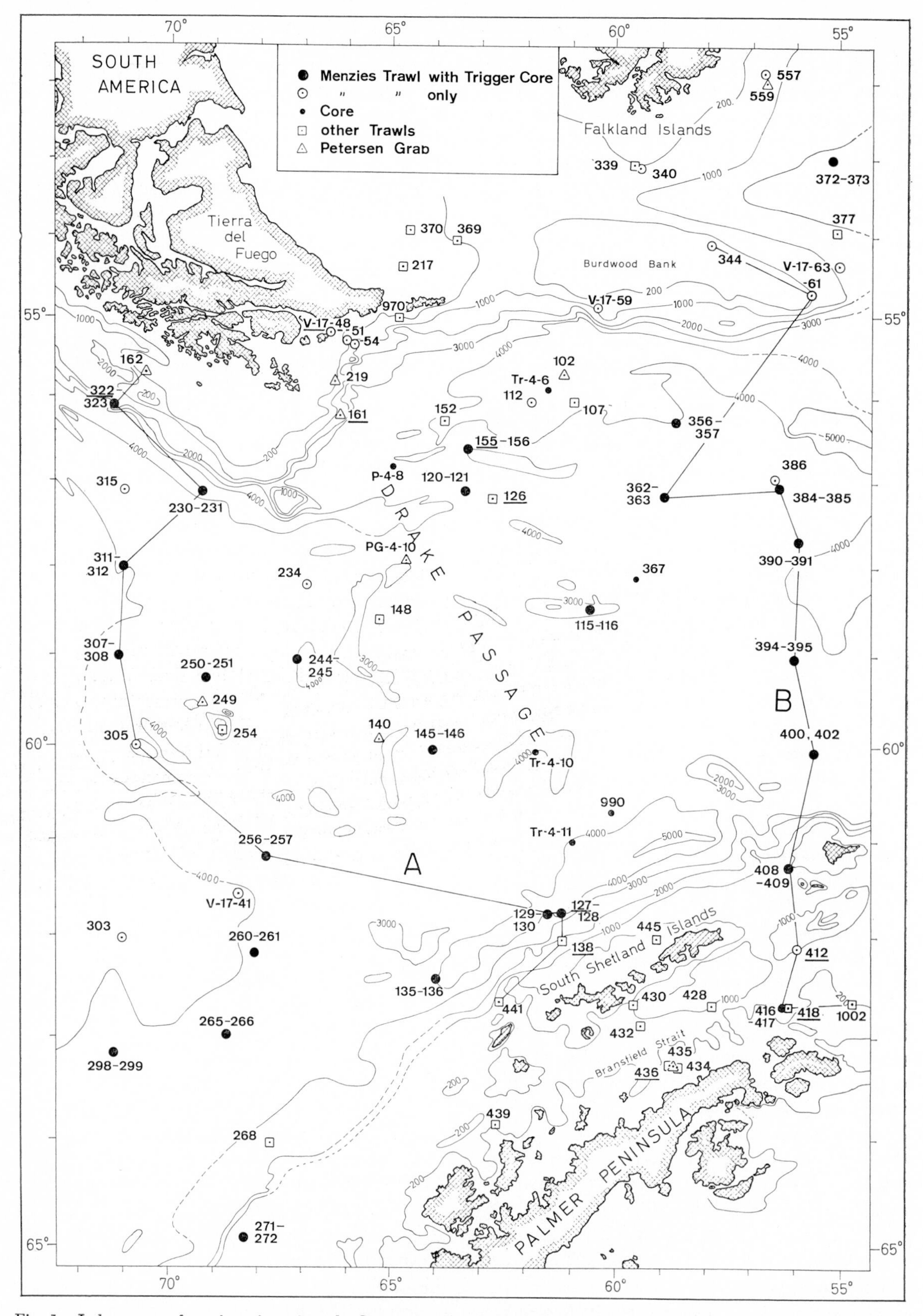

Fig. 1. Index map of stations investigated. Station numbers prefixed 'V' are *Vema* stations, all others *Eltanin*. Underlined are the numbers of those stations, from which typical foraminiferal assemblages are figured on plates 1–8. A and B indicate the location of the faunal profiles 8A and 9B.

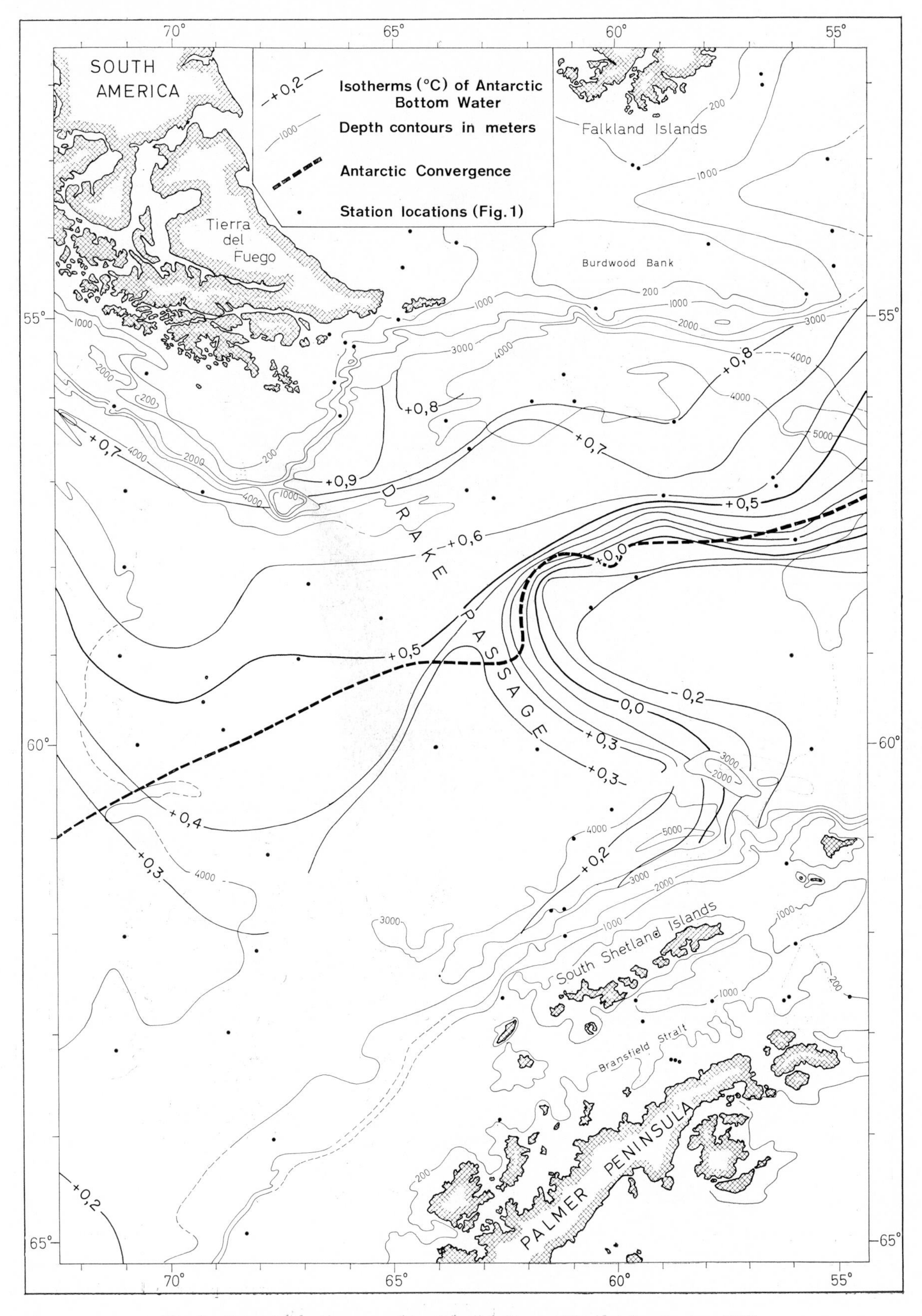

Fig. 2. Potential temperature of the Antarctic Bottom Water (after Gordon, 1967).

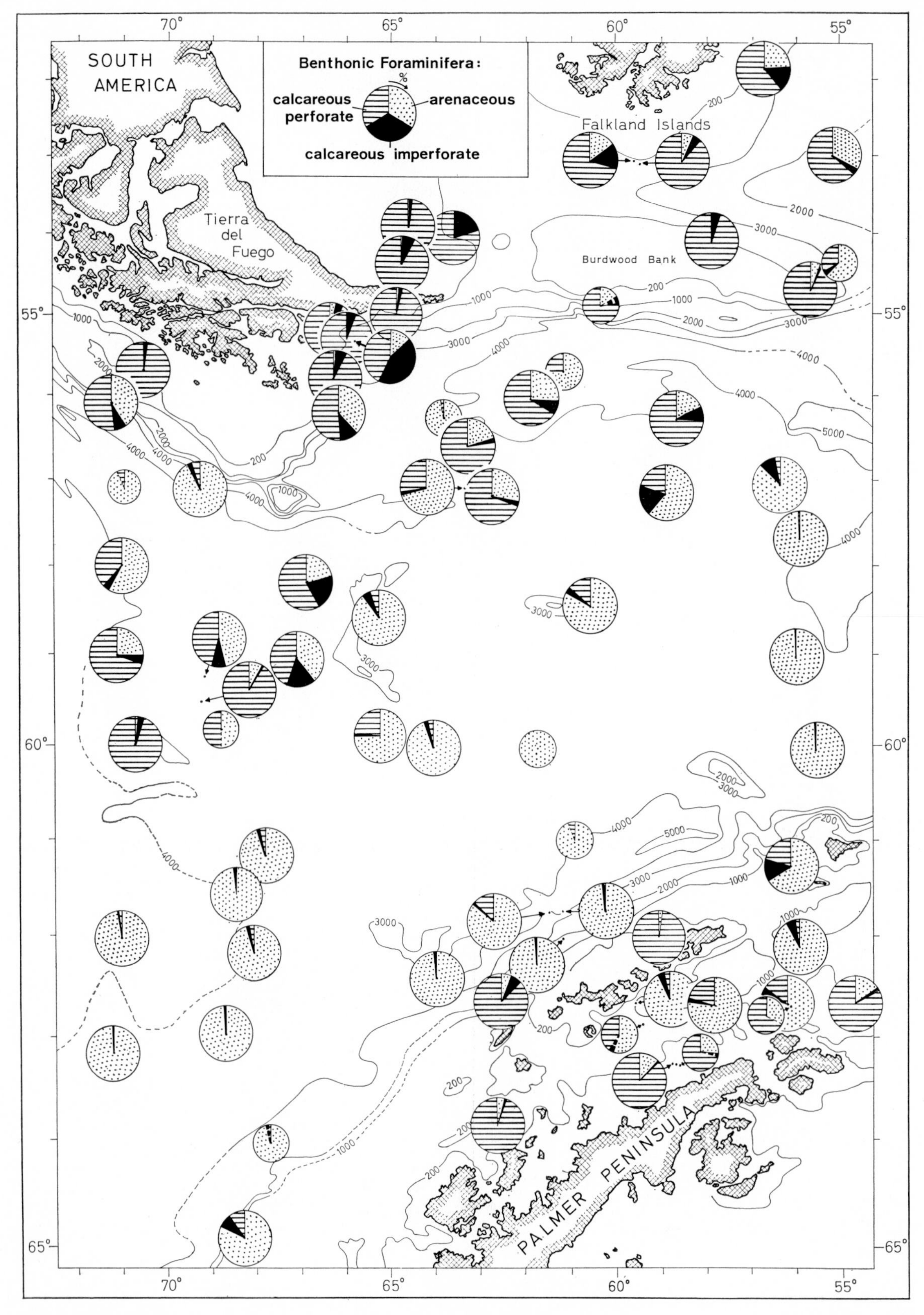

Fig. 3. Map showing percentages of arenaceous, calcareous imperforate, and calcareous perforate Foraminifera. Full circle represents total benthonic Foraminifera.

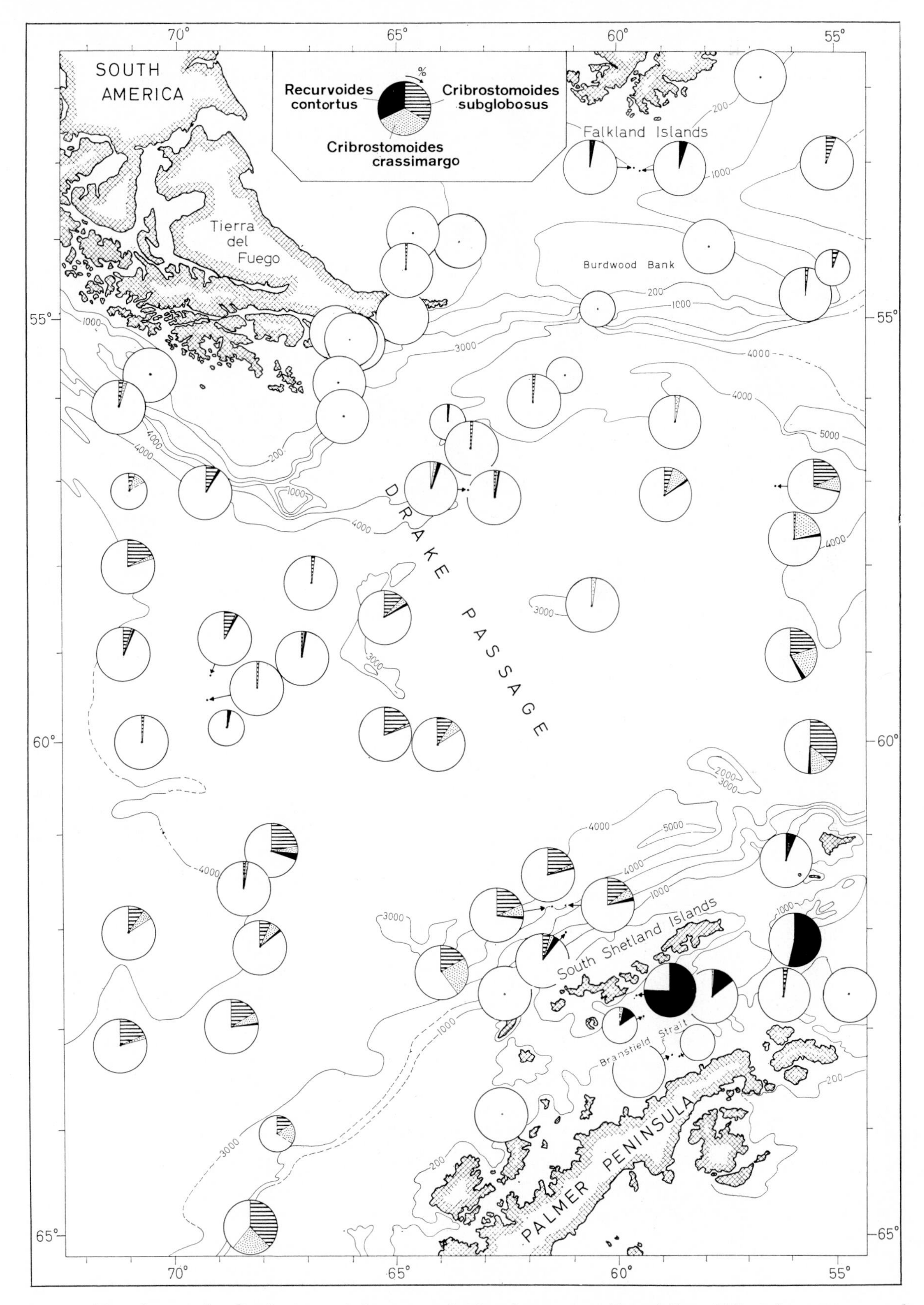

Fig. 4. Map showing the distribution and frequency of selected arenaceous Foraminifera. Full circle represents total benthonic Foraminifera.

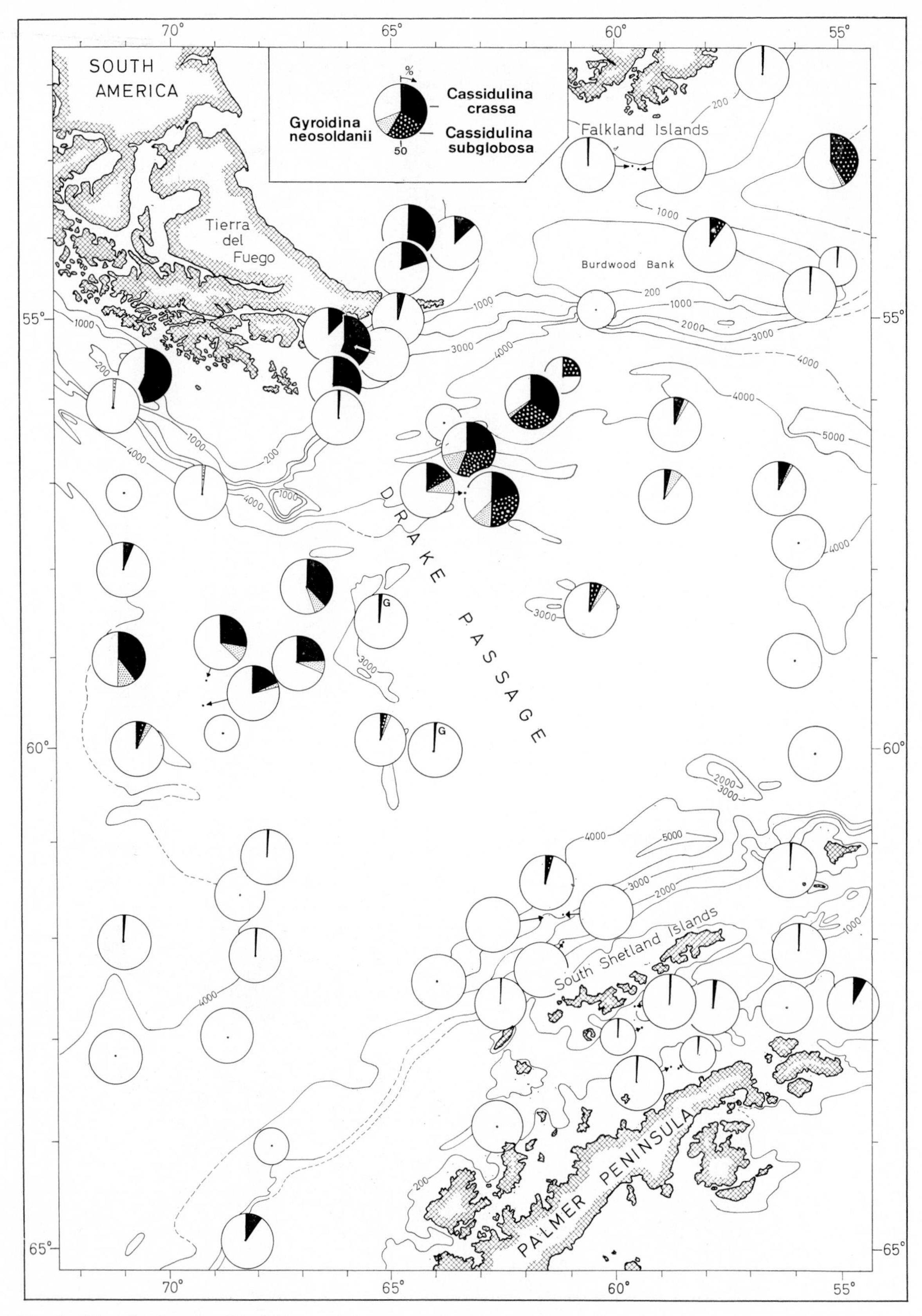

Fig. 5. Map showing the distribution and frequency of selected calcareous perforate Foraminifera. Full circle represents total benthonic Foraminifera. G indicates *Gyroidina neosoldanii* present, but with $<1\%$ frequency.

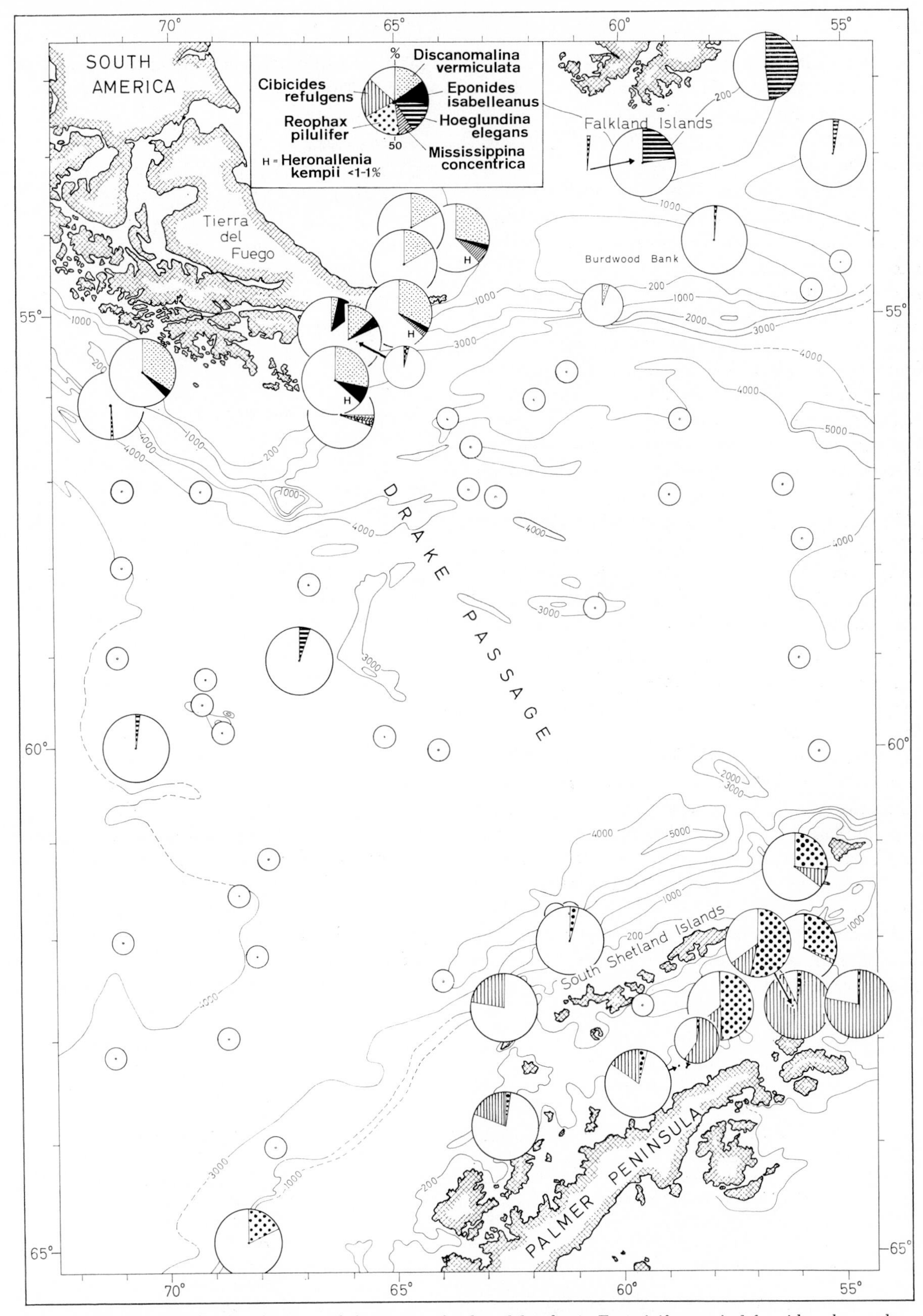

Fig. 6. Map showing the distribution and frequency of selected benthonic Foraminifera typical for either the northern or the southern areas. Full circle represents total benthonic Foraminifera.

A lower depth limit has been observed for *Rotalia* sp. (see Plate 1, Figure 6) at 293 meters. No living specimens of this species have been found in the stained samples, however, and its actual depth range remains uncertain.

Table 2 shows the possibility of a further subdivision of this bathymetric zone with an upper depth limit at 406 meters. Such a distinction is restricted to the South Shetland province and cannot be recognized yet in the Cape Horn province; sample 970 (586 to about 641 meters) does not show essential differences from the shallower stations of this zone. However, at Station 254 (512–622 meters), located on a submarine rise close to the Antarctic Convergence, several specimens of *Textularia* cf. *philippinensis, Rupertia stabilis,* and *Discanomalina vermiculata* were observed together with single specimens of *Psammosphaera parva, Cribrostomoides subglobosus, Recurvoides contortus, Guttulina problema,* and *Cassidulina delicata.* Unfortunately, only 14 specimens of benthonic Foraminifera were recovered here. Additional material would be necessary for supporting a possible further subdivision of zone H 2.

Bathymetric zone H 3, upper depth limit 878 meters. A major faunal change was observed between depths of 641 and 878 meters. *Eponides isabelleanus, Heronallenia kempii,* and *Pyrgo williamsoni* were not found below 641 meters; *Mississippina concentrica* and *Discanomalina vermiculata* show a lower depth limit at 878 meters. Furthermore, zone H 3 is characterized by a significant increase in the frequency of arenaceous Foraminifera at a number of stations (see Figure 3). Although only two samples were available from this zone (161 at 878 meters and V-17-54 at 1274–1362 meters), and although these samples contained quite different assemblages, this trend is clearly visible in both samples. With a greater number of samples at hand a further subdivision of this zone can be expected. The most common species at Station 161 (878 meters) were

Ammolagena clavata
Karreriella novangliae
Textularia mexicana
Textularia cf. *philippinensis*
Dorothia pseudoturris
Pyrgo murrhyna
Pyrgo depressa
Pyrgoella sphaera
Angulogerina carinata bradyana
Ehrenbergina trigona
Ehrenbergina hystrix glabra
Virgulina sp.
Miliolinella subrotunda
Rupertia stabilis
Mississippina concentrica
Cibicides wuellerstorfi
Cibicides bradii
Cibicides tenuimargo
Cibicides lobatulus

The arenaceous species of this list have their upper limit with this sample. The occurrence of *Textularia* and *Dorothia* is interesting here. *Textularia mexicana* has been found at a similar depth at Station 557, east of the Falkland Islands. This species, therefore, might be a good bathymetric indicator for these parts of the southern seas.

The wall of *Karreriella novangliae* is composed of very fine-grained calcareous particles and shows a white color. This is in contrast to the equally fine-grained, but gray tests of this species found in the deep-sea basin of the Drake Passage. The specimens of *Mississippina concentrica* are smaller than the ones from shallower water and show a thicker wall.

Plate 1

Station V-17-48 (55°10′S, 66°23′W, 42 meters), ×30
1. *Eponides isabelleanus*
2. *Cibicides fletcheri*
3. *Cassidulina crassa, rossensis*
4. *Cassidulina pulchella*
5. *Discorbis* aff. *D. nitidus*
6. *Rotalia* sp. A

Station 369 (54°02′S, 63°40′W, 247–293 meters), ×30
7. *Discanomalina vermiculata*
8. *Mississippina concentrica*
9. *Pyrgo bulloides*
10. *Quinqueloculina seminulum*
11. *Ehrenbergina pupa*
12. *Pullenia subcarinata*
13. *Heronallenia kempii*
14. *Cassidulina crassa*
15. *Cibicides fletcheri*
16–17. *Buliminella elegantissima*
18. *Eponides isabelleanus*

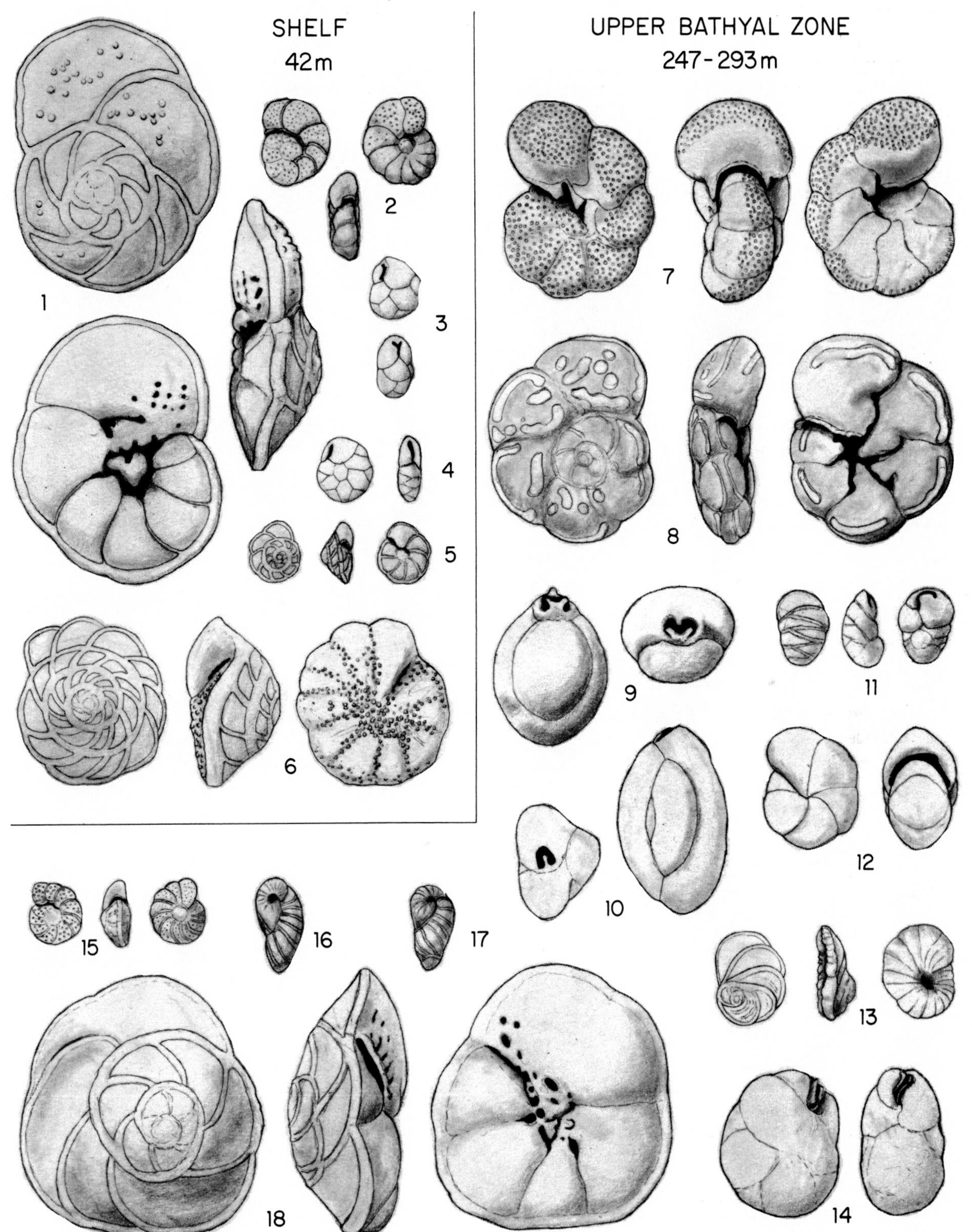
CAPE HORN PROVINCE
SHELF
42m
UPPER BATHYAL ZONE
247-293m
1
2
3
4
5
6
7
8
9
10
11
12
13
14
15
16
17
18

At Station V-17-54 (1274–1362 meters) several species, mostly arenaceous ones, occur for the first time with greater frequency:

Rhabdammina abyssorum—group
Rhabdammina irregularis
Bathysiphon spp.
Ammodiscus sp.
Recurvoides contortus
Cyclammina pusilla
Cruciloculina triangularis
Pullenia bulloides
Hoeglundina elegans

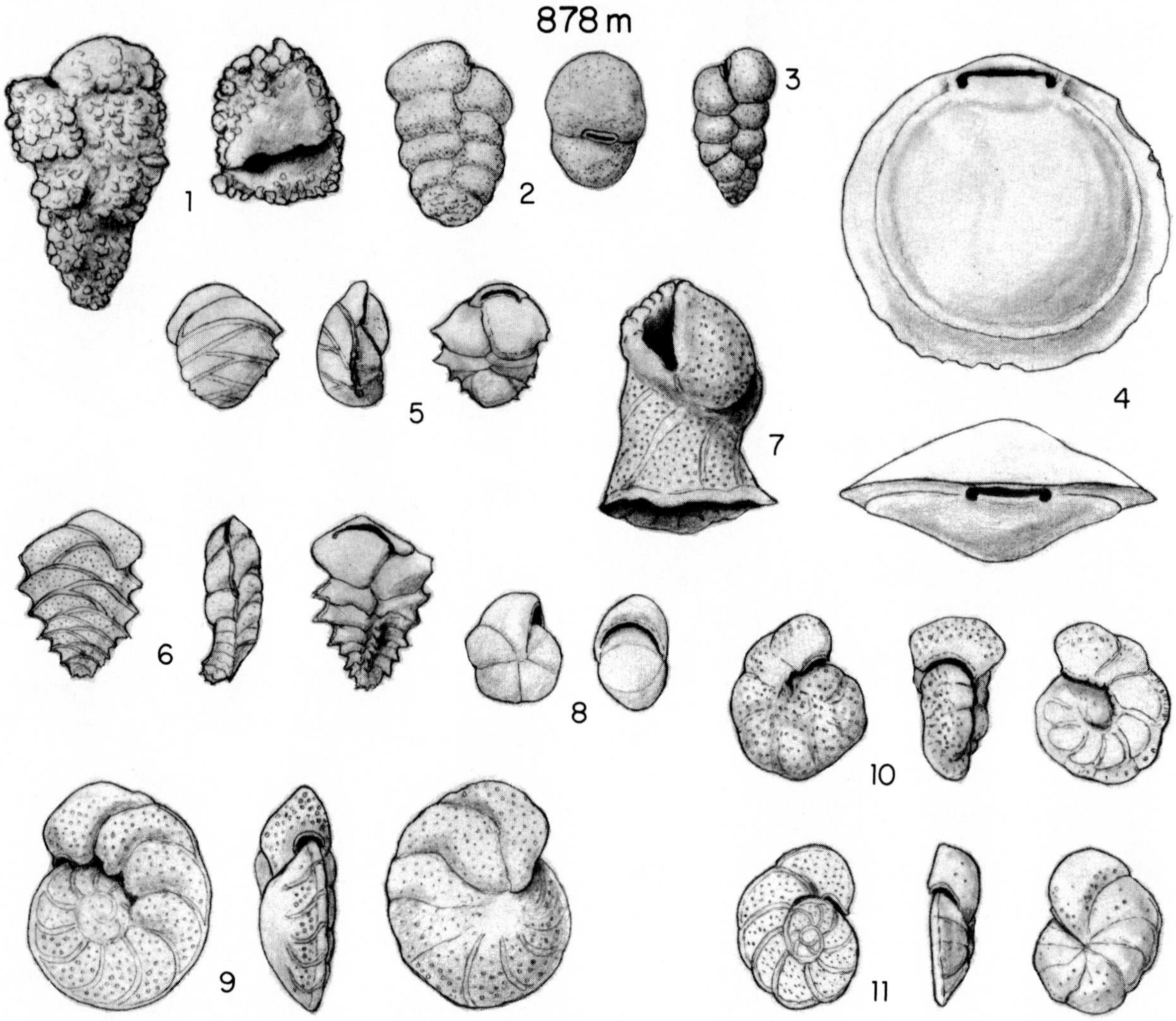

Plate 2

Station 161 (56°12′S, 66°12′W, 878 meters), ×24

1. *Dorothia pseudoturris*
2. *Textularia* cf. *philippinensis*
3. *Karreriella novangliae*
4. *Pyrgo depressa*
5. *Ehrenbergina hystrix glabra*
6. *Ehrenbergina trigona*
7. *Rupertia stabilis*
8. *Pullenia subcarinata*
9. *Cibicides* cf. *wuellerstorfi*
10. *Discanomalina vermiculata*
11. *Cibicides lobatulus*

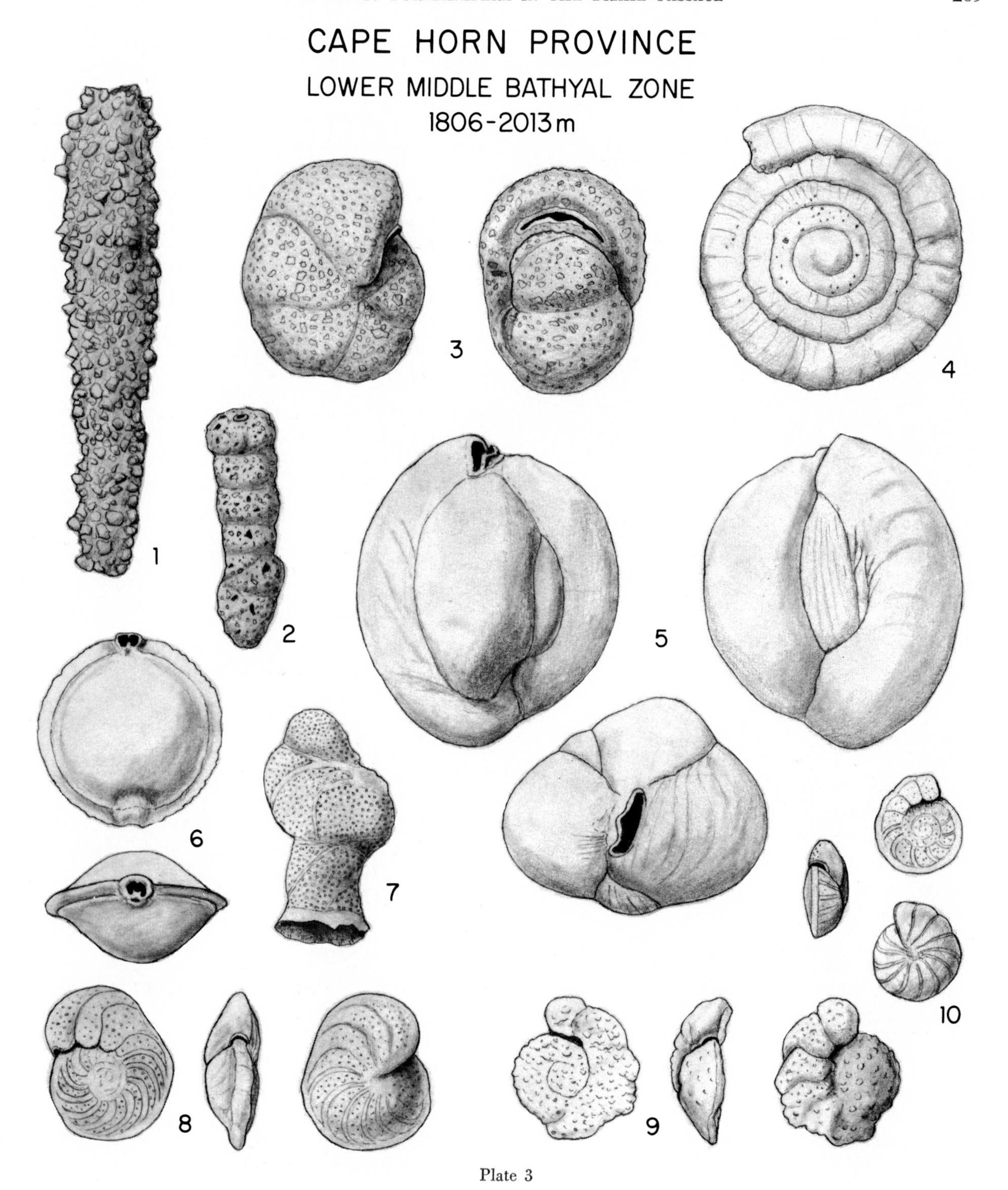

Plate 3

Station 322 (56°04′S, 71°11′W, 1806 to ca. 2013 meters), ×20

1. *Jaculella acuta*
2. *Martinottiella communis*
3. *Cribrostomoides subglobosus*
4. *Ammodiscus pacificus*
5. *Quinqueloculina* sp.
6. *Pyrgo murrhina*
7. *Rupertia stabilis*
8. *Cibicides wuellerstorfi*
9. *Cibicides corticatus*
10. *Cibicides corpulentus*

(In counting foraminiferal populations it is not always clear whether one deals with *Rhabdammina abyssorum* or with particles of broken *Astrorhiza*. 'Rhabdammina *abyssorum*—group' as used here eventually comprises more than one species, although the majority of the individuals should belong to *Rhabdammina abyssorum.*)

In the Falkland subprovince, however, *Hoeglundina elegans* is already abundant at shallower depths (Station 340, 567–578 meters; see also p. 279). At the same station some *Cyclammina pusilla* are found together with the more common *Cyclammina cancellata* (see Table 1).

Bathymetric zone H 4, upper depth limit 1806 meters. From this zone, too, only limited information was available: Stations 322 (1806–2013 meters) southwest of Tierra del Fuego, 372 (1953–1971 meters) and V-17-61 (1814–1919 meters) southeast of the Falklands. A general increase in the average size of benthonic Foraminifera is especially striking at Stations 322 and 372 (see also Plate 3). Species having their upper depth limit in this zone are

Eggerella bradyi bradyi
Karreriella bradyi
Martinottiella communis
Quinqueloculina sp.
Uvigerina bradyana
Angulogerina angulosa
Cibicides corpulentus
Eponides tener stellatus

For *Eggerella bradyi bradyi* and *Karreriella bradyi* the same observations were made as for *Karreriella novangliae* in the previously discussed zone H 3: The tests are composed of calcareous material, resulting in a whitish, smooth appearance; in the deep-sea basin of zone 5 the same species show tests of the typical gray color. *Cibicides corpulentus* is restricted to this zone as well as to those stations of zone 5 which are located north of the Antarctic Convergence.

The following species have their upper depth limit in this zone of the Cape Horn province, but occur in somewhat shallower water in the South Shetland province:

Psammophax consociata
Pelosinella bicaudata
Reophax curtus

Cribrostomoides subglobosus, one of the most frequent deep-water species in the area, becomes common for the first time at Station 322. Its areal distribution is shown in Figure 4, and the relation between frequency and depth in Figure 10.

Shelf and Continental Slope of the South Shetland Province

Bathymetric zone S 1a. Two stations in the shallow water of the Bransfield Strait (435 and 436, each at a depth of 73 meters) furnished a foraminiferal population very limited in number of species. Most important (arranged according to their frequency) are

Cibicides refulgens
Cibicides grossepunctatus
Reophax pilulifer
Rosalina vilardeboana
Cassidulina crassa rossensis

Cribrostomoides jeffreysi is significant at Station 436, and *Reophax nodulosus* at Station 435.

Bathymetric zone S 1b, upper depth limit 128 meters. The assemblage of this zone is essentially the same as in S 1a. *Cibicides refulgens* is by far the most common species. It is always accompanied by *Cibicides grossepunctatus.* In addition, there are a few less common species that were not found in the shallow-water samples mentioned above:

Pyrgo bulloides
Miliolinella subrotunda
Angulogerina carinata carinata
Pullenia subcarinata

The bathymetric zones S 1a and S 1b are thus characterized by a preponderance of the following:

Cibicides refulgens
Cibicides grossepunctatus
Rosalina vilardeboana
Reophax pilulifer

The first three were usually found attached to algae, *Cibicides* at the dorsal side, *Rosalina* at the ventral side. As a consequence of this, many specimens have somewhat distorted tests or some chambers of unusual shape.

Bathymetric zone S 2a, upper depth limit 225 meters. The four species mentioned above, together with the now more common *Cribrostomoides jeffreysi* are predominant. In addition, the following important species have their upper depth limit within this zone:

Psammosphaera fusca
Reophax dentaliniformis
Recurvoides contortus
Quinqueloculina vulgaris
Pyrgo pisum
Pyrgo murrhyna
Glandulina laevigata
Pseudobulimina chapmani
Thurammina papillata Brady

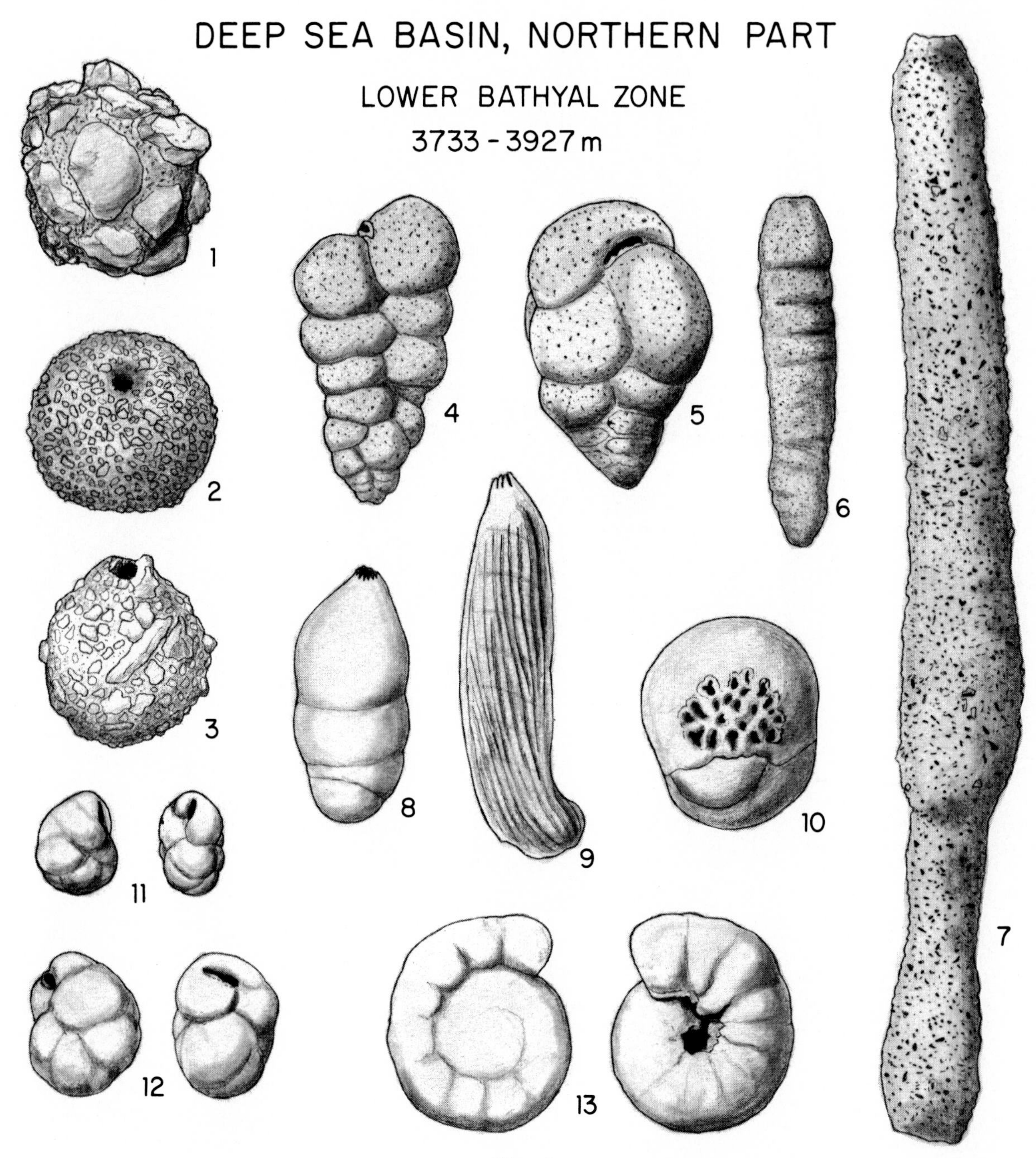

Plate 4

Station 155 (56°31′S, 63°15′W, 3927 meters), ×20
1. *Psammosphaera fuscac*
2. *Saccammina* sp.
3. *Saccammina sphaerica*

Station 126 (57°13′S, 62°48′W, 3733–3806 meters), ×20
4. *Karreriella novangliae*
5. *Eggerella bradyi bradyi*

Station 155 (56°31′S, 63°15′W, 3927 meters), ×20
6. *Martinottiella nodulosa*
7. *Bathysiphon filiformis*
8. *Marginulina obesa*
9. *Marginulina* cf. *planata*
10. *Involvohauerina globularis*
11. *Cassidulina crassa*
12. *Cassidulina subglobosa*
13. *Gyroidina neosoldanii*

DEEP SEA BASIN, SOUTHERN PART

ABYSSAL ZONE, 4758 m

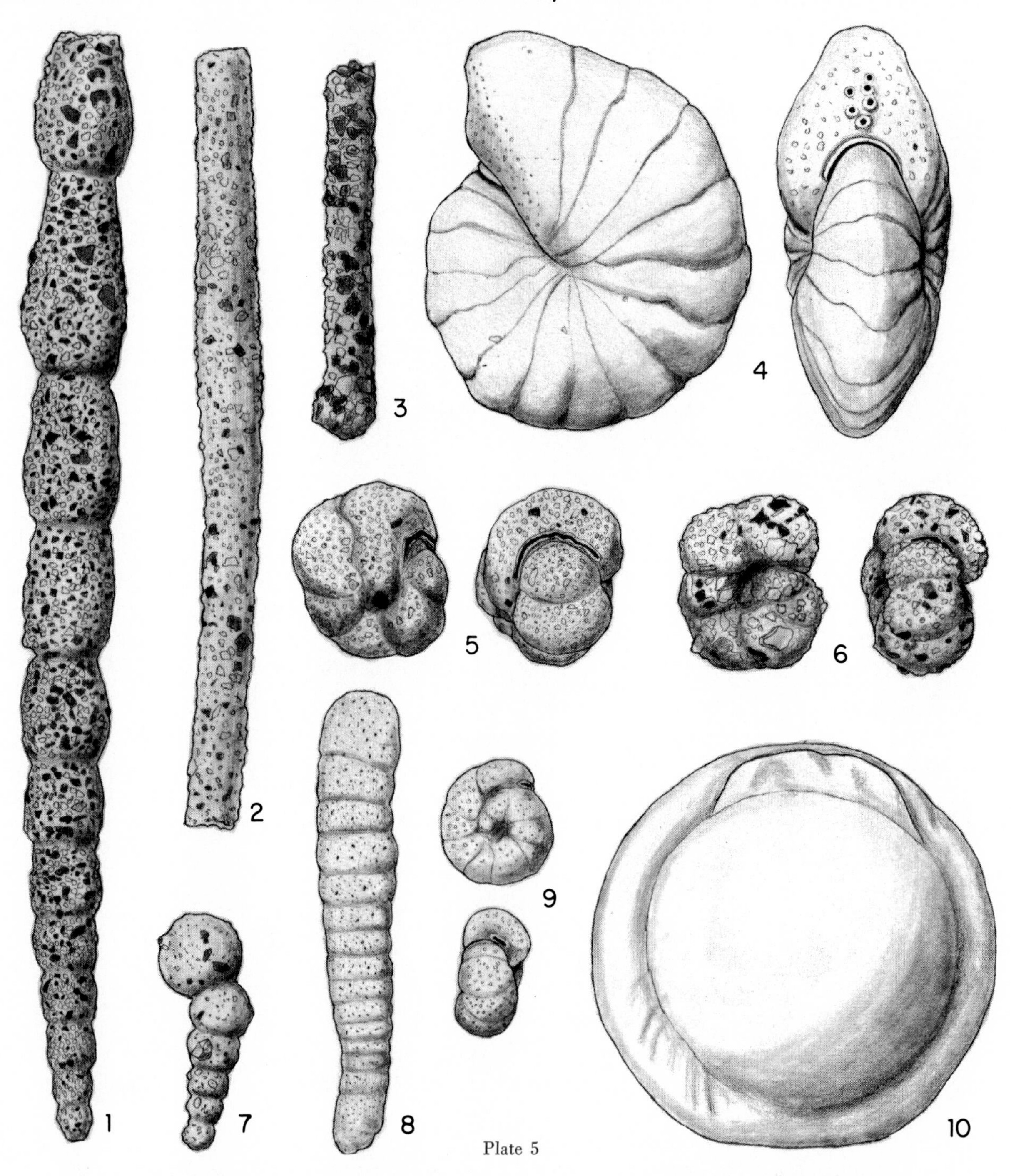

Plate 5

Station 127 (61°45′S, 61°14′W, 4758 meters), ×20

1. *Reophax nodulosus*
2. *Rhabdammina* sp.
3. *Hyperammina cylindrica*
4. *Cyclammina pusilla*
5. *Cribrostomoides subglobosus*
6. *Cribrostomoides crassimargo*
7. *Hormosina globulifera*
8. *Martinottiella nodulosa*
9. *Recurvoides contortus*
10. *Pyrgo wiesneri*

Hormosina normani, the largest and dominant species of this population, is not shown here; see Plate 10, Figures 8–9.

Reophax dentaliniformis and *Pseudobulimina chapmani* are restricted to the South Shetland province. Several specimens of *Thurammina papillata* were found at Station 408, but otherwise this species is found more frequently in deeper waters.

Bathymetric zone S 2b, upper depth limit 406 meters. Of the species mentioned in zones 1 and 2 *Rosalina vilardeboana* and *Pyrgo bulloides* were not observed in samples below about 400 meters. The most common species are usually

Reophax pilulifer
Cibicides refulgens
Cibicides grossepunctatus
Cribrostomoides jeffreysi
Cassidulina crassa
Psammosphaera fusca

Of these *Reophax pilulifer* has been observed with frequencies up to 54% in this zone. The following species are among those whose upper depth limit is within this zone:

Rhabdammina abyssorum
Bathysiphon arenaceus
Pelosinella bicaudata
Cribrostomoides subglobosus
Cyclammina orbicularis

The latter two species, however, are more typical in depths below 1400 meters.

As shown on Figure 7, right side, a progressively higher percentage of arenaceous Foraminifera can generally be observed with increasing depth. Populations from zones S 1a and S 1b (0–225 meters) show a high predominance of calcareous perforate forms; arenaceous Foraminifera were not observed in frequencies higher than 10% of the total of benthonic Foraminifera. In zones S 2a and S 2b (225–662 meters), however, even neighboring stations, collected at similar depths, may show either low or high percentages of arenaceous Foraminifera. This contrast is especially striking at Stations 416 and 418. *Reophax pilulifer, Cibicides refulgens,* and *Cibicides grossepunctatus* are the species that are mostly affected by these fluctuations, for which no explanation can yet be given. Local environmental parameters might influence their being most greatly affected.

Stations of the deeper bathymetric zones consistently show high percentages of arenaceous Foraminifera (see Figures 4 and 7).

Bathymetric zone S 3, upper depth limit 662 meters. In the Cape Horn province a major faunal change has been observed between zones H 2 and H 3. In the South Shetland province, however, changes below zone S 2b concern rather the frequency of a number of species than the generic and specific composition of the foraminiferal populations.

Reophax pilulifer is still prominent in zone S 3, but only at Station 428 (662–1120 meters) is it most common. The two species of *Cibicides* mentioned decrease in number of individuals (probably because of the decrease in the amount of algae), while the arenaceous *Recurvoides contortus* comprises a high percentage of the samples. *Reophax nodulosus* is another ubiquitous and easily recognizable species, although its frequency never exceeds 5%. Upper depth limits of the following species were observed in this zone:

Hormosina globulifera
Cribrostomoides crassimargo
Miliolinella subvalvularis
Laticarinina pauperata

The occurrence of *Cribrostomoides crassimargo,* however, is atypical here. It is normally a deep-water species.

A bathymetric zone corresponding to H 4 (Cape Horn province) cannot be distinguished in the South Shetland province because of the lack of samples within a range of 1437–2763 meters.

Deep-Sea Basin of the Drake Passage
(Lower Bathyal and Abyssal Zone)

The great majority of the deep-sea species occurs throughout the deep-sea basin of the area studied. However, based upon the ratio between frequencies of arenaceous and of calcareous benthonic Foraminifera, a southern area can be distinguished from a northern one (Figure 3). These are separated along a line that coincides roughly with the track of the Antarctic Convergence (Figure 2). The significance of this boundary will be discussed below (p. 290).

Bathymetric zone 5, upper depth limit 2763 meters. At stations with a high predominance of arenaceous Foraminifera the following species become a major constituent of the population:

Cribrostomoides subglobosus
Cribrostomoides crassimargo
Cyclammina pusilla

Station 305 (2780–2826 meters) is located on a submarine rise in the western Drake Passage and shows an unusually high percentage of calcareous perforate Foraminifera (95%). *Rupertia stabilis* Wallich, *Eponides tener stellatus* Silvestri, *Cibicides corpulentus* Phleger and Parker, *Cibicides wuellerstorfi, Cibi-*

cides bradii (Tolmachoff), [not *C. bradyi* (Trauth)], and *Cassidulina crassa* D'Orbigny are the predominant forms.

Upper depth limits of

Reophax insectus
Nodosinum gaussicum
Cibicides sp.

are found in this zone.

Bathymetric zone 6, upper depth limit 3138 meters. The most common and significant species of this zone are

Rhabdammina abyssorum
Rhabdammina irregularis
Bathysiphon filiformis
Hyperammina cylindrica
Hyperammina elongata
Ammolagena clavata
Psammophax consociata
Reophax nodulosus
Hormosina globulifera
Hormosina robusta
Hormosina normani
Cribrostomoides subglobosus
Cribrostomoides crassimargo
Ammobaculites agglutinans
Cyclammina pusilla
Cyclammina orbicularis
Recurvoides contortus

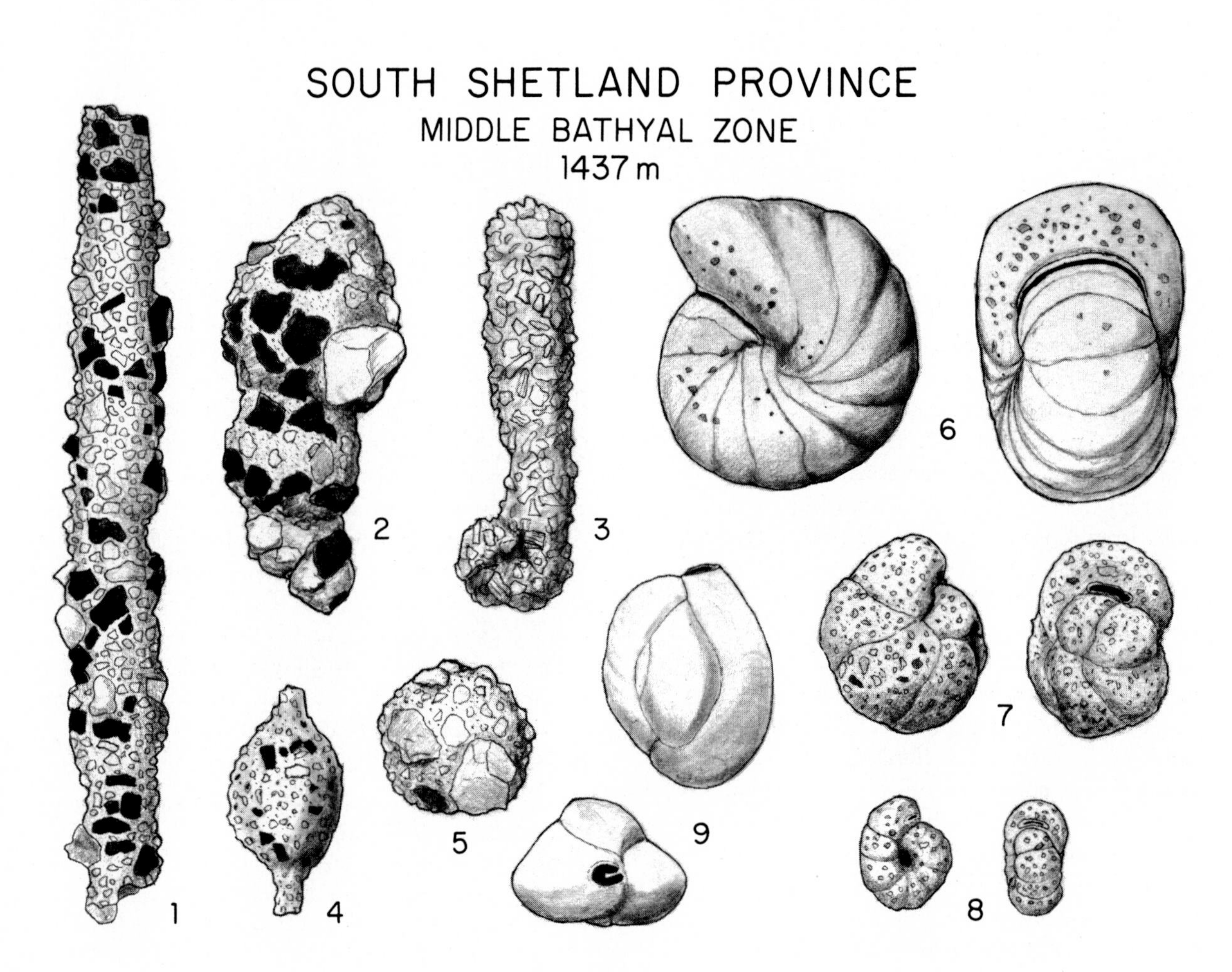

Plate 6

Station 138 (62°02′S, 61°08′W, 1437 meters), ×20

1. *Rhabdammina* sp.
2. *Reophax pilulifer*
3. *Ammobaculites agglutinans*
4. *Pelosinella bicaudata*
5. *Psammosphaera fusca*
6. *Cyclammina orbicularis*
7. *Cribrostomoides subglobosus*
8. *Recurvoides contortus*
9. *Quinqueloculina vulgaris*

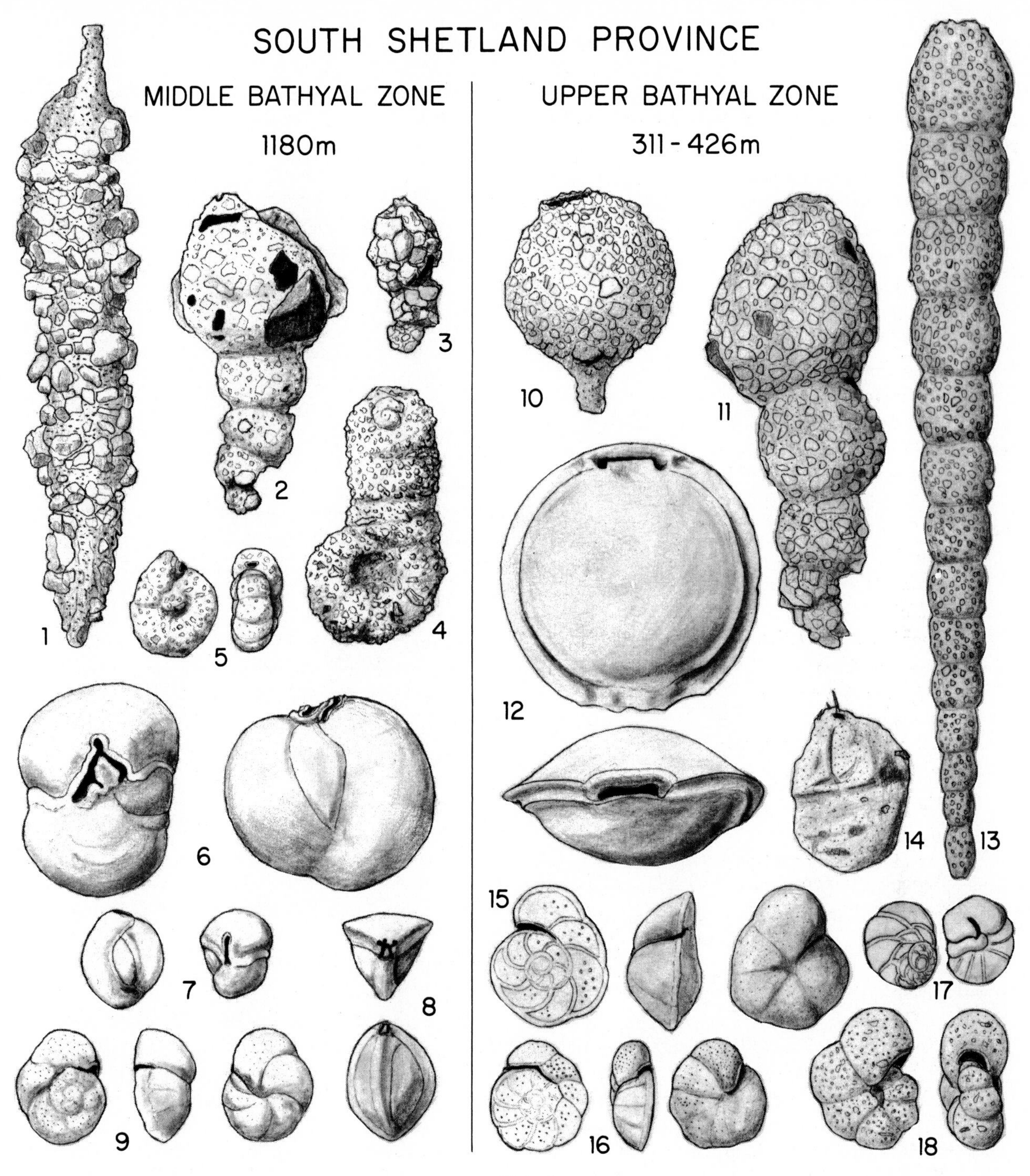

Plate 7

Station 412 (62°06′S, 56°00′W, 1180 meters), ×20

1. *Rhizammina indivisa*
2–3. *Reophax pilulifer*
4. *Ammobaculites agglutinans*
5. *Recurvoides contortus*
6. *Miliolinella* cf. *subvalvularis*
7. *Triloculina* sp.
8. *Cruciloculina triangularis*
9. *Cibicides refulgens*

Station 418 (62°40′S, 56°09′W, 311–426 meters), ×20

10. *Pelosina* sp. aff. *P. rotundata*
11. *Reophax pilulifer*
12. *Pyrgo depressa*
13. *Reophax nodulosus*
14. *Crithionina* sp.
15. *Cibicides refulgens*
16. *Cibicides grossepunctatus*
17. *Pseudobulimina chapmani*
18. *Cribrostomoides jeffreysi*

Eggerella bradyi bradyi }
Eggerella bradyi nitens } showing typically
Karreriella bradyi } gray tests
Karreriella novangliae }
Martinottiella nodulosa
Martinottiella communis
Trochammina rugosa
Pyrgo murrhina
Cassidulina crassa
Cassidulina subglobosa
Gyroidina neosoldanii

The calcareous forms in this list are common or abundant at many stations north of the Antarctic Convergence (see Figure 5), but are rather rare south of this line, as will be discussed below.

Further species observed at a number of stations are

Pyrulina fusiformis
Cibicides wuellerstorfi
Cibicides bradii
Eponides tener stellatus

Confined to zone 6 are

Tolypammina vagans
Pyrgo wiesneri
Involvohauerina globularis
Keramosphaera murrayi
Marginulina cf. *planata*
Marginulina obesa

Stations of the shallowest parts of the abyssal zone, between 4000 and 4200 meters, do not show an essential difference in the composition of the foraminiferal population compared with those from the lower bathyal zone. Unfortunately only one sample was recovered from a depth greater than 4200 meters: from Station 127 at 4758 meters. The difference from the nearby Station 129 (3678–3816 meters) is striking: *Hormosina normani* Brady, with many specimens over 1 cm in length, forms over 50% of the total foraminiferal population. *Cribrostomoides subglobosus, Cribrostomoides crassimargo* (both with varieties of possibly subspecific rank), *Cyclammina pusilla, Cyclammina orbicularis, Reophax nodulosus,* and other arenaceous species are of importance, too. Large miliolids, such as *Pyrgo wiesneri* (with diameters up to 3.5 mm), *Pyrgo murrhyna, Miliolinella subvalvularis, Miliolinella cryptella,* are typical for this population. Most of these forms show an excellent state of preservation, which is not expected for calcareous Foraminifera at this depth. Traces of dissolution can be observed in some of the calcareous perforate species, however, especially in specimens of *Cibicides grossepunctatus.*

Depth Ranges of Selected Foraminiferal Species

As shown in the preceding section, many of the benthonic species have a distinct depth range, although most of them can be found in more than one of the bathymetric zones distinguished here. A compilation of depth ranges of the most common species of the area is given in Figure 10. For practical reasons, the bathymetric distribution of each species has been separated for areas north and south of the convergence. Many species show their greatest frequency in a restricted depth range, but are found less frequently, or very rarely, in greater or shallower depths. Examples of the first case (occurrence at greater depth) can be explained by displacement from a shallow-water environment. More interesting, however, are extensions into shallower water; examples are *Cyclammina pusilla* and *Cribrostomoides subglobosus,* both of which are distributed mainly at depths greater than 1200–1400 meters (see Table 2), but are occasionally found as single specimens up to 110 meters (Station 217) in case of *Cyclammina pusilla* and 73 meters (Station 436) in case of *Cribrostomoides subglobosus.*

Many species are entirely restricted either to shallow or to deep water. A species typical for medium deep water (middle bathyal to uppermost part of lower bathyal zone) is *Rupertia stabilis,* found most frequently between 878 and 2827 meters with only single occurrences up to 225 meters. Its abundance at Station 305, on the flank of a submarine rise, is typical (see cross section, Figure 8B).

Some representatives of the genus *Cibicides* show specific bathymetric distributions. *Cibicides lobatulus,* although not rare in waters shallower than 200 meters, is found primarily at depths similar to those of *Rupertia stabilis. Cibicides wuellerstorfi* is found at all depths between 878 and 4099 meters in the Drake Passage area. It is much more common in the Falkland subprovince, however. At a depth of 512 meters the species was most abundant, making up 76% of the total benthonic Foraminifera at Station 339. This is somewhat contradictory to the results of Bandy and Echols [1964], who found by plotting cumulative percentage curves that in the Gulf of Mexico and in the Gulf of California more than 70% of this species occurred at depths greater than 3000 meters. In the Mediterranean and off California the species has an upper depth limit between 1000 and 1100 meters

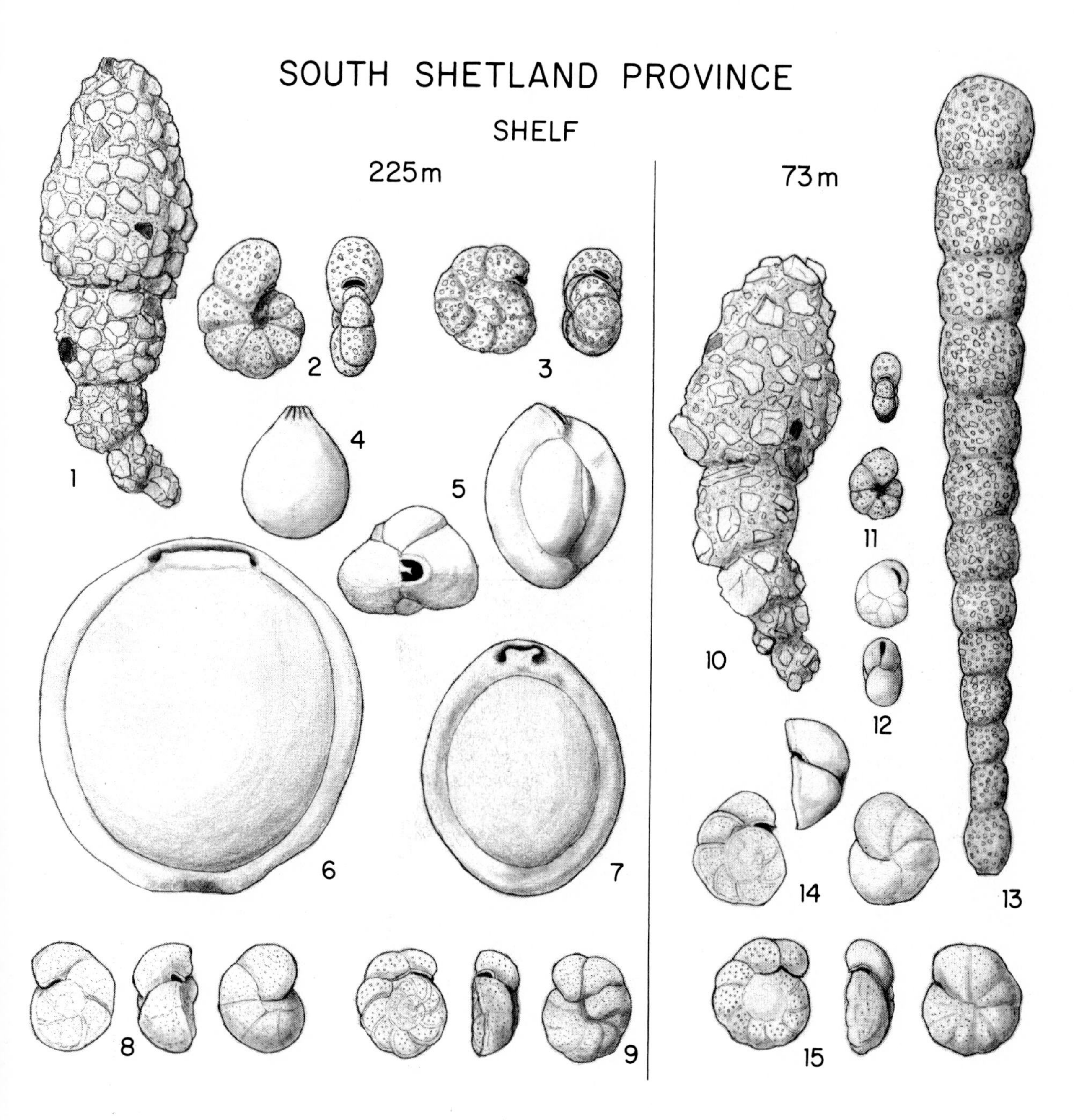

Plate 8

Station 408 (61°16′S, 56°11′W, 225 meters), ×20

1. *Reophax pilulifer*
2. *Cribrostomoides jeffreysi*
3. *Recurvoides contortus*
4. *Glandulina laevigata*
5. *Quinqueloculina* cf. *vulgaris*
6. *Pyrgo depressa*
7. *Pyrgo bulloides*
8. *Cibicides refulgens*
9. *Cibicides grossepunctatus*

Station 436 (63°13′S, 58°47′W, 73 meters), ×20

10. *Reophax pilulifer*
11. *Cribrostomoides jeffreysi*
12. *Cassidulina crassa*
13. *Reophax nodulosus*
14. *Cibicides refulgens*
15. *Cibicides grossepunctatus*

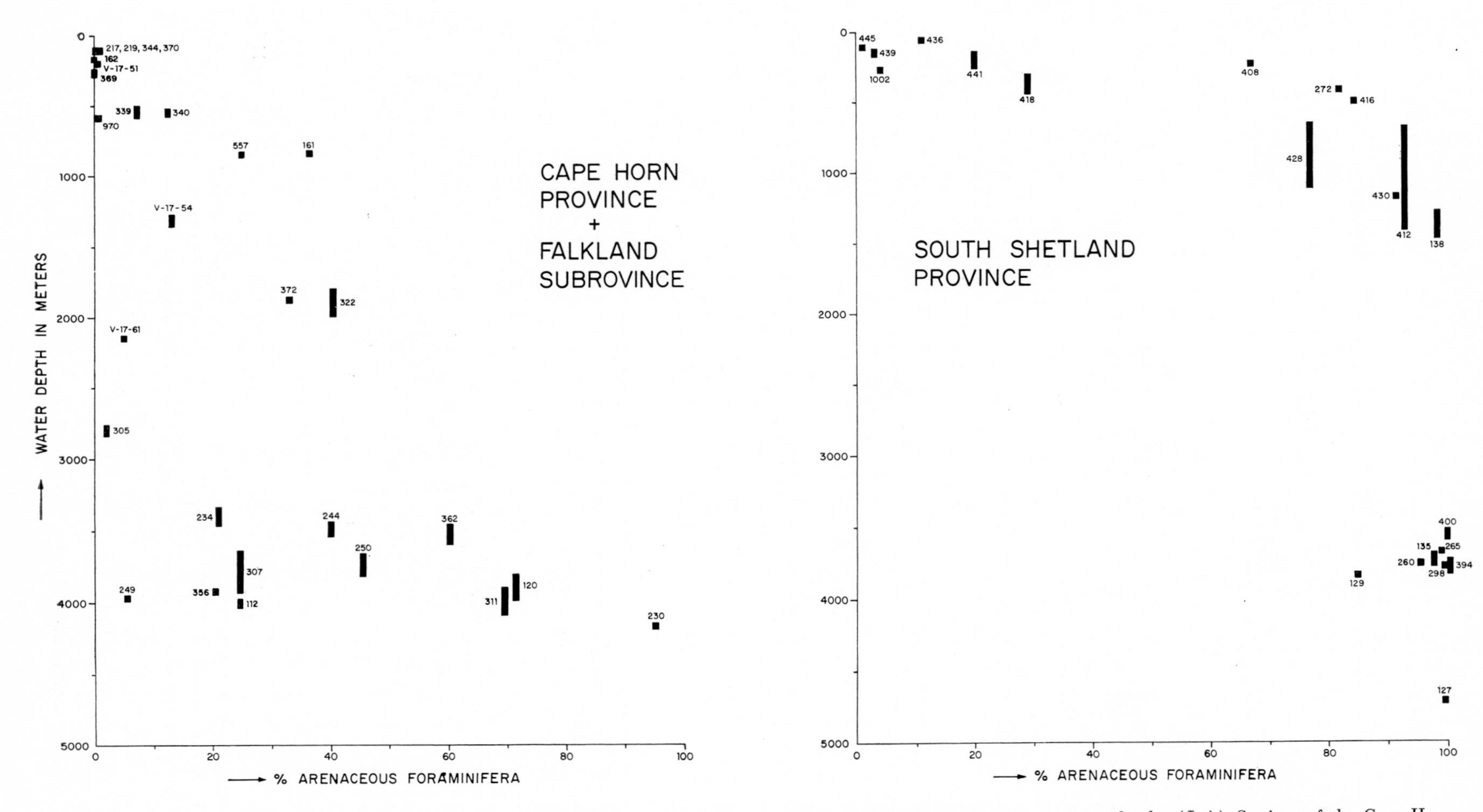

Fig. 7. Diagrams showing the percentage of arenaceous Foraminifera among total benthonic Foraminifera in relation to water depth. (*Left*) Stations of the Cape Horn province and the Falkland subprovince. (*Right*) Stations of the South Shetland province. Stations at or near the Antarctic Convergence are not included here.

[Bandy and Chierici, 1966]. At most stations it is accompanied by *Cibicides bradii*, a species which resembles *Cibicides wuellerstorfi* but has a more convex umbilical side and is usually less common. Both species have been placed in the genus *Planulina* by most authors. However, they show a distinct trochospiral coiling and therefore have to be placed in the genus *Cibicides* rather than in the nearly planispirally coiled *Planulina*. *Cibicides bradii* is somewhat more common at stations of greater depth.

Cibicides fletcheri is a typical form of the shelf and the upper bathyal zone of the South Shetland province, where it was found abundantly and with typical forms down to Station 970 (586–641 meters). *Cibicides corpulentus*, on the other hand, is restricted to deeper waters. It has a first occurrence at Station V-17-54 (1274–1362 meters), but is more typical for depths between 1800 and 2800 meters approximately. Confined to depths below 3000 meters, but mostly rare, is *Cibicides* cf. *pseudoungerianus* (Cushman).

Hoeglundina elegans shows maximum percentage values southeast of the Falklands between 550 and 860 meters but is not rare at considerably greater and lesser depths.

Pullenia subcarinata and *Pullenia bulloides* are two similar species of which the first one is most common in waters not less than 400 meters' depth, the second (more spherical in form) in deeper water. Depth ranges of the two species overlap, however, as may be seen in Table 2.

The name *Gyroidina neosoldanii* is used here for the large forms with an open umbilicus showing a diameter greater than 1 mm. This species is restricted to depths greater than 2782 meters. Considerably smaller forms were found, although rather infrequently, at shallower depths. These might be identified as *Gyroidina soldanii* (D'Orbigny).

All these species are isobathyal within the area covered by this report. A typically heterobathyal species is *Reophax nodulosus*. It is restricted to depths greater than 3294 meters north of the Antarctic Convergence, but it occurs at all depths between 73 and 4758 meters south of this line. Brady [1884] recorded this species from temperate areas at depths greater than 1900 meters, but he observed shallow- as well as deep-water occurrences in arctic and antarctic cold waters. It is therefore possible that the habitat of *Reophax nodulosus* is primarily defined by relatively cold water, a condition encountered in temperate and tropical areas only in deep water, but in arctic and antarctic regions throughout the water column.

A third group of benthonic Foraminifera occurs at stations of all bathymetric zones. *Cassidulina crassa* with its variety *porrecta* Heron-Allen and Earland is probably the most prominent representative of this group. It has been found at all depth zones, abundant at many stations of the shelf, the upper and lower bathyal zones north of the convergence, but is considerably less common or missing at stations of the middle bathyal zone. No explanation is yet available for this kind of distribution. The relative scarcity of samples available from the middle bathyal zone must be considered, however.

From the present foraminiferal records it is difficult to locate the position of a possible dissolution boundary for calcium carbonate in the area of the Drake Passage. At all stations where over 100 individuals of benthonic Foraminifera were recovered, at least a few calcareous imperforate forms were found. This is true also for the deepest sample (4758 meters), where many well-preserved large *Pyrgo* occurred. At a number of deep-water stations south of the convergence, however, benthonic calcareous perforate forms are completely missing (see Figures 1 and 3). These stations are located either in the southwestern (south of 63°S) or the southeastern part of the Drake Passage (south of 59°S) and in a depth greater than 3500 meters. In samples from the deep sea north of the South Shetland Islands some calcareous perforate forms were found, but they are probably displaced, as noted above. It seems, therefore, that in the southern and southeastern part of the Drake Passage living conditions for calcareous perforate forms are unfavorable below 3500 meters of depth, as noted on p. 00. At Station 276, located outside of our map area (about 67°00′S, 75°00′W), even at a depth of 3022–3043 meters no calcareous perforate forms were found.

Displaced Foraminifera

A deep-water occurrence of benthonic Foraminifera known to be restricted otherwise to shallow water has been observed at Station 356 (3678–4136 meters) in the eastern Drake Passage. Among typical deep-water forms, such as *Gyroidina neosoldanii*, *Cribrostomoides crassimargo*, *Cyclammina pusilla*, or *Karreriella novangliae*, the following species were encountered:

Quinqueloculina seminulum
Ehrenbergina pupa
Elphidium crispum
Elphidium lessonii
Bucella frigida
Cassidulina pulchella
Cibicides fletcheri

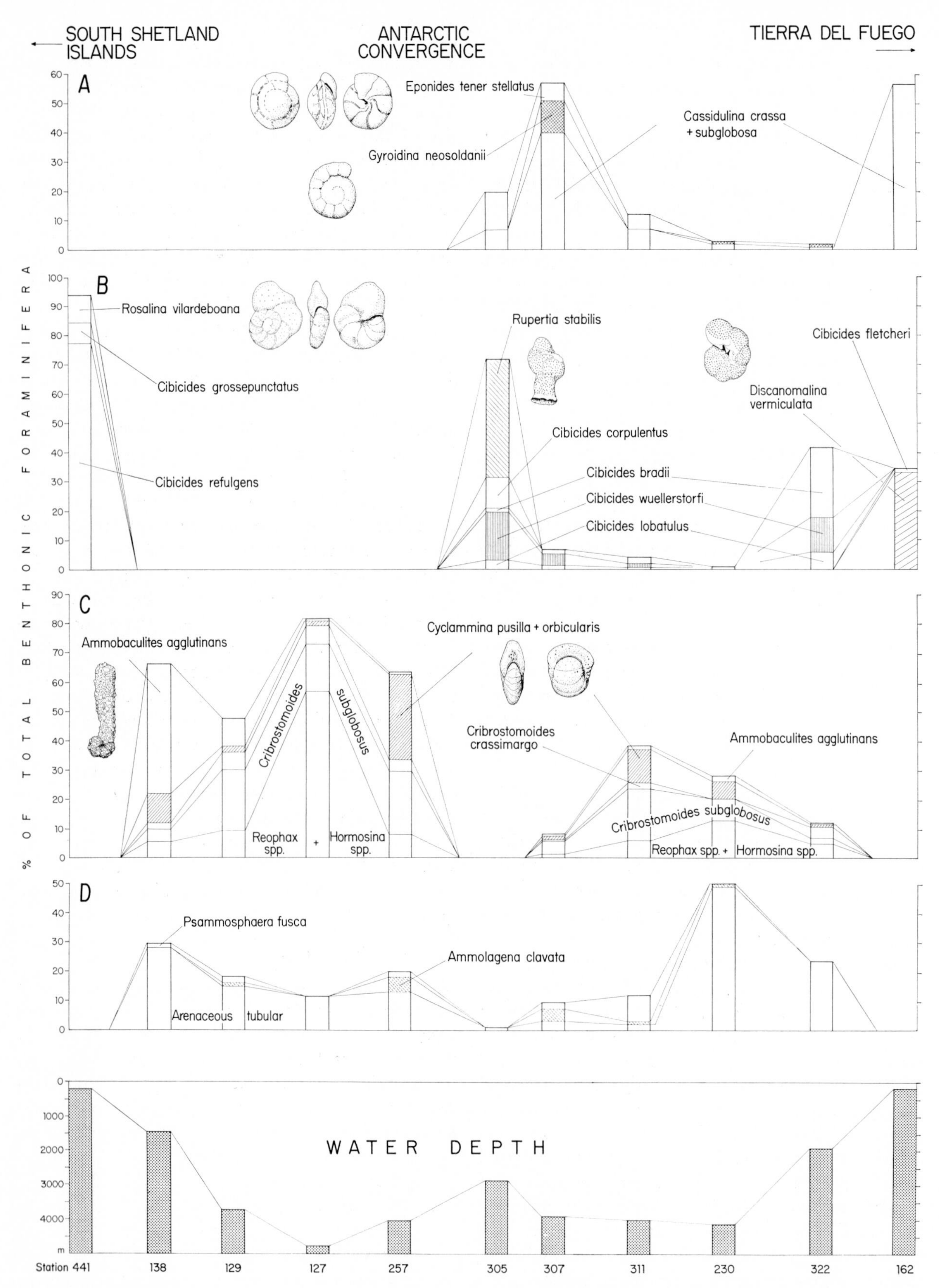

Fig. 8. Faunal profile across the western Drake Passage, showing the frequency of species or selected faunal groups in relation to depth. For locations see Figure 1.

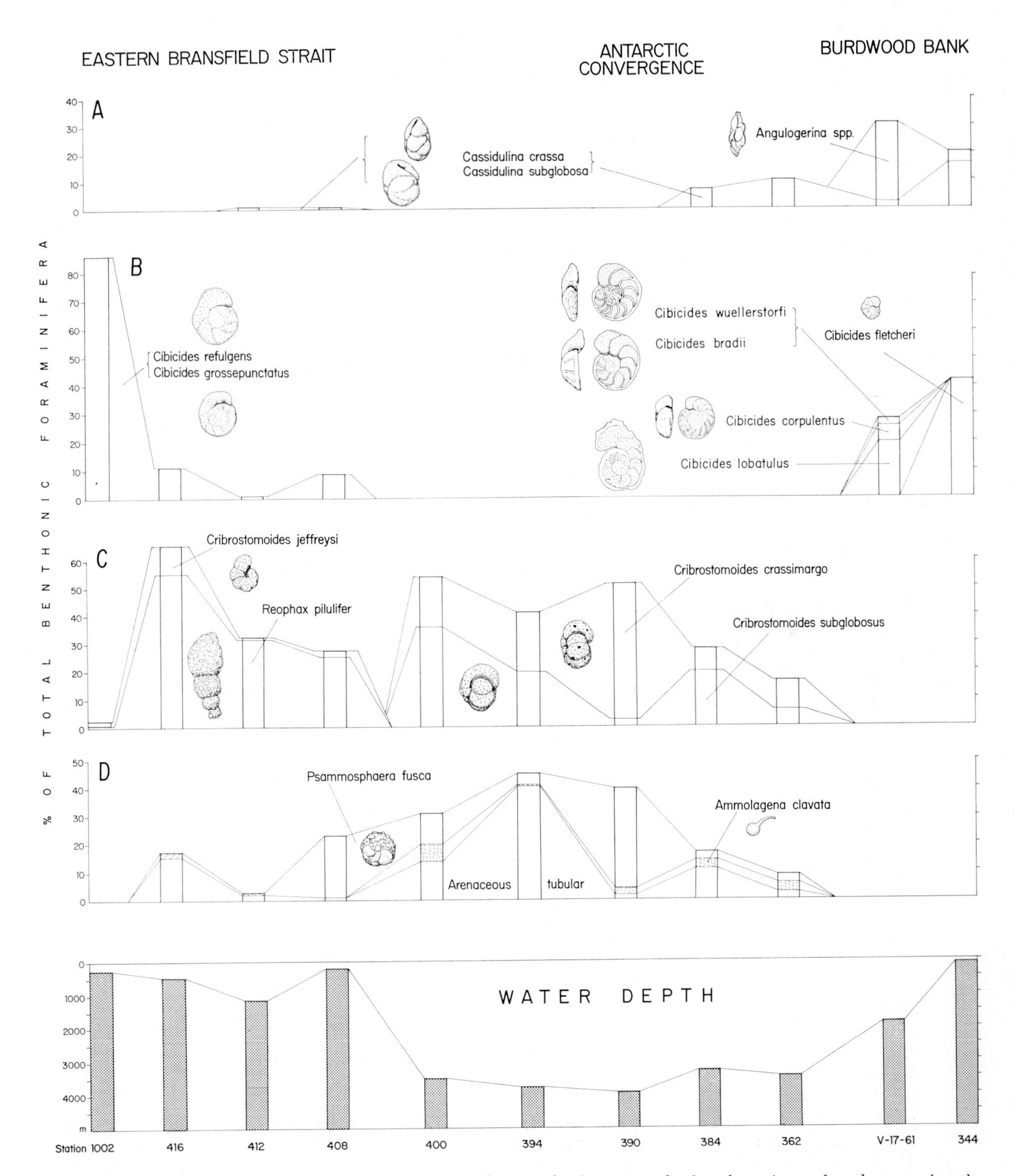

Fig. 9. Faunal profile across the eastern Drake Passage, showing the frequency of selected species or faunal groups in relation to water depth. For location see Figure 1.

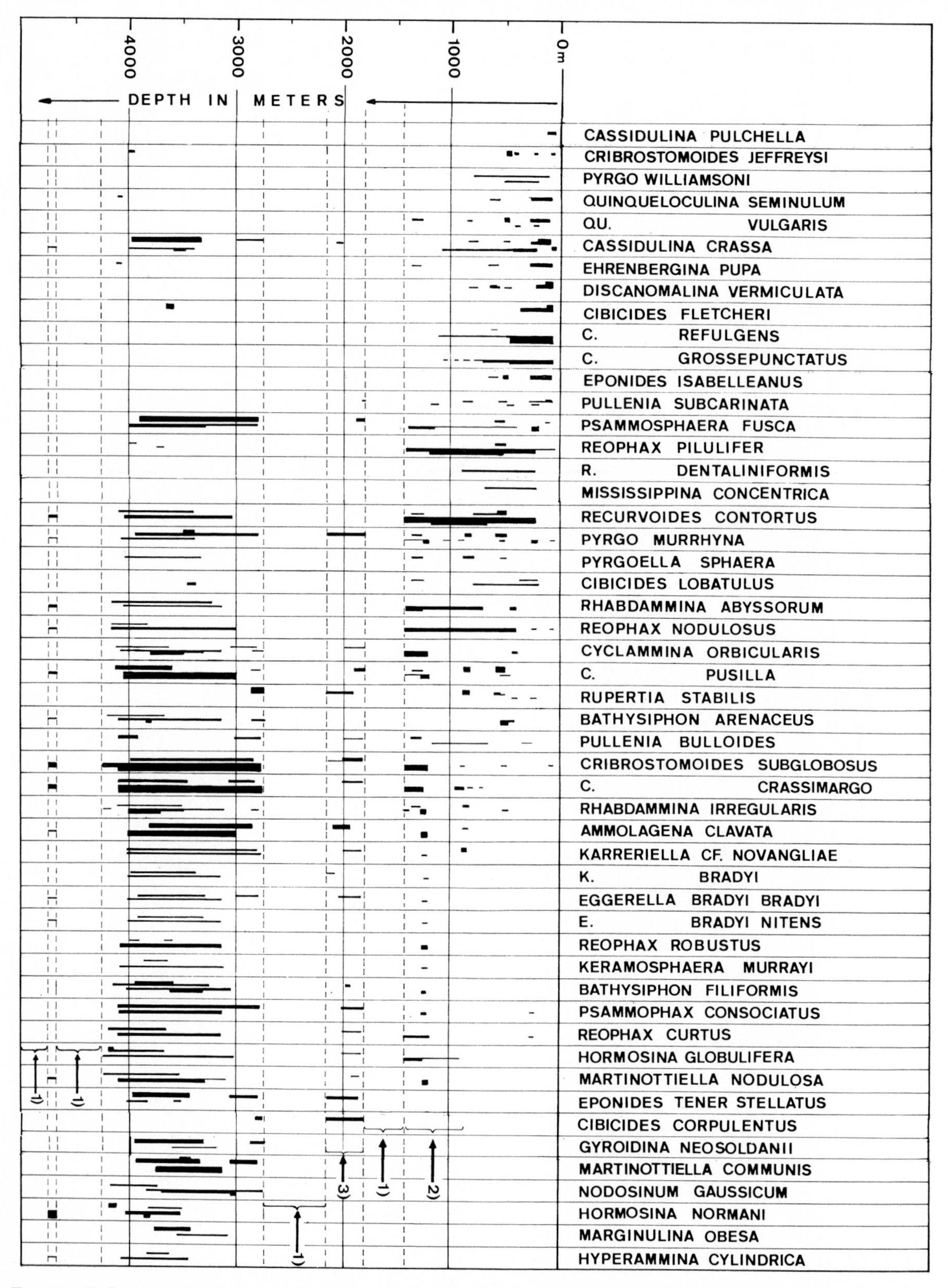

Fig. 10. Bathymetric distribution of selected benthonic Foraminifera in the Drake Passage. For each species the distribution north of the Antarctic Convergence is shown by the upper bar, south of the Antarctic Convergence by the lower bar. (1) No sample available for this depth range. (2) Few records north of the Antarctic Convergence. (3) Few records south of the Antarctic Convergence.

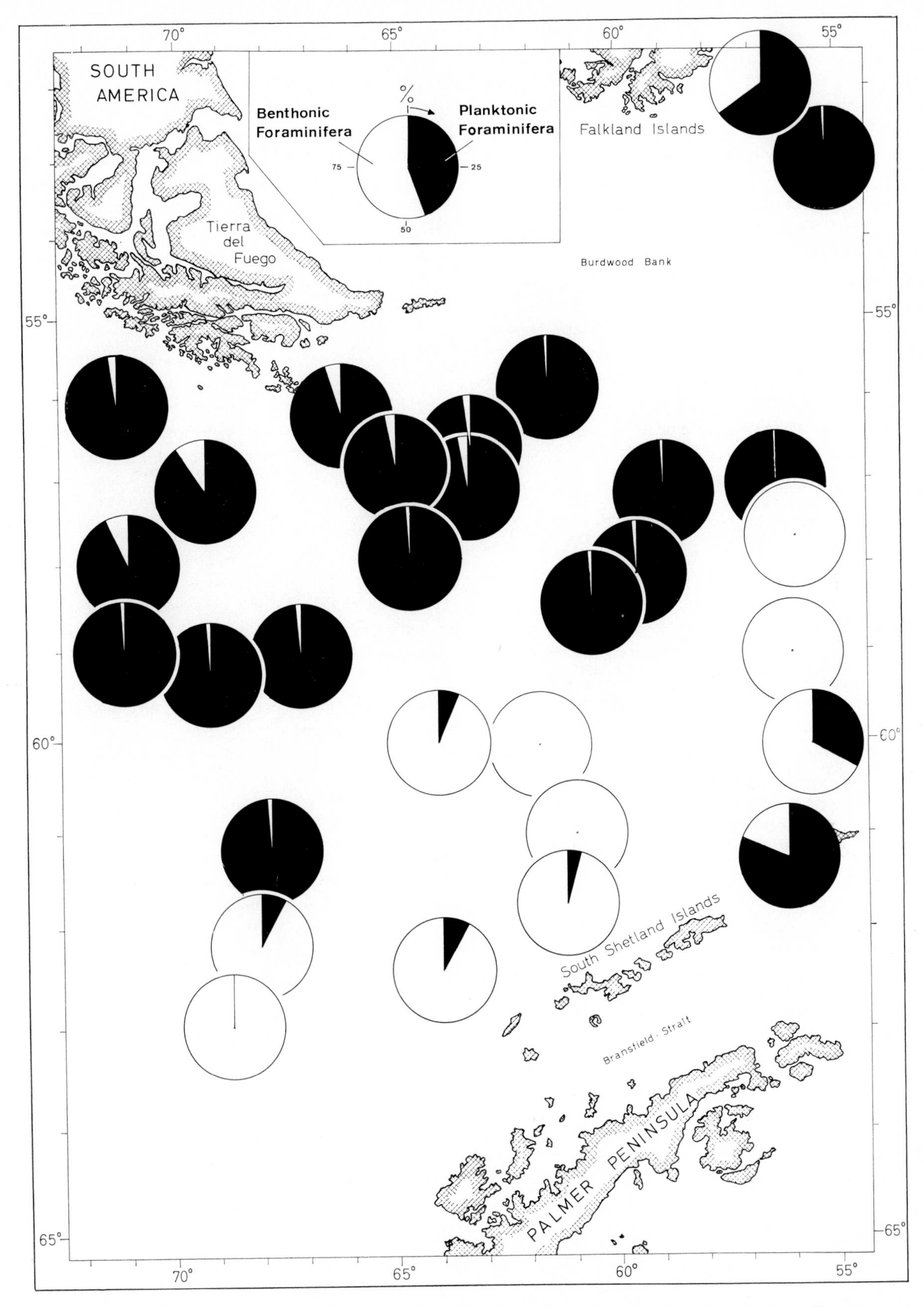

Fig. 11. The ratio between benthonic and planktonic Foraminifera in surface sediments of the Drake Passage.

The last of these species forms 34% of the total benthonic Foraminifera. This case is interesting, inasmuch as the station is distant from any shallow-water area. In any case, the shallow-water assemblage must be derived from the northern shelf areas, where a benthonic population very similar in composition occurs on the Burdwood Bank (Station 344) at a depth of 119 meters. Transport by turbidity current seems to be the most probable explanation, since the shortest distance to the edge of the Burdwood Bank is at least 170 km.

In the deep-sea trench north of the South Shetland Islands, in depths between 3678 and 4758 meters, several calcareous perforate species were found, most probably displaced from shallower water. At Station 129 the following species were observed:

Bolivina interjuncta
Angulogerina carinata carinata
Cancris auriculus
Valvulineria inflata
Valvulineria inaequalis
Cassidulina limbata
Lenticulina sp.

Some of these species were also noted at Stations 127 and 135. Most of the specimens are worn and the tests partly filled with sediment. These populations are almost identical to the ones described by Bandy and Rodolfo [1964] from shallow-water stations along the Peru-Chile trench. Curiously enough, similar assemblages did not occur in the shallow water along the edge of the antarctic shelf, however; all of the species noted above are restricted in our materials to the three deep-water samples mentioned.

A little more than half of the samples has been stained with rose bengal in order to distinguish the living from the dead specimens. This method is especially applicable to samples from the shelf and the upper bathyal zone, in which Foraminifera are often translucent, and staining of protoplasm in the last chamber and around the aperture may easily be seen. In populations of larger deep-sea Foraminifera, especially among the arenaceous ones, a staining effect can be observed only around the aperture, and in many cases is not clearly visible without breaking the last chamber of the specimen. Partial staining effects, probably due to remaining organic matter, make it even more difficult to decide whether a specimen was living or dead at the time of collection. Therefore it was not possible to determine exact ratios between living and dead populations. In reviewing shallow-water as well as deep-water populations, it becomes evident, however, that many more living specimens were recovered in shallow water than in deep water. The density of living populations, therefore, seems to be much greater in the shallower waters of the shelf and the upper bathyal zone than in the deeper parts of the bathyal and in the abyssal zone.

Despite the difficulties, it was possible to detect in each of the stained samples at least one or a few specimens that were probably alive at the time of collection. This is true also for the areas of possible transport of sediment and Foraminifera mentioned above. An exception is Station 249 (3989–4099 meters), where no living Foraminifera were found. The sample contained 94% calcareous perforate specimens, most of which are probably displaced from the middle bathyal zone of the nearby submarine rise. Although it would be desirable to base a distributional study on living specimens only, as postulated by Phleger [1964], it cannot be achieved in this case because of the general scarcity of living specimens in the deep sea. In this connection it is interesting to note the good correspondence of 2 independent distributional studies of Foraminifera in the Gulf of California, the first based on unstained samples [Bandy, 1961], the second on living populations only [Phleger, 1964].

Plate 9

1. *Rhabdammina abyssorum*, ×9, *Eltanin* 412 (62°06′S, 56°00′W, 1180 meters).
2. *Ammolagena clavata*, ×23.5, *Eltanin* 250 (59°14′S, 69°08′W, 3678–3803 meters).
3. *Rhabdammina irregularis*, ×35, *Eltanin* 140 (59°56′S, 65°18′W, 3687 meters).
4. *Hyperammina cylindrica*, ×26.5, *Eltanin* 140 (59°56′S, 65°18′W, 3687 meters).
5. *Pelosinella bicaudata*, ×35, *Eltanin* 140 (59°56′S, 65°18′W, 3687 meters).
6. *Saccammina* sp., ×44, *Eltanin* 126 (57°13′S, 62°48′W, 3733–3806 meters).
7. *Saccammina* sp., broken specimen, showing sponge spicules inside of the test, ×44, *Eltanin* 126 (57°13′S, 62°48′W, 3733–3806 meters).
8. *Tolypammina vagans*, ×23.5, *Eltanin* 140 (59°56′S, 65°18′W, 3687 meters).
9. *Crithionina pisum*, ×23.5, *Eltanin* 250 (59°14′S, 69°08′W, 3678–3803 meters).
10. *Thurammina papillata*, ×19, *Eltanin* 384 (57°03′S, 56°28′W, 3138–3426 meters).

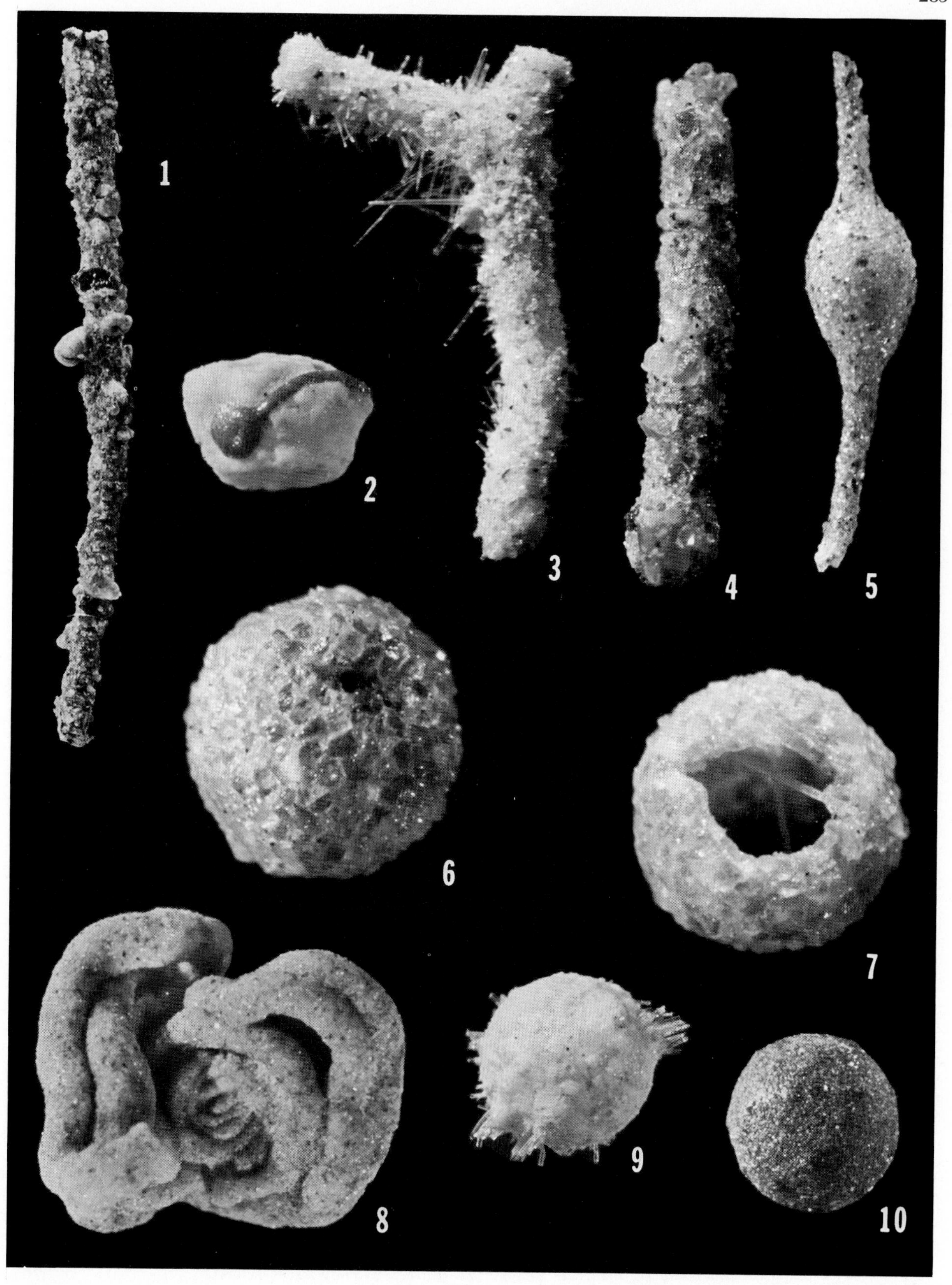
1
2
3
4
5
6
7
8
9
10

AREAL DISTRIBUTION OF BENTHONIC FORAMINIFERA

A distinction of foraminiferal provinces in the Drake Passage area is possible in two different ways: (*a*) by recognizing species that are restricted in their distribution to 1 or another area; (*b*) by recognizing faunal groups that in different provinces represent different proportions of the total population.

An example for the first possibility is the distinction of two main provinces in the shallow water and on the continental slope north and south of the Drake Passage: the Cape Horn province (with the Falkland subprovince) and the South Shetland province, as defined above.

Cape Horn Province (See Plates 1–3)

The following species of benthonic Foraminifera are restricted to this province:

Discorbis aff. *nitidus*
Patellina corrugata
Heronallenia kempii
Polystomellina patagonica
Mississippina concentrica
Rotalia sp.

Other species occur in the Cape Horn province as well as in the Falkland subprovince, but were not found in the South Shetland province. These are

Ehrenbergina pupa
Discanomalina vermiculata
Eponides isabelleanus
Elphidium crispum
Elphidium lessonii

A small, subglobular type of *Ehrenbergina pupa* was found by Earland [1934] at a number of stations in the Drake Passage as well as in the South Shetland area, whereas the typical, slightly elongated form is restricted to the northern shelf areas, according to this author.

Falkland Subprovince

Within the northern shelf areas, samples from south and southeast of the Falkland Islands showed a foraminiferal population somewhat different from the one found around Tierra del Fuego. However, the number of samples available for study from this area was limited, as mentioned above. The foraminifera encountered therein are shown in percentages of the total benthonic Foraminifera in Table 1.

The widespread abundance of *Discanomalina vermiculata* found in the Cape Horn province down to a depth of 878 meters was not observed here. Only one sample from the Falkland subprovince contained this species. On the other hand, *Hoeglundina elegans* was present in all samples shallower than 2000 meters and was especially abundant at Stations 340 (567–578 meters) and 557 (855–866 meters). Sample 344 (119 meters), from the Burdwood Bank, furnished exceptionally numerous *Elphidium crispum*, a species also found around Cape Horn, but infrequently and as dead specimens only.

The distinction of a Falkland subprovince based upon these trends among calcareous perforate Foraminifera is certainly not as clearly defined as the difference between the Cape Horn and the South Shetland provinces. Further investigations based on a greater number of samples will have to be made to show whether the difference observed in the present study is consistent or accidental.

A notable feature of the Falkland subprovince is the occurrence of *Cyclammina cancellata* at Stations 340 and 557, in both cases associated with few specimens of *Cyclammina pusilla*, which is very common throughout the rest of the area. *Cyclammina cancellata*, in contrast, has not been found south of the two stations mentioned. At Station 372, *C. pusilla* alone was present. The opinion of Bandy and Echols [1964] that *Cyclammina orbicularis* is the antarctic counterpart of the tropical and temperate *Cyclammina cancellata* seems to be applicable also to the relation between the latter and the closely related *C. pusilla*. A boundary between the two species can be traced between Stations 340/557 (with bottom temperatures between 3.1 and 4.4°C) and Station 372 (with a bottom temperature around 2.2°C [Friedman, 1964]). Along the Chilean coast a similar boundary must

Plate 10

1. *Nodosinum gaussicum*, ×17.5, *Eltanin* 384 (58°03′S, 56°28′W, 3138–3426 meters).
2. *Reophax nodulosus*, ×13, *Eltanin* 138 (62°05′S, 61°09′W, 1437 meters).
3–5. *Reophax pilulifer*, ×17.5, *Eltanin* 408 (61°16′S, 56°11′W, 223–225 meters).
6. *Hormosina robusta*, ×20, *Eltanin* 303 (62°03′S, 70°55′W, 4077–4176 meters).
7. *Reophax brevis*, ×23.5, *Eltanin* 384 (57°03′S, 56°28′W, 3138–3426 meters).
8–9. *Hormosina normani*, ×6, *Eltanin* 127 (61°45′S, 61°14′W, 4758 meters).

1
2
3
4
5
6
7
8
9

occur between Stations 216 (outside the map area at 52°53′S and 75°36′W) with *C. cancellata* only and 315 with *C. pusilla* only.

South Shetland Province

Foraminiferal populations around the South Shetland Islands and in the Bransfield Strait are distinguished from those of the Cape Horn province by a number of characteristic species as well as by a lower number of total foraminiferal species. *Reophax pilulifer* and *Cibicides refulgens* are especially common forms, in many cases forming the bulk of the population. *Cibicides refulgens* is accompanied by *Cibicides grossepunctatus* and *Rosalina vilardeboana*. Prominent members of this population are relatively large specimens of *Cribrostomoides jeffreysi*. All of these species are found in the Cape Horn province also, although in low percentages only. Two species are restricted entirely to the South Shetland province:

Reophax dentaliniformis
Pseudobulimina chapmani

Other records by Earland [1934], Wiesner [1931], and McKnight [1962] show that these two forms are typical of cold antarctic waters.

Deep-Sea Basin of the Drake Passage

In contrast to the striking difference between populations of the northern and the southern shelf areas, deep-sea assemblages are more uniform with respect to the lists of species present, and there are a great number of species which occur in areas north and south of the convergence (see Table 2). However, a boundary between predominantly arenaceous assemblages in the south (97–100% arenaceous foraminifera) and an apparently random distribution of predominantly calcareous or arenaceous assemblages in the north can be observed and roughly coincides with the position of the Antarctic Convergence, as will be discussed below.

Most of the arenaceous Foraminifera encountered in the deep sea of the Drake Passage are present north as well as south of the convergence, but in considerably higher percentages south of it. Exceptions are

Bathysiphon filiformis
Martinottiella communis
Karrerriella bradyi
Eggerella bradyi bradyi

All are found more frequently north of the convergence. Among the most common forms equally distributed throughout the area are

Psammosphaera fusca
Psammophax consociata
Pelosinella bicaudata
Reophax curtus
Recurvoides contortus (lower bathyal zone)
Eggerella bradyi nitens

Many of the significant arenaceous deep-water forms show, as mentioned, much greater abundance south of the convergence, although they are frequently found north of it. Among these are

Hyperammina cyclindrica
Cribrostomoides subglobosus
Cribrostomoides crassimargo
Reophax nodulosus
Cyclammina pusilla
Cyclammina orbicularis
Ammobaculites aff. *agglutinans*
Hormosina robusta
Hormosina normani
Trochammina rugosa
Ammolagena clavata

The last species is very frequent at selected stations north of the convergence also. A species restricted to areas south of the convergence is *Hormosina globulifera.*

Most of the calcareous Foraminifera are, as noted above, more frequent in areas north of the convergence. Among the large calcareous imperforate forms the most common species, *Pyrgo murrhina,* shows the same trend. Other porcellaneous species, such as *Involvohauerina globularis,* have been found in northern areas only, whereas *Miliolinella subvalvularis* is restricted to southern areas. *Keramosphaera murrayi* was found at 9 stations equally distributed north and south of the convergence. These are the first records of this interesting species outside the type area

Plate 11

1*a*–1*c*. *Cribrostomoides crassimargo*, ×23.5, *Eltanin* 384 (57°03′S, 56°28′W, 3138–3426 meters).
2*a*–2*c*. *Cribrostomoides subglobosus*, ×23.5, *Eltanin* 283 (66°26′S, 74°46′W, 3550–3693 meters).
3*a*–3*c*. *Cribrostomoides subglobosus*, ×23.5, *Eltanin* 276 (67°00′S, 75°01′W, 3020–3041 meters).
4*a*–4*b*. *Cribrostomoides jeffreysi*, ×30, *Eltanin* 408 (61°16′S, 56°11′W, 223–225 meters).
5*a*–5*b*. *Karreriella bradyi*, ×23.5, *Eltanin* 140 (59°56′S, 65°18′W, 3687 meters).

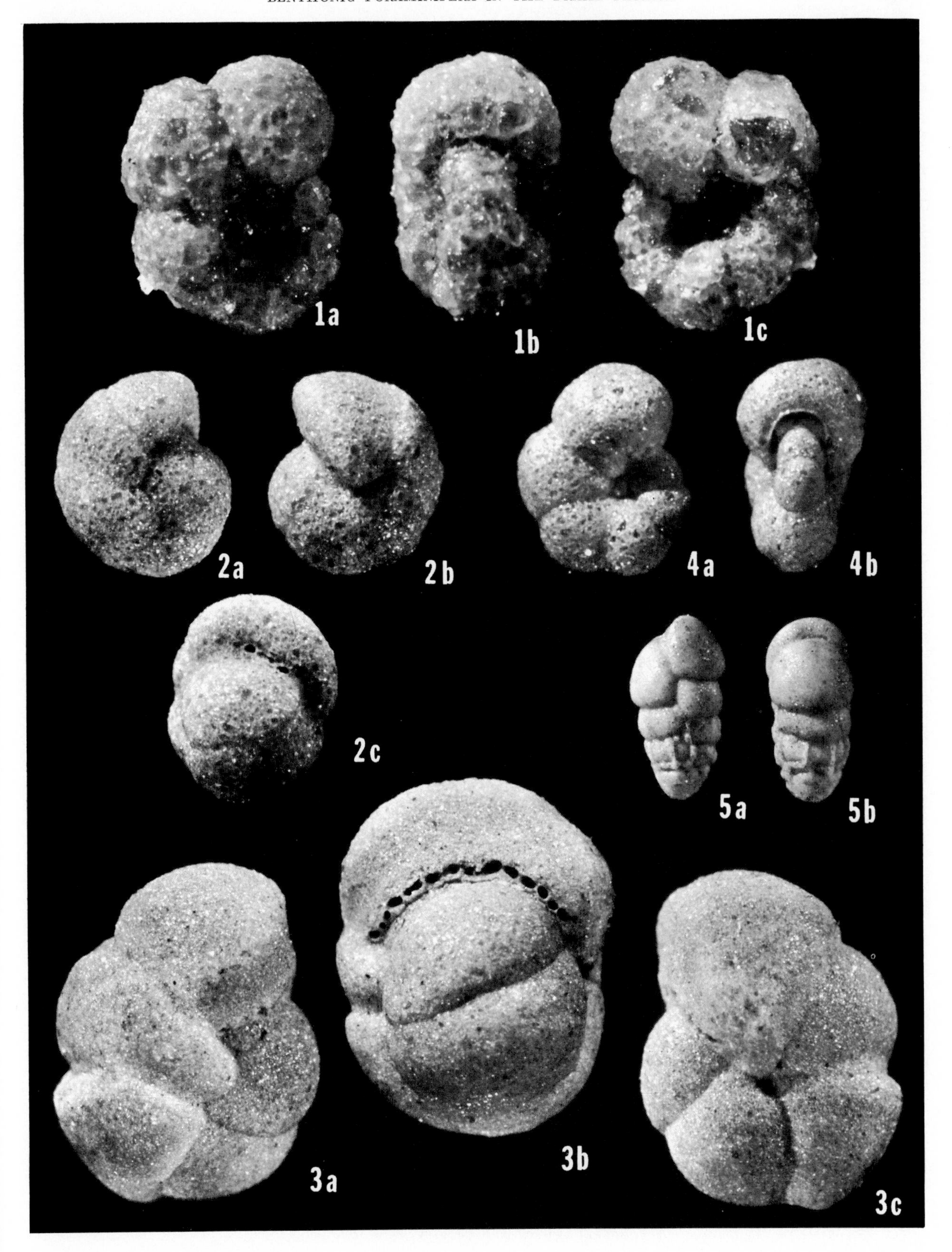
1a
1b
1c
2a
2b
4a
4b
2c
5a
5b
3a
3b
3c

southwest of Australia, from which location it was described by Brady [1884] and Wiesner [1931].

Most of the important calcareous perforate Foraminifera are found predominantly north of the convergence. These are

Cassidulina crassa
Cassidulina subglobosa
Gyroidina neosoldanii
Eponides tener stellatus
Cibicides wuellerstorfi
Cibicides bradii
Pyrulina fusiformis

An inferred influence of the Antarctic Convergence upon the distribution of benthonic Foraminifera seems contradictory, since this line is a feature restricted to surface and subsurface waters. However, in the eastern Drake Passage a strong temperature gradient exists also in the cold Antarctic Bottom Water [Gordon, 1967] and partly coincides with the geographic position of the Antarctic Convergence (see Figure 2). This might explain the rapid change observed across the convergence (Figures 3 and 7). The low bottom temperatures in the deep-sea basin may raise the solubility of calcium carbonate and hamper the development of a substantial population of calcareous Foraminifera in the southeastern Drake Passage. Planktonic Foraminifera were not found at most locations of this area (Figure 11) [Herb, 1968]. Their tests are probably dissolved at depths greater than 3500 meters.

In the southern and southwestern Drake Passage a similar distributional pattern of benthonic Foraminifera was observed. It can be explained in a similar way, although the temperature gradient in the Antarctic Bottom Water is not as great as in the eastern Drake Passage. The exceptionally high percentage of calcareous Foraminifera at Station 129 is due to faunal displacement, as shown above.

In the deep-sea basin north of the Antarctic Convergence the frequency pattern of calcareous perforate Foraminifera shows a more complicated picture than in southern areas; values may vary between 5% and 95%. The highest figures, however, are reached either at stations of the bathymetric zone H 5, described below (Station 305), or at stations where, again, at least part of the calcareous perforate population has been displaced from shallower areas (Stations 356 and 249). In other instances, however, samples which were recovered from equal depth ranges and which show no trace of possible faunal displacement, furnish very different percentage figures. Relatively high values of calcareous Foraminifera (although still lower than in zones H 1 and H 2) were observed especially in the middle part of the western Drake Passage. This area is also characterized by high numbers of planktonic Foraminifera in the sediment (up to 120,000 specimens per gram dry sediment [Herb, 1968]). Areas with a low amount of calcareous perforate forms are the deep-sea trough adjoining the Tierra del Fuego shelf to the south as well as the eastern Drake Passage just north of the Antarctic Convergence (see Figure 3). Some of these stations are close to the level of $CaCO_3$ dissolution, especially Station 230 (4185–4195 meters). The sandy character of the sediment in these areas might also influence the development of a predominantly arenaceous population.

The Antarctic Convergence itself controls some distributional patterns of planktonic Foraminifera [Herb, 1968]. The ratio between planktonic and benthonic Foraminifera, as observed in the Phleger cores, is shown in Figure 11. In areas where planktonic Foraminifera are abundant, their tests furnish a high amount of calcium carbonate to the sediments, assuming that we are not below the level of $CaCO_3$ dissolution. It cannot be proved, however, that the $CaCO_3$ content of the sediment is of any influence upon the frequency of calcareous benthonic Foraminifera observed in our samples. The distribution of these tests in the deep-sea sediments probably depends

Plate 12

1. *Eggerella bradyi bradyi*, ×23.5, *Eltanin* 384 (57°03′S, 56°28′W, 3138–3426 meters).
2a–2b. *Karreriella novangliae*, ×23.5, *Eltanin* 126 (57°13′S, 62°48′W, 3733–3806 meters).
3. *Martinottiella communis*, ×17.5, *Eltanin* 384 (57°03′S, 56°28′W, 3138–3426 meters).
4. *Ammodiscus pacificus*, microspheric form, ×23.5, *Eltanin* 322 (56°05′, 71°11′W, 1806 to ca. 2013 meters).
5. *Ammodiscus pacificus*, megalospheric form, ×23.5, *Eltanin* 322 (56°05′S, 71°11′W, 1806 to ca. 2013 meters).
6a–6b. *Cyclammina cancellata*, ×17.5, *Eltanin* 340 (53°08′S, 59°22′W, 567–578 meters).
7a–7b. *Cyclammina pusilla*, ×17.5, *Eltanin* 127 (61°45′S, 61°14′W, 4758 meters).
8a–8b. *Cyclammina orbicularis*, ×23.5, *Eltanin* 283 (66°26′S, 74°47′W, 3550–3693 meters).
9. *Martinottiella nodulosa*, ×30, *Eltanin* 140 (59°56′S, 65°18′W, 3687 meters).

1
2a
2b
3
4
5
6a
6b
7a
7b
8a
8b
9

on several parameters. Specific oceanographic conditions (water depth, temperature, salinity, CO_3 content), which are responsible for the position of the dissolution boundary of $CaCO_3$, are of primary importance; but the supply of $CaCO_3$ by tests and debris of various planktonic organisms is possibly another important parameter in levels above the dissolution of $CaCO_3$.

A shallow $CaCO_3$ solution boundary at depths between 500 and 550 meters has been observed by Kennett [1966, 1968] in the Ross Sea. Although local conditions of that basin are responsible for this, the general considerations on $CaCO_3$ saturation in antarctic waters made by this author can also be applied, at least in part, for the Drake Passage area. Additional oceanographic data would be needed, however, if a well-founded interpretation for many of the distributional patterns of benthonic Foraminifera described in the present report should be given. Furthermore, the effects of displacement are, in some cases, difficult to detect, especially if one deals with such processes within the lower bathyal and abyssal zone.

Acknowledgments. The author wishes to express his gratitude to Dr. Orville L. Bandy, who kindly offered the opportunity for studying the samples collected by the *Eltanin* in the Drake Passage. Dr. Bandy, as well as Drs. James P. Kennett, Ronald L. Kolpack, and Franz Allemann critically read the manuscript and furnished helpful suggestions. A number of samples collected in this area by the *Vema* was made available to the author by Dr. L. Kolpack. Dr. E. Boltovskoy kindly compared some of our material with the collections at the Smithsonian Institution in Washington. Appreciation is extended to the Allan Hancock Foundation, University of Southern California, for the use of its facilities.

As a part of the U.S. Antarctic Research Program, this investigation received financial support from National Science Foundation grant GA-238. Further financial support was provided by the Swiss National Science Foundation.

REFERENCES

Bandy, O. L.
1960 General correlation of foraminiferal structure with environment. Rep. Session, 21st Int. Geol. Congr., Copenhagen, 1960. Part 22, pp. 7–19. Norden, Copenhagen.
1961 Distribution of Foraminifera, Radiolaria, and diatoms in sediments of the Gulf of California. Micropaleontology, *7:* 1–26.
1963 Larger living Foraminifera of the continental borderland of southern California. Contr. Cushman Fdn Foramin. Res., *14*(4): 121–126.

Bandy, O. L., and R. E. Arnal
1960 Concepts of foraminiferal paleoecology. Bull. Am. Ass. Petrol. Geol., *44:* 1921–1932.

Bandy, O. L., and M. A. Chierici
1966 Depth-temperature evaluation of selected California and Mediterranean bathyal Foraminifera. Mar. Geol., *4:* 259–271.

Bandy, O. L., and R. J. Echols
1964 Antarctic foraminiferal zonation. Antarct. Res. Ser., *1:* 73–91. AGU, Washington, D. C.

Bandy, O. L., and K. S. Rodolfo
1964 Distribution of Foraminifera and sediments, Peru-Chile trench area. Deep Sea Res., *11:* 817–837.

Brady, H. B.
1884 Report on the Foraminifera dredged by HMS *Challenger* during the years 1873–1876. *In* Rep. Scient. Results Voyage HMS *Challenger, 9:* viii–xxi, 1–814, pls. 1–115 (in Atlas).

Chapman, F., and W. J. Parr
1937 Foraminifera. Australasian Antarctic Expedition, 1911–1914, ser. C, *1*(2): 1–190, 4 pls.

Deacon, G. E. R.
1937 The hydrology of the Southern Ocean. 'Discovery' Rep., *15:* 1–124, pls. 1–44.

D'Orbigny, A.
1839 Voyage dans l'Amérique méridionale. Vol. 5, pt. 5. Foraminifères. Pp. 1–86. Bertrand, Paris.

Earland, A.
1933 Foraminifera. 2. South Georgia. 'Discovery' Rep., *7:* 27–138, pls. 1–7.
1934 Foraminifera. 3. The Falklands sector of the Antarctic (excluding South Georgia). 'Discovery' Rep., *10:* 1–208, pls. 1–10.
1936 Foraminifera. 4. Additional records from the Weddell Sea sector from material obtained by the S. Y. Scotia. 'Discovery' Rep., *13:* 1–76, pls. 1, 2, 2A.

Friedman, S. B.
1964 Physical oceanographic data obtained during Eltanin cruises 4, 5 and 6 in the Drake Passage, along the Chilean coast and in the Bransfield Strait. Lamont Geol. Obs. Tech. Rep. 1 cu-1-64: 1–55.

Plate 13

1a–1b. *Pyrgo wiesneri*, ×23.5, *Eltanin* 127 (61°45′S, 61°14′W, 4758 meters).
2a–2b. *Pyrgo murrhina*, ×23.5, *Eltanin* 384 (57°03′S, 56°28′W, 3138–3426 meters).
3a–3b. *Pyrgo depressa*, ×23.5, *Eltanin* 250 (59°14′S, 69°05′W, 3678–3803 meters).
4a–4b. *Pyrgo williamsoni*, ×17.5, *Eltanin* 340 (53°08′S, 59°22′W, 567–578 meters).
5. *Keramosphaera murrayi*, ×23.5, *Eltanin* 394 (58°58′S, 56°03′W, 3724-3825 meters).

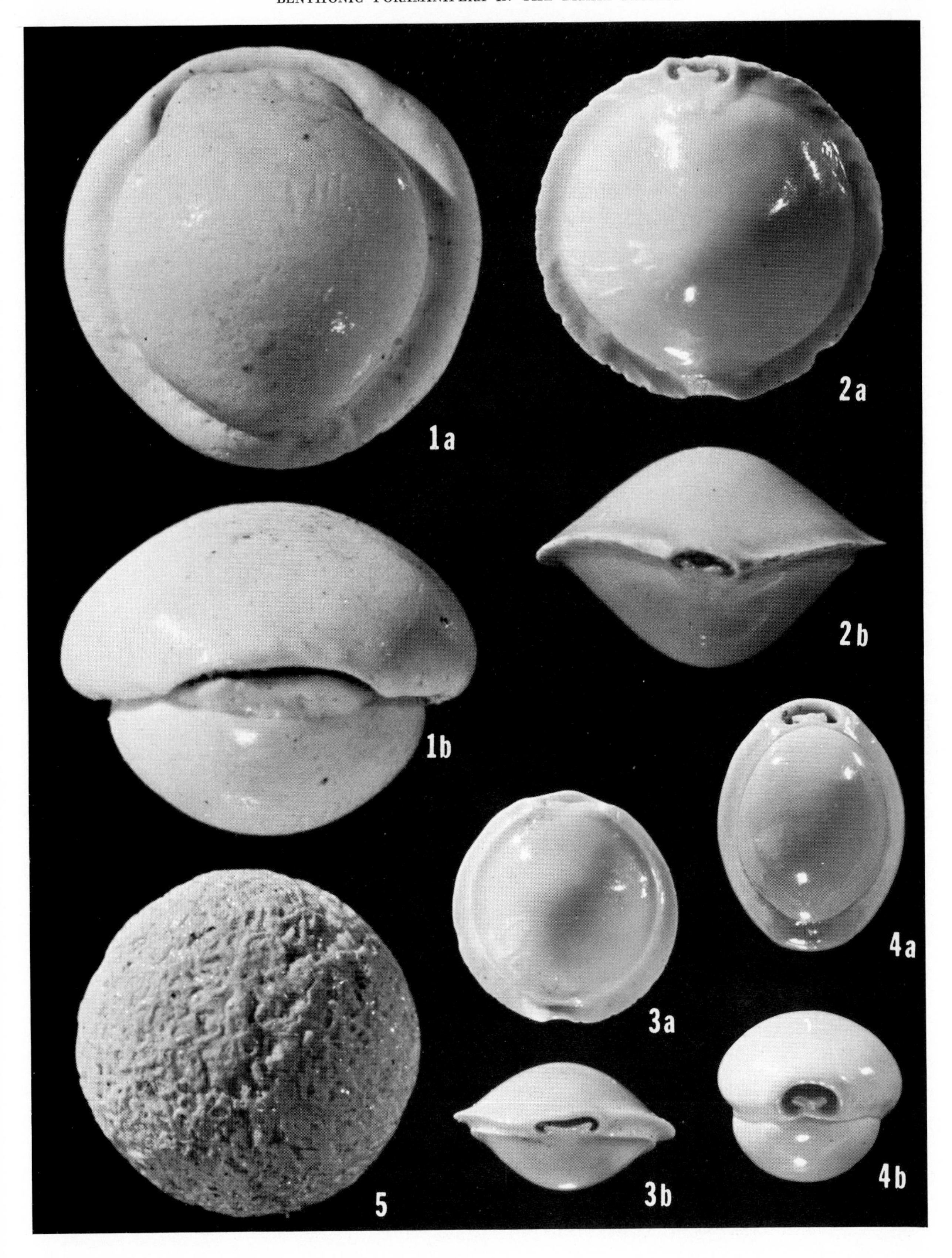
1a
1b
2a
2b
3a
3b
4a
4b
5

Heezen, B. M., and M. Tharp
1961 Physiographic diagram of the South Atlantic. Geol. Soc. Amer., Rochester, N. Y.
Herb, R.
1968 Recent planktonic Foraminifera from sediments of the Drake Passage, Southern Ocean. Eclog. Geol. Helv., *61*(2): 467–480.
Heron-Allen, E., and A. Earland
1922 Protozoa. 2. Foraminifera. *In* Nat. Hist. Rep. Br. Antarct. 'Terra Nova' Exped., 1910. Zool., *6*(2): 25–268, 8 pls.
1932 Foraminifera. 1. The icefree area of the Falkland Islands and adjacent seas. 'Discovery' Rep., *4:* 291–460, pls. 6–17.
Kennett, J. P.
1966 Foraminiferal evidence of a shallow calcium carbonate solution boundary, Ross Sea, Antarctica. Science, N. Y., *153*(3732): 191–193.
1968 The fauna of the Ross Sea. Part 6. Ecology and distribution of Foraminifera. N. Z. Dep. Scient. Ind. Res. Bull., no. 186, 48 pp.
McKnight, W. M., Jr.
1962 The distribution of Foraminifera off parts of the Antarctic coast. Bull. Am. Paleont., *44*(201): 65–158.
Menzies, R. J.
1964 Improved techniques for benthic trawling at depths greater than 2000 meters. Antarct. Res. Ser., *1:* 93–109. AGU, Washington, D. C.
Parr, W. J.
1950 Foraminifera. Rep. BANZ Antarct. Res. Exped., 1929–1931, ser. B, *5*(6): 235–392, pls. 3–15.
Pearcey, F. G.
1914 Foraminifera of the Scottish National Antarctic Expedition. Trans. R. Soc. Edinb., *49:* 991–1044, pls. i, ii.
Pflum, C. E.
1966 The distribution of foraminifera in the eastern Ross Sea, Amundsen Sea, and Bellingshausen Sea, Antarctica. Bull. Am. Paleont., *50*(226); 1–197, pls. 13–18.
Phleger, F. B.
1964 Patterns of living benthonic foraminifera, Gulf of California. A.A.P.G. Mem. 3, pp. 376–394.
Saidova, Kh. M.
1961 The quantitative distribution of bottom Foraminifera in Antarctica. Dokl. Akad. Nauk SSSR, *139*(4): 967–969.
Savage, J. M., and M. C. Caldwell
1965 Biological stations occupiel by the USNS *Eltanin*: Data summary, cruises 1–13. Stud. Antarct. Oceanol. Pp. 1–87. Dep. Biol. Sci. Univ. Sth. Calif., Los Angeles.
Savage, J. M., and S. R. Geiger
1965 Biological stations occupied by the USNS *Eltanin*: A synoptic atlas. Stud. Antarct. Oceanol. Pp. 1–3, 76 pls. Dep. Biol. Sci. Univ. Sth. Calif., Los Angeles.
Shishkevish, L. J.
1964 The distribution of Foraminifera in cores from the northern Scotia Arc, Scotia Sea. M.S. thesis, N. Y. Univ.
Todd, Ruth, and Doris Low
1967 Recent Foraminifera from the Gulf of Alaska and southeastern Alaska. U.S. Geol. Surv. Prof. Pap. 573-A, pp. 1–46, 5 pls.
Uchio, T.
1960 Benthonic Foraminifera of the Antarctic Ocean. Spec. Publs Seto Mar. Biol. Lab., *12:* 3–20.
University of Southern California
1964 USNS *Eltanin* Antarctic Cruise tracks: Cruises 4–13, July 1962–July 1964. Stud. Antarct. Oceanol. 10 pls. Univ. Sth. Calif., Los Angeles.
Wiesner, H.
1931 Die Foraminiferen, *In* E. von Drygalski, Dt. Südpol. Exped., 1901–1903, *20* (Zool., 12): 53–165.

FORAMINIFERAL SPECIES MENTIONED IN TEXT

For complete reference to the original description see Ellis and Messina catalog of foraminifera. The italic numbers in parentheses following references to Tables 1 and 2 correspond to the numerical order of the species in those tables.

Ammobaculites agglutinans (D'Orbigny) = *Spirolina agglutinans* D'Orbigny, pp. 274, 288, Fig. 8, Pls. 6, 7, Table 2(*69*)
Ammodiscus pacificus Cushman and Valentine, Pl. 3, Table 2(*72*)
Ammolagena clavata (Parker and Jones) = *Trochammina irregularis* var. *clavata* Parker and Jones, pp. 266, 274, 288, Figs. 8, 9, 10, Pl. 9, Table 2(*70*)
Angulogerina angulosa (Williamson) = *Uvigerina angulosa* Williamson, p. 270, Table 2(*95*)
Angulogerina carinata bradyana Cushman, p. 266, Table 2(*82*)
Angulogerina carinata carinata Cushman, pp. 270, 284
Angulogerina occidentalis (Cushman) = *Uvigerina occidentalis* Cushman, p. 258, Table 2(*7*)

Plate 14

1*a*–1*c*. *Quinqueloculina* sp., ×23.5, *Eltanin* 340 (53°08′S, 59°22′W, 567–578 meters).
2*a*–2*b*. *Quinqueloculina vulgaris*, ×35, *Eltanin* 340 (53°08′S, 59°22′W, 567–578 meters).
3*a*–3*b*. *Miliolinella* sp., ×23.5, *Eltanin* 412 (62°06′S, 56°00′W, 1180 meters).
4*a*–4*b*. *Pyrgoella spaera*, ×23.5, *Eltanin* 339 (53°07′S, 59°28′W, 512–586 meters).
5. *Pyrgoella sphaera*, large specimen, ×23.5, *Eltanin* 161 (56°12′S, 66°12′W, 878 meters).
6*a*–6*b*. *Involvohauerina globularis*, ×23.5, *Eltanin* 155 (56°32′S, 63°16′W, 3927 meters).
7*a*–7*b*. *Cornuspira* sp., ×23.5, *Eltanin* 408 (61°16′S, 56°11′W, 223–225 meters).

1a
1b
1c
2a
2b
3a
4a
4b
3b
6a
6b
5
7a
7b

Plate 15

1*a*–1*c*. *Eponides isabelleanus,* abnormally high and somewhat distorted specimen, ×26, *Eltanin* 162 (55°42′S, 70°33′W, 174 meters).
2*a*–2*c*. *Eponides isabelleanus,* ×23.5, *Eltanin* 339 (53°07′S, 59°28′W, 512–586 meters).
3. *Marginulina obesa,* ×23.5, *Eltanin* 384 (57°03′S, 56°28′W, 3138–3426 meters).
4. *Marginulina* cf. *planata,* ×17.5, *Eltanin* 155 (56°32′S, 63° 16′W, 3927 meters).
5*a*–5*c*. *Hoeglundina elegans,* ×44, *Eltanin* 557 (51°57′S, 56°39′W, 855–866 meters).
6*a*–6*c*. *Gyroidina neosoldanii,* ×23.5, *Eltanin* 126 (57°13′S, 62°48′W, 3733–3806 meters).
7–9. *Rupertia stabilis,* ×29.5, *Eltanin* 305 (59°59′S, 70°38′W, 2782–2827 meters).

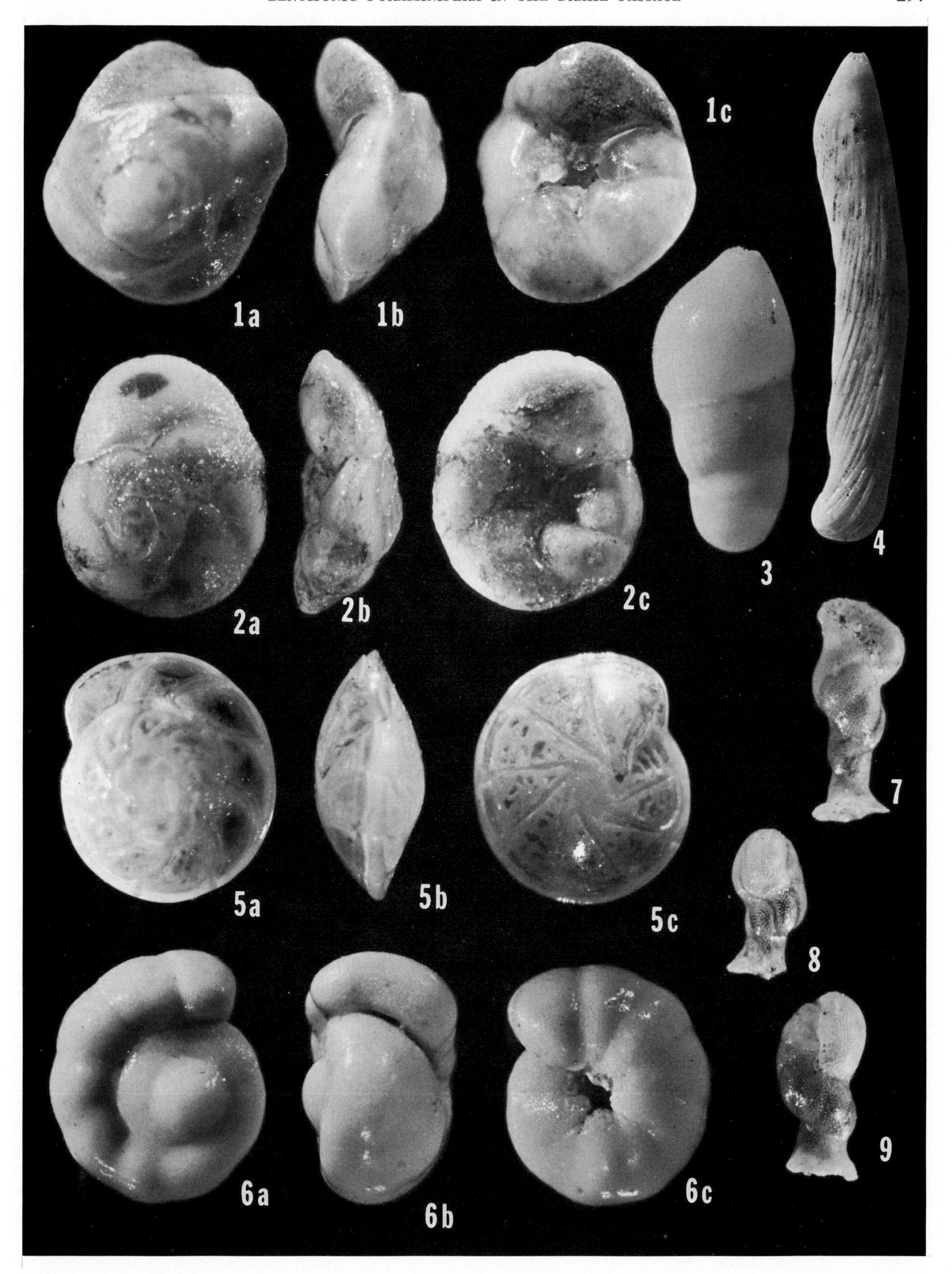
1a
1b
1c
2a
2b
2c
3
4
5a
5b
5c
6a
6b
6c
7
8
9

Plate 16

1*a*–1*c*. *Discanomalina vermiculata,* ×40, *Eltanin* 970 (55°01′S, 64°51′W, 586 to ca. 641 meters).
2*a*–2*c*. *Cibicides wuellerstorfi,* ×29, *Eltanin* 340 (53°08′S, 59°22′W, 567–578 meters).
3. *Dentalin pauperata,* ×17.5, *Eltanin* 412 (62°06′S, 56°00′W, 1180 meters).
4. *Dentalina* cf. *pauperata,* ×17.5, *Eltanin* 340 (53°08′S, 59°22′W, 567–578 meters).
5*a*–5*c*. *Mississippina concentrica,* ×40, *Eltanin* 369 (54°03′S, 63°40′W, 247–293 meters).
6. *Pyrulina* cf. *fusiformis,* ×23.5, *Eltanin* 155 (56°32′S, 63°16′W, 3927 meters).
7*a*–7*b*. *Pullenia bulloides,* ×35, *Eltanin* 322 (56°04′S, 71°11′W, 1803 to ca. 2013 meters).
8*a*–8*b*. *Pullenia subcarinata,* ×45, *Eltanin* 217 (54°22′S, 64°47′W, 106–110 meters).

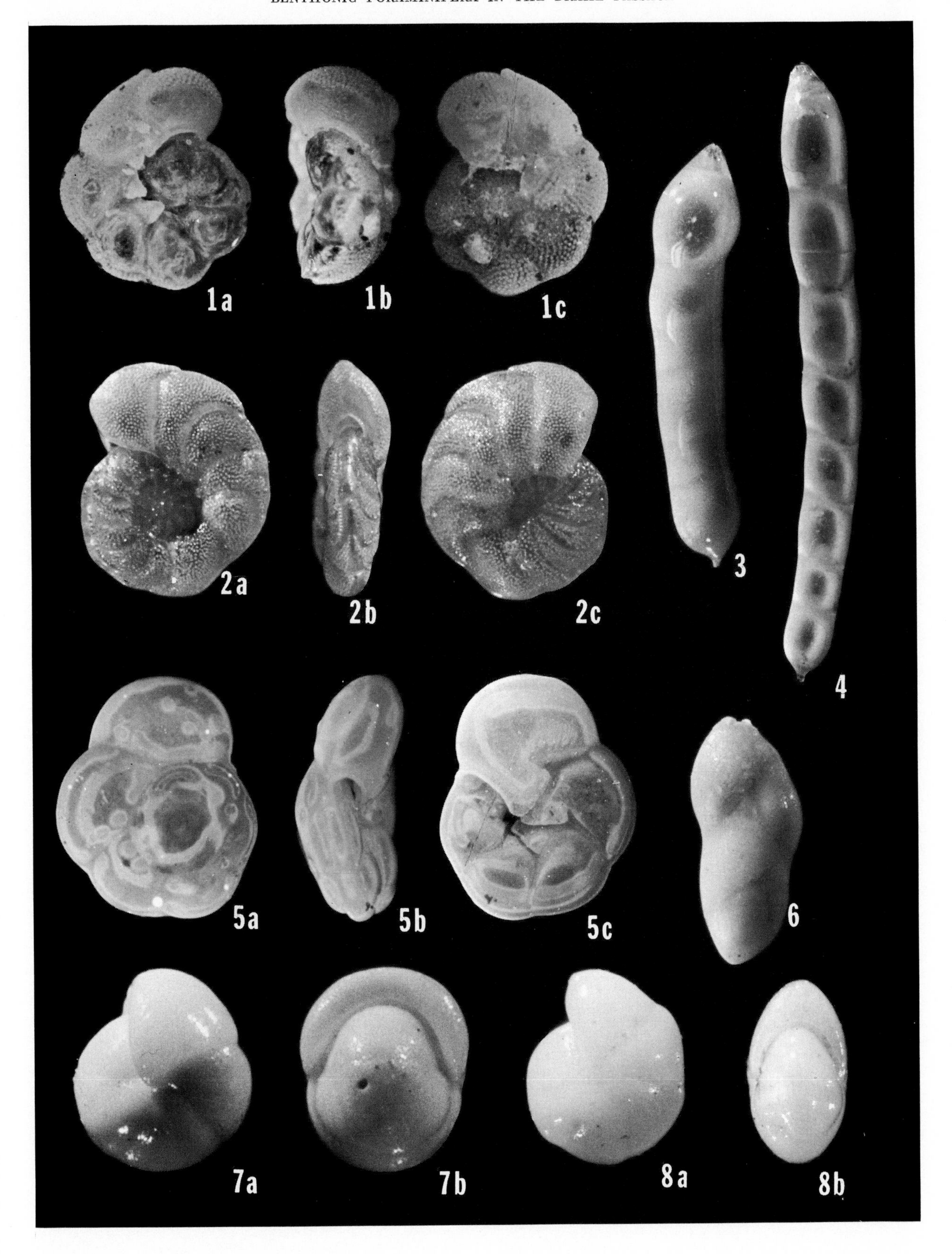
1a
1b
1c
2a
2b
2c
3
4
5a
5b
5c
6
7a
7b
8a
8b

OBSERVATIONS ON PHYTOPLANKTON BLOOM IN THE WEDDELL SEA

SAYED Z. EL-SAYED

Department of Oceanography, Texas A & M University, College Station, Texas 77843

Abstract. In the course of a biological investigation carried out during the International Weddell Sea Oceanographic Expedition (January–March, 1968), aboard the icebreaker *Glacier*, a very extensive bloom of phytoplankton was encountered on February 10, 1968, off the Ronne Ice Shelf. The bloom, which was highly developed at 74°59′S, 60°57′W, covered an area estimated to be at least 15,500 km² (6100 mi²). Dense patches of distinctly yellow-brownish color were visible for miles on end; these patches were interspersed with long, narrow bands of relatively clear water. Very dense concentrations of phytoplankton were also found on the underside of the sea ice and pancake ice.

Chlorophyll a in the bloom region was as high as 190 mg/m³ compared to an average of 0.37 mg/m³ for the rest of the stations occupied during that expedition. Photosynthetic activity of the phytoplankton was also high at the surface (82 mg C/m³/hr); however, since the euphotic zone was shallow (10 meters), daily primary production in the photosynthetic zone was only 1.56 g C/m². Microscopic examination of the phytoplankton samples collected showed the overwhelming abundance of a centric diatom (*Thalassiosira tumida*); however, other diatoms, many dinoflagellates, and tintinnids are also reported in the bloom region. Observations in the bloom region also included temperature, salinity, light penetration, and nutrient salts. The effects of these on the development and sustainment of the bloom are discussed.

INTRODUCTION

Blooms of phytoplankton are of frequent occurrence in many parts of the world ocean and in lakes. Conspicuous among these worldwide phenomena are the blooms of the blue-green alga *Trichodesmium* and of the so-called red tide attributed to toxic and nontoxic dinoflagellates. The vernal outburst of phytoplankton in temperate and high latitudes is usually accompanied by blooms of diatoms and coccolithophorids, which may be extremely patchy. Fogg [1965] describes marine plankton patches as usually elliptical in shape and varying from a few feet across to as much as 50 or 65 km by 200 or 300 km; the mean is about 15 by 65 km.

Despite the intensive investigation carried out by several oceanographic expeditions in the Antarctic, reports on phytoplankton blooms in the oceans bordering the continent are surprisingly few. Murray [1889] gave an early account of the abundance of *Thalassiothrix longissima* var. *antarctica* in antarctic waters. Hardy and Gunther [1935] reported dense zones of *Thalassiothrix antarctica* in February 1926 in the vicinity of South Georgia. Using the continuous plankton recorder, they recorded this species for 125 km, followed by patches of young krill *(Euphausia superba)*, then by another patch of *T. antarctica* 18 km across. Although Hart [1934] preferred to use 'main phytoplankton increase' instead of 'bloom' to refer to the period of maximum production, and 'secondary autumnal increase' for productivity pulse late in the season, he mentioned several observations that could be considered local blooms in the present context. In Deception Island harbor, *Chaetoceros sociale* discolored the surface waters in concentrations estimated by Hart at 25 million cells/l. Other species he found in dense local concentrations include: *Chaetoceros criophilum, Corethron criophilum, Thalassiothrix antarctica,* and *Biddulphia striata* (= *B. weissflogii*). There is no mention in his carefully made observations of the occurrence of phytoplankton abundance on the scale reported in the present contribution. Walsh [1969] has recently reported another bloom of *Biddulphia weissflogii* in the middle of volcanic Port Foster, Deception Island, where 28.4×10^{12} fluorescing cells were found in the water column under 1 m².

In the course of a biological investigation carried out during the International Weddell Sea Oceano-

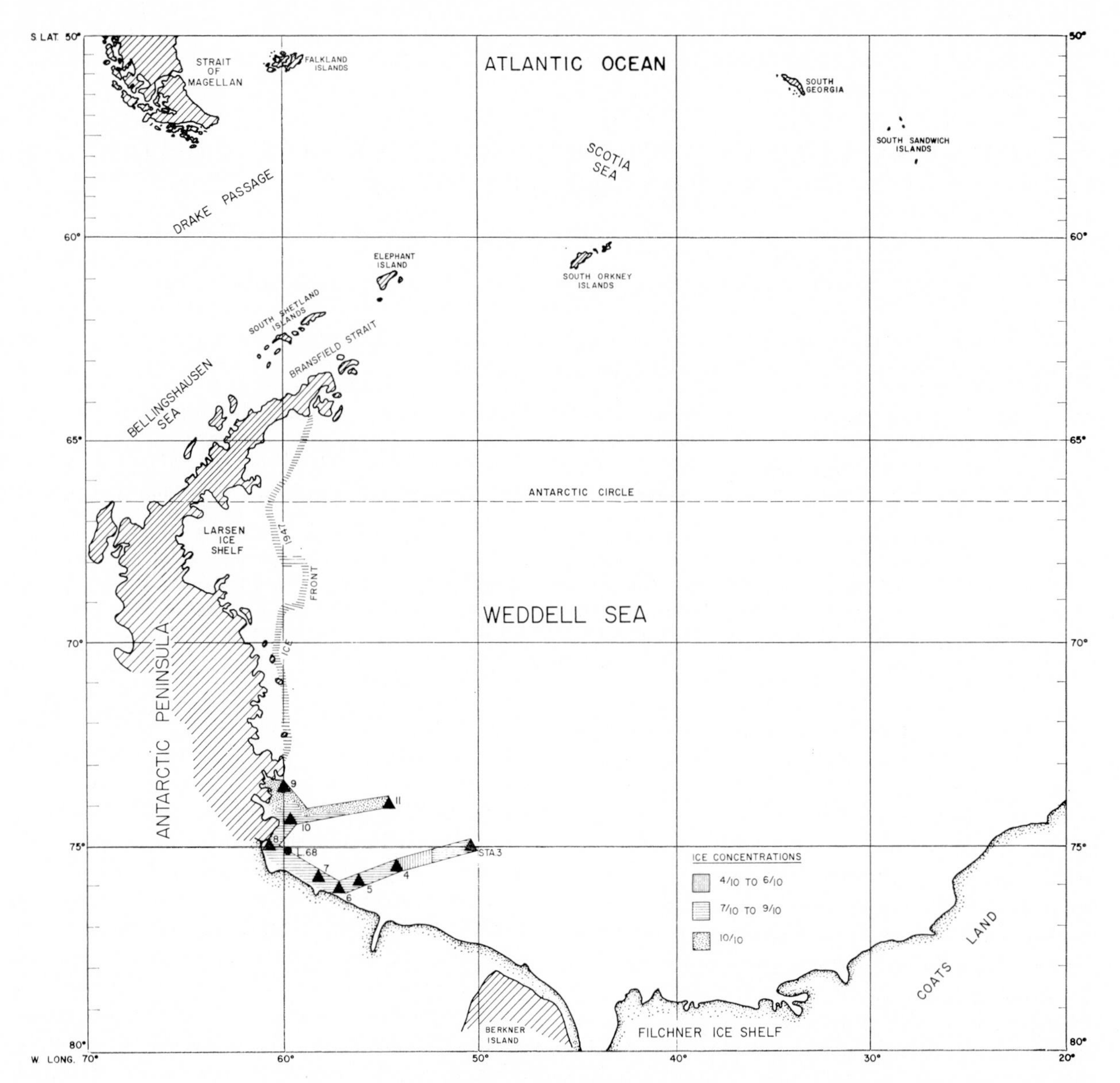

Fig. 1. Stations occupied in southwestern Weddell Sea February 8–13, 1968. Ice concentrations in the area investigated by USCGC *Glacier* are shown.

graphic Expedition (IWSOE, January–March, 1968) aboard the icebreaker USCGC *Glacier*, the author encountered a very extensive bloom of phytoplankton in the southwestern Weddell Sea. Located off the Ronne Ice Shelf (Figure 1), this algal outburst may well be considered the most extensive high-latitude diatom bloom described to date. Because there is little information on marine diatom blooms in general, and in the antarctic seas in particular, the physical, chemical, and biological observations reported herein should be of considerable scientific interest.

METHODS

The physical, chemical, and biological observations were made at Stations 3 through 11 in the southwestern Weddell Sea between February 8 and 13, 1968. The bloom was highly developed at Station 8 (74°58.5′S, 60°57.3′W), which was occupied at local apparent noon (LAN) on February 10, 1968. In addition to observations at the stations occupied, collections of surface phytoplankton samples were made with a plastic bucket at short intervals while the icebreaker plowed her way through the bloom.

The standing crop of phytoplankton was estimated by determining the concentration of chlorophyll a spectrophotometrically according to the modified method of Richards and Thompson [1952] and using the revised equations given by Parsons and Strickland [1963]. Water samples (500 ml) were collected at the surface and at 10 meters for the quantitative estimates of the phytoplankton, using the Utermöhl sedimentation method (with Zeiss inverted microscope). Primary production (simulated in situ) was measured with the use of the carbon-14 uptake method according to Steemann Nielsen [1952], modified by Strickland and Parsons [1968]. Analyses of nutrient salts were made according to the methods outlined in Strickland and Parsons [1968]. Phytoplankton was also collected by a 35-μ mesh net and examined aboard the ship and at Texas A & M University.

RESULTS

The vertical distribution of the physical and chemical parameters collected at the stations occupied in the southwestern Weddell Sea are shown in Figure 2.

The bloom was first sighted at about 0500 LT and continued till about 2300 on February 10, 1968. The extent of the bloom could not be determined with certainty, however, because of a rather dense fog early in the morning. When the fog lifted (between 1000 and 1400), the bloom could be sighted as far as visibility permitted. The recurrence of foggy conditions late in the afternoon and in the evening once more curtailed visibility. The areas affected by the phytoplankton bloom were plotted by the ship's navigation officer, who kept a detailed account of the areal extent of the bloom, in addition to the type and concentration of ice associated with it (Figure 1). The total ice coverage on February 10 was 4/10 to 6/10. The predominant form of ice was slush ice (6/10) south of 75°S and pancake ice (4/10) north of this latitude. Tabular icebergs and several leads in the ice were also observed. The very heavy, impenetrable pack ice encountered on February 11 (Station 9) forced the icebreaker to abandon its northerly pursuit and to backtrack to Station 10. It was estimated that the area affected by bloom was at least 15,500 km^2 (6100 mi^2). This estimate should be regarded as a conservative one; the exact area could have been far greater than the figure reported here.

Dense patches, like thick soup, of distinctly yellow-brownish color, were visible for miles on end; these patches were interspersed with long, narrow bands of relatively clear water (Figure 3). This pattern was typical of the slush-ice region. In the region where pancake ice (20–25 cm in diameter) predominated, very dense concentrations of phytoplankton were found on the underside of the sea ice. The accumulation of phytoplankton underneath the ice and in the interstitial ice crystals imparted a distinct brownish hue to the ice. The water between the pancake ice was relatively clear. The area affected by the bloom was noticeably teeming with Adélie penguins, crabeater seals, and other marine vertebrate life (Siniff et al., 1968). Their abundance seemed to reflect the high productivity of the water in the bloom region.

DISTRIBUTION OF CHLOROPHYLL A AND C^{14} UPTAKE IN SOUTHWESTERN WEDDELL SEA

The distribution of chlorophyll a and C^{14} uptake at various depths at the stations occupied in the southwestern Weddell Sea are plotted in Figure 4. At Stations 8 and 10, phytoplankton standing crop was highest at the surface; the other stations occupied showed maximum chlorophyll-a values at about 10 meters. Surface phytoplankton standing crop at Station 8 was 123 mg/m^3, and at locality 68 (position: 74°59′S, 59°50′W) was 190.39 mg/m^3. The latter value can be regarded as the highest value recorded to date for this pigment in antarctic waters. The photosynthetic pigments at Station 8 and locality 68 are several orders of magnitude higher than the average value of chlorophyll a (0.37 mg/m^3) for the rest of the stations occupied during the entire International Weddell Sea Expedition. Although the preponderance of the phytoplankton population at Station 8 was concentrated at the surface, an exponential reduction in the algal population was noted at a depth of 1 meter; at the latter depth the amount of chlorophyll a was only 2.83 mg/m^3. Between 1 meter and 100 meters, chlorophyll-a concentration showed a less precipitous decrease. Diatom cell counts of the water samples collected at Station 8 also showed a similar drastic decrease from, i.e., 455,590 cells/1 at the surface to 82,000 cells/1 at 10 meters.

The uptake of C^{14} at Station 8, in the above figure, showed a distribution that closely resembles that of the standing crop. The surface water sample exhibited very high C^{14} uptake (82.41 mg C/m^3/hr) compared to values collected at lower depths (2.27 and 2.04 mg C/m^3/hr at 1 meter and 10 meters, respectively). Despite the very high surface C^{14} uptake at Station 8, the integrated values in the euphotic zone (i.e., the zone between the sea surface and the depth to which

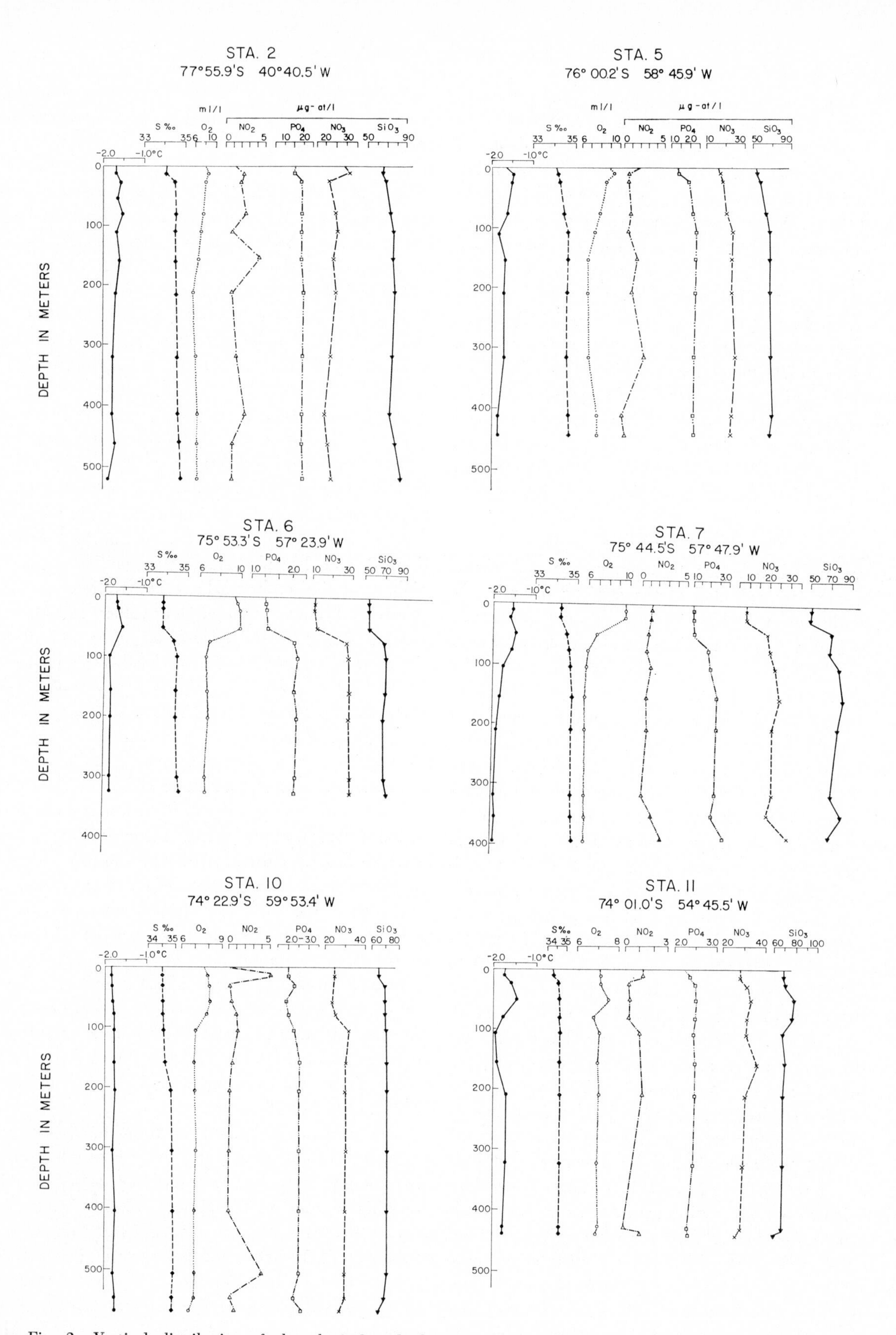

Fig. 2. Vertical distribution of the physical and chemical data collected in southwestern Weddell Sea in February 1968.

Fig. 3. (*Top*) Photograph of bloom of the diatom *Thalassiosira tumida* observed on February 10, 1968. (*Bottom*) Pancake ice underneath which heavy concentrations of diatoms were found.

1% of light penetrates) yielded a moderately high value (65.21 mg C/m^2/hr). On a daily basis, this is equivalent to 1.56 g C/m^2/day. This value is lower than the 3.2 g C/m^2/day recorded by El-Sayed [1968] in February 1965 in the Gerlache Strait and the 3.62 g C/m^2/day reported from Deception Island by Mandelli and Burkholder [1966].

Primary production at the other stations occupied in the southwestern Weddell Sea showed much lower surface values than those reported from either Station 8 or Station 10. In general, they follow a rather familiar pattern, i.e., high photosynthetic activity at or near the surface and minimal photosynthesis at the bottom of the euphotic layer. The rate of photosynthesis per unit of chlorophyll (sometimes referred to as photosynthetic index or assimilation number) at Station 8 was rather low. This rate, which ranged between 0.67 at the surface and 1.42 at 10 meters, seems to corroborate the findings of other investigators [Burkholder and Mandelli, 1965; Horne et al., 1969], who showed that low assimilation numbers usually accompany heavy phytoplankton standing crops.

SPECIES COMPOSITION AND RELATIVE ABUNDANCE OF THE PHYTOPLANKTON IN THE AREA INVESTIGATED

Species composition and number of cells per liter of the phytoplankton organisms at the stations occupied in the southwestern Weddell Sea are given in Table 1.

Microscopic examination of the phytoplankton samples showed the overwhelming abundance in the bloom of a centric diatom that was tentatively identified in the field as belonging to the genus *Coscinodiscus*. Later, when this diatom was studied by Hasle et al. [1971], using light and scanning electron microscopes (Figures 5 and 6), this diatom was transferred from *Coscinodiscus (C. tumidus* as reported by Hustedt [1958]) to the genus *Thalassiosira (T. tumida)*. The taxonomic position and morphological characteristics of this species are discussed by Hasle et al. in a separate paper in this volume. Very few diatoms, other than *Thalassiosira tumida,* were found in the surface sample during this bloom; these included *Eucampia balaustium* and *Charcotia actinochilus.* Among the dinoflagellates, a few *Peridinium mediocre* and *Prorocentrum balticum* were found. Also included in the bloom were crysophycean cysts and the tintinnid *Laackmanniella naviculaefera* in small numbers. It is interesting to note in Table 1 that whereas the surface phytoplankton at Station 8 was overwhelmingly composed of *Thalassiosira tumida* (about 98% of the diatom population), at a depth of 10 meters, *Fragilariopsis* spp. contributed about 70% and *T. tumida* less than 25% of the phytoplankton population at that depth.

It is rather difficult to establish with any degree of certainty the time of the initiation of the bloom. Study of species composition and relative abundance of the phytoplankton at the stations occupied in the southwestern Weddell Sea, however, could be revealing. If one assumes that *Thalassiosira tumida* is a good indicator of the bloom, then from the distribution and abundance of this species in the area investigated, one can speculate on the development and subsequent decline of the bloom. Using the water samples collected at 10 meters (Table 1), the cell counts per liter of *T. tumida* were as follows: Station 3—380; Station 5—9,680; Station 8—19,000; Station 10—18,500; Station 11—0. However, since the water sample collected at Station 8 was, unfortunately, not well preserved, it is very possible that the cell count of this sample was much higher than that reported here. It is likely, then, that the bloom could have originated somewhere in the vicinity of Station 3; and because of the clockwise movement of the surface currents in the Weddell Sea, one can further assume that Stations 5 and 8 could have been seeded from Station 3. The period of vigorous growth that followed the initiation of the bloom could have resulted in increased buoyancy leading to the massive concentration of the *Thalassiosira tumida* at the surface at station 8 (440,000 cells/l). This explanation finds support in experimental observations by Smayda and Boleyn [1965] on flotation and sinking rates of *Thalassiosira* cf. *nana, T. rotula,* and *Nitzschia seriata.* Their data suggest that 'during vigorous growth the observed increased buoyancy adequate to ensure sufficient retention within the euphotic zone occurs and is independent of the species' morphological characteristics' (p. 508). The moderately high surface chlorophyll-a value and the high cell count of *T. tumida* at Station 10 mark the geographical extent of our observations of the bloom. At Station 11 the phytoplankton had fallen off rather rapidly, and hardly any trace of the bloom was discernible.

Judging from the dates at which Stations 3 through 11 were occupied, it would seem plausible that the rise and fall of the bloom in question took place in a matter of a few days—possibly a week. It was unfortunate that the tight icebreaker schedule did not permit us to take advantage of this unique opportunity to make observations during the period of the decline

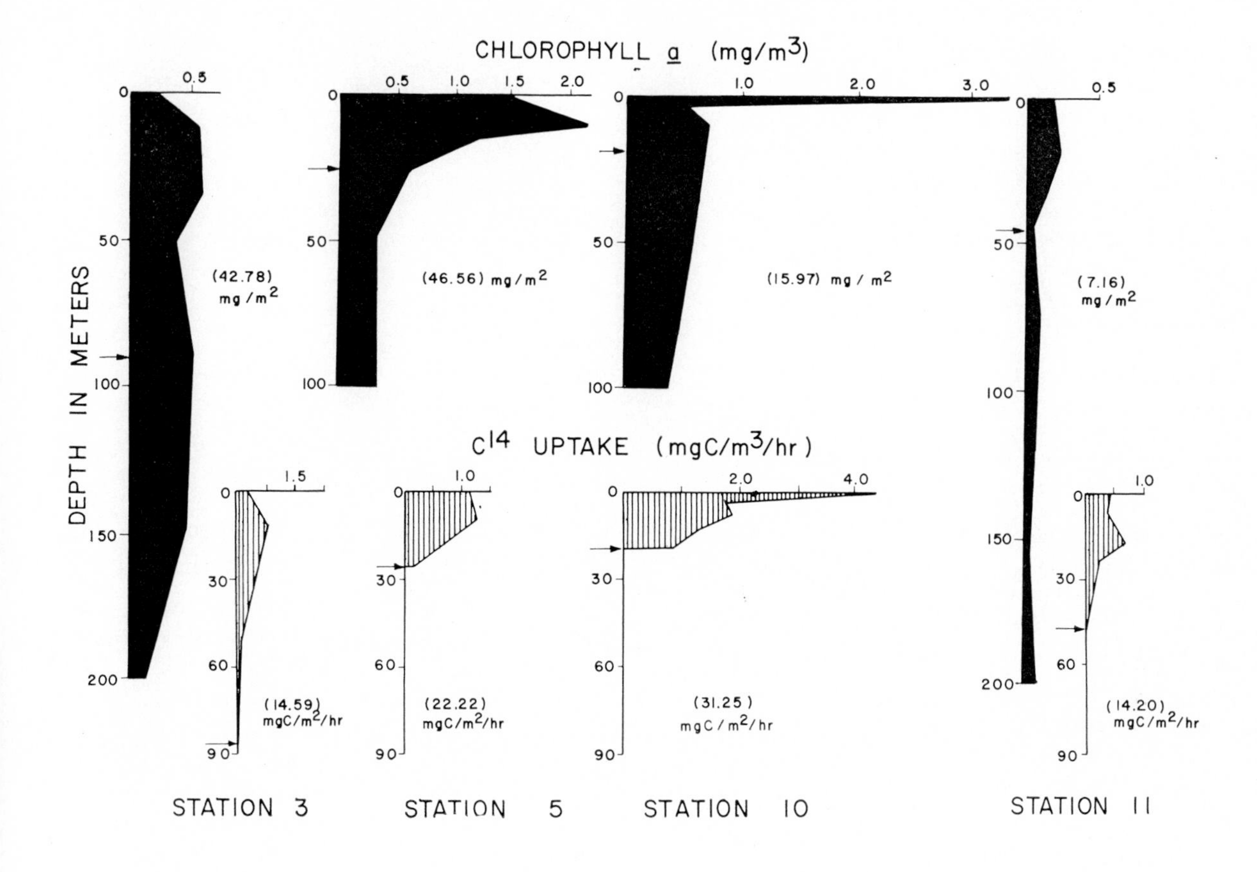

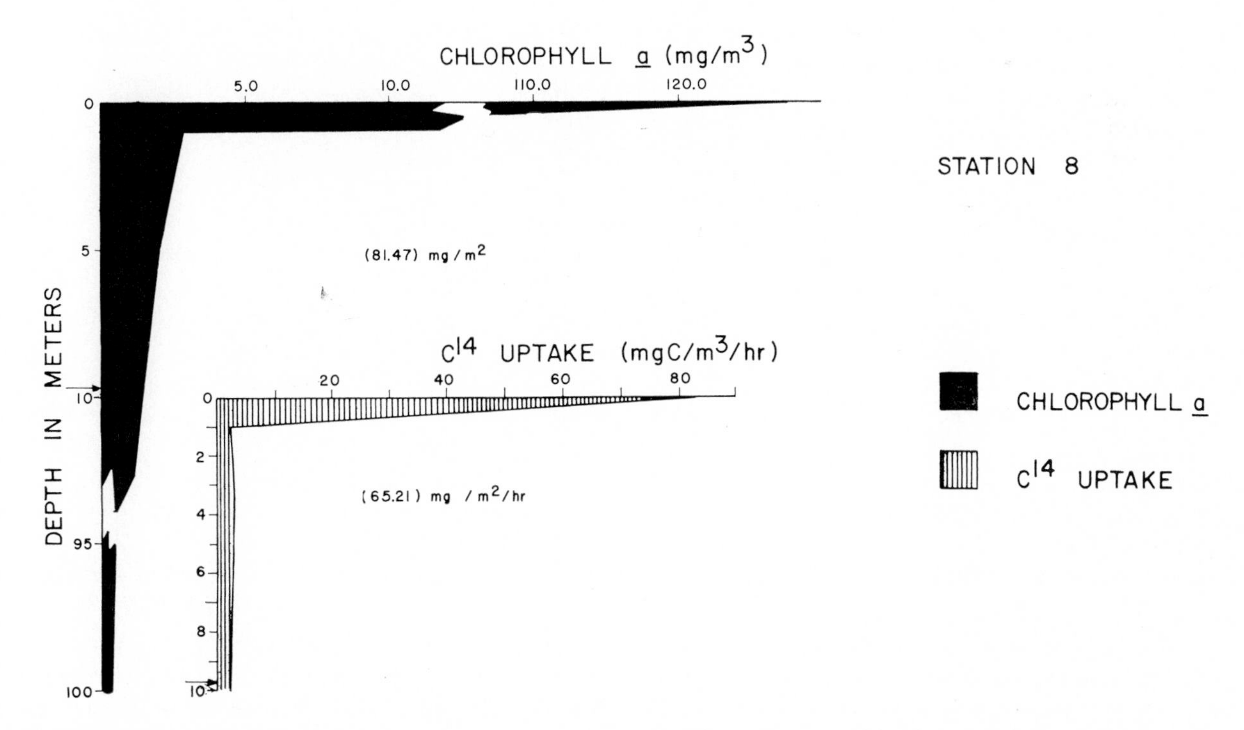

Fig. 4. Distribution of chlorophyll a and C^{14} uptake at various depths at the stations occupied in southwestern Weddell Sea. Integrated values throughout the euphotic zones are given in parentheses. Position of arrow indicates depth of euphotic zone.

TABLE 1. Species Composition and Number of Cells per Liter of Phytoplankton Organisms Collected in Water Samples at the Stations Occupied by USCGC *Glacier* in the Southwestern Weddell Sea*

Species	Station 1, Feb. 6–7, 1968, 10 m	Station 3, Feb. 8, 1968, 10 m	Station 5, Feb. 9, 1968, 10 m	Station 8, Feb. 10, 1968		Station 10, Feb. 11, 1968, 11 m	Station 11, Feb. 12, 1968, 6 m
				0 m	10 m		
Diatoms							
Actinocyclus curvatulus					20		
Amphiprora cf. *kufferathii*			120		80		
Asteromphalus hyalinus			160				
Biddulphia weissflogii			80				
Chaetoceros atlanticus			420				
C. flexuosus			600				
? *C. neglectus*			480				
C. spp.		220					60
Charcotia actinochilus		60		20			
Corethron criophilum	42,500	6,080	12,500		160	40	140
Coscinodiscus furcatus					20		
C. spp.			180				
Dactyliosolen antarcticus			40			20	
Eucampia balaustium			320	110	370	60	
Fragilariopsis curta	2,420	480	140,000				200
F. cylindrus	460	80	500				40
F. spp.	7,060	5,100	342,000	15,380	62,000	20,500	1,280
Melosira sphaerica					10		
Navicula sp.	20						
Nitzschia closterium	1,140	40	40				
N. prolongatoides	380						
N. turgiduloides			4,800				
N. spp.	2,560	20				140	
Pleurosigma sp.			20				
Rhizosolenia alata			40				
R. chunii			80				
R. styliformis					20		
R. var. *oceanica*			40				
R. var. *semispina*	20						
Synedra reinboldii	60		? 20				
Thalassiosira tumida		380	9,680	440,000	19,000	18,500	
T. sp.		1,840	1,680				20
Thalassiothrix antarctica		20	100		40	60	
Tropidoreis cf. *fusiformis*	1,000		520				
T. cf. *glacialis*			80				
T. sp.		60				260	
Unidentified centric							20
Unidentified pennate	20		80				
Dinoflagellates							
Dinoflagellate cysts		120	1,440	80	60		
Dinoflagellates, unidentified		40	160		60		
Dinophysis sp.		100	20				
cf. *Diplopeltopsis minor*			260				
cf. *Gonyaulax tamarensis*			20				
Gymnodinium lohmanni		40	240			40	20
Gymnodiniaceae		20					280
Gymnodineaceae or *Corethron* spores		420				2,500	

* Positions of stations:
1, 74°07.5′S, 39°38.5′W
3, 75°03.1′S, 50°12.2′W
5, 75°00.2′S, 56°45.9′W
8, 74°58.5′S, 60°57.3′W
10, 74°22.9′S, 59°53.4′W
11, 74°01.0′S, 54°45.5′W

TABLE 1. (continued)

Species	Station 1, Feb. 6–7, 1968, 10 m	Station 3, Feb. 8, 1968, 10 m	Station 5, Feb. 9, 1968, 10 m	Station 8, Feb. 10, 1968		Station 10, Feb. 11, 1968, 11 m	Station 11, Feb. 12, 1968, 6 m
				0 m	10 m		
Peridinium cf. *affine*			20				
P. antarcticum			20		20		
P. applanatum					20	40	
P. archiovatum			300				
P. concavum			60		20	? 20	
P. defectum		20	20			20	
P. incertum			20				
P. incognitum			40				
P. mediocre				60		40	
P. nanum					20	20	
P. pyriforme			20				
P. turbinatum					20		
P. variegatum					40		
P. spp.	20		40			140	
Prorocentrum balticum	4,500	240	20	? 20		? 500	120
P. truncatum		200				100	80
Other Forms							
Amphorellopsis sp.		40	20				
Chrysophyceae cysts			500	40	20		
Copepods						20	
Cymatocylis cf. *labiosa*		?	20				
Epiphyte not identified	500	320	200				
Laackmanniella naviculaefera			420	20	140	100	
Monads and flagellates not identified	15,000†	7,880				1,000	2,040
Nauplii						40	
Phaeocystis sp.			1,700,000			810,000	
Pterosperma parallelum	20	20					20
Strombidium elongatum			40				300
S. spp.		40					

* Approximate.
† May contain *Corethron* microspores.

of the bloom. After Station 11, our investigation had to be severely curtailed to allow the ship to rush an invalid crew member to Halley Bay Station.

DISCUSSION

Despite the short duration of the present study, and the fortuitous way in which the bloom was encountered, sufficient observations were made of the biological productivity of antarctic waters; these are worth commenting on.

Hart [1942] showed that the peak of the phytoplankton production in his biogeographical 'Southern Region' (in which he included all the seas south of the Antarctic Circle, excluding the immediate coastal area) is reached during February. Our observations in the Weddell Sea are in good agreement with those of Hart. However, in view of the fact that our observations were made off the Ronne Ice Shelf, the data suggest that the coastal waters should also be included in Hart's aforementioned definition of biogeographical 'Southern Region.'

Further, the low temperatures encountered in the bloom region did not seem to have a deleterious effect on phytoplankton production. This is clearly indicated by the high photosynthetic activity of the phytoplankton in surface water samples at Station 8, where the recorded temperature was −1.69°C. This observation corroborates Gran's [1932] finding, also in the Weddell Sea, that a rich phytoplankton growth can take place at the lowest (−1.5°C) as well as at the highest temperature observed. Our present observations, however, are at variance with the explanation given by Saijo and Kawashima [1964], who attributed their low primary productivity values in the antarctic waters to the effect of the near-freezing water on the photosynthetic rate.

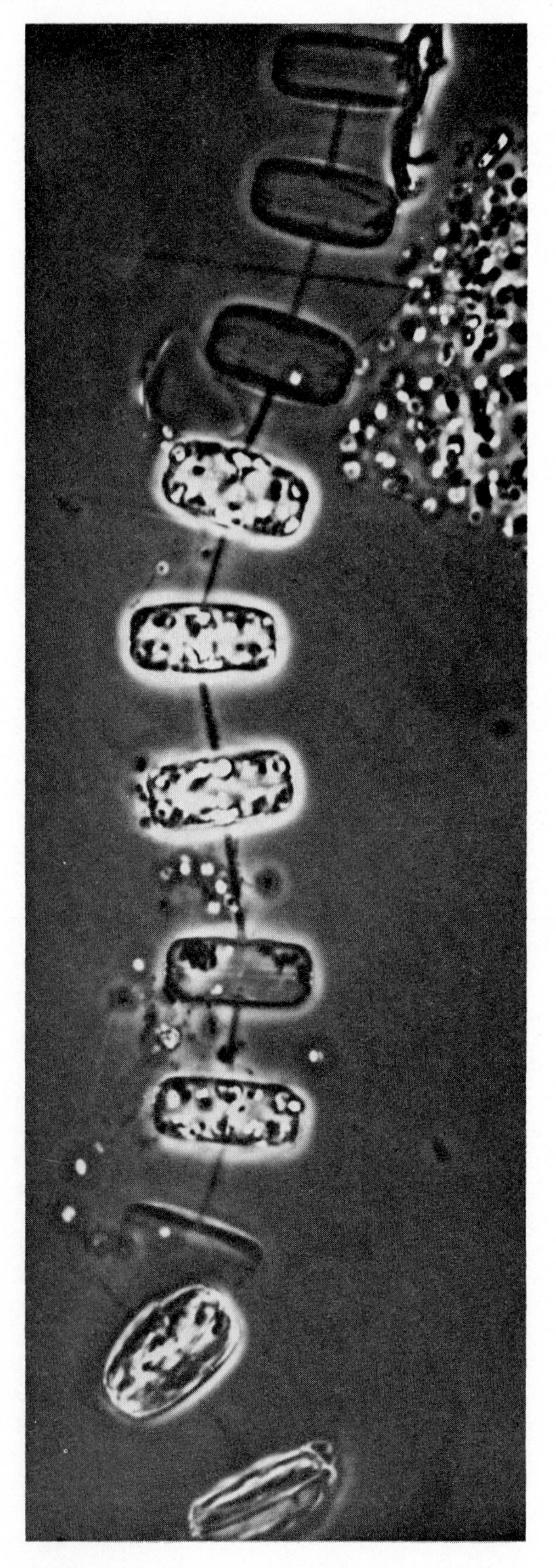

With regard to the effect of light, in view of the high photosynthetic activity of surface phytoplankton at Station 8, it seems unlikely that light intensity was a limiting factor there. However, the low C^{14} uptake values of subsurface samples at the same station could be attributed to low light intensity due to self-shading of phytoplankton. This self-shading had caused the shallowness of the euphotic zone (10 meters) observed at this station.

Study of the concentration of the macronutrient elements at the stations occupied in the southwestern Weddell Sea show their unlikelihood of being limiting factors for phytoplankton growth. Further, examination of the vertical distribution of the phosphates, nitrates, and silicates (Figure 2) at the stations occupied in the bloom show that, despite the reduced quantities of these nutrients at the surface compared to those at lower depths, there seems to be no evidence of nutrient exhaustion in the bloom region. This observation substantiates our belief that it is highly unlikely that the concentration of the nutrient elements studied in the Weddell Sea is a limiting factor in phytoplankton production in that sea. Moreover, the low values of the photosynthetic index are characteristic of cells with ample supplies of mineral nutrients [Fogg, 1965]. On the other hand, the proximity of the bloom to the inshore areas of the southwestern Weddell Sea would suggest that perhaps micronutrients and other trace elements from the neighboring land masses could have contributed to the development of the bloom.

Still remaining unanswered are questions such as: What are the mechanisms that triggered the present bloom? What are the factors that contributed to its decline? It is interesting to mention here that Rao [1969] has attributed the decrease and disappearance of the *Asterionella japonica* bloom in the Bay of Bengal to the exhaustion of thermostable nutrients in the environment, which according to Kain and Fogg [1958] are required by *A. japonica* in natural waters.

Finally, the fortuitous encounter of this enormous bloom in the ice-infested, and until recently inaccessible, southwestern Weddell Sea should force us to reevaluate the potential productivity of antarctic waters. For until the advent of icebreakers as research vessels in the Antarctic, estimates of the pro-

Fig. 5. *Thalassiosira tumida*, chain form, taken with light microscope, ×400. (Photo credit: G. A. Fryxell, Texas A & M University.)

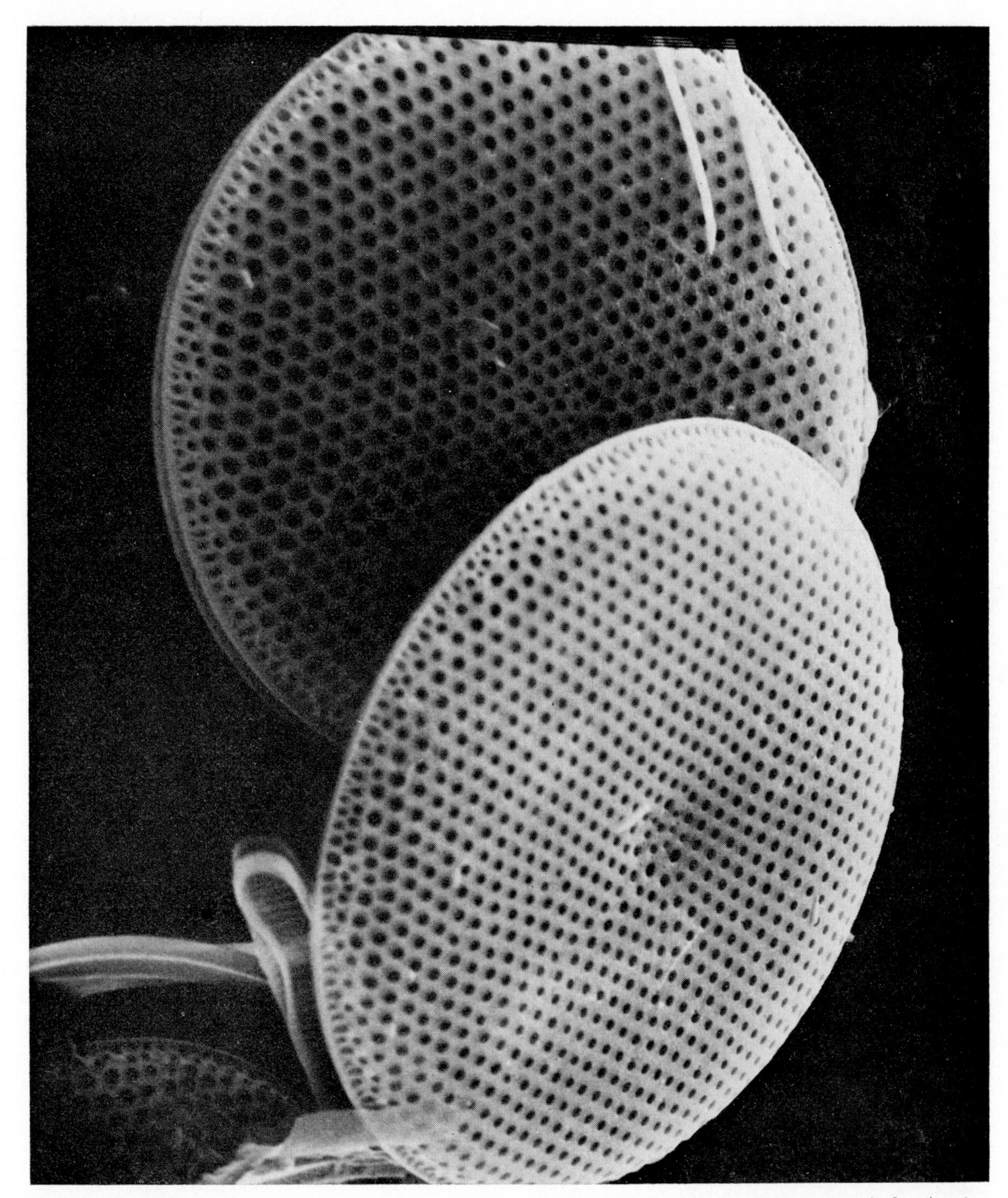

Fig. 6. *Thalassiosira tumida*, scanning electron microscope, ×2000. (Reprinted from *Transactions of the American Microscopical Society*, *89*(4), 272, 1970.)

ductivity of these waters were provided by ships operating in relatively ice-free and easily accessible waters. Thus, judging from the extent and size of the present bloom and the high productivity level reflected in the bloom region, teeming with penguins, winged birds, and seals, it is very possible that the productivity of the antarctic coastal waters could have been underestimated.

Acknowledgments. The author wishes to express his sincere gratitude to Dr. Grethe R. Hasle, Oslo University, for her analysis of the phytoplankton samples included in Table 1. Grateful appreciation is extended to Robert B. Elder and the U. S. Coast Guard Oceanographic Unit for providing the physical and chemical data collected during the IWSOE-1968. This study was supported by National Science Foundation grant G-1487.

REFERENCES

Burkholder, Paul R., and Enrique F. Mandelli
1965 Carbon assimilation of marine phytoplankton in Antarctica. Proc. Nat. Acad. Sci. U.S.A., *54*(2): 437–444.

El-Sayed, Sayed Z.
1968 On the productivity of the southwest Atlantic Ocean and the waters west of the Antarctic Peninsula. *In* G. A. Llano and W. L. Schmitt (eds.), Biology of the Antarctic Seas III. Antarct. Res. Ser., *11:* 15–47. AGU, Washington, D. C.

Fogg, G. E.
1965 Algal Cultures and Phytoplankton Ecology. 126 pp. University of Wisconsin Press, Madison, Wisconsin.

Gran, H. H.
1932 Phytoplankton: Methods and problems. J. Cons. Int. Explor. Mer., *7*(3): 343–358.

Hardy, A. C., and E. R. Gunther
1935 The plankton of the South Georgia whaling grounds and adjacent waters, 1926–1927. 'Discovery' Rep., *11:* 1–456.

Hart, T. J.
1934 On the phytoplankton of the southwest Atlantic and the Bellingshausen Sea, 1929–31. 'Discovery' Rep., *11:* 1–268.
1942 Phytoplankton periodicity in antarctic surface waters. 'Discovery' Rep., *21:* 261–365.

Hasle, G. R., B. R. Heimdal, and G. A. Fryxell
1971 Morphologic variability in fasciculated diatoms as exemplified by *Thalassiosira tumida* (Janisch) Hasle, comb. nov. *In* George A. Llano and I. E. Wallen (eds.), Biology of the Antarctic Seas IV. Antarct. Res. Ser., *17.* AGU, Washington, D.C.

Horne, A. J., G. E. Fogg, and D. J. Eagle
1969 Studies in situ of the primary production of an area of inshore Antarctic Sea. J. Mar. Biol. Ass. U.K., *49:* 393–405.

Hustedt, F.
1958 Diatomeen aus der Antarktis und dem Südatlantik. Dt. Antarkt.-Expedition, 1938–1939, *2:* 103–191.

Kain, J. M., and G. E. Fogg
1958 Studies on the growth of marine phytoplankton. *I. Asterionella japonica* Gran. J. Mar. Biol. Ass. U.K., *37:* 397–413.

Mandelli, E. F., and P. R. Burkholder
1966 Primary productivity in the Gerlache and Bransfield straits of Antarctica. J. Mar. Res., *24:* 15–27.

Murray, John
1889 On marine deposits in the Indian, Southern, and Antarctic oceans. Scott. Geogr. Mag., *5*(11): 405–436, pls. 1–12.

Parsons, T. R., and J. D. H. Strickland
1963 Discussion of spectrophotometric determination of marine-plant pigments, with revised equations for ascertaining chlorophylls and carotenoids. J. Mar. Res., *21*(3): 155–163.

Rao, D. V. Subba
1969 *Asterionella japonica* bloom and discoloration off Waltair, Bay of Bengal. Limnol. Oceanogr., *14*(4): 632–634.

Richards, F. A., and T. G. Thompson
1952 The estimation and characterization of plankton populations by pigment analysis. 2. A spectrophotometric method for estimation of plankton pigments. J. Mar. Res., *11:* 156–172.

Saijo, Yatsuka, and Takuji Kawashima
1964 Primary production in the Antarctic Ocean. J. Oceanogr. Soc. Japan, *19*(4): 22–28.

Siniff, D. B., D. R. Cline, and A. W. Erickson
1968 Population dynamics of antarctic seals and birds (IWSOE-1968). Antarct. J. U.S., *3*(4): 86–87.

Smayda, T. J., and Brenda J. Boleyn
1965 Experimental observations on the floatation of marine diatoms. I. *Thalassiosira* cf. *nana*, *Thalassiosira rotula* and *Nitzschia seriata.* Limnol. Oceanogr., *10*(4): 499–509.

Steemann Nielsen, E.
1952 The use of radioactive carbon (C^{14}) for measuring organic production in the sea. J. Cons. Int. Explor. Mer, *18:* 117–140.

Strickland, J. D. H., and T. R. Parsons
1968 A practical handbook of seawater analysis. Bull. Fish. Res. Bd Can., 167, 311 pp.

Walsh, John J.
1969 Vertical distribution of Antarctic phytoplankton. II. A comparison of phytoplankton standing crops in the Southern Ocean with that of the Florida Strait. Limnol. Oceanogr., *14*(1): 86–94.

MORPHOLOGIC VARIABILITY IN FASCICULATED DIATOMS AS EXEMPLIFIED BY *THALASSIOSIRA TUMIDA* (JANISCH) HASLE, COMB. NOV.

GRETHE R. HASLE AND BERIT R. HEIMDAL

Institute of Marine Biology, B, University of Oslo, Oslo 3, Norway

GRETA A. FRYXELL

Department of Oceanography, Texas A & M University
College Station, Texas 77843

Abstract. *Coscinodiscus inflatus* Karsten is an antarctic, weakly silicified species with valve areolae arranged in sectors; *Coscinodiscus tumidus* Janisch is another antarctic, coarsely silicified species with valve areolae in straight tangential or eccentric curved rows. Cells that could be classified as these two species and also cells having one valve from each of them were found mixed in colonies of *Thalassiosira* type. Light- and electron-microscope examinations of valves of both species found in the same samples demonstrated the presence of many central apiculi, numerous apiculi located both close to the valve margin and in the rest of the valve, and 3–5 labiate processes in the weakly as well as in the coarsely silicified valves. *Coscinodiscus inflatus* and *C. tumidus* are therefore regarded as conspecific and, because of the appearance in colonies produced by long threads emerging from central apiculi and connecting the cells, are transferred to the genus *Thalassiosira*. According to the rule of priority, the name should be *Thalassiosira tumida* (Janisch) Hasle comb. nov. A diagnosis including previous studies and our observations is presented.

Our own observations on materials collected in antarctic and subantarctic waters and the records of species conspecific with *T. tumida*, either according to literature or our own opinion, indicate *T. tumida* as a southern circumpolar species of a wide latitudinal range.

A possible general relationship between degree of silicification and valve areola pattern is discussed with reference to observations of other *Thalassiosira* species grown in unialgal cultures.

INTRODUCTION

Fasciculati is a group of *Coscinodiscus* species with the valve areolae arranged in radial sectors. Hustedt [1958] pointed out the great taxonomic problems within this group and its apparent predominance compared with the rest of the genus in subantarctic and antarctic waters.

The taxonomic problems originated in the pleomorphism (the occurrence of two or more patterns of valve structure in one species) present in the *Fasciculati* group. The pleomorphism was found to be confined to species in which the radial rows of areolae were running parallel to the median row. The sectors varied in width, and instead of being arranged in straight, radial rows the areolae were frequently in eccentric curved tangential rows and also in straight tangential rows. The size of the areolae was more or less variable, and the pleomorphism seemed to appear more frequently in finely structured species than in those with larger areolae.

The pleomorphism could be related neither to cell division nor to seasonal variation; the apparent lack of causality was demonstrated by the presence of various types of areola pattern in one and the same valve. It seemed to be questionable whether regularly fasciculated species really existed. The interspecific distinction was greatly impeded by the pleomorphism. New species may have been described in cases when mere modifications of one species were represented. Hustedt referred in this connection to the many species of this group described from the Antarctic by Karsten [1905].

Coscinodiscus inflatus Karsten is one of the species whose existence was questioned by Hustedt. He suggested that *C. inflatus* was a fasciculated form of *C. tumidus* Janisch, especially since a variety, *C. tumidus* var. *fasciculatus*, had been described by Rattray

[1890]. The fairly large marginal apiculi, which varied in number and also in position inside the sectors, were present in *C. inflatus* as well as in *C. tumidus* var. *fasciculatus.* The valve surface of *C. inflatus* had numerous spines in addition to the valve areolae, apparently in contrast to *C. tumidus* var. *fasciculatus.* Hustedt, however, doubted if this morphologic difference justified the rank of *C. inflatus* as a separate species distinct from *C. tumidus.* Two other species described by Karsten [1905], *C. quinquies-marcatus* and *C. incurvus,* were according to Hustedt's opinion conspecific with *C. inflatus.*

Our examination of species with the same morphologic characteristics as *C. inflatus* demonstrated the presence of many central pores (also noticeable in Hustedt, 1958, fig. 34) and connecting threads emerging from the pores (Figures 11–16). *Coscinodiscus inflatus* should therefore be transferred to the genus *Thalassiosira.* Unlike most *Thalassiosira* species examined, *T. inflata* had more than one labiate process (solitary apiculus or additional process), in addition to the numerous smaller marginal apiculi [Hasle, 1968, figs. 14–16]. The labiate processes had evidently been observed by Karsten [1905, p. 85], who designated them '5 Marken am Rande,' and by Hustedt [1958, p. 115], who regarded them as 'kleine randständige Prozesse' or 'Fortsätze,' while the smaller marginal apiculi had apparently not been observed by these authors.

The morphology of specimens of the *Thalassiosira inflata* type as revealed in light microscopes (LM), as well as in transmission electron microscopes (TEM) and scanning electron microscopes (SEM), and its affinity to *Coscinodiscus tumidus* is the main topic of this paper.

MATERIALS AND METHODS

The samples used are parts of collections previously studied by Cassie [1963], Balech and El-Sayed [1965], Kozlova [1966], and Hasle [1969], in addition to samples collected by the International Weddell Sea Oceanographic Expedition, 1968 [Dale, 1968]. The last-mentioned collection turned out to be the most important part of the material, containing a diversity of morphologic types of the forms in question. The collections originated in the Ross Sea, the eastern Weddell Sea, the Indian and Pacific sectors of the Antarctic, the Pacific sector on both sides of the Antarctic Convergence, and the ice-covered western Weddell Sea, respectively. All samples were collected during the antarctic summer, from December through March. Some of them are water samples; others are net hauls. The water samples were examined under the inverted microscope; the net hauls were examined as water, Hyrax or Coumarone mounts under Leitz Ortholux and Laborlux microscopes with different types of illumination, and under R.C.A. EMU 2A and 2D, JEM 7-5392, and Siemens Elmiskop I transmission electron

Figs. 1–12. Light micrographs of *Thalassiosira tumida* in water mounts. Labiate processes indicated by arrows in this plate and the following plates.

1. Drawing made from a light micrograph showing three cells in colony formation, the connecting strand consisting of many threads, the chromatophores in two cells, and threads protruding from the cells, girdle view, × about 250, Weddell Sea (76°00.2′S, 56°45.9′W, February 9, 1968).
2. Part of colony, girdle view, diameter (d) 40 μm, length of connecting strand 60 μm, phase-contrast illumination, ×300, Weddell Sea (76°00.2′S, 56°45.9′W, February 9, 1968).
3. Part of colony, girdle view, phase-contrast illumination, ×200, Weddell Sea (76°00.2′S, 56°45.9′W, February 9, 1968).

4–6. Colonies consisting of weakly and coarsely silicified cells, dividing cells in Figures 4 and 6, girdle view, bright-field illumination, ×500, Weddell Sea (76°00.2′S, 56°45.9′W, February 9, 1968).

7. Colony of shallow cells, connecting strand distinctly divided into many threads, girdle view, d 70 μm, pervalvar axis 20 μm, bright-field illumination, ×500, Ross Sea (76°35′S, 169°28′E, December 29, 1957).
8. Colony consisting of weakly silicified cells, girdle view, length of connecting strand 25 μm, bright-field illumination, ×500, Ross Sea (76°35′S, 169°28′E, December 29, 1957).
9. One weakly and one coarsely silicified specimen from the same sample, girdle view, bright-field illumination, ×500, Weddell Sea (74°58.5′S, 59°57.3′W, February 10, 1968).
10. A coarsely silicified specimen focused at the girdle with the three bands, bright-field illumination, ×500, Weddell Sea 74°58.5′S, 59°57.3′W, February 10, 1968).
11. Single weakly silicified valve with fasciculated structure, many central apiculi, numerous valve apiculi, one ring of small marginal apiculi, and two of the several labiate processes, valve view, bright-field illumination, ×1500, Ross Sea (77°10′S, 165°40′E, December 29, 1957).
12. Single weakly silicified valve with fasciculated structure, many central apiculi, numerous valve apiculi, one ring of small marginal apiculi, and two of the several labiate processes, valve view, phase-contrast illumination, ×750, Weddell Sea (76°00.2′S, 56°45.9′W, February 9, 1968).

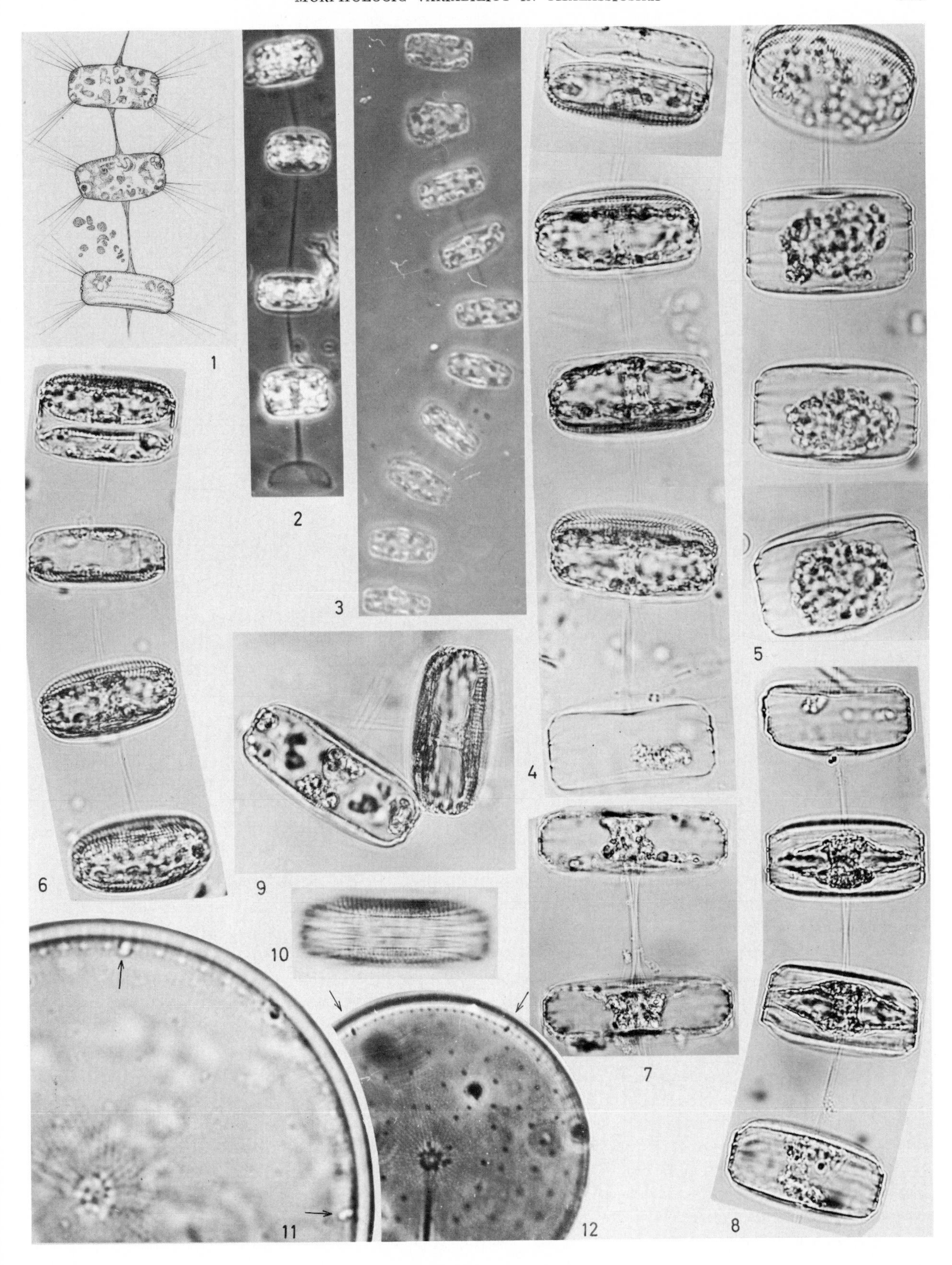
1
2
3
4
5
6
7
8
9
10
11
12

TABLE 1. Records from Material Collected by IWSOE-1968 of ***Thalassiosira tumida***

Station	Date	Position	Ice Cover in 10th	Depth, m	Number of Cells per Liter in Water Samples
	February				
1	6–7	74°07.5′S, 39°36.5′W	6	10	. . ., N*†
3	8	75°03.1′S, 50°12.2′W	9	10	380, N
5	9	76°00.2′S, 56°45.7′W	0	10	9,680, N
8	10	74°58.5′S, 59°57.3′W	6	0	440,000
8	10	74°58.5′S, 59°57.3′W	6	10	19,000, N
10	11	74°22.9′S, 59°53.4′W	8	11	18,500, N
11	12	74°01.0′S, 54°45.5′W	10	6	. . ., N
13	15–16	75°26.7′S, 26°32.6′W	5	7	. . .
14	16	75°31.1′S, 27°14.4′W	5	7	. . ., N
16	17	74°50.6′S, 39°04.9′W	0	6	?, N
18	18–19	72°46.0′S, 42°45.5′W	4	6	120
19	19	71°53.8′S, 48°20.1′W	8	7	. . .
21	20	71°31.7′S, 52°43.4′W	10	7	. . .
24	22	70°51.4′S, 52°12.8′W	10	7	. . ., N
26	23–24	70°55.1′S, 47°45.6′W	4	14	. . ., N
28	25	70°14.3′S, 50°10.3′W	9	10	. . .
30	26–27	69°33.3′S, 48°47.3′W	9	7	. . .
34	29	68°14.9′S, 47°51.3′W	9	12	. . ., N
	March				
36	1	69°01.6′S, 46°00.2′W	9	11	?
38	2	70°09.8′S, 43°41.9′W	10	16	120
40	3	71°47.6′S, 40°41.3′W	9	12	. . .
42	4	70°21.5′S, 37°44.8′W	1	5	440
44	5	69°33.7′S, 39°33.9′W	0	15	20
48	7	68°24.8′S, 43°03.7′W	0	7	160
52	9–10	67°22.0′S, 47°22.3′W	9	10	. . ., N
53	10	67°19.0′S, 48°59.8′W	9	14	. . .
56	12	66°20.1′S, 50°52.9′W	9	8	120
57	12	65°48.1′S, 50°50.5′W	9	10	100
59	13	64°36.0′S, 50°35.0′W	9	7	120

* . . . indicates no cells in 50 ml.
† N indicates collection by net haul.

microscopes and Stereoscan Mk IIa and K Square Ultrascan SM-2 scanning electron microscopes. The specimens examined in Hyrax and Coumarone mounts, as well as those examined under the electron microscopes, were single silica valves cleaned by the method used by Hasle and Fryxell [1970]. The morphologic terms to be used in this paper were defined by Hasle and Heimdal [1970]. (See also Hasle [1971].)

MORPHOLOGY

Light-Microscope Observations

Short chains of specimens of *T. inflata* were observed in samples from the Ross Sea (Figures 7, 8). The identification of the species was based on the presence of weakly silicified valves, valve areolae arranged in sectors in which the rows were parallel to the median row, numerous small marginal apiculi, more than one labiate process, and many central apiculi (Figures 11, 12). In samples collected in the ice-covered western Weddell Sea differently structured cells in colonies were found, particularly in those from Stations 5 and 8 (see Table 1 for position). Some chains were made up either of cells of *T. inflata* exclusively, as in the Ross Sea samples, or of cells of *C. tumidus* only. The latter species was characterized mainly by heavily silicified valves (Figure 9, to the right) and by areolae in tangential or eccentric rows or in more or less distinct sectors. In other cases the two types of cells appeared in the same chains together with cells with one *T. inflata* valve and one *C. tumidus* valve (Figures 4–6). The chains were slightly curved (Figures 1–6, 8), the cells being connected by a fairly thick but flexible strand. The strand consisted of many single threads (Figures 4–9), each of them emerging from one of the many central apiculi (Figures 7, 12). The threads were distinctly separated near the valve surface and twisted midway between two adjacent cells (Figures 7, 8). The intervals between cells in chains varied in length from about 25 to 60 μm, which is about equal to the length of the pervalvar axis to twice its length (Figures 1–8).

The single cells had a rectangular outline in girdle view, the valve surface being flat except for the depressed area around the central apiculi (Figures 4, 5, 8). The valve mantle seemed to consist of two zones, one less slanting than the other (Figures 5–9). The nearly horizontal zone of the valve mantle was attached to the girdle, which apparently consisted of three bands (Figure 10). The ratio of the length of the pervalvar axis to the diameter of the cell varied from about one-third to one-half; in two extremely shallow cells the length of the pervalvar axis was only two-sevenths of that of the diameter (Figure 7).

Cells with unequally structured valves were observed in the dividing stages. In one case the new valves were like those in *C. tumidus* (Figure 6), as shown in the diagram below (heavy line indicates coarsely silicified valve, the *Coscinodiscus tumidus* type; thin line indicates weakly silicified valve, the *Thalassiosira inflata* type):

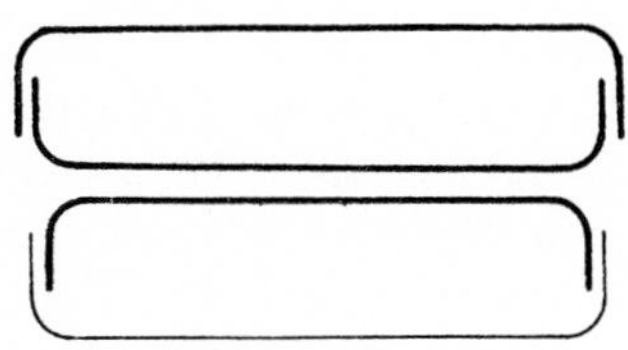

In another case the newly formed valves were like those of *T. inflata* (Figure 4), as shown in the diagram below:

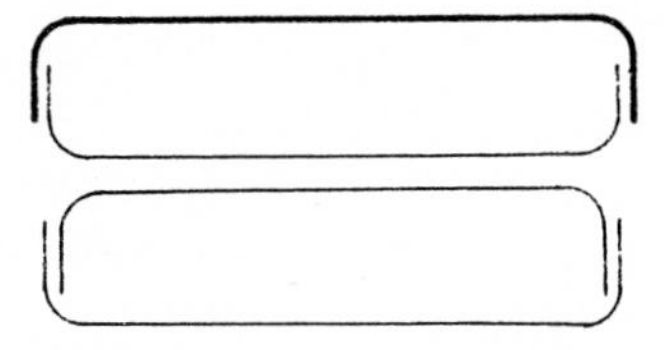

In a dividing mother cell with both valves of the *C. tumidus* type the daughter valves were both like those in *T. inflata,* as shown below:

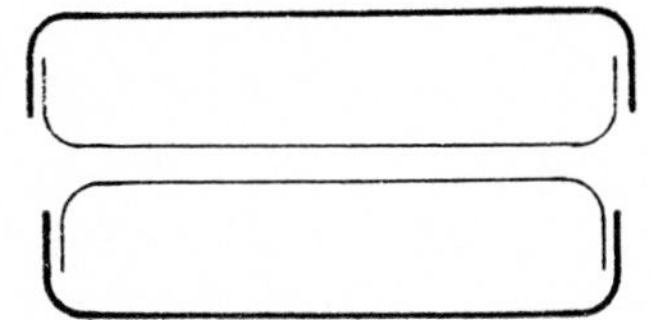

In two chains of weakly silicified cells, auxospore formation was observed. The striking feature was the coarse structure of the auxospore valve.

At first glance it seems reasonable to assume that the heavily silicified cells, the cells of the *C. tumidus* type, were resting spores in the same way as observed in *Thalassiosira nordenskioeldii* Cleve [e.g., Cupp, 1943, fig. 8e] and *T. antarctica* Comber [Hasle and Heimdal, 1968, fig. 4]. The appearance of heavily silicified cells in chains, the coarse structure of the auxospore as well as the formation of two heavily silicified daughter valves inside a mother cell with unequally structured valves evidently contradict this assumption. The explanation left must be that we are dealing with the pleomorphism referred to above.

The examination of single valves in Coumarone mounts prepared particularly from samples collected at Weddell Sea Stations 5 and 8 but also from other stations from this area, as well as examination of Coumarone and Hyrax mounts of specimens from other parts of the Antarctic, greatly supported this assumption. All sorts of transitional forms from the strictly tangentially and coarsely structured *C. tumidus* cells to fasciculated, finely structured *T. inflata* specimens were present. Central apiculi and apiculi scattered over the valve surface as well as labiate processes were more or less readily observed in the various forms, depending upon the thickness of the silica wall. The smaller marginal apiculi were less frequently perceived in the thick-walled specimens, while the labiate processes were seen as protuberances penetrating horizontally from the inner surface of the valve mantle into the cell cavity. The central apiculi and those on the rest of the valve were observed as small dots, in the depressed central part and in the zone between the center and the margin, respectively.

Variability with respect to the curvature of the valve surface was noticeable. The depression around the central apiculi was a generally occurring characteristic (Figures 5, 8, 18); however, its extension varied considerably, and in some specimens this central depression was surrounded by an elevation concentric with the valve margin, giving an inflated appearance to the valve (Figures 22–25). The weakly silicified specimens were so thin-walled that their valves easily changed shape or broke up; thus their shapes in permanent mounts and sometimes also in water mounts (Figure 1) were not adequate evidence of curvatures in intact cells.

The diameter of the specimens examined was 40–72 μm. The average diameter of 50 specimens in a water sample from Station 8 was 62 μm. The corresponding pervalvar axis was 18–29 μm with 22 μm as the average.

The weakly silicified specimens with fasciculated structure, the *T. inflata* type, had 9–14 areolae in 10 μm, 4–5 marginal apiculi in 10 μm, and at least 4 labiate processes (Figures 13–15, 20, 21). The tangential areola walls were partly missing in a few specimens of this type (Figure 13). Specimens with slightly eccentric rows of areolae or slightly fasciculated structure had 6–14 areolae in 10 μm and 4–5 labiate processes (Figures 19, 22, 24, 25), while the coarsely silicified specimens with *lineatus* structure, the *C. tumidus* type, had 5–12 areolae in 10 μm and 4–5 labiate processes (Figures 17, 18, 23). There seemed to be some correspondence between type of structure and size of areola, most of the *lineatus* structured valves having a thicker cell wall and larger areolae, mostly 5–6 in 10 μm, than the valves with areolae arranged in sectors.

In this way we are dealing with a species presenting so great a variability in silicification, arrangement, and size of areola and shape of valve surface that the individual valves might have been referred to more than one species. The chromatophores were of the same type in cells of different valve morphology, namely numerous small plates close to the cell wall.

Electron-Microscope Observations of Thalassiosira inflata

The observations of these thin-walled specimens supported the observations made in a light microscope in phase-contrast–dark-field illumination (Figures 14, 20, 21) of thin-walled, hexagonal areolae; many central apiculi; small, numerous apiculi scattered over the whole valve; numerous, small marginal apiculi in one ring; and 3–5 labiate processes. Moreover, the finer structure was revealed.

The degree of silicification of the diatom valve manifested itself in the thickness of the areola walls, which in some specimens were extremely thin (Figure 27) and in others fairly thick (Figure 28a), and evidently also in the structure of the central part of the valve; in weakly silicified valves the central apiculi were surrounded by apparently homogeneous membranes (Figure 14 in LM; Figures 26, 29a), while in more silicified valves regular areolae were present between the central apiculi (Figure 20 in LM; Figure 28a). The sieve membrane with 12–15 pores in 1 μm was readily observed on electron micrographs (Figures 27, 28, 29, 32c, 36, 37). A pore membrane with one large pore was apparently also present (Figure 28b). As already noticed in the light microscope, the tangential areola walls were occasionally missing in lightly silicified valves (Figures 26, 28a, 29), and scanning micrographs showed that the areola walls extended interiorly as well as exteriorly from the sieve membrane (Figures 31, 32). Areolae of a more irregular shape, sometimes also smaller than the majority, were present at the inner end of most of the areola rows, which formed boundaries between sectors (Figure 14 in LM; Figures 26, 28a, 29). The number of central apiculi varied considerably, the smallest observed being 6 and the greatest 23. The apiculi in the zone between center and valve margin were irregularly located, in contrast to those close to the margin, which appeared in one

Figs. 13–19. Light micrographs of *Thalassiosira tumida* valves in Hyrax and Coumarone mounts.

13. Weakly silicified specimen, fasciculated structure, many central apiculi, numerous valve apiculi, one ring of small marginal apiculi, labiate process visible, parts of valve without tangential areola walls, bright-field illumination, ×1500, Ross Sea (76°35′S, 169°28′E, December 29, 1957).
14. Weakly silicified specimen, fasciculated structure, many central apiculi, numerous valve apiculi, one ring of small marginal apiculi, phase-contrast–dark-field illumination, ×1500, Weddell Sea (74°58.5′S, 59°57.3′W, February 10, 1968).
15. Weakly silicified specimen, fasciculated structure, many central apiculi, numerous valve apiculi, labiate process visible, bright-field illumination, ×1500, Ross Sea (76°35′S, 169°28′E, December 29, 1957).
16. Moderately silicified specimen (d = 66 μm, 9–10 areolae in 10 μm, fasciculated structure poorly developed in central part of valve, apiculi as in Figure 13, labiate processes visible, bright-field illumination, ×1500, Ross Sea (76°35′S, 169°28′E, December 29, 1957).
17. Coarsely silicified specimen (about 6 areolae in 10 μm), straight tangential rows of areolae, central apiculi visible as small, dark openings, valve apiculi visible as white spots, areola sieve membranes just discernible, phase-contrast–dark-field illumination, ×1500, Weddell Sea (74°58.5′S, 59°57.3′W, February 10, 1968).
18. Coarsely silicified specimen (d = 64 μm about 6 areolae in 10 μm), straight tangential rows of areolae, bright-field illumination, ×1500, Weddell Sea (74°58.5′S, 59°57.3′W, February 10, 1968).
19. Coarsely silicified specimen (d = 52 μm, about 6 areolae in 10 μm), straight and curved tangential rows of areolae, central apiculi visible as small, dark openings, valve apiculi visible as dark openings, areola sieve membranes just discernible, phase-contrast–dark-field illumination, ×1500, Weddell Sea (74°58.5′S, 59°57.3′W, February 10, 1968).

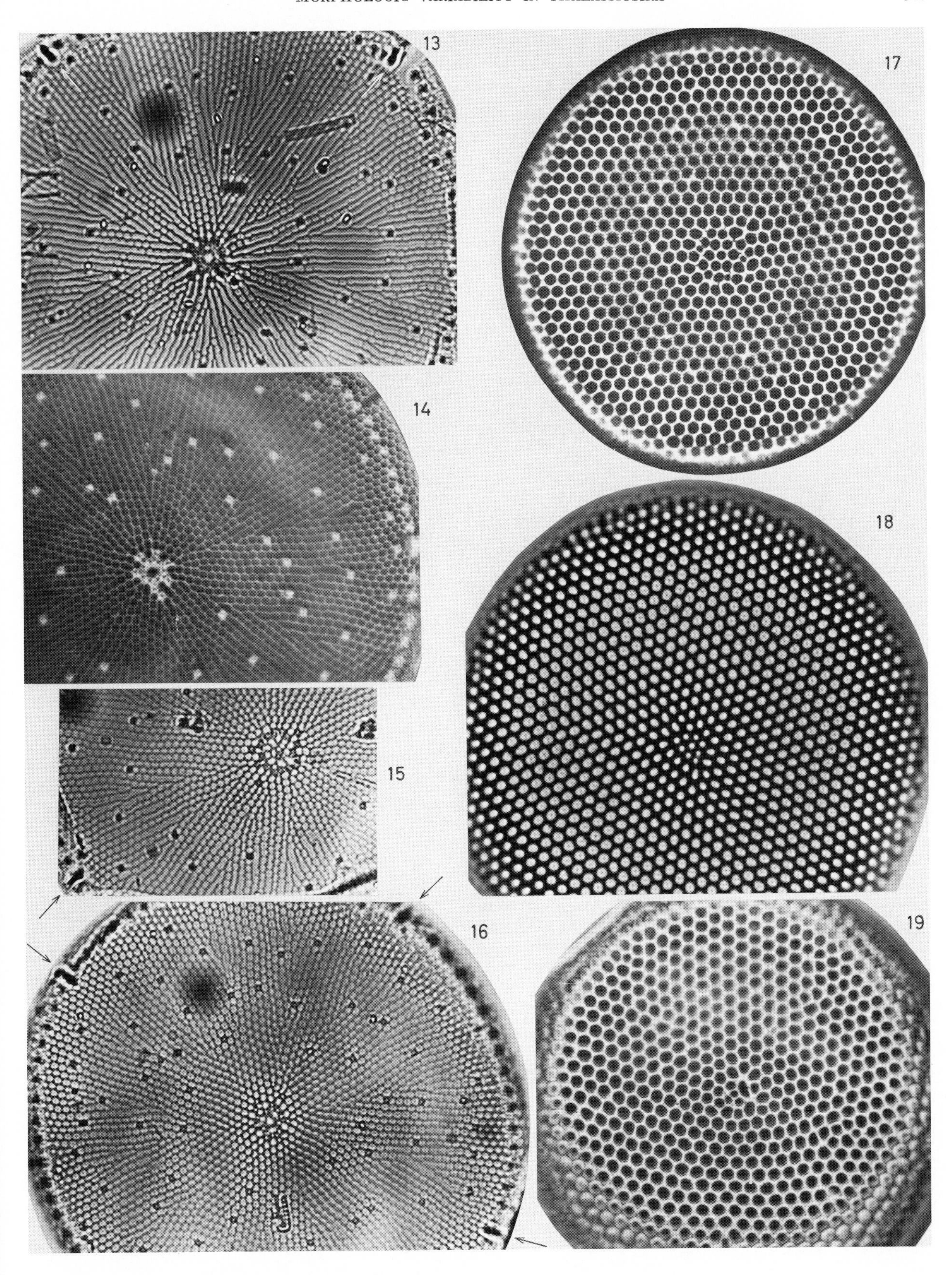
13
14
15
16
17
18
19

ring. There were 4–5 marginal apiculi in 10 μm, and they were separated by 1–3 valve areolae (Figure 14 in LM; Figures 26, 28b). All apiculi mentioned so far were built in the same way. They were tubes open at both ends; their shorter parts penetrated into the cell interior, and the longer parts were outside the cell (Figures 31, 38). The small pores, usually 4 in number, located on the inner surface of the valve wall and surrounding the vertical tube, were readily observed in these thin-walled specimens (Figures 26–28, 29–31, 34–38). Their location on the inside of the valve was convincingly demonstrated by SEM observations (Figures 30, 31, inside view; Figures 32, 33, outside view). Each small pore was partly covered by an arched outgrowth from the inner central tube of the single apiculous. The outgrowths seemed to be club-shaped with the wider portion inserted into the pores (Figures 27, 28, 29).

The labiate processes were particularly conspicuous in *T. inflata* specimens. Their inner part could be seen in a broken area of valve margin studied in TEM (Figure 27). But the scanning microscope was evidently the appropriate instrument for the study of these structures. Micrographs from SEM showed the whole length of the labiate process, the outer slightly trumpet-shaped part, the breakthrough of the cell wall, and the inner purse- or anvil-shaped part (Figures 32, 38). The inner portion of a labiate process obviously had a far more complex structure than the external part. It seemed to consist of a fairly stout stalk flaring into a much wider part. This enlarged part had a longitudinal slit surrounded by two lips (Figures 34–37). There was some evidence that the stalk, or the wide part, or both were bent or curved in some way in these prepared specimens (Figure 38). Although the present materials offered excellent opportunities for revealing the structure of the labiate process, the examination left the authors in the same or even greater doubt concerning its function and dynamics than after previous investigations of the same type [Hasle, 1968; Hasle and Heimdal, 1970]. The intricate structure of its internal portion, so different from that of the central and marginal apiculi, seems to signify a special function that neither this nor the previous studies indicated.

In one of the specimens studied, a part of an intercalary band was still attached to the valve mantle (Figure 27). The structure of this intercalary band consisted of longitudinally running silicified bars traversed by less silicified ribs.

Scanning-Microscope Observations of Coscinodiscus tumidus

Because of the heavy silicification of the valves, direct examination in TEM would give little information about the presence and structure of valve processes. The scanning micrographs verified the presence of numerous central apiculi, one ring of small marginal apiculi close to the valve margin, and apiculi also in other parts of the valve, as well as more than one labiate process.

The internal parts of the protuberances were those readily discernible (Figures 39, 41). They were arranged and built in the same way as in the *Thalassiosira inflata* specimens discussed above.

Figs. 20–25. Light micrographs of *Thalassiosira tumida* valves in Hyrax and Coumarone mounts.

20. Weakly silicified specimen (d = 35 μm, 13 areolae in 10 μm), structure as in Figures 13–15, phase-contrast-dark-field illumination, ×2250, Weddell Sea (58°52.4′S, 37°57.9′W, December 18, 1963).
21. Weakly silicified specimen (d = 47, 10–11 areolae in 10 μm), structure as in Figures 13–15, phase-contrast–dark-field illumination, ×2250, South Shetland Is., 1896, British Museum mount 59677.
22. Moderately silicified specimen (d = 41 μm, 12 areolae in 10 μm, 4 labiate processes), valve and marginal apiculi visible as brighter spots, central area depressed and surrounded by an elevation, slight irregularities in areola pattern close to valve margin, phase-contrast illumination, ×1500, Weddell Sea (66°20.1′S, 50°52.9′W, March 12, 1968).
23. Heavily silicified specimen (d = 47 μm, 8–9 areolae in 10 μm, 5 labiate processes), valve and marginal apiculi visible as brighter spots, central area depressed and surrounded by an elevation, straight and slightly curved tangential rows of areolae, phase-contrast illumination, ×1500, Weddell Sea (66°20.1′S, 50°52.9′W, March 12, 1968).
24. Heavily silicified specimen (d = 48 μm, 8–9 areolae in 10 μm, 5 labiate processes), valve and marginal apiculi visible as brighter spots, central area depressed and surrounded by an elevation, slight irregularities in areola pattern close to valve margin, phase-contrast illumination, ×1500, Weddell Sea (66°20.1′S, 50°52.9′W, March 12, 1968); *a* and *b* indicate micrographs of the same specimen.
25. Moderately silicified specimen (d = 45 μm, 11 areolae in 10 μm, 4 labiate processes), valve and marginal apiculi visible as brighter spots, central area depressed and surrounded by an elevation, a few wide sectors of radial areola rows, rest of valve with straight or curved tangential rows, phase-contrast illumination, ×1500, Weddell Sea (66°20.1′S, 50°52.9′W, March 12, 1968).

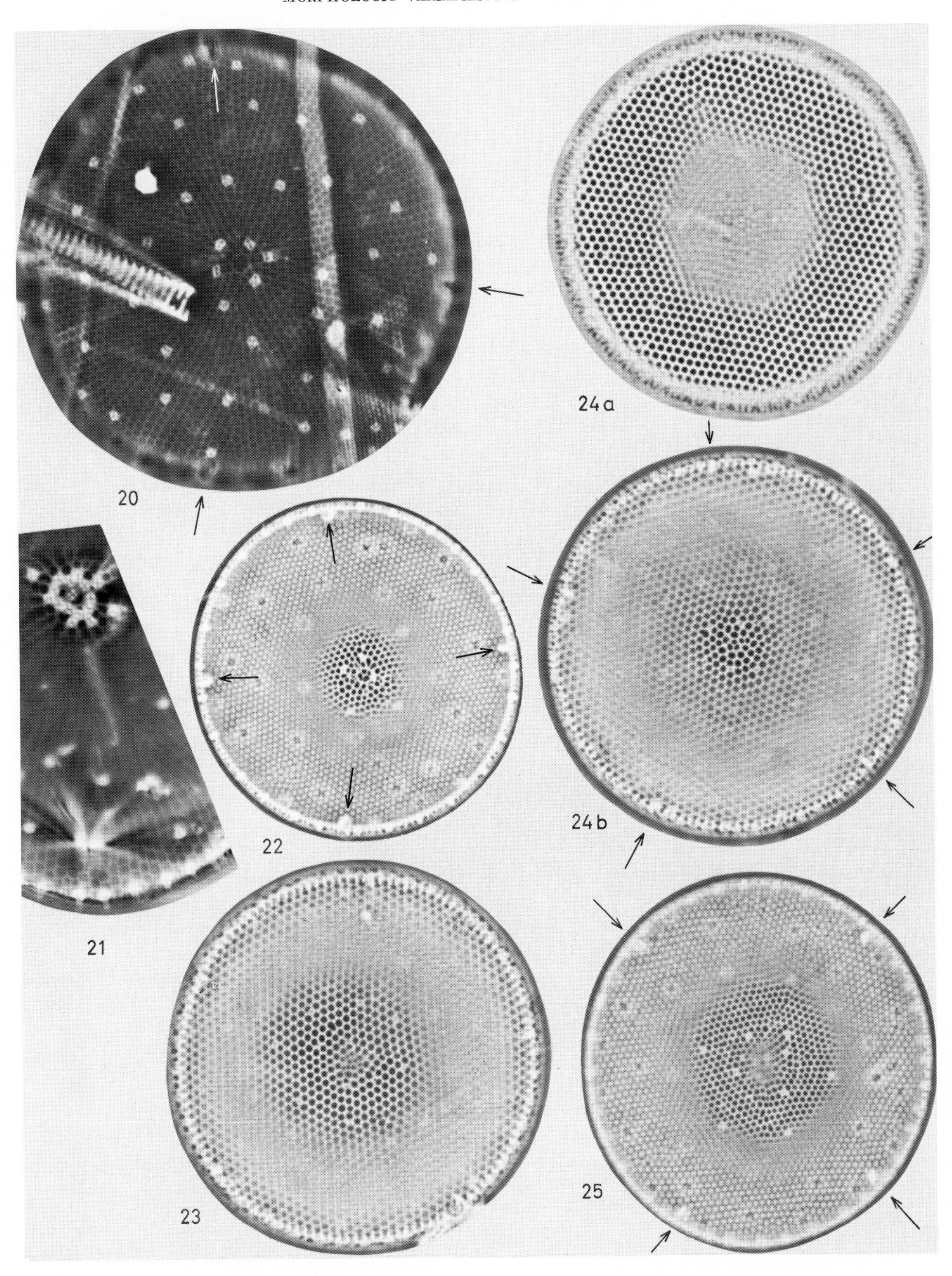

20
21
22
23
24a
24b
25

TABLE 2. Data on *Coscinodiscus tumidus, C. tumidus* var. *fasciculatus*, and *C. inflatus* from the Literature

References	Diameter, μm	Areolae in 10 μm	Number of 'Spines'*
Coscinodiscus tumidus Janisch			
Schmidt et al., 1874–1959, pl. 59, figs. 38, 39	110, 120†	4–6†	
Rattray, 1890, p. 475	164–200	3–4½	
Karsten, 1905, p. 80, pl. 6, fig. 1	80–130	4–5†	
Hendey, 1937, p. 251	100–140		
Hustedt, 1958, p. 120, fig. 29	60†	7†	
Zhuse et al., 1962, p. 77, pl. 1, figs. 4, 5	40–240	7–8, 10–12	
Coscinodiscus tumidus var. *fasciculatus* Rattray			
Rattray, 1890, p. 476	200–280		
Hustedt, 1958, p. 120, figs. 30, 31	74, 96	6–7†	
Manguin, 1960, p. 252, pl. 24, figs. 291, 292	70–188	3–4	
Kozlova, 1966, pl. 4, figs. 10, 11	48, 63†	6–7, 9†	
Coscinodiscus inflatus Karsten			
Karsten, 1905, p. 85, pl. 7, figs. 7, 7a	68–106	10–12	5
Hendey, 1937, p. 253	90–160		5 (usually)
Hustedt, 1958, p. 115, figs. 34, 35	88,100†	8–9†	6, 9†
Zhuse et al., 1962, p. 72, pl. 1, figs. 3, 6	45–80	14–16	
Kozlova, 1966, pl. 2, figs. 3, 4	58, 73†		

* Labiate processes.

† The data were not given by the respective authors, but estimated from the illustrations.

From these observations it was obvious that the sieve membranes were located on the inner surface of the valve. They were finely perforated (in the specimens examined, with 10 pores in 1 μm) and seemed to bulge slightly deeper into the cell than the walls of the areolae (Figure 41). The outside view of these thick-walled specimens did not afford the same possibility for discerning the structure of the protuberances. The apiculi appeared only as open holes surrounded by a slightly thickened ring, while the long external protrusions seen in *T. inflata* in SEM were missing (Figure 43). This was the case also in specimens in which other thin-walled structures like sieve membranes were intact (Figure 40).

Few SEM observations have been made: *T. inflata* was examined from 1 net haul and *C. tumida* from 2. No real conclusion should be drawn concerning a possible difference between the two forms in external length of the protuberances.

The marginal apiculi and the labiate processes were located where the more or less flattened part of the valve turned into the fairly steep valve mantle (Figures 39, 40a). The valve mantle seemed to end in a nearly horizontal flange to which an intercalary band is attached (Figure 39c). The intercalary band was perforated by small pores arranged in fairly regular rows perpendicular to the valvar plane. The girdle band joined to the intercalary band was not perforated except for a row of fine pores close to the junction between the two types of band (Figures 40a, 46).

TAXONOMY

The summarized data on *Coscinodiscus tumidus, C. tumidus* var. *fasciculatus*, and *C. inflatus* available in the literature give a wide range in diameter as well as in number of areolae in 10 μm (Table 2). As in the present material, the most coarsely structured specimens belonged to the *C. tumidus* type.

Coscinodiscus tumidus was first published in Schmidt [1874–1959, pl. 59, figs. 38, 39] by Janisch, while *C.*

Figs. 26–28. Transmission electron micrographs of weakly silicified valves of *Thalassiosira tumida.*

26. Part of a valve, showing closely spaced central apiculi, valve and marginal apiculi, and one large labiate process; radial rows of areolae in sectors, ×2900, Ross Sea (76°35′S, 169°28′E, December 29, 1957).
27. Part of valve margin with valve and marginal apiculi and the inner anvil-shaped part of a labiate process, quadrangular to hexagonal thin-walled areolae with finely perforated sieve membranes, and a part of an intercalary band with longitudinally running silicified bars traversed by less silicified ribs, ×14,600, Ross Sea (76°35′S, 169°28′E, December 29, 1957).
28. Specimen with central apiculi separated by ordinary areolae and with valve and marginal apiculi; Weddell Sea (58°52.4′S, 37°57.9′W, December 18, 1963). (*a*) Central part, ×9800. (*b*) Marginal part, sieve membranes and pore membranes of areolae present, ×8400.

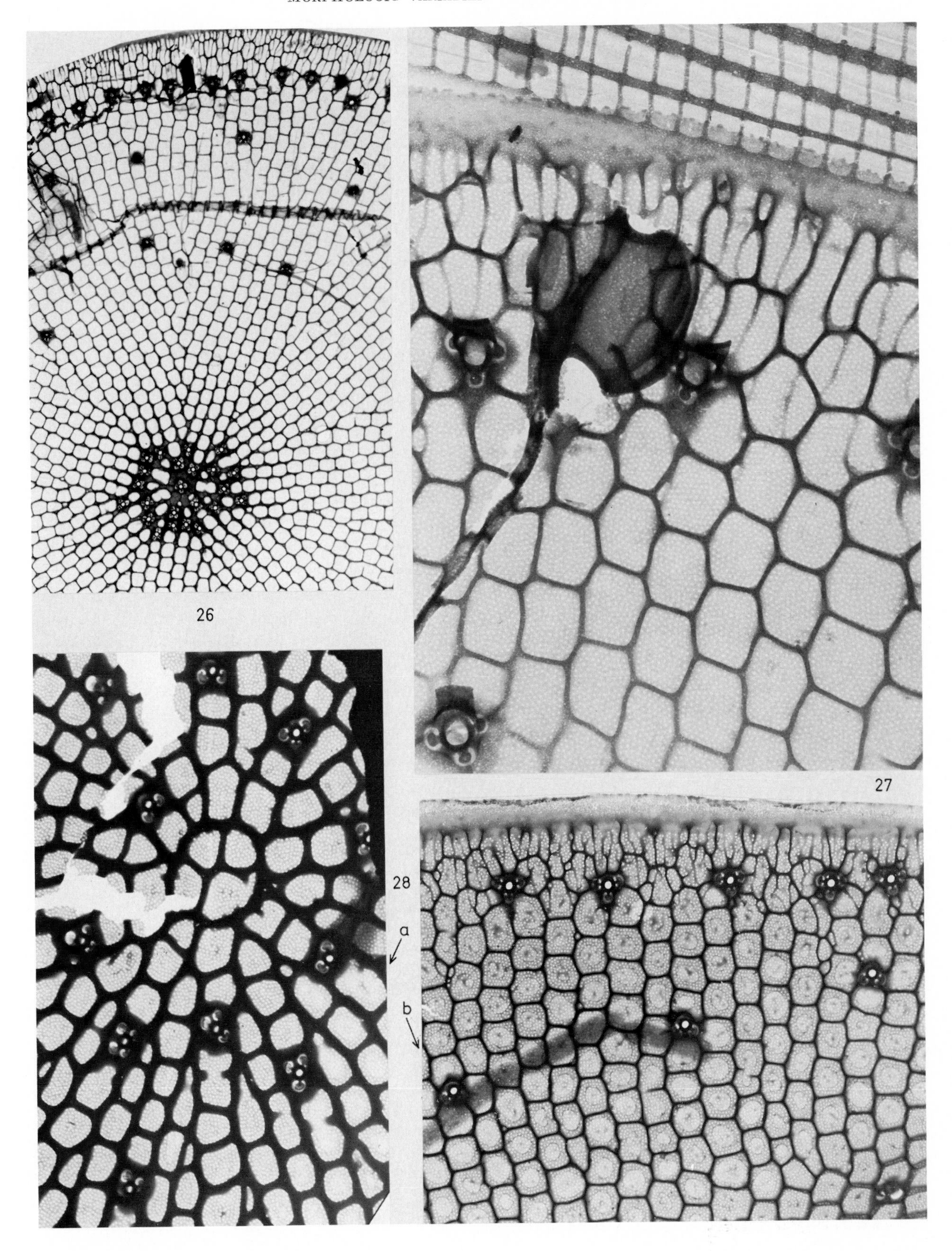
26
27
28
a
b

TABLE 3. *Coscinodiscus* Species with the Same Type of Valve Structure as *Coscinodiscus inflatus*

Species and References	Diameter, μm	Areolae in 10 μm	Number of 'Spines'*
C. chunii, Karsten, 1905, p. 86, pl. 7, figs. 10, 10a	62–128	about 7	6 or more
C. incurvus, Karsten, 1905, p. 85, pl. 7, figs. 8, 8a	56–60		
C. quinquies-marcatus, Karsten, 1905, p. 85, pl. 7, figs. 6, 6a	64–78	10–12	5 (rare, 6)
C. subtilis Ehrenb., Karsten, 1905, p. 86, pl. 7, figs. 11, 11a	82–160		
C. trigonus, Karsten, 1905, p. 84, pl. 5, fig. 10	70–74	12	3
C. quadrifarius, Manguin, 1960, p. 251, pl. 24, figs. 284, 285	47–50	8–10	4

* Labiate processes.

inflatus was described by Karsten [1905]. The valid name for the species under discussion is therefore *Thalassiosira tumida* (Janisch) comb. nov., since it appears in chain-formed colonies and also in other respects agrees with this genus. *Coscinodiscus tumidus* is the basionym of *T. tumida,* while *C. tumidus* var. *fasciculatus, C. inflatus, C. incurvus, C. quinquies-marcatus* (the two latter suggested by Hustedt [1958]), *C. quadrifarius,* and probably also *C. chunii* and *C. trigonus,* are its synonyms (see Table 3). According to Hustedt [1958], *C. subtilis* Ehrenb. in Karsten [1905] might also belong to the same species. The species identified by Manguin [1960] as *Thalassiosira antarctica* Comber positively does not belong to that species (see Hasle and Heimdal [1968]). The excellent illustrations in Manguin [1960], figs. 19–25, show the morphologic features characterizing *T. tumida,* like the many central apiculi, the many central threads, the apiculi scattered over the valve surface, the several labiate processes, and the central flattened, convex, or concave part of the valve surface.

Thalassiosira tumida belongs to the group of *Thalassiosira* species in which the greater length of the processes extends out of the cell and the labiate processes are located at the valve margin [Hasle, 1968]. The presence of more than one labiate process is also found in *Thalassiosira baltica* (Grun.) Ostenf. Up to this time at least a few labiate processes have been found in all *Coscinodiscus* species [Hasle, 1971]. There is also evidence that the marginal spines of *Coscinodiscus crenulatus* Grunow are built in the same way (TEM observations in Hasle [1960], figs. 14, 18).

In colonies in water mounts *T. tumida,* particularly the weakly silicified form, has a certain similarity to *T. gravida* Cleve, *T. rotula* Meunier, *T. antarctica,* and *T. hyalina* (Grun.) Gran. Of these, *T. antarctica* and apparently also *T. gravida* sometimes occur in the same areas as *T. tumida.* The similarity in aspects of

Figs. 29–33. Electron micrographs of weakly silicified valves of *Thalassiosira tumida.*

29. (*a*) A central area with central apiculi separated by an irregular large areola and with valve apiculi and areola sieve membranes; some tangential areola walls are lacking while the radial ones, particularly those close to the center, are well developed; the innermost part of the radial areola row forming the boundary between sectors ending in a triangular areola, transmission electron micrograph, ×6400, Pacific sector of the Southern Ocean (68°13′S, 120°16′W, January 31, 1948). (*b*) Part of *a* photographed under higher magnification, showing perforation of sieve membranes and club-shaped outgrowths inserted into surrounding pores of apiculi, the pores located in inner surface of valve wall, ×13,800.
30. Central area of a specimen, inside view, short tubular parts of central apiculi with club-shaped outgrowths, irregularly shaped areolae with sieve membranes just discernible, contamination by a particle of unknown nature in the center, scanning electron micrograph, ×8400, Ross Sea (76°35′S, 169°28′E, December 29, 1957).
31. Part of slightly broken margin, showing inner short and outer long portions of the tubular apiculi, and inner part of areola walls, scanning electron micrograph, ×8000, Ross Sea (76°35′S, 169°28′E, December 29, 1957).
32. (*a*) View of outer tubular portions of four marginal apiculi, apparently connected with areola walls at their base; one detached labiate process, partition between outer tubular or slightly trumpet-shaped portion and inner anvil- or purse-shaped portion marked by valve wall still attached to the process, scanning electron micrograph, ×8900, Ross Sea (76°35′S, 169°28′E, December 29, 1957). (*b*) Outside view of same marginal area photographed under different magnification and rotated, also showing the longitudinal slit surrounded by two lips, scanning electron micrograph, ×16,000. (*c*) Outside view of another area of the same specimen, outer portions of regularly shaped areolae with sieve membranes, scanning electron micrograph, ×44,700.
33. Outside view of central area, outer portions of the central tubular apiculi, and irregularly shaped areolae, scanning electron micrograph, ×8700, Ross Sea (76°35′S, 169°28′E, December 29, 1957).

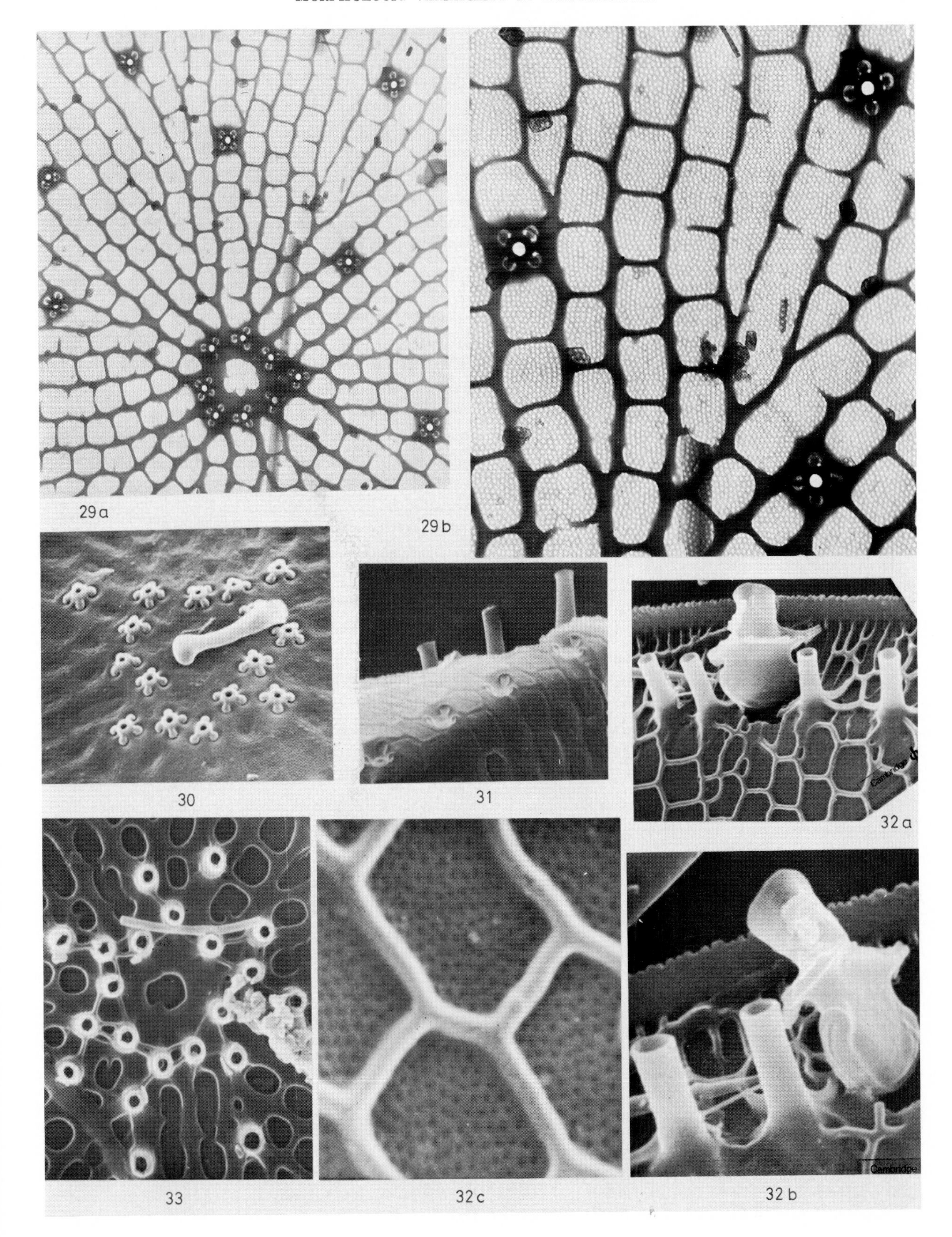
29a
29b
30
31
32a
33
32c
32b
Cambridge
Cambridge

the colonies lies in the shape of the cells in girdle view and the presence of many central pores and many connecting threads. *Thalassiosira hyalina* should be readily distinguished from the other species mentioned by the irregular intervals between the cells, if it appears in not too short chains. The coarsely silicified form seems to be comparable only with *T. decipiens* (Grun.) Jørg. of all species described. The similarity is found in the coarseness of the structure and partly also in the arrangement of the areolae. *Thalassiosira decipiens* has, on the other hand, one central pore and consequently one thin connecting thread between cells in chains. Furthermore, the intervals between cells in chains are greater than in *T. tumida* colonies. The marginal apiculi of *T. decipiens* are larger and more conspicuous than those of the species in question (Figures 44, 45), and they are discernible also in girdle view of entire cells.

DIAGNOSIS

The following diagnosis of *Thalassiosira tumida* (Janisch) Hasle, comb. nov., is based on previous observations of *Coscinodiscus tumidus* Janisch, *C. inflatus* Karsten, and species now considered conspecific with them, as well as on our observations through light and electron microscopes.

Cells are in slightly curved chains formed by a moderately thick connecting strand consisting of many threads. Each thread emerges from one of many central apiculi. Threads are distinctly separated close to the valve surface, twisted midway between adjacent cells in colonies. There are various morphologic types of cells in a single chain.

Intervals between cells in colonies are approximately 25–60 μm (the same length to twice the length of the pervalvar axis). Cell diameter is 40–280 μm; in materials examined by us, it is 40–72 μm, with corresponding pervalvar axis 18–29 μm. Valves are depressed in central parts. Hexagonal valve areolae: 3–16 in 10 μm, 6–14 in present materials, arranged in straight tangential rows or in slightly curved eccentric rows or in radial sectors made up of rows of areolae parallel to the median row or in more than one of these patterns in one valve. Areolae sieve membrane is perforated by 10–15 pores in 1 μm and located on the internal valve surface. Outer pore membrane was observed in a few specimens. There are 6–23 central apiculi in one or more irregular concentric rings; areolae separating central apiculi are variable in size and shape; in some thin-walled valves central apiculi are separated by nonperforated areas. Marginal apiculi: 4–5 in 10 μm and separated by 1–3 areolae; in one row close to valve margin; not observable in girdle view of whole cells. There are 3–9 (3–5 in the present materials) labiate processes located fairly regularly between the small marginal apiculi. Apiculi are built in the same way as central and marginal apiculi and are regularly distributed on remaining part of valve. Internal part of apiculi have club-shaped outgrowths, each one bending into each of the small holes in the valve surface at the base of the apiculus. Outer parts of labiate processes are wider than that of marginal apiculi and slightly trumpet-shaped. Inner part consists of a flattened, fairly wide stalk, almost vertically penetrating from

---

Figs. 34–39. Scanning electron micrographs of *Thalassiosira tumida* valves.

34. (*a*) Weakly silicified valve, inner portion of labiate process, ×15,300 Ross Sea (76°35′S, 169°28′E, December 29, 1957). (*b*) Same specimen viewed from a different angle, with two marginal apiculi, ×17,300.
35. Weakly silicified valve, inner portion of labiate process and marginal apiculus, ×18,700, Ross Sea (76°35′S, 169°28′E, December 29, 1957).
36. (*a*) Weakly silicified valve, inner portion of labiate process and marginal apiculi, ×16,000, Ross Sea (76°35′S, 169°28′E, December 29, 1957). (*b*) Same specimen viewed from a different angle, ×17,300.
37. Weakly silified valve, inner portion of labiate process and marginal apiculi, ×19,300, Ross Sea (76°35′S, 169°28′E, December 29, 1957).
38. Weakly silicified valve, area of valve margin with one marginal apiculus and one labiate process viewed from different angles; the apiculus is open at both ends and tubular in its whole length; its inner part is short; its outer part is long; two outgrowths from its inner part are visible; the outer part of the labiate process is slightly trumpet-shaped and open; the inner part consists of a short stalk flaring into a wider part ending in a labiate portion with a narrow slit, ×17,300, Ross Sea (76°35′S, 169°28′E, December 29, 1957).
39. (*a*) Inside view of a coarsely silified valve; central, valve, and marginal apiculi, two labiate processes, and areola sieve membranes visible, ×1300, Weddell Sea (76°00.2′S, 56°45.9′W, February 9, 1968). (*b, c*) Same specimen, the two labiate processes photographed under higher magnification, × 6700.

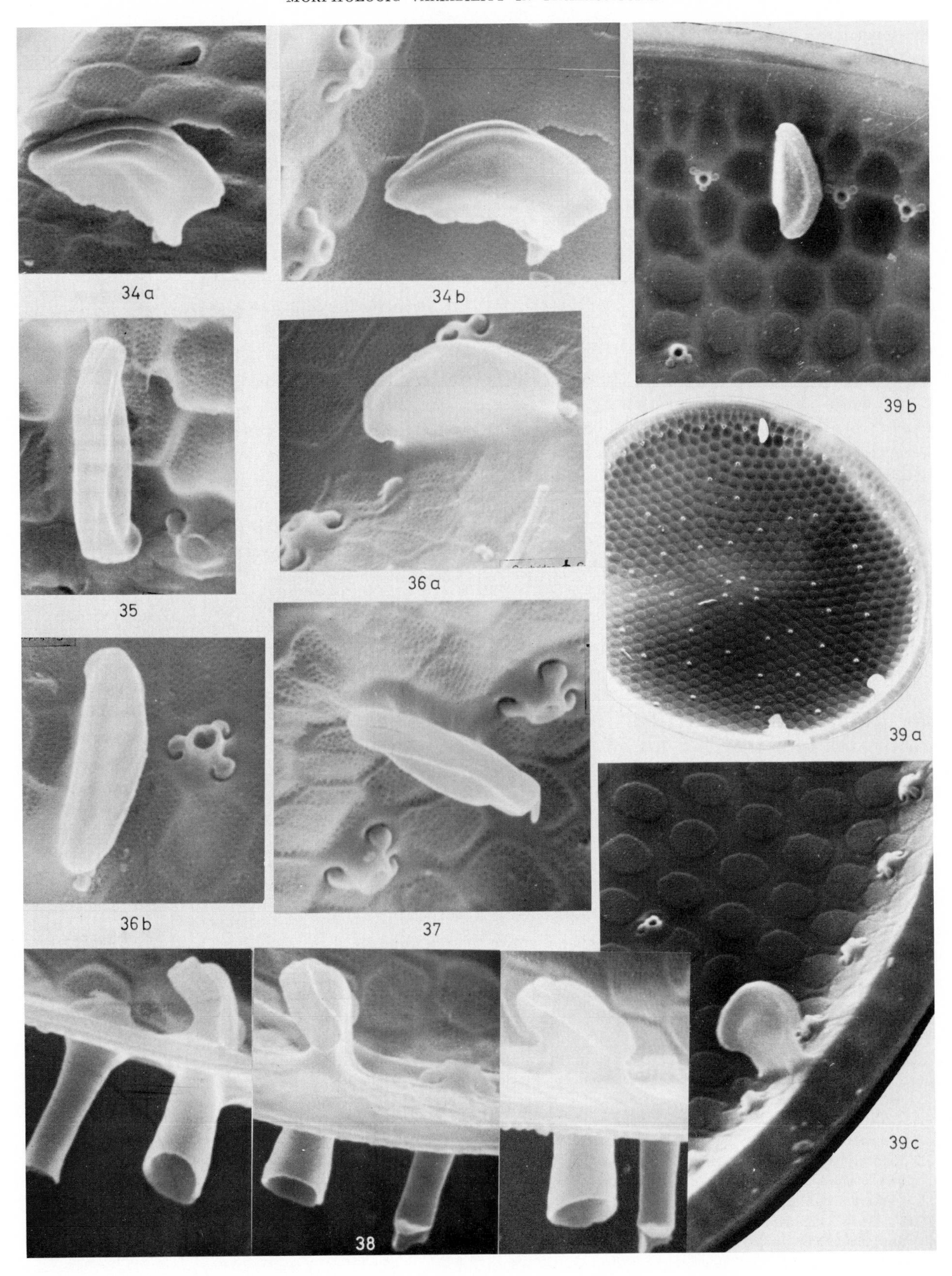
34 a
34 b
39 b
35
36 a
39 a
36 b
37
38
39 c

the inner valve surface, and a wider anvil-shaped or purse-shaped part with a longitudinal slit surrounded by broad lips and facing the cell interior.

Valve mantle is low and is joined to the single intercalary band of each cell half by a homogeneous, well-silicified brim. Intercalary band is perforated by small pores in more or less regular transverse rows. Girdle band is homogeneously silicified except for a single longitudinal row of small pores close to junction between intercalary and girdle bands. Chromatophores are numerous small plates close to the cell wall.

DISTRIBUTION

The records of *Thalassiosira tumida* quoted in Tables 2 and 3, in addition to those made by the present authors on the materials mentioned above, support evidence for a circumpolar distribution of the species within antarctic and subantarctic waters. It has been recorded from the South Atlantic sector [Karsten, 1905; Hendey, 1937; Hustedt, 1958], from the Indian sector [Karsten, 1905; Kozlova, 1962; Zhuse et al., 1962; Manguin, 1960], from the South Pacific sector [Van Heurck, 1909; Hasle, 1969], and from the Weddell Sea, as well as from the Ross Sea. Very few of the records were made north of the Antarctic Convergence.

Hustedt [1958] reported the presence of coarsely structured specimens of *Coscinodiscus tumidus* in his plankton samples from 53°43′S, 0°20′W, while in our samples from the Weddell Sea diatoms of this type were the predominant part of the population, forming a bloom as far south as 74°58.5′S, 59°57.3′W near the Ronne Ice Shelf [El-Sayed, 1968]. Water samples collected at this locality were conservatively estimated to contain between 400,000 and 900,000 cells per liter of *T. tumida* (estimation of the cell number was impeded by the presence of single valves and entire cells). The sea was partly covered by ice [Dale, 1968]. Coarsely silicified *T. tumida* specimens also made up the bulk of diatoms in a sample of floating brown pancake ice collected at 74°00′S, 45°27′W, February 10, 1968. There seemed to be no close relationship between the degree of the ice cover and the presence of *T. tumida* in the Weddell Sea collection; approximately the same abundance was observed in open water between the ice as in waters partly or almost totally covered by ice of various types (Table 1).

Manguin [1960] designated his species *Coscinodiscus quadrifarius* as a meroplanktonic form, appearing at the stations close to the Adélie Coast. According to the description and the illustrations, *C. quadrifarius* seems to be a fasciculated, moderately well silicified form of *T. tumida* like our specimen (Figure 16). Since the species identified by the same author as *Thalassiosira antarctica* Comber is considered in this paper to be a weakly silicified form of *T. tumida*, its distribution is also of interest. Manguin recorded it from the stations close to Adélie Coast except for one station located midway between the coast and the Antarctic Convergence. *Coscinodiscus tumidus* var. *fasciculatus*, which should be a better silicified form, was, on the other hand, more frequently recorded midway between the Antarctic Convergence and the coast than close to the coast. Manguin's material was collected in the first half of February, that is, at the same time of year as the samples collected farthest south in the Weddell Sea. In the present material the weakly silicified form appeared more abundantly in the samples from the Ross Sea than in any of the other samples examined, particularly in two samples collected as far south as 76°35′S and 77°10′S, that is, close to the ice shelf. The heavily silicified forms seemed to occur sparsely if at all. This material was collected at the end of December, which is fairly early in the summer at these southern localities.

Figs. 40–43. Scanning electron micrographs of *Thalassiosira tumida* valves.

40. (*a*) Marginal part of coarsely silicified specimen, outside view, with one intercalary and one girdle band, the former more perforated than the latter; thick-walled areolae, outer opening of valve and marginal apiculi visible, ×6700, Weddell Sea (76°00.2′S, 56°45.9′W, February 9, 1968). (*b*) Broken part of same valve (see Figure 46), photographed to show the presence of areola sieve membranes, ×13,300.
41. Central part of coarsely silicified specimen, inside view, short interior parts of central apiculi with three outgrowths, areola sieve membranes, ×13,300, Weddell Sea (76°00.2′S, 56°45.9′W, February 9, 1968).
42. Central part of coarsely silicified specimen, outside view, exterior parts of central apiculi, ×6700, Weddell Sea (76°00.2′S, 56°45.9′W, February 9, 1968).
43. Marginal area of weakly silicified specimen, outside view, one large labiate process, two smaller marginal apiculi, ×16,000, Ross Sea (76°35′S, 169°28′E, December 29, 1957).

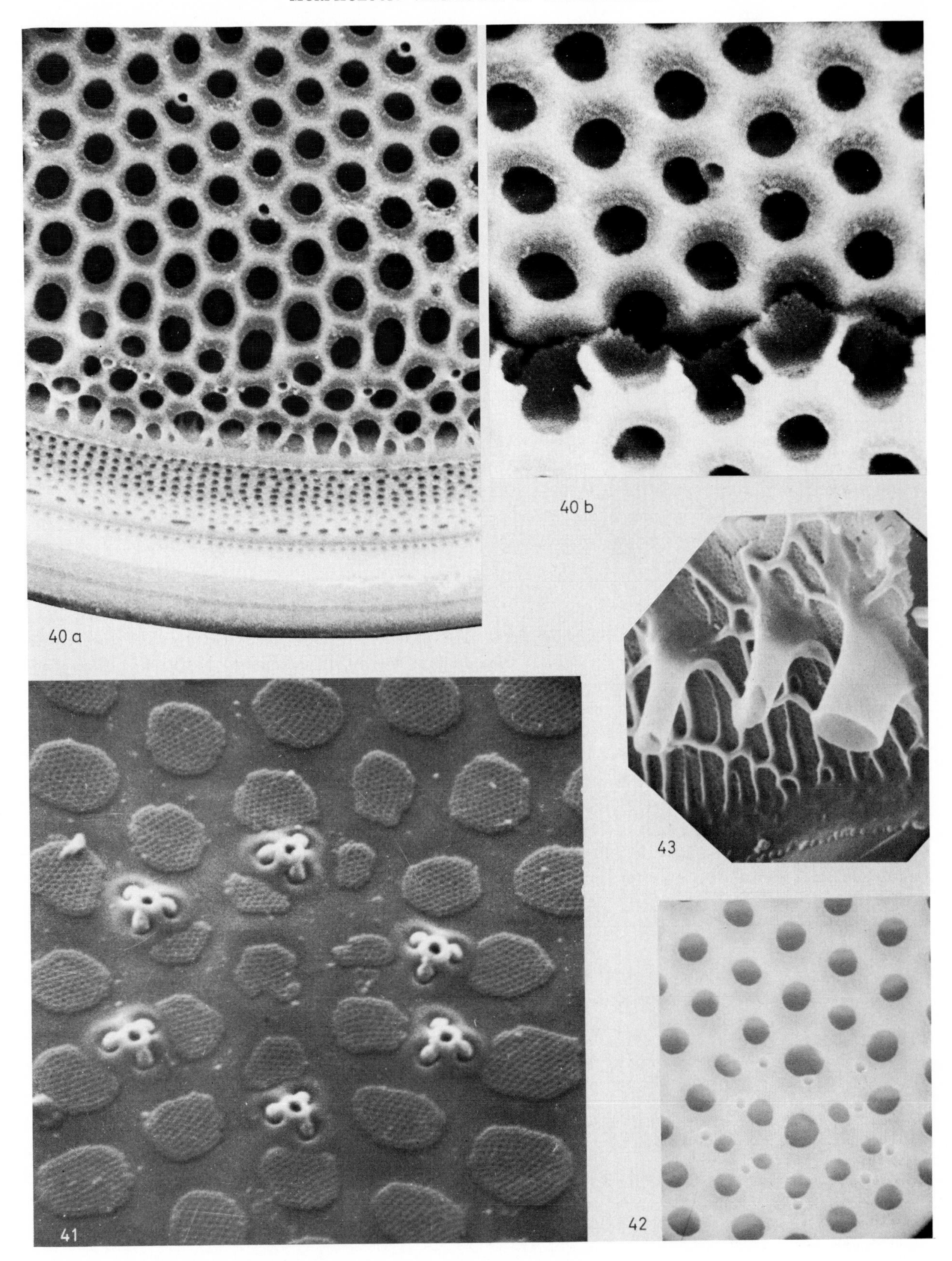
40 a
40 b
41
42
43

Kozlova [1962] considered *C. inflatus* an antarctic species, which in her papers indicates a distribution south of the Antarctic Divergence, and *C. tumidus* as being distributed between the Antarctic Divergence and the Antarctic Convergence. Kozlova [1966] added 'neritic' to the distribution of *C. inflatus* and 'oceanic' to that of *C. tumidus* var. *fasciculatus.* Zhuse et al. [1962], studying diatoms in sediments, listed *C. inflatus* as well as *C. tumidus* as subantarctic (distributed between the Antarctic Convergence and the Antarctic Divergence) and 'oceanic.'

Neither the observations cited from the literature nor those based on the present material seem to be the proper ones to give information on the spatial and seasonal distribution of the various forms of *T. tumida.* There seemed to be a slight indication of greatest abundance of heavily silicified cells near the ice at the end of the season and a predominance of weakly silicified cells earlier in the season. A possible ecological reason for such a change through the season is difficult to figure out; it seems reasonable to assume that more silica would be available at the beginning of the season than at its end.

DISCUSSION

A main asset of the present observations is the demonstration of differently structured cells within the same *Thalassiosira tumida* chain. The apparent correlation between the degree of silicification and the arrangement of the valve areolae may also be regarded as an interesting phenomenon. The latter feature was demonstrated in the tendency of areolae to be in tangential and curved rows in coarsely silicified valves and in sectors in those weakly silicified. Some evidence for different location of the sieve membrane in thin-walled and thick-walled valves was present, the sieve membrane being between the outer and inner surfaces of the cell wall in the former and on the inside of the cell wall in the latter type. The material also offered excellent opportunity for examinations of the valve processes, particularly of the labiate processes, which in this species were considerable in size. The mode of junction between apiculi and inner valve surface was also well illustrated in this species; the location close to the inner valve surface of the small pores surrounding the base of the apiculi was evident. A new element of the apiculi was brought to our knowledge by the study of this species, namely the arched outgrowth from the cylindric wall of the internal apiculi into each single small pore. This morphologic element may be present in other centric diatoms as well, e.g., *Thalassiosira coronata* Gaarder [Hasle, 1968, fig. 11] and *Cyclotella temperei* Per. et Héribaud examined in SEM by Ehrlich [1969].

The structure of the intercalary band and the girdle band of the thick-walled form was as in *Thalassiosira chilensis* Krasske [Helmcke and Krieger, 1953, pl. 14], *T. nordenskioeldii* [Hasle, 1964, pl. 4, fig. 1], and *Coenobidiscus muriformis* [Loeblich et al., 1968, fig. 17].

Although Hustedt [1958] so strongly emphasized the pleomorphism of fasciculated structured centric diatom species, this seems to be the first time pleomorphism has been demonstrated inside *Thalassiosira* colonies. Further investigations will prove whether this is a common feature in the genus, or whether it is specific to *T. tumida.* The correlation between degree of silicification and valve structure is probably a more general phenomenon.

Observations on cultures started from a weakly silicified species identified as *Thalassiosira rotula* showed that areola pattern, in this case formation of

---

Figs. 44–47. Micrographs of *Thalassiosira* spp.

44. *Thalassiosira decipiens* valve, light micrograph, phase-contrast–dark-field illumination, ×2250, North Cape, Norway (71°13′N, 25°44′E, May 1, 1962).

45. (*a*) *Thalassiosira decipiens* valve, partly in side view; one row of comparatively widely spaced long marginal apiculi, one labiate process, transmission electron micrograph, about ×2100, Drøbak, Oslofjord, Norway (about 59°40′N, March 1953). (*b*) Same specimen in valve view one central apiculus and valve areolae in curved tangential rows, transmission electron micrograph, about ×1350.

46. *Thalassiosira tumida* valve and girdle; areola pattern irregular (one sector, straight and slightly curved tangential rows; same specimen as Figures 40 and 42), scanning electron micrograph, ×1500, Weddell Sea (76°00.2′S, 56°45.9′W, February 9, 1968).

47. Two valves from a culture started from a species identified as *Thalassiosira rotula* and grown in a Si-enriched medium for 34 days; the specimen to the left has the valve structure characterizing *T. rotula* with few tangential areola walls, whereas the specimen to the right has the structure of *T. gravida* with well-developed areolae, transmission electron micrograph, about ×6000, inner Oslofjord.

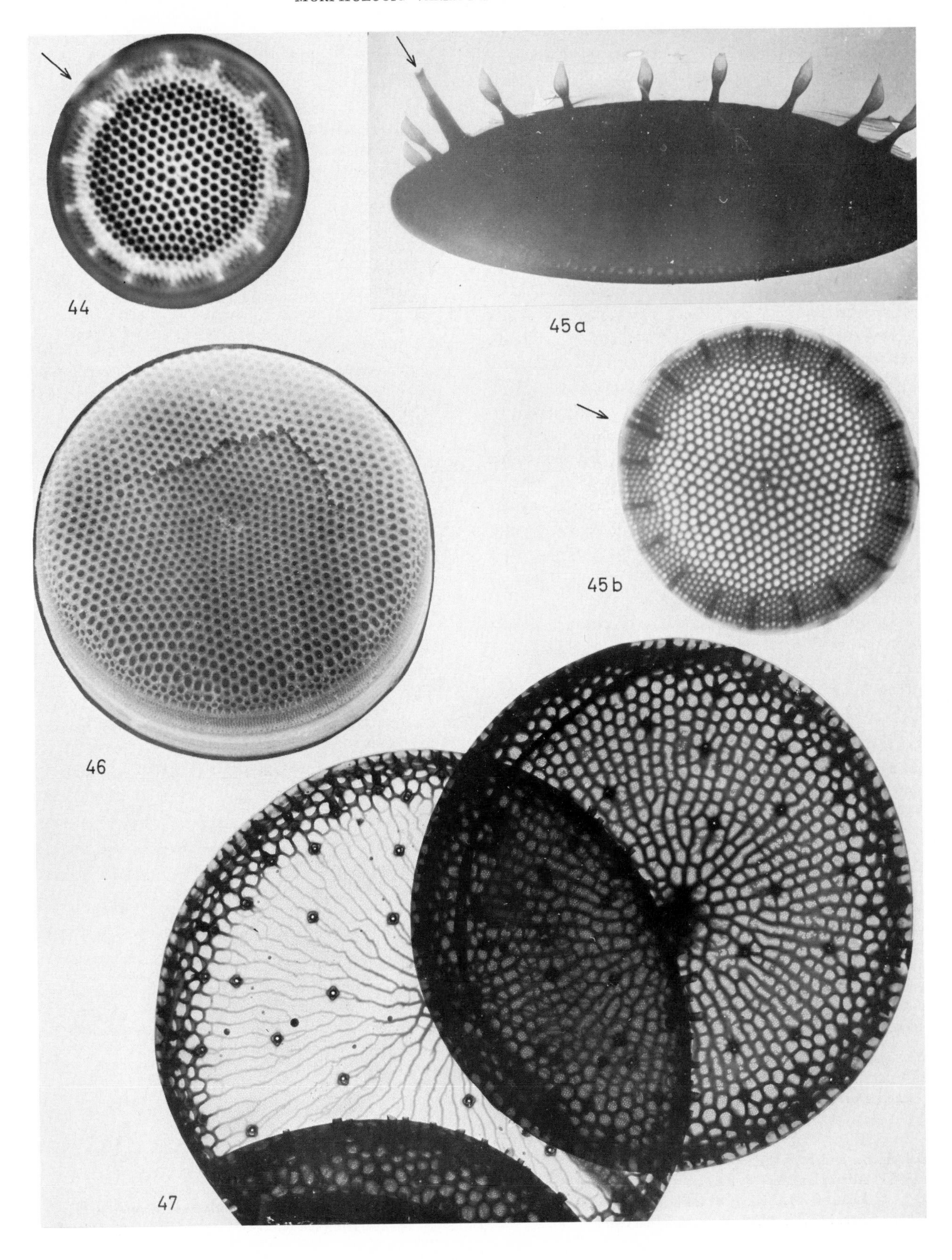
44
45a
45b
46
47

tangential areola walls crossing the radial ones, to some extent seemed to be regulated by the amount of silica available. When grown in a silica-enriched medium, the culture after some time contained a considerable number of cells with coarsely silicified valves. In contrast to the original valves, the coarse valves had well-developed areolae, and because of this a valve structure identical to that of *T. gravida* (Figure 47). (Corresponding observations have been made on two *Cyclotella* species by Belcher et al. [1966].)

Preliminary observations of the distribution of *T. rotula* and *T. gravida* indicated a difference also in their distribution patterns. *Thalassiosira rotula* seemed to have a fairly wide distribution, while *T. gravida* was restricted to higher latitudes of both hemispheres, that is, to waters generally considered rich in silica. The findings of weakly and coarsely silicified *T. tumida* cells in the same colonies complicate a possible theory of a relationship in general between silica supply and variation in valve structure.

The present observations support previous suggestions [Hasle, 1968; Hasle and Heimdal, 1970] that valve processes are more stable morphologic and consequently taxonomic elements than shape, number, and arrangement of areolae, since up to now no variation similar to that found in valve structure has been observed in valve processes.

Acknowledgments. This investigation was carried out partly at the University of Oslo and partly at Texas A & M University. Support for the senior author's work in the United States was provided by Texas A & M University through an invitation extended by Dr. S. Z. El-Sayed, Department of Oceanography. Special optical equipment not provided by the university was obtained from the National Science Foundation under grant GA-1487.

We are grateful to the persons and institutions who placed the materials used for this examination at our disposal, particularly to Dr. S. Z. El-Sayed, for the samples from the Weddell Sea and for providing working facilities in his laboratory. We are also grateful to the Advisory Service, Cambridge Scientific Instruments Limited, Cambridge, England, and to Shell Development Company, Houston, Texas, for the permission to use the scanning electron micrographs made by their instruments, and particularly to Dr. H. P. Studer for operating the K Square Ultrascan microscope.

Photographic credits. Advisory Service, Cambridge Scientific Instruments Ltd., Cambridge, England: Figures 30–38, 43.

Shell Development Co., Houston, Texas: Figures 39–42, 46.

REFERENCES

Balech, E., and S. Z. El-Sayed
1965 Microplankton of the Weddell Sea. *In* George A. Llano (ed.), Biology of the Antarctic Seas II. Antarct. Res. Ser., *5:* 107–124. AGU, Washington, D. C.

Belcher, J. H., E. M. F. Swale, and J. Heron
1966 Ecological and morphological observations on a population of *Cyclotella pseudostelligera* Hustedt. J. Ecol., *54:* 335–340.

Cassie, V.
1963 Distribution of surface phytoplankton between New Zealand and Antarctica, December 1957. Scient. Rep. Transantarct. Exped., *7:* 1–11.

Cupp, E. E.
1943 Marine plankton diatoms of the west coast of North America. Bull. Scripps Instn. Oceanogr., *5* (1): 1–238.

Dale, R. L.
1968 International Weddell Sea Oceanographic Expedition, 1968. Antarct. J. U. S., *3* (4): 80–84.

Ehrlich, A.
1969 Révision de l'espèce *Cyclotella temperei* Peragallo et Héribaud. Examen comparé aux microscopes: photonique, électronique et électronique à balayage. Cah. Micropal., ser. 1, *11:* 1–11.

El-Sayed, S. Z.
1968 Productivity of the Weddell Sea. Antarct. J. U. S., *3* (4): 87–88.

Hasle, G. R.
1960 Phytoplankton and ciliate species from the tropical Pacific. Skr. Norske Vidensk.-Akad. I. Mat.-Naturv. Kl. 1960, *2:* 1–50.
1964 *Nitzschia* and *Fragilariopsis* species studied in the light and electron microscopes. I. Some marine species of the groups *Nitzschiella* and *Lanceolatae.* Skr. Norske Vidensk.-Akad. I. Mat.-Naturv. Kl. N.S., *16:* 1–48.
1968 The valve processes of the centric diatom genus *Thalassiosira.* Nytt Mag. Bot., *15:* 193–201.
1969 An analysis of the phytoplankton of the Pacific Southern Ocean: Abundance, composition, and distribution during the "Brategg" Expedition, 1947–1948. Hvalråd. Skr., *52:* 1–168.
1971 Two types of valve processes in centric diatoms. Nova Hedwigia, in press.

Hasle, G. R., and G. A. Fryxell
1970 Diatoms: Cleaning and mounting for light and electron microscopy. Trans. Am. Microsc. Soc., *89*(4): 469–474.

Hasle, G. R., and B. R. Heimdal
1968 Morphology and distribution of the marine centric diatom *Thalassiosira antarctica* Comber. J. R. Microsc. Soc., *88:* 357–369.
1970 Some species of the centric diatom genus *Thalassiosira* studied in the light and electron microscopes. *In* J. Gerloff and B. J. Cholnoky (eds.), Diaomaceae II. (Beiheft 31 zur Nova Hedwigia.) Pp. 543–581. J. Cramer, Weinheim.

Helmcke, J.-G., and W. Krieger
1953 Diatomeenschalen im elektronenmikroskopischen Bild. I. J. Cramer, Weinheim.

Hendey, N. I.
1937 The plankton diatoms of the Southern Seas. 'Discovery' Rep., *16:* 151–364.

Hustedt, F.
1958 Diatomeen aus der Antarktis und dem Südatlantik. Dt. Antarkt. Exped. 1938/39, *2:* 103–191.

Karsten, G.
1905 Das Phytoplankton des Antarktischen Meeres nach dem Material der deutschen Tiefsee-Expedition 1898–1899. Dt. Tiefsee-Exped. 1898/99, *2* (2): 1–136.

Kozlova, O. G.
1962 Vidovoi sostav diatomovykh vodoroslei v vodakh Indiïskogo sektora Antarktiki (Specific composition of diatoms in the waters of the Indian sector of the Antarctic). (In Russian; English abstract.) Trudy Inst. Okeanol., *61:* 3–18.
1966 Diatoms of the Indian and Pacific sectors of the Antarctic. (English translation of Russian original.) 191 pp. Published for the National Science Foundation, Washington, D. C., by the Israel Program for Scientific Translations, Jerusalem.

Loeblich, A. R., III, W. W. Wight, and W. M. Darley
1968 A unique colonial marine centric diatom, *Coenobiodiscus muriformis* gen. et sp. nov. J. Phycol., *4:* 23–29.

Manguin, E.
1960 Les Diatomées de la Terre Adélie. Campagne du 'Commandant Charcot,' 1949–1950. Annls Sci. Nat., ser. a (Bot.), *12:* 223–363.

Rattray, J.
1890 A revision of the genus *Coscinodiscus* and some allied genera. Proc. R. Soc. Edinb., ser. B, session 1888–89, *16:* 449–692.

Schmidt, A., et al.
1874–1959 Atlas der Diatomaceenkunde. Heft 1–120, pls. 1–480. Munich.

Van Heurck, H.
1909 Diatomées. 127 pp. J. E. Buschmann, Antwerp.

Zhuse, A. P., G. S. Koroleva, and G. A. Nagaeva
1962 Diatomovye vodorosli v poverkhnostnom sloe osadkov Indiïskogo sektora Antarktiki (Diatoms in the surface layer of sediment in the Indian sector of the Antarctic). (In Russian; English abstract.) Trudy Inst. Okeanol., *61:* 19–92.

CORETHRON CRIOPHILUM CASTRACANE: ITS DISTRIBUTION AND STRUCTURE

GRETA A. FRYXELL

Texas A & M University, College Station, Texas 77843

GRETHE R. HASLE

Institute of Marine Biology, B, University of Oslo, Oslo 3, Norway

Abstract. Corethron criophilum Castracane is a cosmopolitan diatom with its largest populations in antarctic waters, often in those waters that are poor in other phytoplankton species. Its distribution and the fine structure of the siliceous parts of the cell are discussed. Two different valve types are pictured, along with three kinds of cell processes.

INTRODUCTION

Corethron criophilum Castracane is a cosmopolitan diatom species that has its largest populations in the antarctic inshore waters [Hasle, 1969]. Hart [1942] considered it essentially oceanic, although abundant in neritic areas and found living in pack ice. It is a cylindrical diatom with its greatest length along the pervalvar axis. Collar-shaped intercalary bands fit together to make the walls of the cylinder; the ends of the cylinder are capped by an upper and a lower hat-shaped valve. Long barbed spines from the lower valve point down and out, as do those from the upper valve. Thus the spines in corresponding positions on the two valves are almost parallel. Shorter, erect, more lightly silicified spines, terminating with a heavier 'claw' (Klammerborsten), are sometimes observed on the upper valve. They were observed routinely on material both from the Antarctic and from a culture isolated from near Plymouth, England. In this study, the 'clawed' spines have not been seen on both valves, in contrast to Hendey's report [1937].

Because of its shape and numerous processes, *C. criophilum* is easily caught in phytoplankton nets, even though the cell diameter may be smaller than the net meshes; it is, therefore, not likely to be overlooked when present in waters through which nets have been towed. In fact, its abundance may well be overestimated.

C. criophilum is most often reported as individual cells, but it is also found in at least two kinds of chains [Hart, 1934; Hendey, 1937] and in colonies embedded in mucus [Hendey, 1937]. Auxospores and microspore-like cellular inclusions have often been observed [Hart, 1934; Hendey, 1937; Hasle, 1969], as well as 'microspores' found embedded in a mucoid matrix [Hendey, 1937]. The numerous chromatophores are round to oval.

TAXONOMY

Corethron exists in many phases and sizes, all of which Hendey [1937] included in a single species after studying enormous quantities of material from the *Discovery* Expedition, which covered both the Indian and Atlantic oceans from tropical waters to the Antarctic. He stated that the type phase, which was described by Castracane in 1886, is probably a summer form or one produced by microspores. Hendey concluded, 'This monotypic genus presents a perfect example of what I mean by a poly-phasic species-system, and can only be understood correctly if the species is conceived as an orbital system in a space-time continuum' [Hendey, 1937, p. 325]. Following Hendey [1964], the classification of *Corethron* is as follows:

Division Chrysophyta
- Class Bacillariophyceae
  - Order Bacillariales
    - Suborder Rhizosoleniineae
      - Family Corethronaceae
        - *Corethron criophilum* Castracane

As will be shown, *C. criophilum* has a wide geographical distribution, and thus the species (sensu

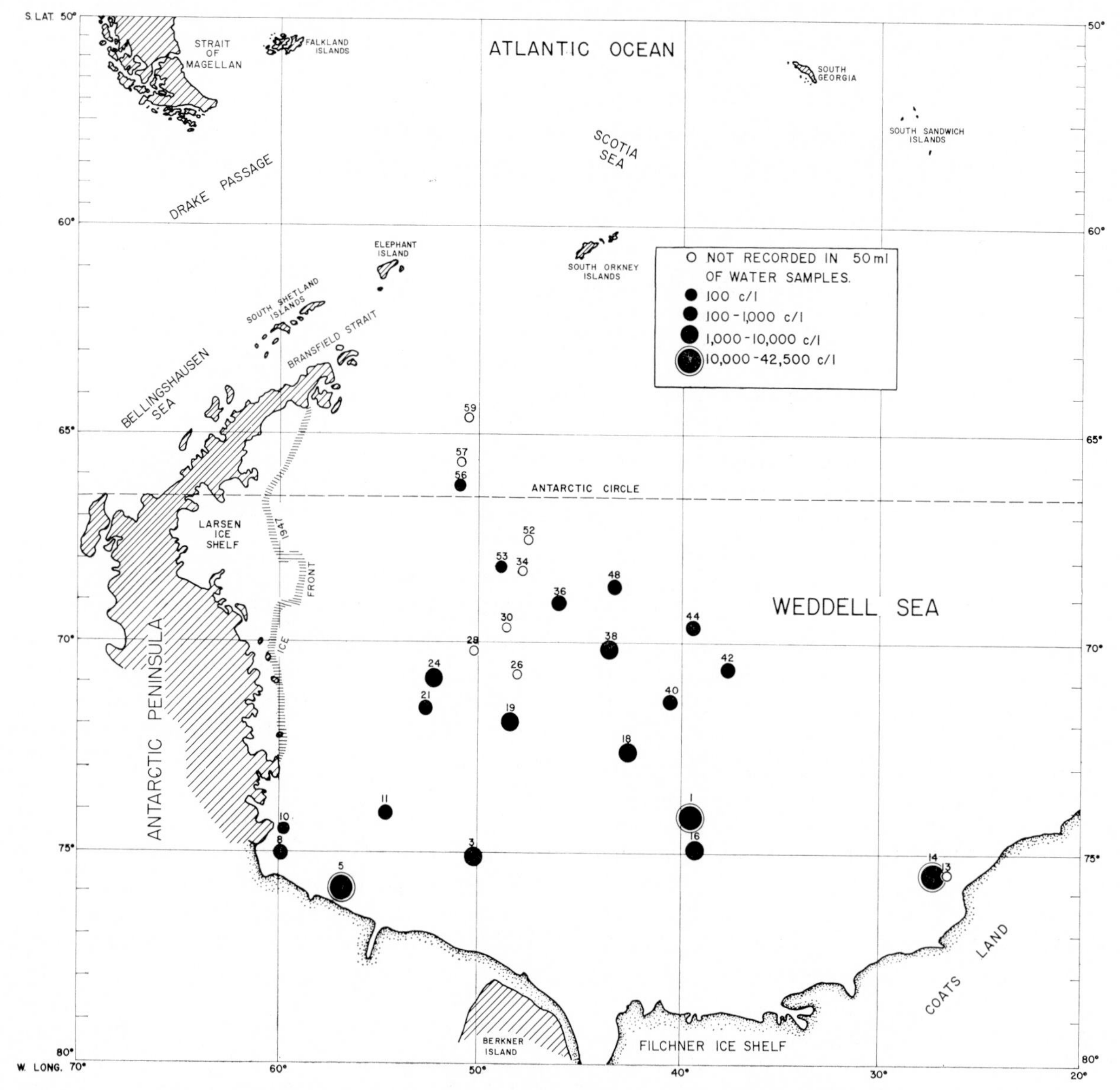

Fig. 1. Quantitative distribution of *Corethron criophilum* during IWSOE-1968.

Hendey) must be able to endure a wide variation of factors affecting growth, such as light, temperature, salinity, dissolved oxygen, minerals, and trace elements. These factors must affect structure, for the usual tubular form commonly pictured in the literature approaches the spherical in the tropics. Hendey failed to find even the presence or absence of spines on one or both valves of specific importance. The degree of silicification varies greatly, apparently geographically.

Hasle [1969], in analyzing material from the *Brategg* Expedition, found that the population showed two distribution patterns and so retained two taxa: *Corethron criophilum* and *C. criophilum* f. *inerme* (Karsten) Hasle. The former is described in the introductory paragraph. Hendey described the latter as robust, strongly siliceous, and forming very long, straight chains. The valves are weakly convex, as is often the case for cells with the larger diameters; and spines are noted only on the terminal cells. Hendey [1937] stated that only rarely were coronas of hairs seen on the terminal cells.

MATERIALS AND METHODS

Net hauls collected during the International Weddell Sea Oceanographic Expedition in 1968 (IWSOE-1968) with a 35-μm phytoplankton net and preserved in neutralized formalin have been studied in water mounts with phase-contrast lighting, using a Leitz Laborlux microscope. Permanent mounts have been made using both Coumarone and Hyrax media [Hasle and Fryxell, 1970] and examined under the same conditions. Net samples, cleaned of organic material, have been studied in the scanning electron microscope (SEM) and the transmission electron microscope (TEM) to augment the information from the light microscope. Phytoplankton in settled-water samples of 50 ml were counted in a Zeiss inverted microscope and cells/liter calculated for the quantitative estimates.

DISTRIBUTION AND VARIABILITY

Hart [1934] found *Corethron valdiviae* Karsten (= *C. criophilum*) frequently dominant in collections of the *Discovery* Expedition. In the February 1929 survey of the Bransfield Strait, *Corethron* was found at all 18 stations and comprised 98% of the phytoplankton collected, which included a few dinoflagellates and silicoflagellates. The following November, *Corethron* was again found at all 18 stations, comprising 89% of the estimated 140,945,800 diatoms in the samples. These samples were taken with a net and allowed to settle; aliquots were then counted. In December 1930, *Corethron* was found at all of the 17 stations sampled and accounted for 68% of the diatoms. When the phytoplankton crop was poor, *Corethron* often dominated [Hart, 1934]. Similar observations were made by Hasle [1969], although not for the Bransfield Strait alone. Considering the year as a whole, Hart estimated that *Corethron* accounted for over 90% of the diatoms in this area. It maintained a high level of abundance throughout the growing season with a maximum in the late spring.

Hart [1934] also found this species to be important in the Bellingshausen Sea. In 1929–30, *Corethron* was found to make up 40% of the phytoplankton taken from 20 stations. The following season it was found in samples of 46 of the 47 stations and made up about 25% of the phytoplankton.

Hendey [1937, p. 326] considered that 'the *Corethron* population is without question the most important constituent of the South Atlantic phytoplankton.' He noted that in very cold waters from 44°S toward the Antarctic, cells are small and narrow, being 10–14 times as long as broad, with no barbs observed on the long spines. The cells observed in the North Atlantic were about twice as long as broad, and the cells found off the coast of South Africa and in the Mediterranean were almost spherical. He noted the corona of shorter clawed spines on organisms collected from about 57°S to the south. Fukase and El-Sayed [1965] found that *C. criophilum* made up more than 50% of net hauls taken just off the continental shelf of northern Argentina and in the Bransfield Strait just off the Antarctic Peninsula.

Cupp [1943] considered *Corethron hystrix* Hensen (= *C. criophilum*) to be a north temperate species, found in small numbers off California, in the Gulf of California, and north to Scotch Cap, Alaska. The diameters of the cells she observed ranged from 12 to 38 μm, but only those cells from 13 to 24 μm were seen to have auxospores.

This species has been recorded from arctic waters in the Atlantic area (net hauls collected between 70° 30′N and 76°36′N, 6°30′E and 31°15′E in August and September, 1967; Heimdal, personal communication, 1969).

Kozlova [1966] stated that the Antarctic Convergence forms the northern limit of this species, with the exception of a section at 20°E where it is found north of the Antarctic Convergence. Kozlova found it planktonic at depths of 0–300 meters. Only rarely were valves found in the sediments, and those seen usually were without spines. The lightly silicified valves were classed as having a partial capacity for preservation in sediments. In one suspended sample from the Pacific Ocean north of Victoria Land, Kozlova found that *Corethron* made up 63% of all the diatoms. The temperature of this sample was −1°C and its salinity 33.8‰. Nets were not used for these collections; water samples were centrifuged or filtered.

Wood [1964] considered two sources of *C. criophilum* in Pacific Australasian waters, one equatorial and the other antarctic. Its occurrence in the Subantarctic was sporadic. The antarctic population was pleomorphic, in contrast to the monomorphic tropical population. He stated that one must 'accept the idea of a single cosmopolitan population divided into local populations by water masses' (p. 33). Hendey [1937] reported that Hustedt also had observed the species in the Pacific from the Sea of Japan.

Hart and Currie [1960] found *C. criophilum* present both inshore and offshore in their investigations of the Benguela Current. It was not dominant in any of their 39 repeated stations. In their comprehensive review of species distribution as recorded in the literature (p. 264), they recorded this diatom as having been found in South Africa's waters, the north and tropical Atlantic, the Indian Ocean, the east and west

TABLE 1. Cells per Liter of Phytoplankton Collected during IWSOE-1968

Station	Depth, meters	*Corethron criophilum*	Diatoms	Total Phytoplankton
1	10	42,500	57,640	77,700
3	10	6,080	14,380	23,800
5	10	12,500	224,720	1,928,140*
8	0	160	81,720	455,590
10	11	40	39,580	854,020
11	6	140	1,760	4,320
13	7	...	3,520	71,880
14	7	14,540	128,400	239,840
16	6	3,800	206,400	234,440
18	6	1,400	13,860	16,840
19	7	3,040	47,140	80,520
21	7	780	3,360	4,360
24	7	2,240	3,440	3,920
26	14	...	5,280	7,600
28	10	...	140	420
30	7	...	300	600
34	12	...	140	420
36	11	260	15,290	15,610
38	16	1,080	357,350	370,650
40	12	420	41,480	42,740
42	5	690	1,410,180	1,426,310
44	15	540	472,150	577,950
48	7	340	416,360	442,920
52	10	...	2,270	2,510
53	14	60	5,890	6,330
56	8	60	11,080	13,940
57	10	...	5,860	9,560
59	7	...	9,900	13,260

**Phaeocystis* sp., 1,700,000 of this total.

Pacific, and the oceans surrounding Antarctica.

In the Indian Ocean, Thorrington-Smith [1969] found *C. criophilum* in small numbers in 21 out of 26 stations both on the edge of and in the Agulhas Current off the Natal coast of Africa.

Hasle [1969] found *C. criophilum* to reach its maximum development in subantarctic waters in December. In January or early February the maximum was farther south. Both net samples and water samples were analyzed. *C. criophilum* f. *inerme* appears to have a more limited distribution.

During the International Weddell Sea Oceanographic Expedition of 1968 (IWSOE-1968), the net collection from Station 1 was predominantly *C. criophilum.* The water sample (reported in Table 1) confirms the predominance of this diatom. It was found in fairly large numbers at Station 5, which had a large population of *Phaeocystis.* It is not uncommon that these two are found together. Although *C. criophilum* was recorded in most other stations, it accounted for only 2½% of the diatoms counted in the water samples from the whole cruise (Figure 1). This figure varies by at least a factor of 10 from the cruises previously reported. No doubt the net hauls of the past have led to overestimation of the preponderance of *C. criophilum.*

The authors see some slight indication that more open waters with less ice cover tend to be richer in *C. criophilum,* although they can find no close correspondence with temperature or salinity from IWSOE-1968 [see hydrographic data, El-Sayed, 1971]. The population of *C. criophilum* does not show a direct or inverse relationship with *Thalassiosira tumida* [Hasle et al., 1971], which was responsible for a bloom observed at its height at Station 8.

STRUCTURE

Intercalary Bands

These bands fit around the cell like a series of collars and give it the cylindrical shape. Adjacent bands overlap in a scalelike manner (Figures 2a, b, c). Okuno [1952a] observed simple 'roundish pores' on the intercalary bands, and this study confirms his findings. In the SEM, the bands look like thin sections of plastic styrofoam, with pores in irregular lines parallel to the pervalvar axis. The TEM shows the pores clearly

Fig. 2. *C. criophilum* structure.

a. Large broken lower valve (above) and an upper valve still attached to several intercalary bands, scanning electron micrograph, ×1000.

b. Two intercalary bands still around the spines of a daughter frustule shown in 2*c*, scanning electron micrograph, ×1300.

c. Two other intercalary bands around daughter lower frustule; note how intercalary bands fit and overlap, scanning electron micrograph, ×1400.

d. Intercalary band folded into the usual collar shape; note the asymmetry, which keeps the bands from joining in a straight line down the length of the cell and not the two sizes of the pores, transmission electron micrograph, ×1800.

e. Outside view of socket of barbed spine on lower valve with pores partly filled, probably with metallic coating used in scanning electron micrograph; note prongs that secure base of spine, scanning electron micrograph, ×13,000.

f. Biological inside view of socket of barbed spine from lower valve (compares with view in 2*e*), scanning electron micrograph, ×13,000.

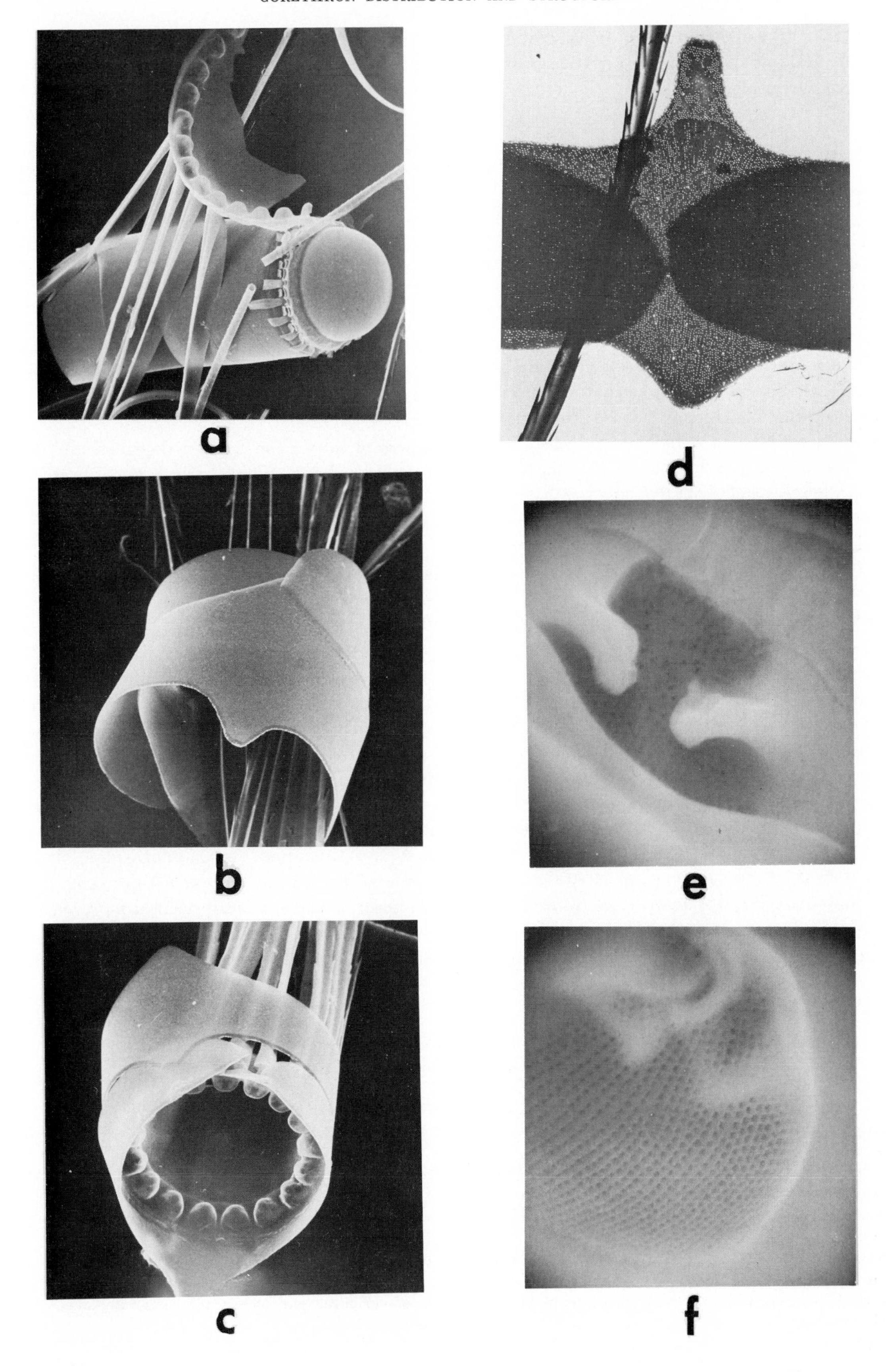
a
b
c
d
e
f

(Figure 2d). Okuno found these simple pores to be about 80 μm; although noting variations, we find similar measurements.

Valves

Two kinds of valves were observed on each cell. For convenience, the valves will be termed 'upper' and 'lower,' as in the introductory paragraph, as Karsten [1905] also considered them. ('Epivalve' and 'hypovalve' are not used because of the possible implication that one is larger than the other. The traditional meanings of these terms are confusing in relation to *Corethron,* since the two valves are differentiated in structure and not just in age.) The valves are more or less hemispherical, with the cells of smallest diameter having the steepest vault (Figures 3a, f). Small pores in irregular lines are found radially on the valve. Okuno [1952b] found the pores scattered irregularly at the center of the valve and about the same size as the pores in lines on the intercalary bands. The 'ring' or 'first intercalary band' of Karsten [1905] is here considered part of the valve, as it does not separate when the structure is cleaned of organic matter.

The lower valve seems to be of simpler construction than the upper, especially in terms of its processes. The long, barbed spines fit into sockets or indentations like modified enarthroses [Okuno, 1954] near the mantle. The mantle is here considered the vertical area that attaches to the intercalary bands, not the major portion of the vault of the valve. The spines are held in place by two prongs extending from the sides of the socket (Figures 2e; 3a, b, c, e, f). The valve mantle is made up of a smooth, vertical flange, more or less scalloped on the valve side and almost straight on the side of the intercalary bands—part of Karsten's 'scalloped crown' (Figures 3a, b, c, e, f).

In this study, some lower valves have been observed with tiny spinules scattered over the valve surface (Figures 3a, d). These were observed by Okuno [1952b] in the TEM and appeared to be hairlike projections that were strongly tapered. In our SEM pictures, there was great variation in the length of the spinules (0.2–1.5 μm), and all appeared blunt, suggesting possible breaks. The thin valve was slightly raised at the spinule area, which indicated that these were not artifacts but structures on the valve itself. Karsten [1905, pl. XIII, fig. 16] and Hendey [1937, pl. VIII, fig. 8] noted spinules on both valves.

The long spines on the lower valve were barbed, both on the antarctic samples and on those from British waters. In the main, the spine is trigonal (Figures 2a, 3a), being approximately T-shaped in cross section. Broken spines that were found in the field of the SEM appeared solid. The barbs on a spine are antrorsely oriented. Pores on the barbed spines were seen by Okuno [1952a] but were not confirmed in our study nor by Kolbe [1948] or Okuno [1952b].

The lining to the socket in which the long, barbed spine fits is like an open network of pores arranged in lines, appearing even more porous than the valve or the intercalary band (Figures 2e, f). There was no evidence that the long, barbed spine itself was involved in the active exchange of material between the cell and the sea around it. The base of the spine is flattened, semicircular, and curved convexly towards the center of the valve. A flaring is sometimes noted just above the base (Figure 4c).

The upper valve is a complex structure indeed. Karsten [1905] found it was given priority and was the first frustule formed by an auxospore. Its general shape is like that of the lower valve, but the sockets for securing the spines are located more on the vault of the frustule with the mantle dropping away and outward (Figures 6a–e). The scalloped vertical flange

---

Fig. 3. Lower valves of *C. criophilum.* Note scalloped edge of mantle, projecting prongs in the sockets of the barbed spine, the spinules on some valves, and the nature of the long, barbed spines.

a. Note broken spine, with trigonal structure, still attached in its socket, scanning electron micrograph, ×1300.

b. Closeup of frustule in *3a,* showing barbed spine in its socket and broken remnants of other spines deep in their sockets, scanning electron micrograph, ×3300.

c. Biological inside view of lower valve with some spines still attached; note break in margin, scanning electron micrograph, ×2000.

d. Closeup of the same valve as in *3a* showing surface spinules scattered over the center of frustule, scanning electron micrograph, ×3300.

e. Lower valve with high vault, showing empty sockets with broken bases of spines and one barbed spine still attached, scanning electron micrograph, ×5000.

f. Small lower valve with high vault; note the prongs in the sockets and the one broken portion of a barbed spine still resting in the center socket, scanning electron micrograph, ×1400.

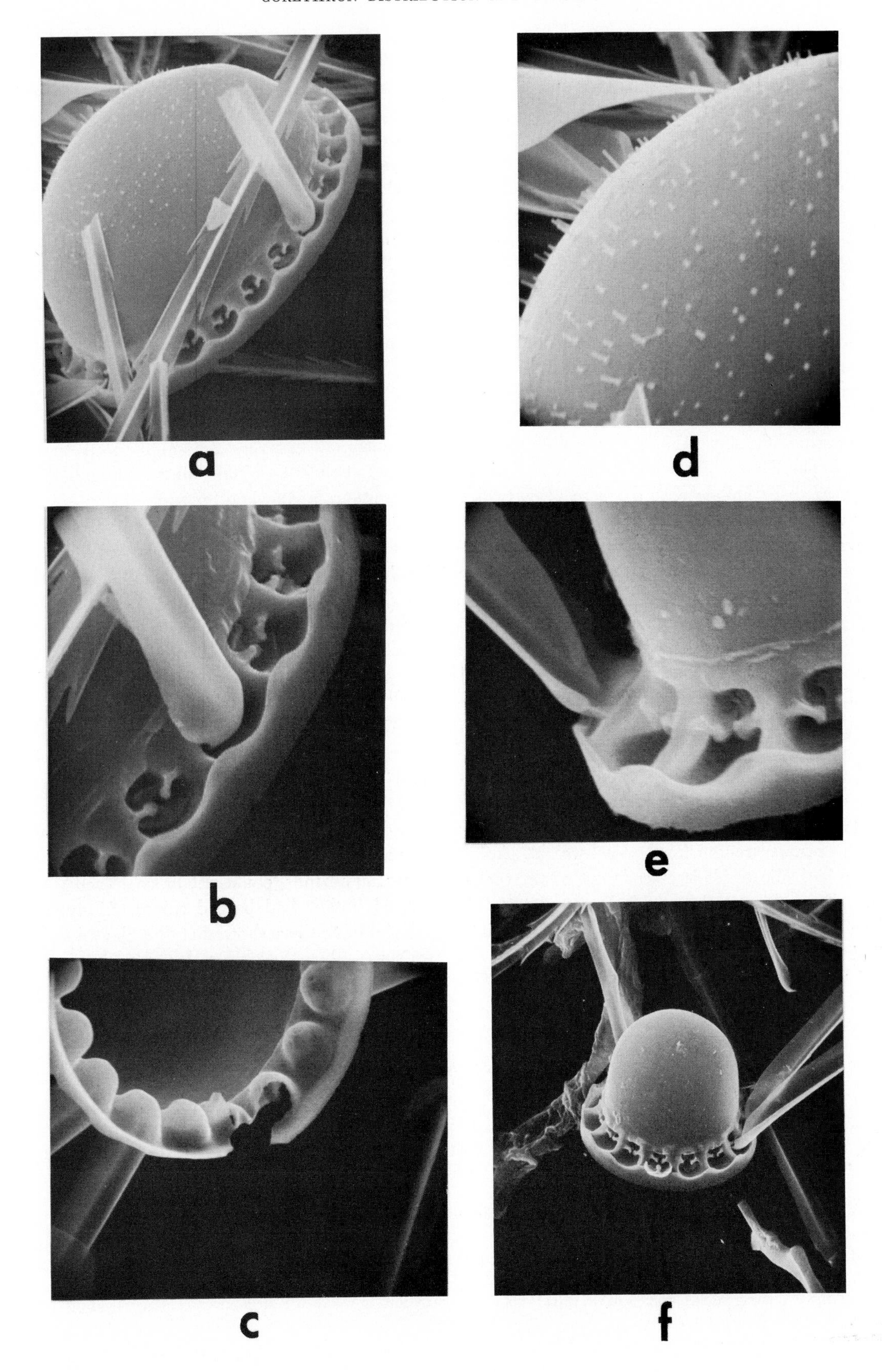
a
d
b
e
c
f

of the lower valve is lacking on the upper valve. The upper valve contains more and smaller sockets than the lower valve; Karsten [1905] indicates that the number is about double. The presence of two kinds of sockets confuses the picture. Open sockets without pronged projections secure the long, tapered, barbed spines, which point downward (Figures 2a, d, 4e) roughly parallel to the larger barbed spines from the lower valve. Barbed spines from the upper valve are about ⅔ the width of those from the lower valve, measured on whole cells near the frustule. The upper barbed spines are not held as firmly as the lower; and in preserved samples, especially in cleaned samples, the barbed spines are likely to be almost entirely missing from the upper valve, apparently because they have been pulled out in handling. The field of vision is often strewn with them. Their bases, usually broken, show a shepherd's crook or an S-shaped terminus. They appear to be hollow near the base (Figure 6d) possibly with perforated walls.

Between one barbed spine and the next, one or sometimes two sockets of a different nature occur where the more lightly silicified shorter spines are secured. These spines terminate in a claw-shaped (sometimes called club-shaped) process, which is well known in the literature (Figures 4d, e, f, 5). Karsten [1905] observed that these claws were formed first in the reproducing cells. A petal-like flap at the outside base of this socket seems to hold the clawed spine erect (Figures 4f, 6a–f). In spite of its lightly silicified nature, it usually maintains an erect posture, and as Kolbe [1948] pointed out, the supporting process shows the greatest strength for the least weight. A flat, monomorphic, silicate blade extends out of the socket on the valves. A central line of denser material is noted on TEM micrographs. The blade is curved into a cylinder within the socket and appears to have a crooked terminus that is more heavily silicified. The blade also recurved into a cylinder 4–6 μm beyond the socket and becomes a hollow rachis-type structure. The rachis has antrorse spicules, one of which flares out to support a single tissue that terminates in the bottom one of two horizontal 'jaws' that are more heavily silicified than the rest of the structure. The cylinder itself expands and terminates in the top 'jaw.' Clawed spines seen in the SEM are often burst at a weakly silicified spot behind the jaws (Figure 4d); it seems possible that trapped gas or fluid ruptured the structure when the sample was subjected to extremely low pressure in preparation. Samples cleaned the same way but mounted in permanent mounts and not subject to low pressures show much less damage. The jaws are fixed structures, and the upper (and possibly the lower) is hollow. Toward the bottom of one side of the upper jaw is a row of comblike teeth that are free at least at their terminal points, although they follow the contours of the jaw for some distance.

Although the base of the clawed spine is in cylindrical form, it does not seem much more likely to be involved in cell-sea exchange than the barbed spine, since, in unbroken valves, the flap over the socket covers the cylindrical part quite completely. The fine structure of the flap is not apparent. The jaws themselves could be flotation devices (as some authors also call the long barbed spines). According to Hendey [1937], the clawed spines are sometimes found on both valves and can act as coupling hooks in the second kind of chain referred to in the introduction. Cells so joined were not observed in this study. It seems that daughter cells, held apart as they are during the formation of the long barbed spines within the parental intercalary bands, could not be so joined while being formed if they had barbed spines. Many more published illustrations show that the clawed spines were seen on only one valve rather than on both valves; in this study clawed spines were seen on only one valve

Fig. 4. Spines found on *C. criophilum*.

a. Barbed spines from lower valve; note the intercalary band draped over and between the spines and note trigonal structure and antrorsely oriented barbs, scanning electron micrograph, ×1300.

b. Young barbed spines still held together in intercalary bands of 2*b* and 2*c*, scanning electron micrograph, ×1400.

c. Base of barbed spine; note flair above portion that fits in socket and beginning of trigonal structure, scanning electron micrograph, ×3300.

d. Barbed spine above lying on portion of claw; below, claw with burst hollow structure on the right; note the comblike teeth on one side of the claw, scanning electron micrograph, ×3300.

e. Barbed spine from upper valve lying across three clawed spines with barbed spine from lower valve across upper left corner, scanning electron micrograph, ×1300.

f. Bases of three clawed spines near broken valve; note curvature into a rachislike structure, scanning electron micrograph, ×7300.

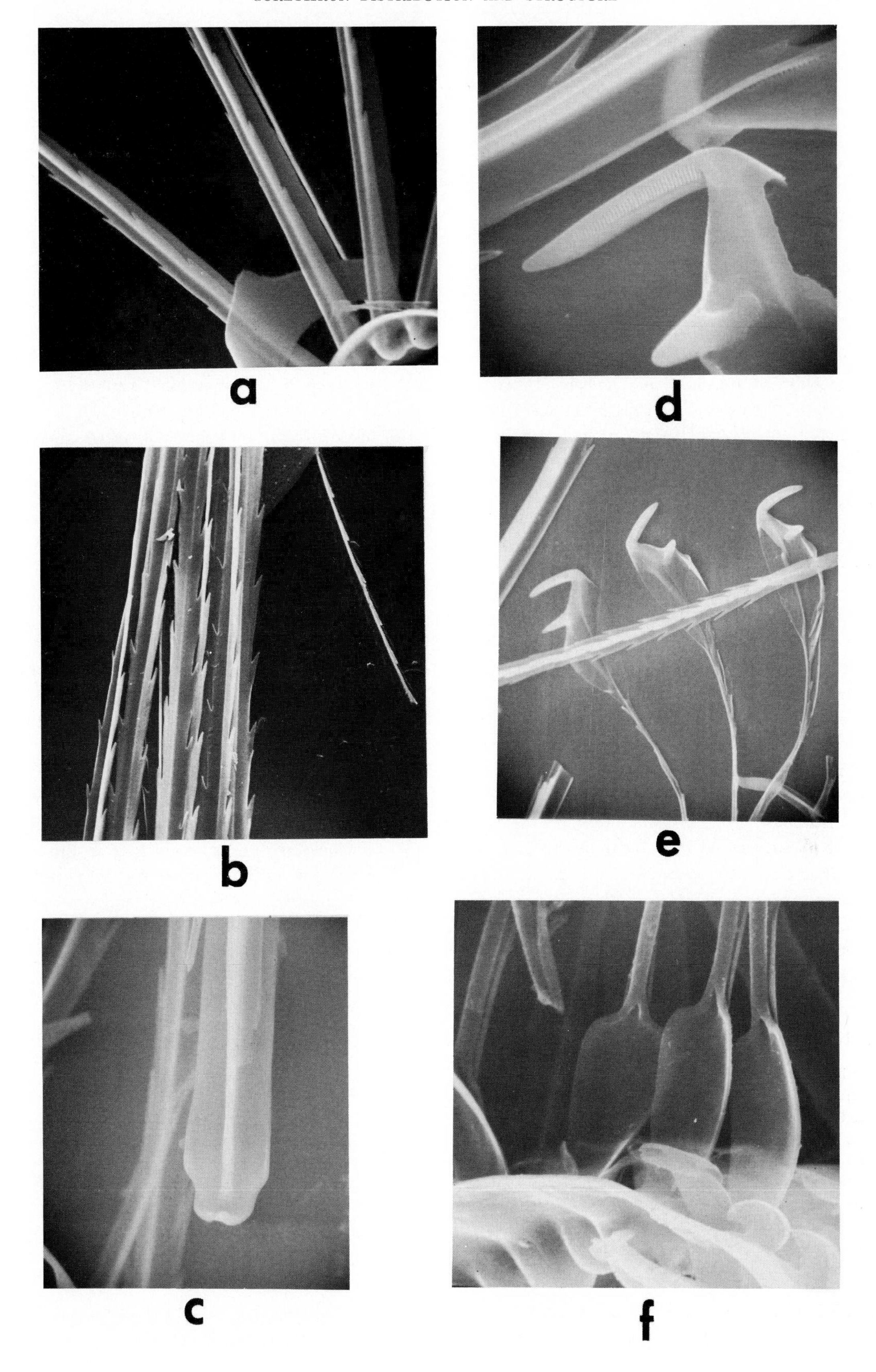
a
d
b
e
c
f

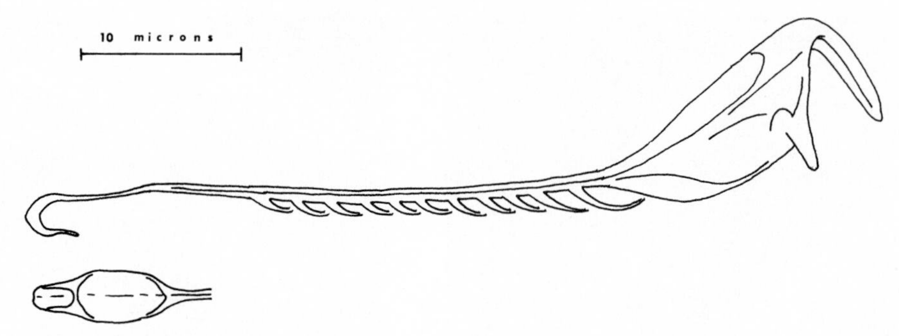

Fig. 5. Artist's reconstruction of the entire clawed spine of the upper valve. Note discussion in text. The number of spicules seems to vary geographically.

of a cell. Under such conditions the new daughter valves are different and only one has clawed spines, so that coupling of the daughter cells by claws would be impossible. If the clawed spines do not usually join two daughter cells, it seems possible that sexual reproduction might be involved in cells held together by claws.

CONCLUSIONS AND QUESTIONS RAISED

This study was undertaken in part to determine whether the cell processes of the solenoid diatom *C. criophilum* might be related to the processes recently noted from certain centric diatoms [Hasle, 1962, 1968], such as the central and marginal tubuli on *Cyclotella* and *Thalassiosira* and now also found on *Detonula, Bacterosira, Skeletonema, Lauderia, Porosira, Schroederella,* and *Planktoniella,* or to the widely distributed labiate process so far noted on the genera mentioned plus *Coscinodiscus.* Diatom processes may prove to be conservative characters that are helpful in establishing taxonomic relationships. The barbed spines and the clawed spines of *C. criophilum* do not seem similar to the labiate processes or the tubuli of the centric diatoms, nor do they show marked similarities to the setae of *Chaetoceros* in fine structure, at this point in the investigation. Thus they provide no evidence in support of an affinity of *C. criophilum* with these groups. Its general structure would indicate that it is properly placed in Rhizosoleniineae.

Furthermore, it is difficult to see how either sort of spine could be significantly involved in metabolic exchange of nutrients or waste with the sea water. The small spinules that are sometimes observed scattered over the center of a valve appear solid, in contrast to the tubuli (the strutted centric processes).

In the cells studied, the upper and lower valves of *C. criophilum* differ in structure, the shape of the mantle, the number and the nature of the sockets, and the kinds of processes. The fine structure of the clawed spine is quite constant, although this species is known for its great variation in other characters. The barbed spines from the upper and lower valves differ in width, size of barbs, and the nature of bases. Spinules were observed on some lower valves only in this study.

The SEM and TEM are tools to augment the light microscope and have confirmed and extended the excellent work of such scientists as Karsten and Hendey.

Much can be learned by combining culture techniques with ecological studies and laboratory examination.

Fig. 6. Upper valves of *C. criophilum.* Note lack of prongs in empty sockets and complex smaller sockets for clawed spines.

a. View into the biological inside of an upper valve, with the outside of the valve showing at the top of the photograph; note the empty sockets that housed the narrow, barbed spines of this valve and the broken ribbonlike blade portions from clawed spines, scanning electron micrograph, ×1300.

b. Closer view of another upper valve; note lack of prongs in sockets and flaps over the broken bases of the clawed spines, scanning electron micrograph, ×3300.

c. View from edge of upper valve; note lack of scalloped edge and the opened flaps over the clawed spine bases, scanning electron micrograph, ×5000.

d. Edge of upper valve with broken narrow barbed spines; note hollow, perforated appearance, scanning electron micrograph, ×3600.

e. View into upper valve with some clawed spines still attached and broken barbed spines extending over the mantle at the bottom left, scanning electron micrograph, ×1400.

f. Two kinds of sockets on upper valve with base of clawed spine near center partly pulled out of its socket; note the downward curving of the barbed spine base on the right, scanning electron micrograph, ×3300.

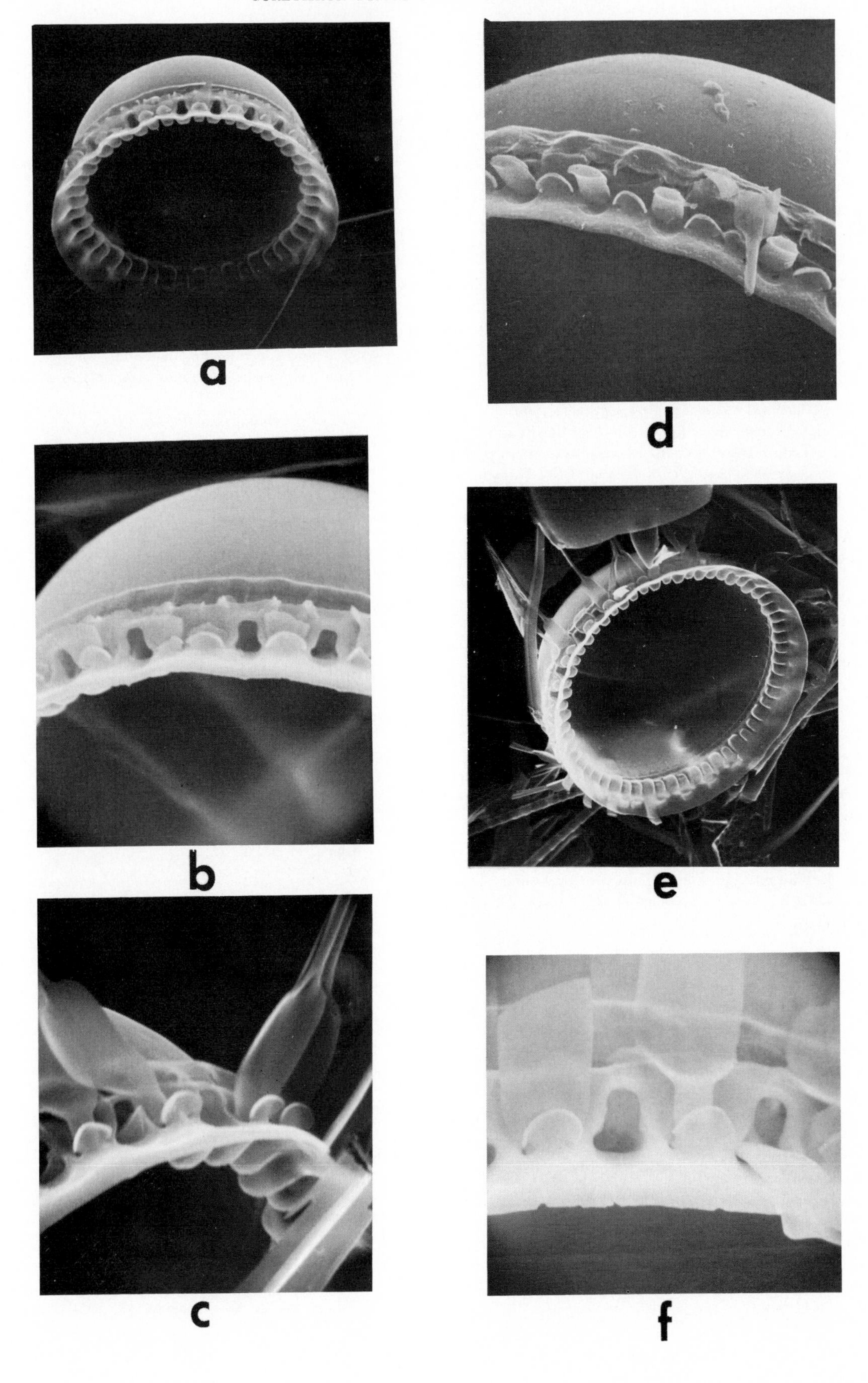
a
d
b
e
c
f

Although large populations of *C. criophilum* have been recorded in the literature from waters of fairly low salinity, its range of salinity tolerance is not known. Sexual reproduction has not been studied, although it undoubtedly occurs. Although the literature might overestimate the abundance of *Corethron* in phytoplankton-poor areas because of the predominance of net samples, the species is widely distributed and worthy of further study.

Acknowledgments. The authors wish to express gratitude to Dr. Sayed Z. El-Sayed, who collected the antarctic samples and aided with their study; to Dr. E. R. Cox for help and encouragement; to the Southwest Center for Advanced Studies, Dallas, Texas, and to Walter R. Brown for the use of the Jeolco JSM-1 SEM and for technical help and encouragement; to the Dow Chemical Company, Freeport, Texas, for the generous use of the Cambridge SEM and to Lee Martin and O. C. Benefiel for skilled assistance; to the Electron Microscopical Unit for Biological Sciences, University of Oslo, Norway; to Dr. G. T. Boalch, Plymouth Lab., England, for the cultures used; to Albert J. Whipple, III, Electron Microscopy Center, Texas A & M University, College Station, Texas, for excellent photography and interest; and especially to Dr. Paul A. Fryxell for Figure 5 and for help throughout the preparation of this manuscript.

This research was supported in part by grant GA 13836 under the U. S. Antarctic Research Program and the National Science Foundation.

Photographic credits. Walter R. Brown, Southwest Center for Advanced Studies, Dallas, Texas: Figures 3a–e, 4a, 6a, b, c. Jeolco JSM-1, SEM.

Electron Microscopical Unit for Biological Sciences, University of Oslo, Norway: Figures 2a–c, 3f, 4b, f, 6d, e. Stereoscan II SEM.

Lee Martin, Dow Chemical Company, Freeport, Texas: Figures 2e, f, 4c–e, 6f. Cambridge Stereoscan Mark IIA SEM.

Albert J. Whipple, III, Electron Microscopy Center, Texas A & M University, College Station, Texas: Figure 2d. RCA EMU-2E TEM.

REFERENCES

Cupp, E. E.
1943 Marine plankton diatoms of the west coast of North America. Bull. Scripps Instn Ocean., *5*(1): 1–238.

El-Sayed, S. Z.
1971 Observations on phytoplankton bloom in the Weddell Sea. *In* G. A. Llano and I. E. Wallen (eds.), Biology of the Antarctic Seas IV, Antarct. Res. Ser., *17:* 000–000. AGU, Washington, D. C.

Fukase, S. and S. Z. El-Sayed
1965 Studies on diatoms of the Argentine coast, the Drake Passage, and the Bransfield Strait. Oceanogr. Mag., *17:* 1–10.

Hart, T. J.
1934 On the phytoplankton of the southwest Atlantic and the Bellingshausen Sea, 1929–31. 'Discovery' Rep., *8:* 1–268.
1942 Phytoplankton periodicity in Antarctic surface waters. 'Discovery' Rep., *21:* 261–356.

Hart, T. J., and R. I. Currie
1960 The Benguela Current. 'Discovery' Rep., *31:* 127–297.

Hasle, G. R.
1962 Three *Cyclotella* species from marine localities studied in the light and electron microscopes. Nova Hedwigia, *4* (3–4): 299–307.
1968 The valve processes of the centric diatom genus *Thalassiosira.* Nytt Mag. Bot., *15:*193–201.
1969 An analysis of the phytoplankton of the Pacific Southern Ocean: Abundance, composition, and distribution during the Brategg Expedition, 1947–48. Hvalråd. Skr., no. 52: 1–168.

Hasle, G. R., and G. A. Fryxell
1970 Diatoms: Cleaning and mounting for light and electron microscopy. Trans. Am. Microsc. Soc., *89:* 469–474.

Hasle, G. R., B. R. Heimdal, and G. A. Fryxell
1971 Morphologic variability in fasciculated diatoms as exemplified by *Thalassiosira tumida* (Janisch) Hasle, comb. nov. *In* G. A. Llano and I. E. Wallen (eds.), Biology of the Antarctic Seas IV, Antarct. Res. Ser., *17:* 000–000. AGU, Washington, D. C.

Hendey, N. I.
1937 The plankton diatoms of the southern seas. 'Discovery' Rep., *16:* 151–364.
1964 An introductory account of the smaller algae of the British coastal waters. Part V. *Bacillariophyceae* (diatoms). Fish. Invest. Min. Agr. Fish. Food (Gt. Brit.), ser. IV, *5:* 1–317.

Karsten, G.
1905 Das Phytoplankton des Antarktischen Meeres nach dem Material der deutschen Tiefsee-Expedition 1898-1899. Dtsch. Tiefsee-Exped., 1898–1899, *2*(2): 1–136.

Kolbe, R. W.
1948 Elektronenmikroskopische Untersuchungen von Diatomeenmembranen I. Ark. Bot., *33A:* 1–21 (Engl. transl. of pp. 12–13).

Kozlova, O. G.
1966 Diatoms of the Indian and Pacific sectors of the Antarctic (English translation of Russian original). 191 pp. Published for the National Science Foundation, Washington, D. C., by the Israel Program for Scientific Translations, Jerusalem.

Okuno, H.
1952a Electron microscopical study on Antarctic diatoms, 2. J. Jap. Bot., *27:* 46–62.
1952b Electron microscopical study on Antarctic diatoms, 3. J. Jap. Bot., *27:* 347–356.
1954 Electron microscopical study on Antarctic diatoms, 5. J. Jap. Bot., *29:* 18–32.

Thorrington-Smith, M.
1969 Phytoplankton studies in the Algulhas Current region off the Natal coast. S. Afr. Assoc. Marine Biol. Res., Invest. Rep. Oceanogr. Res. Inst., no. 23: 1–24.

Wood, E. J. F.
1964 Studies in microbial ecology of the Australasian region. II. Ecological relations of oceanic and neritic diatom species. Nova Hedwigia, *8:* 20–35.

ZOOPLANKTON STANDING CROP IN THE PACIFIC SECTOR OF THE ANTARCTIC

THOMAS L. HOPKINS

Marine Science Institute, University of South Florida, St. Petersburg, Florida 33701

Abstract. Zooplankton standing crop was estimated in the Pacific sector of the Antarctic between 75°W and 175°W from collections made in 1963 to 1965 on eight cruises of the USNS *Eltanin.* Samples were taken with 202-μ mesh closing nets to depths of 2000 meters. Copepods, chaetognaths, and euphausiids made up most of the catch, with copepods averaging 67.3%, 68.8%, and 70.1% of the biomass in the antarctic, subantarctic, and convergence zone waters sampled, respectively. The most important contributors to standing crop were *Rhincalanus, Calanoides, Calanus, Metridia, Eukrohnia, Thysanoessa, Euchaeta, Sagitta, Euphausia, Oithona,* and *Pleuromamma.* Total biomass in the upper 1000 meters averaged 2.67, 2.58, and 2.96 g dry wt (weight)/m^2 in the Antarctic, the Subantarctic, and the convergence zone, respectively.

Seasonal vertical migration noted by other investigators was also evident from *Eltanin* material. The highest percentages (48–71%) of zooplankton biomass in the top 250 meters of the 0- to 1000-meter zone were recorded for the spring and summer months of November through January (no February data). In the late fall, winter, and early spring months of May through October, the percentages declined to 4–33%. This seasonal vertical migration pattern was recorded as well for individual species *(Rhincalanus gigas, Calanoides acutus, Calanus propinquus,* and *Metridia gerlachei).* Deep closing-net tows to 2000 meters indicate that 27–38% of the biomass of the 0- to 2000-meter layer was in the lower half of the water column from March through October.

Biomass in the Pacific sector of the Antarctic is compared to zooplankton abundance in other oceanic regions, and phytoplankton–zooplankton and herbivore–carnivore trophic relationships are discussed.

INTRODUCTION

This paper reports on the standing crop of zooplankton in the upper 2000 meters of the Pacific sector of the Antarctic from 1963 to 1965 during USNS *Eltanin* cruises, 10, 11, 13–15, and 17–19. Data are presented on the seasonal, vertical, and regional distribution of standing crop as well as on the taxonomic composition of biomass. Standing crop in the Pacific sector, in terms of dry weight, is compared to that reported for other oceanic regions; and plankton trophic relationships are discussed. In addition to contributing new quantitative information on the principal taxonomic components of biomass, these cruises fill an important gap in Foxton's [1956] biomass data, since he shows only two complete series of hauls to 1000 meters between 85°W and 170°W, a sector which includes the majority of *Eltanin* collections treated in this paper.

Much of the earlier literature on zooplankton standing crop in the Antarctic has been reviewed by Foxton [1956]. The most extensive plankton work since this review has been that of the Soviet Antarctic Expeditions of the *Ob'* from 1955 to 1958 [Brodskii and Vinogradov, 1957; Korotkevich, 1958; Brodskii 1964; Vinogradov and Naumov, 1964; Voronina, 1967a, b] and the *Eltanin* cruises. Most of the plankton sampling of the Soviet Antarctic Expeditions was to only 500 meters, whereas *Discovery* closing-net hauls, on which Foxton based his report, and the *Eltanin* collections were usually made to at least 1000 meters. It is essential to sample at least the upper 1000 meters in the Antarctic to assay zooplankton biomass, since this zone encompasses most of the biomass of zooplankton that undergoes the annual vertical migration cycle characteristic of plankton in polar waters [Beyer, 1962; Foxton, 1956].

In the present study the samples from the Pacific sector of the Antarctic have been divided into three groups, those collected in antarctic, subantarctic, and convergence zone waters. Samples from the convergence zone have been considered separately, since it is in this zone that faunas of the Antarctic and Subantarctic evidence the most mixing. The convergence

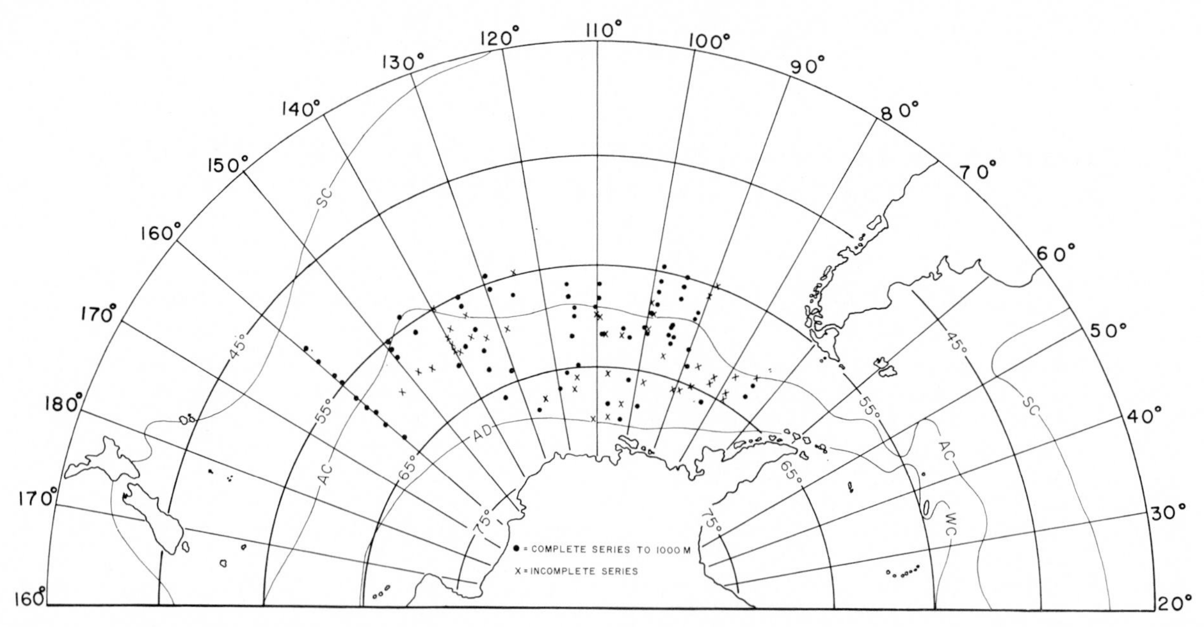

Fig. 1. Positions of stations, made on *Eltanin* cruises 10, 11, 13–15, and 17–19 (1963–1965), from which plankton biomass data were obtained. AC, Antarctic Convergence; AD, Antarctic Divergence; SC, Subtropical Convergence; WC, northern boundary of Weddell Current. Positions of hydrographic boundaries from Deacon [1964].

zone is also where some species occurring in both antarctic and subantarctic regions achieve either their peak or their minimum abundance [David, 1958; Baker, 1959; Kane, 1966].

METHODS

The samples that serve as a basis for this paper were taken with 70- × 70-cm square frame Bé closing nets [Bé, 1962] fitted with 202-μ mesh nets. The depth horizons routinely sampled were 0–100 meters, 100–250 meters, 250–500 meters, and 500–1000 meters. In addition, the 1000- to 2000-meter zone was sampled on tracks subsequent to cruise 14. Table 1 summarizes the number of samples from each depth zone, and Figure 1 shows the sampling locations. Data on season, position, depth, and biomass for each of the collections are in the appendix.

Each net was fitted with a calibrated flowmeter, and in instances of meter malfunction (9 of 375 hauls) an average meter value was assigned.

Samples, preserved in hexamine-buffered 5–10% formalin, were subdivided with a Folsom plankton splitter at the Smithsonian Oceanographic Sorting Center, and one-fourth of each sample was shipped to the Allan Hancock Foundation, University of Southern California, for biomass analysis. Plankton in an aliquot fraction of each sample was sorted into family or genus, with the exception of foraminiferans, radiolarians, medusae, and siphonophores. Large- and intermediate-size plankton were rinsed with distilled water, dried to a constant weight at 50–60°C, then weighed on microcoverslips on an electronic balance. Rinsing with distilled water was necessary to remove hexamine, since comparisons of rinsed and unrinsed material stored in buffered formalin revealed that hexamine added 10–20% to plankton dry weight.

Smaller plankton, such as minute copepods and

TABLE 1. Bé Net Samples from Each Depth Zone in the Antarctic, the Convergence Zone, and Subantarctic Waters

	0–100 meters	100–250 meters	250–500 meters	500–1000 meters	1000–2000 meters
Antarctic	43	44	45	40	11
Convergence zone*	21	19	20	20	8
Subantarctic	23	25	25	23	8

* *Eltanin* crossings of the convergence on the cruises included in this report indicated that the steepest temperature gradient with respect to distance frequently occurred between 2 and 4°C, and stations with surface temperatures within this range were considered in the convergence zone. Mackintosh [1946] shows that the 2°C isotherm is well below the convergence zone in the antarctic summer; consequently, a few of the series included in convergence zone results were possibly from 'true' antarctic water. However, the error would hardly seem significant in view of the similarity of the total biomass values for the three regions.

TABLE 2. A Comparison of the Catch of Three Plankton Nets Based on Six Series of Tows Made in the Antarctic Convergence Zone

	50-cm (diam.), 76-μ mesh net		50-×50-cm, 202-μ mesh net		100-cm (diam.), 330-μ mesh net	
	no./1000 m³	g dry wt/1000 m³	no./1000 m³	g dry wt/1000 m³	no./1000 m³	g dry wt/1000 m³
Plankton <1-mm maximum dimension						
Copepods*	573,000	0.895	140,000	0.735	27,770	0.254
Radiolarians	167,000	0.167	5,930	0.020	1,260	0.014
Invertebrate eggs	44,800	0.090	8,050	0.016	8,670	0.017
Foraminiferans	31,200	0.038	2,460	0.011	59	...
Pteropods†	21,960	0.022	...	...	...	...
Tintinnids	10,950	0.002	...	...	78	...
Polychaetes§	5,980	0.024	132	0.010	71	0.008
Total	854,890	1.238	156,572	0.792	37,908	0.293
Plankton <1-mm maximum dimension	50,815	3.219	59,097	3.960	47,813	4.636
Total zooplankton	905,705	4.457	215,669	4.752	85,721	4.929

* <1-mm metasome length.
† Minute *Limacina*.
§ Phyllodocidae <1-mm maximum width.

nauplii, were not weighed directly because of the large number of individuals required to obtain a significant weight (5×10^{-2} mg). Their biomass was estimated from size–dry-weight curves established for each type of plankton [Hopkins, 1966b].

Bé Closing Nets and Biomass Estimates

In the present study it is certain that very small plankton such as radiolarians, foraminiferans, and copepod nauplii escaped through the meshes of the nets used and that large mobile organisms avoided the 70- × 70-cm mouth opening. Two sets of comparative tows made on *Eltanin* cruises 15 and 23 gave some indication of the quantity of plankton Bé nets collected in relation to nets of other mesh sizes and mouth areas. On cruise 15 it was found that a 76-μ mesh net 50 cm in diameter caught approximately two times more plankton on a volume-filtered basis than the Bé nets in the top 250 meters [Hopkins, 1966a]. However, if those samples from the 76-μ mesh net (which were rich in phytoplankton, since they were taken in the spring bloom toward the end of the cruise) are not included in the calculations, the zooplankton biomass averages for the two nets are about the same (0.045 cm^3/m^3 for the 202-μ net and 0.043 cm^3/m^3 for the 76-μ net).

On cruise 23 another series of comparative tows was made to a depth of 1000 meters with 76-μ mesh, 50 cm in diameter, 50- × 50-cm 202-μ mesh, and 330-μ mesh, 100 cm in diameter, at six stations near the Antarctic Convergence between 94°W and 109°W. (One 50- × 50-cm net sample was lost in shipment, and two 100-cm net hauls were not completed successfully at 2 of the 6 stations because of high winds.) As might be expected, the number of plankters caught increased with decreasing mesh aperture (Table 2). However, the total biomass values for the three nets show relatively little variation. The 76-μ mesh net captured considerably more of the minute plankton; but these organisms, mostly small copepods and radiolarians, contributed far less to total biomass than the larger but less numerous plankton. On the basis of cruise 23 data it appears that the 202-μ mesh Bé nets seriously underestimated total zooplankton numbers while providing as good an estimate of zooplankton biomass as any of the other nets tested on cruises 15 and 23. The similarity in biomass of zooplankton caught by nets of different meshes and mouth apertures has also been reported by Menzel and Ryther [1961], Tranter [1963], and Wickstead [1963] for oceanic tropical and subtropical areas. The foregoing is not to suggest that the plankton escaping through 202-μ meshes can be ignored; this fraction ultimately must be included in energy-flow considerations, since small planktonic organisms have relatively high metabolic rates per unit of biomass [Marshall and Orr, 1962; Johannes, 1964; Pomeroy and Johannes, 1966].

On cruise 23, hauls to a depth of 1000 meters were made at the six comparative tow stations with a

TABLE 3. A Comparison of Midwater Fishes and Scyphomedusae Caught by Various Plankton Nets and a 3-Meter Isaacs-Kidd Midwater Trawl

Catches were made in the upper 1000 meters of the Antarctic Convergence Zone during cruises of the *Eltanin.*

	50-cm, 76-μ mesh net	50-×50-cm, 202-μ mesh net	70-×70-cm, 202-μ mesh net	100-cm, 330-μ mesh net	3-meter IKMT, 4-mm mesh liner	
	Cr.23	Cr.23	Cr.15	Cr.23	Cr.15	Cr.23
Fishes, g dry wt/m²	0.033	0.023	0.001	0.032	0.391	0.546
Scyphomedusae, g dry wt/m²	...	...	0.126	0.026	0.411	0.106

3-meter Isaacs-Kidd midwater trawl (4-mm mesh liner) to estimate the quantity of larger animals that, because of their scarcity or avoidance reactions, were missed by the plankton nets. These collections indicate (Table 3) that micronektonic fishes (1–17 cm) and megaplankton such as the coronate scyphomedusae *Periphylla* and *Atolla* were poorly sampled by even the 100-cm net. These organisms, though relatively sparsely distributed, contributed significantly to animal standing crop because of their large individual size. Fish and scyphomedusa biomass caught by the 3-meter IKMT (0.65 g dry wt/m^2) was 14% of the average zooplankton biomass value (4.71 g dry wt/m^2) for the three plankton nets tested on cruise 23.

Biomass Estimates and Organic Content of Zooplankton

Prior to dry-weight analysis, the zooplankton collected was rinsed with distilled water, which removed hexamine, inorganic salts, and formalin not chemically bound to denatured protoplasm. Work on the euphausiid *Nematoscelis difficilis* has indicated that there is a net loss of organic matter from zooplankton remaining for a period in formalin [Hopkins, 1968]. Chemical analyses of fresh and preserved *Nematoscelis* revealed 20.0% and 29.7% losses of carbon and nitrogen, presumably mostly in the form of proteinaceous compounds. Some liquid fat loss from *Eltanin* material also has to be postulated, since lipids can escape through ruptures in exoskeletons of preserved crustaceans. It would be difficult to correct for this leakage, since fat content varies from species to species [Nakai, 1955] and from season to season

TABLE 4. Contribution in Percent of Types of Zooplankton to Zooplankton Biomass in the Pacific Sector of the Antarctic

	Antarctic	Convergence Zone	Subantarctic
Copepods	67.3	70.1	68.8
Chaetognaths	10.0	14.4	12.3
Euphausiids	7.5	11.0	8.2

in the same species [Fisher, 1962; Littlepage, 1964]. There can be little doubt, then, that the biomass values reported in this paper are underestimates of the organic content of living antarctic plankton. It should be mentioned that the ash content of rinsed *Eltanin* material was probably less than 10% of dry weight, since combustion at 550°C of specimens of 9 of the important genera gave results ranging from 2.4% to 7.2% ash.

Taxonomic Components of Biomass

The principal contributors to zooplankton biomass in the samples from the Pacific sector of the Antarctic were copepods, chaetognaths, and euphausiids (Table 4). Copepods were the largest share of the biomass, averaging 67.3%, 68.8%, and 70.1% of the standing crop in the Antarctic, the Subantarctic, and the convergence zone. Voronina [1967a] found similar percentages for the Antarctic (72.8%) and the Subantarctic (61.5%) in the Indian Ocean sector.

The most important genera in waters south of the convergence zone, for which taxonomic diagnoses are the most reliable, are listed in Table 5 together with their percentage representation of total biomass. These 11 genera constituted almost two-thirds (66.3%) of the zooplankton standing crop. The two genera that individually contributed more than 10% of the biomass were *Rhincalanus* (13.6%) and *Calanoides* (10.6%).

The present estimate of the most important genera in the antarctic plankton was undoubtedly influenced by the type of collecting gear, the locality, the season,

TABLE 5. Contribution of Individual Genera to Zooplankton Biomass, Upper 1000 Meters of Waters South of the Antarctic Convergence Zone

	% of Biomass		% of Biomass
Rhincalanus	13.6	*Thysanoessa*	4.6
Calanoides	10.6	*Euchaeta*	4.4
Calanus	8.2	*Sagitta*	3.9
Metridia	7.9	*Euphausia*	2.9
Eukrohnia	6.1	*Oithona*	2.3
		Pleuromamma	1.8

and the depths sampled. The enormous masses of euphausiids so frequently reported for the Antarctic are concentrated in the East Wind and Weddell drifts [Marr, 1962], whereas the *Eltanin* Bé net collections were farther north in the West Wind Drift. Had a larger portion of the *Eltanin* stations been nearer to the continent, the euphausiid contribution to zooplankton biomass would have been greater. Support for this comes from 8 series of hauls taken in the Weddell Current on *Eltanin* cruise 12 [Hopkins, 1966c] in which juvenile *Euphausia* averaged 30% of the total biomass in the top 1000 meters. Mackintosh [1934] estimated that at least one-half the biomass in the krill zone is *Euphausia superba*, and the *Euphausia* fraction of cruise 12 collections perhaps would have been closer to Mackintosh's estimate if more adult specimens had been captured by the Bé nets. Foxton [1956] admits that *Discovery* N70V nets did not adequately sample adult euphausiids, which possibly led to his underestimation of zooplankton biomass south of the convergence. Naumov [1964] reports, too, that the Juday nets used on Soviet antarctic expeditions were poorly suited for capturing adult krill.

The seasonal influence on the composition of plankton biomass in the upper 500 meters has been studied by Voronina [1967a], who observed that while *Calanus propinquus* and *Calanoides acutus* have overlapping geographical distributions, *C. acutus* is the first to descend to the Warm Deep Water in the fall, leaving *C. propinquus* to contribute proportionately more to the remaining biomass in the upper 500 meters. Voronina indicated that tows to greater depths would be necessary for a more valid comparison of the standing crop of the two species. *Eltanin* collections were made routinely to a depth of 1000 meters, but Mackintosh [1964] and Andrews [1966] have recorded *C. acutus* even below 1000 meters in the winter. Mackintosh [1964] mentions, too, that moderate numbers of *Rhincalanus* were found below 1000 meters during this season. Six *Eltanin* hauls made at 1000–2000 meters in July and August 1965 also revealed significant concentrations of *C. acutus* (23% of its biomass) below 1000 meters, though only small fractions of the *R. gigas* (3.1%) and *C. propinquus* (4.1%) biomasses were in the 1000- to 2000-meter hauls.

Seasonal Vertical Distribution of Zooplankton Biomass

The vertical distribution of zooplankton standing crop by month in the upper 1000 meters is shown in Figure 2. Changes in the fraction of biomass in the upper 250 meters from season to season further substantiate the annual cycle of vertical migration described by a number of investigators [Mackintosh, 1937; Foxton, 1956; Vinogradov and Naumov, 1964; and Seno et al., 1966] for antarctic and subantarctic waters. A similar pattern of migration has also been observed in the subpolar waters of the Norwegian Sea [Østved, 1955; Beyer, 1962]. As can be seen in Figures 3 and 4, a large fraction of the biomass (48–71%) of the upper 1000 meters is in the top 250 meters of antarctic, subantarctic and convergence zone waters in the late spring and summer months of November through January (no February data). Relatively high concentrations of zooplankton in the 0- to 250-meter zone persist in March and April in the Antarctic, though in these months in the subantarctic and the convergence zone, *Eltanin* samples evidence a decline in plankton concentrations to winter levels. In all three regions in the late fall, winter, and early spring months of May through October, the percentages of zooplankton in the upper 250 meters range from no more than 33% to less than 4%.

The annual vertical migration cycle has been recorded as well for a number of individual copepod species south of the convergence, such as *C. propinquus* [Seno et al., 1966; Voronina, 1967a], *C. acutus* [Mackintosh, 1964; Andrews, 1966], *R. gigas* [Ommanney, 1936], and *Metridia gerlachei* [Seno et al., 1966]. *Eltanin* collections suggest, too, that these species were more concentrated in the 0- to 250-meter layer from November to April than from May to October (Figure 5). As Mackintosh [1937] has pointed out, the concentration of these species at different depths in winter and summer ensures the maintenance of at least some of the breeding population south of the convergence, since the Antarctic Surface Current flows in a northeasterly direction toward the convergence, while the underlying Warm Deep Water flows south toward the continent.

Some indication of the biomass of zooplankton below 1000 meters was obtained from 27 collections from the 1000- to 2000-meter depth zone. In the antarctic, subantarctic, and convergence regions sampled, the average percentages of biomass in the lower half of the 2000-meter water column were 27.1%, 26.8%, and 37.8%, respectively. These hauls were all taken between March and October, which spans the period when the annually migrating plankton is in the deeper water. For this reason the values

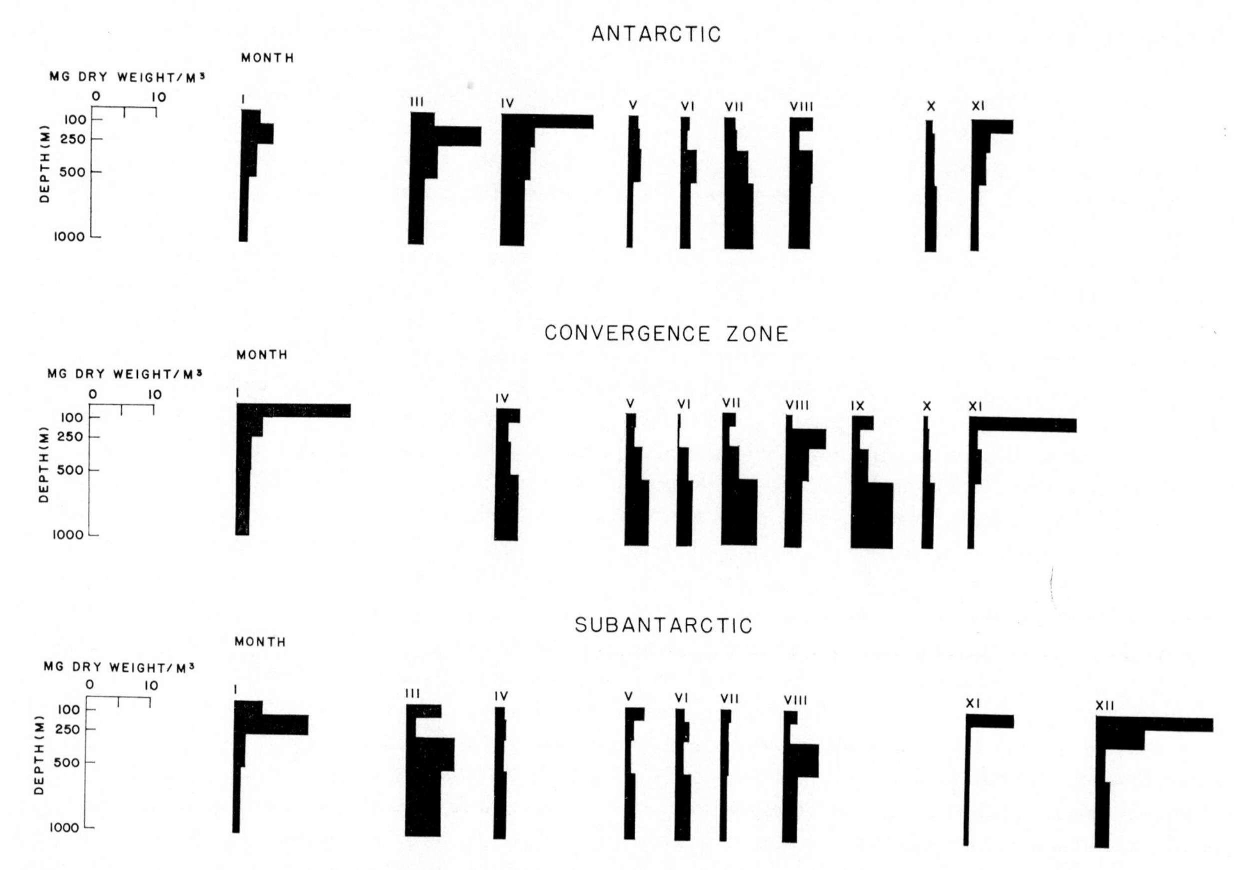

Fig. 2. The vertical distribution of zooplankton biomass by month in the upper 1000 meters of the Antarctic, the Subantarctic, and the convergence zone in the Pacific sector.

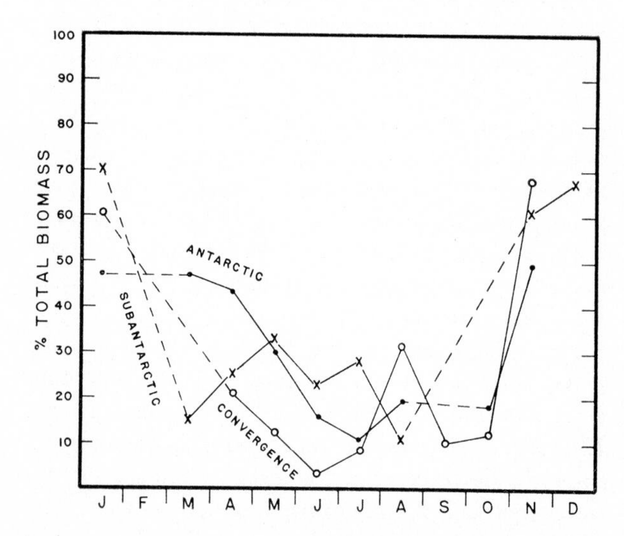

Fig. 3. Percent of zooplankton biomass of the upper 1000 meters found in the 0- to 250-meter zone throughout the year in antarctic, subantarctic and convergence zone waters of the Pacific sector.

may represent maximum percentages and therefore may be somewhat higher than the annual average for the 1000- to 2000-meter layer. Some evidence for this comes from the six samples taken from this depth in April in the Antarctic south of the convergence zone when 44% of the zooplankton captured was taken from the upper 250 meters (see Figure 4). During this month only 12% of the biomass was in the 1000- to 2000-meter layer. Beyer [1962], reporting on collections from the Norwegian Sea, presents figures showing that 30–47% of the zooplankton biomass was in the 1000- to 2000-meter layer from April through August, while 51–72% was in this zone during the remaining fall, winter, and early spring months.

Geographic Distribution of Zooplankton Biomass

Zooplankton biomass in the upper 1000 meters of the Pacific sector of the Antarctic (Figure 6) does

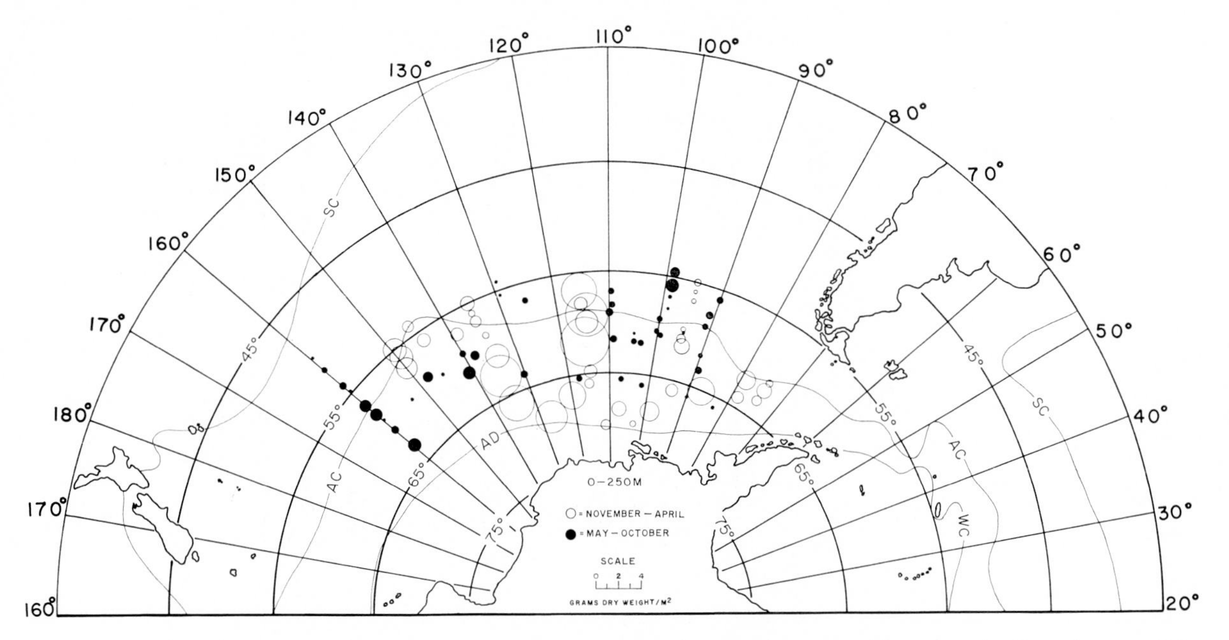

Fig. 4. Zooplankton biomass found in the 0- to 250-meter zone in the austral summer (November–April) and winter (May–October) halves of the year. Hydrographic boundaries as in Figure 1.

not evidence discernible regional trends: stations showing plankton in great abundance are immediately adjacent to stations where negligible quantities were taken. A portion of this variation undoubtedly results from patchiness in antarctic plankton and possibly from seasonal fluctuations in biomass as well. The essentially patchy nature of antarctic plankton has been reported in detail by Hardy and Gunther [1935] and Hardy [1936] in their study of the South Georgia whaling grounds, and by Sheard [1947] for the Australian sector of the Antarctic. Foxton [1956] was unable to discern seasonal variations in zooplankton biomass in the upper 1000 meters in his analyses of *Discovery* collections (Figure 7). Beyer [1962], however, suggests that by using displacement volume as a measure of biomass rather than dry weight, Foxton would not necessarily have detected actual seasonal variations in organic matter. By removing water, which constitutes such a large but variable fraction of displacement volume or wet weight, Beyer was able to note a distinct seasonal trend in the biomass of zooplankton collected from weather station M in the Norwegian Sea. In the present study, 31 complete series of hauls to 1000 meters, taken in the Antarctic, the Subantarctic, and the convergence zone from November to April, gave a mean value of 2.96 g dry wt/m^2, while 36 series taken in the winter months of May to October gave a slightly lower average of 2.45 g/m^2. A *t*-test on log-transformed data shows the means to be significantly different at the 0.01 level. Since there is the possibility of some seasonal fluctuation in biomass, the winter and the summer data points in Figure 6 are differentiated.

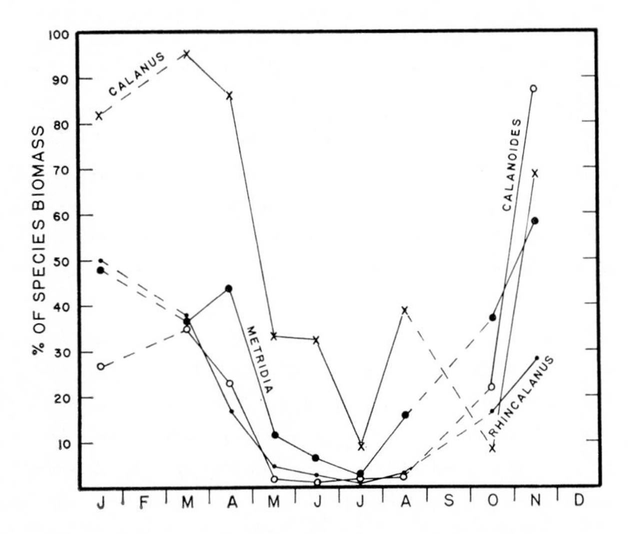

Fig. 5. Percent of *Rhincalanus gigas*, *Calanoides acutus*, *Calanus propinquus*, and *Metridia gerlachei* biomass in the 0- to 250-meter zone of the upper 1000 meters south of the convergence zone in the Pacific sector.

While the *t*-test does support the concept of a seasonal variation in biomass in the upper 1000 meters, wide fluctuations in *Eltanin* results both in

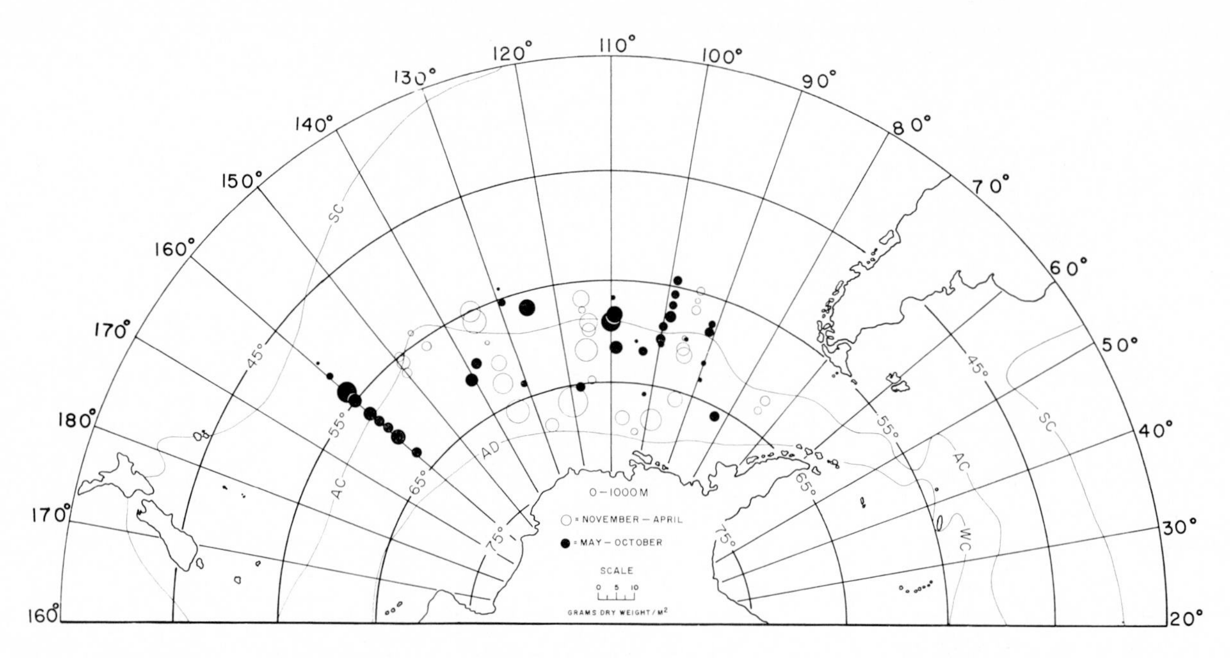

Fig. 6. Distribution of zooplankton biomass in the upper 1000 meters of the Pacific sector of the Antarctic. Hydrographic boundaries as in Figure 1.

winter and summer are apparent, especially when comparisons are made with *Discovery* data (Figure 7). In the absence of possible statistical proof, it is suggested that patchiness may be the largest cause of variation in the present results. Only 375 *Eltanin* collections from different depths, months, and regions were examined, whereas Foxton obtained biomass data from over 2100 *Discovery* samples.

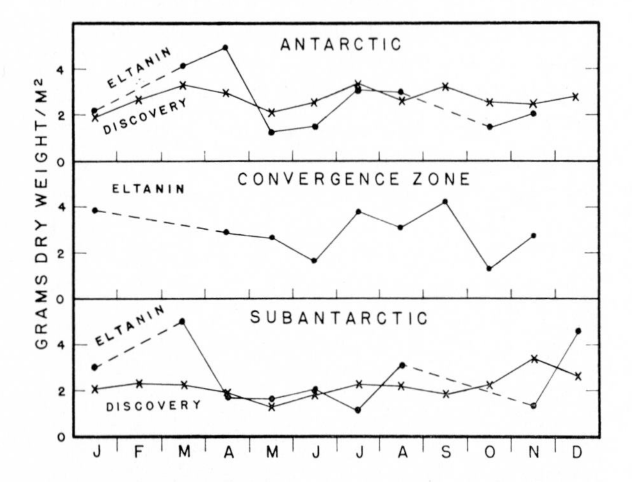

Fig. 7. Seasonal distribution of zooplankton biomass in the upper 1000 meters as determined from *Eltanin* collections from the Pacific sector (*Eltanin* cruises 10, 11, 13–15, 17–19) and *Discovery* material [Foxton, 1956] from all sectors of the Antarctic.

The average standing-crop values for the upper 1000 meters of the Antarctic and Subantarctic were 2.67 and 2.58 g dry wt/m^2, respectively. Foxton's averages [1956] for the Antarctic and the Subantarctic in all sectors about the continent, after converting displacement volume to dry weight, were 2.52 and 2.05 g/m^2, and in the Pacific sector alone (60–180°W), 2.10 and 1.80 g/m^2. The mean value for the convergence zone in the present study was 2.96 g/m^2, which is slightly higher than those for the Antarctic and the Subantarctic. The *Discovery* results also show an increase in standing crop in the vicinity of the convergence. Foxton [1956], however, cautions that the observed increase may be more apparent than real, since *Discovery* N70V nets fished poorly for the strongly swimming euphausiids, which become increasingly abundant south of the convergence.

While there is general agreement on the level of net-caught plankton in the antarctic West Wind Drift in the Pacific sector, *Discovery* results are somewhat lower than those based on *Eltanin* collections. This may stem, in part (assuming the displacement volume to dry-weight conversion factor is valid [see footnote 1, Table 6]) from variations in preparing the samples for biomass analysis. Phytoplankton was removed from samples prior to biomass determination in both studies, but Foxton also removed gelatinous

TABLE 6. Comparison of Zooplankton Standing Crop in the Upper 1000 Meters of the Antarctic with Biomass in the Upper 1000 Meters of Other Oceanic Regions

Region	Biomass, g dry wt/m²*	Net Mesh Aperture, μ	Source
Kuril trench	16.9	170	Jashnov [1961]
Norwegian Sea, weather station M, 66°N, 2°E	6.4	202-569	Beyer† [1962]
East Greenland Current	5.6	73	Hopkins (unpublished data)
Continental slope, N. America, 38–41°N	3.7	170	Jashnov [1961]
Antarctic, West Wind Drift, Pacific sector			
Antarctic	2.1	6.3 mm-569-239	Foxton [1956]
	2.7	202	Hopkins [present study]
Convergence zone	3.0	202	Hopkins [present study]
Subantarctic	1.8	6.3 mm-569-239	Foxton [1956]
	2.6	202	Hopkins [present study]
Gulf Stream, 35–43°N	1.3–2.4	170	Jashnov [1961]
N. Equatorial Current, 16–19°N	1.9	170	Jashnov [1961]
Canary Current	1.8	170	Jashnov [1961]
Sargasso Sea			
Off Bermuda	1.5	366	Menzel and Ryther [1961]
Southwest area, 20–37°N	1.3	170	Jashnov [1961]
Indian Ocean			
Subtropical, 90°E, 2–23°S	0.8	6.3 mm-569-239	Foxton [1956]
Tropical, 29–115°E, 32°S	0.8	6.3 mm-569-239	Foxton [1956]
Mariana deep	0.6	170	Jashnov [1961]
Arctic basin	0.2	73	Hopkins [1969b]

* All displacement volumes have been converted to dry weight assuming that (*a*) 1 cm³ displacement volume equals 1 gram wet weight and (*b*) dry weight equals 0.1 × wet weight. Only the Hopkins and the Menzel and Ryther data were reported directly as dry weight.

† The Nansen net used at weather station M and the Discovery N70V net [Foxton, 1956] have several sizes of mesh with the finest cloth at the cod end.

organisms (siphonophores, medusae, and salps) and large crustaceans (primarily euphausiids). These organisms were included in the dry-weight estimations of *Eltanin* collections.

Standing-Crop Comparisons

Zooplankton standing crop in the upper 1000 meters of the West Wind Drift in the Pacific sector of the Antarctic appears to be only moderately high in relation to other regions of the world ocean (Table 6). Though biomass based on Bé net collections is similar to that recorded for slope waters off North America, the average *Eltanin* values are only one-half those for the rich East Greenland Current and the Norwegian Sea at weather station M and one-eighth the biomass value reported for the eutrophic waters over the Kuril trench. Antarctic and subantarctic values, on the other hand, are somewhat greater than those for the relatively poor waters of the Canary and North Equatorial currents and the Gulf Stream, and considerably greater than those for Sargasso Sea, tropical and subtropical Indian Ocean, Mariana trench, and Arctic basin waters. These comparisons are weakened to some extent because different collecting methods and techniques for biomass analysis were used by different investigators. There is a general trend, though, as Foxton [1956] found, to a decrease in biomass toward the low latitudes.

Zooplankton Standing Crop and Phytoplankton Production

Annual primary production in the Pacific sector of the Antarctic and Subantarctic has been estimated by Burkholder and Burkholder [1967] to be 29–32 g C/m². Zooplankton biomass in the upper 2000 meters of the Pacific sector of the Antarctic and Subantarctic is estimated from the present study at 4.06 g and 4.14 g dry wt/m². From Burkholder and Burkholder's estimate of the carbon fixed annually in photosynthesis and the present figures for zooplankton biomass, it was calculated (see note at end for calculations) that the average annual respiration level supportable by primary production in the West Wind

Drift on either side of the convergence was 36–40 μl O_2/mg dry wt/day. As a point of reference, this is close to the daily respiration rate, measured at 6–7°C, of 2- to 3-mm *Calanus finmarchicus:* 42.4 μl O_2/mg dry wt [Marshall and Orr, 1962]. It must be assumed, however, that animal biomass has been underestimated, since depths below 2000 meters were not sampled and losses occurred through avoidance, escapement, preservation, and handling. These calculations, too, assume an improbable 100% efficiency in ingestion and assimilation of the entire phytoplankton production by the zooplankton. Also assumed is maximum efficiency in converting the assimilated phytoplankton food produced in the summer months to metabolic storage products for zooplankton that must survive the winter. At the present time, then, information is still too limited to state unequivocally that phytoplankton production, as measured by the C^{14} uptake method, can entirely support the pelagic animal population in the Antarctic or that supplemental nutritional sources, such as dissolved organic compounds or detritus formed in situ [Riley et al., 1965], are required. Additional biomass sampling to depths even greater than 2000 meters with various types of nets, and studies of seasonal respiration of different-size groups of plankton, from the minute zooplankton escaping the meshes of the Bé nets to the micronekton avoiding the net mouth, are needed to test this concept. The estimate of annual primary production also should not go unchallenged. A more recent figure (S. Z. El-Sayed, Texas A & M University, personal communication, 1968) for annual primary production in the Antarctic and Subantarctic in the Pacific sector is 163 g C/m^2, a value over 5 times that published by Burkholder and Burkholder. Pomeroy and Johannes [1968] raise the fundamental question of the reliability of the C^{14} method as a measure of primary production and cite evidence that C^{14} productivity values may be 2 to 3 times too low. It is uncertain, further, how directly comparable are values of C^{14} uptake to actual phytoplankton cell production, the measure of primary production perhaps most useful in studying zooplankton–phytoplankton trophic relationships.

TABLE 7. Carnivore/Herbivore Ratio of Biomass in the Upper 1000 Meters of Antarctic, Convergence Zone, and Subantarctic Areas in the Pacific Sector

	Antarctic	Convergence Zone	Subantarctic
January	0.17(1)	0.24(3)	0.24(2)
February	...	...	...
March	0.22(1)	...	0.19(2)
April	0.32(7)	0.30(2)	0.43(4)
May	0.20(2)	0.38(1)	0.47(1)
June	0.31(3)	0.52(1)	0.60(5)
July	0.30(1)	0.40(5)	1.04(1)
August	0.31(6)	0.38(1)	0.66(4)
September	...	0.24(1)	...
October	0.46(2)	0.42(2)	...
November	0.41(5)	0.15(3)	0.72(1)
December	...	...	0.49(1)

Numbers of complete series of hauls are given in parentheses.

Zooplankton Trophic Relationships

Cushing [1959] classifies high-latitude regions of the sea as unbalanced rather than steady-state systems because most of the plankton production is restricted to the summer months. It would be expected, then, that the herbivore contribution to biomass would be highest during the summer months. There is considerable variation from month to month, partially because of the limited number (67) of complete series of hauls; but comparatively low carnivore/herbivore ratios (Table 7) do occur in the austral summer months of January and March in the Antarctic and the Subantarctic and in January and November in the convergence zone. (The carnivore fraction collected consisted mainly of chaetognaths, the copepod genus *Euchaeta*, medusae, and siphonophores. Amphipods and copepods of the families Heterorhabdidae, Phaennidae, and Aetideidae also contributed significantly to predator biomass on occasion. Organisms of uncertain trophic classification were included in the herbivore fraction.)

Slobodkin [1960] states that net trophic or ecological efficiency, that is, the percentage of organic matter of one trophic level incorporated in the next higher trophic level, does not exceed 10–15%. In the present study the carnivore/herbivore ratio exceeds 0.15 in every month in which samples were taken except November (0.149) in the convergence zone. The carnivore fraction has probably been underestimated, since predaceous micronekton (mostly fishes, squids, and decapods), and megaplankton (e.g., coronate scyphomedusae), and the predominately carnivorous animals living below 1000 meters [Vinogradov, 1962] were not included in the calculations. In this investigation, then, it would seem that the carnivore/herbivore ratio is too high throughout the year to mirror net trophic efficiency.

Hedgpeth [1957] suggests that it is possible for

organisms with high rates of turnover to support a relatively large biomass of organisms with lower rates of growth and reproduction. It is postulated that this is the case in the Antarctic and could apply as well to steady-state tropical and subtropical environments. Grice and Hart [1962], for instance, found carnivores to constitute at least 39% of the zooplankton biomass in the upper 200 meters of the Sargasso Sea. Blackburn [1966], on the other hand, found carnivorous standing crop to be less than 10% of herbivore biomass in the eastern tropical Pacific. The area he chose for analysis evidences little annual variation in environmental parameters, and this constancy presumably allows a steady-state relationship between herbivore and carnivore production to evolve so that the carnivore/herbivore biomass ratio reflects net trophic efficiency. It is possible, however, that carnivore standing crop in the eastern tropical Pacific is much higher than Blackburn's estimate, since he took into account only the zooplankton in the upper 300 meters and the micronekton in the top 100 meters and considered all copepods herbivorous. On the basis of the limited information available on carnivore/herbivore ratios, it seems doubtful that biomass data alone can be used with confidence to calculate net trophic efficiency; studies of rates of turnover are necessary as well. Except for McWhinnie and Marciniak's respiration studies on *Euphausia superba* [1964], virtually nothing is known concerning the metabolic rate of the zooplankton and micronekton in antarctic waters; and research is urgently needed to obtain a realistic estimate of energy flow in antarctic pelagic communities. Pomeroy and Johannes [1968] have shown that most of the energy turnover in the pelagic environment is accomplished by ultraplankton passing through the meshes of plankton nets. It is essential, though, to investigate the metabolic rates of the micro- and macroplankton caught with nets as well, since these organisms are important links in the antarctic food chain. The present study gives some indication of the relative importance in terms of biomass of the various genera of net-caught zooplankton found in the Antarctic and suggests the most important taxa for research on metabolic rates.

NOTE

Given 32 g $C/m^2/yr$ fixed by net photosynthesis in the Subantarctic and 4.14 g dry wt/m^2 of zooplankton in the 0- to 2000-meter zone of the Subantarctic:

32 g $C/m^2/yr \times 1/365$ days/yr $= 0.0875$ g $C/m^2/day$ ($= 0.32$ g $CO_2/m^2/day$) fixed in net photosynthesis

0.32 g $CO_2/m^2/day \times 1/44$ g CO_2/mole $CO_2 \times 22.4 \times 10^6$ μl CO_2/mole $CO_2 = 0.163$ μl $CO_2/m^2/day$

0.163×10^6 μl $CO_2/m^2/day$ requires 0.163×10^6 μl O_2 for its respiration, assuming R.Q. = 1

0.163×10^6 μl $O_2/m^2/day \times 1/4.14 \times 10^3$ mg zooplankton dry $wt/m^2 = 39.4$ μl O_2/mg zooplankton dry $wt/m^2/day$ = average annual zooplankton respiration level supportable by annual net phytoplankton production

Acknowledgments. The samples used in this study were collected by Arthur Rothenstein, James Hubbard, and Michael Smiles, who participated in the antarctic skeletal plankton program (NSF grants GA-118 and GA-212) directed by Dr. Allan W. H. Bé (Lamont Geological Observatory). The laboratory portion of the research (NSF grants G-19497, GA-238, and GA-448) was completed at the Allan Hancock Foundation, University of Southern California. I am grateful to Dr. J. M. Savage and Dr. J. L. Mohr for the many courtesies they extended to me during my association with the Antarctic Project at USC and to Sharon Walker, Diana Kennett, Ann Maria Thayer, and Joan Stapleton for their patient assistance in plankton sorting and data processing. I would like to express my thanks to Dr. Allan W. H. Bé, Dr. George D. Grice (Woods Hole Oceanographic Institution), Dr. Peter Foxton (National Institute of Oceanography), and Dr. Sayed Z. El-Sayed (Texas A & M University) for offering their comments on this paper.

This work was supported by NSF grants GA-118 and GA-212, G-19497, GA-238, and GA-448. The paper is contribution 10 of the Marine Science Institute.

REFERENCES

Andrews, K. J. H.
1966 The distribution and life-history of *Calanus acutus* (Giesbrecht). 'Discovery' Rep., *34:* 117–162.

Baker, A. De C.
1959 Distribution and life history of *Euphausia triacantha* Holt and Tattersall. 'Discovery' Rep., *29:* 309–340.

Bé, A. W. H.
1962 Quantitative multiple opening-and-closing plankton samplers. Deep Sea Res., *9:* 144–151.

Beyer, F.
1962 Absorption of water in crustaceans, and the standing crop of zooplankton. Rapp. P.-v. Réun. Cons. Perm. Int. Explor. Mer, *153:* 79–85.

Blackburn, M.
1966 Relationships between standing crop at successive trophic levels in the eastern Pacific. Pacif. Sci., *20:* 36–59.

Brodskii, K. A.
1964 Plankton studies by the Soviet Antarctic Expedition (1955–1958). Inf. Bull. Sov. Antarct. Exped., *1:* 105–110.

Brodskii, K. A., and M. E. Vinogradov
1957 Plankton distribution in the Indian sector of the Antarctic (Engl. transl.). Dokl. Akad. Nauk SSSR, *112:* 120–122.

Burkholder, P. R., and L. M. Burkholder
1967 Primary productivity in surface waters of the South Pacific Ocean. Limnol. Oceanogr., *12:* 606–617.

Cushing, D. H.
1959 The seasonal variation in oceanic production as a

problem in population dynamics. J. Cons. Perm. Int. Explor. Mer, *24:* 455–464.

David, P. M.

1958 The distribution of the Chaetognatha of the Southern Ocean. 'Discovery' Rep., *29:* 199–228.

Deacon, G. E. R.

1964 Antarctic oceanography: the physical environment. *In* R. Carrick et al., (eds.), Biologie Antarctique. Herman, Paris, pp. 81–86.

Fisher, L. R.

1962 The total lipid material in some species of marine zooplankton. Rapp. P.-V. Réun. Cons. Perm. Int. Explor. Mer, *153:* 129–136.

Foxton, P.

1956 The standing crop of zooplankton in the Southern Ocean. 'Discovery' Rep., *28:* 193–235.

Grice, G. D., and A. D. Hart

1962 The abundance, seasonal occurrence and distribution of the epizooplankton between New York and Bermuda. Ecol. Monogr., *32:* 287–309.

Hardy, A. C.

1936 Observations on the uneven distribution of oceanic plankton. 'Discovery' Rep., *11:* 511–538.

Hardy, A. C., and E. R. Gunther

1935 The plankton of the South Georgia whaling grounds and adjacent waters, 1926–1927. 'Discovery' Rep., *11:* 1–456.

Hedgpeth, J. W.

1957 Concepts in marine ecology. *In* J. W. Hedgpeth, (ed.), Treatise on Marine Ecology and Paleoecology. Mem. Geol. Soc. Amer., 67, *2:* 29–52.

Hopkins, T. L.

1966a A volumetric analysis of the catch of the Isaacs-Kidd midwater trawl and two types of plankton nets in the Antarctic. Aust. J. Mar. Freshwat. Res., *17:* 147–154.

1966b The plankton of the St. Andrew Bay System, Florida. Publs Inst. Mar. Sci. Univ. Tex., *11:* 12–64.

1966c Zooplankton standing crop in the Atlantic sector of the Antarctic Ocean, Second International Oceanographic Congress, Moscow, 30 May–9 June, 1966. Abstracts of Papers, p. 160.

1968 Carbon and nitrogen content of fresh and preserved *Nematoscelis difficilis,* a euphausiid crustacean. J. Cons. Perm. Int. Explor. Mer, *31:* 300–304.

1969a Zooplankton biomass related to hydrography along the drift track of *Arlis* II in the Arctic Basin and the East Greenland Current. J. Fish. Res. Bd Can., *26:* 305–310.

1969b Zooplankton standing crop in the Arctic Basin, Limnol. Oceanogr., *14:* 80–85.

Jashnov, V. A.

1961 Vertical distribution of the mass of zooplankton in the tropical region of the Atlantic Ocean. Dokl. Akad. Nauk SSSR, *136:* 705–708.

Johannes, R. E.

1964 Phosphorus excretion and body size in marine animals: microzooplankton and nutrient regeneration. Science, N. Y., *146:* 923–924.

Kane, J. E.

1966 The distribution of *Parathemisto gaudichaudii* (Guer.), with observations on its life history in the 0° to 20°E sector of the Southern Ocean. 'Discovery' Rep., *34:* 163–198.

Korotkevich, V. S.

1958 Distribution of the plankton in the Indian sector of the Antarctic. Dokl. Akad. Nauk SSSR, *122:* 578–581.

Littlepage, J. L.

1964 Seasonal variation in lipid content of two Antarctic marine crustacea. *In* R. Carrick et al., (eds.), Biologie Antarctique. Herman, Paris, pp. 463–470.

Mackintosh, N. A.

1934 Distribution of the macroplankton in the Atlantic sector of the Antarctic. 'Discovery' Rep., *9:* 67–158.

1937 Seasonal circulation of the Antarctic macroplankton. 'Discovery' Rep., *16:* 367–412.

1946 The Antarctic Convergence and the distribution of surface temperatures in Antarctic waters. 'Discovery' Rep., *23:* 177–212.

1964 Distribution of the plankton in relation to the Antarctic Convergence. Proc. R. Soc. (London), Ser. A (physical), *281:* 21–37.

Marr, J. W. S.

1962 The natural history and geography of the Antarctic krill (*Euphausia superba* Dana). 'Discovery' Rep., *32:* 33–464.

Marshall, S. M., and A. P. Orr

1962 Food and feeding in copepods. Rapp. P.-v. Réun. Cons. Perm. Int. Explor. Mer, *153:* 92–98.

McWhinnie, M. A., and P. L. Marciniak

1964 Temperature responses and tissue respiration in Antarctic crustacea with particular reference to the krill *Euphausia superba.* Antarctic Res. Ser., *1:* 63–72. AGU, Washington, D. C.

Menzel, D. W., and J. H. Ryther

1961 Zooplankton in the Sargasso Sea off Bermuda and its relation to organic production. J. Cons. Perm. Int. Explor. Mer, *26:* 250–258.

Nakai, Z.

1955 The chemical composition, volume, weight, and size of the important marine plankton. Spec. Publs Tokai Fish. Res. Lab., *5:* 12–24.

Naumov, A. G.

1964 Some features of the distribution and biology of *Euphausia superba* Dana. Inf. Bull. Sov. Antarct. Exped. (Engl. transl.), *4:* 191–194.

Ommanney, F. D.

1936 *Rhincalanus gigas* (Brady), a copepod of the southern macroplankton. 'Discovery' Rep., *13:* 277–384.

Østvedt, O. J.

1955 Zooplankton investigations from weather ship 'M' in the Norwegian Sea, 1948–49. Hvalråd. Skr., *40:* 1–93.

Pomeroy, L. R., and R. E. Johannes

1966 Total plankton respiration. Deep Sea Res., *13:* 971–973.

1968 Occurrence and respiration of ultraplankton in the upper 500 meters of the ocean. Deep Sea Res., *15:* 381–391.

Riley, G. A., D. van Hemert, and P. J. Wangersky

1965 Organic aggregate in tropical and subtropical surface

waters of the North Atlantic Ocean. Limnol. Oceanogr., *9:* 546–550.

Seno, J., Y. Komaki, and A. Takeda
1966 Report on the biology of the 'Umitaka-Maru' Expedition. Plankton collected in the Antarctic Ocean and adjacent waters by the closing net, with special references to the copepods. J. Tokyo Univ. Fish., *52:* 1–16.

Sheard, K.
1947 Plankton of the Australian-Antarctic quadrant. Part I. Net-plankton volume determination. Rep. BANZ Antarct. Res. Exped., ser. B, *6:* 1–120.

Slobodkin, L. B.
1960 Ecological energy relationships at the population level. Am. Nat., *94:* 213–236.

Tranter, D. J.
1963 Comparison of zooplankton biomass determinations by Indian Ocean Standard Net, Juday Net and Clarke-Bumpus Sampler. Nature, London, *198:* 1179–1180.

Vinogradov, M. E.
1962 Feeding of the deep-sea zooplankton. Rapp. P.-V. Réun. Cons. Perm. Int. Explor. Mer, *153:* 114–120.

Vinogradov, M. E., and A. G. Naumov
1964 Quantitative distribution of plankton in Antarctic waters of the Indian and Pacific Oceans. Inf. Bull. Sov. Antarct. Exped. (Engl. transl.), *1:* 110–112.

Voronina, N. M.
1967a The zooplankton of the Southern Ocean: some study results. Oceanology (Engl. transl.), *6:* 557–563.
1967b Distribution of the zooplankton biomass in the Southern Ocean. Oceanology (Engl. transl.), *6:* 836–846.

Wickstead, J. H.
1963 Estimates of total zooplankton in the Zanzibar area of the Indian Ocean with a comparison of the results with two different nets. Proc. Zool. Soc. Lond., *141:* 577–608.

APPENDIX TABLE

Plankton Biomass Values for *Eltanin* Stations

Sample	Cruise	Date	Lat., °S	Long., °W	Depth, meters	Biomass, g dry wt/m²
			1963 Data			
143	10	Oct. 23	63° 03′	82° 48′	0–100	0.157
145	10		63° 02′	82° 46′	500–1080	1.230
147	10	Oct. 25	63° 53′	83° 04′	100–250	0.319
148	10		63° 53′	83° 04′	250–528	0.161
154	10	Oct. 28	65° 57′	82° 56	0–100	0.058
155	10		65° 57′	82° 56′	100–250	0.182
156	10		65° 57′	82° 56′	250–493	0.720
157	10		65° 58′	82° 54′	500–1010	1.865
160	10	Oct. 29	64° 55′	78° 50′	250–492	0.570
161	10	Oct. 30	64° 52′	78° 42′	500–1120	0.985
168	10	Nov. 3	62° 59′	74° 58′	0–100	0.037
169	10		62° 59′	74° 58′	100–250	0.725
170	10		62° 59′	74° 58′	250–555	0.287
171	10		62° 53′	74° 55′	500–1005	0.835

APPENDIX TABLE (continued)

Sample	Cruise	Date	Lat., °S	Long., °W	Depth, meters	Biomass, g dry wt/m²
			1963 Data			
173	10	Nov. 5	61° 56′	75° 01′	0–100	1.110
174	10		61° 56′	75° 01′	100–250	0.134
175	10		61° 56′	75° 01′	250–525	0.580
176	10		61° 56′	75° 01′	500–932	0.740
178	10	Nov. 6	61° 04′	75° 08′	0–100	0.173
179	10		61° 04′	75° 08′	100–250	0.345
180	10		61° 04′	75° 08′	250–526	0.668
198	10	Nov. 15	64° 01′	79° 06′	0–100	0.837
199	10		64° 01′	79° 06′	100–250	0.127
200	10		64° 01′	79° 06′	250–508	0.191
210	10	Nov. 18	62° 37′	78° 55′	0–100	0.910
211	10		62° 37′	78° 55′	100–250	0.730
212	10		62° 37′	78° 55′	250–515	0.423
236	11	Dec. 31	56° 54′	115° 18′	0–100	1.858
237	11		56° 54′	115° 18′	100–250	1.190
238	11		56° 54′	115° 18′	250–525	0.339
239	11		56° 54′	115° 18′	500–1000	1.160
			1964 Data			
240	11	Jan. 1	57° 47′	115° 14′	0–100	0.587
241	11		57° 47′	115° 14′	100–250	0.436
242	11		57° 47′	115° 14′	250–500	0.365
243	11		57° 47′	115° 14′	500–1000	0.345
246	11	Jan. 2	58° 58′	114° 47′	0–100	0.424
247	11		58° 58′	114° 47′	100–250	3.140
248	11		58° 58′	114° 47′	250–475	0.505
249	11	Jan. 3	58° 58′	114° 47′	500–1000	0.649
251	11	Jan. 4	59° 57′	114° 57′	0–100	1.196
252	11		59° 57′	114° 57′	100–250	0.608
253	11		59° 57′	114° 57′	250–480	0.805
254	11		59° 57′	114° 57′	250(?)–1000	0.985
260	11	Jan. 7	61° 59′	115° 12′	0–100	3.274
261	11		61° 59′	115° 12′	100–250	0.920
262	11		61° 59′	115° 12′	250–500	0.575
263	11		61° 59′	115° 12′	500–1000	1.275
273	11	Jan. 11	64° 59′	114° 50′	0–100	0.900
274	11		64° 59′	114° 50′	100–250	0.176
275	11		64° 59′	114° 50′	250–500	0.250
276	11		64° 59′	114° 50′	500–1000	0.745
280	11	Jan. 13	65° 52′	115° 05′	0–100	0.297
281	11		65° 52′	115° 05′	100–250	0.318
282	11		65° 52′	115° 05′	250–475	0.212
287	11	Jan. 14	66° 53′	115° 31′	500–1000	0.410
298	11	Jan. 18	70° 12′	110° 58′	0–100	0.135
299	11		70° 12′	110° 58′	100–250	0.629
300	11		70° 12′	110° 58′	250–490	0.782
303	11	Jan. 19	70° 09′	106° 40′	500–1000	0.765
306	11	Jan. 20	70° 07′	102° 55′	0–100	0.124
307	11		70° 07′	102° 55′	100–250	0.311
308	11		70° 07′	102° 55′	250–500	0.918
309	11		70° 07′	102° 55′	500–1000	0.505
326	11	Jan. 28	65° 57′	89° 10′	500–1000	0.400

APPENDIX TABLE
(continued)

Sample	Cruise	Date	Lat., °S	Long., °W	Depth, meters	Biomass, g dry wt/m²
			1964 Data			
328	11	Jan. 29	65° 08′	86° 53′	0–100	0.620
329	11		65° 08′	86° 53′	100–250	1.722
330	11		65° 08′	86° 53′	250–500	0.600
336	11	Jan. 31	63° 07′	86° 53′	0–100	0.336
444	13	May 19	54° 59′	89° 49′	100–250	0.187
445	13		54° 59′	89° 49′	250–490	0.148
446	13		54° 59′	89° 49′	500–1000	0.600
449	13	May 20	56° 05′	90° 10′	0–100	0.310
450	13		56° 05′	90° 10′	100–250	0.142
451	13		56° 05′	90° 10′	250–510	0.248
458	13	May 23	57° 49′	90° 46′	0–100	0.288
459	13		57° 49′	90° 46′	100–250	0.247
460	13		57° 49′	90° 46′	250–480	0.235
461	13		57° 49′	90° 46′	500–955	0.960
465	13	May 24	58° 45′	91° 05′	0–100	0.140
466	13	May 24	58° 45′	91° 05′	100–250	0.207
467	13		58° 45′	91° 05′	250–500	0.580
468	13		58° 45′	91° 05′	500–1000	1.865
480	13	May 29	62° 00′	90° 20′	0–100	0.042
481	13		62° 00′	90° 20′	100–250	0.213
482	13		62° 00′	90° 20′	250–500	0.415
483	13		62° 00′	90° 20′	500–1000	0.800
486	13	May 30	63° 03′	89° 54′	0–100	0.243
487	13		63° 03′	89° 54′	100–250	0.232
488	13		63° 03′	89° 54′	250–520	0.472
489	13		63° 03′	89° 54′	500–980	0.020
497	13	June 4	66° 10′	90° 20′	0–100	0.030
498	13		66° 10′	90° 20′	100–250	0.041
499	13		66° 10′	90° 20′	250–500	0.133
507	13	June 8	66° 12′	97° 55′	500–1030	1.110
510	13	June 10	66° 09′	102° 14′	0–100	0.121
511	13		66° 09′	102° 14′	100–250	0.176
512	13		66° 09′	102° 14′	250–480	0.265
513	13		66° 09′	102° 14′	500–1000	0.535
516	13	June 11	65° 30′	107° 11′	0–100	0.238
517	13		65° 30′	107° 11′	100–250	0.041
518	13		65° 30′	107° 11′	250–475	0.780
528	13	June 15	65° 17′	117° 35′	0–100	0.111
529	13		65° 17′	117° 35′	100–250	0.239
530	13		65° 17′	117° 35′	250–520	1.050
531	13		65° 17′	117° 35′	500–1050	0.925
543	13	June 22	64° 08′	130° 11′	0–100	0.095
544	13		64° 08′	130° 11′	100–250	0.248
545	13		64° 08′	130° 11′	250–480	0.820
546	13		64° 08′	130° 11′	500–1000	0.575
558	13	June 28	55° 43′	129° 37′	0–100	0.031
559	13		55° 43′	129° 37′	100–250	0.038
560	13		55° 43′	129° 37′	250–500	0.455
561	13		55° 43′	129° 37′	500–1000	1.250
564	13	June 29	54° 31′	129° 39′	0–100	0.044
565	13		54° 31′	129° 39′	100–250	0.064
566	13		54° 31′	129° 39′	250–500	0.222

APPENDIX TABLE
(continued)

Sample	Cruise	Date	Lat., °S	Long., °W	Depth, meters	Biomass, g dry wt/m²
			1964 Data			
567	13		54° 31′	129° 39′	500–1000	0.350
570	14	Aug. 2	50° 01′	159° 41′	0–100	0.028
571	14		50° 01′	159° 41′	100–250	0.114
572	14		50° 01′	159° 41′	250–460	0.220
573	14		50° 00′	159° 39′	500–1000	0.435
576	14	Aug. 3	51° 59′	159° 57′	0–100	0.249
577	14		51° 59′	159° 57′	100–250	0.128
578	14		51° 59′	159° 57′	250–480	0.321
579	14		51° 59′	159° 57′	500–1000	1.185
582	14	Aug. 5	53° 59′	159° 59′	0–100	0.441
583	14		53° 59′	159° 59′	100–250	0.165
584	14		53° 59′	159° 59′	250–460	3.500
585	14		53° 59′	159° 58′	500–970	1.280
588	14	Aug. 6	54° 59′	159° 52′	0–100	0.039
589	14		54° 59′	159° 52′	100–250	0.098
590	14		54° 59′	159° 52′	250–500	1.525
591	14		54° 59′	159° 52′	500–1000	1.685
602	14	Aug. 8	57° 00′	160° 03′	0–100	0.083
603	14		57° 00′	160° 03′	100–250	0.908
604	14		57° 00′	160° 03′	250–460	0.980
605	14		57° 00′	160° 03′	500–1000	1.495
608	14	Aug. 11	58° 03′	160° 04′	0–100	0.830
609	14		58° 03′	160° 04′	100–250	0.160
610	14		58° 03′	160° 04′	250–580	0.815
611	14		58° 03′	160° 04′	500–1040	0.895
615	14	Aug. 12	59° 08′	159° 47′	0–100	0.046
616	14		59° 08′	159° 47′	100–250	0.146
617	14		59° 08′	159° 47′	250–500	0.985
618	14		59° 08′	159° 47′	500–1010	1.355
629	14	Aug. 14	60° 53′	160° 09′	0–100	0.381
630	14		60° 53′	160° 09′	100–250	0.113
631	14		60° 53′	160° 09′	250–520	2.120
632	14		60° 53′	160° 09′	500–1000	1.185
642	14	Aug. 17	62° 55′	159° 56′	0–100	0.842
643	14		62° 55′	159° 56′	100–250	0.250
644	14		62° 55′	159° 56′	250–520	0.708
645	14		62° 55′	159° 56′	500–1000	0.860
655	14	Aug. 20	59° 52′	152° 45′	0–100	0.004
656	14		59° 52′	152° 45′	100–250	0.072
657	14		59° 52′	152° 45′	250–548	0.651
665	14	Aug. 22	59° 59′	145° 12′	0–100	0.042
666	14		59° 59′	145° 12′	100–250	0.072
667	14		59° 59′	145° 12′	250–500	0.900
678	14	Aug. 24	59° 58′	136° 56′	0–100	0.166
679	14		59° 58′	136° 56′	100–250	0.540
680	14		59° 58′	136° 56′	250–520	0.670
681	14		59° 58′	136° 56′	500–1000	1.480
686	14	Aug. 25	59° 59′	132° 41′	500–1000	1.445
691	14	Aug. 27	59° 59′	128° 56′	100–250	0.244
692	14		59° 59′	128° 56′	250–520	0.737
693	14		59° 59′	128° 56′	500–1140	2.575
713	14	Sept. 1	57° 07′	125° 23′	0–100	0.302

APPENDIX TABLE
(continued)

Sample	Cruise	Date	Lat., °S	Long., °W	Depth, meters	Biomass, g dry wt/m²
			1964 Data			
714	14		57° 07′	125° 23′	100–250	0.136
715	14		57° 07′	125° 23′	250–500	0.630
716	14		57° 07′	125° 23′	500–1000	3.210
723	14	Sept. 5	54° 51′	124° 51′	100–250	0.116
724	14		54° 51′	124° 51′	250–500	0.262
725	14		54° 51′	124° 51′	500–1085	0.745
732	15	Oct. 12	60° 12′	95° 04′	0–100	0.002
733	15		60° 12′	95° 04′	100–250	0.042
734	15		60° 12′	95° 04′	250–500	0.063
735	15		60° 12′	95° 02′	500–1000(?)	0.585
736	15		60° 13′	94° 58′	1000–2000	0.350
746	15	Oct. 16	59° 01′	99° 39′	1000–2000	0.150
755	15	Oct. 17	58° 03′	99° 54′	1000–2000	0.070
757	15	Oct. 19	61° 03′	99° 56′	0–100	0.097
758	15		61° 03′	99° 56′	100–250	0.177
759	15		61° 03′	99° 56′	250–500	0.425
760	15		61° 03′	99° 55′	400–620	1.120
762	15	Oct. 21	61° 02′	104° 58′	0–100	0.070
763	15		61° 02′	104° 58′	100–250	0.075
764	15		61° 02′	104° 58′	250–500	0.188
765	15		61° 03′	104° 58′	500–900	0.400
777	15	Oct. 26	60° 09′	109° 51′	0–100	0.059
779	15		60° 09′	109° 51′	250–500	0.092
780	15		60° 09′	109° 49′	500–1000	0.045
821	15	Nov. 11	58° 24′	134° 36′	500–1000	0.290
822	15		58° 25′	134° 36′	0–1000	0.173
823	15		58° 25′	134° 36′	100–250	0.218
824	15		58° 25′	134° 36′	250–500	0.453
827	15	Nov. 12	57° 32′	138° 45′	0–100	0.810
828	15		57° 32′	138° 45′	100–250	0.231
829	15		57° 32′	138° 45′	250–500	0.645
846	15	Nov. 16	56° 03′	144° 45′	0–100	0.649
847	15		56° 03′	144° 45′	100–250	0.390
848	15		56° 03′	144° 45′	250–500	1.085
849	15		56° 02′	144° 42′	500–1000	0.615
861	15	Nov. 18	53° 59′	145° 16′	0–100	0.733
862	15		53° 59′	145° 16′	100–250	0.100
863	15		53° 59′	145° 16′	250–500	0.210
864	15		54° 00′	145° 17′	500–1000	0.325
867	15	Nov. 19	55° 06′	149° 47′	0–100	1.548
868	15		55° 06′	149° 47′	100–250	0.303
869	15		55° 06′	149° 47′	250–500	0.290
871	15		55° 06′	149° 47′	500–1000	0.210
873	15	Nov. 20	56° 02′	149° 44′	0–100	1.793
874	15		56° 02′	149° 44′	100–250	0.090
875	15		56° 02′	149° 44′	250–500	0.678
877	15		56° 02′	149° 43′	500–1000	0.640
879	15	Nov. 21	57° 00′	150° 08′	0–100	1.183
880	15		57° 00′	150° 08′	100–250	0.488
881	15		57° 00′	150° 08′	250–500	0.420
882a	15		57° 01′	150° 07′	500–1000	0.265

APPENDIX TABLE
(continued)

Sample	Cruise	Date	Lat., °S	Long., °W	Depth, meters	Biomass, g dry wt/m²
			1965 Data			
929	17	Mar. 25	55° 01′	135° 00′	0–100	0.839
930	17		55° 01′	135° 00′	100–250	0.234
931	17		55° 01′	135° 00′	250–500	2.060
932	17		55° 04′	135° 00′	500–1000	2.340
933	17		55° 10′	135° 00′	1000–2000	1.120
936	17	Mar. 26	55° 59′	135° 04′	500–1000	2.810
937	17		56° 00′	135° 03′	0–100	0.263
938	17		56° 00′	135° 03′	100–250	0.276
939	17		56° 00′	135° 03′	250–500	2.820
942	17	Mar. 27	56° 55′	135° 00′	1000–2000	2.160
943	17		56° 55′	135° 00′	0–100	0.650
944	17		56° 55′	135° 00′	100–250	0.192
945	17		56° 55′	135° 00′	250–500	0.680
952	17	Mar. 28	59° 00′	135° 31′	1000–2000	1.910
956	17	Mar. 30	61° 08′	134° 20′	0–100	0.237
957	17		61° 08′	134° 20′	100–250	1.665
958	17		61° 08′	134° 20′	250–500	1.050
961	17		61° 10′	134° 18′	500–1000	1.060
972	17	Apr. 1	63° 04′	135° 03′	500–1000	0.760
973	17		63° 04′	135° 02′	0–100	2.419
974	17		63° 04′	135° 02′	100–250	0.930
975	17		63° 04′	135° 02′	250–500	1.105
986	17	Apr. 3	65° 02′	134° 52′	1000–2000	0.750
988	17		65° 59′	134° 41′	0–100	1.859
989	17		65° 59′	134° 41′	100–250	1.050
990	17		65° 59′	134° 41′	250–500	1.650
991	17		65° 59′	134° 41′	500–1000	1.630
993	17		66° 00′	134° 41′	1000–2000	1.430
1001	17	Apr. 7	68° 05′	126° 47′	0–100	2.201
1002	17		68° 05′	126° 47′	100–250	0.422
1003	17		68° 05′	126° 47′	250–500	0.417
1004	17	Apr. 8	68° 06′	126° 46′	500–1000	0.555
1006	17		68° 07′	126° 46′	1000–2000	0.050
1012	17	Apr. 9	67° 29′	124° 26′	1000–2000	0.520
1014	17	Apr. 10	66° 59′	120° 08′	0–100	1.381
1015	17		66° 59′	120° 08′	100–250	0.882
1016	17		66° 59′	120° 08′	250–500	1.340
1018	17		66° 59′	120° 07′	500–1000	4.215
1027	17	Apr. 12	67° 59′	106° 51′	0–100	0.397
1028	17		67° 59′	106° 51′	100–250	0.770
1029	17		67° 59′	106° 51′	250–500	0.815
1031	17	Apr. 13	67° 59′	106° 52′	1000–2000	0.590
1032	17		68° 01′	106° 51′	500–1000	1.430
1039	17	Apr. 14	67° 56′	103° 02′	1000–2000	0.800
1041	17		67° 56′	98° 52′	0–100	1.077
1042	17		67° 56′	98° 52′	100–250	0.622
1043	17		67° 56′	98° 52′	250–500	1.040
1045	17	Apr. 15	67° 56′	98° 51′	500–1000	2.820
1062	17	Apr. 19	66° 01′	94° 32′	0–100	0.562
1063	17		66° 01′	94° 32′	100–250	0.494
1064	17		66° 01′	94° 32′	250–500	1.205
1066	17		66° 00′	94° 31′	500–1000	1.530

APPENDIX TABLE (continued)

Sample	Cruise	Date	Lat., °S	Long., °W	Depth, meters	Biomass, g dry wt/m²
			1965 Data			
1081	17	Apr. 22	63° 00′	95° 08′	1000–2000	1.050
1083	17	Apr. 23	62° 07′	94° 43′	500–1000	2.020
1086	17		62° 09′	94° 44′	0–100	0.762
1087	17		62° 09′	94° 44′	100–250	0.463
1088	17		62° 09′	94° 44′	250–500	0.893
1091	17	Apr. 24	61° 10′	95° 02′	0–100	0.470
1092	17		61° 10′	95° 02′	100–250	0.215
1093	17		61° 10′	95° 02′	250–500	0.911
1094	17		61° 12′	95° 00′	500–1000	1.810
1097	17	Apr. 25	60° 04′	95° 10′	0–100	0.113
1098	17		60° 04′	95° 10′	100–250	0.242
1099	17		60° 04′	95° 10′	250–500	0.213
1101	17		60° 06′	95° 10′	500–1000	0.475
1112	17	Apr. 28	57° 05′	94° 49′	0–100	0.114
1113	17		57° 05′	94° 49′	100–250	0.237
1114	17		57° 05′	94° 49′	250–500	0.152
1115	17		57° 04′	94° 48′	500–1000	2.050
1117	17		57° 03′	94° 46′	1000–2000	0.490
1119	17	Apr. 29	56° 04′	94° 52′	0–100	0.037
1120	17		56° 04′	94° 52′	100–250	0.145
1121	17		56° 04′	94° 52′	250–500	0.073
1122	17		56° 04′	94° 52′	500–1000	1.290
1124	17		56° 04′	94° 50′	1000–2000	1.400
1126	17	Apr. 30	55° 01′	94° 51′	0–100	0.168
1127	17		55° 01′	94° 51′	100–250	0.292
1128	17		55° 01′	94° 51′	250–500	0.355
1129	17		55° 01′	94° 50′	500–1000	1.380
1131	17	Apr. 30	55° 01′	94° 48′	1000–2000	4.170
1132	18	June 5	54° 42′	99° 06′	0–100	0.120
1133	18		54° 42′	99° 06′	100–250	0.650
1134	18		54° 42′	99° 06′	250–500	0.375
1135	18		54° 43′	99° 08′	500–1000	1.550
1138	18	June 6	56° 00′	99° 22′	0–100	0.404
1139	18		56° 00′	99° 22′	100–250	0.660
1140	18		56° 00′	99° 22′	250–500	0.173
1142	18		56° 01′	99° 22′	500–1000	0.965
1144	18	June 8	57° 04′	99° 22′	0–100	0.065
1145	18		57° 04′	99° 22′	100–250	0.138
1146	18		57° 04′	99° 22′	250–500	0.117
1147	18		57° 03′	99° 22′	500–1000	1.595
1149	18		57° 02′	99° 23′	1000–2000	1.230
1152	18	June 10	58° 06′	99° 12′	0–100	0.027
1153	18		58° 06′	99° 12′	100–250	0.100
1154	18		58° 06′	99° 12′	250–500	0.542
1156	18		58° 05′	99° 13′	500–1000	1.650
1157	18		58° 05′	99° 14′	1000–2000	0.780
1158	19	July 13	58° 52′	99° 56′	1000–2000	1.540
1160	19		59° 00′	100° 00′	0–100	0.065
1161	19		59° 00′	100° 00′	100–250	0.213
1162	19		59° 00′	100° 00′	250–500	0.292
1163	19		59° 00′	100° 00′	500–1000	2.075
1165	19	July 15	60° 40′	100° 18′	500–1000	1.435

APPENDIX TABLE (continued)

Sample	Cruise	Date	Lat., °S	Long., °W	Depth, meters	Biomass, g dry wt/m²
			1965 Data			
1166	19		60° 40′	100° 18′	0–100	0.116
1167	19		60° 40′	100° 18′	100–250	0.103
1168	19		60° 40′	100° 18′	250–500	0.868
1172	19	July 16	61° 36′	99° 44′	1000–2000	1.490
1177	19	July 17	61° 58′	102° 48′	0–100	0.251
1178	19		61° 58′	102° 48′	100–250	0.182
1179	19		61° 58′	102° 48′	250–500	0.710
1183	19		61° 59′	102° 51′	500–1000	1.120
1186	19	July 19	62° 06′	104° 51′	0–100	0.092
1187	19		62° 06′	104° 51′	100–250	0.216
1188	19		62° 06′	104° 51′	250–500	1.170
1191	19		62° 07′	104° 50′	1000–2000	2.140
1193	19	July 20	61° 57′	108° 00′	100–250	0.242
1194	19		61° 57′	108° 00′	250–500	0.760
1195	19		61° 57′	108° 00′	500–1000	3.160
1197	19		61° 57′	108° 00′	1000–2000	2.670
1200	19	July 21	62° 05′	109° 29′	0–100	0.196
1201	19		62° 05′	109° 29′	100–250	0.142
1202	19		62° 05′	109° 29′	250–500	0.670
1203	19		62° 05′	109° 25′	500–1000	2.325
1205	19	July 24	62° 05′	109° 25′	1000–2000	2.280
1208	19		59° 52′	110° 06′	500–1000	3.745
1209	19		59° 52′	110° 06′	0–100	0.056
1211	19		59° 52′	110° 06′	250–500	0.319
1213	19	July 25	58° 53′	110° 08′	0–100	0.556
1214	19		58° 53′	110° 08′	100–250	0.039
1215	19		58° 53′	110° 08′	250–500	0.782
1216	19		58° 53′	110° 08′	500–1000	3.750
1219	19	July 26	58° 04′	109° 46′	0–100	0.142
1220	19		58° 04′	109° 46′	100–250	0.245
1221	19		58° 04′	109° 46′	250–500	0.730
1222	19		58° 04′	109° 45′	500–1000	2.970
1225	19	July 27	57° 08′	110° 01′	500–1000	0.520
1227	19	July 28	57° 08′	109° 59′	0–100	0.143
1228	19		57° 08′	109° 59′	100–250	0.168
1229	19		57° 08′	109° 59′	250–500	0.282
1245	19	Aug. 9	54° 58′	140° 06′	250–500	0.780
1254	19	Aug. 11	56° 59′	140° 12′	500–1000	1.010
1255	19		57° 00′	140° 12′	1000–2000	0.740
1259	19	Aug. 12	58° 11′	139° 59′	1000–2000	0.980
1262	19	Aug. 13	59° 03′	139° 59′	0–100	0.258
1263	19		59° 03′	139° 59′	100–250	0.239
1265	19		59° 03′	139° 58′	500–1000	1.850
1270	19	Aug. 15	59° 56′	139° 20′	500–1000	2.400
1272	19		59° 56′	139° 19′	1000–2000	1.750
1275	19	Aug. 16	61° 05′	140° 34′	0–100	0.940
1276	19		61° 05′	140° 34′	100–250	0.138
1277	19		61° 05′	140° 34′	250–500	0.658
1278	19		61° 06′	140° 32′	500–1000	1.805
1291	19	Aug. 20	59° 02′	147° 49′	0–100	0.218
1292	19		59° 02′	147° 49′	100–250	0.390
1293	19		59° 02′	147° 49′	250–500	0.195